建筑施工技术（中英文对照版）
Building Construction Technology (Chinese and English Version)

主 编 李 辉 黄 敏
　　　 戴明元 吕 颖
主 审 张义琢

北京理工大学出版社
BEIJING INSTITUTE OF TECHNOLOGY PRESS

内 容 提 要

本书共分为9章，主要内容包括绪论、土方工程施工、地基与基础工程施工、砌体工程施工、钢筋混凝土结构工程施工、预应力混凝土结构工程施工、结构安装工程施工、防水工程施工、装饰装修工程施工等。

本书可作为高等职业院校建筑工程技术、工程管理、工程监理等专业的教材，也可以作为建筑行业工程技术人员培训及自学用书，尤其适用于建筑类中外合作办学相关专业和涉外建设工程技术人员培训使用。本书可供相关院校研究生教学及工程技术人员参考。

版权专有　侵权必究

图书在版编目（CIP）数据

建筑施工技术：汉英对照/李辉等主编. --北京：北京理工大学出版社，2021.8
ISBN 978-7-5763-0253-0

Ⅰ.①建… Ⅱ.①李… Ⅲ.①建筑施工－施工技术－高等职业教育－教材－汉、英　Ⅳ.①TU74

中国版本图书馆CIP数据核字（2021）第176746号

出版发行 /	北京理工大学出版社有限责任公司
社　　址 /	北京市海淀区中关村南大街5号
邮　　编 /	100081
电　　话 /	（010）68914775（总编室）
	（010）82562903（教材售后服务热线）
	（010）68944723（其他图书服务热线）
网　　址 /	http://www.bitpress.com.cn
经　　销 /	全国各地新华书店
印　　刷 /	北京紫瑞利印刷有限公司
开　　本 /	787毫米×1092毫米　1/16
印　　张 /	48
字　　数 /	1257千字
版　　次 /	2021年8月第1版　2021年8月第1次印刷
定　　价 /	120.00元

责任编辑／江　立
责任校对／周瑞红
责任印制／李志强

图书出现印装质量问题，请拨打售后服务热线，本社负责调换

前 言 Foreword

本书依据《高等职业学校专业教学标准（试行）》中"高等职业学校建筑工程技术专业教学标准"和"实训导则"编写，并满足双语教学的语言环境，紧贴当前高职教育的教学要求。全书以施工现场的实际工作过程为线索，实际工作内容和现场理论相结合，以满足高等职业教育培养目标的需要。本书在内容编排上，做到了从简单到复杂、从单一到综合，结合工种培训实践要求，既便于教师讲授，也适合学生自学。

本书依托李辉、黄敏主编的优秀中文教材"十三五"国家职业教育规划教材《建筑施工技术》内容，针对中外合作办学、海外教学双语特点进行优化和升级编写、翻译而成。全书共计9个章节，中英文对照。第一章为绪论、第二章为土方工程施工、第三章为地基与基础工程施工、第四章为砌体工程施工、第五章为钢筋混凝土结构工程施工、第六章为预应力混凝土结构工程施工、第七章为结构安装工程施工、第八章为防水工程施工、第九章为装饰装修工程施工。

在教学过程中，建议使用多种教学方法相结合的方式进行教学，强调对学生独立收集信息、独立计划、独立实施、独立检查、独立工作能力的培养；可采用多样化的教学手段，如施工现场教学、教学模型、教学多媒体、教学录像、实地参观等方式相结合，有效调动学生的主动性。

本书由四川建筑职业技术学院李辉、黄敏主编；戴明元翻译1~5章、吕颖翻译6~9章。多年在海外从事国际工程设计和施工管理的高级工程师张义琢老师主审了英文部分并对中英文表述提出了许多宝贵意见，李育枢协助审阅了部分英文翻译。

中英文教材的编写是我们随着"双高"建设，中外合作办学，职业教育"走出去"对教材建设的尝试。由于时间仓促和作者水平有限,加上新内容的不断添加,书中难免存在不妥之处,敬请读者批评指正。

本书在编写过程中得到了四川建筑职业技术学院、华西集团第七建筑公司的大力支持,在此表示真诚的感谢!

编 者

目 录 Contents

第 1 章 绪论 ······ 2
Chapter 1　Introduction ······ 3

第 2 章 土方工程施工 ······ 16
Chapter 2　Earthwork Construction ······ 17

第 3 章 地基与基础工程施工 ······ 82
Chapter 3　Subbase and Foundation Construction ······ 83

第 4 章 砌体工程施工 ······ 202
Chapter 4　Masonry Works Construction ······ 203

第 5 章 钢筋混凝土结构工程施工 ······ 272
Chapter 5　Construction of Reinforced Concrete Structure ······ 273

第 6 章 预应力混凝土结构工程施工 ······ 430
Chapter 6　Prestressed Concrete Structure Construction ······ 431

第 7 章 结构安装工程施工 ······ 506
Chapter 7　Installation of Prefabricated Structural Members Construction ······ 507

第 8 章 防水工程施工 ······ 598
Chapter 8　Waterproofing Construction ······ 599

第 9 章 装饰装修工程施工 ······ 688
Chapter 9　Finishing and Decoration Work Construction ······ 689

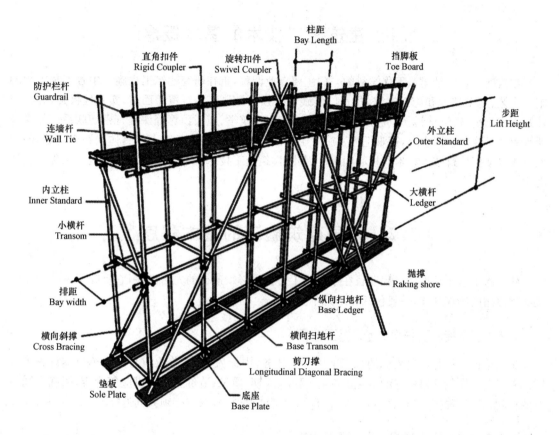

第1章 绪论

1.1 建筑施工技术的基本概念

建筑施工技术是建筑工程学科的一门主要专业课程,重在研究建筑工程施工中各工种工程的施工工艺、施工技术和施工方法,是一门实践性很强的学科。内容覆盖面广,实践性极强,并且与建筑工程测量、建筑材料、建筑力学、建筑结构、房屋建筑学、施工预算等课程有密切联系。如果该课程教学只停留在课堂上,很难把涉及施工工艺、施工工序及施工关键点等具体问题讲清楚,同学们也难以掌握建筑施工技术的核心问题。因此,建筑施工技术教学与实践相结合是本课程特点的要求。

1.2 我国建筑施工的基本程序

建筑施工程序是拟建工程项目在整个施工过程中必须遵循的先后次序,它反映了整个施工阶段必须遵循的客观规律,一般包括以下几个阶段:

1.2.1 承接施工任务,签订施工合同

施工单位一般是通过投标的方式承接施工任务,中标后,施工单位应与建设单位签订施工合同。施工合同应包括协商好的承包的内容、要求、工期、质量、造价及材料供应等内容,明确合同双方的权利、义务、责任。施工合同一经签订后,具有法律效力,双方必须共同遵守。

1.2.2 做好施工准备,提出开工报告

签订施工合同后,施工单位应全面做好施工准备工作。施工准备工作包括调查、研究、收集、资料,技术资料准备,物资准备,施工人员准备,施工现场准备,季节性施工准备。

工程具备开工条件后,施工单位向监理单位提出工程开工报告,经审查批准后,工程即可正式开工。

Chapter1　Introduction

1.1　The Basic Ideas of Building Construction Technology

The building construction technology is a major professional course of building engineering, which focuses on the study of construction skills, construction technology and construction method for various types of trades in the construction projects. It covers a wide range of contents and has very strong practicality, and it is closely related to courses such as architectural engineering surveying, building materials, architectural mechanics, building structure, building architecture and construction budget. If the teaching of this course only stays in the classroom, it is difficult to explain clearly the construction technology, construction process and key points of construction, and students are also difficult to understand the core problems of construction technology. Therefore, it is the characteristic of this course to combine the teaching of construction technology with practice.

1.2　Basic Procedures of Building Construction in China

The construction procedure is the sequence that must be followed in the whole construction process of the proposed project, which reflects the objective law that must be followed in the whole construction stage, and generally includes the following stages:

1.2.1　Undertake Construction Work and Sign Construction Contract

Generally, the Contractor undertakes construction works through tendering. After acceptance of the tender, the Contractor signs construction contract with the Employer. The contract shall include the contents, requirements, duration, quality, cost and material supply that have been negotiated, and specify the rights, obligations and responsibilities of the both parties. Once the contract is signed, it shall have legal effect and the both parties shall be abided by it.

1.2.2　Prepare for Construction and Put Forward Work Report

After signing the construction contract, the Contractor shall make full preparations for commencing the work, which include investigation and research, information collection, technical data preparation, material preparation, construction personnel preparation, construction site preparation and seasonal construction preparation.

After the project meets the commencing conditions, the Contractor shall submit the project commencement report to the Consultants, and the project can be officially started after the review and approval.

1.2.3 组织施工

施工单位应按照施工组织设计精心组织施工。在施工中做好动态控制工作,保证质量目标、进度目标、造价目标、安全目标、现场目标的实现。严格履行施工合同,处理好内外关系,处理好合同变更,搞好索赔。编制好施工技术资料。

1.2.4 竣工验收,交付使用

竣工验收是施工的最后阶段,在竣工验收前,施工企业内部进行自检,检查各分部分项工程的施工质量,整理工程竣工资料,进行竣工结算。自检不合格的项目应进行整改,达到合格才能交付验收。

在施工单位自检合格的基础上,由建设单位(项目)负责人组织施工单位(含分包单位)、设计、监理等单位(项目)负责人进行竣工验收。

1.2.5 回访保修阶段

工程交工后保修是我国一项基本法律制度,回访保修的责任应由施工单位承担。施工单位应建立施工项目交工后的回访与保修制度,提高工作质量,听取用户意见,改进服务方式。

1.3 我国建筑施工发展

建筑业是国民经济建设中的支柱产业之一,是相关行业赖以发展的基础性先导产业,建筑业在促进我国国民经济和社会发展中起着重要作用。我国建筑业经过几十年的发展,尤其是近20年来,建成了一大批规模宏大、结构新颖、技术难度大的建筑物,取得了显著的成绩和突破性进展,也充分显示了我国建筑技术的实力。特别是超高层建(构)筑物和新型钢结构建筑的兴起对我国建设工程技术进步产生了巨大的推动力,促使我国建筑施工水平再上新台阶,有些已达到国际先进水平。

1.2.3 Organizing Construction

The Contractor shall carefully organize the construction according to the construction organization design. Do a good job of dynamic control in the construction to ensure the realization of quality objectives, progress objectives, cost objectives, safety objectives and site objectives. Strictly perform the construction contract, deal with the internal and external relations, deal with the contract changes, and make good claims. Prepare the construction technical data.

1.2.4 Completion Acceptance and Delivery

Completion acceptance is the final stage of construction. Before the completion acceptance, the Contractor shall conduct internal self-inspection to check the construction quality of each subdivision and items, put project file in order and settle final account for the completed project. The unqualified items in the self-inspection shall be rectified, and only when they are qualified can they be delivered for acceptance.

On the basis of the self-inspection of the Contractor, the person in charge of the project on behalf of the Employer (project manager) shall organize the persons in charge of the project on behalf of the Contractor (including the subcontractor), the designer, the Consultant and other units to carry out the completion acceptance.

1.2.5 Return Visit and Maintenance Period

The maintenance after project delivery is a basic legal system in China. The responsibility of return visit and maintenance should be borne by the Contractor. The Contractor should establish the return visit and maintenance system after the completion of the construction project, improve the work quality, listen to the opinions of users and improve the service mode.

1.3 Development of Building Construction in China

The building industry is one of the pillar industries in the national economic construction, and it is the basic and leading industry on which also the related industries rely for development. It plays an important role in promoting China's national economy and social development. After decades of development, especially in the past 20 years, China's building industry has built a large number of buildings with large scale, novel structure and difficult technology, and has made remarkable achievements and breakthroughs, which also fully shows the strength of China's building technology. In particular, the rise of super high-rise buildings and new steel structure buildings has become a driving force to the technical progress of China's construction, and pushed China's construction level to a new stage, some of which have reached the international advanced level.

1.3.1 地基与基础工程施工技术

1. 地基加固技术

在地基处理方面,我国根据土质条件,加固材料和工艺特点,结合国外软土地基加固的新工艺,研究开发出具有中国特色的多种复合地基加固方法。按照加固机理大体分为 4 类:第 1 类是压密固结法,如强夯、降水压密、真空预压等,适用于大面积松软地基处理;第 2 类是加筋体复合地基处理,如砂桩、碎石桩、水泥粉煤灰碎石桩、夯实水泥土桩、水泥土搅拌桩等,应用范围广,已成为地基加固的主体;第 3 类是换填垫层法,如砂石垫层、灰土垫层等,适用范围较小;第 4 类是浆液加固法,如水泥注浆、化学注浆等,主要用于既有建筑地基的加固处理。上述加固方法有不少已形成系列,集中反映在《建筑地基处理技术规范》(JGJ 79—2012)中,有的处理技术已接近或达到国际先进水平。

2. 桩基技术

混凝土灌注桩具有适用于任何土层、承载力大、对周围环境影响小等特点,因而发展最快。目前已施工的混凝土灌注桩桩径达 3 m、孔深达 104 m。在灌注桩施工中国内还研究应用了后压浆技术,即成桩后通过预埋的注浆管用一定压力将水泥浆压入桩底和桩侧,使之对桩侧底泥皮、桩身和桩端底沉渣、桩周底土层产生充填胶结、加筋、固化效应。采用后压浆技术,可减少桩体积 40%,成本降低效果显著。沉管灌注桩在振动、锤击沉管灌注桩基础上,研究了新的桩型,如新工艺的沉管桩、沉管扩底桩(静压沉管夯扩灌注桩和锤击振动沉管扩底灌注桩)、直径 500 mm 以上的大直径沉管桩等。先张法预应力混凝土管桩逐步扩大应用范围,在针对起吊不当、偏打、打桩应力过高、挤土、超静水压力等原因而产生的施工裂缝方面,研究出了有效的预防措施。挖孔桩近年来已可开挖直径 3.4 m、扩大头直径达 6 m 的超大直径挖孔桩。在一些复杂地质条件下,亦可施工深 60 m 的超深人工挖孔桩。大直径钢管桩在建筑物密集地区高层建筑中应用较多,在减少挤土桩沉桩时对周围环境影响的技术方面达到了较高的水平。CFG 桩复合地基是一种采用长螺旋钻成孔管内泵压水泥粉煤灰碎石桩、桩间土和褥垫层组成的一种新型复合地基形式,适用于饱和及非饱和的粉土、黏性土、砂土、淤泥质土等地质条件。桩检测技术包括成孔后检测和成桩后检测,后者主要是动力检测,我国桩基动力检测的软硬件系统正在赶超国际水平。

1.3.1 Construction Technology of Sub-base and Foundation Works

1. Sub-base Reinforcing Technology

In terms of sub-base treatment, according to the soil conditions, reinforcing materials and technological characteristics, China has fully adopted the new technology of soft soil foundation reinforcement in foreign countries, studied and developed a variety of composite foundation reinforcing methods with Chinese characteristics. According to the reinforcement mechanism, it can be divided into four categories: The first is compaction consolidation method, such as dynamic compaction, dewatering compaction and vacuum preloading, which is suitable for large area soft foundation treatment; The second is reinforced composite foundation treatment, such as sand pile, gravel pile, cement fly ash gravel pile, rammed cement soil pile, cement soil mixing pile, etc. , which has been widely used and has become the main body of foundation reinforcement; The third is the replacement of filling layer method, such as sand cushion, lime soil cushion, etc. , which has a small scope of application; The fourth is the slurry reinforcement method, such as cement grouting, chemical grouting, etc. , which is mainly used for the reinforcement of the existing building foundation. Many of the above reinforcement methods have formed a series, which are mainly reflected in *Technical Code for Ground Treatment of Buildings* (JGJ 79-2012), and some of them have reached the international advanced level.

2. Pile Foundation Technology

The cast-in-place concrete pile has the characteristics of being suitable for any soil layer, large bearing capacity and little impact on the surrounding environment, so it develops the fastest. At present, the diameter of the cast-in-place concrete pile is up to 3m and the hole depth is up to 104m. In the construction of cast-in-place pile, the post grouting technology is also studied and applied in China, that is, after the pile is completed, the cement slurry is pressed into the bottom and side of the pile with a certain pressure through the embedded grouting pipe, so as to have the effect of filling, cementation, reinforcement and solidification on the bottom mud skin of the pile side, the sediment of the pile body and the bottom of the pile end, and the soil layer around the pile. After using the post grouting technology, the pile volume can be reduced by 40%, and the cost can be reduced significantly. Based on the vibration and hammering of the cast-in-place pipe pile, the new pile types are studied, such as the new technology of the cast-in-place pipe pile, the cast-in-place pipe belled pile (static pressure cast-in-place pipe ramming pile and hammering vibration cast-in-place pipe belled pile), the large diameter cast-in-place pipe pile with diameter more than 500mm, etc. The application scope of pretensioned prestressed concrete pipe pile is gradually expanded, and effective measures are developed to prevent construction cracks caused by improper lifting, eccentric driving, excessive piling stress, soil compaction, and excess hydrostatic pressure. In recent years, the extra large diameter bored pile with diameter of 3.4m and enlarged head diameter of 6m can be excavated. Under some complicated geological conditions, it is also possible to construct super deep man dug piles with a depth of 60m. Large diameter steel pipe pile is widely used in high-rise buildings in densely populated areas, and it has reached a high level in technology to prevent the impact of soil compaction pile on the surrounding environment. CFG pile composite foundation is a new type of composite foundation, which is composed of long auger drilling hole pipe pumping cement fly ash gravel pile, soil between piles and cushion. It is suitable for saturated and unsaturated silt, cohesive soil, sandy soil, muddy soil and other geological conditions. Pile detection technology includes detection after hole forming and detection after pile forming, the latter is mainly dynamic testing. The software and hardware system of pile foundation dynamic testing in China is catching up with or reaching the international level.

3. 深基坑支护技术

为适应不同坑深和环境保护要求,在支护墙方面发展了土钉墙、水泥土墙、排桩和地下连续墙等。土钉墙费用低、施工方便,适宜于深度不大于 15 m,周围环境保护要求不十分严格的工程,因此,土钉墙和复合土钉墙近年来发展十分迅速,在软土地区得到应用。地下连续墙宜用于基坑较深,环境保护要求严格的深基坑工程。近年在北京中银大厦施工中,基础外墙采用封闭式三合一型(防水、护坡、承重)800 mm 厚的地下连续墙,深度达 30 米,在施工中要采取可拆式锚杆等特殊措施,与锚杆,降水,土方同步进行,解决了地下连续墙的锚固问题。预应力地下连续墙作为一个新趋势,也得到了研究与应用。预应力地下连续墙可提高支护墙的刚度 30% 以上,墙厚度可减薄,内支撑的数量可减少。由于曲线布筋张拉后产生反拱作用,可减少支护墙地变形,支护墙裂缝少,提高了抗渗性。因此,在解决了设计和施工工艺之后,预应力地下连续墙会得到一定的发展。内支撑 H 型钢、钢管、混凝土支撑皆有应用,布置方式根据基坑形状有对撑、角撑、桁(框)架式、圆环式等,还可多种布置方式混合使用。圆环式支撑受力合理、能为挖土提高较大的空间。深、大基坑土方开挖目前多采用反铲挖土机下坑,以分层、分块、对称、限时的方式开挖土方,以减少时空效应的影响,限制支护墙的变形。逆作法施工工艺在有多层地下室的深基坑工程中应用。逆作法或半逆作法能有效地降低施工费用、加快整个工程的施工进度,还能较好地控制周围环境的变形,可用于地铁车站、高层建筑多层地下室、构筑物的深基础和车站广场人防工程等地的施工。逆作法施工工艺在软土地区解决了中柱桩承载不足、中柱桩过多的问题。

1.3.2 混凝土工程施工技术

1. 预拌混凝土和混凝土泵送技术

(1)预拌混凝土技术。商品混凝土的应用数量和比例标志着一个国家的混凝土工业生产的水平。随着预拌混凝土的发展,我国的混凝土泵送技术提高很快,泵送高度在上海金茂大厦达到 382.5 m,在世界上已名列前茅。

3. Supporting Technology of Deep Foundation Pit

In order to meet the requirements of different pit depth and environmental protection, soil nailing wall, cement wall, row pile and underground continuous wall are developed. Soil nailing wall has the advantages of low cost and convenient construction, which is suitable for projects with a depth of less than 15m and less stringent environmental protection requirements. Therefore, soil nailing wall and composite soil nailing wall have developed rapidly in recent years, and have been applied in soft soil areas. The underground continuous wall is suitable for deep foundation pit with strict environmental protection requirements. In recent years, in the construction of Bank of China Tower in Beijing, the closed three in one (waterproof, slope protection, load-bearing) 800mm thick diaphragm walls is used for the foundation external wall, with a depth of 30m. Special measures such as detachable anchor rod should be taken in the construction, which is carried out simultaneously with anchor rod, dewatering and earthwork, so as to solve the anchoring problem of diaphragm wall. As a new trend, prestressed continuous wall has been studied and applied. Prestressed continuous wall can increase the stiffness of retaining wall by more than 30%, the thickness of the wall can be thinner, and the number of internal support can be reduced. Due to the anti-arch effect after the stretching of the curved tendons, the deformation of the supporting wall can be reduced, the supporting wall has fewer cracks, and the anti-permeability can be improved. Therefore, after solving the design and construction process, the prestressed continuous wall will be further developed. Internal support H-steel, steel pipe, concrete support are all used. According to the shape of the foundation pit, the layout can be divided into diagonal brace, angle brace, truss (frame) type, ring type, etc. it can also be mixed with a variety of layout. The ring brace is reasonable in force and can improve the space for excavation. At present, backhoe excavators are mostly used in the excavation of deep and large foundation pits to excavate the earth in a layered, partitioned, symmetrical and time-limited manner, so as to reduce the influence of time and space effect and limit the deformation of retaining walls. The application of reverse construction technology in deep foundation pit works with multi-storey basement. The application of reverse method or semi reverse method can effectively reduce the construction cost, speed up the construction progress of the whole project, and better control the deformation of surrounding environment. Reverse construction technology can be used in the construction of subway station, multi-storey basement of high-rise building, deep foundation of structure and civil air-defense construction of station square. In soft soil area, reverse construction technology solves the problem of insufficient bearing capacity of center pillar pile and prevents too many center pillar piles.

1.3.2 Construction Technology of Concrete Works

1. Ready Mixed Concrete and Concrete Pumping Technology

(1) Ready Mixed Concrete Technology. The application quantity and proportion of commercial concrete mark the level of a country's concrete industrial production. With the development of ready mixed concrete, the pumping technology of concrete in China has been improved rapidly. The pumping height of Jin Mao Tower in Shanghai has reached 382.5m, which is among the best in the world.

(2)混凝土外加剂技术。商品混凝土产量的增大,极大地推动了混凝土外加剂(特别是各种减水剂)的发展。如在水下混凝土、喷射混凝土、商品混凝土和泵送混凝土中的应用和发展。

(3)预防混凝土碱集料反应的措施。我国许多地方存在混凝土碱集料反应,给结构带来严重危害。必须采用相应的技术措施,保证混凝土安全,延长结构使用寿命。要解决混凝土碱集料反应,重点在选用的低碱水泥、砂石料、外加剂和低碱活性集料等,选用高品质减水剂、膨胀剂,严格控制砂石料的含泥量及其级配,混凝土试配时首先考虑使用低碱活性集料以及优选低碱水泥(碱含当量0.6%以下),掺加矿粉掺和料及低碱、无碱外加剂。

2. 高强高性能混凝土

目前我国已利用多种地方材料(磨细砂渣、无机超细粉、粉煤灰、硅粉等)和超塑化剂在工业化生产水平C60的高强混凝土,C80高强混凝土在一些大城市开始用于工程实践,也已基本掌握了配置C100高强混凝土的技术,并在国家大剧院工程中应用。此外,一些特种混凝土如纤维混凝土、水下不分散混凝土、特细砂混凝土等,亦成功配制和应用。

(1)大体积混凝土浇筑。

我国在高层建筑的桩基承台或箱基底板大体积混凝土浇筑方面达到很高水平。一般可采取以下措施保证大体积混凝土施工质量:

①进行混凝土试配。

②根据混凝土用量,组织商品混凝土供应站、现场泵车、备用电源、混凝土罐车确保现场混凝土供应的连续性。

③混凝土采用斜面推进、大斜面分层下料,分层振筑。

④现场测温设备采用"大体积混凝土温度微机自动测试仪",对混凝土内外温差进行适时监控。

(2)预应力混凝土技术。新Ⅲ级钢筋和低松弛高强度钢绞线的推广,以及开发研究的新型预应力锚夹具的应用,都为推广预应力混凝土创造了条件。目前大跨度预应力框架和高层建筑大开间的无粘结预应力楼板应用较为普遍,后者能减少板厚、减低高度、减轻建筑物自重,优越性显著。在构筑物中,如压力管道、水池、贮罐、核电站、电视塔应用更普遍,如天津电视塔采用了最长束达310 m的竖向预应力筋,其预应力束长度为国内之最。

(2) Concrete admixture technology. The increase of commercial concrete output greatly promotes the development of concrete admixtures(especially various water reducing agents), such as application and development of underwater concrete construction technology, shotcrete, commercial concrete and pumping concrete.

(3) Measures to prevent concrete alkali-aggregate reaction. Concrete alkali-aggregate reaction exists in many areas in our country and cause severe harm to the structure. Corresponding technical measures must be taken to ensure the safety of concrete and extend the service life of the structure. To solve the alkali-aggregate reaction of concrete, low alkali cement, aggregate, admixture and low alkali active aggregate should be selected. High quality water reducing agent and expansion agent should be selected. The mud content and gradation of aggregate should be strictly controlled. Low alkali active aggregate, low alkali cement(alkali equivalent less than 0.6%), mineral powder admixture, low alkali and non alkali active aggregate should be considered first in concrete trial Alkali admixture.

2. High Strength and High Performance Concrete

At present our country has used a variety of local materials(fine grinding slag, inorganic ultra-fine powder, fly ash, silicon powder, etc.)and super plasticizer to produce C60 high strength concrete in the industrialized level. C80 high-strength concrete began to use in engineering practice in some big cities. Configuration of C100 high strength concrete technology has been basically grasped, and used in the national grand theatre engineering. In addition, some special concrete such as fiber concrete, underwater non-dispersible concrete, ultra-fine sand concrete, etc., has also been successfully prepared and applied.

(1) Mass concrete pouring.

In China, the mass concrete pouring of pile cap or box base slab of high-rise buildings has reached a very high level. Generally, the following measures can be taken to ensure the construction quality of mass concrete:

①The concrete is tested.

②According to the amount of concrete, commercial concrete supply station, on-site pump truck, standby power supply and concrete tank truck shall be organized to ensure the continuity of on-site concrete supply.

③Concrete adopts slop advancing, large slop layered unloading and layered vibrating construction.

④The on-site temperature measuring equipment adopts the "mass concrete temperature microcomputer automatic tester" to timely monitor the temperature difference between the inside and outside of the concrete.

(2) Prestressed concrete technology. The promotions of new Ⅲ grade steel and low relaxation high strength steel strands, as well as the application of new types of prestressed anchorage fixture researched and developed, create conditions for the promotion of prestressed concrete. At present, large-span prestressed frame and larger bay unbounded prestressed floor slab in high-rise buildings are widely used. The latter has obvious advantages by reducing floor thickness, height and weight of the building. In the structures, such as penstock, pool, storage tank, nuclear power station and TV Tower, the application is more common. For example, Tianjin TV tower adopts vertical prestressed tendons with the longest tendons up to 310m, and the length of prestressed tendons is the largest in China.

3. 钢筋技术

在粗钢筋连接方面，除广泛应用的电渣压力焊外，机械连接（套筒挤压连接、锥螺纹连接、直螺纹连接）不受钢筋化学成分、可焊性及气候影响，质量稳定，无明火，操作简单，施工速度快。尤其是直螺纹连接，可确保接头强度不低于母材强度，连接套筒通用Ⅱ、Ⅲ级钢筋，该技术正得到国内广泛推广。

4. 模板工程施工技术

(1) 模板脚手架体系的发展

近20年来，竖向模板经历了小钢模→钢框竹胶合板→全钢组合大模板，目前市场的主流体系除组合钢模板外，木胶合板模板使用量也比较大。水平模板体系一直难以工具化，国内主要采用木胶合板模板和竹胶合板模板体系（欧美多采用铝木结合）。全钢大模板具有拼缝少，施工过程中混凝土不易漏浆；刚度大，构件不易变形、鼓肚；周转次数多；模板表面平整光洁，成型质量好，能很好保证清水混凝土质量的优点。

(2) 模板脚手架技术。随着经济飞速发展，国内许多专利系统模板被应用，很多新型模板技术及工法已经使用。如：墙体模板体系；柱模体系；井筒模板体系；早拆体系；滑模、爬升模板体系；预应力圆孔、大型屋面、异型（楼梯模、门窗洞口模等）多向新型模板系统；路、桥梁、隧道模板体系；饰面混凝土模板系统；竹胶合板及高强人造板模板；钢框胶合板模板及其支撑系统；铝制、玻璃钢模壳及其他材质的新型结构模板系统等。在脚手架技术方面，扣件式钢管脚手架、碗扣式钢管脚手架、门式钢管脚手架以及爬、挑、挂脚手架得到广泛应用。此外还有一些特殊脚手架，如：吊脚手架（吊篮）桥式脚手架、塔式脚手架，而木、竹脚手架则因为成本低廉，在高度较低建筑物施工中使用。超高层建筑的发展，促进了高层建筑模板体系的系统研究，目前已有模板CAD辅助设计软件。高层建筑施工配套升降式脚手架亦日益完善。

1.3.3 钢结构安装技术

除原钢板箱形柱焊接技术、高强螺栓施工技术和钢结构安装技术继续发展、提高外，钢结构预应力技术方面发展很快。

3. Reinforcement Technology

In the connection of thick reinforcement, in addition to the widely used electroslag pressure welding, the mechanical connection (sleeve extrusion connection, taper thread connection, straight thread connection) is not affected by the chemical composition, weldability and climate of reinforcement, with stable quality, no open fire, simple operation and fast construction speed. In particular, the straight thread connection can ensure that the strength of the joint is not lower than that of the base metal, and the connection sleeve is generally used for grade II and III steel bars. This technology is being widely promoted in China.

4. Formwork Construction Technology

(1) Development of formwork and scaffolding system

In the past 20 years, vertical formwork experienced small steel formwork → steel frame bamboo plywood → all steel composite large formwork. At present, the mainstream system in the market is not only composite steel formwork, but also wood plywood formwork. The horizontal formwork system has always been difficult to implement. Wood plywood formwork and bamboo plywood formwork system are mainly used in China (aluminum wood combination is mostly used in Europe and America). The full steel large formwork has the advantages of less joint, less leakage of concrete in the construction process, large rigidity, less deformation and bulging of components, more turnover times, smooth and smooth formwork surface, good molding quality, which can ensure the quality of fair faced concrete.

(2) Formwork and scaffolding technology. With the rapid development of economy, many domestic patent system formworks have been applied, and many new formwork technologies and construction methods have been used, such as wall formwork system, column formwork system, shaft formwork system; early dismantling system; slip form, climbing formwork system; prestressed circular hole, large roof, special-shaped (stair formwork, door and window opening formwork, etc.) multi-directional new formwork system; road, bridge, tunnel formwork system; facing concrete formwork system; bamboo plywood and high-strength wood-based panel formwork; steel frame plywood formwork and its support system, new structural formwork system made of aluminum, FRP and other materials. In scaffold technology, coupler connected steel tube scaffolding, steel tube scaffolding with bowl type coupler, portal steel tube scaffolding and climbing, outrigger and suspended scaffolding are widely used. In addition, there are some special scaffold, such as flying scaffolding (basket), bridge-type scaffolding, tower scaffolding, etc. Wood and bamboo scaffolding for their lower cost are mainly used in lower building construction. The development of super high-rise buildings has promoted the systematic research on the formwork system of high-rise buildings. At present, there is CAD software for formwork. The attached lifting scaffold for high-rise building construction is becoming more and more perfect.

1.3.3 Steel Structure Installation Technology

In addition to the continuous development and improvement of the original steel box column welding technology, high-strength bolt construction technology and steel structure installation technology, the steel structure prestress technology has developed rapidly.

20世纪90年代以后我国大跨度公共建筑兴建较多,预应力技术在空间钢结构中得到较广泛的应用,创造出多种空间钢结构的新体系,如预应力网架与网壳、索网、索拱、索膜、斜拉体系等,充分发挥受拉杆件的强度潜力,结构轻盈,时代感强。在空间钢结构预应力施工中也创造了许多新颖的施加预应力的方法,有张拉、整体下压、整体顶升等多种,工艺简易、经济而且可靠。

1.3.4 建筑防水技术

近年,我国建筑防水材料应用量稳步增长,特别是新型防水材料增长很快。到2010年,按原国家建材局"新型建材及制品导向目录"要求及市场走势,SBS、APP改性沥青防水卷材仍是主导产品,将大力发展;高分子防水卷材重点发展EPDM、PVC(P型)两种产品,并积极开发TPO产品;防水涂料着重发展前景看好的聚氨酯、丙烯酸酯类防水涂料;密封材料仍重点发展硅酮、聚氨酯、丙烯酸酯密封膏,尽快开发防水保温一体材料;刚性防水材料、渗透结晶型防水材料、金属屋面材料、沥青油毡瓦、水泥瓦、土工材料也有一定的发展。

1.3.5 建筑装饰施工技术

我国建筑装饰行业兴起是改革开放政策带来的成果,这一行业保持了20年高速持续发展,其施工技术、制品制造技术也有了很大的进步,尤其幕墙专业已经接近国际水平。有的工种已经进行了彻底的改变,建筑装饰行业常用的各种电动工具已经在全行业得到了普及。有的企业已经开始走装饰配件生产工厂化、现场施工装配化的路子。这种应用全新生产方式的示范工程已经显示出工期短、质量好、无污染等特点,是当前通常施工方式无法比拟的。背栓系列、石材干挂技术、组合式单体幕墙技术、点式幕墙技术、金属幕墙技术、微晶玻璃与陶瓷复合技术、石材毛面铺设整体研磨等有较大发展。

1.3.6 信息化管理技术

随着我国改革开放的深入发展、加入WTO以及国际建筑业投资的加大,建筑市场竞争日趋激烈,对建筑安装企业本身的管理技术的要求也越来越高,这就要求建筑安装企业运用现代管理手段,提高企业的竞争力。对施工企业来讲,实现办公自动化,不仅可以提高工作效率,更为重要的是营造管理信息化建设所需要的氛围,提高人们对管理信息化的认识,初步感受管理信息化所带来的好处,为更高层次的管理信息化奠定基础。

Since the 1990s, there have been many large-span public buildings in China. Prestressed technology has been widely used in space steel structure, creating a variety of new systems of space steel structure, such as the system of prestressed space truss and shell, cable net, cable arch, cable membrane, cable-stayed, etc. , which give full play to the strength potential of the tension bar, light structure, and strong sense of the times. In the construction of prestressed spatial steel structure, many new methods of applying prestress are also created, such as tension, overall lowering, overall jacking and so on, which are simple in process, economical and reliable.

1.3.4 Building Waterproof Technology

In recent years, the application of building waterproof materials in China is growing steadily, especially the new waterproof materials. By 2010, SBS and APP modified asphalt waterproofing membrane will still be the leading products and will be vigorously developed according to the requirements and market trend of "new building materials and products guide catalogue" of the former National Building Materials Bureau; high-molecular waterproof rolling material focuses mainly on developing the two products of EPDM and PVC(P type), and actively developing TPO products; waterproofing coatings are mainly developing polyurethane and acrylate which will have promising prospects; sealing materials still focus on the development of silicone, polyurethane, acrylate sealing paste, and waterproof insulation integrated materials will be developed as soon as possible; rigid waterproof material, permeable crystalline waterproof material, metal roofing material, asphalt felt tile, cement tile and geotechnical materials should be developed as well.

1.3.5 Construction Technology of Building Decoration

The rise of China's building decoration industry is brought about by the reform and opening-up policy and has maintained a high-speed and sustainable development for 20 years. The construction technology and products manufacturing technology of building decoration industry have also made great progress, especially the professional curtain wall which is close to international level. Some trades have thorough changed, and commonly used electric tools for building decoration are very popular in the industry. Some enterprises have begun to take the way of factory production of decorative accessories and assembly of on-site construction. This demonstration project with new production mode has shown the characteristics of short construction period, good quality and no pollution, which is unmatched by the current common construction mode. Great development has been made in back bolt series, stone dry hanging technology, combined single curtain wall technology, point curtain wall technology, metal curtain wall technology, glass ceramics and ceramic composite technology, stone rough surface laying and overall grinding technology, etc.

1.3.6 Information Management Technology

With the further development of China's reform and opening up, China's accession to the WTO and the increasing investment in the international construction industry; the competition in the construction market is becoming increasingly fierce; which requires the construction and installation enterprises to use modern management means to improve their competitiveness. For construction enterprises, the realization of office automation can not only improve work efficiency, but also create the atmosphere needed for the construction of management informatization, improve people's understanding of management informatization, and preliminarily feel the benefits brought by management informatization, so as to lay a foundation for higher level management informatization.

第 2 章 土方工程施工

2.1 土方工程的基本概念

2.1.1 土方工程的施工特点

1. 土方工程概念

土方工程是建筑工程施工中的重要工作,它包括土的开挖、运输和填筑等主要施工过程,以及排水、降水和土壁支撑等准备工作与辅助工作。

土方工程按开挖和填筑的几何特征不同,可分为场地平整、挖基槽、挖基坑、挖土方、回填土等工程项目。

(1)场地平整系指厚度在 300 mm 以内的挖填和找平工作。

(2)挖基槽系指挖土宽度在 3 m 以内,且长度等于或大于宽度 3 倍者。

(3)挖基坑系指挖土底面积在 20 m^2 以内,且底长为底宽 3 倍以内者。

(4)挖土方系指山坡挖土或基槽宽度大于 3 m,坑底面积大于 20 m^2 或场地平整挖填厚度超过 300 mm 者。

(5)回填土分夯填和松填。

2. 土方工程的施工特点

土方工程的工程量大,劳动强度大。建筑工地的场地平整,土方工程量有时可达数百万立方米以上,施工面积达数平方公里;大型基坑的开挖,有的深达 20 多米。因此土方工程应尽可能采用机械化施工以节约工期、降低劳动强度。

土方工程施工条件复杂,且多为露天作业,受气候、水文、地质等影响较大,难以确定的因素较多。因此在组织土方工程施工前,必须做好施工组织设计,选择好施工方法和施工机械,制订合理的调配方案,以保证工程质量,并取得较好的经济效果。

Chapter 2　Earthwork Construction

1.1　Basic Concept of Earthwork

2.1.1　Construction Characteristics of Earthwork

1. Earthwork Concept

Earthwork is an important work in the building construction, which includes the main construction process of soil excavation, transportation and filling; as well as the preparation and auxiliary work of drainage, dewatering and soil wall support.

According to the different geometric characteristics of excavation and filling, earthwork can be divided into site leveling, foundation groove, foundation pit, earth excavation, backfilling and other engineering items.

(1) Site leveling means excavation, filling and leveling with a thickness of less than 300mm.

(2) Excavation of foundation groove means that the excavation width is within 3m, and the length is equal to or greater than 3 times the width.

(3) Excavation of foundation pit refers to the excavation of the bottom area within $20m^2$, and the bottom length is 3 times the bottom width.

(4) Earthwork means that the width of slope excavation or foundation groove is greater than 3m, and the area of pit bottom is greater than $20m^2$, or site leveling of excavation and filling thickness exceeds 300mm.

(5) Backfill is divided into ramming and loose filling.

2. Construction Characteristics of Earthwork

The quantity of earthwork is large and the labor intensity is high. The quantity of earthwork in site leveling sometimes may amount to several million cubic meters above, the construction area amounts to several square kilometers, and some of large foundation pits are more than 20 meters deep. Therefore, earthwork excavation should be mechanized as far as possible to save time and reduce labor intensity.

The conditions of earthwork construction are complex; and most of the construction is in the open air; which is greatly affected by the climate, hydrology, geology, etc. ; and many factors are difficult to determine. Therefore, before the earthwork starting, the construction organization design must be well prepared, choosing good construction method and construction machinery, working out reasonable allocation, to ensure the engineering quality and achieve better economic results.

2.1.2 土的基本性质

1. 土的组成

土一般由土颗粒(固相)、水(液相)和空气(气相)三部分组成,如图 2.1 所示。这三部分间的比例关系随着周围条件的变化而变化,表现出土的不同物理状态,如干燥与潮湿、密实与松散等等。

图中符号:
m——土的总质量($m=m_s+m_w$)(kg)
m_s——土中固体颗粒的质量(kg)
m_w——土中水的质量(kg)
V——土的总体积($V=V_a+V_s+V_w$)(m³)
V_a——土中空气体积(m³)
V_s——土中固体颗粒体积(m³)
V_w——土中水所占的体积(m³)
V_v——土中孔隙体积($V_v=V_a+V_w$)(m³)

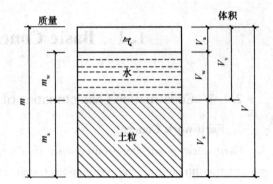

图 2.1 土的三相示意图

2. 土的物理性质

(1)土的天然密度和干密度。

土在天然状态下单位体积的质量,叫土的天然密度(简称密度)。一般黏土的密度为 1 800~2 000 kg/m³,砂土为 1 600~2 000 kg/m³。土的密度按下式计算:

$$\rho = \frac{m}{V} \tag{2.1}$$

干密度是土的固体颗粒质量与总体积的比值,用下式表示:

$$\rho_d = \frac{m_s}{V} \tag{2.2}$$

式中 $\rho、\rho_d$——分别为土的天然密度和干密度;
　　　m——土的总质量(kg);
　　　m_s——土中固体颗粒的质量(kg);
　　　V——土的体积(m³)。

2.1.2 Basic Properties of Soil

1. Soil Composition

Soil generally consists of soil particles (solid phase), water (liquid phase) and air (air phase), as shown in Fig. 2.1. The proportional relationship between the three parts varies with the surrounding conditions, showing the different physical states of the excavations, such as dryness and dampness, compactness and looseness, and so on.

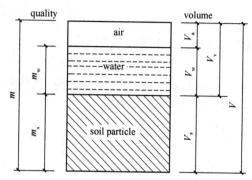

Fig. 2.1 Three-phase Diagram of Soil

Symbols in the figure:

m——Total mass of soil ($m = m_s + m_w$)(kg)
m_s——Mass of solid particles in soil(kg)
m_w——Mass of water in soil(kg)
V——Total volume of soil ($V = V_a + V_s + V_w$)(m³)
V_a——Air volume in soil(m³)
V_s——Volume of solid particles in soil(m³)
V_w——Volume of water in soil(m³)
V_v——Pore volume in soil ($V_v = V_a + V_w$)(m³)

2. Physical Properties of Soil

(1) Natural density and dry density of soil.

The mass per unit volume of a soil in its natural state is called its natural density (shortened as density). Generally, the density of clay is about 1800—2000kg/m³, and sandy soil is about 1600—2000kg/m³. The density of soil is calculated as follows:

$$\rho = \frac{m}{V} \tag{2.1}$$

Dry density is the ratio of the solid particle mass of soil to the total volume, expressed as follows:

$$\rho_d = \frac{m_s}{V} \tag{2.2}$$

In the formula:

ρ、ρ_d——are the natural density and dry density of soil respectively;
m——total mass of soil(kg);
m_s——mass of solid particles in soil(kg);
V——volume of soil(m³).

(2)土的天然含水量。在天然状态下,土中水的质量与固体颗粒质量之比的百分率叫土的天然含水量,反映了土的干湿程度,用下式表示:

$$w = \frac{m_w}{m_s} \times 100\% \tag{2.3}$$

式中 m_w——土中水的质量(kg);
m_s——土中固体颗粒的质量(kg)。

(3)土的孔隙比和孔隙率。孔隙比和孔隙率是土中孔隙的比率,它反映了土的密实程度。孔隙比和孔隙率越小土越密实。

孔隙比
$$e = \frac{V_v}{V_s} \tag{2.4}$$

孔隙率
$$n = \frac{V_v}{V} \times 100\% \tag{2.5}$$

式中 V——土的总体积(m^3),$V = V_s + V_v$;
V_s——土的固体体积(m^3);
V_v——土的孔隙体积(m^3)。

(4)土的可松性。天然土经开挖后,其体积因松散而增加,虽经振动夯实,仍然不能完全复原,这种现象称为土的可松性。土的可松性用可松性系数表示:

土的最初可松性系数
$$K_s = \frac{V_2}{V_1} \tag{2.6}$$

土的最后可松性系数
$$K'_s = \frac{V_3}{V_1} \tag{2.7}$$

式中 K_s、K'_s——土的最初、最后可松性系数;
V_1——土在天然状态下的体积(m^3);
V_2——土挖后松散状态下的体积(m^3);
V_3——土经压(夯)实后的体积(m^3)。

(5)土的透水性。土的透水性是指水流通过土中孔隙的难易程度,用渗透系数 K(单位时间内水穿透土层的能力,单位 m/d)表示。根据土的渗透系数不同,可分为透水性土(如砂土)和不透水性土(如黏土)。土的透水性影响施工降水与排水的速度,一般土的渗透系数见表2.1。

Chapter 2 Earthwork Construction

(2) Natural moisture content of soil. In natural state, the percentage of the mass of water in the soil to the mass of solid particles is called the natural moisture content of the soil, which reflects the dry and wet degree of the soil, and is expressed as follows:

$$w = \frac{m_w}{m_s} \times 100\% \tag{2.3}$$

In the formula:

m_w—the mass of water in soil(kg);

m_s—mass of solid particles in soil(kg).

(3) Void ratio and porosity of soil. The void ratio and porosity are the ratios of pores in the soil, which reflect the degree of compactness of the soil. The smaller the void ratio and porosity is, the denser the soil will be.

Void ratio
$$e = \frac{V_v}{V_s} \tag{2.4}$$

Porosity
$$n = \frac{V_v}{V} \times 100\% \tag{2.5}$$

In the formula:

V—total volume of soil(m³), $V = V_s + V_v$;

V_s—solid volume of soil(m³);

V_v—pore volume of soil(m³).

(4) Looseness of soil. After excavation, the volume of natural soil increases due to its looseness. Although it is compacted by vibration, it cannot be completely restored. This phenomenon is called soil looseness. The looseness of soil is expressed by the looseness coefficient, i. e.

Initial coefficient of looseness of soil
$$K_s = \frac{V_2}{V_1} \tag{2.6}$$

Final coefficient of looseness of soil
$$K'_s = \frac{V_3}{V_1} \tag{2.7}$$

In the formula:

K_s, K'_s—initial and final looseness coefficient of soil;

V_1—the volume of soil in its natural state(m³);

V_2—volume of soil in loose state(m³) after excavation;

V_3—compacted volume of soil(m³).

(5) Water permeability of soil. The water permeability of soil refers to the difficulty of water flowing through the pores in the soil, expressed by the permeability coefficient K(the ability of water penetrating the soil in unit time, unit m/d). Depending on the difference of permeability coefficient, soil can be divided into permeable soil(such as sand) and impervious soil(such as clay). The permeability of soil affects the speed of construction precipitation and drainage, and the permeability coefficient of ordinary soil is shown in Table 2.1.

表 2.1 土的渗透系数参考表

土的名称	渗透系数/(m·d^{-1})	土的名称	渗透系数/(m·d^{-1})
黏土、粉质黏土	<0.1	含黏土的中砂及纯细砂	20~25
粉质砂土	0.1~0.5	含黏土的细砂及纯中砂	35~50
含黏土的粉砂	0.5~1.0	纯粗砂	50~75
纯粉砂	1.5~5.0	粗砂夹卵石	50~100
含黏土的细砂	10~15	卵石	100~200

2.1.3 土的分类与现场鉴别方法

在建筑施工中,根据土的坚硬程度及开挖的难易程度,将土分为松软土、普通土、坚土、砂砾坚土、软石、次坚石、坚石、特坚石等八类。前四类属一般土,后四类属岩石。土的类别对土方工程施工方法的选择、劳动量和机械台班的消耗及工程费用都有较大的影响,应高度重视。土的这种八类分类法及其现场鉴别方法见表 2.2。

表 2.2 土的工程分类与现场鉴别方法

土的分类	土的名称	可松性系数 K_s	可松性系数 K'_s	现场鉴别方法
一类土（松软土）	砂；粉质砂土；冲积砂土层；种植土；泥炭（淤泥）	1.08~1.17	1.01~1.03	能用锹、锄头挖掘
二类土（普通土）	粉质黏土；潮湿的黄土；夹有碎石、卵石的砂；种植土；填筑土及粉质砂土	1.14~1.28	1.02~1.05	用锹、条锄挖掘,少许用镐翻松
三类土（坚土）	软及中等密度黏土；重粉质黏土；粗砾石；干黄土及含碎石、卵石的黄土、粉质黏土；压实的填筑土	1.24~1.30	1.05~1.07	主要用镐,少许用锹、条锄挖掘
四类土（砂砾坚土）	重黏土及含碎石、卵石的黏土；粗卵石；密实的黄土；天然级配砂石；软泥灰岩及蛋白石	1.26~1.35	1.06~1.09	整个用镐、条锄挖掘,少许用撬棍挖掘
五类土（软石）	硬石灰纪黏土；中等密度的页岩、泥灰岩、白垩土；胶结不紧的砾岩；软的石灰岩	1.30~1.40	1.10~1.15	用镐或撬棍、大锤挖掘,部分用爆破方法
六类土（次坚石）	泥岩；砂岩；砾岩；坚实的页岩；泥灰岩；密实的石灰岩；风化花岗岩、片麻岩	1.35~1.45	1.11~1.20	用爆破方法开挖,部分用风镐

Table 2.1 Reference Table of Soil Permeability Coefficient

Name of Soil	Permeability Coefficient/(m·d^{-1})	Name of Soil	Permeability Coefficient/(m·d^{-1})
clay, loam	<0.1	medium sand and pure fine sand with clay	20—25
sandy loam	0.1—0.5	fine sand with clay and pure medium sand	35—50
silt with clay	0.5—1.0	pure coarse sand	50—75
pure silt	1.5—5.0	coarse sand with pebble	50—100
fine sand with clay	10—15	pebble	100—200

2.1.3 Soil Classification and Field Identification Method

In construction, according to the degree of hardness and difficulty of excavation, the soil is divided into eight categories of soft soil, ordinary soil, hard soil, gravel hard soil, soft rock, secondary hard rock, hard rock, and special hard rock. The first four are general soil, and the last four are rocks. The type of soil has great influence on the selection of earthwork construction methods, the amount of labor, the consumption of machinery and engineering cost. The eight classifications and field identification methods of soil are shown in Table 2.2.

Table 2.2 Engineering Classification and Field Identification Method of Soil

Classification of Soil	Name of Soil	Coefficient Looseness		Field Identification Method
		K_s	K_s'	
Type I (soft soil)	sand, sandy silt, alluvial sandy silt, planting soil, peat(silt)	1.08—1.17	1.01—1.03	Can be excavated by spade and hoe
Type II (ordinary soil)	silty clay, moist loess, sand with gravel, planting soil, fill soil and sandy silt	1.14—1.28	1.02—1.05	Can be excavated by spade and mattock, and some of them can be loosen by pick
Type III (hard soil)	soft and medium-density; heavy silty clay, coarse gravel, dry loess and loess with gravel, silty clay, compacted filled soil	1.24—1.30	1.05—1.07	Be excavated by pick, and a few of them by spade and mattock
Type IV (gravel hard soil)	heavy clay and clay containing gravel and pebble, coarse pebble, dense loess, natural grade sand-stone, soft rock and opal	1.26—1.35	1.06—1.09	All are excavated by pick and mattock, and a few of them by crowbar
Type V (soft rock)	hard lime clay, medium density shale, marl, chalky soil, unconsolidated conglomerate, soft limestone	1.30—1.40	1.10—1.15	Excavated by pick or crowbar, or sledgehammer, part of them by blasting
Type VI (secondary hard rock)	mudstone, sandstone, conglomerate, solid shale, marl, density limestone, weathered granite, gneiss	1.35—1.45	1.11—1.20	Excavated by blasting, part of them by compressed air pick

续表

土的分类	土的名称	可松性系数 K_s	可松性系数 K'_s	现场鉴别方法
七类土（坚石）	大理岩；辉绿岩；玢岩；粗、中粒花岗岩；坚实的白云岩、砂岩、砾岩、片麻岩、石灰岩、风化痕迹的安山岩、玄武岩	1.40～1.45	1.15～1.20	用爆破方法开挖
八类土（特坚石）	安山岩；玄武岩；花岗片麻岩；坚实的细粒花岗岩、闪长岩、石英岩、辉长岩、辉绿岩、玢岩	1.45～1.50	1.20～1.30	用爆破方法开挖

2.2 土方量的计算与土方调配

2.2.1 基坑、基槽土方量的计算

1. 基坑土方量的计算

基坑土方量可按立体几何中的拟柱体（由两个平行的平面做底的一种多面体）体积公式计算（图2.2）。即：

$$V = \frac{H}{6}(A_1 + 4A_0 + A_2) \tag{2.8}$$

式中　H——基坑深度(m)；
　　　A_1、A_2——基坑上、下的底面积(m^3)；
　　　A_0——基坑中截面的面积(m^3)。

2. 基槽土方量的计算

基槽和路堤的土方量可以沿长度方向分段后，再用同样方法计算（图2.3）。

$$V_1 = \frac{L_1}{6}(A_1 + 4A_0 + A_2) \tag{2.9}$$

式中　V_1——第一段的土方量(m^3)；
　　　L_1——第一段的长度(m)。

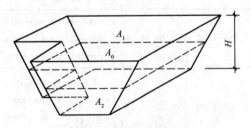

图 2.2　基坑土方量计算

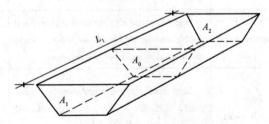

图 2.3　基槽土方量的计算

续表

Classification of Soil	Name of Soil	Coefficient Looseness		Field Identification Method
		K_s	K'_s	
Type Ⅶ (hard rock)	marble; diabase; porphyrite; coarse and medium grained granite; solid dolomite, sandstone, conglomerate, gneiss, limestone, andesite and basalt with weathering trace	1.40—1.45	1.15—1.20	Excavated by blasting
Type Ⅷ (extra hard rock)	andesite; basalt; granite gneiss; solid fine grained granite, diorite, quartzite, gabbro, diabase, porphyrite	1.45—1.50	1.20—1.30	Excavated by blasting

2.2 Calculation of Earthwork Volume and Allocation of Earthwork

2.2.1 Calculation of Earthwork Volume of Foundation Pit and Groove

1. Calculation of Earthwork Quantity of Foundation Pit

The earthwork volume of foundation pit can be calculated according to the volume formula of pseudo column(a polyhedron with two parallel planes as the bottom) in solid geometry(Fig. 2.2). Namely:

$$V = \frac{H}{6}(A_1 + 4A_0 + A_2) \tag{2.8}$$

In the formula:
H—Foundation pit depth(m);
A_1, A_2—bottom area above and below the foundation pit(m³);
A_0—sectional areas of foundation pit(m³).

2. Calculation of Earthwork Volume of Foundation Trench

The earthwork volume of foundation trench and embankment can be segmented along the length direction, and then calculated by the same method(see Fig. 2.3).

$$V_1 = \frac{L_1}{6}(A_1 + 4A_0 + A_2) \tag{2.9}$$

In the formula:
V_1—the earthwork volume in the first section(m³);
L_1—length of the first section(m).

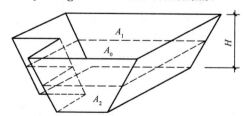

Fig. 2.2 Calculation of Earthwork Volume of Foundation Pit

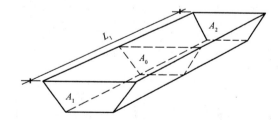

Fig. 2.3 Calculation of Earthwork Volume of Foundation Trench

将各段土方量相加即得总土方量

$$V = V_1 + V_2 + \cdots + V_n \tag{2.10}$$

式中　$V_1、V_2、\cdots V_n$——各分段的土方量(m^3)。

2.2.2　场地平整土方量的计算

场地平整就是将天然地面改造成我们所要求的设计平面。场地设计平面通常由设计单位在总图竖向设计中确定。由设计平面的标高和天然地面的标高之差,可以得到场地各点的填挖高度,由此可计算场地平整的土方量。

场地平整的土方量计算通常采用方格网法。其计算步骤为:

1. 划分方格网

根据已有地形图(一般用 1/500 的地形图),将地面划分成若干个方格网,尽量与测量的纵、横坐标网对应,方格一般采用 20 m×20 m～40 m×40 m。

2. 在方格网上标注填挖高度

(1)场地设计标高值的确定。

在进行场地设计时,设计单位应综合各方面因素确定场地的设计标高,这个设计标高可以是一个固定值,但实际上由于排水的要求,场地表面均应有一定的泄水坡度。因此,应根据场地泄水坡度的要求(单向泄水或双向泄水),计算出场地内各方格角点实际施工时所采用的设计标高。

①单向泄水时,场地各点设计标高的求法。场地用单向泄水时,以设计给定的场地中心线(与排水方向垂直的中心线)标高 H_0 作为原始标高(图 2.4),场地内任意一点的设计标高为:

$$H_n = H_0 \pm l \cdot i \tag{2.11}$$

式中　H_0——场地内设计确定的标高;

　　　l——该点至场地中心线的距离;

　　　i——场地泄水坡度(不小于 2‰)。

例如:图 2.4 中 H_{52} 点的设计标高为:

$$H_{52} = H_0 - l \cdot i = H_0 - 1.5ai$$

②双向泄水时,场地各点设计标高的求法。场地用双向泄水时,设计给定的场地中心点标高 H_0 作为原始标高(图 2.5),场地内任意一点的设计标高为:

Add the earthwork volume of each section to get the total earthwork volume.
$$V = V_1 + V_2 + \cdots + V_n \tag{2.10}$$
In the formula:
$V_1, V_2, \cdots, V_n$—earthwork volume of each section(m^3).

2.2.2 Calculation of Earthwork Volume in Site Leveling

Site leveling is to transform the natural ground into the required design plane. The site design plane is usually determined by the design unit in the vertical design of the general drawing. Based on the difference between the elevation of the design plane and that of the natural ground, the filling and excavation height of each point of the site can be obtained, and the earthwork volume of site leveling can be calculated accordingly.

The square grid method is usually used to calculate the earth volume of site leveling. The calculation steps are as follows:

1. Partition Grid

According to the existing topographic map(generally 1/500 topographic map), the ground is divided into several square grids, which correspond to the vertical and horizontal coordinates of the survey as far as possible. The square grids are generally 20 m×20 m—40 m×40 m.

2. Mark the Filling and Excavation Height on the Grid

(1) Determination of site design elevation value.

In site design, the design unit shall determine the design elevation of the site based on various factors. The design elevation may be a fixed value, but in fact, due to the requirements of drainage, the site surface shall have a certain drainage slope. Therefore, the design elevation used in the actual construction of each grid point on the site should be calculated according to the requirements of the site drainage slope(one-way drainage or two-way drainage).

①Method of calculating the design elevation of each point of the site in the case of one-way water discharge. When one-way drainage on the site, the elevation H_0 of the given site center line (the center line perpendicular to the drainage direction) is used as the original elevation(Fig. 2.4). The design elevation of any point in the site is:
$$H_n = H_0 \pm l \cdot i \tag{2.11}$$
In the formula:

H_0—the elevation determined by the design on the site;

l—the distance from this point to the center line of the site;

i—drainage slope of the site(not less than 2‰).

For example, the design elevation of point H_{52} in Fig. 2.4 is:
$$H_{52} = H_0 - l \cdot i = H_0 - 1.5ai$$

②Method of calculating the design elevation of each point in the site during two-way water discharge. When two-way drainage on the site is adopted, the elevation H_0 of the site center point given by the design is taken as the original elevation(Fig. 2.5). The design elevation of any point on the site is:

$$H_n = H_0 \pm l_x \cdot i_x \pm l_y \cdot i_y \tag{2.12}$$

式中　l_x、l_y——该点对场地中心线 $x-x$、$y-y$ 的距离；

　　　i_x、i_y——$x-x$、$y-y$ 方向的泄水坡度。

例如：图 3.4 中场地内 H_{42} 点的设计标高为：

$$H_{42} = H_0 \pm l_x \cdot i_x \pm l_y \cdot i_y = H_0 - 1.5a \cdot i_x - 0.5a \cdot i_y$$

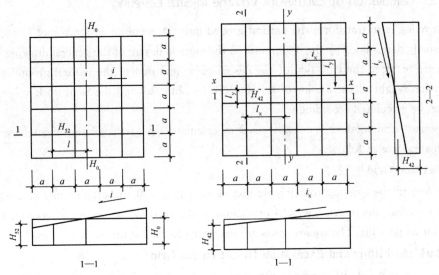

图 2.4　单向泄水坡度的场地　　　图 2.5　双向泄水坡度的场地

(2)场地填挖高度的标注。将设计标高和自然地面标高分别标注在方格点的右上角和右下角。设计地面标高与自然地面标高的差值，即各角点的填挖高度，填在方格网的左上角，挖方为(－)，填方为(＋)。

3. 计算方格网的零点位置

在一个方格网内同时有填方或挖方时，要先算出方格网的零点位置，并标注于方格网上，连接零点就得零线，它是填方区与挖方区的分界线(图 2.6)。零点位置的确定可用计算法(图 2.6)或图解法(图 2.7)。

零点的位置按下式计算：

$$x_1 = \frac{h_1}{h_1+h_2} \times a; \quad x_2 = \frac{h_2}{h_1+h_2} \times a \tag{2.13}$$

式中　x_1、x_2——角点至零点的距离(m)；

　　　h_1、h_2——相邻两角点的施工高度(m)，均用绝对值；

　　　a——方格网的边长(m)。

$$H_n = H_0 \pm l_x \cdot i_x \pm l_y \cdot i_y \qquad (2.12)$$

In the formula:

l_x, l_y—the distance between this point and the center line of the site $x-x, y-y$;

i_x, i_y—$x-x, y-y$ drain slope of y direction.

For example, the design elevation of H_{42} in Fig. 2.5 is:

$$H_{42} = H_0 \pm l_x \cdot i_x \pm l_y \cdot i_y = H_0 - 1.5a \cdot i_x - 0.5a \cdot i_y$$

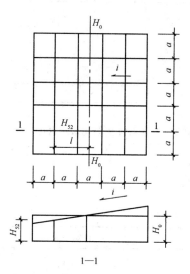

Fig. 2.4 Site of One-way Drainage Slope

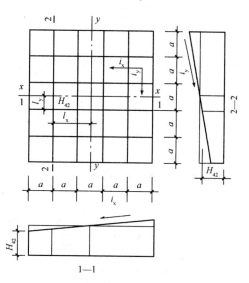

Fig. 2.5 Site of Two-way Drainage Slope

(2) Marking of site filling and excavation height

Mark the design elevation and natural ground elevation at the upper right and lower right corner of the grid point respectively. The difference between the designed ground level and the natural ground level, i. e. , the filling and excavation height of each corner point is filled in the upper left corner of the grid. The excavation is ($-$), the filling is ($+$).

3. Calculating the Zero Point Position of Grid

When there is filling or excavation in a grid at the same time, the zero point of the grid should be calculated and marked on the grid. Connecting the zero point, the zero line is the boundary between the filling area and the excavation area (Fig. 2.6). The zero point position can be determined by calculation method (Fig. 2.6) or graphic method (Fig. 2.7).

The position of the zero point is calculated as follows:

$$x_1 = \frac{h_1}{h_1 + h_2} \times a; \quad x_2 = \frac{h_2}{h_1 + h_2} \times a \qquad (2.13)$$

In the formula:

x_1, x_2—the distance from the corner point to the zero point (m);

h_1, h_2—construction height of two adjacent corner points (m), both in absolute value;

a—length of the grid (m).

在实际工作中,为省略计算,常采用图解法直接求出零点,如图 2.7 所示。方法是用尺子在各角上标出相应比例,再用尺子相连,与方格相交点即为零点位置,非常方便,同时可避免计算或查表出错。

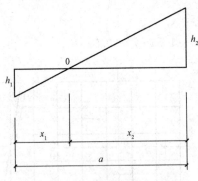

图 2.6 零点位置计算法示意图

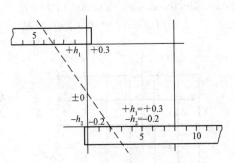

图 2.7 零点位置图解法示意图

4. 计算方格网中各方格的土方量

按照表 2.3 所列公式,分别计算每个方格内的挖方或填方量。

表 2.3 常用方格网计算公式

项目	图式	计算公式
一点填方或挖方 （三角形）		$V = \dfrac{1}{2}bc\dfrac{\sum h}{3} = \dfrac{bch_3}{6}$ 当 $b = c = a$ 时,$V = \dfrac{a^2 h_3}{6}$
二点填方或挖方 （梯形）		$V_+ = \dfrac{b+c}{2}a\dfrac{\sum h}{4} = \dfrac{a}{8}(b+c)(h_1+h_3)$ $V_- = \dfrac{d+e}{2}a\dfrac{\sum h}{4} = \dfrac{a}{8}(d+e)(h_2+h_4)$
三点填方或挖方 （五角形）		$V = \left(a^2 - \dfrac{bc}{2}\right)\dfrac{\sum h}{5} = \left(a^2 - \dfrac{bc}{2}\right)\dfrac{h_1+h_2+h_4}{5}$
四点填方或挖方 （正方形）		$V = \dfrac{a^2}{4}\sum h = \dfrac{a^2}{4}(h_1+h_2+h_3+h_4)$

注:1. a——方格网的边长(m);b、c——零点到一角的边长(m);h_1、h_2、h_3、h_4——方格网四角点的填挖高度(m),用绝对值代入;$\sum h$——填方或挖方高度的总和(m),用绝对值代入;V——挖方或填方体积(m³);

2. 本表公式是按各计算图形底面积乘以平均施工高程而得出的。

In practical work, in order to omit the calculation, the zero point is often obtained directly by graphic method, as shown in Fig. 2.7. The method is to mark the corresponding proportion on each corner with a ruler and connect it with a ruler. The intersection point with the square grid is the position of the zero point, which is very convenient. At the same time, it can avoid mistakes in calculation or table checking.

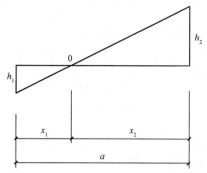

Fig. 2.6 Schematic Diagram of Calculation Method of Zero Point Position

Fig. 2.7 Schematic Diagram of Graphic Method of Zero Point Position

4. Calculating the Earthwork Volume of Each Grid in the Grid

According to the formula listed in Table 2.3, calculate the volume of excavation or filling in each grid.

Table 2.3 Commonly used grid calculation formula

Project	Schematism	Calculation Formula
One point fill or excavation (triangle)		$V = \dfrac{1}{2}bc\dfrac{\sum h}{3} = \dfrac{bch_3}{6}$ 当 $b = c = a$ 时, $V = \dfrac{a^2 h_3}{6}$
Two points fill or excavation (trapezoid)		$V_+ = \dfrac{b+c}{2}a\dfrac{\sum h}{4} = \dfrac{a}{8}(b+c)(h_1 + h_3)$ $V_- = \dfrac{d+e}{2}a\dfrac{\sum h}{4} = \dfrac{a}{8}(d+e)(h_2 + h_4)$
Three points fill or excavation (five-pointed star)		$V = \left(a^2 - \dfrac{bc}{2}\right)\dfrac{\sum h}{5} = \left(a^2 - \dfrac{bc}{2}\right)\dfrac{h_1 + h_2 + h_4}{5}$
Four points fill or excavation (square)		$V = \dfrac{a^2}{4}\sum h = \dfrac{a^2}{4}(h_1 + h_2 + h_3 + h_4)$

Notes: 1. a—side length of grid (m); b, c—side length from zero to one angle (m); h_1, h_2, h_3, h_4—height of fill or excavation in four single points of grid (m), brought in by absolute value; $\sum h$—total of fill or excavation height (m), brought in by absolute value; V—volume of excavation or fill (m³);

2. The formula in this table is obtained by multiplying the bottom area of each calculation figure by the average construction elevation.

5. 计算边坡土方量

图 2.8 是一场地边坡的平面示意图。从图中可看出：边坡的土方量可以划分为两种近似的几何形体进行计算，一种为三角棱锥体（如体积 1~3,5~11），另一种为三角棱柱体（如体积 4）。

(1) 三角棱锥体边坡体积。例如图 2.8 中的①，其体积计算公式为：

$$V_1 = \frac{1}{3} A_1 l_1 \qquad (2.14)$$

式中 l_1——边坡①的长度；

A_1——边坡①的端面积，即：

$$A_1 = \frac{h_2(mh_2)}{2} = \frac{mh_2^2}{2}$$

h_2——角点的挖土高度；

m——边坡①的坡度系数，$m = \dfrac{宽}{高}$。

注：在计算 A_1 时，图 2.8 剖面系近似表示，实际上，地表面不完全是水平的。

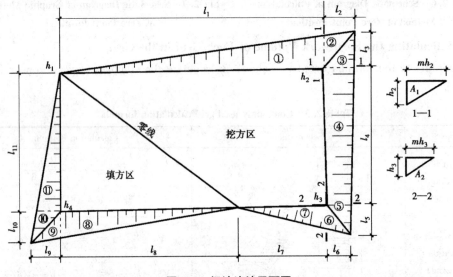

图 2.8　场地边坡平面图

(2) 三角棱柱体边坡体积。例如图 2.8 中的④计算公式如下：

$$V_4 = \frac{A_1 + A_2}{2} l_4 \qquad (2.15)$$

当两端横断面面积相差很大的情况下，则应按下式计算：

$$V_4 = \frac{l_4}{6}(A_1 + 4A_0 + A_2) \qquad (2.16)$$

式中 l_4——边坡④的长度；

A_1、A_2、A_0——边坡④的两端及中部的横截面面积，算法同上。

Chapter 2　Earthwork Construction

5. Calculating Sloping Earthwork

Fig. 2.8 is a schematic plan of a site slope. It can be seen from the figure that the soil volume of the slope can be divided into two approximate geometric shapes for calculation. One is a triangular pyramid(e. g. volume 1—3,5—11), and the other is a triangular prism(e. g. volume 4).

(1) Volume of triangular pyramid slope. For example, in Fig. 2.8①, its volume calculation formula is:

$$V_1 = \frac{1}{3} A_1 l_1 \tag{2.14}$$

In the formula：

l_1— the length of the slope ①;
A_1—the end area of slope ①, i. e.

$$A_1 = \frac{h_2(mh_2)}{2} = \frac{mh_2^2}{2}$$

h_2—the height of the corner;
m—the slope coefficient of slope ①, m＝width/height.

Note：In the calculation of A_1, the section in Fig. 2.8 is approximate. In fact, the ground surface is not completely horizontal.

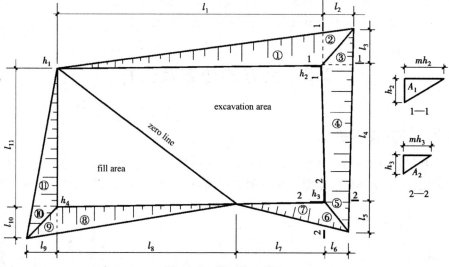

Fig. 2.8　Site Slope Plan

(2) Volume of triangular prism slope. For example, the calculation formula of ④ in Fig. 2.8 is as follows：

$$V_4 = \frac{A_1 + A_2}{2} l_4 \tag{2.15}$$

When the cross-sectional areas at both ends differ greatly, the calculation shall be carried out according to the following formula：

$$V_4 = \frac{l_4}{6}(A_1 + 4A_0 + A_2) \tag{2.16}$$

In the formula：

l_4—Length of the slope ④;
A_1, A_2, A_0—Both ends of the slope ④ and the middle of the cross-sectional area, the algorithm is the same as above.

6. 计算土方总量

将挖方区（或填方区）的所有方格土方量和边坡土方量汇总后可获得场地平整挖（填）方的工程量。

2.2.4 土方调配

土方量计算完成后，即可着手土方的调配工作。土方调配，就是对挖土的利用、土方的堆弃和填土的取得这三者之间的关系进行综合协调的处理。好的土方调配方案，应该使土方运输量或运输费用最小，而且又能方便施工。土方调配应按以下原则进行：

（1）应力求达到挖方与填方基本平衡和就近调配，使挖方量与运距的乘积之和尽可能为最小，即使土方运输量或费用最小；

（2）土方调配应考虑近期施工与后期利用相结合的原则，考虑分区与全场相结合的原则，还应尽可能与大型地下建筑物的施工相结合，以避免重复挖运和场地混乱；

（3）合理布置挖、填方分区线，选择恰当的调配方向、运输线路，使土方机械和运输车辆的性能得到充分发挥；

（4）好土用在回填质量要求高的地区。

总之，进行土方调配，必须根据现场具体情况、有关技术资料、工期要求、土方施工方法与运输方法，综合考虑上述原则，并经计算比较，选择经济合理的调配方案。

2.3 土方的边坡及支护

2.3.1 土方边坡保护

土方开挖过程中及开挖完毕后，基坑（槽）边坡土体由于自重产生的下滑力在土体中产生剪应力，该剪应力主要靠土体的内摩阻力和内聚力平衡，一旦土体中力的体系失去平衡，边坡就会塌方。

6. Calculating the Total Amount of Earthwork

After the earthwork volume of all squares and slopes in the excavation area(or filling area) is summarized, the excavation(filling) volume of site leveling can be obtained.

2.2.4 Earthwork Allocation

After the calculation of earthwork volume is completed, earthwork allocation can be started. Earthwork allocation is a comprehensive and coordinated treatment of the relationship among the utilization of excavated soil, the stacking of earthwork and the acquisition of filled soil. A good earthwork allocation plan should make transport amount of earthwork be smallest or transport cost be lowest, and make construction work be convenient. Earthwork allocation shall be carried out according to the following principles:

(1) Trying to reach the basic balance between the excavation and filling and the nearest allocation, and making the sum of the product of the excavation amount and the transportation distance as short as possible, even if the transport volume is the smallest or the transport cost is the lowest.

(2) Earthwork allocation should consider the combination principles of the recent construction and late utilization, the combination principle of section and the whole site, and combination with the construction of large underground buildings as far as possible to avoid repeated excavation and site disorder.

(3) Reasonably arrange section line of excavation and fill, choose the appropriate allocation direction and transportation lines, so that the earth moving machinery and transport vehicles can get into full play.

(4) Good soil is used in areas with high backfill quality requirements.

In a word, the allocation of earthwork must be based on the specific situation of the site, relevant technical data, construction period requirements, earthwork construction method and transportation method, comprehensive consideration of the above principles, and through calculation and comparison, the selection of economic and reasonable allocation scheme.

2.3 Side Slope and Support of Earthwork

2.3.1 Earthwork Slope Protection

During and after the earth excavation, the sliding force caused by the self weight of the foundation pit(trench) slope produces shear stress in the soil. The shear stress is mainly balanced by the internal friction resistance and cohesive force of the soil. Once the force system in the soil is out of balance, the slope will collapse.

造成边坡塌方的原因有两方面。其一是土体剪应力增加。如坡顶堆物、行车等荷载；基坑边坡太陡；开挖深度较大；雨水或地面水渗入土中，使土的含水量增加从而使土的自重增加；地下水的渗流产生一定的动水压力；土体竖向裂缝中的积水产生侧向静水压力等。其二是土的抗剪强度（土体的内摩阻力和内聚力）降低。如本身土质较差或因气候影响使土质变软；土体内含水量增加而产生润滑作用；饱和的细砂、粉砂受振动而液化等。

为了防止塌方保证施工安全，在基坑（槽）开挖深度超过一定限度时，土壁应做成有斜率的边坡以减少土体自重 P，或者增加临时的土壁支撑形成外加压力来保持土体的稳定（图 2.9）。

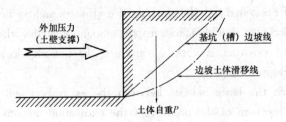

图 2.9　防止边坡土体塌方原理示意图

土方边坡的坡度以土方挖方深度 H 与底宽 B 之比表示。即：

$$土方边坡坡度 = \frac{H}{B} = \frac{1}{m} （图 2.10）$$

式中 $m = \dfrac{B}{H}$ 称为坡度系数

土方边坡的大小主要与土质、开挖深度、开挖方法、边坡留置时间的长短、边坡附近的各种荷载状况及排水情况有关。根据施工经验，当地质条件良好，土质均匀且地下水位低于基坑（槽）或管沟底面标高时，挖方边坡可作成直立壁不加支撑，但深度不宜超过下列规定：

密实、中密的砂土和碎石类土（充填物为砂土）——1.0 m；
硬塑、可塑的粉土及粉质黏土——1.25 m；
硬塑、可塑的黏土和碎石类土（充填物为黏性土）——1.5 m；
坚硬的黏土——2 m。

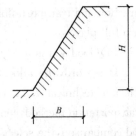

图 2.10　边坡坡度示意图

Chapter 2 Earthwork Construction

There are two reasons for slope collapse. One is the increase of soil shear stress, such as stocking on top of the slope and traffic load, too steep of the foundation pit, greater depth in excavation, increased water content of the soil causing the increase of its own weight because of rainwater or surface water infiltrating into the soil, hydrodynamic pressure produced by seeping of groundwater, accumulated water in the vertical crack of soil producing lateral hydrostatic pressure and so on. The other is the decrease of the shear strength of soil (internal friction resistance and cohesion), such as the poor quality of the soil itself or the soil getting soft because of the influence of climate, lubrication effect caused by the increase of water content in soil, and liquefaction of saturated fine sand and silt because of vibration, etc.

In order to prevent the collapse and ensure the safety of construction, when the excavation depth of foundation pit(groove) exceeds a certain limit, the soil wall should be made into a sloping side slope to reduce the dead weight of soil mass P, or temporary soil wall support should be used to maintain the stability of soil mass with external pressure(Fig. 2. 9).

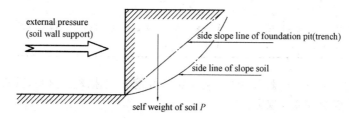

Fig. 2. 9 Schematic Diagram of Preventing Side Slope Earth Collapse

The gradient of side slope is expressed as the ratio of earthwork excavation depth H to bottom width B. That is:

The gradient of side slope $=H/B=1/m$. (Fig. 2. 10).

In the formula:

$m=B/H$ is gradient coefficient.

The size of the earthwork slope is mainly related to the soil quality, excavation depth, excavation method, length of retaining time of the slope, various load conditions near the slope and drainage conditions. According to the construction experience, when the geological conditions are good, the soil quality is uniform and the underground water level is lower than the foundation pit(groove) or the bottom elevation of the pipe trench, the excavation slope can be made into an upright wall without support, but the depth should not exceed the following stipulations.

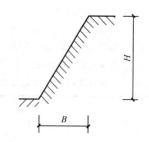

Fig. 2. 10 Schematic Diagram of Side Slope Gradient

Dense, medium-dense sand and gravel soil(the filling material is sand)—1. 0m;

Hard and malleable silt and silty clay—1. 25m;

Hard plastic, plastic clay and gravel soil(filling material is clay)—1. 5m;

Hard clay—2m.

挖方深度超过上述规定时,应考虑放坡或做成直立壁加支撑。

当地质条件良好,土质均匀且地下水位低于基坑(槽)或管沟底面标高时,挖方深度在5 m以内,根据施工经验,不加支撑的边坡的最陡坡度应符合表2.4规定。

表2.4 挖方深度在5 m以内不加支撑的边坡的最陡坡度

土的类别	边坡坡度(高∶宽)		
	坡顶无荷载	坡顶有静载	坡顶有动载
中密的砂土	1∶1.00	1∶1.25	1∶1.50
中密的碎石类土(充填物为砂土)	1∶0.75	1∶1.00	1∶1.25
硬塑的粉土	1∶0.67	1∶0.75	1∶1.00
中密的碎石类土(充填物为黏土)	1∶0.50	1∶0.67	1∶0.75
硬塑的粉质黏土、黏土	1∶0.33	1∶0.50	1∶0.67
老黄土	1∶0.1	1∶0.25	1∶0.33
软土(经井点降水后)	1∶1.00	—	—

注:静载指堆土或材料等;动载指机械挖土或汽车运输作业等。静载或动载距挖方边缘的距离应保证边坡和直立壁的稳定;堆土或材料应距挖方边缘0.8 m以外,高度不超过1.5 m。

永久性挖方边坡应按设计要求放坡。对于临时性挖方,根据现行规范,其边坡的挖方深度及边坡的最陡坡度应符合表2.5规定。

表2.5 临时性挖方边坡值

土的类别		边坡值(高∶宽)
砂土(不包括细砂、粉砂)		1∶1.25～1∶1.50
一般性黏土	硬	1∶0.75～1∶1.00
	硬、塑	1∶1.00～1∶1.25
	软	1∶1.50 或更缓
碎石类土	充填坚硬、硬塑黏土	1∶0.50～1∶1.00
	充填砂土	1∶1.00～1∶1.50

注:1. 设计有要求时,应符合设计要求;
  2. 如采用降水措施或其他加固措施,可不受本表限制,但应计算复核;
  3. 开挖深度,对软土不应超过4 m,对硬土不应超过8 m。

When the excavation depth exceeds the above stipulations, slope or vertical wall with support shall be considered.

When the geological conditions are good, the soil quality is uniform and the underground water level is lower than the bottom elevation of the foundation pit(groove) or pipe trench, the excavation depth should be within 5m. According to the construction experience, the steepest gradient of side slope without support should meet the requirements in Table 2.4.

Table 2.4 The Steepest Gradient of Side Slope without Support within 5m Excavation Depth

Types of Soil	Gradient of Side Slope(depth : width)		
	Without Load on Top of the Slope	With Dead Load on Top of the Slope	With Live Load on Top of the Slope
Medium-dense sand	1 : 1.00	1 : 1.25	1 : 1.50
Medium-dense gravel(filling material is sand)	1 : 0.75	1 : 1.00	1 : 1.25
Hard plastic silt	1 : 0.67	1 : 0.75	1 : 1.00
Medium-dense gravel(filling material is clay)	1 : 0.50	1 : 0.67	1 : 0.75
Hard plastic silt clay, clay	1 : 0.33	1 : 0.50	1 : 0.67
Old loess	1 : 0.1	1 : 0.25	1 : 0.33
Soft soil(after dewatering)	1 : 1.00	—	—

Note: Dead load refers to stocking or materials. Live load refers to mechanical excavation or vehicle transportation. The distance of dead load or live load to the side of the excavation should ensure the stability of side slope and vertical wall. Stocking or materials should keep 0.8m away from the side of excavation, and the hight should not exceed 1.5m.

The permanent excavated slope shall be sloped according to the design requirements. For temporary excavation, the excavation depth and steepest gradient of the side slope shall comply with the stipulations of Table 2.5 according to the current specifications.

Table 2.5 Side Slope Value of Temporary Excavation

Types of Soil		Slope Value(height : width)
Sand soil(excluding fine sand and silt)		1 : 1.25—1 : 1.50
Ordinary clay	hard	1 : 0.75—1 : 1.00
	Hard, plastic	1 : 1.00—1 : 1.25
	soft	1 : 1.50 or softer
Soil of gravel types	Filling with hard and hard-plastic soil	1 : 0.50—1 : 1.00
	Filling with sandy soil	1 : 1.00—1 : 1.50

Note: 1. Design requirement should be followed when required;
2. If dewatering measure or other reinforcing measures are adopted, it will not restricted by this table but it should be calculated and reviewed;
3. As for the depth of excavation, soft soil should not exceed 4m and hard soil should not exceed 8m.

2.3.2 土壁支撑

开挖基坑(槽)时,如地质和周围条件允许,可放坡开挖。但在建筑稠密地区施工或基坑深度较大时,无法按放坡的宽度要求开挖;或者施工时有防止地下水渗入基坑要求,这时就需要用土壁支撑支撑土体,以保证施工的顺利和安全,并减少对相邻已有建筑物等的不利影响。土壁支撑的种类甚多,如用于较窄沟槽的横撑式支撑;用于深基坑的支护结构:板桩、灌注桩、深层搅拌桩、地下连续墙等。横撑式支撑根据挡土板的不同,分为水平挡土板[图 2.11(a)]和垂直挡土板[图 2.11(b)]两类,前者挡土板的布置又分断续式和连续式两种。湿度小的黏性土挖土深度小于 3 m 时,可用断续式水平挡土板支撑;松散、湿度大的土可用连续式水平挡土板支撑;挖土深度可达 5 m。对松散和湿度很高的土可用垂直挡土板式支撑,挖土深度不限。

采用横撑式支撑时,应随挖随撑,支撑牢固。施工中应经常检查,如有松动、变形等现象时,应及时加固或更换。支撑的拆除应按回填顺序依次进行,多层支撑应自下而上逐层拆除,随拆随填。

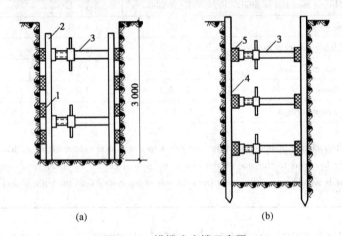

图 2.11 横撑式支撑示意图
(a)断续式水平挡土板支撑;(b)垂直挡土板支撑
1—水平挡土板;2—竖棱木;3—工具式横撑;
4—竖直挡土板;5—横棱木

2.3.2 Soil Wall Support

When excavating foundation pits(trenches), slope excavation may be carried out if geological and surrounding conditions are allowed. However, it is impossible to excavate in accordance with the required slope width in densely built areas or when the foundation pit is deep, or when the construction is required to prevent groundwater seepage into the foundation pit, then it is necessary to support the soil with soil wall to ensure the smooth and safe construction, and reduce the adverse impact on the adjacent existing buildings. There are many kinds of soil wall supports, such as transverse support for narrow trenches, supporting structure for deep foundation pit, sheet pile, cast-in-place pile, deep mixing pile, and underground continuous wall, etc. According to the different types of retaining plate, transverse bracing can be divided into two types: horizontal retaining plate [Fig. 2. 11(a)] and vertical retaining plate [Fig. 2. 11(b)]. The arrangement of the former retaining plate can be divided into intermittent type and continuous type. When the excavation depth of clay soil with low humidity is less than 3m, it can be supported by intermittent horizontal retaining plate. Loose and humid soil can be supported by continuous horizontal retaining plate, and the soil depth can reach 5m. For loose and humid soil, vertical retaining plate can be used to support the excavation depth which is unlimited.

When using the transverse brace type support, it should be digging along with the brace, and the support should be firm. Construction should be checked very often. If loose, deformation and other phenomena are found. They should be timely reinforced or replaced. The removal of support should be carried out in sequence according to the backfill sequence, and the multi-layer support should be removed layer by layer from bottom to top, removing while backfilling.

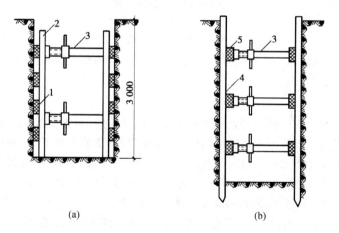

Fig. 2. 11 Schematic Diagram of Transverse Brace Support
(a)intermittent horizontal retaining plate; (b)vertical retaining plate suppors
1—horizontal retaining plate; 2—vertical joist; 3—tool-type transverse brace;
4—vertical retaining plate; 5—horizontal joist

2.4 降、排水施工

2.4.1 常见控制地下水的方法

在开挖基坑、地槽、管沟或其他土方时,土的含水层常被切断,地下水将会不断地渗入坑内。雨季施工时,地面水也会流入坑内。为了保证施工的正常进行,防止边坡塌方和地基被水浸泡导致承载能力的下降,必须做好基坑降水工作。降水方法可分集水井降水和井点降水两类。

1. 集水井降水法

这种方法是在基坑或沟槽开挖时,在坑底设置集水井,沿坑底的周围或中央开挖排水沟,使水由排水沟流入集水井内,然后用水泵抽出坑外(图 2.12)。

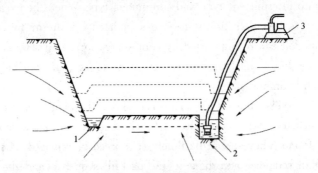

图 2.12 集水井降水示意图
1—排水沟;2—集水坑;3—水泵

从基坑中直接抽出地下水的方法比较简单,应用也较广,但若土质为细砂或粉砂,地下水渗出时会产生流砂现象,使边坡塌方,坑底冒砂,工作条件恶化,并有引起附近建筑物下沉的危险,此时常用井点降水的方法进行施工。

2. 井点降水法

(1)井点降水的概念与作用。井点降水就是在基坑开挖前,预先在基坑四周埋设一定数量的滤水管(井),在基坑开挖前和开挖过程中,利用真空原理,不断抽出地下水,使地下水位降低到坑底以下,从根本上解决地下水涌入坑内的问题[图 2.13(a)];防止边坡由于受地下水流的冲刷而引起的塌方[图 2.13(b)];使坑底的土层消除了地下水位差引起的压力,因此防止了坑底土的上冒[图 2.13(c)];由于没有水压力,使板桩减少了横向荷载[图 2.13(d)];由于没有地下水的渗流,也就消除了流砂现象[图 2.13(e)]。降低地下水位后,由于土体固结,还能使土层密实,增加地基土的承载能力。其中,防治流砂现象是井点降水的主要目的。

Chapter 2 Earthwork Construction

2.4 Dewatering and Drainage Construction

2.4.1 Common Methods of Controlling Groundwater

In the excavation of foundation pits, trenches or other earthworks, the aquifer of the soil is often cut off, and the groundwater will continue to seep into the pit. During the rainy season, the ground water will also flow into the pit. In order to ensure the normal operation of construction and prevent the decline of bearing capacity caused by slope collapse and foundation soaked by water, it is necessary to do a good job in foundation pit dewatering. The dewatering method can be divided into collective well dewatering and well point dewatering.

1. Collective Well Dewatering Method

In this method, when the foundation pit or trench is excavated, a collective well is set at the bottom of the pit, and a drainage ditch is dug around or in the center of the pit, so that the water flows into the collective well from the drainage ditch, and then the water is pumped out of the pit(Fig. 2.12).

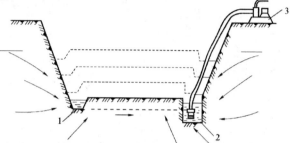

Fig. 2.12 Schematic Diagram of Collective Well Dewatering
1—drainage ditch; 2—catch pit; 3—pump

The method of pumping ground water directly from foundation pit is simple and widely used. However, if the soil is fine sand or silt, the phenomenon of flowing sand will occur when the groundwater seeps out, causing slope collapse, sand arisen in pit bottom, poor working condition, and the danger of the nearby buildings sunk down. In this case, well point dewatering is often used in the construction.

2. Well Point Dewatering Method

(1) Concept and Function of Well Point Dewatering. Well point dewatering is that before excavating foundation pit, a certain number of filter tubes(well) will be buried around the foundation pit in advance. Before or during the excavation of foundation pit, the vacuum principle is used constantly to pump underground water, which will lower the underground water level below the bottom of the pit, and fundamentally solve the problem of groundwater flowing into the pit [Fig. 2.13(a)], to prevent the landslide of the slope caused by scouring of underground water [Fig. 2.13(b)]; and to make the bottom soil layer eliminate the pressure caused by of the ground water level difference, so as to prevent soil in the pit bottom uplifting [Fig. 2.13(c)]. Since there is no water pressure, lateral load of the plate pile is reduced [Fig. 2.13(d)]. As there is no leakage of underground water, the phenomenon of flowing sand is also eliminated [Fig. 2.13(e)]. After lowering the groundwater level, the density of soil layer can be achieved and the bearing capacity of the foundation soil can be increased because of the consolidation of the soil. Among them, preventing the phenomenon of flowing sand is the main purpose of well point dewatering.

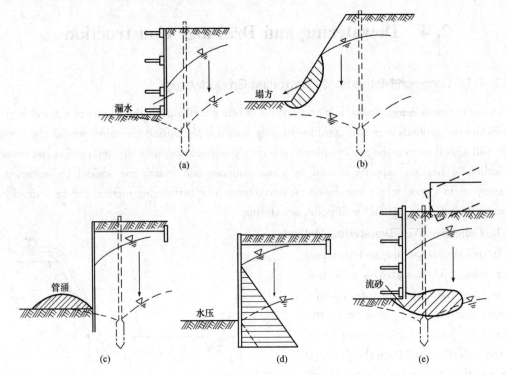

图 2.13 井点降水的概念与作用
(a)防止涌水；(b)使边坡稳定；(c)防止土体上冒；
(d)减小横向荷载；(e)防止流砂

(2)流砂现象产生的原因。

图 2.14 所示的试验说明：由于高水位的左端(水头为 h_1)与低水位的右端(水头为 h_2)之间存在压力差，水经过长度为 l，断面积为 F 的土体由左端向右端渗流[图 2.14(a)]。

土颗粒处于悬浮状态，土的抗剪强度等于零，土颗粒能随着渗流的水一起流动，这种现象就叫"流砂现象"。

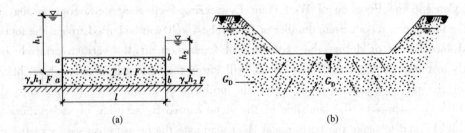

图 2.14 动水压力原理图
(a)水在土中渗流时的力学现象；(b)动水压力对地基土的影响

Chapter 2　Earthwork Construction

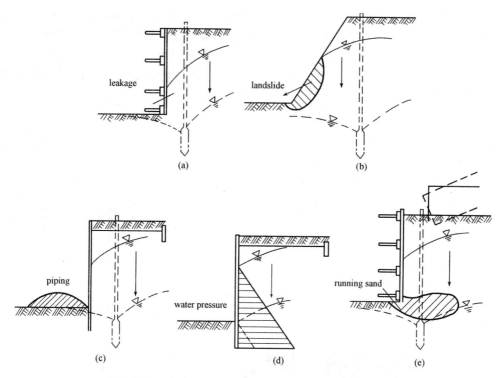

Fig. 2.13　Concept and Function of Well Point Dewatering
(a) prevent water flow; (b) make the side slope stable; (c) prevent soil from rising;
(d) reduce the lateral load; (e) prevent running sand

(2) Reason of running sand phenomenon

The test shown in Fig. 2.14 shows that due to the pressure difference between the left end of the high water level (head h_1) and the right end of the low water level (head h_2), water flows from the left end to the right end through the soil with length l and sectional area F [Fig. 2.14(a)].

When the soil particles are suspended, the shear strength of the soil is equal to zero, and the soil particles can flow with the seepage water. This phenomenon is called running sand phenomenon.

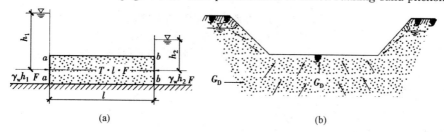

Fig. 2.14　Schematic Diagram of Dynamic Water Pressure
(a) mechanical phenomenon of water seepage in soil;
(b) influence of hydrodynamic pressure on foundation soil

· 45 ·

(3)管涌现象产生的原因。当基坑坑底位于不透水土层内,而不透水土层下面为承压蓄水层,坑底不透水层的覆盖厚度的重量小于承压水的顶托力时,基坑底部即可能发生涌冒现象。

(4)井点降水的方法。井点降水有两类:一类为轻型井点(包括一般轻型井点、电渗井点与喷射井点),一类为管井井点(包括深井泵)。各种井点降水方法一般根据土的渗透系数、降水深度、设备条件及经济性选用(表2.6)。以下以轻型井点为例做重点阐述。

表 2.6 各种井点的适用范围

井点类型		土层渗透系数/(m·d^{-1})	降低水位深度/m
轻型井点	一级轻型井点	0.1~50	3~6
	二级轻型井点	0.1~50	6~12
	喷射井点	0.1~5	8~20
	电渗井点	<0.1	根据选用的井点确定
管井类	管井井点	20~200	3~5
	深井井点	10~250	>15

①一般轻型井点设备。轻型井点设备由管路系统和抽水设备组成(图2.15)。

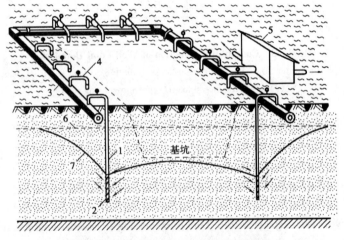

图 2.15 轻型井点降低地下水位示意图
1—井点管;2—滤管;3—总管;4—弯联管;
5—水泵房;6—原有地下水位线;7—降低后地下水位线

(3) Reason of piping phenomenon. When bottom of foundation pit is within non-permeable layer, further, under which is bearing storage layer, and the weight of cover thickness at the pit bottom of non-permeable layer is smaller than the top supporting force of bearing water, piping phenomenon may occur at the bottom of foundation pit.

(4) Well point dewatering method. There are two types of dewatering methods: one is light well point(including general light well point, electro percolation well point and injection well point). The other is the pipe well point (including the deep well pump). All well point dewatering methods are generally selected according to soil permeability coefficient, dewatering depth, equipment conditions and economy(Table 2.6). The following light wells are taken as an example to focus on.

Table 2.6 Scope of Application of Different Well Points

Types of Well Point		Coefficient of Soil Permeability/(m·d^{-1})	Depth of Reducing Water Level/m
Light well point	Type I light well point	0.1—50	3—6
	Type II light well point	0.1—50	6—12
	Injection well point	0.1—5	8—20
	Electro percolation well point	<0.1	Determined according to selected well point
Types of pipe well	Tubular well point	20—200	3—5
	Deep well point	10—250	>15

②General light well point. Equipment: Light well point equipment consists of pipe system and pump system(Fig. 2.15).

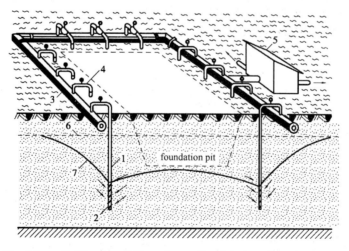

Fig. 2.15 Schematic Diagram of Groundwater Level Reduction by Light Well Points

1—well point pipe; 2—filter pipe; 3—main pipe; 4—bending connection;
5—punping house; 6—original ground water level; 7—reduced ground water level

管路系统包括：滤管、井点管、弯联管及总管等。

滤管(图 2.16)为进水设备，通常采用长 1.0～1.2 m，直径 38 mm 或 51 mm 的无缝钢管，管壁钻有直径为 12～19 mm 的呈星棋状排列的滤孔，滤孔面积为滤管表面积的 20%～25%。骨架管外面包以两层孔径不同的铜丝布或塑料布滤网。滤管上端与井点管连接。井点管为直径 38 mm 或 51 mm、长 5～7 m 的钢管，可整根或分节组成。井点管的上端用弯联管与总管相连。集水总管为直径 100～127 mm 的无缝钢管，每段长 4 m，其上装有与井点管联结的短接头，间距 0.8 m 或 1.2 m。

抽水设备是由真空泵、离心泵和水气分离器(又叫集水箱)等组成，其工作原理如图 2.17 所示。抽水时先开动真空泵 19，将水气分离器 10 内部抽成一定程度的真空，使土中的水分和空气受真空吸力作用而吸出，经管路系统，再经过滤箱 8(防止水流中的细砂进入离心泵引起磨损)进入水气分离器 10。水气分离器内有一浮筒 11，能沿中间导杆升降。当进入水气分离器内的水多起来时，浮筒即上升，此时即可开动离心泵 24，将在水气分离器内的水和空气向两个方向排去，水经离心泵排出，空气集中在上部由真空泵排出。为防止水进入真空泵(因为真空泵为干式)，水气分离器顶装有阀门 12，并在真空泵与进气管之间装一副水气分离器 16。为对真空泵进行冷却，特设一个冷却循环水泵 23。

一套抽水设备的负荷长度(即集水总管长度)，采用 W5 型真空泵时，不大于 100 m；采用 W6 型真空泵时，不大于 120 m。

②轻型井点的布置。井点系统的布置，包括平面布置与高程布置，应根据基坑大小与深度、土质、地下水位高低与流向、降水深度要求等而确定。

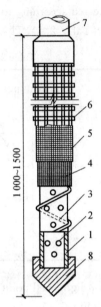

图 2.16　滤管构造

1—钢管；2—管壁上的小孔；3—缠绕的塑料管；
4—细滤网；5—粗滤网；6—粗铁丝保护网；
7—井点管；8—铸铁头

Pipeline system includes: filter pipe, well point pipe, bend pipe and main pipe.

The filter tube (Fig. 2.16) is the inlet device, usually 1.0—1.2m seamless steel tube with a diameter of 38mm or 51mm. The pipe surface has been drilled star-shaped filter holes with diameter of 12—19mm, and the area is 20%—25% of the surface area of the filter tube. The skeleton tube is coated with two layers of copper wire cloth or plastic cloth with different pore sizes. The upper end of the filter pipe is connected with the well point pipe. The well point pipe is a steel pipe with a diameter of 38mm or 51mm and a length of 5—7m. The upper end of the wellpoint pipe is connected to the main pipe by a bent union. The collecting water main is a seamless steel tube with a diameter of 100—127mm, each section of which is 4m long, and a short connector connected with the well point pipe with a spacing of 0.8m or 1.2m.

The pumping equipment is composed of a vacuum pump, a centrifugal pump and a water vapor separator (also known as a collecting water tank), etc. Its working principle is shown in Fig. 2.17. When pumping water, first start the vacuum pump 19, then pump the internal of water vapor separator 10 to a certain degree of vacuum, so that the soil moisture and air are sucked out by vacuum suction, which enter into the water vapor separator 10 through the pipeline system and the filter tank 8 (to prevent fine sand flowing into the centrifugal pump which may cause wear). There is a float 11 in the water vapor separator, which can be lifted and lowered along the middle guide rod. When there is more water in the water-air separator, the float will rise and the centrifugal pump will be started at this time. The water and air in the water-air separator will be discharged in two directions. The water will be discharged by the centrifugal pump and the air will be concentrated in the upper part and discharged by the vacuum pump. To prevent water from entering the vacuum pump (because the vacuum pump is dry), valve 12 is installed at the top of the water-air separator, and a water-air separator 16 is installed between the vacuum pump and the intake pipe. For the cooling of the vacuum pump, a cooling circulation pump 23 is specially designed.

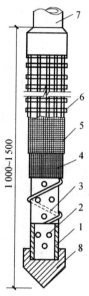

Fig. 2.16 Filter Pipe Structure
1—steel pipe; 2—samll holes on pipe surface;
3—rapped plastic pipe; 4—fine filter net;
5—coarse filter net; 6—heavy iron wire net;
7—well point pipe; 8—cast iron head

The load length of a set of pumping equipment (i.e., the length of the collecting water main) shall not be greater than 100m when 5W vacuum pump is adopted, and shall not be greater than 120m when 6W vacuum pump is used.

②Layout of light well point. Well point system layout, including plane layout and elevation layout, should be determined according to the size and depth of foundation pit, soil quality, level and flow direction of underground water, requirements of dewatering depth, etc.

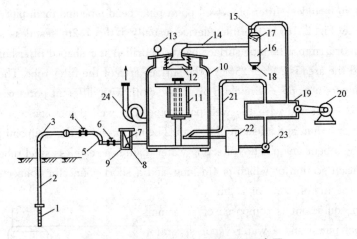

图 2.17 轻型井点设备工作原理示意图
1—滤管；2—井点管；3—弯管；4—阀门；5—集水总管；6—闸门；
7—滤管；8—过滤箱；9—淘沙孔；10—水气分离器；11—浮筒；
12—阀门；13—真空计；14—进水管；15—真空计；16—副水气分离器；
17—挡水板；18—放水口；19—真空泵；20—电动机；21—冷却水管；
22—冷却水箱；23—循环水泵；24—离心水泵

平面布置：当基坑或沟槽宽度小于 6 m，且降水深度不超过 5 m 时，可用单排线状井点，布置在地下水流的上游一侧，两端延伸长度以不小于槽宽为宜(图 2.18)；如宽度大于 6 m 或土质不良，则用双排线状井点(图 2.19)；面积较大的基坑宜用环状井点(图 2.20)。有时亦可布置成 U 形，以利挖土机和运土车辆出入基坑。井点管距离基坑壁一般可取 0.7~1.0 m，以防局部发生漏气。井点管间距一般为 0.8 m、1.2 m、1.6 m，由计算或经验确定。井点管在总管四角部位应适当加密。

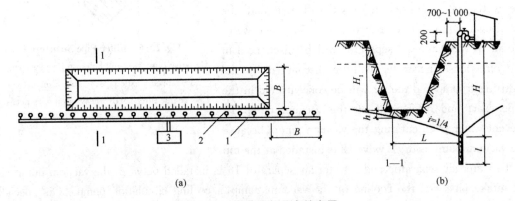

图 2.18 单排线状井点的布置
(a)平面布置；(b)高程布置
1—总管；2—井点管；3—抽水设备

Chapter 2   Earthwork Construction

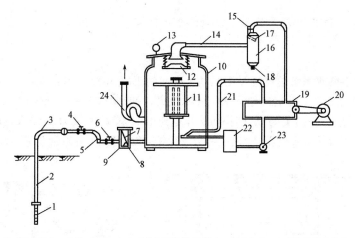

Fig. 2.17   Schematic Diagram of Light Well Point Equipment Working Principle
1—filter pipe; 2—well point pipe; 3—bend pipe; 4—valve; 5—collecting water main; 6—lock-gate hatch;
7—filter pipe; 8—filter tank; 9—sand washing hole; 10—water-air separator; 11—float; 12—valve;
13—vacuum gauge; 14—inlet pipe; 15—vacuum gauge; 16—auxiliary water-air separator;
17—eliminator; 18—dewatering outlet; 19—vacuum pump; 20—electric motor;
21—cooling water pipe; 22—cooling water tank; 23—circulating pump; 24—centrifugal pump

Plane Layout: When the width of foundation pit or trench is less than 6m, and the dewatering depth is no more than 5m, a single row of linear parallel points can be used to arrange on the upstream side of the groundwater flow, and the extension length at both ends should be no less than the slot width(Fig. 2.18). If the width is greater than 6m or the soil quality is poor, double-row linear well points are used(Fig. 2.19). Annular well points are recommended for foundation pits with large area(Fig. 2.20). Sometimes it can be arranged in a u-shape to facilitate the excavators and earth moving vehicles in and out of the foundation pit. The distance between the well point pipe and the foundation pit wall can be 0.7 − 1.0m to prevent local air leakage. Well point pipe spacing is generally 0.8m, 1.2m, 1.6m, determined by calculation or experience. The well point pipe shall be properly encrypted at the four corners of the main pipe.

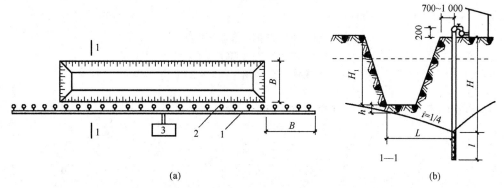

Fig. 2.18   Layout of Single Row Linear Well Point
(a) plane layout; (b) elevation layout
1—main pipe; 2—well point pipe; 3—pumping equipment

· 51 ·

高程布置:井点降水深度,考虑抽水设备的水头损失以后,一般不超过 6 m。井点管埋设深度按下式计算:

$$H \geqslant H_1 + h + iL \tag{2.17}$$

式中 H_1——井点降水系统总管埋设面至基坑底面的距离(m);

h——基坑底面至降低后的地下水位线的距离,一般取 0.5~1.0 m;

i——水力坡度,根据实测:双排和环状井点为 1/10,单排井点为 1/4~1/5;

L——双排井点为井点管至基坑中心的水平距离,单排井点为井点管至基坑另一边的距离(m)。

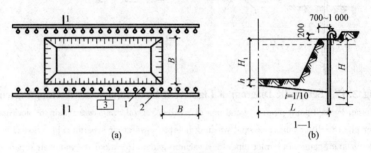

图 2.19 双排线状井点布置图
(a)平面布置;(b)高程布置
1—井点管;2—总管;3—抽水设备

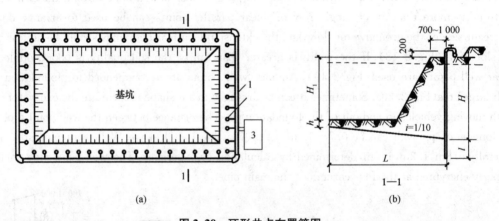

图 2.20 环形井点布置简图
(a)平面布置;(b)高程布置
1—总管,2—井点管;3—抽水设备

此外,在确定井点管埋深时,还要考虑井点管一般要露出地面 0.2 m 左右。

Elevation Layout: Well point dewatering depth is generally not exceeding 6m after considering the head loss of pumping equipment. The buried depth of well point pipe is calculated as follows:

$$H \geqslant H_1 + h + iL \qquad (2.17)$$

In the formula:

H_1—The distance from the buried surface of the well point dewatering system main pipe to the bottom of the foundation pit(m);

H—The distance from the foundation pit bottom to the lowered underground water level, generally 0.5—1.0m;

i—According to the measured hydraulic gradient, the double-row and annular well points are 1/10 and the single-row well points are 1/4 to 1/5.

L—The double-row well point is the horizontal distance from the well point pipe to the center of the foundation pit, while the single-row well point is the distance from the well point pipe to the other side of the foundation pit(m).

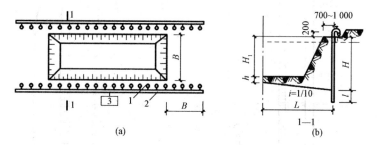

Fig. 2.19 Layout of Double Row Linear Well Point

(a) plane layout; (b) elevation layout

1—main pipe; 2—well point pipe; 3—pumping equipment

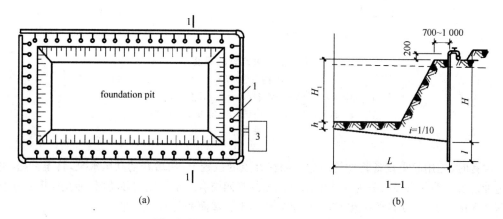

Fig. 2.20 Layout of Annular Well Point

(a) plane layout; (b) elevation layout

1—main pipe; 2—well point pipe; 3—pumping equipment

In addition, when determining the buried depth of well point pipe, it is also necessary to consider that the well point pipe is generally about 0.2m above the ground.

根据式(2-17)算出的 H 值,如大于 6 m,则应降低井点管抽水设备的埋置面,以适应降水深度要求。即将井点系统的埋置面(布置标高)接近原有地下水位线(要事先挖槽),个别情况下甚至稍低于地下水位(当上层土的土质较好时,先用集水井排水法挖去一层土,再布置井点系统),就能充分利用抽吸能力,使降水深度增加。井点管露出地面的长度一般为 0.2 m。当一级井点系统达不到降水深度要求时,可采用二级井点,即先挖去第一级井点所疏干的土,然后再在其底部装设第二级井点。

③井点管埋设。一般用水冲法,分为冲孔[图 2.21(a)]与埋管[图 2.21(b)]两个过程。

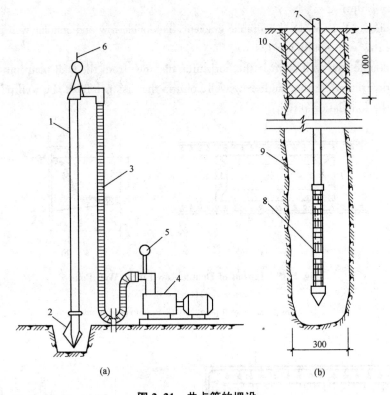

图 2.21　井点管的埋设
(a)冲孔;(b)埋管
1—冲管;2—冲嘴;3—胶皮管;4—高压水泵;5—压力表;
6—起重机吊钩;7—井点管;8—滤管;9—填砂;
10—黏土封口

冲孔时,先用起重设备将冲管吊起并插在井点的位置上,然后开动高压水泵,将土冲松,冲管则边冲边沉。冲孔直径一般为 300 mm,以保证井管四周有一定厚度的砂滤层,冲孔深度宜比滤管底深 0.5 m 左右,以防冲管拔出时,部分土颗粒沉于底部而触及滤管底部。

If the H value calculated according to equation 2-17 is greater than 6m, the buried surface of well point pipe pumping equipment should be reduced to meet the requirements of dewatering depth. When the buried surface(layout elevation) of the well point system is close to the original underground water level line(grooves should be dug in advance), or even slightly lower than the underground water level in some cases(when the soil quality of the upper soil is better, the layer of soil should be excavated by the water-collecting well drainage method first, and then the well point system should be arranged), the pumping capacity can be fully utilized to increase the dewatering depth. The length of the well point pipe above the ground is generally 0.2m. When the first-level well point system fails to meet the requirement of dewatering depth, the second-level well point can be adopted, that is, the soil dredged by the first-level well point can be excavated first, and then the second-level well point can be installed at the bottom.

③Embedment of well point pipe. Generally water flushing method is used, which is divided into two processes: punching hole[Fig. 2.21(a)]and burying pipe[Fig. 2.21(b)].

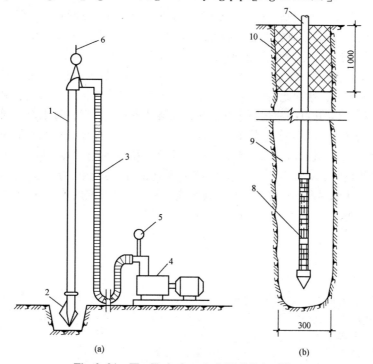

Fig. 2.21 The Embedment of Well Point Pipe

(a)punching hole; (b)burying pipe

1—flushing pipe; 2—flushing nozzle; 3—rubber pipe; 4—high pressure water pump;
5—pressure gauge; 6—crane hook; 7—well point pipe; 8—filter pipe; 9—sand filling; 10—clay sealing

When punching, first use the lifting equipment to lift the flushing pipe and insert it in the position of the well point, and then start the high-pressure water pump. The punching diameter is generally 300mm to ensure that there is a certain thickness of sand filter layer around the well pipe. The punching depth should be about 0.5m deeper than the bottom of the filter pipe to prevent some soil particles from sinking to the bottom of the filter pipe when the punch pipe is pulled out.

井孔冲成后,立即拔出冲管,插入井点管,并在井点管与孔壁之间迅速填灌砂滤层,以防孔壁塌方。砂滤层的填灌质量是保证轻型井点顺利抽水的关键。一般宜选用干净粗砂,填灌均匀,并填至滤管顶上 1～1.5 m,以保证水流畅通。

井点填砂后,在地面以下 0.5～1.0 m 范围内须用黏土封口,以防漏气。

井点管埋设完毕,应接通总管与抽水设备进行试抽水,检查有无漏水、漏气,出水是否正常,有无淤塞等现象,如有异常情况,应检修好后方可使用。

④井点管使用。井点管使用时,应保证连续不断地抽水,并准备双电源。正常出水规律是"先大后小,先混后清"。抽水时需要经常观测真空度以判断井点系统工作是否正常,真空度一般应不低于 55.3～66.7 kPa,并检查观测井中水位下降情况,如果有较多井点管发生堵塞,影响降水效果时,应逐根用高压水反向冲洗或拔出重埋。井点降水工作结束后所留的井孔,必须用砂砾或黏土填实。

2.4.2 降水对周围的影响及防止措施

在弱透水层和压缩性大的黏土层中降水时,由于地下水流失导致的地下水位下降、地基自重应力增加和土层压缩等,会造成较大的地面沉降,又由于土层的不均匀性和降水后地下水位呈漏斗曲线,四周土层的自重应力变化不一而导致不均匀沉降,使周围建筑物基础下沉或房屋开裂。因此,在建筑物附近进行井点降水时,为防止降水影响或损害区域内的建筑物,就必须阻止原有建筑物下地下水的流失。为达到此目的,除可在降水区域和原有建筑物之间的土层中设置一道固体抗渗帷幕外,还可用回灌井点补充地下水的办法来保持地下水位。即在降水井点和原有建筑物之间打一排井点,向土层灌入足够数量的水,以形成一道隔水帷幕,使原有建筑物下的地下水位保持不变或降低较少。这样,就能防止因降水而造成的地面沉降,或者至少可以降低沉降率。

回灌井点是防止井点降水损害周围建筑物的一种经济、简便、有效的办法,它能将井点降水对周围建筑物的影响降低到最小程度。为确保基坑施工的安全和回灌的效果,回灌井点与降水井点之间应保持一定的距离,一般不宜小于 6 m。

After the hole is punched, the punch pipe should be pulled out immediately and inserted into the well point pipe, and the sand filter layer should be filled quickly between the well point pipe and the hole wall to prevent the hole wall from collapsing. The filling quality of sand filter is the key to ensure the pumping of light well. Generally, it is advisable to select clean coarse sand, fill and fill evenly, and fill to the top of the filter tube 1—1.5m to ensure smooth flow of water.

After the well point is filled with sand, it shall be sealed with clay within 0.5—1.0m below the ground to prevent air leakage.

The well point pipe shall be connected to the main pipe and pumping equipment for pumping test after completion. Check whether there is water leakage and air leakage, and whether the water outlet is normal, or whether there is blockage and other phenomena. If there is any abnormal situation, it shall be repaired before use.

④Use of well point pipe. When using well point pipe, continuous pumping shall be ensured and dual power supply be prepared. The rule of normal outflow is "first big then small, first muddy then clear". Vacuum pumping needs often observing to determine that the well point system works normal when pumping. Generally vacuum degree should be not less than 55.3—66.7kPa, and the reducing of water level in the well should be examined the observed. If there are more well point pipes clogging and influencing dewatering effect, they should be flushed with high-pressure water one by one or pulled out and re-buried. After the completion of well point dewatering, the well hole must be filled with gravel or clay.

2.4.2　Influence of Dewatering on Surroundings and Preventive Measures

When dewatering in aquitard and clay layer with high compressibility, due to the groundwater loss, the groundwater level will drop, the self weight stress of foundation will increase and the soil layer will be compressed, which will cause large land subsidence. Moreover, due to the non-uniformity of the soil layer and the funnel curve of the groundwater level after the dewatering, the self weight stress of the surrounding soil layer varies, which leads to the uneven settlement and makes the foundation of the surrounding buildings sink or crack. Therefore, when carrying out well point dewatering in nearby building structure, the groundwater running under the existing buildings must be stopped in order to prevent affection or damage to the building structure in the area because of the dewatering. For this purpose, apart from building a solid penetration resistant partition between dewatering area and existing building, a method of recharge well point to make-up groundwater can be adopted to maintain the groundwater level, i.e., a row of well point can be excavated between dewatering well point and existing building, recharging enough water to the soil layer to form a diaphragm curtain, so as to make the groundwater level of the existing building unchanged or less reduced. In this way, the ground settlement can be prevented due to dewatering, or at least, settlement ratio can be reduced.

Recharge well point is an economical, simple and effective way to prevent well point dewatering from damaging surrounding buildings. It can minimize the impact of well point dewatering on surrounding buildings. To ensure the safety of pit construction and effect of recharge, distance between recharge well point and dewatering well point must be kept, generally, not less than 6m.

为了观测降水及回灌后四周建筑物、管线的沉降情况及地下水位的变化情况，必须设置沉降观测点及水位观测井，并定时测量记录，以便及时调节灌、抽量，使灌、抽基本达到平衡，确保周围建筑物或管线等的安全。

2.5 土方机械化施工

土方工程的施工过程主要包括：土方开挖、运输、填筑与压实等。常用的施工机械有：推土机、铲运机、单斗挖土机、装载机等，施工时应按照施工机械的特点正确选用，并按照现场条件选择正确的施工方法，以加快施工进度。

2.5.1 推土机

推土机操纵灵活，运转方便，所需工作面较小、行驶速度快、易于转移，能爬30°左右的缓坡，因此应用较广。多用于道路工程、场地清理和平整、开挖深度1.5 m以内的基坑，填平沟坑，以及配合铲运机、挖土机工作等。此外，在推土机后面可安装松土装置，破、松硬土和冻土，也可拖挂羊足碾进行土方压实工作。推土机可以推挖一～三类土，经济运距100 m以内，效率最高为60 m。

2.5.2 铲运机

铲运机的特点是能综合完成挖土、运土、平土和填土等全部土方施工工序，对行驶道路要求较低，操纵灵活、运转方便，生产率高，在土方工程中常应用于大面积场地平整，开挖大基坑、沟槽以及填筑路基、堤坝等工程。适宜于铲运含水量不大于27%的松土和普通土，不适于在砾石层和冻土地带及沼泽区工作，当铲运三、四类较坚硬的土时，宜用推土机助铲或用松土机配合将土翻松0.2～0.4 m，以减少机械磨损，提高生产率。

在选定铲运机斗容量之后，其生产率的高低主要取决于机械的开行路线和施工方法。

2.5.3 单斗挖土机

单斗挖土机在土方工程中应用较广，种类很多，按其行走装置的不同，分为履带式和轮胎式两类。单斗挖土机还可根据工作的需要，更换其工作装置。按其工作装置的不同，分为正铲、反铲、拉铲和抓铲等。按其操纵机械的不同，可分为机械式和液压式两类，如图2.22所示。

Settlement view point and water level inspection well must be established to inspect the settlement of the surrounding buildings, pipelines, and the change of groundwater level after dewatering and recharging. Measurement and record must be done at regular intervals so as to adjust the volume of recharge and pumping, and keep the balance of recharge and pumping to ensure the safety of surrounding buildings or pipelines, etc.

2.5 Mechanized Construction of Earthwork

Construction process of earthwork mainly includes excavation, transportation, backfill and compaction, etc. The commonly used construction machines are bulldozer, scraper, single-bucket excavator, and shovel loader, etc. When construction, proper choice of construction machines must be according to their features and the choice of proper construction method must be according to the site conditions, so as to speed up construction progress.

2.5.1 Bulldozer

Bulldozer is flexible in operation, easy in running, smaller working face required, faster in speed, convenient in transfer, and being able to climb a gentle slope of about 30 degrees, so it is widely used. Bulldozer is often used in road construction, site clearing and leveling, excavation depth within 1.5m of foundation pit, filling trench and pit, as well as working together with scraper and excavator. Besides, loosen soil devices can fixed in the back of bulldozer to break and loose hard soil and frozen soil. It can also pull sheep-food roller to compact earthwork. The bulldozer can push and excavate from the first to third kind of soil, the economic transportation distance is less than 100m, and the highest efficiency is 60m.

2.5.2 Scraper

The characteristic of scraper is that it can comprehensively complete all the earthwork construction process of excavation, transportation, soil leveling and backfilling, and requires lower level of traffic road. It is also flexible in operation, easy in running and higher in production efficiency. It is often used in large area site leveling, excavation of large foundation pit, trench and filling of subgrade, embankment and other projects in earthwork. It is suitable to shovel and transport loosen and ordinary soil with water content not more than 27%. It is unsuitable to work in gravel layer and frozen area, as well as marshland. When shoveling harder soil of type III and type IV, it is better to use bulldozer to help shoveling, and use loosener to work together to loose soil 0.2—0.4m, so as to reduce wear of the machine and improve productivity.

After selecting the bucket capacity of scraper, its productivity mainly depends on the route and construction methods of the machine.

2.5.3 Single-bucket Excavator

Single-bucket excavator is widely used in earthwork and has many different types. According to walking device, single-bucket excavator can be divided into two types, crawler type and tire type. The working device of single bucket excavator can also be changed according to the need of work. According to the different working devices, it can be divided into forward shovel, backhoe, dragline and clamshell, etc. According to its different mechanical operation, it can be divided into two types: mechanical and hydraulic (Fig. 2.22).

1. 正铲挖土机施工

(1)特点及适用范围。正铲挖土机外形如图2.22所示。正铲挖土机的挖土特点是："向前向上，强制切土"。其挖掘能力大，生产率高，适用于开挖停机面以上的一～三类土，它与运土汽车配合能完成整个挖运任务。可用于开挖大型干燥基坑以及土丘等。

(2)开挖方式。根据挖土机的开挖路线与运输工具的相对位置不同，可分为正向挖土侧向卸土和正向挖土后方卸土两种。

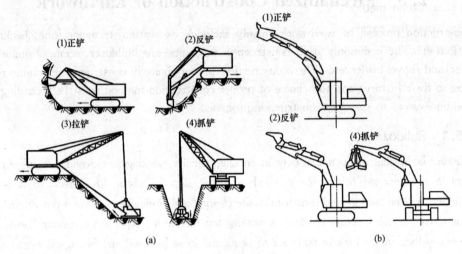

图 2.22 单斗挖土机
(a)机械式；(b)液压式

正向挖土侧向卸土：挖土机沿前进方向挖土，运输工具停在侧面装土[图2.23(a)]。采用这种作业方式，挖土机卸土时动臂回转角度小，运输工具行驶方便，生产率高，使用广泛。

正向挖土后方卸土：挖土机沿前进方向挖土，运输工具停在挖土机后方装土[图2.23(b)]。这种作业方式所开挖的工作面较大，但挖土机卸土时动臂回转角大，生产率低，运输车辆要倒车开入，一般只宜用来开挖工作面较狭小且较深的基坑。

Chapter 2 Earthwork Construction

1. Construction of Forward Shovel Excavator

(1) Characteristic and scope. The shape of forward shovel excavator is just like Fig. 2. 22. The characteristic of forward shovel excavator is forward and upward, conforcement cutting soil. It is powerful in excavation, higher in production efficiency, and suitable for excavating type Ⅰ —type Ⅲ soil above floor level. It is working together with soil transfer truck to complete the whole excavation. It can be used to excavate large dry foundation pit and hillock.

(2) Excavation method. According to the excavation route and the relative position of the transport tools, the excavator can be divided into two types: forward excavation and lateral unloading and forward excavation and rear unloading.

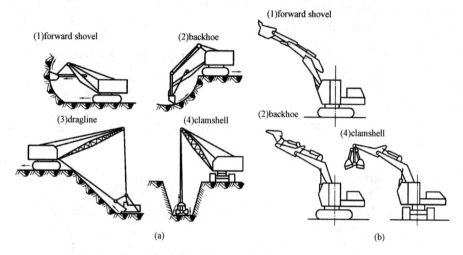

Fig. 2. 22 Single-bucket Excavator
(a) mechanical type; (b) hydraulic type

Forward excavation and lateral unloading: Excavator excavates soil along the forward direction, and the transportation vehicle stays at the side to load soil [Fig. 2. 23 (a)]. With this operation mode, the boom rotation angle of the excavator is small, the transportation tool is convenient to drive, the productivity is high, and it is widely used.

Forward excavation and rear unloading: Excavator excavates soil along the forward direction, and the transportation vehicle stops behind the excavator to load the soil [Fig. 2. 23 (b)]. The working face excavated by this operation mode is large, but when the excavator unloads the soil, the turning angle of the boom is large, and the productivity is low. The transportation vehicle has to reverse, so it is generally only suitable for excavating the narrow and deep foundation pit of the working face.

2. 反铲挖土机施工

(1)特点及适用范围。反铲挖土机的挖土特点是："后退向下,强制切土"。其挖掘力比正铲小,能开挖停机面以下的一～三类土(索式反铲只宜挖一、二类土),适用于挖基坑、基槽和管沟、有地下水的土壤或泥泞土壤。

(2)开挖方式。反铲挖土机挖土时可采用沟端开挖和沟侧开挖两种方式。

沟端开挖:挖土机停在基槽(坑)的端部,向后侧边退边挖土,汽车停在基槽两侧装土[图 2.24(a)]。沟端开挖工作面宽度为:单面装土时为 $1.3R$,双面装土时为 $1.7R$。基坑较宽时,可多次开行开挖或按 Z 字形路线开挖。为了能很好地控制所挖边坡的坡度或直立的边坡,反铲的一侧履带应靠近边线向后移动挖土。

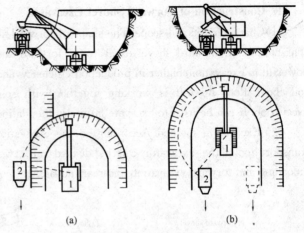

图 2.23 正铲挖土机开挖方式
(a)侧向卸土;(b)后方卸土
1—正铲挖土机;2—自卸汽车

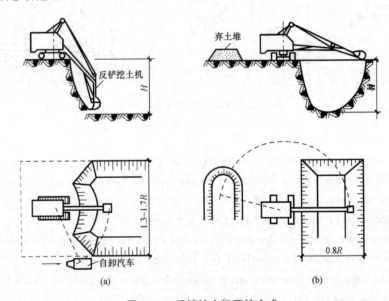

图 2.24 反铲挖土机开挖方式
(a)沟端开挖;(b)沟侧开挖

2. Construction of Backhoe Excavator

(1) Characteristics and scope of application. The characteristics of the backhoe excavator are retreat and downward, conforcement cutting soil. Its digging force is smaller than that of the foreword shovel, and it can excavate class Ⅰ—Ⅲ soil below the parking surface (only class Ⅰ、Ⅱ soil is suitable for cable backhoe). It is suitable for digging foundation pit, foundation trench, and pipe trench, soil with underground water or muddy soil.

(2) Excavating method. When backhoe excavates soil, it can adopt two ways: Trench end excavation and trench side excavation.

Trench end excavation: Excavator is stopped at the end of the foundation trench (pit), moving back side while excavating,

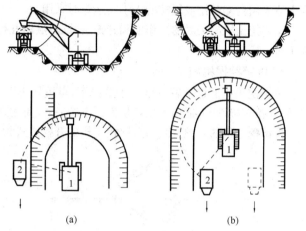

Fig. 2.23 Excavation Method of Forward Excavator
(a) lateral unloading; (b) rear unloading
1—forward excavator; 2—self-unloading vehicle

and the vehicle is stopped at both sides of the foundation trench loading soil [Fig. 2.24(a)]. The width of working face for trench end excavation is 1.3R for one side loading soil and 1.7R for both sides loading soil. When foundation pit is wider, repeated moving and excavation can be adopted or follow the Z road to excavate. In order to control the slope of the excavated slope or the vertical slope, the track on one side of the backhoe should move backward to excavate soil near the sideline.

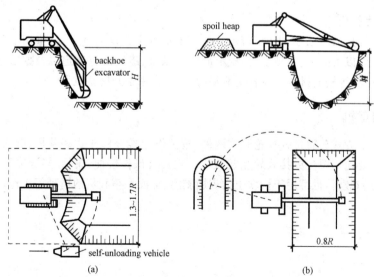

Fig. 2.24 Excavation Method of Back Shovel Excavator
(a) trench end excavation; (b) trench side excavation

沟侧开挖：挖土机沿基槽的一侧移动挖土[图 2.24(b)]。沟侧开挖能将土弃于距基槽边较远处，但开挖宽度受限制（一般为 0.8R），且不能很好地控制边坡，机身停在沟边稳定性较差，因此只在无法采用沟端开挖或所挖的土不需运走时采用。

反铲挖土机施工时的工作面和开挖层数、每层的开行次数以及开行次序的确定与正铲挖土机施工时一样。

3. 拉铲挖掘机施工

拉铲挖土机的挖土特点是："后退向下，自重切土"，其挖土半径和挖土深度较大，但不如反铲灵活，开挖精确性差，适用于挖停机面以下的一、二类土，可用于开挖大而深的基坑或水下挖土。

拉铲挖土机的开挖方式与反铲挖土机的开挖方式相似，可沟侧开挖也可沟端开挖（图 2.25）。

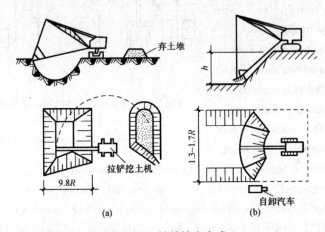

图 2.25　拉铲挖土方式

4. 抓铲挖土机施工

抓铲挖土机外形如图 2.22 中(4)所示。其挖土特点是："直上直下，自重切土"，挖掘力较小，适用于开挖停机面以下的一、二类土，如挖窄而深的基坑、疏通旧有渠道以及挖取水中淤泥等，或用于装卸碎石、矿渣等松散材料。在软土地基的地区，常用于开挖基坑等。

2.5.4　装载机

装载机按行走方式分履带式和轮胎式两种，按工作方式分单斗式装载机、链式和轮斗式装载机。土方工程主要使用单斗铰接式轮胎装载机。它具有操作轻便、灵活、转运方便、快速等特点。适用于装卸土方和散料，也可用于松软土的表层剥离、地面平整和场地清理等工作。

Trench side excavation: Excavator is moving to excavate along one side of foundation trench [Fig. 2.24(b)]. Trench side excavation can abandon soil farther away from trench side, but the width of excavation is limited(generally 0.8R), nor the side slope is well controlled. The stability of the machine body stopping at the trench side is poor, so it is only used when the trench end excavation cannot be used or the excavated soil does not need to be transported.

The working face, the number of excavation layers, the operation times of each layer and the operation sequence of backhoe are the same as those of forward shovel.

3. Construction of Dragline Excavator

The excavation characteristic of dragline excavator is back moving and downward, self-load cutting soil. Its excavation radius and depth are larger, but it is not as flexible as backhoe, and its accuracy of excavation is poor. It suits to excavate I and II types of soil below the machine. It can be used for excavation of large and deep foundation pit or underwater excavation.

The excavation method of the dragline excavator is similar to that of the backhoe excavator, which can be excavated at the trench side or at the trench end(Fig. 2.25)

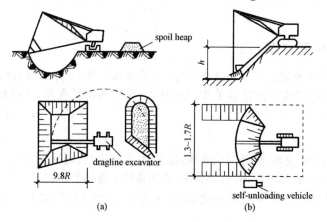

Fig. 2.25 Dragline Excavation Method
(a)trench side excavation; (b)trench end excavation

4. Construction of Clamshell Excavator

The outline of clamshell excavator is as shown in Fig. 2.22(4). Its excavation characteristic is straight up and straight down, cutting soil with self weight. The power of excavation is smaller, and it suits to excavate I and II types of soil below the machine, such as digging narrow and deep foundation pit, dredging waterway, digging mud in the water, etc. , or loading and unloading loose materials such as gravel, slag, etc. In the soft soil foundation area, it is often used to excavate foundation pit.

2.5.4 Shovel Loader

Shovel loader is divided into crawler type and tire type according to the walking mode, and single bucket loader, chain loader and wheel bucket loader according to the working mode. Single bucket articulated tire loader is mainly used in earthwork. It has the characteristics of easy operation, flexibility, convenient and fast transportation. It is suitable for loading and unloading earthwork and bulk materials, as well as surface stripping of soft soil, ground leveling and site cleaning.

2.5.5 土方施工机械的选择

土方机械的选择,通常先根据工程特点和技术条件提出几种可行方案,然后进行技术经济比较,选择效率高、费用低的机械进行施工,一般可选用土方单价最小的机械。开挖基坑时根据下述原则选择机械:

(1)土的含水量较小,可结合运距长短、挖掘深浅,分别采用推土机、铲运机或正铲挖土机配合自卸汽车进行施工。当基坑深度在1~2 m,基坑不太长时可采用推土机;当基坑深度在2 m以内,长度较大的线状基坑,宜由铲运机开挖;当基坑较大,工程量集中时,可选用正铲挖土机挖土。

(2)如地下水位较高,又不采用降水措施,或土质松软,可能造成正铲挖土机和铲运机陷车时,则采用反铲、拉铲或抓铲挖土机配合自卸汽车较为合适。

(3)移挖回填以及基坑和管沟的回填,运距在60~100 m以内可用推土机。

2.6 土方开挖与填筑

2.6.1 施工准备及辅助工作

土方工程施工前通常需完成下列准备工作:施工现场准备,土方工程的测量放线和编制施工组织设计等;有时尚需完成下列辅助工作,如:基坑、沟槽的边坡保护,土壁的支撑,降低地下水位等。

1. 场地清理

(1)地面以上的场地清理:拆迁或改建通信、电力设备(电杆、高压线等)。

(2)地面的场地清理:拆除房屋,迁移树木,去除耕植土及河塘淤泥等。

(3)地面以下的场地清理:拆迁或改建通信、电力设备(地下埋设部分),拆除或改建上下水管道及其他各种管道。

2. 排除地面水

场地内低洼地区的积水必须排除,同时应注意雨水的排除,使场地保持干燥,以利土方施工。地面水的排除一般采用排水沟、截水沟、挡水土坝等措施。

2.5.5 Selection of Earthwork Construction Machinery

In the selection of earthwork machinery, several feasible schemes are usually put forward according to the project characteristics and technical conditions, and then the technical and economic comparison is carried out. The machinery with high efficiency and low cost is selected for construction, and the machinery with the lowest earthwork unit price is generally selected. When excavating foundation pit, the machinery shall be selected according to the following principles.

(1) When water content of soil is less, bulldozer, scraper, or forward shovel excavator can be separately used in combination with self-load vehicles to carry out excavation according to transportation distance and excavation depth. When depth of foundation pit is in 1—2m and is not very long, bulldozer can be sued. The linear foundation pit with a larger length within 2m in depth should be excavated by scraper. When foundation pit is larger and quantity is concentrated, forward shovel excavator can be selected.

(2) If the groundwater level is higher and dewatering measures are not adopted, or soil is loose and soft, which may cause forward shovel excavator and scraper to be trapped, it is more appropriate to use backhoe, dragline or clamshell excavator together with self-unloading vehicle.

(3) Bulldozer can be used for moving excavation and backfilling as well as for foundation pit and pipe trench with transportation distance within 60—100m.

2.6 Earthwork Excavation and Backfilling

2.6.1 Construction Preparation and Ancillary Works

Before earthwork construction, the following preparations are usually completed: preparation of construction site, surveying and setting out of earthwork and preparation of construction organization design, etc. Sometimes it is also required to complete the following ancillary works, such as side slope protection of foundation pit and trench, supporting of soil wall, reducing groundwater level, etc.

1. Site Clearance

(1) Site clearance above the ground: Removal or reconstruction of communication and power equipments(poles, cables, etc.);

(2) Site clearance on the ground: Demolishing houses, removal of trees, eliminating top soil and pound silt, etc.

(3) Site clearance below the ground: Removal or reconstruction of communication and power equipments(buried below the ground), removal or reconstruction of water supply and sewer pipes, as well as other various pipes.

2. Eliminating Surface Water

Surface water in low areas of the site must be eliminated. At the same time, attention must also be paid to eliminate rain water so as to keep the site dry and suit for earthwork construction. Eliminating surface water is generally by measures of drainage ditch, intercepting drain, retaining dam, etc.

应尽量利用自然地形来设置排水沟,使水直接排至场外,或流向低洼处再用水泵抽走。主排水沟最好设置在施工区域的边缘或道路的两旁,其横断面和纵向坡度应根据最大流量确定。一般排水沟的横断面不小于 0.5 m×0.5 m,纵向坡度一般不小于 3‰。平坦地区,如出水困难,其纵向坡度不应小于 2‰,沼泽地区可减至 1‰。场地平整过程中,要注意排水沟保持畅通,必要时应设置涵洞。

3. 修建临时设施

根据施工现场的实际需要,修建临时办公室、宿舍、食堂、厕所、仓库等;修筑临时道路;布置临时水电管线及安全防护设施。

2.6.2 土方开挖

1. 放线

土方开挖时的放线工作主要是在天然地面放出开挖范围线,俗称放灰线。放灰线时,可用装有石灰粉末的长柄勺靠着木质板侧面,边撒、边走,在地上撒出灰线。

(1)基槽放线:根据房屋主轴线控制点,首先将外墙轴线的交点用木桩测设在地面上,并在桩顶钉上铁钉作为标志。房屋外墙轴线测定以后,再根据建筑物平面图,将内部开间所有轴线都一一测出。最后根据中心轴线,再考虑基础宽度、放坡、基础施工工作面等,确定实际开挖的范围,用石灰在地面上撒出基槽开挖边线(图 2.26)。同时在房屋四周设置龙门板或轴线控制桩,以便于基础施工时复核轴线位置。

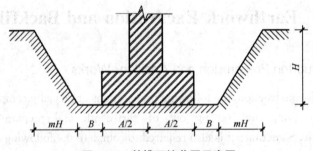

图 2.26 基槽开挖范围示意图

本图中:
基槽开挖范围 $= 2mH + 2B + A$
其中:A——基础宽度;B——基础施工工作面;H——基坑高度;m——坡度系数。

Try to set up drainage ditch by natural topography so that water can be discharged directly from the site, or flow to the lower area and pump out. It is better to set up main drainage ditch at the edge of the construction area or both sides of the road. Its cross section and vertical slop will be decided according to maximum flow rate. The cross section of drainage ditch is generally not less than 0.5m×0.5m, and vertical slop not less than 3‰. In flat area, if it is difficult to drain off water, its vertical slop should not be less than 2‰. In marshland, vertical slop can be reduced to 1‰. During the grading, attention must be paid to keep drainage ditch through. When necessary, discharge culvert should be set up.

3. Construction of Temporary Facilities

According to the actual needs of the construction site, build temporary offices, dormitories, canteens, toilets, warehouses, etc. ; build temporary roads; arrange temporary water and electricity pipelines and safety protection facilities.

2.6.2 Earthwork Excavation

1. Setting Out

The setting out work of earthwork excavation is mainly to set out the excavation scope line on the natural ground, commonly known as lime line. When placing lime line, the long handled spoon with lime powder can be used to lean against the side of the wooden board, scattering powder while you are walking, and the lime line can be sprinkled on the ground.

(1) Foundation Trench Setting-out: According to the control point of the main axis of the building, the intersection of the external wall axis is measured on the ground with wooden stake, and iron nails are nailed on the top of the stake as a sign. After measuring the axis of the external wall of the building, all the axes of the internal bay are measured one by one according to the building plan. Finally, according to the central axis, considering the foundation width, slope, working face of foundation construction, etc., determine the actual excavation scope, and use lime to scatter the excavation sideline of foundation trench on the ground (Fig. 2.26). At the same time, the gantry plate or axis control stake is set around the building, so as to check the axis position during the foundation construction.

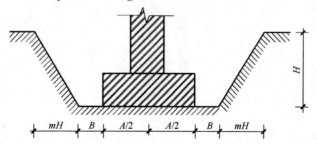

Fig. 2.26 Schematic Diagram of Foundation Trench Excavation

In Fig. 2.26:

Scope of foundation trench excavation $= 2mH + 2B + A$

Among them: A— width of foundation; B—working area for foundation construction; H—height of foundation pit; m—slope coefficient

(2)柱基放线:在基坑开挖前,从设计图上查对基础的纵横轴线编号和基础施工详图,根据柱子的纵横轴线,用经纬仪测定基础中心线的端点,同时在每个柱基中心线上,测定基础定位桩,每个基础的中心线上设置四个定位木桩,其桩位离基础开挖线的距离为0.5~1.0 m。若基础之间的距离不大,可每隔1~2个或几个基础打一定位桩,但两定位桩的间距以不超过20 m为宜,以便拉线恢复中间柱基的中线。定位桩桩顶用钉子标明中心线的位置。然后按柱基的尺寸再考虑放坡、基础施工工作面等,确定实际开挖的范围,放出基坑上口挖土灰线。

大基坑开挖,根据房屋的控制点用经纬仪放出基坑四周的挖土边线。

2. 基坑(槽)开挖

开挖基坑(槽)应按规定的尺寸合理确定开挖顺序和分层开挖深度,连续进行施工,尽快完成。挖出的土除预留一部分用作回填外,不得在场地内任意堆放,应把多余的土运到弃土地区,以免妨碍施工。为防止坑壁滑坍,根据土质情况及坑(槽)深度,在坑顶两边一定距离(一般为0.8)内不得堆放弃土,在此距离外堆土高度不得超过1.5 m。为防止基底土(特别是软土)受到水浸泡或其他原因的扰动,基坑(槽)挖好后,应立即做垫层或浇筑基础,否则,挖土时应在基底标高以上保留150~300 mm厚的土层,待基础施工时再行挖去。如用机械挖土,为防止基底土被扰动,结构被破坏,不应直接挖到坑(槽)底,应根据机械种类,在基底标高以上留出200~300 mm,待基础施工前用人工铲平修整。挖土不得挖至基坑(槽)的设计标高以下,如个别处超挖,应用与基土相同的土料填补,并夯实到要求的密实度。如用当地土填补不能达到要求的密实度时,应用碎石类土填补,并仔细夯实到要求的密实度。如在重要部位超挖时,可用低强度等级的混凝土(如C10)填补。

在软土地区开挖基坑(槽)时,尚应符合下列规定:

(2) Column Base Setting-out: Before the excavation of foundation pit, check the number of vertical and horizontal axis of foundation and the construction details of foundation from the design drawing. According to the vertical and horizontal axis of the column, the end point of the foundation center line is measured with a theodolite, and the foundation alignment stake is measured on the center line of each column base. Four alignment wooden stakes are set on the center line of each foundation, and the distance between the stake position and the foundation excavation line is 0.5—1.0m. If the distance between the foundations is not large, a certain alignment stake can be driven every 1—2 or several foundations, but the distance between the two alignment stakes should not exceed 20m, so that the middle line of the middle column foundation can be restored by pulling wires. Nail should be fixed on the top of the alignment stake to show the position of the central line. Then, according to the size of the column base on the construction drawing, considering the slope and foundation construction working face, the actual excavation scope is determined, and the excavation lime line of the foundation pit is set out.

In the excavation of large foundation pit, according to the control point of the building, the side line of excavation around the foundation pit is set out with theodolite.

2. Foundation Pit(Trench)Excavation

The excavation sequence and layered excavation depth of foundation pit (trench) shall be reasonably determined according to the specified size, and the construction shall be carried out continuously and completed as soon as possible. In addition to reserving a part of the excavated soil for backfilling, the soil dug up shall not be piled up arbitrarily in the site, and the surplus soil shall be transported to the spoil area, so as not to interfere construction. In order to prevent the pit wall from sliding, according to the soil conditions and the depth of the pit(trench), the soil shall not be piled up and abandoned within a certain distance(generally 0.8m)on both sides of the pit top, and the height of the soil piled outside the distance shall not exceed 1.5m. In order to prevent the base soil(especially the soft soil) from being disturbed by water immersion or other reasons, the cushion or pouring foundation should be made immediately after the foundation pit(trench) is excavated. Otherwise, the 150—300mm thick soil layer should be reserved above the base elevation during the excavation, and then it should be excavated during the foundation construction. If mechanical excavation is used, in order to prevent the base soil from being disturbed and the structure from being damaged, it should not be directly excavated to the bottom of the pit(trench). According to the type of machinery, 200—300mm should be reserved above the base elevation, and it should be leveled and trimmed manually before the foundation construction. Soil excavation shall not be carried out below the design elevation of foundation pit(trench). In case of over excavation in other places, it shall be filled with the same soil material as the foundation soil and compacted to the required compactness. If the required compactness cannot be achieved by filling with local soil, gravel soil shall be used for filling, and carefully tamped to the required compactness. In case of over excavation in important parts, low strength concrete(such as C10)can be used to fill.

When excavating foundation pit (trench) in soft soil area, the following requirements shall be met:

(1)施工前必须做好地面排水和降低地下水位工作,地下水位应降低至基坑底以下 0.5~1.0 m 后,方可开挖。降水工作应持续到回填完毕;

(2)施工机械行驶道路应填筑适当厚度的碎石或砾石,必要时应铺设工具式路基箱(板)等;

(3)相邻基坑(槽)开挖时,应遵循先深后浅或同时进行的施工顺序,并应及时作好基础;

(4)在密集群桩上开挖基坑时,应在打桩完成后间隔一段时间,再对称挖土。在密集群桩附近开挖基坑(槽)时,应采取措施防止桩基位移;

(5)挖出的土不得堆放在坡顶上或建筑物(构筑物)附近。

基坑(槽)开挖有人工开挖和机械开挖,对于大型基坑应优先考虑选用机械化施工,以加快施工进度。

2.6.4 土方回填

1. 土方回填的要求

(1)对回填土料的选择。

选择回填土料应符合设计要求。如设计无要求时,应符合下列规定:

碎石类土、砂土(使用细、粉砂时应取得设计单位同意)和爆破石碴,可用作表层以下的填料;含水量符合压实要求的黏性土,可用作各层填料;碎块草皮和有机质含量大于8%的土,仅用于无压实要求的填方;淤泥和淤泥质土一般不能用作填料,但在软土或沼泽地区,经过处理含水量符合压实要求后,可用于填方中的次要部位;含盐量符合规定的盐渍土,一般可以使用,但填料中不得含有盐晶、盐块或含盐植物的根茎。

对碎石类土或爆破石碴用作填料时,其最大粒径不得超过每层铺填厚度的 2/3(当使用振动辗时,不得超过每层铺填厚度的 3/4)。铺填时,大块料不应集中,且不得填在分段接头处或填方与山坡连接处。填方内有打桩或其他特殊工程时,块(漂)石填料的最大粒径不应超过设计要求。

(2)土方回填施工要求。

土方回填前,应根据工程特点、填料种类、设计压实系数、施工条件等合理选择压实机具,并确定填料含水量控制范围、铺土厚度和压实遍数等参数。

(1) Before construction, the ground drainage and lowering of groundwater level must be done well. The excavation can be started only after the groundwater level is lowered to 0.5—1.0m below the bottom of foundation pit. The dewatering shall continue until the backfill is completed;

(2) The road for construction machinery should be filled with crushed stone or gravel of appropriate thickness, and if necessary, the tool type subgrade box(plate) should be laid;

(3) When excavating the adjacent foundation pit(trench), the construction sequence should be from deep to shallow or carried out at the same time and the foundation should be done in time;

(4) When excavating the foundation pit on the dense pile group, it is necessary to excavate the soil symmetrically at intervals after piling. Measures should be taken to prevent the displacement of pile foundation when excavating the foundation pit(groove) near the dense pile group;

(5) The excavated soil shall not be stacked on the top of the slope or near the buildings (structures).

The excavation of foundation pit (trench) includes manual excavation and mechanical excavation. For large foundation pit, mechanized construction should be preferred to speed up the construction progress.

2.6.4 Backfilling

1. Backfilling Requirement

(1) Selection of backfill materials.

The selection of backfill materials should be in line with the design requirement. If there is no requirement in design, the following requirements shall be met.

Gravel soil, sandy soil(fine and silty sand shall be approved by the designer) and blasting ballast can be used as the filling below the surface. The cohesive soil with water content meeting the compaction requirements can be used as filling materials for each layer. The crushed turf and soil with organic matter content more than 8% are only used for filling without compaction requirements. Generally, silt and muddy soil can not be used as filling materials, but in soft soil or swamp areas, after treatment, the water content can meet the compaction requirements, it can be used in the secondary parts of the fill. Generally, the saline soil with salt content meeting the requirements can be used, but the filler shall not contain salt crystals, salt blocks or rhizomes of salt plants.

When gravel soil or blasting ballast is used as filler, its maximum particle size shall not exceed 2/3 of the paving thickness of each layer(when vibratory roller is used, it shall not exceed 3/4 of the paving thickness of each layer). The bulk materials shall not be concentrated and shall not be filled in the joint of sections or the joint of filling and hillside. When there are piling or other special projects in the fill, the maximum particle size of the block(floating) stone filler shall not exceed the design requirements.

(2) Construction requirement of earthwork backfilling.

Before earthwork backfilling, compaction machines and tools should be reasonably selected according to the engineering characteristics, filler types, design compaction coefficient and construction conditions, and parameters such as filler water content control range, soil thickness and compaction times should be determined.

土方回填施工应接近水平状态，施工过程中应分层填土、分层压实和分层测定压实后土的干密度，在每层的压实系数和压实范围符合设计要求后，才能填筑上层。

填土应尽量采用同类土填筑。如采用不同填料分层填筑时，为防止填方内形成水囊，上层宜填筑透水性较小的填料，下层宜填筑透水性较大的填料，填方基底表面应作成适当的排水坡度，边坡不得用透水性较小的填料封闭。因施工条件限制，上层必须填筑透水性较大的填料时，应将下层透水性较小的土层表面作成适当的排水坡度或设置盲沟。

分段填筑时，每层接缝处应作成斜坡形，辗迹重叠0.5～1.0 m。上、下层接缝应错缝布置且距离不应小于1 m。

回填基坑和管沟时，应从四周或两侧均匀地分层进行，以防基础和管道在土压力作用下产生偏移或变形。

2. 填土压实的方法

填土压实方法有碾压、夯实和振动三种。

(1)碾压法。碾压法是利用机械滚轮的压力压实土壤，使之达到所需的密实度，此法多用于大面积填土工程如场地平整、大型车间的室内填土等工程。碾压机械有光面碾(压路机)、羊足碾和气胎碾。光面碾对砂土、黏性土均可压实，羊足碾需要较大的牵引力，且只宜压实黏性土，因在砂土中使用羊足碾会使土颗粒受到"羊足"较大的单位压力后会向四周移动，从而使土的结构遭到破坏，气胎碾在工作时是弹性体，其压力均匀，填土质量较好。还可利用运土机械进行碾压，也是较经济合理的压实方案，施工时使运土机械行驶路线能大体均匀地分布在填土面积上，并达到一定重复行驶遍数，使其满足填土压实质量的要求。

碾压机械压实填方时，行驶速度不宜过快，一般平碾控制在2 km/h，羊足碾控制在3 km/h。否则会影响压实效果。

用碾压法压实填土时，铺土应均匀一致，碾压遍数要一样，碾压方向以从填土区的两边逐渐压向中心，每次碾压应有150～200 mm的重叠。

The earthwork backfill construction should be close to the horizontal state. During the construction process, fill the soil layer by layer, compact the soil layer by layer and measure the dry density of the compacted soil layer by layer. The upper layer can be filled only after the compaction coefficient and compaction range of each layer meet the design requirements.

Similar soil shall be used for filling. If different fillers are used for layered filling, in order to prevent the formation of water pockets in the filling, the upper layer should be filled with less permeable fillers, and the lower layer should be filled with more permeable fillers. The surface of the filling base should be made into an appropriate drainage slope, and the slope should not be closed with less permeable fillers. Due to the limitation of construction conditions, when the upper layer must be filled with filler with high permeability, the surface of the lower layer with low permeability should be made into an appropriate drainage slope or a blind ditch should be set.

When filling in sections, the joints of each layer shall be made into slope shape, and the rolling trace shall overlap by 0.5~1.0m. The upper and lower joints shall be staggered and the distance shall not be less than 1m.

When the foundation pit and pipe trench are backfilled, it should be carried out in layers around or on both sides to prevent the foundation and pipeline from deflection or deformation under the action of earth pressure.

2. Compaction Method of Filling Soil

There are three compaction methods: rolling, tamping and vibrating.

(1) Rolling method. Rolling method is to use the pressure of mechanical roller to compact the soil to achieve the required compactness. This method is commonly used for large area filling works such as site leveling, indoor fill of large workshop and other works. Rolling machinery is divided into smooth roller, sheep foot roller and pneumatic tire roller. Smooth roller can compact sandy soil and cohesive soil evenly. Sheep foot roller needs larger traction force and is only suitable for compaction of cohesive soil. Because the use of sheep foot roller in sandy soil will make the soil particles move around under the larger unit pressure of "sheep foot", which will damage the soil structure. The pneumatic tire roller is an elastic body in working, its pressure is even, and the filling quality is good. It is also an economical and reasonable compaction scheme to use the earth moving machinery for rolling. During the construction, the driving route of the earth moving machinery can be roughly evenly distributed on the filling area, and reach a certain number of repeated driving times, so as to meet the requirements of filling compaction quality.

When the roller compaction machine is used to compact the fill, the driving speed should not be too fast. Generally, the flat roller should be controlled at 2km/h, and the sheep foot roller should be controlled at 3km/h. Otherwise, the compaction effect will be affected.

When using the rolling method to compact the fill, the soil shall be paved evenly and uniformly, and the rolling times shall be the same. The rolling direction shall be gradually from both sides of the fill area to the center, and each rolling shall be overlapped by 150—200mm.

(2)夯实法。夯实法是利用夯锤自由下落的冲击力来夯实土壤,主要用于小面积回填。夯实法分人工夯实和机械夯实两种。夯实机械有夯锤、内燃夯土机和蛙式打夯机。人工夯土用的工具有木夯、石夯等。其中蛙式打夯机轻巧灵活,构造简单,在小型土方工程中应用最广。

夯实法的优点是,可以夯实较厚的土层。采用重型夯土机(如1 t以上的重锤)时,其夯实厚度可达1～1.5 m。但对木夯、石夯或蛙式打夯机等夯土工具,其夯实厚度则较小,一般均在200 mm以内。

(3)振动法。振动法是将重锤放在土层的表面或内部,借助于振动设备使重锤振动,土壤颗粒即发生相对位移达到紧密状态。此法用于振实非黏性土效果较好。

3. 填土压实的影响因素

填土压实的影响因素较多,主要有压实功、土的含水量以及每层铺土厚度。

(1)压实功的影响。填土压实后的密度与压实机械在其上所施加的功有一定的关系。土的密度与所耗的功的关系如图2.27所示。当土的含水量一定时,一开始压实,土的密度急剧增加,待到接近土的最大密度时,压实功虽然增加许多,而土的密度则变化甚小。此外,松土不宜用重型碾压机械直接滚压,否则土层有强烈起伏现象,效率不高。如果先用轻碾压实,再用重碾压实就会取得较好效果。

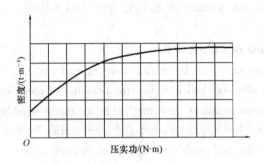

图2.27 土的密度与压实功的关系示意图

(2)含水量的影响。在同一压实功条件下,填土的含水量对压实质量有直接影响。较为干燥的土颗粒之间的摩阻力较大,因而不易压实;当含水量超过一定限度时,土颗粒之间孔隙由水填充而呈饱和状态,也不能压实;当土的含水量适当时,水起了润滑作用,土颗粒之间的摩阻力减少,压实效果最好。每种土都有其最佳含水量。土在这种含水量的条件下,使用同样的压实功进行压实,所得到的密度最大(图2.28),各种土的最佳含水量和最大干密度可参考表2.7。

(2) Tamping method.

Tamping method takes advantage of impact of free falling tamper to compact soil. It is commonly used for small area filling. Tamping method is divided into manual tamping and mechanical tamping. Tamping machineries are of tamper, internal combustion compacter and frog ramming machine. Tools for artificial ramming are wood ram, stone ram, etc. Among them, frog ramming machine is most widely used in small earthworks for its lightweight, flexible and simple structure.

The advantage of tamping method is that thicker soil layer can be compacted. When using heavy ram compactor (heavy hammer more than 1t), tamping thickness can reach 1—1.5m. However, tamping tools such as wood ram, stone ram and frog ramming machine are of less tamping thickness, generally within 200mm.

(3) Vibrating method.

Vibration method is to place the heavy hammer on the surface or inside of the soil layer. With the help of vibration equipment, the heavy hammer vibrates, and the soil particles move relatively to a compact state. This method can be used to vibrate non cohesive soil effectively.

3. Influencing Factors of Filling Compaction

There are many influencing factors of filling compaction, such as compaction work, soil moisture content and thickness of each layer.

(1) Influence of compaction work. The density of compacted soil has a certain relationship with the work exerted on it by compaction machinery. The relationship between soil density and work consumed is just as shown in Fig. 2.27. When the moisture content of soil is constant, the density of soil increases sharply at the beginning of compaction. When it is close to the maximum density of soil, the density of soil changes little although the compaction work increases a lot. In

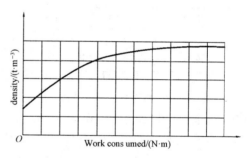

Fig. 2.27 Schematic Diagram of Soil Density and Compaction Work

addition, it is not suitable to use heavy rolling machinery to directly roll loose soil, otherwise the soil layer has strong undulation phenomenon and the efficiency is not high. If the light roller is used first, then the heavy roller will get better effect.

(2) Influence of soil moisture. Under the condition of the same compaction work, the moisture content of fill has a direct impact on the compaction quality. Frictional resistance between dryer soil particles is larger, so it is not easy to compact. When soil moisture exceeds a certain limit, porosity between soil particles is saturated for filling with water, and it is also not easy to compact. When soil moisture is appropriate, water takes as lubrication action, and frictional resistance between soil particles reduces, then compaction result will be the best. Every kind of soil has its best moisture content. Soil under this condition of moisture content and using the same compaction work to compact will get maximum density(Fig. 2.28). The optimum moisture content and maximum dry density of various soils can be referred to Table 2.7.

工地简单检验黏性土含水量的方法一般是以手握成团落地开花为适宜。为了保证填土在压实过程中处于最佳含水量状态,当土过湿时,应予翻松晾干,也可掺入同类干土或吸水性土料;当土过干时,则应预先洒水润湿。

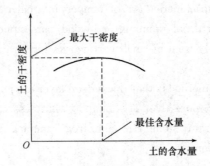

图 2.28　土的干密度与含水量的关系示意

表 2.7　土的最优含水量和最大干密度参考表

项次	土的种类	变动范围		项次	土的种类	变动范围	
		最佳含水量/%（质量比）	最大干密度/(kN·m^{-3})			最佳含水量/%（质量比）	最大干密度/(kN·m^{-3})
1	砂土	8～12	18.0～18.8	3	粉质黏土	12～15	18.5～19.5
2	黏土	19～23	15.8～17.0	4	粉土	16～22	16.1～18.0

注:1. 表中土的最大干密度应根据现场实际达到的数字为准。
　　2. 一般性的回填可不作此项测定。

(3)铺土厚度的影响。土在压实功的作用下,土壤内的应力随深度增加而逐渐减小(图 2.29),其影响深度与压实机械、土的性质和含水量等有关。铺土厚度应小于压实机械压土时的作用深度。同时还有最优土层厚度问题,铺得过厚,要压很多遍才能达到规定的密实度;铺得过薄,虽可减少每层压实遍数,但却要增加机械的总压实遍数。最优的铺土厚度应能使土方压实而机械的功耗费最少,可按照表 2.8 选用。在表中规定压实遍数范围内,轻型压实机械取大值,重型的取小值。

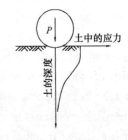

图 2.29　压实作用沿深度的变化示意图

The method of simple examining clay soil moisture content on site is generally to clump soil in hands and drop it on earth, and if it is loose enough, it is suitable to compact. In order to ensure that the filling soil is in the best moisture content state in the compaction process, when the soil is too wet, it should be turned loose and dried, or it can be mixed with the same kind of dry soil or water absorbing soil material. When the soil is too dry, it should be watered in advance.

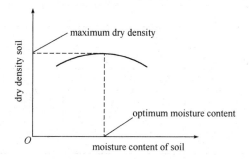

Fig 2.28 Schematic Diagram of Dry Density and Water Content

Table 2.7 Reference Table of Optimal Water Content and Maximum Dry Density of Soil

Items	Types of Soil	Range of Change		Items	Types of Soil	Range of Change	
		Optimal Water Content/% (Quality Ratio)	Maximum Dry Density /(kN·m^{-3})			Optimal Water Content/% (Quality Ratio)	Maximum Dry Density /(kN·m^{-3})
1	Sandy Soil	8—12	18.0—18.8	3	Silty Clay	12—15	18.5—19.5
2	Clay	19—23	15.8—17.0	4	Silt	16—22	16.1—18.0

Notes: 1. Maximum dry density in the table should be decided according to the actual figure reached on site.
2. General fill may not refer to this table.

(3) Influence of soil thickness. Under the effect of compaction work, the stress in the soil decreases gradually with the increase of depth (Fig. 2.29), and the influence depth is related to compaction machinery, soil properties and moisture content. Soil thickness should be less than the effective depth of rolling machinery. At the same time, there is the problem of optimal soil layer thickness. If it is too thick, it needs to be pressed many times to achieve the specified compactness; if it is too thin, it can reduce the compaction times of each layer, but it needs to increase the total compaction times of machinery. The optimal paving thickness should be able to compact the earthwork with the least mechanical power consumption, which can be selected according to Table 2.8. Within range of specified compaction passes in the table, light rolling machinery takes large value and heavy rolling machinery takes small value.

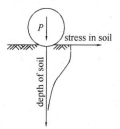

Fig 2.29 Change Diagram of Compaction along the Depth

上述三方面因素之间是互相影响的。为了保证压实质量,提高压实机械的生产率,重要工程应根据土质和所选用的压实机械在施工现场进行压实试验,以确定达到规定密实度所需的压实遍数、铺土厚度及最优含水量。

表 2.8 填方每层的铺土厚度和压实遍数参考表

压实机具	每层铺土厚度/mm	每层压实遍数/遍
平碾	200～300	6～8
羊足碾	200～350	8～16
蛙式打夯机	200～250	3～4
振动压路机	120～150	10
推土机	200～300	6～8
拖拉机	200～300	8～16
人工打夯	不大于200	3～4

习 题

1. 土方工程有什么施工特点?
2. 场地平整土方量的计算步骤是怎样的?
3. 土方边坡及工作面大小如何确定? 与开挖有何影响?
4. 如何确定基坑(槽)开挖线? 简述其步骤。
5. 常用的土方施工机械有哪些? 各种施工机械适用什么条件?
6. 何谓流砂现象? 产生的主要因素有哪些? 如何防治?
7. 常用的降水方法有哪些? 各适用什么条件?
8. 降水对周围有什么影响? 采取什么防止措施?

The above three factors influence each other. To ensure rolling quality and increase productivity of rolling machinery, the important project should have compaction test on site according to soil quality and rolling machinery selected, so as to determine rolling passes, soil thickness and optimal moisture content required by specified density.

Table 2.8 Reference Table of Soil Thickness for Each Layer and Rolling Passes

Rolling Machinery	Soil Thickness for Each Layer/mm	Rolling Passes for Each Layer/pass
Flat Roller	200—300	6—8
Sheep-foot Roller	200—350	8—16
Frog Ramming Machine	200—250	3—4
Vibratory Roller	120—150	10
Bulldozer	200—300	6—8
Tractor	200—300	8—16
Manual Ramming	Not more than 200	3—4

Exercises

1. What are the construction characteristics of earthwork?

2. What is the calculation procedure of the earthwork volume for site leveling?

3. How to determine the size of earthwork slope and working face? What is the impact of excavation?

4. How to determine the excavation line of foundation pit(trench)? Describe the steps.

5. What is the commonly used earthwork construction machinery? What are the applicable conditions of construction machinery?

6. What is quicksand? What are the main factors and how to prevent it?

7. What are the commonly used dewatering methods? What are the applicable conditions of each method?

8. What is the impact of dewatering on the surrounding area? What preventive measures should be taken?

第 3 章 地基与基础工程施工

3.1 地基处理与加固

在施工过程中如发现局部地基土质过软或过硬,不符合设计要求时,应本着使建筑物各部位沉降尽量趋于一致,减小地基不均匀沉降的原则对地基进行处理,即地基的局部处理。

在软弱地基上建造建筑物或构筑物,利用天然地基有时不能满足设计要求,需要对地基进行人工处理,提高其承载力、减小沉降,以满足结构对地基的要求,即地基的整体加固。

3.1.1 地基整体加固

地基处理与加固是本着改善地基土的不良特性,提高地基承载力,改变其变形性质或渗透性质,满足工程设计要求为目的而采取的人工处理地基的方法。欧美国家称地基处理,亦有称地基加固。常见的人工地基处理方法有换土地基、重锤夯实地基、强夯地基、水泥土搅拌地基、振冲地基、水泥粉煤灰碎石桩复合地基、堆载预压地基、化学加固等。

1. 换土地基

当建筑物基础下的持力层比较软弱,不能满足上部荷载对地基的要求时,常采用换土回填来处理软弱地基。这时先将基础下一定深度、宽度范围内承载力低的软土层挖去,然后回填强度较大的砂、石或灰土等,并夯至密实。地基换填按其回填的材料可分为砂地基、砂石地基、灰土地基等。

(1)砂地基和碎(砂)石地基。砂地基和碎(砂)石地基是将基础下一定深度、宽度范围内的土层挖去,然后用强度较大的砂或碎石等回填,并经分层夯实至密实,以起到提高地基承载力、减少沉降、加速软弱土层的排水固结,防止冻胀和消除膨胀土的胀缩等作用。

Chapter 3 Subbase and Foundation Construction

3.1 Subbase Treatment and Reinforcement

In the process of construction, if local subbase soil is found too soft or too hard, which does not meet design requirements, the subbase should be treated according to the principle of making settlement of all parts of the building as consistent as possible and reducing differential settlement of subbase, i. e. local treatment of subbase.

When the building or the structure is constructed on soft subbase, the natural subbase sometimes cannot meet design requirements. Therefore, the subbase should be artificially treated to improve its bearing capacity and reduce the settlement, so as to meet the requirements of the structure on the subbase, and it is called overall reinforcement of subbase.

3.1.1 Integral Subbase Reinforcement

Subbase treatment and reinforcement is a method of artificial subbase treatment, which aims to improve bad characteristics of subbase soil, improve bearing capacity of subbase, change deformation property or permeability property and meet the requirements of engineering design. It is called subbase treatment in European and American countries, and it is also called subbase reinforcement. There are some common artificial subbase treatment methods: displaced subbase, heavy tamping subbase, dynamic compaction subbase, cement soil mixing subbase, vibro-compacted subbase, cement pulverized fuel ash gravel pile composite subbase, preloading subbase, chemical consolidation and so on.

1. Displaced Subbase

When the bearing stratum under the building foundation is weak and cannot meet the requirement of upper load on the subbase, the soft subbase is often treated by backfilling with replacement soil. At this time, within a certain range of depth and width, the soft layer with low bearing capacity under the foundation should be dug out, and then the sand, stone, lime treated soil and so on with high strength should be backfilled and tamped. According to the backfill materials, the subbase replacement can be divided into sand subbase, sand and gravel subbase, lime soil subbase, etc.

(1) Sand subbase and sand-gravel subbase. Sand subbase and gravel (sand) subbase is to excavate the soil layer within a certain depth and width under the subbase, and then backfill it with sand or gravel with high strength, and tamp it to dense layer by layer, so as to improve the bearing capacity of subbase, reduce settlement, accelerate the drainage consolidation of soft soil layer, prevent frost heave and eliminate the expansion and contraction of expansive soil. The subbase has the advantages of simple construction technology, short construction period and low cost.

该地基具有施工工艺简单、工期短、造价低等优点。适用于处理透水性强的软弱黏性土地基，但不宜用于湿陷性黄土地基和不透水的黏性土地基，以免聚水而引起地基下沉和降低承载力。

质量检验：

①环刀取样法。在夯实振捣后的砂垫层中用容积不小于 200 cm³ 的环刀取样，取样点应位于每层厚度的 2/3 深度处。检验点数量：每个单体工程不少于 3 点，对大基坑每 50～100 m² 不应于 1 个检验点；对于基槽每 10～20 延米不应少于 1 个点；每个独立柱基不应少于 1 个点。

②贯入测定法。当采用贯入法检查时，先将垫层表面的砂刮去 3 cm 左右，以不大于通过试验所确定的贯入度值为合格。钢筋贯入测定法，用直径 20 mm，长 1 250 mm 的平头钢筋，距离砂层面 700 mm 自由下落，记录贯入深度。钢叉贯入测定法，用水撼法使用的钢叉，距离砂层面 500 mm 处自由下落，记录插入深度。以上钢筋或钢叉的插入深度，可根据砂的控制干密度预先进行小型试验确定。

另外，土体原位测试的一些方法，如载荷试验、标准贯入试验、静力触探试验和旁压试验等，也可以用来进行垫层的质量检验。这些内容可参考有关的文献和资料。

(2)灰土地基。灰土地基是用石灰和黏性土拌和均匀，然后分层夯实而成。采用的体积配合比一般为 2∶8 或 3∶7(石灰∶土)，其承载能力可达 300 kPa。适用于一般黏性土地基加固，施工简单，取材方便，费用较低。

质量检验：灰土地基的质量检查，可以采用环刀法或钢筋贯入法检验。灰土地基的检验必须分层进行，每夯实一层，均应检查该层的平均压实系数 λ_c，一般为 0.93～0.95 为土在施工时实际达到的干密度与室内采用击实试验得到的最大干密度的比值。当压实系数满足设计要求时，才能铺筑上层。如果无设计规定时，也可以按表 3.1 的要求执行。如果用贯入仪检查灰土质量时，应先进行现场试验以确定贯入度的具体要求。

表 3.1 灰土质量标准

土料种类	黏土/(t·m⁻³)	粉质黏土/(t·m⁻³)	粉土/(t·m⁻³)
灰土最小干密度	1.45	4.50	1.55

Chapter 3 Subbase and Foundation Construction

It is suitable for the treatment of soft clay foundation with strong permeability, but it is not suitable for collapsible loess subbase and impermeable clay subbase, so as to avoid water accumulation causing subbase subsidence and reducing bearing capacity.

Quality inspection:

①Ring Knife Sampling Method: In the tamped and vibrated sand cushion, a ring knife with a volume of not less than 200cm³ shall be used for sampling and the sampling point shall be at the depth of 2/3 of the thickness of each layer. Number of inspection points: No less than 3 points for each single project, no less than 1 inspection point for every 50—100m² of large foundation pit, no less than 1 inspection point for every 10—20 linear meters of foundation trench and no less than 1 inspection point for each independent column base.

②Penetration Resistance Method: When the penetration method is used for inspection, the sand on the cushion surface shall be scraped off for about 3cm, and it shall be regarded as qualified if it is not greater than the penetration value determined by the test. In the steel bar penetration test method, the flat head steel bar with diameter of 20mm and length of 1250mm is used to fall freely 700mm away from the sand layer, and the penetration depth is recorded. The steel fork used in the water shaking method falls freely 500mm away from the sand layer, and the insertion depth is recorded. The insertion depth of the above steel bar or steel fork can be determined by small-scale test in advance according to the controlled dry density of sand.

In addition, some methods of soil in-situ test, such as load test, standard penetration test, static cone penetration test and pressure meter test, can also be used for quality inspection of cushion. These contents can be referred to relevant literature and materials.

(2)Lime soil subbase.

Lime soil subbase is mixed evenly with lime and clay soil, then layered and compacted. The volume mix ratio used is generally 2 : 8 or 3 : 7(lime: soil). Its load capacity can reach 300kPa. It is suitable for general clay soil subbase reinforcement, simple in construction, convenient in material acquisition and lower in cost.

Quality inspection. The quality inspection of lime soil foundation can be carried out by ring knife method or steel bar penetration method. The inspection of lime soil subbase must be carried out in layers, and the average compaction coefficient λ_c of each compacted layer should be checked. Generally, it is 0.93—0.95, which is the ratio of the actual dry density of soil during construction and the maximum dry density obtained by indoor compaction test. When compaction factors meet the design requirement, the upper layer can be paved. If there are no design regulations, Table 3.1 can be taken as reference. If penetration test is used to inspect the quality, in-situ test should be done first to determine the actual requirement of penetration degree.

Table 3.1 Lime Soil Quality Standard

Type of Soil	Clay t/m³	Silty Clay t/m³	Silt t/m³
Minimum Density of Lime Soil	1.45	4.50	1.55

2. 重锤夯实地基

重锤夯实地基是用起重机械将夯锤(1.5~3 t)提升到一定高度(1.5~4.5 m)后,利用夯锤自由下落时的冲击能来夯实基土表面,使其形成一层较为均匀的硬壳层,从而使地基得到加固。加固深度一般为 1.2 m。该法具有施工简单,费用较低,但布点较密,夯击遍数多,施工期相对较长,同时夯击能量小,孔隙水难以消散,加固深度有限,当土的含水量稍高,易形成"橡皮土",处理较困难等特点。

重锤夯实地基适用于处理地下水位以上稍湿的黏性土、砂土、湿陷性黄土、杂填土和分层填土地基,是一种浅层的地基加固方法。为了提高地基强度,减少其压缩性及不均匀性,也可用于消除湿陷性黄土的表层湿陷性。但当夯击振动对邻近的建筑物、设备以及施工中的砌筑工程或浇筑混凝土等产生有害影响时,或地下水位高于有效夯实深度以及在有效深度内存在软黏土层时,不宜采用。

3. 强夯地基

强夯法是用起重机械(起重机或起重机配三脚架、龙门架)将夯锤(一般为 8~30 t,最重达 200 t)吊起,从高处(一般为 6~30 m,最高达 40 m)自由下落,对土体进行强力夯实的地基加固方法。强夯法是在重锤夯实法的基础上发展起来的,但在作用机理上,又与它有区别。强夯法属高能量夯击,是用巨大的冲击能(一般为 500~800 kJ),使土中出现冲击波和很大的应力,迫使土颗粒重新排列,排除孔隙中的气和水,从而提高地基强度,降低其压缩性。地基经强夯加固后,承载能力可以提高 2~5 倍,压缩性可降低 200%~1 000%,其影响深度在 10 m 以上,国外加固影响深度已达 40 m。是一种效果好、速度快、节省材料,施工简便的地基加固方法。其缺点是施工时噪音和振动很大,离建筑物小于 10 m 时,应挖防震沟,沟深要超过建筑物基础深。

强夯法适用于处理碎石土、砂土、低饱和度的粉土与黏性土、素填土和杂填土等地基。高饱和度的粉土、软塑及流塑的黏性土等地基上可用块石、碎石或其他粗颗粒材料进行强夯置换。强夯置换应通过现场试验确定其适用性和处理效果。当强夯所产生的振动,对周围已有或正在施工的建(构)筑物有影响时不得采用,如应采用应采取隔振、防振措施并设置监测点。

4. 注浆地基

注浆地基常用的有水泥注浆地基和硅化注浆地基。水泥注浆适用于孔隙比较大的砂土、卵石土及人工填土。

2. Heavy Hammer Tamping Subbase

The heavy hammer tamping subbase is to use the lifting machinery to lift the hammer(1.5—3T) to a certain height(1.5—4.5m), and then use the impact energy when the hammer falls freely to tamp the subbase soil surface, so as to form a more uniform hard shell layer, and strengthen the subbase. The depth of reinforcement is generally 1.2m. This method is easy in construction and lower in cost, but compacting spots are denser, tamping times are more and construction period is relatively longer. At the same time, there are features of small tamping energy, not easy dissipating pore water, limitation of reinforcement depth, easy forming *rubber soil* when water content of soil is higher and difficulty in treatment, etc.

Heavy hammer tamping subbase is suitable for dealing with slightly damp clay, sandy soil, collapsible loess, miscellaneous fill and layered fill above groundwater level. It is a shallow subbase reinforcement method. To improve subbase strength, reduce its compressibility and unevenness, it can also use to eliminate surface collapsibility of collapsible loess. However, when the vibration of ramming has harmful effects on the adjacent buildings, equipment, masonry works or concrete pouring, or when the groundwater level is higher than the effective tamping depth and there is soft clay layer in the effective depth, it should not be used.

3. Dynamic Compaction Subbase

The dynamic compaction method is a subbase reinforcement method that uses the lifting machinery(crane, or crane with tripod, gantry) to lift the rammer(generally 8—30t, the heaviest up to 200t) and fall freely from the height(generally 6—30m, the heaviest up to 40m) to tamp the soil strongly. The method is developed based on heavy tamping method, but they are different in mechanism of action. Dynamic compaction belongs to compaction of high energy. It uses huge impacting energy(generally 500—800 kJ) to make shock wave and high stress in the soil, to force the soil particles to rearrange and eliminate air and water in pores, so as to increase the strength of subbase and reduce its compressibility. Bearing capacity of subbase can be increased by 2—5 times and compressibility can be reduced by 200%—1000% after the subbase is reinforced by dynamic compaction, and the influenced depth is more than 10m. The influenced depth of reinforcement in foreign countries has reached 40m. Dynamic compaction is a subbase reinforcement method with good effect, fast speed, saving material and simple construction. Its disadvantage is loud noise and vibration in construction. When the distance is less than 10m from the building, earthquake-proof trench should be dug and the depth of trench should exceed the depth of building foundation.

The dynamic compaction method is suitable for the treatment of gravel soil, sandy soil, silt and cohesive soil with low saturation, plain fill and miscellaneous fill. The subbase of silty soil with high saturation, soft plastic and fluid-plastic clay soil can be compacted and replaced by rock block, crushed stone or other coarse granular material. The applicability and treatment effect of dynamic tamping replacement should be determined by site test. Dynamic compaction should not be used when its vibration has an impact on the surrounding buildings(structures) which are existed or under construction. If the method is adopted, vibration isolation and anti-vibration measures should be taken and monitoring points should be set up.

4. Grouting Subbase

The commonly used grouting subbases are cement grouting subbase and silicified grouting subbase. Cement grouting subbase is suitable for larger pores sandy soil, pebble soil and artificial earth fill.

(1)机具设备。

成孔机械主要采用钻机、振动打拔管机。压浆泵主要采用泥浆泵、砂浆泵、齿轮泵、手摇泵。其他配套机具主要有注浆花管、搅拌机、灌浆管、阀门、压力表、磅秤、贮液罐、倒链等。

(2)施工要点。

①水泥注浆地基施工要点。

a. 地基注浆加固前,应通过试验确定灌浆段长度、灌浆孔距、灌浆压力等有关技术参数;灌浆段长度根据土的裂隙、松散情况、渗透性以及灌浆设备能力等条件选定。在一般地质条件下,段长多控制在5~6 m;在土质严重松散、裂隙发育、渗透性强的情况下,宜为2~4 m;灌浆孔距一般不宜大于2.0 m,单孔加固的直径范围可按1~2 m考虑;孔深视土层加固深度而定;灌浆压力是指灌浆段所受的全压力,即孔口处压力表上指示的压力,所用压力大小视钻孔深度、土的渗透性以及水泥浆的稠度等而定,一般为0.3~0.6 MPa。

b. 灌浆一般用净水泥浆,水灰比变化范围为0.6~2.0,常用水灰比从0.8∶1~1∶1。要求快凝时,可采用快硬水泥或在水中掺入水泥用量1%~2%的氯化钙;要求缓凝时,可掺加水泥用量0.1%~0.5%的木质素磺酸钙;亦可掺加其他外加剂以调节水泥浆性能。在裂隙或孔隙较大、可灌性好的地层,可在浆液中掺入适量细砂或粉煤灰,比例为1∶0.5~1∶3。对不以提高固结强度为主的松散土层,亦可在水泥浆中掺加细粉质黏土配成水泥黏土浆,灰泥比为1∶3~1∶8(水泥∶土,体积比),可以提高浆液的稳定性,防止沉淀和析水,使填充更加密实。

c. 灌浆施工方法是先在加固地基中按规定位置用钻机或手钻钻孔到要求的深度,孔径一般为55~100 mm,并探测地质情况,然后在孔内插入直径38~50 mm的注浆射管。管底部1.0~1.5 m管壁上钻有注浆孔,在射管之外设有套管,在射管与套管之间用砂填塞。地基表面空隙用1∶3水泥砂浆或黏土、麻丝填塞,而后拔出套管,用压浆泵将水泥浆压入射管而透入土层孔隙中,水泥浆应连续一次压入,不得中断。灌浆先从稀浆开始,逐渐加浓。灌浆次序一般把射管一次沉入整个深度后,自下而上分段连续进行,分段拔管直至孔口为止。灌浆孔宜分组间隔灌浆,第1组孔灌浆结束后,再灌第2组、第3组。

Chapter 3 Subbase and Foundation Construction

(1) Machinery and equipment.

Hole forming machinery mainly uses driller and vibro-driver extractor. Slurry pumps are mainly mud pump, mortar pump, gear pump and hand pump.

Other supporting machinery mainly include grouting pipe with small holes, mixer, grouting pipe, valve, pressure gauge, weighing machine, liquid tank, chain block and so on.

(2) Main Points of construction.

①Construction points of cement grouting subbase.

a. Before the reinforcement of grouting subbase, the relevant technical parameters such as the length of grouting section, grouting hole distance and grouting pressure should be determined by test. The length of grouting section is determined by the conditions of the cracks, looseness, and permeability of soil and the ability of grouting equipment. Under general geological conditions, the length of section is mainly controlled at 5—6m. In the case of serious loose soil, developed cracks and strong permeability, the length should be controlled at 2—4m. Generally, grouting hole distance should not be more than 2.0m. The diameter range of single-hole reinforcement can be considered as 1—2m. The depth of hole depends on the depth of soil reinforcement. The grouting pressure means all pressure on the grouting section and the pressure is indicated on the pressure gauge at the orifice. The pressure depends on drilling depth, permeability of soil and consistency of cement slurry, and the pressure is generally 0.3—0.6MPa.

b. Grounting is generally to use neat cement slurry. The range of water cement ratio is 0.6—2.0, and commonly used water cement ratio is from 0.8 : 1—1 : 1. When quick setting is required, fast hardening cement can be used or 1%—2% calcium oxide of cement dosage can be mixed in water. When retarding is required, Calcium Lignosulfonate with cement content of 0.1%—0.5% can be added or other admixtures can also be added to adjust the performance of cement slurry. For layers with large cracks or pores and good groutability, appropriate amount of fine sand or pulverized fuel ash can be mixed into the slurry, and the ratio is 1 : 0.5—1 : 3. For loose soil layer which is not mainly to improve the degree of consolidation, it can also be mixed with fine silty clay in cement slurry to form cement clay slurry, and the ratio of mortar is 1 : 3—1 : 8 (cement : soil, volume ratio), so as to improve the stability of slurry, prevent precipitation and water separation, and make the fill more dense.

c. Grouting construction method is firstly using drill or hand drill to drill holes to the required depth in the reinforced subbase at the specified position. The aperture is generally 55—100mm and geological conditions are detected. Then the grouting shooting tube of 38—50mm in diameter is inserted in the hole. Grouting holes are drilled on pipe walls of 1.0—1.5m in the bottom tube and there is a sleeve outside the shooting tube. Sand is used to fill between the shooting tube and the sleeve. The subbase surface voids are filled with 1 : 3 cement mortar or clay and hemp thread. Then the sleeve is pulled out, and the cement slurry is pressed into shooting tube by slurry pump and penetrated into the pores of the soil layer. The cement slurry should be pressed in one time continuously without interruption. Grouting starts with thin slurry and thickens gradually. Grouting sequence is generally after sinking shooting tube into the entire depth in one time, piecewise continues from the bottom to the top until the tube is pulled out in sections to the orifice. The grouting holes should be grouted with group intervals. After the first group of holes is filled, the second and third groups are grouted.

d. 灌浆完成后,拔出灌浆管,留下的孔用 1∶2 水泥砂浆或细砂砾石填塞密实;亦可用原浆压浆堵孔。

e. 注浆充填率应根据加固土要求达到的强度指标、加固深度、注浆流量、土体的孔隙率和渗透系数等因素确定。饱和软黏土的一次注浆充填率不宜大于 15%～17%。

f. 注浆加固土的强度具有较大的离散性,加固土的质量检验宜采用静力触探法,检测点数应满足有关规范要求。检测结果的分析方法可采用面积积分平均法。

②硅化注浆地基施工要点。

a. 施工前,应先在现场进行灌浆试验,确定各项技术参数。

b. 灌注溶液的钢管可采用内径为 20～50 mm、壁厚大于 5 mm 的无缝钢管。它由管尖、有孔管、无孔接长管及管头等组成。管尖作成 25°～30°圆锥体,尾部带有丝扣与有孔管连接;有孔管长一般为 0.4～1.0 m,每米长度内有 60～80 个直径为 1～3 mm 向外扩大成喇叭形的孔眼、分 4 排交错排列;无孔接长管一般长 1.5～2.0 m,两端有丝扣。电极采用直径不小于 $\phi 22$ mm 的钢筋或 $\phi 33$ mm 钢管。通过不加固土层的注浆管和电极表面,须涂沥青绝缘,以防电流的损耗和作防腐。灌浆管网系统包括输送溶液和输送压缩空气的软管、泵、软管与注浆管的连接部分、阀等,其规格应能适应灌注溶液所采用的压力。泵或空气压缩设备应能以 0.2～0.6 MPa 的压力,向每个灌浆管供应 1～5 L/min 的溶液压入土中。灌浆管间距 1.73R,各行间距为 1.5R(R 为一根灌浆管的加固半径,其数值见表 3.2);电极沿每行注浆管设置,间距与灌浆管相同。土的加固可分层进行,砂类土每一加固层的厚度为灌浆管有孔部分的长度加 0.5R,湿陷性黄土及黏土类土按试验确定。

表 3.2 土的压力硅化加固半径

项次	土的类别	加固方法	土的渗透系数/(m·d^{-1})	土的加固半径/m
1	砂土	压力双液硅化法	2～10	0.3～0.4
			10～20	0.4～0.6
			20～50	0.6～0.8
			50～80	0.8～1.0
2	粉砂	压力单液硅化法	0.3～0.5	0.3～0.4
			0.5～1.0	0.4～0.6
			1.0～2.0	0.6～0.8
			2.0～5.0	0.8～1.0

d. After grouting, the grouting pipe shall be pulled out, and the hole left shall be filled with 1 : 2 cement mortar or fine sand gravel. The hole can also be plugged with original slurry grouting.

e. The filling rate of grouting should be determined by the required index of strength, reinforcement depth, grouting flow rate, porosity and permeability coefficient of soil according to the reinforced soil. The primary grouting filling rate of saturated soft clay should not be greater than 15%—17%.

f. The strength of reinforced soil by grouting has great discreteness. The quality inspection of the reinforced soil should be determined by the cone penetration method, and the number of check points should meet the related code requirement. The analysis of inspection results can be determined by the area average integration method.

②Main points of silicified grouting subbase construction.

a. Before the construction, grouting test should be carried out on the site to determine the technical parameters.

b. Seamless steel pipe with inner diameter of 20—50mm and wall thickness greater than 5mm can be used for steel pipe filled with solution. It is composed of pipe tip, perforated pipe, non-perforated long pipe and pipe head. The pipe tip is a cone of 25°—30°, and the tail has screw thread to connect with perforated pipe. The length of perforated pipe is generally 0.4—1.0m, and there are 60—80 flared holes with diameter of 1—3mm in each meter, which are staggered in four rows. The length of non-perforated long pipe is generally 1.5—2.0m with screw thread at both ends. The electrode uses rebar with a diameter of not less than 22mm or steel pipe with a diameter of 33mm. When passing through the grouting pipe and the electrode surface of non- reinforced soil layer, asphalt for insulation must be coated to prevent the current loss and corrosion. The grouting pipe network system includes hose, pump, and connection part between the hose and grouting pipe, valve and so on for transporting solutions and compressed air. Its specification should be able to suit the pressure used in the perfusion solution. The pump or air compressor should be able to supply 1—5L/min solution to each grouting pipe for compacting into the soil at a pressure of 0.2—0.6MPa. The spacing of grout pipe is 1.73 R, and the spacing of each line is 1.5 R (R is reinforcement radius of a grouting pipe, and its value is shown in Table 3.2). The electrodes are arranged along each row of the grouting pipe, and the spacing is the same as the grouting pipe. The reinforcement of soil can be carried out in layers. The thickness of each reinforcement layer of sand soil is the length of perforated part of grouting pipe and plus 0.5R. Collapsible loess and clay soil are determined by tests.

Table 3.2 Pressure Silicification Reinforcement Radius of soil

Items	Type of Soil	Reinforcement Method	Soil Permeability Coefficient/(m · d^{-1})	Soil Reinforcement Radius/m
1	Sandy Soil	Pressure Double Liquid Silicification	2—10	0.3—0.4
			10—20	0.4—0.6
			20—50	0.6—0.8
			50—80	0.8—1.0
2	Silty Sand	Pressure Single Liquid Silicification	0.3—0.5	0.3—0.4
			0.5—1.0	0.4—0.6
			1.0—2.0	0.6—0.8
			2.0—5.0	0.8—1.0

c. 灌浆管的设置,用打入法或钻孔法(振动打拔管机、振动钻或三脚架穿心锤)沉入土中,保持垂直和距离正确,管子四周孔隙用土填塞夯实。电极可用打入法或先钻孔 2~3 m 深再打入。

d. 硅化加固的土层以上应保留 1 m 厚的不加固土层,以防溶液上冒,必要时须夯填素土或灰土层。

e. 灌注溶液的压力一般在 0.2~0.4 MPa(始)和 0.8~1.0 MPa(终)范围内,采用电动硅化法时,不超过 0.3 MPa(表压)。

f. 土的加固程序,一般自上而下进行,如土的渗透系数随深度而增大时,则应自下而上进行。如相邻土层的土质不同时,渗透系数较大的土层应先进行加固。

灌注溶液的次序,根据地下水的流速而定。当地下水流速在 1 m/d 时,向每个加固层自上而下灌注水玻璃,然后再自下而上灌注氯化钙溶液,每层厚 0.6~1.0 m;当地下水流速为 1~3 m/d 时,轮流将水玻璃和氯化钙溶液均匀地注入每个加固层中;当地下水流速大于 3 m/d 时,应同时将水玻璃和氯化钙溶液注入,以减低地下水流速,然后再轮流将两种溶液注入每个加固层。采用双液硅化法灌注,先从单数排的灌浆管压入,然后从双数排的灌浆管压入;采用单液硅化法时,溶液应逐排灌注。

(3)质量检查。

①施工前应掌握有关技术文件(注浆点位置、浆液配比、注浆施工技术参数、检测要求等)。浆液组成材料的性能应符合设计要求,注浆设备应确保正常运转。

②施工中应经常抽查浆液的配比及主要性能指标,注浆的顺序、注浆过程中的压力控制等。

③施工结束后,应检查注浆体强度、承载力等。检查孔数为总量的 2%~5%,不合格率大于或等于 20%时应进行二次注浆。检验应在注浆后 15 d(砂土、黄土)或 60 d(黏性土)进行。

c. The grouting pipe is arranged to sink into the soil by means of driving or drilling(vibro-driver extractor, vibratory drill or tripod core hammer) and keep vertical and correct distance. Pores around pipe are filled and tamped with earth. The electrode can be driven or drilled 2—3m deep before driving.

d. 1m thick unreinforced soil layer should be reserved above the silicified soil layer to prevent the solution from rising. If necessary, plain soil or lime soil layer should be filled and tamped.

e. The pressure of perfusion solution is generally within the range of 0.2—0.4MPa(beginning) and 0.8—1.0MPa(end), and when electric silicification method is adopted, it shall not exceed 0.3MPa(gauge pressure).

f. Soil reinforcement sequence is generally from top to bottom. If the permeability coefficient of soil increases with depth, the reinforcement should be carried out from bottom to top. If the soil quality of adjacent soil layer is different, the soil layer with large permeability coefficient should be reinforced first.

The sequence of perfusion solution is determined by underground water flow velocity. When underground water flow velocity is at 1m/d, the sodium silicate is poured from top to bottom into each reinforcement layer, and then the calcium chloride solution is poured from bottom to top. The thickness of each layer is at 0.6—1.0m. When underground water flow velocity is at 1—3m/d, the sodium silicate and calcium chloride solution are in turn injected evenly into each reinforcement layer. When underground water flow velocity is greater than 3m/d, the sodium silicate and calcium chloride solutions should be injected at the same time to reduce the underground water flow velocity, and then the two solutions are in turn injected into each reinforcement layer. When the double liquid silicification method is used for grouting, first press in from odd rows of the grouting pipe, and then press in from double rows of the grouting pipe. When using single liquid silicification method, the solution should be grouted row by row.

(3) Quality inspection.

①The relevant technical documents(position of grouting point, grouting proportion, technical parameters of grouting construction, testing requirements, etc.) should be mastered before the construction. The properties of slurry composition material should meet the design requirements, and the grouting equipment should be ensured in normal operation.

②Grouting proportion and its main performance index, grouting sequences and the pressure control during grouting should be checked frequently during the construction.

③After the completion of construction, the strength and bearing capacity of the grouting body and so on should be checked. The number of holes for inspection is 2%—5% of the total quantity, and the second grouting should be carried out when the unqualified rate is more than 20%. Inspection should be carried out at 15d(sandy soil, loess) or 60d(clay soil) after grouting.

5. 振冲地基

利用振动和水冲加固土体的方法称为振冲法。振冲法分为振冲挤密法和振冲置换法两类，用于振密松砂地基时，称为"振冲挤密"。用于黏性土地基，在黏性土中制造一群以碎石、卵石或砂砾材料组成的桩体，从而构成复合地基。这种方法称之为"振冲置换"。

振冲法适用于处理砂土、粉土、粉质黏土、松散与稍密卵石土、素填土和杂填土等地基。对于处理不排水抗剪强度≥20 kPa的饱和黏性土和饱和黄土地基，应在施工前通过现场试验确定其适用性。不加填料振冲密实法适用于处理粘粒含量不大于10%的中砂、粗砂地基。

(1)机具设备、材料。

①主要机具有振冲器、起重机械、水泵及供水管道、加料设备和控制设备等。

②主要材料为粒径20~100 mm的卵、碎石等粗骨料。

(2)施工工艺：清理平整场地→定桩位→成孔→清孔(对置换法)→填料→振实(重复填料→振实至桩顶)→铺设褥垫层。

(3)施工要点。振冲施工过程如图3.1所示。

①清理平整施工场地，对置换法宜超挖200~300 mm，对密实法应根据土层松散情况适当预留100~200 mm。

②根据施工图纸布置桩位，对置换法宜人工取定位孔，定位孔直径宜与设计桩径一致，深度不小于500 mm。

③组织泥浆排放系统，设置沉淀池(对密实法可不设沉淀池)。

④施工机具就位，使振冲器对准桩位

⑤启动供水泵，水压可用200~600 kPa，水量可用200~400 L/min，将振冲器徐徐沉入土中，造孔速度宜为0.5~2.0 m/min，直至达到设计深度。记录振冲器经过深度的水压、电流和时间。

⑥造孔后边提升振冲器边冲水直至孔口，再放至孔底，重复两三次扩大孔径并使孔内泥浆变稀，开始填料制桩。

⑦大功率振冲器投料可不提出孔口，小功率振冲器下料困难时，可将振冲器提出孔口填料，每次填料厚度不宜大于500 mm。将振冲器沉入填料中进行振密制桩，当电流达到规定的密实电流值和规定的留振时间后，将振冲器提升300~500 mm。

5. Vibroflotation Subbase

The method of using vibration and water flushing to reinforce soil is called vibroflotation method. Vibroflotation method is divided into vibroflotation compaction method and vibroflotation replacement method. When it is used to vibroflotation loose sand foundation, it is called "vibroflotation compaction". The method used in clay soil subbase is called "vibroflotation replacement", which makes a group of piles composed of crushed stone, pebbles or sand and gravel inclayey soil subbase to form a composite subbase.

The vibroflotation method is suitable for the treatment of sandy soil, silt, silty clay, loose and slightly dense pebble soil, plain fill and miscellaneous fill, etc. The applicability of saturated clay and saturated loess subbase with undrained shear strength greater than or equal to 20kPa should be determined through site tests before the construction. The vibro-compaction without filler is suitable for treating medium and coarse sand subbase with clay content less than 10%.

(1) Machinery, equipment and materials.

①The main machinery and equipment are vibroflot, hoisting machinery, water pump and water supply pipeline, feeding equipment and control equipment, etc.

②The main materials are coarse aggregates such as pebbles and gravel with grain diameter of 20—100mm.

(2) Construction technology: site cleaning and leveling → pile position determination → hole forming → hole cleaning(pair replacement method)→ filling → tamping(repeated filling → tamping to pile top)→ bedding laying.

(3) Construction points.

The vibroflotation construction process is shown in Fig. 3.1.

①Site cleaning and leveling: The replacement method is appropriate to over-dig 200—300mm, and the compaction method should properly reserve 100—200mm according to loose soil condition.

②Pile layout should be done according to the construction drawing. For the replacement method, the positioning hole should be manually taken, the diameter of positioning hole should be in accordance with the diameter of designed pile, and the depth should not be less than 500mm.

③Organize mud discharge system and set up sedimentation tank (not required for the compaction method).

④The construction machine is in place and the vibroflot has alignment with the pile position.

⑤Start the water supply pump. Available water pressure can be 200—600kPa and water volume can be 200—400L/min. The vibroflot is sunk into the soil slowly, and the speed of hole-forming should be at 0.5—2.0m/min until the designed depth is reached. Record the water pressure, electric current and time when the vibroflot goes through the depth.

⑥After the hole is formed, lift the vibroflot while flush water until the orifice, and then put it to the bottom of hole. Repeat two or three times to enlarge aperture, make the mud thin in the hole, and then start to fill and make the pile.

⑦High-power vibroflot feeding may not lift out of orifice. When low-power vibroflot feeding is difficult, the vibroflot can be lifted out of the orifice to fill, and the thickness of each filling should not be more than 500mm. Put the vibroflot into the filler to compact and form pile. When current reaches the specified density current value and the specified holding time, lift the vibroflot 300—500mm.

⑧重复以上步骤,自下而上逐段制作桩体直至孔口,记录各段深度的填料量、最终电流值和留振时间,并均应符合设计规定。

⑨桩体施工完毕后宜及时铺设垫层。

⑩关闭振冲器和水泵,场地内振冲桩施工完毕后,应对垫层压实找平处理。

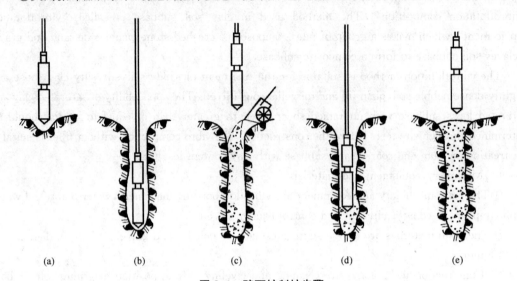

图 3.1　碎石桩制桩步骤
(a)定位;(b)振冲下沉;(c)加填料;(d)振密;(e)成桩

(4)质量控制要点。

①施工前,应检查振冲器的性能及电流表,电压表的准确度及填料的性能。

②施工中,应检查供水压力、供水量、振冲点位置;严格控制密实电流、填料量、留振时间。

③施工结束后,应在有代表性的地段做地基强度(动力触探)及地基承载力(单桩复合地基静载)检验。检验的时间除砂土地基外,应间隔一定时间方可进行,对黏性土地基,间隔时间,为3～4周;对粉土地基,为2～3周。

6. 水泥土搅拌地基

水泥土搅拌法简称 MIP 法,是利用水泥或石灰等材料作为固化剂,通过特制的深层搅拌机械,在地基深处就地将固化剂和地基土强制搅拌,使软土硬结成具有整体性、水稳定性和一定强度的桩体的地基处理方法。根据施工方法的不同,水泥土搅拌法分为深层搅拌或水泥浆搅拌(简称湿法)和粉体喷射搅拌(简称干法)两种。

⑧Repeat the above steps, the pile is made up to the orifice section by section from the bottom to the top. Record filling quantity of each section, final current value and the holding time of vibration, and all should meet the design requirements.

⑨Cushion should be laid in time after the completion of pile construction.

⑩Turn off the vibroflot and water pump, and the cushion should be treated with compaction and leveling after the completion of vibroflotation pile construction in the site.

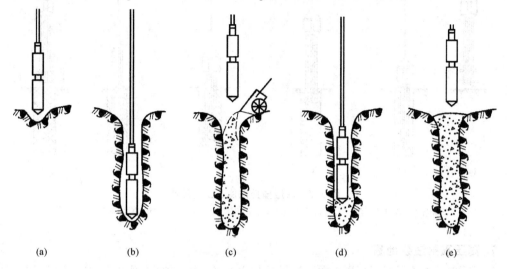

Fig. 3.1 Steps of Making Gravel Pile
(a)positioning; (b)vibro-sinking; (c)filling; (d)vibrating densification; (e)pile formed

(4) Main Points of quality control.

① Before construction, the performance of vibroflot, the accuracy of ampere meter and voltmeter and performance of filling should be checked.

②In construction, the pressure of water supply, the volume of water supply and the position of vibration point should be checked. The compaction current, the amount of filler and the vibration time are strictly controlled.

③After construction, the subbase strength(dynamic penetration) and subbase bearing capacity (single pile composite subbase dead load) should be tested in representative sections. Except for sandy soil subbase, the test should be carried out at a certain interval. For clayey soil subbase, the interval is 3—4 weeks and 2—3 weeks for silty subbase.

6. Cement Soil Mixing Subbase

Cement soil mixing method is MIP method for short. It is a subbase treatment method which uses cement or lime and other materials as curing agent, through the special deep mixing machine, the curing agent and the foundation soil are forced to mix in locally in the deep subbase, so as to make the soft soil hard into a pile with integrity, water stability and certain strength. According to the different construction methods, cement soil mixing method is divided into deep mixing or cement slurry mixing method(wet method) and powder jet mixing method(dry method).

深层搅拌法是用于加固饱和软黏土地基的一种方法，它是利用水泥、石灰等材料作为固化剂，通过特制的深层搅拌机械，如图3.2所示。在地基深处就地将软土和固化剂（浆液）强制搅拌，利用固化剂和软土之间所产生的一系列物理化学反应，使软土硬结成具有整体性、水稳定性和一定强度的地基。深层搅拌法还常作为重力式支护结构用来挡土、挡水（图3.2）。

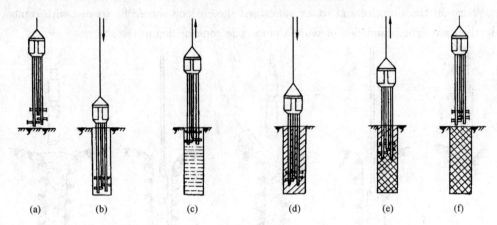

图3.2 深层搅拌法施工工艺流程
(a)定位；(b)预拌下沉；(c)喷浆搅拌机上升；(d)重复搅拌下沉；(e)重复搅拌上升；(f)完毕

7. 高压喷射注浆地基

高压喷射注浆法是指用高压水泥浆通过钻杆由水平方向的喷嘴喷出，形成喷射流，以此切割土体并与土拌和形成水泥土加固体的地基处理方法。高压喷射注浆地基主要用于地基加固，提高地基的抗剪强度，改善土体的变形性质，使地基在荷载下不产生破坏或过大的变形。

高压喷射注浆地基适用于处理淤泥、淤泥质土、流塑、软塑或可塑黏性土、粉土、砂土、黄土、素填土和碎石土等地基。但对含有较多大粒块石、坚硬黏性土、大量植物根茎或含过多有机质的土以及地下水流过大、喷射浆液无法在注浆管周围凝聚的情况下，不宜采用。

高压喷射注浆法所形成的固结体形状与喷射流移动方向有关。一般分为旋转喷射（简称旋喷）、定向喷射（简称定喷）和摆动喷射（简称摆喷）三种形式，如图3.3所示。

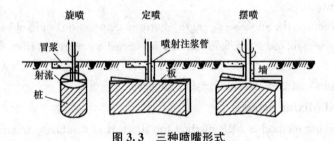

图3.3 三种喷嘴形式

Chapter 3 Subbase and Foundation Construction

Deep mixing method is a method used for the reinforcement of saturated soft clay subbase. It uses cement, lime and other materials as curing agent, through the special deep mixing machine, the soft soil and curing agent (slurry) are forced to mix locally in the deep subbase, and use a series of physical and chemical reaction between curing agent and soft soil to make soft soil hard into the subbase with integrity, water stability and a certain strength. Deep mixing method is often used as gravity supporting structure for retaining soil and water. The construction process of deep mixing method is shown in Fig. 3. 2.

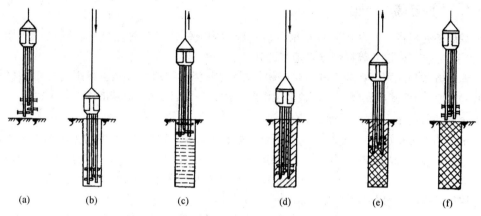

Fig. 3. 2 Construction Process of Deep Mixing Method
(a) positioning; (b) premix and sink; (c) shotcrete mixer rising; (d) remix and sink; (e) remix and rise; (f) end

7. High Pressure Jet Grouting Subbase

High Pressure jet grouting method is a method for the subbase treatment, which means high-pressure cement slurry is ejected from the horizontal nozzle by the drill stem to form a jet stream, so as to cut the soil and mix with soil to form cement soil solid. High Pressure jet grouting subbase is mainly used to reinforce subbase, improve the shear strength of subbase and improve the deformation properties of soil, so that the subbase under load does not produce damage or excessive deformation.

High Pressure jet grouting subbase applies to treat the subbase of mud, mucky soil, flow plastic soil, soft plastic or plastic clay, silt soil, sandy soil, loess soil, plain fill and gravel soil, etc. However, it is not suitable for soil with more large stone, hard clay, many plant roots or soil with too much organic matter as well as the situation that groundwater flow is too large and jet slurry cannot condense around grouting pipe.

The shape of consolidating body formed by the high pressure jet grouting method is related to the moving direction of jet stream. The high pressure jet grouting is generally divided into three types of spin jet, directional jet and swing jet, as shown in Fig. 3. 3.

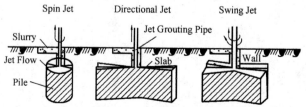

Fig. 3. 3 Three Types of Spray Nozzel

旋喷法施工时,喷嘴一边喷射一边旋转和提升,固结体呈圆柱状。主要用于加固地基,提高地基的抗剪强度,改善地基土的变形性质,也可组成闭合的帷幕,用于截阻地下水流和治理流砂。旋喷法施工后,在地基中形成的圆柱体,简称旋喷桩。定喷法施工时,喷嘴一边喷射一边提升,喷射的方向固定不变,固结体形如板状或壁状。摆喷法施工时,喷嘴一边喷射一边提升,喷嘴的方向呈较小的角度来回摆动,固结体形如较厚的墙板状。定喷和摆喷两种方法通常用于基坑防渗、改善地基土的水流性质及稳定边坡等工程。

3.1.2 地基局部处理

地基局部处理指在浅基础开挖基槽(坑)的施工中或验槽(坑)时,发现基槽(坑)范围内有洞穴、软弱土层或岩基、墙基等局部异常地基的处理。

处理的方法和原则是将局部软弱层或硬物尽可能挖除,回填与天然土压缩性相近的材料,分层夯实;处理后的地基应保证建筑物各部位沉降量趋于一致,以减少地基的不均匀下沉。

1. 松土坑(填土、墓穴、淤泥等)的处理

当坑的范围较小(在基槽范围内),可将坑中松软土挖除,使坑底及四壁均见天然土为止,回填与天然土压缩性相近的材料。当天然土为砂土时,用砂或级配砂石回填;当天然土为较密实的黏性土,则用3∶7灰土分层回填夯实;如为中密可塑的黏性土或新近沉积黏性土,可用1∶9或2∶8灰土分层回填夯实,每层厚度不大于20 cm[图3.4(a)]。

当坑的范围较大(超过基槽边沿)或因条件限制,槽壁挖不到天然土层时,则应将该范围内的基槽适当加宽,加宽部分的宽度可按下述条件确定:当用砂土或砂石回填时,基槽每边均应按1∶1坡度放宽;当用1∶9或2∶8灰土回填时,按0.5∶1坡度放宽;当用3∶7灰土回填时,如坑的长度≤2 m,基槽可不放宽,但灰土与槽壁接触处应夯实[图3.4(b)]。

When using spin jet in the construction, the nozzle is jetting while spinning and lifting, and the consolidating body has formed the shape of cylinder. It is mainly used to reinforce the subbase, increase the shear strength of subbase and improve the deformation properties of subbase soil, and it can also be used to form the closed layer to resist the underflow and treat the flowing sand. After spin jet construction, the cylinder formed in the subbase is called spin pile in short. When using directional jet in the construction, the nozzle is jetting while lifting, and the direction of jetting is fixed. The consolidating body is shaped like the board or the wall. When using the swing jet in the construction, the nozzle is jetting while lifting, the direction of jetting is swinging back and forth at small angle, and the consolidating body is shaped like thicker wallboard. Directional jet and swing jet are usually used to prevent the foundation pit from the seepage, improve the flow properties of subbase soil and stabilize the side slope, etc.

3.1.2 Local Treatment of Subbase

Local treatment of subbase refers to the treatment of local abnormal subbase such as cave, soft soil layer, rock foundation and wall foundation in the scope of foundation trench (foundation pit) during the excavation of foundation trench (foundation pit) or inspection of foundation trench (foundation pit).

Treatment method and principles are to remove local weak layers or hard objects as much as possible, backfill with the materials that are similar to the natural soil in compressibility and tamp in layers. The treated subbase should ensure the settlement of all parts of the building to be consistent to reduce the differential settlement of subbase.

1. Treatment of Loose Soil Pit (Filling, Grave, Mud, etc.)

When the pit is small (within the range of foundation trench), the soft soil in the pit can be removed until the natural soil can be seen at the bottom and walls, and then the material with similar compressibility to the natural soil can be backfilled. When the natural soil is sandy soil, sand or graded sand and gravel can be backfilled. When the natural soil is more dense clay soil, 3 : 7 lime soil is used to backfill and tamp in layers. If the natural soil is medium density plastic clay soil or newly deposited clay soil, 1 : 9 or 2 : 8 lime soil can be used to backfill and tamp in layers, and the thickness of each layer is no more than 20cm [Fig. 3.4(a)].

When the pit is large (beyond the edge of foundation trench) or the natural soil layer cannot be excavated in the trench wall because of the limited conditions, the foundation trench within the scope should be widened appropriately, and the width of the widened part can be determined by the following conditions: When backfilling uses sandy soil or sandstone, each side of foundation trench should be broadened at a gradient of 1 : 1. When backfilling uses 1 : 9 or 2 : 8 lime soil, the foundation trench should be broadened at a gradient of 0.5 : 1. When 3 : 7 lime soil is used to backfill, and if the length of pit is ≤2m, the foundation trench may not be broadened, but the contact part between lime soil and the trench wall should be tamped [Fig. 3.4(b)].

如坑在槽内所占的范围较大(长度在 5 m 以上),且坑底土质与一般槽底天然土质相同,可将此部分基础加深,做 1∶2 踏步与两端相接,踏步多少根据坑深而定,但每步高不大于 0.5 m,长不小于 1.0 m[图 3.4(c)]。

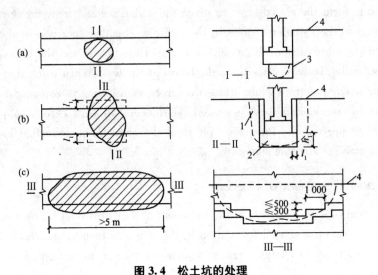

图 3.4 松土坑的处理

1—软弱土;2—2∶8 灰土;3—松土全部挖除然后填以好土;4—天然地面

对于较深的松土坑(如坑深大于槽宽或大于 1.5 m 时),由于回填材料与天然地基密实度相差较大,会造成基础不均匀下沉,所以还要考虑加强上部结构的强度,以抵抗地基不均匀沉降而引起的内力。因此槽底处理后,还应适当考虑加强上部结构的强度,方法是在灰土基础上 1~2 皮砖处(或混凝土基础内)、防潮层下 1~2 皮砖处及首层顶板处,加配 4 根 $\phi 8$~$\phi 12$ 的钢筋跨过该松土坑两端各 1 m 或加钢筋砖过梁,以防产生过大的局部不均匀沉降,如图 3.5 所示。

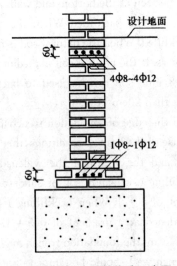

图 3.5 基础内配筋构造示意图

Chapter 3　Subbase and Foundation Construction

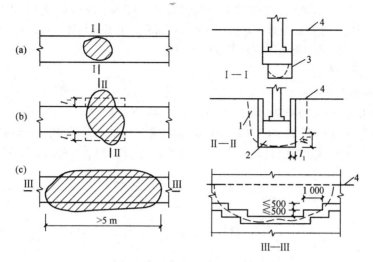

Fig. 3.4　Treatment of Loose Soil Pit
1—soft soil;2—2∶8 lime soil;
3—remove all loose soil and backfill with good soil;4—natural ground

When the scope of pit in the trench is large(The length is more than 5m), and the soil texture at the bottom of pit is the same as the natural soil at the general bottom of trench, this part of foundation can be deepened and 1∶2 steps are made to connect with the both sides. The number of steps depends on the depth of pit, but the height of every step is not more than 0.5m and the length is no less than 1.0m [Fig. 3.4(c)].

For the deeper loose soil pit(for example, the depth of pit is larger than the width of trench or larger than 1.5m), differential settlement may occur to the foundation because of the big difference between the backfilled material and the density of natural subbase. So, strengthening the superstructure should be considered to resist the internal force caused by the differential settlement of subbase. Therefore, after the treatment of trench bottom, strengthening the superstructure should be properly considered. The method is that in the place of 1 or 2 bricks of the lime soil foundation(or concrete foundation), 1 or 2 bricks under the damp-proofing course and at the top of the first floor, 4 rebars with $\phi 8 - \phi 12$ should be added across 1m of the both sides of the loose soil pit or the reinforced brick lintel can be added, so as to prevent the large local differential settlement as shown in Fig. 3.5.

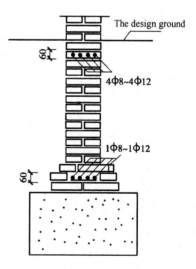

Fig. 3.5　Schematic Diagram of Reinforcing Bars in Foundation

· 103 ·

如遇到地下水位较高,坑内无法夯实时,可将坑(槽)中软弱的松土挖去后,再用砂土、碎石或混凝土代替灰土回填。如坑底在地下水位以下时,回填前先用粗砂与碎石(比例为1∶3)分层回填夯实;地下水位以上用3∶7灰土回填夯实至要求高度。

2. 砖井或土井的处理

(1)砖井或土井在室外。

当砖井或土井在室外,距离基础边缘5 m以外时,不予考虑。

当砖井或土井在室外,距离基础边缘5 m以内时,应先用素土分层夯实,回填到室外地坪以下1.5 m,将井壁四周砖圈拆除或松软部分挖去,之后用素土分层回填并夯实,如图3.6所示。

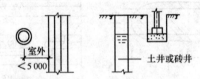

图3.6 室外砖井或土井的处理方法

(2)砖井或土井在室内基础附近。

当井在室内基础附近时,应采用以下措施:

①将地下水降低到最低可能的限度,用中、粗砂及块石、卵石或碎砖等回填到地下水位以上0.5 m。

②砖井应将四周井圈拆至基槽(基坑)底以下1 m以上,然后用素土分层回填夯实。

③若井已回填,但不密实或有软土,可用大块石将下面软土挤紧,再分层回填素土夯实,如图3.7所示。

图3.7 室内基础附近砖井或土井的处理方法

If the groundwater level is higher and the pit cannot be compacted, the weak loose soil in the pit (trench) can be removed, and then sandy soil, crushed stone or concrete can be used to replace the lime soil for backfilling. If the bottom of pit is below the groundwater level, coarse sand and crushed stone(1 : 3) can be filled in layers and compacted before backfilling. If the bottom of pit is above the groundwater level, 3 : 7 lime soils are used to backfill and compact to the required height.

2. Treatment of Brick Well or Earth Well

(1) The brick well or earth well is outdoor.

When the brick well or earth well is outdoor and 5m away from the foundation edge, it will not be considered.

When the brick well or earth well is outdoor and within 5m of the foundation edge, it should be tamped in layers by using plain soil first, and then backfilled to 1.5m below the outdoor floor. The bricks around the well wall should be demolished or the soft part should be removed, and then backfilled and tamped in layers by using plain soil as shown in Fig. 3.6.

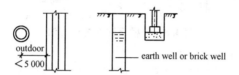

Fig. 3.6 Treatment of Outdoor Brick Well or Earth Well

(2) The brick well or earth well is near the indoor foundation.

When the well is near the indoor foundation, the following measures should be taken:

①The groundwater should be reduced to the lowest possible limit. The medium sand, coarse sand and block stone, pebble or broken bricks and so on are used to backfill to reach 0.5m above the groundwater level.

②For brick well, the well circle around should be demolished to 1m below the bottom of the foundation trench(foundation pit), and then backfilled and tamped in layers with plain soil.

③If the well has been backfilled, but the backfilling is not compacted or there is soft soil, the big block stone can be used to compact the underlying soft soil, and then the plain soil is used to backfill and tamp in layers, as shown in Fig. 3.7.

Fig. 3.7 Treatment of brick well or earth well near the indoor foundation

(3)砖井或土井在基槽(基坑)中间。当砖井或土井在基槽(基坑)中间,应先用素土分层回填夯实至基础底下2m处,将井壁四周松软部分挖去,若有井圈时,则应将井的砖圈拆除至槽(坑)底以下1~1.5m,在此拆除范围内用2∶8或3∶7灰土分层夯实至槽(坑)底,如图3.8所示。若井内有水时,应用中、粗砂及块石、卵石或碎砖回填至水位以上0.5m,然后再按上述方法处理;如井已回填,但不密实,且挖出困难或不能夯填密实时,则可在部分拆出的砖石井圈上加钢筋混凝土盖封口,上部再用素土或用2∶8或3∶7灰土分层夯实至槽(坑)底,如图3.9所示。如井的直径大于1.5m时,则应适当考虑加强上部结构的强度,如在墙内配筋或做地基梁跨越砖井等。

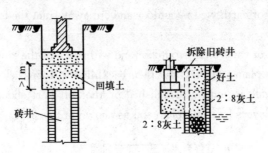

图3.8　基槽下砖井处理方法

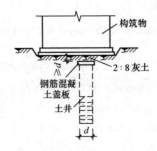

图3.9　基槽下砖井钢筋混凝土盖处理方法

(4)井在基础的转角处。若井在基础的转角处,除采用上述拆除回填办法处理外,还应对基础加强处理。

Chapter 3 Subbase and Foundation Construction

(3) The brick well or earth well is in the middle of the foundation trench(foundation pit).

When the brick well or earth well is in the middle of the foundation trench(foundation pit), the plain soil should be used to backfill and be tamped in layers to 2m below the foundation, and the soft part around the well wall should be dug out. If there is well circle, the brick around the well should be demolished to 1—1.5m below the trench(pit). Within the scope of this demolition, 2:8 or 3:7 lime soil is used to tamp in layers to the bottom of trench(pit), as shown in Fig. 3.8. If there is water in the well, the medium sand, coarse sand and the block stone, pebbles or broken bricks are used to backfill to 0.5m above the water level, and then treated by the above methods. If the well has been backfilled but not dense and it is difficult to dig out or cannot be compacted, the reinforced concrete cover can be added on the demolished masonry well circle and the upper part can be tamped in layers by using plain soil or 2:8 or 3:7 lime soil to the bottom of trench(pit), as shown in Fig. 3.9. If the diameter of well is greater than 1.5m, the strength of superstructure should be properly considered to be strengthened, such as in-wall reinforcement or foundation beam across the brick well.

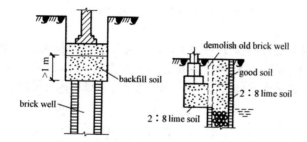

Fig. 3.8 Treatment of brick well below foundation trench

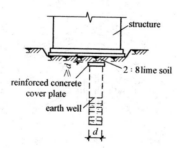

Fig. 3.9 Treatment of Reinforced Concrete Cover of Brick Well under Foundation Trench

(4) The well is at the corner of the foundation. If the well is at the corner of the foundation, the foundation should be reinforced in addition to the demolition and backfilling method mentioned above.

①当井位于房屋转角处,而基础压在井上部分不多,并且在井上部分所损失的承压面积,可由其余基槽承担而不引起过多的沉降时,则可采用从基础中挑钢筋混凝土梁的办法解决,如图 3.10 所示。

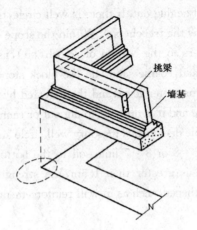

图 3.10 墙角下砖井处理方法(一)

②当井位于墙的转角处,而基础压在井上的面积较大,且采用挑梁办法较困难或不经济时,则可将基础沿墙长方向向外延长出去,使延长部分落在天然土上。落在天然土上的基础总面积,应等于井圈范围内原有基础的面积(即 $A_1+A_2=A$),然后在基础墙内再采用配筋或钢筋混凝土梁来加强,如图 3.11 所示。

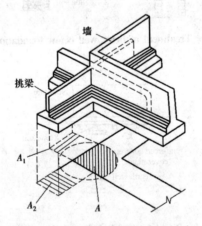

图 3.11 墙角下砖井处理方法(二)

① When the well is located at the corner of the house, but the foundation on the well is not much, and the pressure area lost on the well can be borne by other foundation trench without causing excessive settlement, then the reinforced concrete beam over hanged from the foundation can be used to solve the problem, as shown in Fig. 3.10.

② When the well is located at the corner of the wall, and the foundation pressure has a large area on the well, moreover, it is more difficult or not economic to use cantilever beam, then the foundation can extend outward along the length of the wall to make the extension part fall on the natural soil. The total area of foundation falling on the natural soil should be equal to the area of original foundation within the well circle (That is: $A_1 + A_2 = A$), and then reinforcement or reinforced concrete beams can be used to strengthen in the foundation wall, as shown in Fig. 3.11.

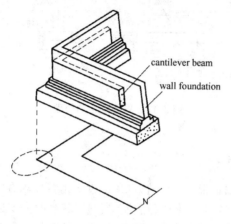

Fig. 3.10 Treatment of Brick Well below the Wall Corner(I)

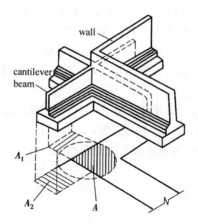

Fig. 3.11 Treatment of Brick Well below the Wall Corner(II)

3.1.3 局部范围内硬土(或其他硬物)的处理

当柱基或部分基槽下,有较其他部分过于坚硬的土质时,例如:基岩、旧墙基、老灰土、化粪池、大树根、砖窑底、压实的路面等,均应尽可能挖除,以防建筑物由于局部落于较硬物上造成不均匀沉降,而使上部建筑物开裂。

局部范围硬土处理方法为:首先在地坑、地槽范围内尽可能地挖除硬土或坚硬物,以免基础局部落在硬物上造成不均匀沉降使上部建筑物开裂。硬土、硬物挖除后,若深度小于 1.5 m 时,可用砂、砂卵石或灰土回填;若长度大于 5 m 时,则将槽底做 1∶2 踏步灰土垫层与两端紧密连接,然后做落深基础,如图 3.12 所示。

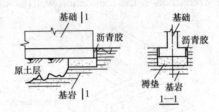

图 3.12 局部硬土的处理

3.1.4 橡皮土的处理

当地基为黏性土,且含水量很大、趋于饱和的黏性土地基回填夯实时,由于原状土被扰动,颗粒之间的毛细孔遭到破坏,水分不易渗透和散发,当气温较高时夯实或碾实,表面会形成硬壳,更阻止了水分的渗透和散发,埋藏深的土水分散发慢,往往长时间不易消失,形成软塑状的橡皮土,踩上去有颤动的感觉。

橡皮土的处理方法:

(1)对含水量很大的黏土、粉质黏土、淤泥质土、腐质土等原状土,暂停一段时间施工,避免再直接拍打,使其含水量逐渐降低,或将土层翻起进行晾槽;

(2)对已成的橡皮土可采取在上面铺一层碎石或碎砖后进行夯实,将表土层挤紧;

(3)严重的橡皮土,可将土蹭翻起并破碎均匀,掺加石灰粉吸水分,并水化改变原土结构成为灰土,使之具有一定强度和水稳性;

(4)当为荷载大的房屋地基,可采用打石桩,将毛石依次打入土中,间距 400~500 mm,直至打不下去为止,最后在上面铺满 50 mm 的碎石在夯实;

(5)采取换土法,挖去橡皮土,重新填好土或级配砂石夯石。

3.1.3 Treatment of Local Hard Soil(or Other Hard Object)

When there is too hard soil below column foundation or part of the foundation trench, such as bedrock, old wall foundation, old lime soil, septic tank, big tree root, brick kiln bottom, compacted road surface, and so on, they should be removed as far as possible, so as to prevent the building from differential settlement because it falls locally on the harder object to make the superstructure crack.

The treatment of local hard soil is first to remove hard soil or hard objects as far as possible within the scope of foundation pit and foundation trench, so as to prevent foundation from falling locally on hard objects to cause differential settlement and make superstructure crack. If the depth is less than 1.5m after removing hard soil and hard objects, the sand, sandy pebble or lime soil can be used to backfill. If the length is greater than 5m, the trench bottom should be made into 1 : 2 step lime soil cushion to closely connect with both ends, and then deep foundation is made as shown in Fig. 3.12.

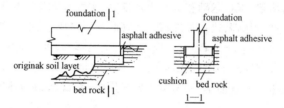

Fig. 3.12 Treatment of Local Hard Soil

3.1.4 Treatment of Rubber Soil

When the subbase is clayey soil and with high water content, backfilling and tamping the tending saturation clayey soil will form a hard shell on the surface because of the disturbance of undisturbed soil, the destruction of pores between particles, difficult penetration and emission of moisture and the compaction and rolling at the higher temperature. Even more, it prevents moisture from permeating and sending out, so that the moisture of deep-buried soil sends out slowly and often not easily disappears for a long time. As a result, it forms soft plastic rubber soil and feels trembling when stepping on it.

Treatment of rubber soil:

(1) For undisturbed soil with high water content, such as clay, silty clay, mucky soil and humus soil, the construction should be suspended for a period of time and direct patting again should be avoided so as to reduce the water content gradually, or the soil layer is turned up to dry the trench.

(2) For the formed rubber soil, a layer of crushed stone or crushed brick can be laid on the soil first, and then be tamped, squeezing the topsoil.

(3) For serious rubber soil, the soil layer can be turned up and broken evenly, mixed with lime to absorb moisture and the original soil structure is changed into lime soil by hydration to make it have certain strength and water stability.

(4) When it is the building subbase with heavy load, stone piles can be used. Ashlars are hit in turn into the soil with 400—500mm spacing until they cannot go down. Finally 50mm crushed stone are covered on the top and tamped.

(5) Take the method of soil replacement, dig out the rubber soil, refill the soil or compact with graded sand.

3.2 浅基础工程施工

浅基础指埋入地层深度较浅,一般埋置深度小于基础宽度,或埋置深度小于 5 m,建造在自然地基上可用敞开挖基坑修筑的基础。一般工业与民用建筑的基础多采用天然浅基础。它技术简单,施工方便,施工设备简便,工期短,工程造价低,在设计计算时可以忽略基础侧面土体对基础的影响,故在保证建筑安全和正常使用的前提下,应尽量采用浅基础。

3.2.1 浅基础的分类

浅基础是应用最广泛的基础,指除桩基础、地下连续墙以外的全部基础。

按受力特点可分为刚性基础和柔性基础。刚性基础又称无筋扩展基础,主要有:砖基础、毛石基础、灰土基础、三合土基础、碎石基础和混凝土基础等;柔性基础又称扩展基础,主要是钢筋混凝土基础、桩(预制桩、灌注桩)基础、独立基础、条形基础、筏板基础、箱形基础等。

按构造形式分为单独基础、条形基础(带形基础)、交梁基础、筏板基础等。单独基础又叫独立基础,多呈柱墩形,截面可做成阶梯形或锥形等;带形基础是指长度远大于其高度和宽度的基础,常见的有墙下条形基础和柱下条形基础,材料主要采用砖、毛石、混凝土和钢筋混凝土等。

3.2.2 刚性基础

刚性基础是用抗压强度大,而抗弯、抗拉强度小的材料建造的基础,其特点是抗压性能好,但抗拉、抗弯、抗剪性能差。适用于基础土较好且均匀,上部荷载较小的多层民用建筑和墙承重的轻型厂房。刚性基础形式如图 3.13 所示。

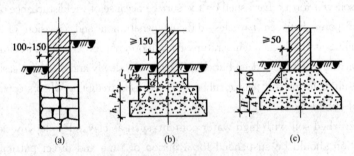

图 3.13 刚性基础形式
(a)矩形;(b)阶梯形;(c)锥形

3.2 Construction of Shallow Foundation

The shallow foundation refers to the foundation which the buried depth is shallow, the embedment depth is generally less than the foundation width or less than 5m, and it is built on natural subbase with open excavation pit. Generally, the foundation of industrial and civil buildings uses the natural shallow foundation which is simple in technology, convenient in construction, easy in construction equipment, short in construction period and low in construction cost. The influence of lateral soil on the foundation can be ignored in design calculation. Therefore, shallow foundation should be applied as far as possible under the premise of ensuring building safety and normal use.

3.2.1 Classification of shallow foundation

The shallow foundation is the most widely used foundation. It refers to all the foundations apart from pile and underground continuous walls.

Shallow foundation is divided into rigid foundation and flexible foundation by stress characteristic. Rigid foundation is also known as non-reinforced extended foundation, mainly brick foundation, rubble foundation, lime-soil foundation, lime/cement-sand-gravel foundation, gravel foundation and concrete foundation, etc. Flexible foundation is known as extended foundation, mainly reinforced concrete foundation, pile(precast pile, cast-in-place pile)foundation, pad foundation, strip foundation, raft foundation and box foundation, etc.

According to structure form, shallow foundation is divided into individual foundation, strip foundation(strip line), cross beam foundation and raft foundation, etc. Individual foundation is also known as pad foundation, mostly in the form of column pier. Its section is ladder shape or cone shape. Strip line is the foundation whose length is much larger than its height and width. The common ones are strip foundation under wall and strip foundation under column. The materials mainly include brick, rubble, concrete, and reinforced concrete, etc.

3.2.2 Rigid Foundation

Rigid foundation is the one made by the materials of high compressive strength but low bending strength and tensile strength. Its characteristic is good at compressive resistance, but poor at tensile resistance, bending resistance and shear resistance. This foundation is good for multistory civil buildings which have better foundation soil, even and low upper load and light factory buildings that wall bears load. The form of rigid foundation is shown in Fig. 3.13.

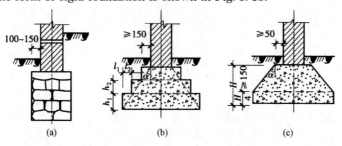

Fig. 3.13 The form of rigid foundation
(a)rectangle; (b)step structure; (c)cone

刚性基础的特点是稳定性好,施工简便,因此只要地基强度能够满足要求,它是房屋、桥梁、涵洞等结构物首先考虑的基础形式。它的主要缺点是用料多,自重大。当基础承受荷载较大,按地基承载力确定的基础底面宽度也较大时,为了满足刚性角的要求,则需要较大的基础高度,导致基础埋深增大。所以刚性基础一般适于六层和六层以下(三合土基础不宜超过四层)的民用建筑和砌体承重的厂房以及荷载较小的桥梁基础。

1. 构造要求

根据这类基础的特点,在构造上要采取措施使基础内的拉应力和剪应力不超过基础材料容许的抗拉和抗剪强度。故此类基础多做成阶梯形和锥形,即限制 α 角,如图 3.14 所示,使其满足刚性角的要求,亦即基础底面宽度应符合下式要求:

$$b \leqslant b_0 + 2H_0 \tan\alpha$$

式中:b 为基础底面宽度;b_0 为基础顶面的墙体宽度或柱脚宽度;H_0 为基础高度;b_2 为基础台阶宽度;$\tan\alpha$ 为基础台阶的宽高比 $b_2:H_0$,见表 3.3。

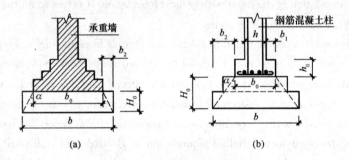

图 3.14 刚性基础截面形式
(a)墙下刚性基础;(b)柱下刚性基础

表 3.3 无筋扩展基础台阶宽高比的允许值

基础材料	质量要求	台阶宽高比的允许值		
		$p_k \leqslant 100$	$100 < p_k \leqslant 200$	$200 < p_k \leqslant 300$
混凝土基础	C15 混凝土	1:1.00	1:1.00	1:1.25
毛石混凝土基础	C15 混凝土	1:1.00	1:1.25	1:1.50
砖基础	砖不低于 MU10、M5 砂浆	1:1.50	1:1.50	1:1.50
毛石基础	M5 砂浆	1:1.25	1:1.50	—

The characteristics of rigid foundation are good at stability and easy in construction. Therefore, only when the strength of subbase meets requirements, it is the foundation form that the structures such as building, bridge and culvert are considered first. Its main drawbacks are more materials to be used and big self-weight. When the foundation bears a large load and the width of foundation determined according to the bearing capacity of subbase is also large, in order to meet the requirements of rigid angle, a larger foundation height is required, which results in an increase of deep bury of foundation. Therefore, the rigid foundation is generally suitable for civil buildings with 6 floors and 6 floors below (lime/cement-sand-gravel foundation should not exceed 4 floors) and factory buildings with masonry bearing as well as bridge foundation with smaller load.

1. Structural Requirements

According to the characteristics of this kind of foundation, measures should be taken structurally to make the tensile stress and shear stress in the foundation not exceed the allowable tensile strength and shear strength of foundation materials. Therefore, this kind of foundation is usually made into stepped form and cone shape, which is to restrict Angle α(as shown in Fig. 3. 14), so that it can meet the requirement of rigid angle, and the width of foundation should meet the requirement of the following equation:

$$b \leqslant b_0 + 2H_0 \tan\alpha$$

In this equation: b is the width of foundation, b_0 is the width of wall at the top of foundation or width of column footing, H_0 is the height of foundation, b_2 is the width of foundation step, and $\tan\alpha$ is the width and height ratio of foundation steps $b_2 : H_0$, as shown in Table. 3. 3.

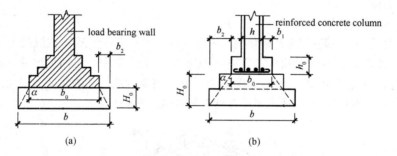

Fig. 3. 14 Section Form of Rigid Foundation

(a) rigid foundation under the wall; (b) rigid foundation under the column

Table 3. 3 Allowable Width and Height Ratio of Steps for Non-reinforced Extended Foundation

Foundation Materials	Quality Requirements	Allowable Width and Height Ratio of Steps		
		$p_k \leqslant 100$	$100 < p_k \leqslant 200$	$200 < p_k \leqslant 300$
Concrete foundation	C15 concrete	1 : 1.00	1 : 1.00	1 : 1.25
Rubble concrete foundation	C15 concrete	1 : 1.00	1 : 1.25	1 : 1.50
Brick foundation	The brick is MU10 at least, M5 mortar	1 : 1.50	1 : 1.50	1 : 1.50
Rubble foundation	M5 mortar	1 : 1.25	1 : 1.50	—

续表

基础材料	质量要求	台阶宽高比的允许值		
		$p_k \leqslant 100$	$100 < p_k \leqslant 200$	$200 < p_k \leqslant 300$
灰土基础	体积比为3:7或2:8的灰土,最小干密度:粉土 1.55 t/m³ 粉质黏土:1.50 t/m³ 黏土 1.45 t/m³	1:1.25	1:1.50	—
三合土基础	体积比 1:2:4～1:3:6 (石灰:砂:骨料),每层约虚铺 220 m,夯至 150 mm	1:1.50	1:2.00	

注:1. p_k 为荷载效应标准组合时基础底面处的平均压力值(kPa);
2. 阶梯形毛石基础的每阶伸出宽度,不宜大于 200 mm;
3. 当基础由不同材料叠合组成时,应对接触部分作抗压验算;
4. 对混凝土基础,当基础底面处的平均压力值超过 300 kPa 时,应进行抗剪验算。

采用无筋扩展基础的钢筋混凝土柱,其柱脚高度 h_0,不得小于 b_1,如图 3.20(b)所示,并不应小于 300 mm 且不小于 $20d$(d 为柱中的纵向受力钢筋的最大直径)。当柱纵向钢筋在柱脚内的竖向锚固长度不满足锚固要求时,可沿水平方向弯折,弯折后的水平锚固长度不应小于 $10d$ 也不应大于 $20d$。

2. 施工要点

(1)砖基础。

①做垫层。在大放脚下面为基础垫层。垫层一般为灰土、碎砖三合土或混凝土等。

②基础弹线。基础开挖、垫层施工完毕后,应根据基础平面图尺寸,用钢尺量出各墙的轴线位置及基础的外边沿线,并用墨斗弹出,如图 3.15 所示。基础放线尺寸的长度和宽度允许偏差应符合有关规定。砖基础砌筑方法、质量要求详见本书第四章相关内容。

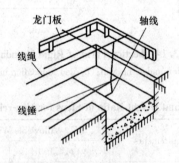

图 3.15 基础弹线

续表

Foundation Materials	Quality Requirements	Allowable Width and Height Ratio of Steps		
		$p_k \leq 100$	$100 < p_k \leq 200$	$200 < p_k \leq 300$
Lime-soil foundation	Volume ratio is 3 : 7 or 2 : 8. The minimum dry density, silt 1.55t/m³, silty clay 1.50t/m³, clay soil 1.45t/m³	1 : 1.25	1 : 1.50	—
Cement-sand-gravel foundation	Volume ratio is 1 : 2 : 4 — 1 : 3 : 6 (lime: sand: aggregate), about 220mm for each layer, tamping to 150 mm.	1 : 1.50	1 : 2.00	—

Notes: 1. p_k is the average pressure value (kPa) at the bottom of the foundation in the standard combination of load effect;

2. The each extended width of stepped rubble foundation should not be more than 200mm;

3. When the foundation is made up of different materials, the contacts should have a compressive test.

4. For the concrete foundation, when the average pressure value at the bottom of the foundation exceeds 300kPa, shear check should be carried out.

For reinforced concrete columns with non-reinforced extended foundation, the height of column footing h_0 should not be less than b_1, see Fig. 3.14(b), and should not be less than 300mm as well as not less than $20d$ (d is the maximum diameter of longitudinal tensile reinforcement in the column). When the vertical anchorage length of column longitudinal tensile reinforcement in the column footing does not meet the requirement of anchorage, it can be bent in the horizontal direction. The horizontal anchorage length after bending should not be less than $10d$, and not be more than $20d$.

2. Main Points of Construction

(1) Brick foundation.

①Bedding. Foundation bedding is under the large stepped footing. The bedding is mainly lime soil, broken brick cement-sand-gravel or concrete, etc.

B. Setting out with ink line snapped on the foundation.

After the foundation excavation and bedding construction are completed, the position of axial line of each wall and the outer line of foundation should be measured by steel tape according to the size of foundation plan, and the ink fountain should be used to snap on them, as shown in Fig. 3.15. The allowable deviation of the length and width of the line setting out on the foundation should comply with the relevant regulations. The detail of construction method of brick foundation and quality requirement see chapter 4 *Masonry Works*.

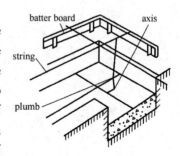

Fig. 3.15 Foundation Snapped with Ink Line

在垫层转角处、交接处及高低处立好基础皮数杆、弹线。

③砌筑砖基础。先检查垫层质量,进行垫层面找平,砌筑时,可依皮数杆先在转角及交接处砌几皮砖,再在其间拉准线砌中间部分。其中第一皮砖应以基础底宽线为准砌筑。内外墙的砖基础应同时砌起。基础底标高不同时,应从低处砌起,并由高处向低处搭接。砖砌大放脚通常采用一顺一丁砌筑方式,最下一皮砖以丁砌为主。水平灰缝和竖向灰缝的厚度应控制在10 mm左右,砂浆饱满度不得小于80%,错缝搭接,在丁字及十字接头处要隔皮砌通。

当地基承载力大于等于 150 kPa 时,采用等高式大放脚,即两皮一收,两边各收进 1/4 砖长;若地基承载力小于 150 kPa 时,采用不等高式大放脚,即两皮一收与一皮一收间隔,两边各收进 1/4 砖长,见图 3.16 所示。

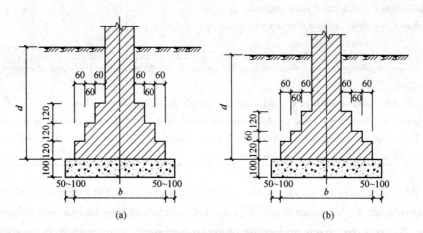

图 3.16 砖基础剖面图
(a)等高式的"两皮一收"砌法;(b)不等高式的"二一间隔收"砌法

砖和砂浆砌筑基础所用砖和砂浆的强度等级,根据地基土的潮湿程度和地区的严寒程度而要求不同。地面以下或防潮层以下的砖砌体,所用材料强度等级不得低于表 3.4 所规定的数值。

The story pole should be set up at the bed corner, the junction, the high and low places, and snap the ink line.

③Masonry brick foundation. Firstly, check the quality of bedding and level its surface. When laying brick, several bricks can be built according to the story pole at the corner and the junction, and then pull a line in between and lay the middle part. Among them, the first brick shall be laid according to the width line of the foundation bottom. The brick foundations of the internal and external walls should be laid at the same time. When the elevation at the bottom of foundation is different, it should be laid from the lower part and overlapped from the higher part to the lower part. Laying the stepped footing is usually by the stretcher bond, the brick at bottom is mainly laid by header bond. The thickness of horizontal mortar joint and vertical mortar joint should be controlled about 10mm. The plumpness of mortar should not be less than 80%. The blocks are overlapped with staggered seams. In the T-shape and cross joints, bricks will be laid through every one brick layer.

When the bearing capacity of subbase is more than or equal to 150kPa, the equal height type of large stepped footing is adopted, that is, two bricks and then setback, both sides should have a setback of 1/4 brick length. If the bearing capacity of the subbase is less than 150kPa, unequal height type of large stepped footing is adopted, that is, the methods of *two bricks and then setback* and *one brick and then setback* are used at interval, both sides should have a setback of 1/4 brick length, as shown in Fig. 3. 16.

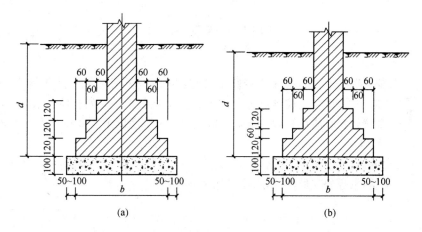

Fig. 3. 16 Sectional Drawing of Brick Foundation
(a) two bricks and one setback masonry method of equal height;
(b) two one interval setback masonry method of unequal height

The strength grade of bricks and mortar used for masonry foundation varies according to the degree of humidity of subbase soil and the cold degree of the region. For brick masonry below the ground or damp-proofing course, the strength grade of materials used on the foundation should not be lower than the value specified in Table 3. 4.

表 3.4　基础用砖、石材及砂浆最低强度等级

地基的潮湿程度	黏土砖		石材	白灰、水泥混合砂浆	水泥砂浆
	严寒地区	一般地区			
稍潮湿的	MU10	MU7.5	MU20	M2.5	M2.5
很潮湿的	MU15	MU15	MU20	M5	M5
含水饱和的	MU20	MU20	MU30	—	M5

(2)毛石基础。毛石基础砌筑前,先清除杂物,打好底夯。不宜采用混合砂浆。水泥砂浆强度等级为 M2.5～M5,在铺砌第一皮毛石时,基底如为素土,可不铺砂浆;基底如为各种垫层,应先铺 4 cm 左右的砂浆。毛石砌到室内地坪以下 5 cm 处,应设置防潮层。每日砌筑高度不宜超过 1.2 m。基础顶面每边比墙宽 100 mm,每阶高度一般为 300～400 mm。毛石基础如图 3.17 所示。

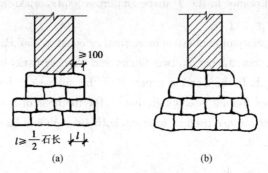

图 3.17　毛石基础
(a)阶梯形;(b)梯形

(3)混凝土基础。混凝土浇筑前应进行验槽,轴线、基坑(槽)尺寸和土质等均应符合设计要求。局部软弱土层应挖去,用灰土或砂砾分层回填夯实至基底相平。如有地下水应挖沟排除;对粉土或细砂地基,基坑较深时应用轻型井点降低地下水位至坑底以下 500 mm 处。基槽(坑)内浮土、积水、淤泥应清除干净。

如地基土质良好,且无地下水,基槽(坑)底部台阶可利用原槽(坑)浇筑,但应保证尺寸正确。上部台阶应支模浇筑,模板要支撑牢固,木模应浇水湿润。

Chapter 3 Subbase and Foundation Construction

Table 3.4 The Lowest Strength Grade of Brick, Stone and Mortar Used in the Foundation

Degree of Humidity of Subbase	Clay Brick		Stone	White Lime and Cement Mixed Mortar	Cement Mortar
	Severe Cold Area	Common Area			
Damp slightly	MU10	MU7.5	MU20	M2.5	M2.5
Very damp	MU15	MU15	MU20	M5	M5
Water-saturated	MU20	MU20	MU30	—	M5

(2) Rubble foundation. Before construction of rubble foundation, firstly remove sundries and tamp the bottom well. Composite mortar should not be used. Cement mortar of grade M2.5—M5 is used. When paving first layer of rubble, if the base is plain soil, mortar will not be used. If the base is variety of cushion, it should be laid about 4cm mortar first. Damp-proofing course should be laid when the rubble is paved to 5cm below the indoor floor. Daily masonry height should not exceed 1.2m. The top surface of foundation should be 100mm wider than the wall on each side, and the height of each step is generally 300—400mm. The rubble foundation is shown in Fig. 3.17.

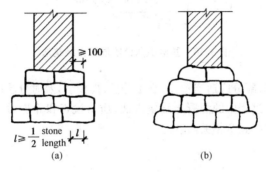

Fig. 3.17 Rubble Foundation
(a) step-structure; (b) trapezoid

(3) Concrete foundation. Groove inspection should be carried out before pouring concrete, and the axial line, size of foundation pit (foundation trench), soil texture and so on should meet the design requirements. The local weak soil layer should be dug out and backfilled with lime soil or gravel in layers to tamp to even with the base. If there is groundwater, it should be removed by digging trench. For silt subbase or fine sand subbase, a light well point should be used to lower the groundwater level to 500mm below the bottom of pit when the foundation pit is deep. The loose soil, stagnant water and mud in the foundation trench (foundation pit) should be cleaned up.

If the soil property of subbase is well and there is no groundwater, the steps at the bottom of foundation trench (foundation pit) can be poured by using original trench (pit), but the correct size should be ensured. The upper steps should be poured by setting up formwork. The formwork should be set up firmly, and wooden formwork should be watered to wet.

基础混凝土浇筑高度在 2 m 以内时，混凝土可直接卸入基槽(坑)内；浇筑高度在 2 m 以上时，应通过漏斗、串筒或溜槽下灰。浇筑阶梯形基础应按台阶分层一次浇筑完成，每层先浇边角，后浇中间。施工时应注意防止上下台阶交接处混凝土出现蜂窝和脱空现象。要使第一台阶混凝土浇完后稍停 0.5~1.0 h，待下部沉实，再浇上一台阶。锥形基础如斜坡较陡，斜面部分应支模浇筑，或随浇随安装模板，并应注意防止模板上浮。斜坡较平时，可不支模，但应注意斜坡部位及边角部位混凝土的捣固密实，振捣完后，再用人工将斜坡表面修正、拍直、拍实。

当基槽(坑)底标高成阶梯形式时，应先从最低处开始浇筑，按每阶高度，其各边搭接长度应不小于 500 mm，如图 3.18 所示。

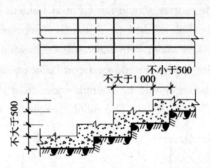

图 3.18 基槽纵向台阶混凝土浇筑法

混凝土浇筑完后，外露部分应适当覆盖，洒水养护；拆模后，及时分层回填土方并夯实。

混凝土主要包括施工过程中的混凝土浇筑方法、质量检查和养护后的质量检查，质量要求应符合有关规定。参见本书第五章相关内容。

3.2.3 柔性基础

当不便于采用刚性基础或采用刚性基础不经济时，可以采用钢筋混凝土基础。柱下钢筋混凝土独立基础和墙下钢筋混凝土条形基础，统称为钢筋混凝土扩展基础或柔性基础。钢筋混凝土扩展基础的抗弯和抗剪性能良好，可在竖向荷载较大、地基承载力不高等情况下使用。该类基础的高度不受台阶宽高比的限制，其高度比刚性基础小，适宜于需要"宽基浅埋"的情况。例如，有些建筑场地浅层土承载力较高，即表层具有一定厚度的所谓"硬壳层"，而在该硬壳层下土层的承载力较低，并拟利用该硬壳层作为持力层时，更可考虑采用此类基础形式。

If the pouring height of concrete on the foundation is within 2m, the concrete can be directly poured into the foundation trench (foundation pit). If the pouring height is more than 2m, the concrete should be poured through a funnel, a tumbling barrel or a chute. The pouring of stepped foundation should be completed in one-time according to step layers. The corners of each layer should be poured first and then poured in the middle. In the construction, attention should be paid to prevent honeycomb and disengaging of concrete at the junction between the upper and lower steps. After completing the pouring of first step concrete, a pause about 0.5—1.0 hour should be kept, waiting the lower part sinking done, then pouring the upper step. If the tapered foundation has a steep slope, the bevel part should be poured by setting up the formwork, or the formwork should be installed with the pouring, and care should be taken to prevent the formwork from floating. When the slope is relatively flat, the formwork support may not be used, but attention should be paid to the consolidation and compaction of concrete in the slope parts and corner parts. After completing the vibration, the surface of slope should be modified, flapped straight and compacted by manpower.

When the elevation at the bottom of foundation trench (foundation pit) is in the form of ladder, it should be poured from the lowest place first. According to the height of each step, the lap length of each side should be no less than 500mm, as shown in Fig. 3.18.

After pouring concrete, exposed parts should be properly covered and sprinkled for curing. After the formwork is removed, the earthwork shall be backfilled in layers and compacted in time.

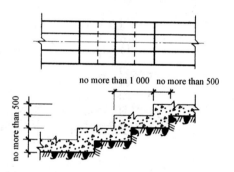

Fig. 3.18 **Pouring Method of Longitudinal Step Concrete for Foundation Trench**

The concrete mainly includes the concrete pouring method in the construction process, quality inspection and quality inspection after curing, and the quality requirements shall meet the relevant provisions. See *Chapter* 5 of this book.

3.2.3 Flexible Foundations

When rigid foundation is inconvenient to adopt or is uneconomical, we can choose reinforced concrete foundation. Reinforced concrete pad foundation under columns and reinforced concrete strip foundation under walls are general called reinforced concrete extend foundation or flexible foundation. Reinforced extend foundation has good bending and shearing resistance, and it can be used in the condition of the large vertical load and low bearing capacity of subbase. The height of this kind of foundation is not limited by the ratio of step width to height. This kind of foundation height is lower than rigid foundation and suitable for the *wide but sallow buried foundation*. For example, on some construction sites shallow soil layer has high bearing capacity, that is, the surface layer has certain thickness so called *hard crust*. The bearing capacity of soil under the hard crust is low. When using the hard crust as bearing layer, this type of foundation can be considered to adopt.

1. 墙下钢筋混凝土条形基础

墙下钢筋混凝土条形基础和柱下钢筋混凝土独立基础由于钢筋混凝土的抗弯性能好,可放大基础底面尺寸,以减小地基应力,也可减小埋深,节省材料和土方开挖量。适用于多层民用建筑和厂房的柱基和墙基。

(1)构造要求。墙下钢筋混凝土条形基础是砌体承重结构墙体及挡土墙、涵管下常用的基础形式,其构造如图 3.19 所示。如果地基不均匀或承受荷载有差异时,为了增强基础的整体性和抗弯能力,可以采用有肋的墙基础[图 3.19(b)],肋部配置足够的纵向钢筋和箍筋。锥形基础的边缘高度不宜小于 200 mm;阶梯形基础的每阶高度,宜为 300～500 mm。垫层的厚度不宜小于 70 mm,工程上常为 100 mm,垫层混凝土强度等级宜取 C10。墙下钢筋混凝土条形基础底板受力钢筋的最小直径不宜小于 10 mm,间距不宜大于 200 mm,也不宜小于 100 mm。墙下钢筋混凝土条形基础纵向分布钢筋的直径不小于 8 mm,间距不大于 300 mm,每延米分布钢筋的面积应不小于受力钢筋面积的 1/10,当有垫层时,钢筋保护层的厚度不小于 40 mm,无垫层时不小于 70 mm。混凝土强度等级不应低于 C20,且应满足耐久性要求。

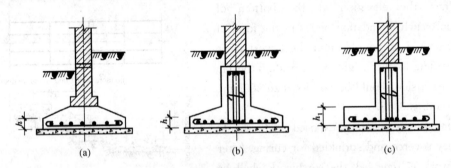

图 3.19 墙下钢筋混凝土独立基础
(a)板式无肋,阶梯形;(b)梁、板式有肋,锥形;(c)梁、板式有肋,阶梯形或矩形

柱下和墙下钢筋混凝土条形基础,在 T 字形与十字形交接处的钢筋应沿一个主要受力方向通长放置。

(2)施工要点。基坑验槽清理同刚性基础。垫层混凝土在基坑验槽后应立即浇筑,以免地基土被扰动。垫层达到一定强度后,在其上划线、支模、铺放钢筋网片。

上下部垂直钢筋应绑扎牢固,将钢筋弯钩朝上,连接柱插筋,下端用 90°弯钩与基础钢筋扎牢。底部钢筋网应用垫块垫起,保证位置正确。

1. Reinforced Concrete Strip Foundation under Walls

The reinforced concrete strip foundation under the wall and the reinforced concrete pad foundation under the column can enlarge the foundation bottom size to reduce the subbase stress, reduce the buried depth and save the amount of material and earthwork excavation due to the good bending performance of reinforced concrete, which are suitable for column foundation and wall foundation of multi-story civil buildings and factory buildings.

(1) Construction requirements. Reinforced concrete strip foundation under walls is a common foundation form for masonry loading-bearing structural wall, retaining wall and culvert pipe. Its structure is shown as Fig. 3.19. If foundation is uneven or the bearing load is difference, to increase the integrity and bending resistance of the foundation, the ribbed wall foundation can be used [Fig. 3.19(b)], and rib should have enough longitudinal reinforcement and stirrup. The edge height of the cone foundation should not be less than 200mm, and stepped foundation height of each step should be at 300—500mm. Cushion thickness should not be less than 70mm, and 100mm are always used on project. Cushion concrete strength grade should be C10. The bottom bearing reinforcement of reinforced concrete strip foundation under walls minimum diameter should not be less than 10mm, spacing should not be more than 200mm and also should not be less than 100mm. Vertical distribution reinforcement diameter for reinforced concrete strip foundation under walls should not be less than 8mm, and space should not be more than 300mm. Per linear meter distribution in reinforced area should not be less than 1/10 of bearing reinforced area. When there is cushion, the thickness of reinforcement cover should not be less than 40mm. When here is not cushion, the thickness should not be less than 70mm. Concrete strength grade should not be less than C20, and should satisfy the requirement of durability.

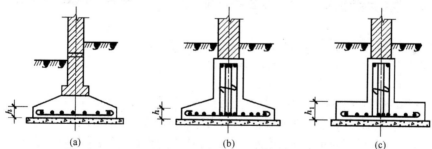

Fig. 3.19 Reinforced Concrete Pad Foundation under the Wall
(a) slab without ribs, stepped form; (b) beam and slab with ribs, cone;
(c) beam and slab with ribs, stepped form or rectangle

For reinforced concrete strip foundation under columns and walls, the reinforcement in T-shape and cross shape should be placed along the length of major force.

(2) Construction points. Foundation pit inspection and cleaning are the same as rigid foundation. Cushion concrete should be immediately poured after inspection of foundation pit to prevent foundation soil from being disturbed. When cushion concrete reaches certain strength, line can be drawn, formwork can be erected and steel mesh can be laid on it.

Top and bottom vertical reinforcement should be firmly tied, putting steel hook up, connecting the column steel dowel, and using 90° hook to fasten the bottom and foundation reinforcement. Bottom steel mesh should be lifted with block and ensure the position is right.

对阶梯形基础,每一台阶应整体分层浇筑,浇筑完一台阶应稍停 0.5～1.0 h,待其沉实后再浇筑上层。每一台阶浇完应原浆表面抹平。锥形基础应保持锥体斜面坡度正确,防止模板上浮。浇筑柱下基础时,应注意保证柱子插筋位置的正确,防止位移。

条形基础应根据高度分段分层连续浇筑,不留施工缝,各段各层间相互衔接,做到逐段逐层呈阶梯状推进。基础上有插筋时,要保证插筋位置正确,防止浇筑混凝土时移位。

2. 柱下钢筋混凝土独立基础

柱下独立基础,当柱荷载的偏心距不大时,常用方形;偏心距大时,则用矩形。

(1)构造要求。工程中,柱下基础底面形状大多采用矩形,因此也称其为柱下钢筋混凝土独立基础。柱下钢筋混凝土独立基础只不过是条形基础的一种特殊形式,有时也统一称为条形基础、带形基础或条式基础。柱下钢筋混凝土独立基础可以做成阶梯形和锥形,如图 3.20 所示。独立基础下一般设有素混凝土垫层,其厚度一般为 100 mm,强度等级一般用 C10、C15;阶梯形基础的每阶高度宜为 300～500 mm;锥形基础边缘高度不宜小于 200 mm。底板受力钢筋的最小直径不宜小于 10 mm,间距不宜大于 200 mm,无垫层时钢筋保护层不宜小于 70 mm,有垫层时钢筋保护层不宜小于 40 mm。基础高度 h 小于等于 350 mm,用一阶;h 大于等于 350 mm 且小于 900 mm,用二阶;h 大于等于 900 mm,用三阶。基础台阶的宽度比不大于 2.5。

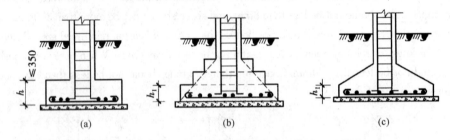

图 3.20　柱下钢筋混凝土独立基础
(a)阶梯形或矩形;(b)阶梯形;(c)锥形

柱基础插筋的数目与直径应与柱内纵向受力钢筋相同。当基础高度在 900 mm 以内时,插筋应伸至基础底部的钢筋网,并在端部做成直弯钩;当基础高度较大时,位于柱子四角的插筋应伸到基础底部,其余的钢筋只需伸至锚固长度即可。插筋伸出基础部分长度应按柱的受力情况及钢筋规格确定。柱子插筋必须与柱子纵向受力钢筋相吻合,其锚固、搭接等必须符合设计要求和规范要求。

For step foundation, each step should be integrally poured in layers. After pouring a step, stop for 0.5—1h, waiting for it to settle before pouring the upper layer. After each pouring, the surface should be smoothed with same material. Cone foundation should keep the slop of the cone side correct to prevent template from floating. When pouring foundation under columns, the position of column steel dowel should be ensured to be correct to prevent displacement.

Strip foundation should according to the height be segmented and layered continuous pouring without leaving construction joint. The sections and layers connect with each other, pushing forward by stages. When there is dowel-bar reinforcement on foundation, correct position should be kept to prevent displacement when pouring.

2. Reinforced Concrete Pad Foundation under Columns

Pad foundation under columns: When the eccentricity of column load is small, the square foundation is usually used; when the eccentricity is large, the rectangle is used.

(1) Construction requirements. In construction, the shape of foundation under columns mostly is rectangle. Therefore it is also called reinforced concrete pad foundation under columns. It is a special form of strip foundation, and sometimes it is also called stripe foundation, band foundation or bar foundation. Reinforced concrete pad foundation under columns can be made rectangle, stepped and cone, as shown in Fig. 3.20. Plain concrete cushion is general designed under the pad foundation, with thickness of 100mm, strength grade of C10 or C15. Every step height of stepped foundation should be 300—500mm. The edge height of cone foundation should not be less than 200mm. The minimum diameter of stress reinforcement on bottom slab should not be less than 10mm, with spacing no more than 200mm. When there is not cushion, the thickness of reinforcement protective cover should not be less than 70mm. when there is cushion, the thickness of reinforcement protective cover should not be less than 40mm. When foundation height is less than or equal to 350mm, one step is used. When height is more than 350mm and less than 900mm, two steps are used. When height is equal to or more than 900mm, three steps are used. The width ratio of foundation step should not be more than 2.5.

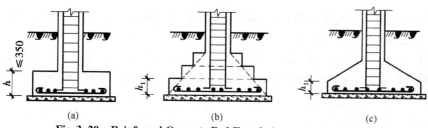

Fig. 3.20 Reinforced Concrete Pad Foundation under the Column
(a) rectangle; (b) stepped form; (c) cone

The number and diameter of insert bars in the column foundation will be the same as longitudinal steel bars in the column. When foundation height is within 900mm, dowel-bar should insert into the steel mesh in foundation bottom and the straight hook is formed at the end part. When foundation height is larger, the dowel-bars at four corners of column should be extended to the foundation bottom, and the rest will just be extended to the anchorage length. The length of dowel-bars stretched out the foundation should be determined according to the stress of the column and specification of reinforcement. Column dowel-bars must be identical with column longitudinal reinforcement, and the anchorage and overlap must comply with requirement of design and specification.

(2)施工要点。

①基础钢筋、柱、墙钢筋安装。根据设计要求的规格、品种和间距安装基础钢筋。应使 HPB300 级钢筋的弯钩朝上。设有避雷带时,应与基础钢筋焊接完好。

柱、墙钢筋插入基础时其位置应正确,并保证在混凝土浇筑时不偏斜。一般底部应与基础钢筋点焊固定(防雷接地处应增强焊接),上部则应绑扎一定的箍筋以增加骨架刚度并间隔一定距离用钢管支撑牢牢夹住。

雨后施工应保证钢筋表面不粘泥。涂刷模板隔离剂时不得污染钢筋。

②基础模板及支撑安装。独立基础的模板主要是侧模板。由于独立基础一般都有两阶或两阶以上的台阶,因此,模板安装中必须解决每一阶的模板组成以及各阶模板之间的连接,以保证尺寸形状以及相互位置的正确。

独立基础的侧模板可以用木模或钢模进行拼装。上阶模板可以支撑在下阶模板上,相互之间位置关系可以统一用钢管支撑解决或各自设置支撑。当土质较好时,也可以利用土壁作最下阶模板,即所谓原槽浇筑。图 3.21 所示是独立基础模板安装的一种形式。

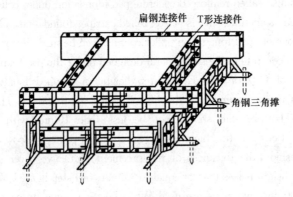

图 3.21 独立基础模板

模板安装完毕后,必须复核轴线、标高以及尺寸等,各项偏差应在允许偏差范围内。自检合格后,再报专业监理工程师进行验收。

③混凝土浇筑。当基础台阶有多阶时,应将下阶混凝土浇筑满后才浇筑上阶混凝土,避免混凝土出现脱节现象。当台阶较高时,混凝土浇筑应分层进行。混凝土浇筑完毕应适时进行混凝土养护。

(2) Construction points.

①Installation of Foundation Reinforcement, Column and Wall Reinforcement

Foundation reinforcement will be installed according to the specification, variety and spacing of design requirement. The HPB300 grade reinforcement should be hooked upward. When there is lightning strip, it should be well welded with foundation reinforcement.

When column reinforcement and wall reinforcement insert into foundation, the position should be correct, and guarantee no skew will occur when pouring concrete. Generally bottom should be fixed by spot welding with foundation reinforcement (lightning protection location should enhance welding), the upper part should tie up certain stirrups to increase skeleton stiffness and at certain distance steel pipes are used to clamp them firmly.

Construction after raining should keep the reinforcement surface without mud. Don't contaminate reinforcement when brushing template isolator.

②Foundation formwork and support installation. Pad foundation formwork is mainly side template. Normally pad foundation has two or more steps, so in template installation, each step template composition and connection of each step template must be solved to keep size, shape and mutual position right.

Side formwork of pad foundation can assemble with wooden and steel formwork. Upper step formwork can be supported on the lower step formwork, and the position relation between each other can be solved by the steel pipe support or set support separately. When soil quality is good, it can also use the soil wall as lower step formwork, calling original trench pouring. Fig. 3.21 is a form of pad foundation formwork installation.

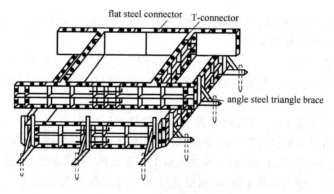

Fig. 3.21　Pad Foundation Formwork

After formwork is installed, the axis, level and size must be rechecked, and any deviation should be within the allowable deviation. After the self check, report to professional consultant for acceptance.

③Pouring concrete. When the foundation steps are multiple steps, the lower step concrete should be fully poured before the upper step concrete is pouring, so as to avoid concrete disconnect. When a step is higher, concrete should be layered pouring. Concrete should be timely cured after pouring.

3. 筏板基础

筏板基础分平板式和梁板式两类(图3.22),像倒置的无梁楼盖和肋形楼盖。这种楼盖整体性好、抗弯刚度大,可调整上部结构的不均匀沉降,多用于高层建筑。适用于土质软弱不均匀而上部荷载又较大时。

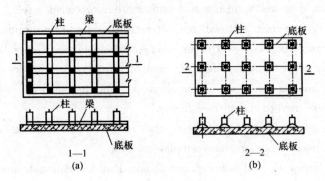

图 3.22　筏板基础
(a)梁板式;(b)平板式

(1)构造要求。

①混凝土强度等级不宜低于C20,钢筋无特殊要求,钢筋保护层厚度不小于35 mm。

②基础平面布置应尽量对称,以减小基础荷载的偏心距。底板厚度不宜小于200 mm,梁截面和板厚按计算确定,梁顶高出底板顶面不小于300 mm,梁宽不小于250 mm。

③底板下宜设厚100 mm的C10混凝土垫层,每边伸出基础底板不小于100 mm。

(2)施工要点。

①施工前,如地下水位较高,可采用人工降低地下水位至基坑底不小于500 mm,以保证在无水情况下进行基坑开挖和基础施工。

②参照独立基础钢筋施工要点。当设有上下层钢筋时,钢筋支架的数量应足够,上层钢筋应支承牢固并保证其位置正确。筏板基础的构造钢筋安装应符合要求。垫块设置应规范,应有足够的强度避免压碎。

③梁板式筏板基础施工时,可以将底板模板和梁模板同时支好,混凝土一次连续浇筑完成。梁模板的支承支架可以与钢筋支架一并考虑,应将其固定牢固,应保证有足够的数量。也可先浇筑底板混凝土,待达到25%设计强度后,再在底板上支梁模板,继续浇筑完梁混凝土。

④混凝土浇筑时一般不留施工缝,必须留设时,应按施工缝要求处理,并应设置止水带。

⑤基础浇筑完毕,应及时覆盖和洒水养护。

3. Raft Foundation

Raft foundation can be divided into two types: flat plate and beam plate(Fig. 3.22), which just like an inverted beamless floor and ribbed floor. This type of floor has good integrity and high bending rigidity, which can adjust the differential settlement of superstructure, and always used for high-rise buildings. It is suitable for the soil which is weak and uneven and the upper load is larger.

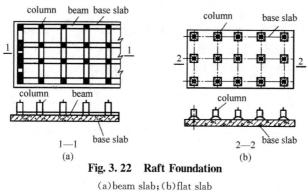

Fig. 3.22 Raft Foundation
(a)beam slab; (b)flat slab

(1) Structural requirements.

①Concrete strength should not be less than C20, reinforcement has no special requirement, and the thickness of reinforcement protective cover should not be less than 35mm.

②The foundation layout should be as symmetry as possible, to reduce the eccentricity of foundation load. The thickness of bottom plate should not be less than 200mm. Girder cross-section and plate thickness are determined by calculation. The top of the beam is not less than 300mm above the bottom plate, and the beam width should not be less than 250mm.

③100mm thickness C10 concrete cushion should be done under the bottom plate. Each side extends foundation bottom plate not less than 100mm.

(2) Construction points.

①Before construction, if the groundwater level is higher, it can be artificially lowered to the bottom of the foundation pit not less than 500mm, to guarantee the foundation excavation and foundation construction without groundwater.

②Referring to pad foundation reinforcement construction, when both the upper layer and lower layer have reinforcement, the amount of reinforcement supports should be enough. The upper reinforcement should be strongly supported and the position be kept correct. The installation of structural reinforcement of raft foundation should comply with the requirement. Cover block should be standard with sufficient strength to avoid the crush.

③When constructions of beam plate raft foundation, the bottom template and beam template can be supported at the same time, and the concrete is poured continuously at one time. The support bracket of beam template can be considered together with reinforcement support, and fixed them firmly. A sufficient quantity should be guaranteed. Also floor concrete can be poured first. After reaching 25% of the design strength, beam template can be erected on floor and continue to pour beam concrete.

④General, no construction joints are left when concrete is poured. If it is must, the construction joints shall be treated according to the requirements, and water stop should be provided.

⑤After foundation is poured, it should be covered and cured with water in time.

⑥属大体积混凝土施工时,应严格执行经批准的施工技术方案。
⑦采用泵送混凝土施工时,管道的移动不应造成模板支承体系的变形变位。

3.3 深基础工程施工

一般的结构当中,常采用在地表浅层地基土上建造浅基础,但当建筑物荷载较大而对变形和稳定性要求较高时,或地基的软弱土层较厚,对软土进行人工处理又不经济时,往往采用桩基础。目前,桩基础是一种常用的深基础形式,主要由桩及桩承台组成。

3.3.1 桩基础

桩基础是深基础应用最多的一种基础形式,它由若干个沉入土中的桩和连接桩顶的承台或承台梁组成。

1. 概述

(1)桩基础的作用。

桩基础是指用承台或梁将沉入土中的桩联系起来承受上部结构的一种基础形式。桩基础中桩的作用是借其自身穿过松软的压缩性土层,将来自上部结构的荷载传递至地下深处具有适当承载力且压缩性较小的土层或岩石上,或者将软弱土层挤压密实,从而提高地基土的承载力,以减少基础的沉降。承台的作用则是将各单桩连成整体,承受并传递上部结构的荷载给群桩。

桩基础具有承载力大、稳定性好、沉降量小等特点,不仅能承受竖向力、水平力、上拔力、振动力等,而且更便于实现机械化施工,尤其当软弱土层较厚,上部结构荷载很大,天然地基的承载能力又不能满足设计要求时,采用桩基础可省去大量土方挖填、支撑装拆及降水排水设施布设等工序,施工速度快、质量好,因而能获得较好的经济效果,已广泛用于房屋地基、桥梁、水利等工程中。

(2)桩基础的分类。

工程中的桩基础,往往由数根桩组成,桩顶设置承台,把各桩连成整体,并将上部结构的荷载均匀传递给桩(图3.23)。承台一般为钢筋混凝土构件,有低承台和高承台之分。桩一般为钢筋混凝土桩或钢管桩。钢筋混凝土桩按照施工方法的不同可以分为预制桩和灌注桩。预制桩是指在地面上(工厂或现场)按设计图纸要求直接制作好桩身;灌注桩是指先用一定的成孔方法形成桩孔,然后放入钢筋笼,最后浇筑混凝土成桩。

⑥In case of mass concrete construction, the approved construction technical scheme shall be strictly implemented.

⑦When pumping concrete is used, the movement of pipe should not cause the deformation and displacement of formwork support system.

3.3 Deep Foundation Construction

In general structure, shallow foundations are always built on shallow ground. However, when building load is larger and higher deformation and stability are required, or the soft layer of subbase is thicker and artificial treatment to the soft soil is uneconomical, pile foundation is often used. At present, pile foundation is a common form of deep foundation, mainly composed of pile and pile cap.

3.3.1 Pile Foundation

Pile foundation is one of the most widely used foundation forms in deep foundation. It is composed of several piles sunk into the soil and pile caps or bearing beams connecting the pile top.

1. Introduction

(1) The role of pile foundation.

Pile foundation is a form of foundation in which the pile sunk in the soil is connected with pile cap or beam to bear superstructure. The role of pile in the pile foundation is to lend itself through the soft compressible soil, and transfer the superstructure load to soil or rock in deep underground that has good bearing capacity and good compressibility, or squeeze the soft soil layer so as to increase the foundation carrying capacity and to reduce foundation settlement. The role of pile cape is to connect each single pile into a whole, bear and transfer superstructure load to group piles.

Pile foundation has the characteristics of large bearing capacity, good stability and small settlement. Not only it can bear the vertical force, horizontal force, upward force, vibration force and so on, but also it is easier to realize mechanical construction. Especially when the soft soil is thicker, the superstructure load is large and the natural subbase bearing capacity does not conform to the design requirements, the pile foundation can save a lot of earthwork excavation and filling, support assembly and disassembly, and the layout of rainfall and drainage facilities. It is also fast in construction and good in quality, and can get better economic results. It is now widely used in building foundation, bridges, water conservancy projects, etc.

(2) Classification of pile foundation.

Pile foundation in project is usually composed of several piles, and the pile top is provided with pile cap to connect each pile as a whole and transfer superstructure load to pile uniformly(Fig. 3.23). Pile cap is generally a reinforced concrete member and is divided into low cap and high cap. Pile general is reinforced concrete pile and steel pipe pile. According to the different construction method, reinforced concrete pile can be divided into precast pile and cast-in-site pile. The precast pile refers to the pile body directly made on the ground(factory or site) according to the design drawings. The cast-in-site pile refers to a certain method is used to form a pile hole first, and put the reinforcement cage into the hole, and finally pour concrete to form pile.

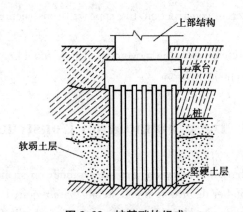

图 3.23 桩基础的组成
1—坚硬土层;2—桩;3—承台;4—上部结构;5—软弱土层

按承载性质不同,桩基础可分为端承桩和摩擦桩。

①端承桩:是指穿过软弱土层并将建筑物的荷载通过桩传递到桩端坚硬土层或岩层上。桩侧较软弱土对桩身的摩擦作用很小,其摩擦力可忽略不计,如图 3.24(a)所示;施工时以控制贯入度为主,桩尖进入持力层深度或桩尖标高可做参考。

②摩擦桩:是指沉入软弱土层一定深度通过桩侧土的摩擦作用,将上部荷载传递扩散于桩周围土中,桩端土也起一定的支承作用,桩尖支承的土不甚密实,桩相对于土有一定的相对位移时,即具有摩擦桩的作用,如图 3.24(b)所示;施工时以控制桩尖设计标高为主,贯入度可做参考。

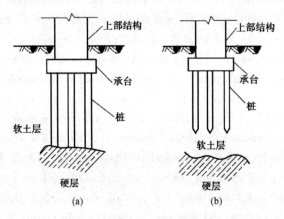

图 3.24 桩基础
(a)端承桩;(b)摩擦桩

2. 钢筋混凝土预制桩

钢筋混凝土预制桩可以在预制构件厂预制,亦可以在施工现场预制,是用沉桩设备将钢筋混凝土预制桩沉入或埋入土中而成。预制桩主要有钢筋混凝土预制桩和钢桩两类。采用预制桩施工,桩身质量易保证,施工机械化程度高,施工速度快,且可不受气候条件变化的影响。但当土层变化复杂时桩长规格较多,桩入土后易被冲压破损、变形而达不到设计标高。其特点为:坚固耐久,不受地下水或潮湿环境影响,能承受较大荷载,施工机械化程度高,进度快,能适应不同土层施工。

Piles are classified end bearing pile and friction pile according to their bearing properties:

①End bearing pile. It refers to the pile passing through soft soil and transferring the load of the building through the pile to the hard soil or rock layer at the end of the pile. Soft soil on pile side has very little friction on pile body, and the frictional force can be ignored, see Fig. 3.24(a). The penetration degree is mainly controlled during construction, and the depth of the pile tip entering the bearing stratum or the elevation of the pile tip can be used as a reference.

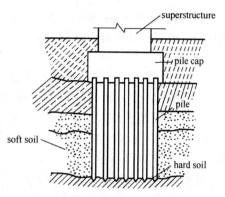

Fig. 3.23 Composition of Pile Foundation

②Friction pile. It means that when the friction pile sinks into a certain depth of soft soil, the upper load is transferred and diffused into the soil around the pile through the frictional action of the soil on the side of the pile. The soil at the end of the pile also plays a supporting role. The soil supported by the tip of the pile is not very compact. When the pile has a certain relative displacement with respect to the soil, it has the function of friction pile, see Fig. 3.24(b). During construction, the design elevation of pile tip is mainly controlled, and the penetration degree can be used as a reference.

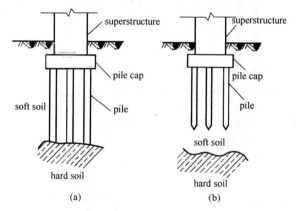

Fig. 3.24 Pile Foundation

(a)end bearing pile; (b)friction pile

2. Precast Reinforced Concrete Pile

Precast reinforced concrete pile can be prefabricated in prefabricated construction plants and also can be prefabricated at the construction site. It is the pile that uses pile sinking equipment to sink or bury into the soil. Precast piles are mainly two types: precast reinforced concrete pile and steel pile. When precast piles are used for construction, the quality of pile body is easy to be guaranteed, the construction is highly mechanized and the construction speed is fast, furthermore, it is not affected by climate change. However, when the soil is complex, pile length specifications will be more. After the pile enters into the soil, it is easily damaged and deformed by stamping, and deformation cannot reach the design elevation. Its characteristics are strong and durable, not affected by groundwater or humid environment, able to bear the larger load, high degree of construction mechanization, fast progress and suitable for different soil layers.

(1)钢筋混凝土预制桩的制作。预制桩主要有预制普通钢筋混凝土方桩、预制预应力混凝土管桩等。预制钢筋混凝土桩分实心桩和空心桩(空腹桩),有钢筋混凝土桩和预应力钢筋混凝土桩。

①钢筋混凝土实心桩。钢筋混凝土实心桩截面有三角形、圆形、矩形、六边形、八边形,桩身截面一般沿桩长不变。为了便于预制一般做成正方形断面,断面一般为 200 mm×200 mm~600 mm×600 mm。单根桩的最大长度,一般根据打桩架的高度而定,限于运输条件,工厂预制时单根桩长一般在 12 m 以内;如在现场预制,长度不宜超过 30 m,如需打设 30 m 以上的桩,则将桩预制成几段,在打桩过程中逐段接桩,桩的接头不宜超过两个。

钢筋混凝土实心桩由桩尖、桩身和桩头组成,如图 3.25 所示。

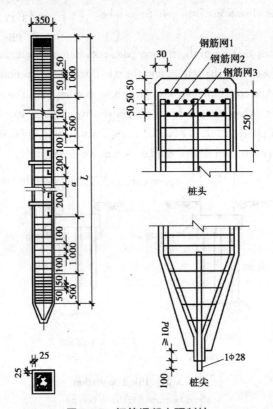

图 3.25　钢筋混凝土预制桩

②钢筋混凝空腹桩。预制预应力混凝土空腹桩有空心正方形、空心三角形和空心圆形(即管桩)。按其截面形式分为管桩、空心方桩;按其混凝土强度等级可分为预应力高强混凝土桩(PHC)和空心方桩(PHS,混凝土等级不低于 C80)、预应力混凝土管桩(PC)和空心方桩(PS,混凝土等级不低于 C50)、预应力混凝土薄壁管桩(PTC);按桩身混凝土有效预压应力值或抗弯性能分为 A、AB、B、C 四种类型,其混凝土有效预压应力值(N/mm²)分别为 4.0、6.0、8.0、10.0。混凝土管桩一般在预制厂用离心法生产,直径多为 400~600 mm,壁厚为 65~125 mm,每节长度为 7~15 m,用法兰连接,桩的接头不宜超过 4 个,下节桩底端可设桩尖,可以使闭口的,即闭口十字型,亦可以是开口的,即开口型。

(1) The production of precast reinforced concrete piles. The precast piles mainly include precast reinforced concrete square pile and precast prestressed concrete pipe pile, etc. Precast reinforced concrete pile can be divided into solid pile and hollow pile(cavity pile), reinforced concrete pile and prestressed reinforced concrete pile.

①Reinforced concrete solid pile. The section of reinforced concrete solid pile has triangle, roundness, rectangle, hexagon, and octagon. The section of pile body is generally invariant along the length of the pile. For easy precast, square sections are usually made. Generally, the section size is 200mm×200mm—600mm×600mm. The maximum length of single pile is usually determined according to the height of pile driving frame. Limited by the transportation, the length of a single pile precast in factory is generally within 12m. If it is precast on site, the length should not be more than 30m. If pile above 30m is required, it should be precast into several segments, and connected piece by piece in the progress of piling. The joint of the pile should not be more than two.

Reinforced concrete solid pile is composed of pile tip, pile body and pile head, see Fig. 3. 25.

②Reinforced concrete hollow pile.

Precast prestressed concrete hollow pile has hollow squre, hollow triangle and hollow circle (i. e. pipe pile). According to its section form, it is divided into pipe pile and hollow square pile. According to its concrete strength grade, it can be divided into prestressed high strength concrete (PHC) pile and hollow square pile (PHS, concrete grade not lower than C80), prestressed concrete pipe pile (PC) and hollow square pile (PS, concrete grade not lower than C50), and prestressed concrete thin-walled pipe pile. According to the effective prepressure stressed value or bending performance of pile body concrete, it is divided into A, AB, B, C, four kinds, and the concrete effective prepressure stressed value is 4.0, 6.0, 8.0, and 10.0 respectively. Concrete pipe piles are generally produced by centrifugal method in prefabrication plants, with diameter mostly 400—600mm, wall thickness 65—125mm, length of each pile 715m, and connected with flange. Pile joints should not exceed 4. Pile tip can be set at the bottom end of the next pile, which may be closed, i. e. closed cross type, or open, i. e. open type.

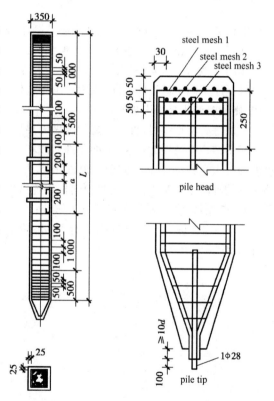

Fig. 3. 25 Precast Reinforced Concrete Pile

预应力混凝土管桩的编号方法如图3.26所示。预应力混凝土管桩应采用强度等级不低于42.5的硅酸盐水泥、普通硅酸盐水泥、矿渣硅酸盐水泥、粉煤灰硅酸盐水泥。其中细骨料宜采用洁净的天然硬质中粗砂,细度模数为2.3～3.4;粗骨料应采用碎石,其最大粒径应不大于25 mm,且应不超过钢筋净距的3/4。预应力钢筋应采用预应力混凝土用钢棒、预应力混凝土用钢丝。预应力混凝土管桩的保护层厚度不得小于25 mm。管桩接头宜采用端板焊接,端板的宽度不得小于管桩的壁厚,接头的端面必须与桩身的轴线垂直。放张预应力筋时,预应力混凝土管桩的混凝土抗压强度不得低于35 MPa。

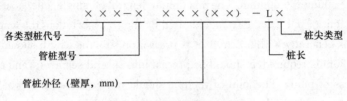

图3.26 管桩的编号图解

③桩的制作。较短的桩一般在预制厂制作,较长的桩一般在施工现场附近露天预制。桩的预制方法有并列法、间隔法、重叠法和翻模法四种。

预制场地的地面要平整、夯实,并防止浸水沉陷。对于两个吊点以上的桩,现场预制时,要根据打桩顺序来确定桩尖的朝向,因为桩吊升就位时,桩架上的滑轮组有左右之分,若桩尖的朝向不恰当,则临时调头是很困难的。

预制桩叠浇预制时,桩与桩之间要做隔离层(可涂皂脚、废机油或黏土石灰膏),以保证起吊时不互相粘结。叠浇层数,应由地面允许荷载和施工要求而定,一般不超过四层,上层桩必须在下层桩的混凝土达到设计强度等级的30%以后,方可进行浇筑。为防止由于混凝土收缩而产生裂缝,预制桩的混凝土浇筑工作应由桩顶向桩尖连续浇筑,浇筑中应保证钢筋位置正确、模板及支撑牢固、混凝土振捣密实,要制作好混凝土试块;严禁中断,不得留施工缝,制作完成后,应洒水养护不少于7d,并应在桩身上标明编号、制作日期和吊点位置并填写制桩记录。如果设计时不考虑埋设吊环,则应标明绑扎点位置。

Chapter 3　Subbase and Foundation Construction

The numbering method of prestressed concrete pipe pile is shown in Fig. 3.26. Prestressed concrete pipe pile should use Portland cement, ordinary Portland cement, slag Portland cement, and fly ash Portland cement with strength grade not less than 42.5. Among them, fine aggregate should be the clean natural hard coarse sand. The fineness is 2.3—3.4. Coarse aggregate shall be crushed stone, its maximum particle size is not greater than 25mm, and should not exceed 3/4 of the net distance of reinforcement. Prestressed reinforcement should be the steel bars for prestressed concrete and steel wire for prestressed concrete. The cover thickness of prestressed concrete pipe pile shall not be smaller than 25mm. The end face of the joint must be perpendicular to the axis of the pile body. When tendons are placed, the compressive strength of presressed concrete pipe pile shall not be less than 35MPa.

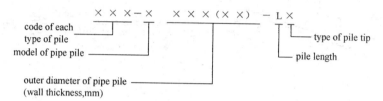

Fig. 3.26　**Numbering Diagram of Pipe Pile**

③Pile production. Shorter piles are usually made in prefabricated plant, and longer piles are usually prefabricated on construction site or nearby open area. Prefabrication methods of piles include parallel method, interval method, overlap method and flip method.

The ground of the prefabricated site should be leveled and tamped to prevent flooding subsidence. For piles with two or more hanging points, the orientation of the pile tip should be determined according to the order of pile driving, because when pile is lifted to the position, the pulley block on the pile frame has left and right parts. If the orientation of the pile tip is not appropriate, it is very difficult to make a temporary u-turn.

When precast piles are stacking prefabrication, isolation layer should be made between piles (soap stock, waste engine oil or clay lime paste can be coated) to ensure that they do not stick to each other when lifting. Number of stacked layers should be determined by the allowable ground load and construction requirements, normally not more than four layers, and upper pile pouring must be after lower pile concrete reaching 30% of design strength. In order to prevent cracks caused by concrete shrinkage, the concrete pouring of precast piles should be continuously poured from the top of piles to the pile tips. During the pouring, the position of the reinforcement should be correct, the formwork and support should be firm, the concrete should be vibrated and compacted, and concrete test block should be made. It is strictly forbidden to interrupt during pouring, and no construction joints are allowed. After the production is completed, it should be sprinkled curing for not less than seven days, and the number, date of making and location of lifting point should be marked on the pile body, and pile making record should be filled in. If the hoisting ring is not considered in the design, the location of the binding point should be indicated.

桩的主筋上端以伸至最上一层钢筋网之下为宜，并应连成"⌐—⌐"形，这样能更好地接受和传递桩锤的冲击力。主筋必须位置正确，桩身混凝土保护层要均匀，不可过厚，否则打桩时容易剥落，桩身保护层厚不宜小于 30 mm，以确保钢筋骨架受力不偏心，使混凝土有良好的抗裂和抗冲击性能。

(2)钢筋混凝土预制桩的起吊。钢筋混凝土预制桩应在混凝土达到设计强度等级的 70% 方可起吊，重叠生产的桩在起吊前应用撬棍或其他工具将桩拨动使其脱开，严禁使用千斤顶顶升桩尖。起吊时，必须合理选择吊点，防止在起吊过程中过弯而损坏。预制桩吊点合理位置图如图 3.27 所示。当吊点少于或等于 3 个时，其位置按正负弯矩相等的原则计算确定。当吊点多于 3 个时，其位置按反力相等的原则计算确定。长度为 20~30 m 的桩，一般采用 3 个吊点。

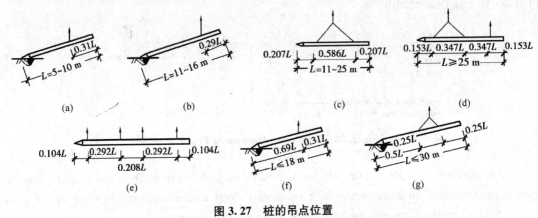

图 3.27 桩的吊点位置

(a)(b)一点起吊；(c)两点起吊；(d)三点起吊；(e)四点起吊；(f)、(g)管桩一点及两点起吊

(3)钢筋混凝土预制桩的运输。运输时桩身的混凝土强度应达到设计强度的 100%。可以用平板车进行运输，运输中应保持平稳。

打桩前需将桩从制作处运至施工现场堆放或直接运至桩架前以备打桩，并应根据打桩顺序随打随运以避免二次搬运。

桩的运输方式，在运距不大时，可用起重机吊运；当运距较大时，可采用轻便轨小平台车运输。严禁现场以拖拉方式代替装车辆运输。装运时桩的支撑应按设计吊钩位置或接近吊钩位置叠放平稳并垫实，支撑或绑扎牢固，以防运输中晃动或滑动。

The upper end of the main rib of the pile is appropriate to extend below the top layer of steel mesh, and should be connected into a " ⌐ ¬ " shape, so that the impact force of the pile hammer can be better accepted and transmitted. The main ribs must be in the correct position. The concrete protective layer of the pile should be even, not too thick. Otherwise, it will be easy to peel off when pile driving. The protective layer of the pile should not be less than 30mm to ensure that the reinforcement of the steel frame is not eccentric, so that the concrete has good crack resistance and impact resistance.

(2) Lifting of precast reinforced concrete piles. Precast reinforced concrete piles should be lifted until the concrete reaches 70% of design strength grade. Overlapping production piles should be removed by crowbar or other tools before being lifted, and jacks are strictly prohibited to raise pile tip. When lifting, the hoisting point must be chosen reasonably to prevent the damage caused by bending in the lifting process. Reasonable location map of precast pile lifting points are shown as Fig. 3. 27. When lifting points are less than or equal to 3, the position is determined according to the principle of equal bending movement. When the lifting points are more than 3, its position is calculated and determined according to the principle of equal reaction. Three lifting points are generally used for lifting piles with a length of 20—30m.

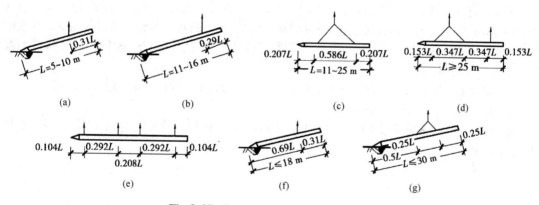

Fig. 3. 27 Location of Pile Lifting Points
(a) one point lifting; (b) one point lifting; (c) two points lifting; (d) three points lifting; (e) four points lifting;
(f) one point lifting of prestressed pipe pile; (g) two points lifting of prestressed pipe pile

(3) Transportation of precast reinforced concrete piles. During transportation, the concrete strength of the piles should reach 100% of the designed strength. They can be transported by flat car and be kept stable during transportation.

Before piling, the piles shall be transported from the fabrication site to the construction site for stacking or directly transported to the pile frame for piling. The piles shall be driven and transported according to the piling sequence to avoid secondary handling.

When the transportation distance is short, the pile can be lifted by crane; when the transportation distance is long, the light rail small platform vehicle can be used for transportation. It is strictly prohibited to transport by dragging instead of loading vehicle on site. At the time of shipment, pile support should be stowed and padded according to the design hook position or close to the hook position, and supported or tied firmly to prevent shaking and sliding during transportation.

(4)钢筋混凝土预制桩的堆放。预制桩在桩堆放时,应按规格、桩号分层叠置在平整、坚实的地面上,支承点应设在吊点处,为防止产生附加弯矩,各层垫木应上下对齐,支撑平稳,最下层的垫木应适当加宽,堆放层数不宜超过4层。运桩和堆放的桩尖方向应符合吊升的要求,以免临时再需将桩调头。

不同规格的桩,应分别堆放。

(5)钢筋混凝土预制桩的沉桩。预制桩进入到土层中可以有锤击沉桩法(打入法)、静力压桩法、振动沉桩法、水冲法等多种施工方法。这里只介绍锤击沉桩法

预制桩的锤击沉桩法施工,就是利用桩锤落到桩顶上的冲击力来克服土对桩的阻力,使桩沉到预定的深度或达到持力层的一种打桩施工方法,是预制桩最常用的沉桩方法。其特点为施工速度快,机械化程度高,适用范围广,但施工时有冲撞噪声和对地表层有振动,在城区和夜间施工有所限制。

a. 打桩设备。

打桩设备主要包括桩锤、桩架和动力装置。桩锤对桩施加冲击力,将桩打入土中。

桩锤有落锤、单动汽锤(单动蒸汽锤)、双动汽锤(双动蒸汽锤)、柴油打桩锤见和液压锤等。桩锤选择首先应根据施工条件选择桩锤的类型,然后决定锤重,一般锤重大于桩重的1.5~2倍时效果较为理想(桩重大于2 t时可采用比桩轻的锤,但不宜小于桩重的75%)。

桩架的作用是将桩吊到打桩位置,支持桩身和桩锤,并在打桩过程中引导桩的方向,保证桩锤沿着所要求的方向冲击。选择桩架时,应考虑桩锤的类型、桩的长度和施工条件等因素。桩架的高度由桩的长度、桩锤高度、桩帽厚度及所用滑轮组的高度来确定。此外,还应留1~3 m的高度作为桩锤的伸缩余地。

桩架高度=桩长+桩锤高度+桩帽高度+滑轮组高度+起锤工作余地(1~3 m)

常用的桩架形式有以下三种:

滚筒式桩架:行走靠两根钢滚筒在垫木上滚动,优点是结构比较简单,制作容易,但在平面转弯、调头方面不够灵活,操作人员较多。适用于预制桩和灌筑桩施工。

(4) Stacking of precast reinforced concrete piles.

Precast piles should be stacked in layers on a level, solid ground according to specification and pile numbers. The support point should be set at lifting point. To prevent additional bending, each layer of cushion wood should be aligned up and down for stable support, the bottom layer of the cushion wood shall be properly widened, and stacking layers should not be more than 4 layers. The tip direction of pile transport and stacking should meet the requirements of lifting, so as not to temporarily need to turn the pile.

Piles of different specifications should be stacked separately.

(5) Pile-sinking of precast reinforced concrete pile.

The precast pile can be placed into soil by hammer sinking pile method(driving method), static pressure pile method, vibration pile sinking method, water flushing method and other construction methods. Here hammer sinking pile method is introduced.

Construction of precast piles by hammer sinking pile method is a piling construction method to use the impact force of the pile hammer falling on pile top to overcome the resistance of soil to the pile, and make the pile sink to the predetermined depth or reach the bearing layer. It is the most commonly used method for pile sinking. Its characteristics are fast in construction, high degree of mechanization and wide range of applications, but there is the collision noise during construction and vibration on the surface layer. The construction is limited in urban area and at night.

a. Piling equipment.

Piling equipment mainly includes pile hammer, pile frame and power equipment. The pile hammer exerts impact force on the pile, driving the pile into the earth.

Pile hammers include drop hammer, single action steam hammer(single action steam hammer), double action steam hammer(double action steam hammer), diesel pile hammer and hydraulic hammer, etc. The selection of pile hammer should firstly choose the type of pile hammer according to the construction condition, and then decide hammer weight. Generally, that the weight of hammer is more than 1.5—2 times the weight of pile will get relatively idea effect(when pile weight is more than 2t, lighter hammer can be chosen but should not be less than 75% of the pile weight).

The role of the pile frame is to hang the pile to the position of the piling, supporting the pile body and pile hammer, and guide the direction of the pile during piling to ensure the pile hammer impacting along the required direction. When choosing pile frame, the type of pile hammer, the length of the pile and the construction conditions should be considered. The height of the pile frame is determined by the length of the pile, the height of the pile hammer, the thickness of the pile cap and the height of the pulley block to be used. In addition, a height of 1—3m should be reserved for expansion of the pile hammer.

Pile frame height = pile length+pile hammer height+pile cap height+pulley block height+ hammer lifting working space(1—3m).

There are three common forms of pile frame.

The drum pile frame: It is walking on two steel rollers rolling on the cushion wood. Its advantage is that the structure is relatively simple and easy to produce, but it is not flexible enough in plane turn and u-turn, and more operators are needed. It is suitable for precast pile and cast-in-place pile construction.

多功能桩架：多功能桩架的机动性和适应性很大，在水平方向可做360°旋转，导架可以伸缩和前后倾斜，底座下装有铁轮，底盘在轨道上行走。这种桩架可适用于各种预制桩和灌筑桩施工。

履带式桩架：以履带起重机为底盘，增加导杆和斜撑组成，用以打桩。移动方便，比多功能桩架更灵活，可用于各种预制桩和灌筑桩施工。

动力设备包括驱动桩锤用的动力设施，如卷扬机、锅炉、空气压缩机和管道、绳索和滑轮等。

打桩机的动力装置，主要根据所选的桩锤性质而定。选用蒸汽锤则需配备蒸汽锅炉；用压缩空气来驱动，则需考虑电动机的或内燃机的空气压缩机；用电源作动力，则应考虑变压器容量和位置、电缆规格及长度、现场供电情况等。

b. 打桩前的准备工作。

Ⅰ. 清理。打桩施工前应认真清除施工现场内所有妨碍打桩施工的障碍物。

Ⅱ. 场地平整。应进行整个施工场地的平整工作，并保证场地排水良好。

Ⅲ. 进行打桩试验。施工前应作数量不少于2根桩的打桩工艺试验，通过试桩可以了解桩的沉入时间、最后贯入度、持力层的强度、桩的承载力等以及发现施工过程中可能发生的问题和意外，并检验所选施工工艺、打桩设备能否完成打桩任务。

Ⅳ. 抄平放线。打桩现场应设置数量不少于2个的控制桩，要测设出基础轴线和每个桩位。

c. 施工工艺：确定桩位和沉桩顺序→打桩机就位→吊桩喂桩→校正→锤击沉桩→接桩→再锤击沉桩→送桩→收锤→切割桩头。

d. 确定桩位和沉桩顺序。

Ⅰ. 确定桩位。根据设计图纸编制工程桩测量定位图，并保证轴线控制点不受打桩时振动和挤土的影响，保证控制点的准确性。根据实际打桩线路图，按施工区域划分测量定位控制网，一般一个区域内根据每天施工进度放样10～20根桩的桩位，在桩位中心点地面上打入一支 φ6 的长度为 30～40 cm 的钢筋，并用红油漆等标示。

The multi-functional pile frame: The multi-functional pile frame has great mobility and adaptability, and can be rotated 360 degrees horizontally. The guide frame can be expanded and tilted backward and forward, iron wheels are set under the pedestal and chassis are walking on the track. It is suitable for all kinds of precast pile and cast-in-place pile construction.

Crawler pile frame: With crawler crane as the chassis, the guide rod and diagonal brace are added to pile driving. It is easy to move, more flexible than multi-functional pile frame, and can be used for various precast piles and cast-in-place piles construction.

Power equipment includes power facilities for driving pile hammers, such as hoist, boiler, air compressor, pipeline, rope, and pulley wheel, etc.

The power plant of the pile driver is mainly determined according to the properties of the selected pile hammer. Steam hammer should be equipped with steam boiler. When using compressed air to drive, an electric motor or an internal combustion engine air compressor should be considered. When using power supply as drive, transformer capacity and position, cable specification and length, on-site power supply situation should be considered.

b. Preparation before piling.

Ⅰ. Cleaning. Before piling construction, all the obstacles that hinder the piling construction in the construction site should be carefully removed.

Ⅱ. Site leveling. The entire construction site should be leveled and the site should be well drained.

Ⅲ. Piling test. Before the construction, the piling process experiment of no less than 2 piles shall be carried out. Through the test piles, the sinking time of the pile, the final penetration degree, the strength of the bearing layer, and the bearing capacity of the pile can be understood, and find out the problems and accidents that may occur in the construction process can be found out, and whether the selected construction technology and piling equipment can complete the piling works can be checked.

Ⅳ. Leveling and setting out. At least 2 control piles shall be set at the piling site, and the foundation axis and each pile position shall be measured and set out.

c. Construction process: Determine the pile position and pile driving sequence → pile driver in place → pile lifting and pile delivering → correction → pile driving with hammer → pile extension → pile driving with hammer again → pile follower → hammer closing → pile head cutting.

d. Determination of pile position and piling sequence.

Ⅰ. Determination of pile position. According to the design drawings, the project pile measurement and positioning map is prepared, and the axis control point is ensured not to be affected by the vibration and soil compaction during piling, and the accuracy of the control point is guaranteed. According to the actual piling route map, the measurement and positioning control network is divided according to the construction area. Generally, in one area, 10 to 20 pile positions are staked according to the daily construction schedule, and a steel bar with a length of 30 to 40cm is inserted into the ground at the center of the pile position, and marked with red paint or the like.

桩机移位后，应进行第二次核样，核样根据轴线控制网点所标示的工程桩位坐标点（X、Y值），采用极坐标法进行核样，保证工程桩位偏差值小于 10 mm，并以工程桩位点中心，用白灰按桩径大小画一个圆圈，以方便插桩和对中。工程桩在施工前，应根据施工桩长在匹配的工程桩身上划出以米为单位的长度标记，并按从下至上的顺序标明桩的长度，以便观察桩入土深度及记录每米沉桩锤击数。

Ⅱ. 打桩顺序。

打桩顺序合理与否，影响打桩速度、打桩质量，及周围环境。当桩的中心距小于 4 倍桩径时，打桩顺序尤为重要，打桩顺序影响挤土方向。打桩向哪个方向推进，则向哪个方向挤土。

根据桩的规格、长短和桩群的密集程度正确选择打桩顺序，以保证施工质量。当桩较稀疏，即桩中心距大于 4 倍桩边长或桩径时，可采用由一侧向单一方向进行，或由两侧向中心打设，土体挤压不是很明显，打桩顺序可相对灵活，易保证质量，如图 3.28(a)、(b)、(e)所示。当桩较紧密，即桩中心距小于等于 4 倍桩边长或桩径时，可采用自中间向两个方向对称进行、分段相对打设、分段打设或自中间向四周进行，土体挤压均匀，易保证质量，如图 3.28(c)、(d)、(f)、(g)所示。第一种打桩顺序，打桩推进方向宜逐排改变，以免土朝一个方向挤压，而导致土壤挤压不均匀，对于同一排桩，必要时还可采用间隔跳打的方式。对于大面积的桩群，宜采用后两种打桩顺序，以免土壤受到严重挤压，使桩难以打入，或使先打入的桩受挤压而倾斜。大面积的桩群，宜分成几个区域，由多台打桩机采用合理的顺序同时进行打设。

当桩的规格、埋深、长度和基础的设计标高不同，宜按先深后浅、先大后小、先长后短的原则确定打桩顺序。

e. 桩机就位。为保证打桩机下地表土受力均匀，防止不均匀沉降，保证打桩机施工安全，采用厚度为 20～30 mm 厚的钢板铺设在桩机履带板下，钢板宽度比桩机宽 2 m 左右，保证桩机行走和打桩的稳定性。桩机行走时，应将桩锤放置于桩架中下部以桩锤导向脚不伸出导杆末端为准。根据打桩机桩架下端的角度计初调桩架的垂直度，并用线坠由桩帽中心点吊下与地上桩位点初对中。

After the pile driver is displaced, the second sample should be checked. According to the coordinate points(X and Y values) of the project piles shown in the axis control points, the polar sample method is used to verify the sample to ensure that the deviation of the project pile position is less than 10mm, and the pile position is centered on the project pile point. Draw a circle with white ash according to the diameter of the pile to facilitate piling and centering. Before construction of project piles, the length mark in meters should be marked on the matched project piles according to the length of the construction piles, and the length of the piles should be marked in order from bottom to top, so as to observe the depth of piles to the soil and record the number of pile hammer strokes per meter.

II. Piling sequence.

Whether the piling sequence is reasonable or not directly affects the piling speed, pile quality and the surrounding environment. When the center distance of pile is less than 4 times of pile diameter, piling sequence is particularly important. Piling sequence affects the direction of compaction. Where the pile is going, the soil is being squeezed there.

According to the size of the pile, length and density of the pile group, the piling sequence is correctly selected to ensure the construction quality. When pile is sparse, that is, when the pile center distance is greater than 4 times the length of the pile or the diameter of the pile, it can be carried out in a single direction from one side, or from both sides to the center. Soil compaction is not obvious, piling sequence is relatively flexible, and it is easy to ensure the quality, shown as Fig. 3.28(a), (b), (e). When the pile is tight, that is, when the pile center distance is less than or equal to 4 times the length of the pile or the diameter of the pile, it can be carried out symmetrically from the center to the two directions, section relatively piling. Section piling or from the center to the around, the soil is compacted evenly. It is easy to ensure the quality, shown as Fig. 3.28 (c), (d), (f), (g). For the first piling sequence, the driving direction of piling should be changed row by row, to prevent the soil from being squeezed in one direction, resulting in uneven soil compaction. For the same row of piles, if necessary, it can also use the interval jump to pile. For large pile groups, it is appropriate to adopt the later two piling sequences to avoid soil being severely squeezed, making it difficult for piles to be driven, and thus making the first driven piles be squeezed and inclined. The large pile groups should be divided into several areas, by multiple pile drivers to use reasonable sequence to pile at the same time.

When the specification, buried depth, length of pile and design elevation of foundation are different, the pile driving sequence should be determined according to the principle of first deep to shallow, first large to small, and first long to short.

e. Piling machine in place. In order to ensure the uniform stress of the surface soil under the pile driver, prevent the uneven settlement and ensure the construction safety of the pile driver, the steel plate with the thickness of about 20—30mm is laid under the track plate of the pile driver, and the width of the steel plate is about 2m wider than that of the pile driver, so as to ensure the stability of the driving and driving of the pile driver. When pile driver is walking, the pile hammer should be placed in the middle and lower part of the pile frame, and the pile hammer guide foot should not extend the end of the guide rod. According to the angle gauge of the lower end of the pile frame of the pile driver, the verticality of the pile frame shall be initially adjusted, and the plumb line shall be used to lift down the pile cap center point and initially align with the pile site on the ground

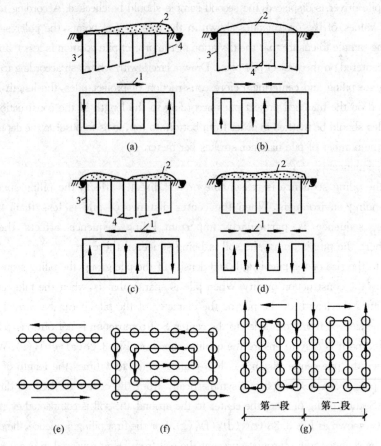

图 3.28 打桩顺序和土的挤密情况
(a)逐排单向打设;(b)两侧向中心打设;(c)中部向两侧打设;(d)分段相对打设;
(e)逐排打设;(f)自中部向边沿打设;(g)分段打设
1—打设方向;2—土的挤密情况;3—沉降量大;4—沉降量小

f. 起吊,对中和调直。

Ⅰ. 桩应由吊车将桩转运至打桩机导轨前,当管桩单节长小于等于 20 m 时,采用专用吊钩勾住两端内壁直接进行水平起吊;当管桩单节长大于 20 m 时,应采用四点吊法转运。

Ⅱ. 桩摆放平稳后,在距管桩端头 $0.21L$ 处,将捆桩钢丝绳套牢,一端拴在打桩机的卷扬机主钩上,另一端钢丝绳挂在吊车主钩,打桩机主卷扬向上先提桩,吊车在后端辅助用力,使管桩与地面基本成 $45°\sim 60°$ 角向上提升,将管桩上口喂入桩帽内,将吊车一端钢丝绳松开取下,将管桩移至桩位中心。

Chapter 3 Subbase and Foundation Construction

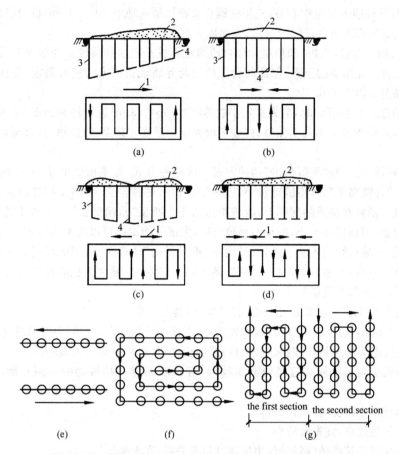

Fig. 3. 28 Piling Sequence and Conditions of Soil Compaction
(a) drive row by row in one direction; (b) drive from both sides to center;
(c) drive from center to both sides; (d) drive by section relatively;
(e) drive row by row; (f) drive from the middle to the edge;
(g) drive by section
1—Pile driving direction; 2—Conditions of soil compaction;
3—Large amount of settlement; 4—Small amount of settlement

f. Lifting, centering and straightening.

I. The pile shall be transported to the front of the guide rail of the pile driver by the crane. When the single section length of the pipe pile is less than or equal to 20m, the special hook shall be used to hook the inner walls of both ends for direct horizontal lifting; when the single section length of the pipe pile is greater than 20m, the four point lifting method shall be adopted for transfer.

II. After the pile is placed stably, the steel wire rope for binding pile shall be fastened at 0.21L from the end of pipe pile, and one end shall be tied to the main hook of hoist of pile driver and the other end shall be hung on the main hook of crane. The main hoist of the pile driver shall lift the pile first, and the crane shall assist in the rear end to lift the pipe pile upward at an angle of 45° to 60° with the ground. The upper opening of the pipe pile shall be fed into the pile cap, the steel wire rope at one end of the crane shall be loosened and removed, and the pipe pile shall be moved to the center of the pile position.

Ⅲ．对中：管桩插入桩位中心后，先用桩锤自重将桩插入地下 30～50 cm，桩身稳定后，调正桩身、桩锤、桩帽的中心线重合，使之与打入方向成一直线。

Ⅳ．调直：用经纬仪（直桩）和角度计（斜桩）测定管桩垂直度和角度。经纬仪应设置在不受打桩机移动和打桩作业影响的位置，保证两台经纬仪与导轨成正交方向进行测定，使插入地面时桩身的垂直度偏差不得大于 0.5%。

g．打桩方法。打桩开始时，应先采用小的落距（0.5～0.8 m）进行轻的锤击，使桩正常沉入土中 1～2 m 后，经检查桩尖不发生偏移，再逐渐增大落距至规定高度，继续锤击，直至把桩打到设计要求的深度。

打桩有"轻锤高击"和"重锤低击"两种方式。这两种方式，如果所做的功相同，而所得到的效果却不相同。"轻锤高击"，所得的动量小，而桩锤对桩头的冲击力大，因而回弹也大，桩头容易损坏大部分能量均消耗在桩锤的回弹上，故桩难以入土。相反"重锤低击"所得的动量大，而桩锤对桩头的冲击力小，因而回弹也小，桩头不易被击碎，大部分能量都可以用来克服桩身与土壤的摩阻力和桩尖的阻力，故桩很快入土。此外，又由于"重锤低击"的落距小，因而可提高锤击频率，打桩效率也高，正因为桩锤频率较高，对于较密实的土层，如砂土或黏性土也能较容易地穿过，所以打桩宜采用"重锤低击"，低垂重打。

打桩顺序应根据桩的密集程度及周围建（构）筑物的关系：

Ⅰ．若桩较密集且据周围建（构）筑物较远，施工场地开阔时宜从中间向四周进行。

Ⅱ．若桩较密集场地狭长，两端距建（构）筑物较远时，宜从中间向两端进行。

Ⅲ．若桩较密集且一侧靠近建（构）筑物时，宜从毗邻建（构）筑物的一侧开始，由近及远地进行。

Ⅳ．根据桩入土深度，宜先长后短。

Ⅴ．根据管桩规格，宜先大后小。

Ⅵ．根据高层建筑塔楼（高层）与裙房（低层）的关系，宜先高后低。

Ⅲ. Centering. After the pipe pile is inserted into the center of the pile position, the pile shall be inserted into the underground 30—50cm by the weight of the pile hammer. After the pile body is stable, the center line of the pile body, pile hammer and pile cap shall be adjusted to coincide with the driving direction.

Ⅳ. Straightening. The verticality and angle of pipe pile are measured by theodolite (straight pile) and angle gauge (inclined pile). The theodolite shall be set at a position not affected by the movement of the pile driver and the piling operation, so as to ensure that the two theodolites are in the orthogonal direction with the guide rail, so that the verticality deviation of the pile body when inserted into the ground shall not be greater than 0.5%.

g. Piling method.

When start piling, first use a small drop distance (0.5—0.8m) for light hammering, and make the pile sink into the soil 1—2m normally. After checking, the pile tip will not send offset, and then gradually increase the drop distance to the specified height, continue hammering until the pile is driven to the depth required by the design.

There are two ways of piling: "light hammer high hit" and "heavy hammer low hit". For these two methods, the works are the same, but the effects are not the same. A "light hammer high hit" gains little momentum, but pile hammer has great impact on the pile head, so it has a great rebound. The pile head is easily damaged, and most of the energy is consumed on the rebound of the pile hammer, so the pile is difficult to be burried. In contrast, "heavy hammer low hit" gains great momentum, and the impact of pile hammer on the pile head is little, so it has little rebound. The pile head is not easy to be damaged, and most of the energy can be used to overcome the frictional resistance of the pile and soil and the resistance of the pile tip, so the pile can be quickly buried. Besides, because the drop distance of "heavy hammer light hit" is small, so the hammer frequency can be improved, and piling is also highly efficientive. Because the pile hammer has high frequency, it can easily pass through the denser soil, such as sand or clay. So piling should use "heavy hammer low hit" and "low hammer heavy hit".

The piling sequence should be based on the density of piles and relationship between the surrounding structures.

Ⅰ. If piles are denser and far from the surrounding structures, and when the site is open, piling can proceed from the center to the perimeter.

Ⅱ. If piles are denser and site is narrow, the two ends are far from structures, piling can proceed from the center to the two ends.

Ⅲ. If piles are denser and one side is close to the structure, piling should start from the side of adjacent structure and proceed from near to far.

Ⅳ. According to the depth of pile into the soil, piling should be first long and then short.

Ⅴ. According to the size of pipe pile, piling should be first big and then small.

Ⅵ. According to the relationship between high-rise tower (high-rise) and podium (low-rise), piling should be first high and then low.

h. 接桩。当桩的长度较大时,由于桩架高度以及制作运输等条件限制,往往需要分段制作和运输,沉桩时,分段之间就需要接头。一般混凝土预制桩接头不宜超过2个,预应力管桩接头不宜超过4个,应避免在桩尖接近硬持力层或桩尖处于硬持力层中时接桩。打桩施工接桩一般在下节桩距地面一定高度处(预制桩一般0.5~0.8 m,预应力管桩一般1~1.2 m)进行。

预制桩的接桩工艺主要有硫磺胶泥浆锚法、焊接法接桩和法兰螺栓接桩法三种。桩的接头应有足够的强度,能传递轴向力、弯矩和剪力,前一种适用于软弱土层,后两种适用于各类土层。预制混凝土方桩接长的方法有硫磺胶泥浆锚法和焊接法。浆锚法仅适用于软弱土层。预应力管桩接长采用焊接法。工程中经常使用的是焊接法接桩。

硫磺胶泥浆锚接桩法:如图3.29所示,图中a为$15d$(锚筋直径)。在上节桩的下端伸出钢筋,长度为15倍的钢筋直径,下节桩的上端设预留锚筋孔,孔径为$2.5d$,孔深大于$15d$,一般取$15d+30$ mm。接桩时,把上节桩伸出的4根锚筋,插入下节桩的预留孔中,孔内灌满硫磺胶泥并热铺于桩的顶面,厚度为1~2 cm,胶泥灌注时间不得超过2 min,然后将两节桩压紧,胶泥很快冷却硬化,只需停5~10 min,就可继续锤击沉桩。

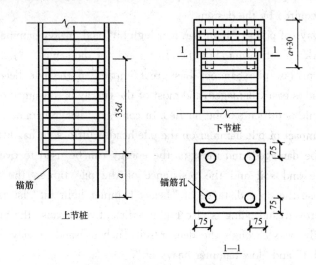

图3.29 浆锚法接桩节点构造

h. Pile extension. When pile length is large, due to the limitation of pile frame height and conditions of production and transportation, it often needs to be made and transported in sections. When piling, connections between sections are required. Generally, concretions of precast concrete pile are not more than 2, and the connections of prestressed pipe pile are not more than 4. Pile extension should be avoided when pile tip is near the hard bearing layer or when it is in the hard bearing layer. In pile driving construction, pile extension is generally carried out at a certain hight of the lower pitch pile from the ground(precast piles are general 0.5—0.8m, and prestressed pipe pile are general 1—1.2m.

There are three kinds of pile extension techniques for precast piles: sulfur glue mud anchor method, welding method and flange bolt method. Pile joints should be strength enough to transfer axial forces, bending moment and shear forces. The former one is suitable for the soft soil layer and the latter two are suitable for all kind of soil layer. The extension methods of precast concrete square piles are sulfur glue mud anchor method and welding method. Mud anchor method is just applicable for the soft soil. The welding method is used for extension of prestressed pipe pile. Welding method is often used for pile extension in project.

Sulfur glue mud anchor method is shown as Fig. 3.29. In the figure, a is $15d$ (anchor bar diameter). The steel bar is extended from the lower end of the upper pitch pile with length of 15 times the diameter of the steel bar. The upper end of lower pitch pile is provided with a reserved anchor bar hole with an aperture of $2.5d$ and a hole depth is great than $15d$, generally $15d + 30$mm. To extend the pile, insert the 4 anchor bars protruding from the upper pile into the reserved hole of the lower pile. Fill the hole with sulfur glue and heat it on the top surface of the pile. The thickness is 1—2cm and the glue filling time should not exceed 2min, then the two piles are pressed tightly, and the glue is cooled and hardened quickly. Just stop for 5—10 minutes and hammering the pile can be continued.

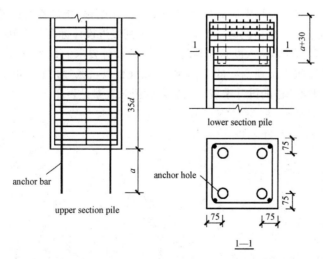

Fig. 3.29 Structure of Pile Extension by Mud Anchor Method

焊接接桩法:如图3.30所示,当桩沉至操作平台时,在下节桩上端部焊接4个63 mm×8 mm、长150 mm的短角钢,这4个短角钢与桩的主筋焊在一起,然后把上节桩吊起,在其下端把4个63 mm×8 mm、长150 mm的短角钢焊在主筋上,最后把上下两桩对准用4根角钢或扁钢焊接,使之成为一个整体。

法兰螺栓接桩法:如图3.31所示。制作预制桩时,在桩的端部设置法兰,需接桩时用螺栓把它们连在一起,这种方法,施工简便,速度快。主要用于混凝土管桩。但法兰盘制作工艺复杂,用钢量大。

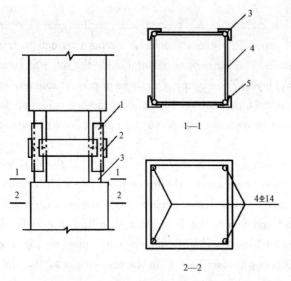

图3.30 焊接法接桩节点构造
1—角钢;2—钢板;3—桩内预埋角钢;4—桩箍筋;5—桩主筋

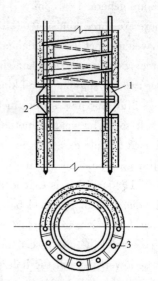

图3.31 管桩法兰接桩节点构造
1—法兰盘;2—螺栓;3—螺栓孔

i. 打桩质量要求。打桩过程中应做好测量和记录。用落锤、单动汽锤或柴油锤打桩时,从开始即需统计桩身每沉落1 m所需的锤击数。当桩下沉到接近设计标高时,则应以一定落距测量其每个阵击(10击)的沉落值(即贯入度),并判断是否符合设计要求的最小贯入度。如用双动汽锤,从开始应记录桩身每下沉1 m所需要的工作时间,以观察其沉入速度。当接近设计标高时,应测量每分钟的下沉值,以保证桩的设计承载力。贯入度记录表参见表3.5。

Welding pile extension method: As shown in Fig. 3.30, when pile is sunk to the operating platform, four short angle steels of 63mm×8mm and 150mm long are welded at top end of the lower pile. These four short angle steels are welded together with the main bars of the pile and then the upper joint pile is lifted up. At the bottom, four short angle steels of 63mm×8mm and 150mm long are welded on the main bar. Finally, align the upper and lower piles with four angle steels or flat steels and welded, so that they become a whole.

Flange bolt method for pile extension is shown as Fig. 3.31. When making precast piles, flanges are set at the end of the piles, and bolts are attached to them when the piles are connected. This method is simple and fast in construction and is mainly used for concrete pipe piles. But flange production process is complex, and large amount of steel is needed.

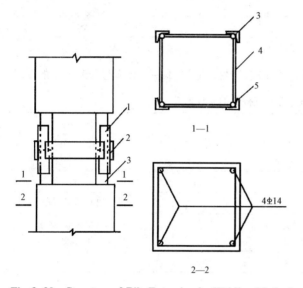

Fig. 3.30 Structure of Pile Extension by Welding Method
1—angle steel; 2—steel plate; 3—embedded angle steel in pile;
4—pile stirrup; 5—pile main bar

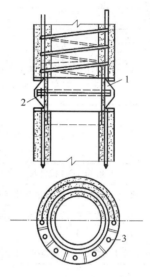

Fig. 3.31 Structure of Pile Extension by Flange Bolt Method
1—flange; 2—bolt; 3—bolt hole

i. Piling quality requirements. Measurement and record shall be made during pile driving. When piling with drop hammer, single acting steam hammer or diesel hammer, it is necessary to count the number of hammers required for each 1m of pile sinking from the beginning. When the pile is sunk close to the design elevation, the subsidence value (penetration) of each array (10 hits) should be measured at a certain pitch, and the minimum penetration degree should be determined whether it meets the design requirements. If double action steam hammer is used, from beginning, it is necessary to record the working time required for each 1m of sinking of pile, so as to observe the sinking speed. When approaching the design elevation, the subsidence value per minute should be measured to ensure the design bearing capacity of the pile. The penetration record is shown in Table 3.5.

表 3.5　钢筋混凝土预制桩施工记录

施工单位：　　　　　　　　　　　　　　工程名称：
施工班组：　　　　　　　　　　　　　　桩的规格：
桩锤类型及冲击部分重量：　　　　　　　自然地面标高：
桩帽重量：　　　　气候：　　　　　　　桩顶设计标高：

编号	打桩日期	桩入土每米锤击次数 1 2 3 4 5……	落距/mm	桩顶高出或低于设计标高/mm	最后贯入度/mm	备注

工程负责人：　　　　　　　　　　　　　记录：

打入桩（预制混凝土方桩、先张法预应力管桩、钢桩）的桩位偏差，必须符合表 3.6 的要求。

表 3.6　打入桩桩位的允许偏差

项	项目	允许偏差
1	盖有基础梁的桩：(1)垂直基础梁的中心线 (2)沿基础梁的中心线	$100+0.01H$ $150+0.01H$
2	桩数为 1～3 根桩基中的桩	100
3	桩数为 4～16 根桩基中的桩	1/2 桩径或边长
4	桩数大于 16 根桩基中的桩：(1)最外边的桩 (2)中间桩	1/3 桩径或边长 1/2 桩径或边长

注：H 为施工现场地面标高与桩顶设计标高的距离。

钢筋混凝土预制桩的检验标准应符合表 3.7 的规定。

Chapter 3 Subbase and Foundation Construction

Table 3.5 Construction Record of Precast Reinforced Concrete Pile

Construction Unit: Project name:
Construction team: Pile Size:
Type of Pile Hammer and Impact Weight: Natural ground level:
Pile Cap Weight: Climate: Design Elevation of Pile Top:

Serial Number	Piling Date	Number of hammer hits per meter of pile into soil	Drop Distance/mm	Pile Top Above or Below Design Level/mm	Final Penetration /mm	Remarks

Project Manager: Record:

The pile positions deviation of driven piles (precast concrete square pile, pretensioned prestressed pipe pile and steel pile) must comply with the requirements in Table 3.6.

Table 3.6 The Allowable Deviation of Driven Pile Position

Item	Project	Allowable Deviation
1	A pile covered with a foundation beam: (1) The center line of the vertical foundation beam (2) Along the center line of the foundation beam	$100+0.01H$ $150+0.01H$
2	The number of piles is 1—3 piles in the pile foundation	100
3	The number of piles is 4—16 piles in the pile foundation	1/2 pile diameter or side length
4	The number of the piles is more than the pile in 16 pile foundation: The outermost pile The middle pile	1/3 pile diameter or side length 1/2 pile diameter or side length

Note: H is the distance between the ground elevation of the construction site and the designed elevation of the pile top.

The inspection standard for precast reinforced concrete pile shall comply with the provisions of Table 3.7.

表 3.7　钢筋混凝土预制桩的质量检验标准

项	序号	检查项目	允许偏差或允许值		检查方法
			单位	数值	
主控项目	1	桩体质量检验	按基桩检测技术规范		按基桩检测技术规范
	2	桩位偏差	见表 3.6		用钢尺量
	3	承载力	按基桩检测技术规范		按基桩检测技术规范
一般项目	1	砂、石、水泥、钢材等原材料(现场预制时)	符合设计要求		查出厂质保文件或抽样送检
	2	混凝土配合比及强度(现场预制时)	符合设计要求		检查称量或查试块记录
	3	成品桩外形	表面平整,颜色均匀,掉角深度<10mm,蜂窝面积小于总面积0.5%		直观
	4	成品桩裂缝(收缩裂缝或起吊、装运、堆放引起的裂缝)	深度小于 20 mm,宽度小于 0.25 mm,横向裂缝不超过边长的一半		裂缝测定仪(在地下水有侵蚀及锤击数超过 500击的长桩地区不适用)
	5	成品桩尺寸:横截面边长	mm	±5	用钢尺量
		桩顶对角线差	mm	<10	用钢尺量
		桩尖中心线	mm	<10	用钢尺量
		桩身弯曲矢高		<l/1 000	用钢尺量,l 为桩长
		桩顶平整度	mm	<2	用水平尺量
	6	电焊接桩:焊缝质量	按地基基础质量验收规范		按地基基础质量验收规范
		电焊结束后停歇时间	min	>1.0	秒表测定
		上下节点平面偏差	mm	<10	用钢尺量
		节点弯曲矢高		<l/1000	用钢尺量,l 为桩长
	7	硫磺胶泥接桩:胶泥浇筑时间	min	<2	秒表测定
		浇筑后后停歇时间	min	>71.0	秒表测定
	8	桩顶标高	mm	±50	水准仪
	9	停锤标准	设计要求		现场实测或查沉桩记录

Chapter 3 Subbase and Foundation Construction

Table 3.7 Quality Inspection Standards for Precast Reinforced Concrete Piles

Item	Serial Number	Inspection Item	Allowable Deviation or Allowable Value		Inspection Method
			Unit	Numerical Value	
Dominant Project	1	Pile quality inspection	According to the technical specification of foundation pile detection		According to the technical specification of foundation pile detection
	2	The pile location deviation	See Table 3.6		use steel tape to measure
	3	Bearing capacity	According to the technical specification of foundation pile detection		According to the technical specification of foundation pile detection
General Project	1	Sand, stone, cement, steel and other raw materials (when prefabricated on site)	Comply with design requirements		Find out the factory quality assurance documents or sample for inspection
	2	Concrete mix ratio and strength (when prefabricated on site)	Comply with design requirements		Check the weighing or check the test block record
	3	Shape of finished pile	Surface is flat, color is uniform, the corner depth is < 10mm and the honeycomb area is less than 0.5% of the total area		Perceptual intuition
	4	Cracks in finished piles (shrinkage cracks or cracks caused by lifting, shipping, stacking)	The depth is less tan 20mm, the width is less than 0.25mm, and the transverse cracks are not more than half the length of the side		Fracture meter (not applicable in areas with groundwater erosion and long piles with more than 500 hammer strikes)
	5	Finished pile size: Cross section length of side	mm	±5	Steel tape to measure
		Diagonal difference of pile top	mm	<10	Steel tape to measure
		Center line of pile tip	mm	<10	Steel tape to measure
		Bending height of pile body		<l/1000	Steel tape to measure, l for pile length
		Flatness of pile top	mm	<2	Measure with horizontal scale
	6	Electric welding pile extension: Weld quality	According to the quality acceptance code of foundation		According to the quality acceptance code of foundation
		Stop time after welding	min	>1.0	Stopwatch measurement
		Plane deviation of upper and lower nodes	mm	<10	Steel tape to measure
		Node bending vector height		<l/1000	Steel tape to measure, l for pile length
	7	Sulfur cement pile extension: Cement pouring time	min	<2	Stopwatch measurement
		Stop time after pouring	min	>71.5	Stopwatch measurement
	8	Pile top elevation	mm	±50	Level
	9	Stop hammer standard	Design requirement		Field measurement or pile sinking record

3. 钢筋混凝土灌注桩

灌注桩基础由灌注桩和承台组成。灌注桩是先以一定的手段（人工或机械）在桩位上形成桩孔，然后放入钢筋笼，最后浇筑混凝土所成的桩。灌注桩按成孔方式的不同可分为泥浆护壁成孔灌注桩、沉管灌注桩、干作业成孔灌注桩、人工挖孔灌注桩和爆扩成孔灌注桩等。由于一个基础下可能有多根桩，以及在成孔时因振动等原因可能造成塌孔或造成成桩质量问题，因此必须采取一定的成孔顺序。施工中对土有挤密作用和振动影响时，可以结合现场情况，采取间隔一或两个桩位成孔、先成中间孔再成周围孔、在临桩混凝土初凝前或终凝后成孔等方式。

灌注桩的优点是灌注桩施工不存在沉桩挤土问题，振动和噪音均很小，对邻近建筑物、构筑物及地下管线、道路等的危害极小。但是，混凝土灌注桩的成桩工艺较复杂，尤其是湿作业成孔时，成桩速度也较预制打入桩慢。且其成桩质量与施工好坏密切有关，成桩质量难以直观的进行检查。

（1）干作业成孔灌注桩。干作业成孔灌注桩是指不用泥浆和套管护壁的情况下，用人工钻具或机械钻成孔，下钢筋笼、浇混凝土成桩。该工作主要优点是不同循环介质，噪音和振动小，对环境影响小；施工速度快，设备简单，易操作；由于干作业成孔，混凝土质量能得到较好的控制。缺点是孔底常留有虚土不易清除干净，影响桩的承载力；螺旋钻具回转阻力较大，对地层的适应性有一定的条件限制。

干作业成孔灌注桩成孔深度为 8~20 m，成孔直径为 300~600 mm，适用于地下水位较低，在成孔深度内无地下水的土质，无须护壁可直接取土成孔；也适用于地下水位以上的人工填土、黏性土、粉土、砂性土、风化软岩及粒径不大的砾砂层。

干作业扩底灌注桩适用于地下水位以上的黏土、粉土、砂土、填土和粒径不大的砾砂层，扩底部宜设在强度较高的持力层中。

除上述两种干作业成孔灌注桩施工方法还有手摇钻成孔灌注桩，它是用人力旋转钻具钻进，成孔直径一般为 200~350 mm，孔深 3~5 m。本法使用机具简单、操作容易，但劳动强度大，适用于缺乏成孔机具设备、桩数不多的情况。

3. Reinforced Concrete Cast-in-Place Pile

Cast-in-place pile foundation consists of cast-in-place pile and pile cap. Cast-in-place pile is a pile that made by certain means(manual or mechanical)to form a pile hole on the pile position, then put steel reinforcement cage, and finally pour concrete. According to the different types of hole forming, the piles can be divided into mud wall cast-in-place pile, sinking pipe cast-in-place pile, dry operation cast-in-place pile, manual digging cast-in-place pile and blasting expansion cast-in-place pile, etc. Because there may be multiple piles under one foundation and the hole collapse or pile quality problems may be caused due to vibration and other reasons during hole forming, a certain sequence of hole forming must be adopted. When there is compaction effect and vibration influence on soil during construction, it can be combined with site conditions to form holes at intervals of one or two pile positions, to form intermediate holes first and then to surrounding holes, and to form holes before initial setting or after final setting of temporary pile concrete.

The advantages of cast-in-place pile construction are that there is no pile sinking and soil squeezing problem, the vibration and noise are very small, and the damage to adjacent buildings, structures, underground pipelines and roads is minimal. However, the pile forming technology of concrete cast-in-place pile is more complicated, especially when wet operation drilling, the speed of pile forming is slower than the precast driven pile, and the pile forming quality is closely related to the construction quality, so it is difficult to inspect the pile quality directly.

(1)Dry operation cast-in-place pile. Dry operation cast-in-place pile refers to drilling holes with manual drilling tools or machines, placing reinforcement cages and pouring concrete without mud and casing wall protection. The main advantages of the operation are different circulating medium, low noise and vibration, little influence to the environment, fast speed in construction, simple equipment and easy operation. Due to hole forming in dry operation, the quality of concrete can be well controlled. The disadvantage is that it is not easy to remove the soil at the bottom of the hole, which affects the bearing capacity of the pile. Auger has great resistance to gyration and has certain limitation to the adaptability of formation.

The hole depth of dry operation cast-in-place pile is 8—20m, and the hole diameter is 300—600mm. It is suitable for the soil where the underground water is lower and no underground water within the hole depth. The hole can be made directly from soil without backing. It is also suitable for artificial filling soil, clay soil, silty soil, sandy soil, weathered soft rock and gravel sand layer with small particle size.

Dry operation belled cast-in-place pile is suitable for clay, silt, sand, fill and gravel sand layer with small particle size above the groundwater level. The enlarged bottom should be set in the bearing stratum with higher strength.

In addition to the above two kinds of dry operation bored pile construction methods, there are also hand drill bored pile, which is drilled by hand rotary drilling tool, the hole diameter is generally 200—350mm, and the hole depth is 3—5m. This method does not need energy, the machine is simple and easy to operate, but the labor intensity is high, so it is suitable for the situation of lacking drilling machines and equipment and few piles.

①成孔机械。成孔机械主要有螺旋钻机,钻孔扩机,洛阳铲等。干作业成孔一般采用螺旋钻机钻孔。螺旋钻头外径一般为 300～600 mm,最大可达 800 mm,常有的直径分别为 400 mm、500 mm、600 mm,钻孔深度相应为 12 m、10 m、8 m。

螺旋钻机由主机、滑轮组、螺旋钻杆、钻头、滑动支架和出土装置等组成,一般采用步履式全螺旋钻孔。这种钻机根据钻杆形式不同可分为:整体式螺旋、装配式长螺旋和短螺旋三种。

②施工工艺:场地清理→测量放线定桩位→桩机就位→钻孔取土成孔→清除孔底沉渣→成孔质量检查验收→吊放钢筋笼→浇筑孔内混凝土,如图 3.32 所示。

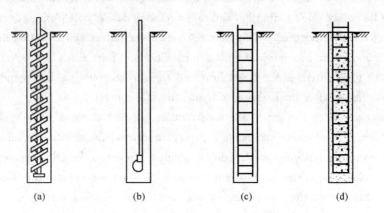

图 3.32 干作业成孔桩
(a)螺旋钻机钻孔;(b)空转清土后掏土;(c)放入钢筋骨架;(d)浇筑混凝土

③施工要点及注意事项。干作业成孔一般采用螺旋钻成孔,还可采用机扩法扩底。为了确保成桩后的质量,施工中应注意以下几点:

a. 螺旋钻进应根据地层情况,合理选择和调配钻进参数,并可通过电流表来控制进尺速度,电流值增大,说明孔内阻力增大,应降低钻进速度。

b. 开始钻进及穿过软硬土层交接处时,应缓慢进尺,保持钻具垂直和位置正确,防止因钻杆晃动引起孔径扩大及增多孔底虚土;钻进含有砖头、石块或卵石的土层时,应控制钻杆跳动与机架摇晃。

c. 钻进中遇蹩车,钻杆摇晃、移动、偏斜,不进尺或进尺缓慢时,应停机检查找出原因,采取措施,避免盲目钻进,导致桩孔严重倾斜、跨孔甚至卡钻、折断钻具等恶性事件发生。

①Hole forming machinery.

Hole forming machinery is mainly auger drilling machine, drilling and expanding machine and Luoyang shovel, etc. Dry operation hole forming generally uses auger drilling machine to drill holes. Auger bit diameter is generally 300—600mm, and maximum 800mm. The diameter commonly used is 400mm, 500mm and 600mm separately. The drilling depth is corresponding to 12m, 10m and 8m.

The auger is composed of main engine, pulley block, spiral drill pipe, drill bit, sliding support and digging device, etc. According to the different types of drill pipe, this kind of drilling rig can be divided into three types: integral screw, assembled long screw and short screw.

②Construction process. site cleaning→measuring and setting pile position→pile machine in place→drilling hole, collecting earth and hole forming→removing sediment from the bottom of the hole→inspection and acceptance of hole forming quality→lifting and putting in steel cage→pouring concrete in the hole. as shown in Fig. 3. 32.

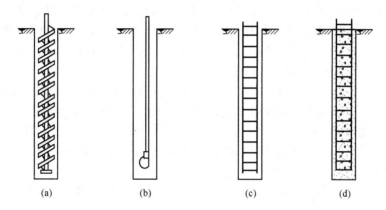

Fig. 3. 32　Dry Operation for Driven Pile

(a)auger drilling; (b)remove soil after idling; (c)put in the steel frame; (d)pour concrete

③Construction points and matters needing attention.

Dry operation hole forming generally uses auger to drill holes, and can use machine expansion method to expand bottom. To ensure formed pile quality, following matters need to be attention during construction.

a. Auger drilling should be based on the soil layer condition, reasonably choosing and adjusting drilling parameters, and penetration speed can be controlled through ampere meter. The increase of current value indicates that internal resistance of the hole increases, and the drilling speed should be reduced.

b. When drilling begins and crosses the junction of soft and hard soil, penetration speed should be slowly. Keep drilling equipment vertical and correct position and avoid hole diameter to be enlarged and bottom soil increased due to drill stem shaking. When drilling into the soil layer containing bricks, stones or pebbles, drill stem jumping and machine frame shaking should be controlled.

c. When suffered a bump, drill stem shaking, movement, deflection, no penetration speed or speed slowly in drilling, it should be stopped to check and find out the reasons, and take measures to avoid blind drilling which leads to vicious incidents such as pile hole badly tilted, cross hole and even stuck drilling, and broken drilling equipment, etc.

d. 遇孔内渗水、塌孔、跨孔和缩径等异常情况时,立即起钻,同有关单位研究处理,采取相应的技术措施;上述情况不严重时,可调整钻具参数,投入适量黏土球,上下活动钻具等,保持钻具顺畅。如塌孔处理,宜钻至塌孔处以下 1~2 m,用低等级混凝土填至塌孔以上 1m,待混凝土初凝后继续钻至设计深度。

e. 冻土层、硬土层施工时,宜采用高转速,小给进量,恒钻压。若钻头进入硬土层时,造成钻孔偏斜,可提起钻头上下反复扫钻几次,以便削去硬土。若纠正无效,可在孔中局部回填黏土至偏孔处 0.5 m 以上,再重新钻进。

f. 短螺旋钻进,每次进尺宜控制在钻头的 2/3 左右,砂层、粉土层可控制在 1.2~1.8 m,黏土、粉质黏土在 0.6 m 以下。

g. 钻至设计深度后,应使钻具在孔内空转数圈清除孔口积土、孔内虚土。清孔目的是将孔内的浮土、虚土取出,减少桩的沉降。方法是钻机在原深处空转清土,然后停止旋转,提钻卸土。

对于摩擦桩,桩底虚土不得大于 300 mm;对于端承桩,桩底虚土不得大于 100 mm。

钻孔时,如遇软塑土层含水量大的情况,可用叶片螺距较大的钻杆,这样工效可高一些;在可塑或硬塑的土层中,或含水量较少的砂土层中,则应采用叶片螺距较小的钻杆,以便能均匀平稳的钻进土中。一节钻杆钻完后,可接上第二节钻杆,直至钻到要求的深度。

h. 清理虚土后,应盖好孔口盖,防止杂物落入,按规定验收,并做好施工记录。

i. 一级建筑桩基,应配置桩顶与承台的连接钢筋笼,其主筋采用 6~10 根直径为 12~14 mm,配筋率不小于 0.2%,锚入承台 30 倍主筋直径,伸入桩身长度不小于 10 倍桩身直径,且不小于承台下软弱土层层底深度。

钢筋笼吊入孔内,不要碰撞孔壁,骨架要扎好保护层垫块,控制保护层厚度,又要保证钢筋骨架垂直度及钢筋笼的标高。

j. 在灌注混凝土之前,必须对孔深、孔径、垂直度和孔底虚土厚度进行复查,不合格应及时处理。

k. 混凝土应连续浇注,分层夯实,分层的高度按采用的振捣工具而定,一般每次浇筑高度不得大 1.5 m。

Chapter 3　Subbase and Foundation Construction

　　d. In case of water seepage in the hole, hole collapse, cross hole and hole shrinkage, drill should be lifted immediately. Study and deal with them with relevant units, and take corresponding technical measures. If the above cases are not serious, parameter of drill equipment can be adjusted, an appropriate amount of clay ball can be put in and drill equipment can be moved up and down to keep it clean. If dealing with collapsed hole, it is advisable to drill down to 1—2m below the collapsed hole and use low grade concrete to fill to 1m above the collapsed hole. Continue drilling down to designed depth after initial setting of concrete.

　　e. When construction in frozen soil layer and hard soil layer, it is advisable to use high speed, small feed and constant drillpressure. If the drill enters into hard soil layer and the borehole deflection is caused, the drill can be lifted up and down, and repeatedly swept several times to remove the hard soil. If the correction is not effective, the hole can be partially backfilled with clay to more than 0.5m away from the hole, and then re-drilling.

　　f. Short auger drilling in, the penetration speed each time is better controlled at about 2/3 of the drill bit. Sand and silt can be controlled at 1.2—1.8m, clay and silt clay below 0.6m.

　　g. After drilling to designed depth, make the drill string for several times into the hole to clear out the accumulated soil and empty soil inside the hole. The purpose of clearing the hole is to remove the floating soil in the hole and reduce the settlement of the pile. The method is to idle the drill in the original depth to clear the soil, and then stop the rotation, lift the drill to unload the soil.

　　As for friction piles, the void soil at the bottom of piles shall not be greater than 300mm; for end bearing piles, the void soil at the bottom of piles shall not be greater than 100mm.

　　When drilling, if soft plastic soil moisture content is large, drill stem with larger pitch can be used, so the effect can be higher. In plastic and hard plastic soil, or sand soil with little moisture content, drill stem with smaller pitch should be used, so as to evenly and stably drill to the soil. After one drill stem is drilled, the second drill stem can be attached until the required depth is reached.

　　h. After clearing the void soil, the hole cover shall be covered to prevent the debris from falling into the hole. The acceptance shall be carried out according to the regulations, and the construction record shall be made..

　　i. The pile foundation of the first-class building should be equipped with the connection reinforcement cage between the pile top and the bearing platform. The main reinforcement should be 6—10, with the diameter of 12—14mm, and the reinforcement ratio should not be less than 0.2%. The anchor should be 30 times of the diameter of the main reinforcement, and the length of the pile body should not be less than 10 times of the diameter of the pile body, and not less than the depth of the soft soil layer under the bearing platform.

　　When reinforcement cage is hoisted into the hole, do not collide with the hole wall. Protective cover block should be tied well on the frame, cover thickness shall be controlled, and verticality of steel frame as well as steel cage elevation should be ensured.

　　j. Before pouring concrete, the hole depth, hole diameter, verticality and the thickness of void soil at the bottom of the hole must be rechecked, and the unqualified shall be treated in time.

　　k. The concrete shall be continuously poured and compacted in layers. The height of each layer shall be determined according to the vibration tools used. Generally, the pouring height shall not be greater than 1.5m each time.

(2)人工挖孔灌注桩。人工挖孔灌注桩是指桩孔采用人工挖掘方法进行成孔,然后安放钢筋笼,浇筑混凝土而成的桩。人工挖孔灌注桩其结构上的特点是单桩的承载能力高,受力性能好,既能承受垂直荷载,又能承受水平荷载;人工挖孔灌注桩,具有机具设备简单,施工操作方便,占用施工场地小,无噪音,无振动,不污染环境,对周围建筑物影响小,施工质量可靠,可全面展开施工,工期缩短,造价低等优点,因此得到广泛应用。

人工挖孔灌注桩适用于土质较好,地下水位较低的黏土、粉质黏土及含少量砂卵石的黏土层等地质条件。可用于高层建筑、公用建筑、水工结构(如泵站、桥墩)作桩基,起支承、抗滑、挡土之用。对软土、流砂及地下水位较高,涌水量大的土层不宜采用。

①施工机具。人工挖孔施工的施工机具较为简单。常用的有:电动或手动葫芦、提土桶、潜水泵、鼓风机和输风管、镐、锹、土筐、照明灯、对讲机、电铃等。

②施工工艺。人工挖孔灌注桩护壁方法有现浇混凝土护壁、喷射混凝土护壁、砖砌体护壁、沉井护壁、套管护壁、型钢或木板桩护壁等。下面以应用较广的现浇混凝土分段护壁为例说明人工挖孔灌注桩的施工工艺流程。

定位放线→挖第一节桩土方、清理、核对中心线→绑第一节护壁钢筋→支第一节护壁模板→浇第一节混凝土、做护圈、拆模→第二次投测标高及核对中心线→安装提升设备→挖第二节桩土方、清理、核对中心线和垂直度→绑第二节护壁钢筋→支第二节护壁模板→浇第二节护壁混凝土、拆模→重复以上施工至设计要求深度,检查持力层情况→扩孔→验收桩孔和持力层→清孔→安放钢筋笼→浇筑混凝土。人工挖孔灌注桩构造示意图如图 3.33 所示。

③施工要点及注意事项。

a. 成孔质量控制。成孔质量包括垂直度和中心线偏差、孔径、孔形等。成孔中应随时检查桩孔的垂直度和中心线偏差并控制在允许偏差范围之内。

(2) Manually excavated cast-in-place pile. Manually excavated cast-in-place pile refers to the pile that hole forming is manual excavated, and then steel cage is placed in and concrete is poured. The structural characteristics of manually excavated cast-in-place pile are high bearing capacity of single pile and good mechanical performance, which can bear both vertical load and horizontal load. Manually excavated cast-in-place pile has the advantages of simple machines and tools, convenient construction operation, small occupation of construction site, no noise, no vibration, no environmental pollution, little influence on surrounding buildings, reliable construction quality, comprehensive construction, shortened construction period, low cost and so on, so it has been widely used.

Manually excavated cast-in-place pile is suitable for the geological conditions of better soil quality, lower underground water clay, silty clay and clay layer with small amount of sand pebble. It can be used in high-rise buildings, public buildings, hydraulic structures(pumping stations, piers)as pile foundations for supporting, anti-sliding and retaining purposes. It is not suitable for soft soil, drift sand, soil layer with high-level underground water and large amount of water inflow.

①Construction machines. The construction machines of manually excavated cast-in-place pile are simpler. Electrical or manual hoist, lifting bucket, submersible pump, air-blower and air conveyance duct, pickax, shovel, soil box, headlamp, interphone, electrical bell and so on are commonly used.

②Construction process.

The wall protection methods of manual hole digging cast-in-place pile include cast-in-place concrete wall protection, shotcrete wall protection, brick masonry wall protection, open caisson wall protection, casing wall protection, steel or wood board pile wall protection, etc. Taking the widely used cast-in-place concrete segmental retaining wall as an example, the construction process of manual digging cast-in-place pile is illustrated.

Positioning and setting out→excavation of the first section of pile earthwork, cleaning and checking of the central line→binding of the first section of retaining wall steel bar→supporting the first section of retaining wall formwork→pouring the first section of concrete, making the retaining ring and removing the formwork→the second time measuring the elevation and checking the center line→installing the lifting equipment→excavating the earthwork of the second section of pile, clearing, checking the center line and verticality→binding the second section of retaining wall reinforcement→supporting the second section of retaining wall formwork→pouring the second section wall protection concrete, formwork removal→repeat the above construction to the design depth, check the bearing layer condition→reaming→acceptance of pile hole and bearing layer→hole cleaning→reinforcement cage placement→concrete pouring. See Fig. 3.33 for the structure of manual hole digging cast-in-place pile.

③Construction points and matters needing attention.

a. Quality control of hole forming. The quality of hole forming includes verticality, center line deviation, hole diameter, hole shape, etc. The verticality and center line deviation of pile hole should be checked at any time during hole forming and controlled within the allowable deviation range.

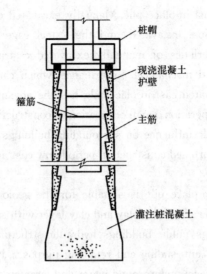

图 3.33 人工挖孔灌注桩构造示意图

b. 防止塌孔。护壁是人工挖孔桩施工中防止塌孔的构造措施。施工中应按照设计要求做好护壁。护壁混凝土强度在达到 1 MPa 后方能拆除模板。

地面水往孔边渗流会造成土的抗剪强度降低,可能造成塌孔,要解决好地面水的排水问题。

地下水对挖孔有着重要影响,尤其是地下水丰富以及地下水位高时。如遇地下水位高或地下水量大时,应先采取降水措施,再进行挖孔施工。如遇局部小的渗流,可以边排水边挖,但应有治理流砂现象发生的预案或直接采取措施防止流砂现象发生。在松软土层或流砂层中挖孔时,可以将施工段高度减小(如 300~500 mm)或采用钢护筒护壁。

c. 混凝土浇筑。安放钢筋笼、浇筑混凝土前要清孔,清孔必须干净,混凝土浇筑前应排除孔内大量的积水。如地下水较多不能及时抽干时,可以采取导管法水下浇筑混凝土。混凝土浇筑应连续进行,一个桩孔一次浇筑完毕。孔深较大时,应采取措施防止混凝土因倾落高度过大而产生离析等质量问题。

④安全措施。必须特别注意操作人员井下作业的安全问题。要制定和采取可靠的安全措施,要严格遵守安全操作规程,要有发生安全事故的处理预案,主要包括:井下人员的安全设施设备;孔口的安全防护;安全教育;井下有害气体(主要为甲烷、二氧化碳、硫化氢等)的测定和防护措施;新鲜空气的输送等。其他事项如下:

a. 施工人员进入孔内必须戴安全帽,孔内有人作业时,孔上必须有人监督防护。

b. 孔内必须设置应急软爬梯供人员上下井;使用的电动葫芦、吊笼等应安全可靠并配有自动卡紧保险装置;不得用麻绳和尼龙绳吊挂或脚踏井壁凸缘上下;电动葫芦使用前必须检验其安全起吊能力。

b. Prevent hole collapse. The retaining wall is a structural measure for preventing the hole collapse during the construction of manually excavated cast-in-place pile. In the construction, the retaining wall should be well done according to the design requirements. The formwork can be removed only after the strength of the retaining wall concrete reaches 1MPa.

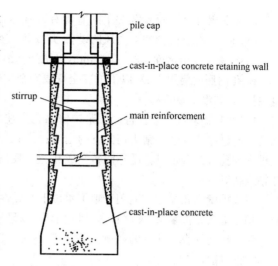

Fig. 3.33 the Schematic Diagram of Manually Digging Cast-in-place Pile

The seepage of the surface water to the edge of the hole will cause the shear strength of the soil to be decreased and may cause the hole collapse. So the drainage problem of the surface water should be solved.

Groundwater has an important impact on digging holes, especially when groundwater is rich and the groundwater level is high. In case of high groundwater level or large groundwater volume, dewatering measures should be taken first, and then digging hole construction can be carried out. In case of local small seepage, digging can be continued while draining. However, there should be a preplan for the occurrence of flowing sand control or direct measures to prevent the occurrence of flowing sand. When digging holes in soft soil layers or flowing sand layers, the height of the construction section can be reduced (for example, 300—500mm) or the steel pile casing retaining wall can be used.

c. Concrete pouring. Place steel cage, clean the hole before pouring concrete and the hole must be clear. A large amount of water should be removed from the hole before the concrete is poured. If there is a large amount of groundwater that cannot be drained in time, the ducting method can be adopted to pour concrete under water. Concrete pouring should be carried out continuously, and one pile hole is poured in one time. When the hole depth is large, measures should be taken to prevent concrete from segregation and other quality problems due to excessive tilt height.

④ Safety measures. Special attention must be paid to the safety of operators working underground. It is necessary to develop and adopt reliable security measures, follow strictly the safe operating procedures, and have a handling preplan for the occurrence of a safety accident. These mainly include safety facilities and equipment for underground personnel, safety protection of openings, safety education, measurement and protective measures for harmful gases (mainly for hospitals, carbon dioxide, hydrogen sulfide, etc.), and delivery of fresh air. Other matters are as follows:

a. The construction personnel must wear helmet into the hole. When there is a person working in the hole, there must be supervision and protection on the hole.

b. An emergency soft ladder shall be provided in the hole for the personnel to go up and down. The electrical hoist and cage used shall be safe and reliable, and is equipped with automatic clamping safety device. The manila rope and nylon rope or hole protruding shall not be used for up and down. Electrical hoist must be tested for safe lifting capacity before using.

c. 每日开工前必须检测井下的有毒有害气体,并有足够的安全防护措施。桩孔开挖深度超过 10 m 时,应有专门向井下送风的设备,风量不宜少于 25 L/s。

d. 护壁应高出地面 200~300 mm,以防杂物滚入孔内;孔周围要设 0.8 m 高的护栏。

e. 孔内照明要用 12 V 以下的安全灯或安全矿灯。使用的电器必须有严格的接地、接零和漏电保护器(如潜水泵等)。

(3)泥浆护壁成孔灌注桩。泥浆护壁钻孔灌注桩是利用原土自然造浆或人工造浆进行护壁,钻孔时通过循环泥浆将钻头切削下的土渣排出孔外而成孔,而后吊放钢筋笼,水下灌注混凝土而成桩。成孔方式有正(反)循环回转钻成孔、正(反)循环潜水钻成孔、冲击钻成孔、冲抓锥成孔、钻斗钻成孔等。

泥浆护壁钻孔灌注桩适用于地下水位下的黏性土、粉土、砂土、人工填土、碎石土及风化岩层,也适用于地质条件复杂、夹层较多、风化不均、软硬变化较大的岩层。

钻扩桩是在直状孔桩的基础上发展起来的。当钻孔达到设计持力层后,换上特殊的扩底钻头,将桩的底部扩大。

①成孔机械。成孔机械主要有回转钻机、潜水钻机、冲击钻等,以回转钻机应用最为广泛。回转钻机配有笼式柱状钻头和扩底钻头。常用的 3 种扩底钻头是:MRS 系列扩底钻头、YKD 系列油压扩底钻头和 MRR 系列基岩扩底钻头。

②成孔方法。回转钻成孔是国内灌注桩施工中最常用的方法之一。按排渣方式不同分为正循环回转钻成孔和反循环回转钻成孔两种。

正循环回转钻成孔由钻机回转装置带动钻杆和钻头回转切削破碎岩土,由泥浆泵往钻杆输进泥浆,泥浆沿孔壁上升,从孔口溢浆孔溢出流入泥浆池,经沉淀处理返回循环池,见图 3.34。正循环成孔泥浆的上返速度低,携带土粒直径小,排渣能力差,岩土重复破碎现象严重,适用于填土、淤泥、黏土、粉土、砂土等地层,对于卵砾石含量不大于 15%、粒径小于 10 mm 的部分砂卵砾石层和软质基岩及较硬基岩也可使用。桩孔直径不宜大于 1 000 mm,钻孔深度不宜超过 40 m。

Chapter 3　Subbase and Foundation Construction

　　c. It is necessary to test toxic and harmful gases under the hole before starting work every day, and there should be sufficient safety protection measures. When the depth of pile hole excavated exceeds 10m, there should be equipment for delivering air to the hole below. The air volume should not be less than 25L/s.

　　d. The retaining wall should be 200—300mm above the ground to prevent sundries from rolling into the hole. A 0.8m high guardrail should be placed around the hole.

　　e. Hole lighting should use safety lamp below 12V or a safety miner's lamp. The electrical appliances used must have strict grounding, zeroing and leakage protectors (such as submersible pumps).

　　(3) Cast-in-place pile with mud retaining wall. The cast-in-place pile with mud retaining wall is to use natural mud of raw soil or manual mud for protection. When drilling, the soil stag cut by the drill bit is discharged out of the hole through circulating mud to form a hole, then the steel cage is lifted and released, and the concrete is poured underwater to form pile. The way of hole forming includes positive(reverse) circular turning drilling, positive(reverse) circulating diving drilling, impact drilling, punching and grasping cones drilling, and drilling bucket drilling, etc.

　　The cast-in-place pile with mud retaining wall is suitable for clayey soil, silt, sandy soil, artificial fill, gravel soil and weathered rock stratum under the groundwater level, as well as the rock stratum with complex geological conditions, many interlayer, uneven weathering and large change of soft and hard.

　　Drilling broaden pile is developed on the basis of straight hole piles. When the hole reaches the designed bearing stratum, a special reaming bit is used to enlarge the bottom of the pile.

　　① Hole forming machinery. The hole forming machinery mainly includes rotary drilling machine, diving drill and impact drill, etc., and rotary drilling machine is the most widely used. The rotary drilling machine is equipped with cage type column bit and bottom expanding bit. The three kinds of bottom expanding bits are MRS series bottom expanding bits, YKD series hydraulic bottom expanding bits and MRR series base rock bottom expanding bits.

　　② Hole forming method.

　　Hole forming by rotary drill is one of the most commonly used methods for cast-in-place pile construction in China. According to the different ways of slag removal, hole forming can be divided into positive circulation rotary drilling and reverse circulation rotary drilling.

　　The hole forming by positive circulating rotary drill is driven by a rotary drilling device to drive the drill pipe and drill bit to rotate and cut the rock and soil. The mud is pumped into the drill pipe by the mud pump. The mud rises along the hole wall, overflows from the overflow hole and flows into the mud pool. After sedimentation, it returns to the circulating pool, as shown in Fig. 3.34. The upward return speed of the positive circulating drilling mud is low, the carrying diameter of soil particles is small, the slag discharge ability is poor, and the phenomenon of repeated rock and soil crushing is serious. It is suitable for filling, mud, clay, silt, sand and other soil layer. It can also be used in part of the sand gravel layer and soft bedrock as well as harder bedrock with pebbles not more than 15% and the particle size less than 10mm. The diameter of the pile hole should not exceed 1000mm, and the drilling depth should not exceed 40m.

反循环回转钻成孔由钻机回转装置带动钻杆和钻头回转切削破碎岩土,利用泵吸、气举、喷射等措施抽吸循环护壁泥浆,挟带钻渣从钻杆内腔抽吸出孔外的成孔方法,如图 3.34 所示。当孔深小于 50 m 时,宜选用泵吸或射流反循环;当孔深大于 50 m 时,宜采用气举反循环。

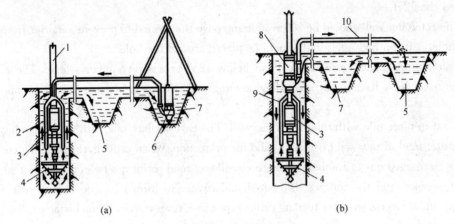

图 3.34　正反循环回转钻机钻孔示意图
(a)正循环回转钻机;(b)反循环回转钻机
1—钻杆;2—送水管;3—主机;4—钻头;5—沉淀池;6—潜水泥浆泵;
7—泥浆池;8—砂石泵;9—抽渣管;10—排渣胶管

③施工工艺。定位放线→埋设护筒→泥浆制备→钻机就位→钻进成孔(泥浆循环排渣)→成孔检测→清孔→安放钢筋笼→下导管→再次清孔→浇筑混凝土成孔。

对扩底桩,还应包括换扩底钻头、扩底成孔等工序。

④施工要点。

a. 埋设护筒。护筒的作用是固定桩孔位置、保护孔口、维持孔内水头、防止塌孔和为钻头导向。护筒用钢板制成,高出地面 0.4~0.6 m,内径应比钻头直径大 100~200 mm,上部开 12 个溢浆孔。护筒埋置深度在黏土中不小于 1.0 m,在砂土中不小于 1.5 m,其高度应满足孔内泥浆液面高度要求。护筒内泥浆面应保持高出地下位 1 m 以上。护筒应埋设准确,允许偏差不大于 50 mm。

The hole forming by reverse circulating rotary drill is driven by a rotary drilling device to drive the drill pipe and drill bit to rotate and cut the rock and soil, and the circulating retaining wall mud is pumped by means of pumping, air lift, and jet, etc. And the drilling slag is sucked out from the inner cavity of the drill pipe, as shown in Fig. 3. 34. When the hole depth is less than 50m, pumping or jet reverse circulation should be used. When the hole depth is greater than 50m, air lift reverse circulation should be adopted.

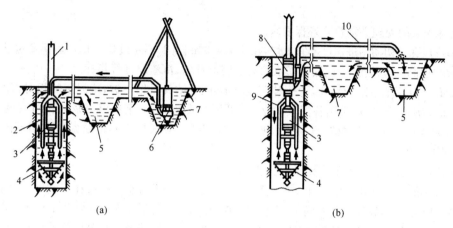

Fig. 3. 34 **Drilling Diagram of Positive and Reverse Circulating Rotary Drilling Rig**
(a)Positive circulation rotary drilling rig; (b)Reverse circulation rotary drilling rig
1—drill pipe; 2—water hose; 3—main drill; 4—drill bit;
5—setting pond; 6—diving mud pump; 7—mud pond;
8—sand and stone pump; 9—slag pipe; 10—slag discharge hose

③Construction process.

Setting out →bury pipe casing → mud preparation → drilling machine in place →drilling and hole forming(mud circulation and slag discharge)→ hole forming inspection →hole cleaning → place steel cage → put in conduit →hole cleaning again →hole forming by pouring concrete.

For the belled pile, the process also includes the replacement of the belled bit, bottom expanding and hole forming.

④Construction points.

a. Bury pile casing. The function of the pile casing is to fix the position of the pile hole, protect the opening, maintain the water head in the hole, avoid the hole collapse and guide the drill bit. The pile casing is made of steel plate, which is 0. 4—0. 6m above the ground surface. The inner diameter should be 100—200mm larger than the drill diameter, and 12 mud overflow holes are opened in the upper part. The depth of buried pile casing is not less than 1. 0m in the clay and not less than 1. 5m in the sand, and its height should meet the requirements of mud level height in the hole. The mud surface in the pile casing should be kept more than 1m above the underground level. The pile casing should be buried accurately, and the allowable deviation is not more than 50mm.

b. 泥浆制备。泥浆的作用是保护孔壁、防止塌陷、排出土渣及冷却与润滑钻头的作用。泥浆可以利用孔内原土造浆(在黏土和粉质黏土中)和选用高塑性的黏土或膨润土制备。泥浆的技术指标主要是泥浆密度、粘度、含砂率和胶体率等。在砂土和较厚夹砂层中成孔时,泥浆密度应控制在 $1.1 \sim 1.3 \text{ g/cm}^3$;在容易塌孔的土层中成孔时,泥浆密度应控制在 $1.3 \sim 1.5 \text{ g/cm}^3$;泥浆粘度为 $18 \sim 22 \text{ s}$;含砂率为 $4 \sim 8\%$;胶体率为不小于 90%。

施工中应经常测定并保持泥浆指标。为保证泥浆质量可以掺入增重剂、增粘剂、分散剂等材料。

对废弃泥浆的处理必须符合环保要求。

c. 钻进成孔。初期钻进速度不宜太快,在孔深 4.0 m 以内,不超过 2 m/h,以后不要超过 3 m/h。

d. 成孔检测。应检测成孔孔径、扩底直径、孔深、孔斜、沉渣厚度等指标。

e. 清孔。以原土造浆的钻孔,清孔时注入清水,同时钻具只钻不进,待泥浆的相对密度降 1.1 左右即可认为清孔合格;对于制备泥浆的钻孔,采用换浆法清孔,至换出泥浆的相对密度 $1.15 \sim 1.25$ 时认为合格。对于扩底桩,宜采用泵吸反循环清孔。应根据不同地质条件选用不同泵量和方法,以利于清孔。

清孔符合要求后应再次验收孔深、沉渣厚度等。

f. 吊放钢筋笼。清孔后应立即安放钢筋笼、浇混凝土。钢筋笼一般都在工地制作,制作时要求主筋环向均匀布置,箍筋直径及间距、主筋保护层、加劲箍的间距等均应符合设计要求。分段制作的钢筋笼,其接头采用焊接且应符合施工及验收规范的规定。钢筋笼主筋净距必须大于 3 倍的骨料粒径,加劲箍宜设在主筋外侧,钢筋保护层厚度不应小于 35 mm(水下混凝土不得小于 50 mm)。可在主筋外侧安设钢筋定位器,以确保保护层厚度。为了防止钢筋笼变形,可在钢筋笼上每隔 2 m 设置一道加强箍,并在钢筋笼内每隔 $3 \sim 4 \text{ m}$ 装一个可拆卸的十字形临时加劲架,在吊放入孔后拆除。吊放钢筋笼时应保持垂直、缓缓放入,防止碰撞孔壁。

b. Mud preparation.

The function of mud is to protect the hole wall, prevent collapse, discharge soil slag and cool and lubricate the bit. Mud can be prepared by using the original soil pulping in the holes(in clay and silty clay) and using highly plastic clay or betonies. The technical parameters of mud are mainly mud density, viscosity, sand content and colloid fraction. When forming holes in sand and thicker sand layers, the mud density should be controlled at $1.1-1.3 g/cm^3$; when forming holes in soil layers that are easy to collapse, the slurry density should be controlled at $1.3-1.5 g/cm^3$. The viscosity of mud is 18 to 22s. The sand content is 4% to 8%. The colloid fraction is not less than 90%.

The mud index should be measured and maintained frequently during construction. In order to ensure the quality of the mud, it can be mixed with weight increasing agent, viscosity increasing agent, dispersant and other materials. The treatment of waste mud must meet the requirements of environmental protection.

c. Drilling and hole forming. The initial drilling speed should not be too fast. The hole within 4.0m deep should be no more than 2m/h, and no more than 3m/h afterwards.

d. Hole forming inspection. The hole diameter, bottom expanding diameter, hole depth, hole inclination and sediment thickness should be inspected.

e. Hole cleaning.

When drilling with mud made from original soil, clean water is injected into the hole, and the drilling tool cannot drill in at the same time. When the relative density of mud drops about 1.1, the hole cleaning is considered qualified. For the preparation of mud drilling, the mud replacement method is used to clean the hole. When the relative density of the replaced mud is $1.15-1.25$, it is considered qualified. For belled pile, pump suction reverse circulation cleaning should be used. Different pump volume and methods should be selected according to different geological conditions to facilitate hole cleaning.

After the hole cleaning meets the requirements, the hole depth and sediment thickness shall be accepted again.

f. Lifting and placing steel cage.

After cleaning the hole, the steel cage should be placed immediately and the concrete should be poured. Steel cages are generally made on the construction site, and the main reinforcements are required to be uniform circumferential arranged during the fabrication. The diameter and distance of the stirrups, the protective layer of the main reinforcements, and the distance of the stiffening hoops should meet the design requirements. When steel cage is made in section, its joint should be welded and complied with the regulations of construction and acceptance specifications. The net distance of the main reinforcement of the steel cage must be greater than 3 times the aggregate size. The stiffening bars should be placed outside the main reinforcement. The thickness of the protective layer of steel cage should not be less than 35mm(the underwater concrete should not be less than 50mm). A steel fixer can be placed on the outside of the main reinforcements to ensure the thickness of the protective layer. In order to prevent the deformation of the steel reinforcement cage, a reinforcing bar can be placed every 2m on the steel reinforcement cage, and a removable cross-shaped temporary stiffening frame is installed every 3—4m in the steel cage and then demolished it after being lifted and released into the hole. When lifting the steel cage, keep it vertical and release it slowly to prevent collision with hole wall.

若造成塌孔或安放钢筋笼时间太长,应进行二次清孔后再浇筑混凝土。

g. 水下混凝土浇筑。钢筋笼定位后,须在 4 h 内浇捣混凝土以防塌孔。因泥浆护壁成孔灌注桩的桩孔内充满了水和泥浆,混凝土浇筑采取水下导管法,混凝土强度等级不低于 C20,坍落度为 18～22 cm,水泥用量不宜少于 360 kg/m³,可掺用减水剂或加气剂。其浇筑方法如图 3.35 所示,所用设备有金属导管、承料漏斗和提升机具等。

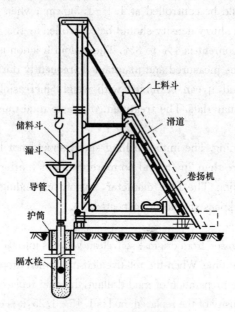

图 3.35 水下灌筑混凝土

导管一般用无缝钢管制作,采用直径 200～300 mm 的钢管,壁厚不宜小于 3 mm,每节长 2～4 m,最下一节为脚管,长度不小于 4 m,各节管用法兰盘和螺栓连接。承料漏斗利用法兰盘安装在导管顶端,其容积应大于保证管内混凝土所必须保持的高度和开始浇筑时导管埋置深度所要求的混凝土的体积。导管使用前应试拼装、过球和进行封闭水压试验,试水压力取为 0.6～1.0 MPa,以 15 min 不漏水为宜。

隔水栓(球塞)用来隔开混凝土与泥浆(或水),可用木球或混凝土圆柱塞等,其直径宜比导管内径小 20～25 mm。用 3～5 mm 厚的橡胶圈密封,其直径宜比导管内径大 5～6 mm。

导管安装时其底部高出孔底 300～500 mm。开始浇筑时管内混凝土应足够,要保证导管一次埋入混凝土中 0.8 m 以上。

If the hole collapse is caused or the reinforcement cage is placed for a long time, the concrete shall be poured after secondary hole cleaning.

g. Underwater concrete pouring. After the reinforcement cage is positioned, concrete shall be poured and tamped within 4 hours to prevent hole collapse. For the hole of mud retaining wall cast-in-place pile is filled with water and slurry, the underwater conduit method is adopted for concrete pouring. The concrete strength grade is not less than C20, the slump is 18—22cm, the cement dosage should not be less than 360kg/m^3, and water reducing agent or aerator can be added. The pouring method is shown in Fig. 3. 35, and the equipment used includes metal conduit, material hopper and lifting machine.

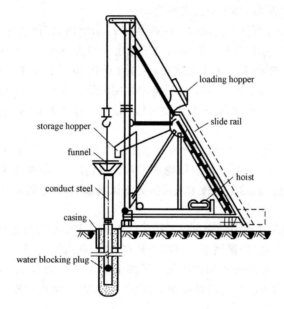

Fig. 3. 35　Underwater Concrete Pouring

The conduit is generally made of seamless steel pipe with diameter of 200—300mm and wall thickness not less than 3mm. The length of each section is 2—4m. The bottom section is the foot pipe and the length is not less than 4m. Flange plates and bolts are used to connect each section. The loading hopper is installed on the top of the conduit by means of a flange plate, and the volume should be greater than the required concrete volume to ensure the necessary height of the concrete in the conduit and the buried depth of the conduit at the beginning of pouring. Before the conduit is used, it should be trial assembled, balled and carried out the closed water pressure test. The test water pressure should be 0. 6—1. 0MPa and it should be no leakage in 15 minutes.

The water blocking plug(ball plug) is used to separate the concrete from the slurry(or water). The wooden ball or concrete cylindrical plug can be used, and its diameter should be 20—25mm smaller than the inner diameter of the conduit. Seal with 3—5mm thick rubber ring, and its diameter should be 5—6mm larger than the inner diameter of the conduit.

When the conduit is installed, its bottom is 300—500mm higher than the bottom of the hole. At the beginning of pouring, the concrete in the pipe shall be sufficient to ensure that the pipe is buried more than 0. 8m in the concrete at one time.

浇筑时，用提升机具将承料漏斗和导管悬吊起来后，沉至孔底，往导管中放隔水栓，隔水栓用绳子或铁丝吊挂，然后向导管内灌一定数量的混凝土，并使其下口距地基面约 300 mm，立即迅速剪断吊绳(水深在 10 m 以内可用此法)、或让球塞下滑至管的中部或接近底部再剪断吊绳，使混凝土靠自重推动球塞下落，冲向基底，并向四周扩散，球塞被推出管后，混凝土则在导管下部包围住导管，形成混凝土堆，这时可把导管再下降至基底 100～200 mm 处，使导管下部能有更多的部分埋入首批浇筑的混凝土中。然后不断地将混凝土通过承料漏斗浇入导管内，管外混凝土面不断被挤压上升。随着管外混凝土面的上升，相应地逐渐提升导管。导管应缓缓提升，每次 200 mm 左右，严防提升过度，务必保证导管下端埋入混凝土中的深度不少于规定的最小埋置深度；一般情况下，在泥浆中浇混凝土时，导管最小埋置深度不能小于 1 m，适宜的埋置深度为 2～6 m，但也不宜过深，以免混凝土的流动阻力太大，易造成堵管。

混凝土浇筑过程应连续进行，间歇时间应控制在 15 分钟内，任何情况不得超过 30 分钟，并控制在 4～6 h 内浇完，上升速度不大于 2 m/h，浇筑速度一般为 30～35 m^3/h，浇注混凝土量不得小于灌注桩的计算体积，每根桩不少于 1 组试块，混凝土强度必须符合设计要求。混凝土浇筑的桩顶最终标高应高出设计标高 0.8～1 m。然后将浮浆层消除 0.6～0.8 m，留 0.2 m 待后凿除。

水下混凝土宜比设计强度提高一个强度等级，必须具备良好的和易性，配合比应通过试验确定。

⑤常见问题的分析和处理。泥浆护壁成孔灌注桩施工时常易发生孔壁坍塌、斜孔、孔底隔层、夹泥、流砂等工程问题，水下混凝土浇筑属隐蔽工程，一旦发生质量事故难以观察和补救，所以应严格遵守操作规程，在有经验的工程技术人员指导下认真施工，并做好隐蔽工程记录，以确保工程质量。

a. 孔壁坍塌。指成孔过程中孔壁土层不同程度坍落。主要原因是提升下落冲击锤、掏渣筒或钢筋骨架时碰撞护筒及孔壁；护筒周围未用黏土紧密填实，孔内泥浆液面下降，孔内水压降低等造成塌孔。塌孔处理方法：一是在孔壁坍塌段用石子黏土投入，重新开钻，并调整泥浆容重和液面高度；二是使用冲孔机时，填入混合料后低锤密击，使孔壁坚固后，再正常冲击。

During pouring, the loading hopper and the conduit are suspended by the lifting equipment and then sunk to the bottom of the hole. The waterproof plug is placed in the conduit and suspended by rope or iron wire, then the conduit is filled with a certain amount of concrete, making its lower end about 300mm from the subbase surface, and the lifting rope is cut off immediately(the water depth within 10m can use this method)or the lifting rope is cut off when the button plug slides down to the middle or near the bottom of the conduit, making the concrete push the button plug dropping by its own weight , rush toward the foundation base and spread to the surroundings. After the button plug is pushed out of the conduit, the concrete surrounds the conduit at the lower part of the conduit to form a concrete heap. At this time, the conduit can be lowered to the foundation base 100—200mm so that more parts of the conduit can be buried in the first batch of concrete. The concrete is then continuously poured into the conduit through the loading hopper, and the concrete surface outside the conduit is continuously squeezed up. As the concrete surface outside the conduit rises, the conduit is gradually raised accordingly. The conduit should be slowly raised about 200mm each time, and the over-lifting should be strictly prevented. It is necessary to ensure that the low end of the conduit depth is buried in the concrete not less than the specified minimum buried depth. In general, when pouring concrete in mud, the minimum buried depth of the conduit should not be less than 1m and the suitable buried depth is 2—6m, but it should not be too deep so as to avoid the concrete flow resistance too large, resulting in blocking of the pipe.

The concrete pouring process should be carried out continuously. The intermission time should be controlled within 15min, and should not exceed 30min under any circumstances. The pouring should be controlled within 4—6 h. The rising speed is not more than 2m/h, and the pouring speed is generally 30—35m^3/h. The amount of concrete poured should not be less than the calculated volume of the cast-in-place piles. Each pile shall be not less than one set of test blocks, and the concrete strength shall meet the design requirements. The final elevation of the pile top of concrete poured should be 0.8—1m higher than the design elevation. Then the floating layer is removed 0.6—0.8m, leaving 0.2m for later chisel removing.

Underwater concrete should be one grade higher than the design strength grade, and must have good workability. The mix proportion should be determined by experiments.

⑤Analysis and treatment of common problems.

In construction, mud retaining wall cast-in-place piles are easy to have engineering problems such as hole wall collapse, hole shift, hole-bottom interlayer, mud, quick sand and so on. Underwater concrete pouring is a hidden engineering. It is difficult to observe and remedy the quality accident once it occurs, so the operation rules should be strictly observed. Under the guidance of experienced engineering and technical personnel, the construction should be carried out seriously, and the hidden works should be recorded to ensure the quality of the project.

a. Hole wall collapse. Hole wall collapse refers to the collapse of soil layer in different degrees during hole forming. The main reason is that when lifting the falling impact hammer, slag bucket or reinforcement framework, it collides with the casing and hole wall. The surrounding of the casing is not tightly filled with clay, the mud level in the hole decreases, and the water pressure in the hole decreases, resulting in hole collapse. The treatment methods of hole collapse are as follows: first, use gravel clay in the collapsed section of hole wall, re-drill, and adjust the mud bulk density and liquid level height; second, when using the punching machine, fill in the mixture and hammer it tightly to make the hole wall firm before normal impact.

b. 偏孔。指成孔过程中出现孔位偏移或孔身倾斜。偏孔的主要原因是桩架不稳固,导杆不垂直或土层软硬不均。对于冲孔成孔,则可能是由于导向不严格或遇到探头石及基岩倾斜所引起的。处理方法为:将桩架重新安装牢固,使其平稳垂直;如孔的偏移过大,应填入石子黏土,重新成孔;如有探头石,可用取岩钻将其除去或低锤密击将石击碎;如遇基岩倾斜,可以投入毛石于低处,再开钻或密打。

c. 孔底隔层。指孔底残留石渣过厚,孔脚涌进泥砂或塌壁泥土落底。造成孔底隔层的主要原因是清孔不彻底,清孔后泥浆浓度减少或浇筑混凝土、安放钢筋骨架时碰撞孔壁造成塌孔落土。主要防治方法为:做好清孔工作,注意泥浆浓度及孔内水位变化,施工时注意保护孔壁。

d. 夹泥或软弱夹层。指桩身混凝土混进泥土或形成浮浆泡沫软弱夹层。其形成的主要原因是浇筑混凝土时孔壁坍塌或导管口埋入混凝土高度太小,泥浆被喷翻,掺入混凝土中。防治措施是经常注意混馄凝土表面标高变化,保持导管下口埋入混凝土表面标高变化,保持导管下口埋入混凝土下的高度,并应在钢筋笼下放孔内 4 h 内浇筑混凝土。

e. 流砂。指成孔时发现大量流砂捅塞孔底。流砂产生的原因是孔外水压力比孔内水压力大,孔壁土松散。流砂严重时可抛入碎砖石、黏土,用锤冲入流砂层,防止流砂涌入。

(4)沉管灌注桩。沉管灌注桩又叫套管成孔灌注桩,是利用锤击打桩设备或振动沉桩设备,将带有钢筋混凝土的桩尖(或钢板靴)或带有活瓣式桩靴的钢管沉入土中(钢管直径应与桩的设计尺寸一致),形成桩孔,然后放入钢筋骨架并浇筑混凝土,随之拔出套管,利用拔管时的振动将混凝土捣实,便形成所需要的灌注桩。又称为套管成孔灌注桩。由于有锤击和振动两种沉管方式,因此,又可分为锤击沉管灌注桩和振动沉管灌注桩;按沉桩振动锤的不同,振动沉管灌注桩分为振动沉管灌注桩和振动冲击沉管灌注桩。沉管桩对周围环境有噪声、振动、挤压等影响。

b. Eccentric hole. Eccentric hole refers to hole position deviation or hole body inclination during hole forming. The main reasons of the eccentric hole are unstable pile frame, non vertical guide rod or uneven soil layer. For punching hole, it may be caused by loose guiding or inclination of probe stone and bedrock. The treatment methods are as follows: re-install the pile frame firmly to make it stable and vertical; if the deviation of the hole is too large, it should be filled with stone clay to re-form the hole; if there is probe stone, it can be removed by rock drill or crushed by low hammer; if the bedrock is inclined, rubble can be put into the low place, and then drill or compact.

c. Hole bottom interlayer. Hole bottom interlayer means that the residual slag at the bottom of the hole is too thick, and the hole foot gushes into the mud and sand or the wall collapses, and the soil falls to the bottom. The main reason for the hole bottom interlayer is that the hole is not completely cleaned, and after the hole is cleaned, the mud concentration is reduced or the hole wall is collided to cause soil falling when the concrete is poured and the steel frame is placed. The main prevention methods are: do a good job in hole cleaning, pay attention to the change of mud concentration and water level in the hole, and pay attention to the protection of hole wall during construction.

d. Mud or weak interlayer. Mud or weak interlayer refers to concrete mixed into soil or foamed foam weak interlayer. The main reason is that when pouring concrete, the hole wall collapses or the height of the conduit embedded in the concrete is too small, the slurry is sprayed over and mixed into the concrete. The prevention and control measures are to pay attention to the surface elevation change of concrete, keep the surface elevation change of the lower opening of the conduit embedded in the concrete, keep the height of the lower opening of the conduit embedded in the concrete, and place the concrete in the lower hole of the reinforcement cage within 4 hours.

e. Quick sand. It means that a large amount of quicksand is found at the bottom of the hole when hole forming. The reason for quicksand is that the water pressure outside the hole is greater than that inside the hole, and the soil on the wall of the hole is loose. When the quicksand is serious, the broken masonry and clay can be thrown into it, and then be hammered into the quicksand layer to prevent the inflow of quicksand.

(4) Sinking pipe cast-in-place pile. Sinking pipe cast-in-place pile is also called casing hole forming cast-in-place pile. It uses hammering pile-driving equipment or vibrating pile-sinking equipment to sink pile tip (or steel plate shoe) with reinforced concrete or steel tube with flapper-type drive shoe into the soil (the diameter of the steel tube should be the same as the design size of the pile), forming the pile hole, then put the steel skeleton and pour concrete, and then pull out the casing, using the vibration of pulling out the pipe to compact the concrete, and the required cast-in-place pile is formed, also called casing cast-in-place pile. Because there are two types of sinking pipes, hammer strike and vibrating, so they can be divided into hammer strike cast-in-place pile and vibration impact cast-in-place pile. According to the difference of pile-sinking vibrating hammer, the vibration sinking pipe cast-in-place pile is divided into a vibration sinking pipe cast-in-place pile and vibratory impact sinking pipe cast-in-pace pile. The sinking pipe pile has influence on the surrounding environment such as noise, vibration and extrusion, etc.

套管成孔灌注桩整个施工过程在套管护壁条件下进行,不受地下水位高低和土质条件好坏的限制,适合于地下水位高,地质条件差的可塑、软塑、流塑以上黏土、淤泥及淤泥质土、稍密和松散的砂土中施工。图 3.36 所示为沉管灌注桩施工过程。

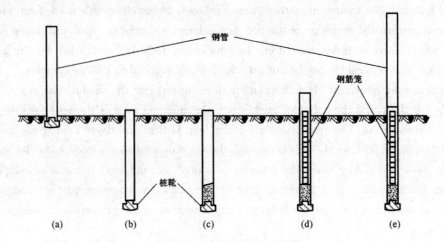

图 3.36　沉管灌注桩施工过程
(a)就位;(b)沉管;(c)开始灌注混凝土;(d)下钢筋骨架继续浇筑混凝土;(e)拔管成桩

①锤击沉管灌注桩。锤击沉管灌注桩又称为打拔管式灌注桩,是用锤击沉桩设备(落锤、汽锤、柴油锤)将桩管打入土中成孔。锤击沉管桩适用于一般黏性土、淤泥质土、砂土和人工填土地基,但不能在密实的砂砾石、漂石层中使用。

a. 成孔机械。锤击沉管灌注桩机械设备示意图如图 3.37 所示。预制混凝土桩靴如图 3.38 所示。

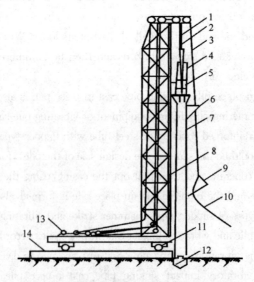

图 3.37　锤击沉管灌注桩机械设备示意图
1—桩帽钢丝绳;2—桩管钢丝绳;3—吊斗钢丝绳;4—桩锤;5—桩帽;6—混凝土漏斗;
7—桩管;8—桩架;9—混凝土吊斗;10—回绳;11—行驶用钢管;12—预制混凝土桩靴;13—卷扬机;14—枕木

The whole construction process of the casing cast-in-place pile is carried out under the casing retaining wall, and is not limited by the height of the groundwater level and the quality of the soil condition. It is suitable for the construction of plasticity, soft plastic, clay above the flow plastic, mud and mucky soil, slightly dense and loose sand with high groundwater level and poor geological conditions. Fig. 3. 36 shows the construction process of the sinking pipe cast-in-place pile.

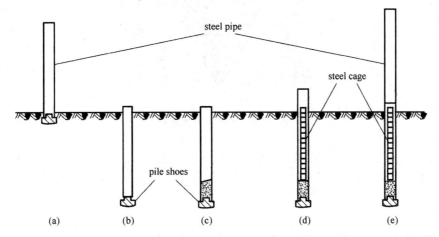

Fig. 3. 36 Construction Process of Sinking Pipe Cast-in-place Pile
(a) pipe in place; (b) sink pipe; (c) start to pour concrete;
(d) put in steel cage and continue to pour concrete; (e) pull out pipe and form pile

①Hammer strike cast-in-place pile.

Hammer strike cast-in-place pile is also known as pull out pipe cast-in-place pile. It uses hammer striking pile equipment (drop hammer, steam hammer, diesel hammer) to drive the pile pipe into the soil for hole forming. Hammer strike cast-in-place pile is suitable for general clay soil, mucky soil, sand soil and artificial fill subbase, but it cannot be used in dense sand gravel and boulder layers.

a. Hole forming machinery. The schematic diagram of mechanical equipment for hammer driven cast-in-place pile is shown in Fig. 3. 37. Precast concrete pile shoes are shown in Fig. 3. 38.

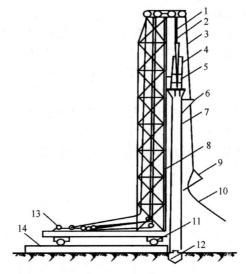

Fig. 3. 37 Schematic Diagram of Mechanical Equipment for Hammer Driven Cast-in-place Pile

1—pile cap wire rope; 2—pile pipe wire rope; 3—bucket wire rope;
4—pile hammer; 5—pile cap; 6—concrete funnel; 7—pile pipe;
8—pile frame; 9—concrete bucket; 10—back rope; 11— steel pipe for driving; 12—precast concrete pile shoes; 13—hoist; 14—sleeper

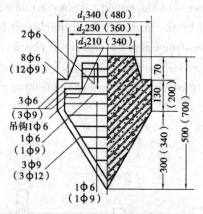

图 3.38 预制混凝土桩靴

b. 施工工艺:定位放线→安放桩尖→桩机就位→桩管安装并套在桩尖上→校正钢管垂直度→锤击沉管至设计要求的贯入度或设计标高→安放钢筋笼→浇筑混凝土→拔管、敲击钢管振实混凝土。

c. 施工要点。

Ⅰ. 安放桩靴、桩管安装和校正。锤击沉管灌注桩施工时,首先将打桩机就位,吊起桩管,对准预先在桩位埋好的预制混凝土桩尖,放置麻、草绳垫于桩管和桩尖连接处,以作缓冲和防止泥水进入桩管,然后缓慢放下桩管,套入桩尖,将桩管压人土中。

然后桩管上部扣上桩帽,检查桩管与桩锤、桩尖是否在一条垂直线上,其垂度偏差应小于0.5%桩管高度。

Ⅱ. 沉管。初打时应低锤轻击,观察桩管有无偏移时,方能正常施打。桩锤施打的冲击频率,视桩锤的类型和土质而定,宜采用低锤密击,即小落距、高频率,尽量控制每分钟击打 70 次以上。直至将桩管打至设计要求贯入度或桩尖标高,并检查管内有无泥、水浆灌入。

当桩距稀疏时(中心距大于 3.5 倍桩径或大于 2 m),可采用连打方法;当桩距较密集时(中心距小于等于 3.5 倍桩距或 2 m),为防止断桩现象应采用跳跃施打的方法,中间空出的桩应待邻桩混凝土达到设计强度等级的 50% 以上方可施打;对于土质较差的饱和淤泥质土,可采用控制时间的连打方法,即必须在邻桩混凝土终凝前,将影响范围内(中距小于等于 3.5 倍桩径或 2 m)的桩全部施工完毕。

b. Construction process: setting out → place pile tip → pile machine in place → pile pipe installation and placed on the pile tip → correct perpendicularity of the steel pipe → hammer the sinking pipe to the penetration degree of design requirement or design elevation → place steel reinforcement cage → pour concrete → pull out the pipe and tap the steel pipe to compact the concrete.

c. Construction points.

Ⅰ. Place the pile shoes, pile pipe installation and correction. During the construction of cast-in-place pile by hammering, firstly, the pile driver is put in place, the pile pipe is lifted, the precast concrete pile tip is aligned with the pre-buried pile position, and hemp and straw ropes are placed at the joint of the pile pipe and the pile tip to cushion and prevent muddy water from entering the pile pipe. Then, the pile pipe is slowly put down, put into the pile tip, and the pile pipe is pressed into the soil.

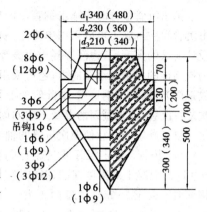

Fig. 3.38 Precast Concrete Pile Shoes

Then fasten the pile cap on the upper part of the pile pipe, check whether the pile pipe is in a vertical line with the pile hammer and pile tip, and the sag deviation shall be less than 0.5% of the pile pipe height.

Ⅱ. Sinking pipe. During the initial driving, it is necessary to strike with *low hammer light blow*, and observe whether the pile pipe is deviation before normal driving. The impact frequency of the pile hammer is determined by the type of the pile hammer and the soil quality. It is better to use the low hammer close blow, that is, small drop distance, high frequency, and try to control more than 70 blows per minute until the pile pipe is driven to the designed penetration or pile tip elevation, and check whether there is mud or water slurry in the pipe.

When the pile spacing is sparse(the center distance is more than 3.5 times of the pile diameter or more than 2m), the continuous driving method can be used; When the pile spacing is dense(the center distance is less than or equal to 3.5 times of the pile spacing or 2m), in order to prevent the pile from breaking, the jumping driving method should be adopted. The pile with empty space in the middle can be driven above 50% of the design strength grade of the adjacent pile concrete. For the saturation mucky soil with poor soil quality, the continuous hitting method of controlling time can be used. That is to say, all piles within the influence range(medium distance less than or equal to 3.5 times the pile diameter or 2m)must be completed before the permanent setting of the adjacent pile concrete.

锤击沉管拔管时应锤击钢管使产生振动而使混凝土密实。锤击次数视桩锤的类型而定。锤击沉管灌注桩施工可以采用单打法、复打法。单打法是一般正常的沉管方法，它是将桩管沉入到设计要求的深度后，边灌混凝土边拔管，最后成桩，适用于含水量较小的土层，也可采用活瓣桩靴。为提高桩的质量和承载能力，还可采用复打法，其施工顺序是：在第一次混凝土浇筑施工完毕（未放钢筋笼）及拔出钢管后，清除管壁上及桩孔周边地面的污泥，立即在原桩位上再安放桩尖，第二次复打沉管，使未凝固的混凝土向四周挤压扩大桩径，然后再放入钢筋笼和进行第二次混凝土浇筑。复打施工应注意使两次沉管的轴线重合，且必须在第一次浇筑的混凝土初凝前进行。

Ⅲ．拔管。锤击沉管桩混凝土强度等级不得低于 C20，每立方米混凝土的水泥用量不宜少于 300 kg。混凝土坍落度在配钢筋时宜为 80～100 mm，无筋时宜为 60～80 mm。碎石粒径在配有钢筋时不大于 25 mm，无筋时不大于 40 mm。预制钢筋混凝土桩尖的强度等级不得低于 C30。混凝土充盈系数（实际灌注混凝土体积与按设计桩身直径计算体积之比）不得小于 1.0，成桩后的桩身混凝土顶面标高应至少高出设计标高 500 mm。

浇筑混凝土以及拔管时应保证混凝土的质量，桩管内应尽量灌满混凝土，应保持管内混凝土高度不少于 2 m 高度。在测得混凝土确已流出桩管后，然后开始拔管。凡灌注配有不到孔底的钢筋笼的桩身混凝土时，第一次混凝土应先灌至笼底标高，然后放置钢筋笼，再灌混凝土至桩顶标高。第一次拔管高度应控制在能容纳第二次所需灌入的混凝土量为限，不宜拔得过高。在拔管过程中应用专用测锤或浮标检查混凝土面的下降情况。

拔管速度不易过快，应均匀，对一般土层以 1 m/min 为宜，在软弱土层及软硬土层交界处宜控制在 0.3～0.8 m/min 为宜。采用倒打拔管时，桩锤的冲击频率为：单动汽锤不得少于 50 次/min，自由落锤轻击不得少于 40 次/min。在管底未拔至桩顶设计标高之前，倒打和轻击不得中断。拔管时应保持对桩管进行低锤密击，使桩管不断受到冲击振动，以振实混凝土。拔管过程中，应用吊铊或浮标测定混凝土下落和扩散情况，注意使桩管内混凝土高度保持略高于地面，这样一直至全管拔出为止。灌入桩管内的混凝土，从拌制到最后拔管结束不得超过混凝土的初凝时间。

When pulling out pipes of hammer strike sinking pipe, the steel pipe should be hammered to cause vibration and make the concrete dense. The number of hammer strokes depends on the type of pile hammer. The construction of hammer strike cast-in-place pile can adopt the single driving method and re-driving method. The single driving method is a normal sinking pipe method. It is to sink the pile pipe to the depth of design requirement, and to put out the pipe while the concrete is being poured, then form the pile finally. The method is suitable for the soil layer with small water content and can also use flapper-type drive shoes. In order to improve the quality and carrying capacity of the pile, the re-driving method can also be adopted. Its construction sequence is that after the first concrete pouring completed(without the steel cage) and after the steel pipe is pulled out, the sludge on the pipe wall and the surrounding ground of the pile hole are removed. The pile tip is immediately placed on the original pile position, and the second time re-drive the sunk pipe, making the unsetting concrete squeeze around to enlarge the pile diameter, and then place the steel cage and pour the second concrete. The re-driving construction should pay attention to make the axes of the two sinking pipes coincidence, and must be carried out before the first pouring of the concrete initial setting.

Ⅲ. Pull out pipe. The concrete strength grade of hammer strike cast-in-place pile shall not be lower than C20, and the cement per cubic meter of concrete shall not be less than 300kg. The concrete slump is suitable for 80—100mm with reinforcement and 60—80mm without reinforcement. The particle size of the gravel is not more than 25mm with reinforcement and not more than 40mm without reinforcement. The strength grade of precast reinforced concrete pile tip shall not be lower than C30. The concrete filling coefficient(the ratio of the actual poured concrete volume to the calculated volume of the designed pile body diameter) shall not be less than 1.0, and the elevation of the concrete top surface of the pile after completion shall be at least 500mm higher than the design elevation.

The quality of the concrete should be ensured when pouring concrete and pulling out the pipe. The pile pipe should be filled with concrete as much as possible and the concrete height inside the pipe should be kept no less than 2m. After the concrete has actually been measured out of the pile pipe, then pipe is pull out. When pouring the pile concrete with steel cage less than the bottom of the hole, the first concrete should be poured to the bottom level of the cage, then the steel cage is placed, and the concrete is poured to the top level of the pile. The height of the first pipe pulling should be limited to the amount of concrete required for the second pulling, and it should not be pulled too high. In the process of pipe pulling, special plumb bob or float should be used to check the drop of the concrete surface.

The speed of pipe pulling should not be too fast but should be uniform. For general soil layer, 1m/min is appropriate, and 0.3—0.8m/min at the junction of the soft weak soil layer and the soft hard soil layer. When using the inverted hit to pull, the impact frequency of the pile hammer is that the single action steam hammer should not be less than 50 times/min, and the free drop hammer light hit should not be less than 40 times/min. The inverted hit and light hit should not be interrupted before the bottom of the pipe is pulled to the design elevation of the pile top. The pile pipe should be kept low-hammered close hit when the pipe is pulled, so that the pile pipe is continuously impacted and vibrate the concrete. During the process of pipe pulling, the drop and spread circumstance of the concrete are measured by using hanging thallium or float. Pay attention to that the concrete height in the pile pipe is kept slightly higher than the ground, until the whole pipe is pulled out. The concrete poured into the pile pipe shall not exceed the initial setting time of the concrete from mixing to the final end of pipe pulling.

②振动沉管灌注桩。振动、振动冲击沉管灌注桩是利用振动桩锤(又称激振器)、振动冲击锤将桩管沉入土中,然后灌注混凝土而成。这两种灌注桩与锤击沉管灌注桩相比,更适合于稍密及中密的砂土地基施工。振动沉管灌注桩和振动冲击沉管桩的施工工艺完全相同,只是前者用振动锤沉桩,后者用振动带冲击的桩锤沉桩。

振动沉管灌注桩适用范围与锤击沉管灌注桩大致相同,还适用于稍密及中密的碎石类土中施工。

a. 成孔机械。振动沉管灌注桩机械设备如图3.39所示。活瓣桩靴如图3.40所示。

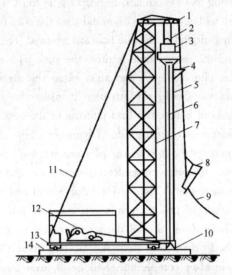

图3.39 振动沉管灌注桩机械设备示意图
1—导向滑轮;2—滑轮组;3—激振器;4—混凝土漏斗;5—桩管;6—加压钢丝绳;
7—桩架;8—混凝土吊斗;9—回绳;10—活瓣桩靴;11—缆风绳;12—卷扬机;
13—行驶用钢管;14—枕木

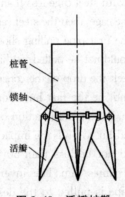

图3.40 活瓣桩靴

b. 施工工艺:定位放线→桩机就位→桩管安装并校正垂直度→振动沉管至设计要求深度→安放钢筋笼→混凝土浇筑→拔管。

②Vibro-sinking cast-in-place pile.

Vibration and vibration impact pipe sinking cast-in-place pile is made by using vibration pile hammer(also known as vibrator)and vibration impact hammer to sink the pile pipe into the soil, and then pouring concrete. These two types of cast-in-pace piles are more suitable for the construction of slightly dense and medium dense sandy soil foundations than hammer striking pipe cast-in-place piles. The construction process of the vibro-sinking cast-in-place pile and the vibratory impact sinking pipe cast-in-place pile is exactly the same, except that the former uses a vibrating hammer to sink the pile, and the latter uses a vibrating impact pile hammer to sink the pile.

The applicable range of the vibro-sinking cast-in-place pile is roughly the same as that of the hammer striking cast-in-place pile, and it is also suitable for the construction of slightly dense and medium dense gravel soil.

a. Hole forming machine: The mechanical equipment of vibro-sinking cast-in-place pile is shown in Fig. 3. 39. The valve shoe is shown in Fig. 3. 40.

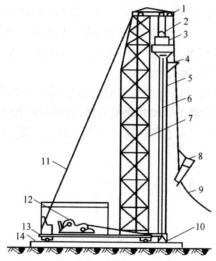

Fig. 3. 39 Schematic Diagram of Mechanical Equipment for Vibro-sinking Cast-in-place Pile

1—guide pulley; 2—pulley block; 3—vibrator; 4—concrete funnel; 5—pile pipe;
6—pressure wire rope; 7—pile frame; 8—concrete bucket; 9—return rope;
10—valve pile shoe; 11—guy rope; 12—hoist; 13—steel pipe for driving; 14—sleeper

b. Construction process: setting out →pile machine in place → pile pipe installation and correction of perpendicularity → vibrating sinking pipe to designed requirement depth → placing steel cage → concrete pouring → pulling out the pipe.

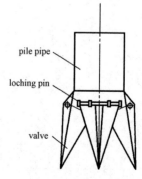

Fig. 3. 40 Valve Pile Shoe

c. 施工要点：

Ⅰ. 安放桩靴、桩管安装和校正。振动沉管施工时，先安好桩机，将桩管下端活瓣闭合或埋好预制桩尖，对准桩位，徐徐放下桩管压入土中。然后较正垂直度，符合要求后即可开动振动器沉管。

Ⅱ. 沉管。沉管时，由电动机带动的偏心块旋转而产生振动，桩管受振后与土体之间的摩擦力减小，当振动频率与土体自振频率相同时（一般黏性土自振频率为 $600\sim750$ r/min，砂土为 $900\sim1\,000$ r/min），土体结构因共振而破坏，同时在桩管上加压，桩管即沉入土中。加压常利用桩架自重，通过收紧加压滑轮组上的钢丝绳把压力传到桩管上，直到桩管沉到要求深度为止。沉管过程中，应经常探测管内有无地下水，若发现进水过多时，应将桩管拔出，检查桩尖缝隙是否过疏；若过疏应加以修理并将桩孔用砂填满再重新沉管。倘若发现有少量水量，可灌入 $0.1\,\mathrm{m}^3$ 左右混凝土或砂浆封堵桩管尖部，然后再继续施工。混凝土浇筑时应将桩管灌满或略高于地面。

振动灌注桩沉管时可采用单振法（单打法）、反插法或复振法（复插法、复打法）施工。

单振法：同锤击沉管的单打法也可采用预制桩靴。

反插法：是在拔管过程中边振边拔，每次拔管 $0.5\sim1.0$ m，再向下反插 $0.3\sim0.5$ m，如此反复并保持振动，直至桩管全部拔出。反插法能增加桩的截面和提高桩的承载能力，如在桩尖处 1.5 m 范围内，宜多次反插以扩大桩的局部断面。穿过淤泥夹层时，应放慢拔管速度，并减少拔管高度和反插深度。在流动性淤泥中不宜使用反插法。

复振法：同锤击沉管的复打法。采用全长复打的目的是提高桩的承载力；局部复打主要是为了处理沉桩过程中所出现的质量缺陷，如发现或怀疑出现缩颈、断桩等缺陷，局部复打深度应超过断桩或缩颈区 1 m 以上。复打必须在第一次灌注的混凝土初凝之前完成。

c. Construction points:

I. Place pile shoes, installation and correction of pile pipe. During the construction of vibrating sinking pipe, the pile machine shall be installed first, the valve at the lower end of the pile pipe shall be closed or the prefabricated pile tip shall be buried, the pile position shall be aligned, and the pile pipe shall be slowly put down and pressed into the soil. Then, after the perpendicularity is corrected and the requirements are met, the vibrator can be started to sink the pipe.

II. Sinking pipe.

When sinking the pipe, the eccentric block driven by the motor rotates and generates vibration, and the friction between the pile pipe and the soil decreases. When the vibration frequency is the same as the natural frequency of soil(generally, the natural frequency of cohesive soil is 600—750r/min, and that of sandy soil is 900—1000r/min), the soil structure will be destroyed due to resonance, and the pile pipe will sink into the soil when it is pressurized on the pile pipe. The self weight of the pile frame is often used to transfer the pressure to the pile pipe by tightening the steel wire rope on the pressure pulley block until the pile pipe is sunk to the required depth. In the process of sinking pipe, it is necessary to frequently detect whether there is groundwater in the pipe. If there is too much water in the pipe, the pile pipe should be pulled out to check whether the gap at the pile tip is too sparse. If it is too sparse, it should be repaired and the pile hole should be filled with sand before re-sinking. If a small amount of water is found, about $0.1m^3$ of concrete or mortar can be poured to plug the tip of the pile pipe, and then the construction can be continued. When pouring concrete, the pile pipe should be filled or slightly higher than the ground.

The single vibration method(single driving method), reverse inserting method or compound vibration method(compound inserting method and compound driving method) can be used in the construction of vibration cast-in-place pile.

Single vibration method: It is the same as the single driving method of sinking pipe by hammering. Precast pile shoes can also be used.

Reverse insertion method: During the process of pulling out the pipe, pull out the pipe 0.5—1.0m each time, and then insert the pipe 0.3—0.5m backward, so repeatedly and keep the vibration until all the piles are pulled out. The reverse insertion method can increase the pile section and improve the bearing capacity of the pile. For example, within 1.5 m of the pile tip, the reverse insertion method should be used for many times to expand the local section of the pile. When passing through the silt interlayer, the speed of pipe drawing should be slowed down, and the height of pipe drawing and the depth of reverse insertion should be reduced. It is not suitable to use the reverse insertion method in the flowing silt.

Compound vibration method: It is the same as that of hammering sinking pipe. The purpose of full-length re-driving is to improve the bearing capacity of the pile. Local re driving is mainly to deal with the quality defects in the process of pile sinking. If defects such as necking and broken pile are found or suspected, the depth of local re-driving should be more than 1m above the broken pile or necking area. The re-driving must be completed before the initial setting of the first pouring concrete.

Ⅲ. 拔管。当桩管沉到设计标高,且最后 30 s 的电流值、电压值符合设计要求后,停止振动,安放钢筋笼,并用吊斗将混凝土灌入桩管内,然后再开动激振器和卷扬机,拔出钢管,边振边拔,从而使桩的混凝土得到振实。

d. 常见问题的分析和处理。沉管灌注桩施工时易发生断桩、缩颈、桩靴进水或进泥、吊脚桩等问题,施工中应加强检查并及时处理。

Ⅰ. 断桩。断桩的裂缝为水平或略带倾斜,一般都贯通整个截面,常常出现于地面以下 1～3 m 软硬土层交接处。

断桩原因主要有:桩距过小,邻桩施打时土的挤压产生的水平推力和隆起上拔力的影响;软硬土层传递水平力不同,对桩产生剪应力;桩身混凝土终凝不久;强度弱,承受不了外力的影响。

避免断桩的措施如下:布桩应坚持少桩疏排的原则,桩与桩之间中心距不宜小于 3.5 倍桩径;桩身混凝土强度较低时,尽量避免振动和外力的干扰,因此要合理确定打桩顺序和桩架行走路线;采用跳打法或控制时间法以减少对邻桩的影响。控制时间法指在邻桩混凝土初凝以前,必须把影响范围内的桩施工完毕。

断桩的检查与处理:在浅层(2～3 m)发生断桩,可用重锤敲击桩头侧面,同时用脚踏在桩头上,如桩已断,会感到浮振;深处断桩目前常用动测或开挖的办法检查。断桩一经发现,应将断桩段拔出,将孔清理后,略增大面积或加上铁箍连接,再重新浇混凝土补做桩身。

Ⅱ. 缩颈桩。缩颈桩又称瓶颈桩,是指部分桩径缩小、桩截面积不符合设计要求。

缩颈桩产生的原因是:拔管过快,管内混凝土存量过少,混凝土本身和易性差,出管扩散困难造成缩颈;在含水量大的黏性土中沉管时,土体受到强烈扰动和挤压,产生很高的孔隙水压力,拔管后,这种水压力便作用到新浇筑的混凝土桩上,使桩身发生不同程度的缩颈现象。

Ⅲ. Pull out pipe. When the pile pipe is sunk to the design elevation, and the current value and voltage value of the last 30s meet the design requirements, stop the vibration, place the reinforcement cage, and pour the concrete into the pile pipe with the bucket, then start the vibrator and hoist, pull out the steel pipe, pulling out while vibrating, so that the concrete of the pile can be compacted.

d. Analysis and treatment of common problems. During the construction of sinking pipe cast-in-place pile, the problems such as pile breaking, necking, water or mud entering into pile shoe and end-suspended pile are easy to occur, so the inspection should be strengthened and the treatment should be carried out in time.

Ⅰ. Broken pile. The cracks of the broken pile are horizontal or slightly inclined, and generally run through the whole section. They often appear at the junction of soft and hard soil layers 1—3m below the ground.

The main reasons for pile breaking are as follows: the pile spacing is too small, the influence of horizontal thrust and uplift force produced by the soil extrusion when the adjacent piles are driven; the shear stress produced by the different horizontal force transmitted by the soft and hard soil layers; the final setting of the concrete in the pile body is not long; the strength is weak, and it cannot bear the influence of external force.

The measures to avoid pile breaking are as follows: pile arrangement should adhere to the principle of few piles and sparse row, and the center distance between piles should not be less than 3.5 times of pile diameter. When the concrete strength of pile body is low, the interference of vibration and external force should be avoided as far as possible, so the driving sequence and driving route of pile frame should be determined reasonably. The jump driving method or the control time method should be adopted to reduce the influence on the adjacent piles. The control time method means that before the initial setting of the adjacent pile concrete, the pile construction within the influence range must be completed.

Inspection and treatment of broken pile: If the pile is broken in the shallow layer(2—3m), the heavy hammer can be used to knock the side of the pile head, and at the same time, the foot can be used on the pile head. If the pile has been broken, it will feel floating vibration; at present, the deep broken pile is often checked by dynamic measurement or excavation. Once the broken pile is found, the broken pile section shall be pulled out, the hole shall be cleaned, the area shall be slightly increased or the iron hoop shall be added for connection, and then the concrete shall be poured again to make up the pile body.

Ⅱ. Necking pile. The necking pile, also known as bottle-neck pile, refers to part of the reduction of pile diameter and the pile cross-sectional area does not meet the design requirements.

The causes of necking pile are that the pipe is pulled out too fast, the amount of concrete in the pipe is too little, the concrete itself is of poor workability, and the pipe diffusion is difficult to cause necking. When sinking pipe in the clay soil with large moisture content, the soil is subjected to strong disturbance and squeezing, resulting in high pore water pressure. After the pipe is pulled out, the water pressure acts on the newly-poured concrete pile, causing the pile body necking to different degrees.

防治措施:在容易产生缩颈的土层中施工时,要严格控制拔管速度,采用"慢拔密击";混凝土坍落度要符合要求且管内混凝土必须略高于地面,以保持足够的压力,使混凝土出管扩散正常。施工时可设专人随时测定混凝土的下落情况,遇有缩颈现象,可采取复打处理。

Ⅲ. 桩尖进水、进泥砂。桩尖进水、进泥砂常见于地下水位高、含水量大的淤泥和粉砂土层,是由于桩管与桩尖接合处的垫层不紧密或桩尖被打破所致。

处理办法:可将桩管拔出,修复改正桩靴缝隙或将桩管与预制桩尖接合处用草绳、麻袋垫紧后,用砂回填桩孔后重打;如果只受地下水的影响,则当桩管沉至接近地下水位时,用水泥砂浆灌入管内约 0.5 m 作封底,并再灌 1 m 高的混凝土,然后继续沉桩。若管内进水不多(小于 200 mm)时,可不作处理,只在灌第一槽混凝土时酌情减少用水量即可。

Ⅳ. 吊脚桩。吊脚桩即桩底部的混凝土隔空,或混凝土中混进了泥砂而形成松软层。形成吊脚桩的原因是由于混凝土桩尖质量差,强度不足,沉管时被打坏而挤入桩管内,且拔管时冲击振动不够,桩尖未及时被混凝土压出或活瓣未及时张开。

防治措施:为了防止出现吊脚桩,要严格检查混凝土桩尖的强度(应不小于 C30),以免桩尖被打坏而挤入管内。沉管时,用吊砣检查桩尖是否有缩入管内的现象。如果有,应及时拔出纠正并将桩孔填砂后重打。

(5)灌注桩质量标准。

①成桩质量检验。

a. 灌注桩成孔施工的允许偏差应满足表 3.8 的要求。

Prevention and control measures are that the speed of the pipe pulling should be strictly controlled when construction is carried out in the soil which is easy to produce necking, and "slow pulling and close striking" should be adopted. The concrete slump should meet the requirements and the concrete in the pipe must be slightly higher than the ground to keep enough pressure, making the concrete out of the pipe diffusion normally. During the construction, specially-assigned person can be appointed to measure the falling of the concrete at any time, in case of necking, the re-driving method can be used.

Ⅲ. Water, mud and sand in pile tip. Water, mud and sand in the pile tip are often found in the silt and silt soil layer with high water level, which is caused by the incompact cushion at the joint between the pile pipe and the pile tip or the broken pile tip.

Treatment method: The pile pipe can be pulled out, the gap between the pile shoes can be repaired or corrected or the joint between the pile pipe and the prefabricated pile shoe should be padded tightly with straw rope and sack, driving it again after the pile hole is backfilled with sand. If it is only affected by groundwater, when the pile pipe sinks close to the groundwater level, the cement mortar is used to fill the pipe with about 0.5m as the bottom seal, and the concrete of 1m high is poured again, and then the pile-sinking is continued. If there is not much water in the pipe (less than 200mm), it cannot be treated. Only when pouring the first slot of concrete, the water consumption can be reduced as appropriate.

Ⅳ. End-suspended pile. The end-suspended pile is the concrete separation at the bottom of the pile, or the concrete is mixed into the sand and forms the soft layer. The reason for the formation of the end-suspended pile is that the quality of the concrete pile tip is poor, the strength is insufficient, pile tip is broken and squeezed into the pile pipe when sinking the pipe, and the impact vibration is not enough when the pipe is pulled out. The pile tip is not pressed out by concrete in time or the valve is not opened in time.

Prevent measures: In order to prevent the occurrence of end-suspended pile, the strength of concrete pile tip(no less than C30) should be strictly checked to avoid the pile tip being damaged and squeezed into the pipe. When sinking the pipe, check whether the pile tip has the phenomenon of shrinking into the pipe with the lifting weight. If any, the pile should be pulled out to correct in time and re-drive after filling sand.

(5) Quality standard for cast-in-place pile.

① Pile quality inspection.

a. The allowable deviation of the cast-in-place pile construction should meet the requirements of Table 3.8.

表 3.8 灌注桩成孔施工的允许偏差

序号	成孔方法		桩径允许偏差/mm	垂直度允许偏差/%	桩位允许偏差/mm 单桩、条形桩基沿垂直轴线方向和群桩基础中的边桩	条形桩基沿垂直轴线方向和群桩基础中间桩
1	泥浆护壁冲钻、挖、冲孔桩	$d \leqslant 1\,000$ mm	±50	<1	$d/6$ 且不大于 100	$d/4$ 且不大于 150
		$d > 1\,000$ mm	±50		$100+0.01H$	$150+0.01H$
2	锤击(振动)沉管、振动冲击沉管成孔	$d \leqslant 500$ mm	−20	<1	70	150
		$d > 500$ mm			100	150
3	螺旋钻、机动洛阳铲干作业成孔		−20	<1	70	150
4	人工挖孔桩	现浇混凝土护壁	±50	<0.5	50	150
		长钢套管护壁	±20	<1	100	200

b. 桩身质量应进行检验。抽检数量不应少于总数的 30% 且不少于 20 根;

c. 桩应进行承载力检验,应采用静载荷试验的方法进行检验。检验桩数不应少于总数的 1% 且不少于 3 根,当总桩数少于 50 根时,不应少于 2 根。

②质量要求(验收)。

a. 钢筋笼质量应符合表 3.9 的要求。

表 3.9 灌注桩成孔施工的允许偏差

项	序	检查项目	允许偏差/mm	检查方法
主控项目	1	主筋间距	±10	用钢尺量
	2	钢筋笼长度	±100	用钢尺量
一般项目	1	钢筋材质检验	设计要求	抽样送检
	2	箍筋间距	±20	用钢尺量
	3	钢筋笼直径	±10	用钢尺量

b. 灌注桩质量应符合表 3.10 的要求。

Table 3.8 the Allowable Deviation of the Cast-in-place Pile Construction

Serial Number	Drilling Method		Allowable Deviation of Pile Diameter /mm	Allowable Deviation of Perpendicularity(%)	Allowable Deviation of Pile Position/mm	
					Single pile, strip pile foundation along vertical axis direction and pile groups in the side of pile foundation	Strip pile foundation along vertical axis direction and pile groups in the middle of pile foundation
1	Drilling, digging, and punching piles with mud retaining wall	$d \leqslant 1000$mm	±50	<1	d/6 and no more than 100	d/4 and no more than 150
		$d > 1000$mm	±50		$100 + 0.01H$	$150 + 0.01H$
2	Hammer(vibration)sinking pipe, vibration impact sinking pipe	$d \leqslant 500$mm	−20	<1	70	150
		$d > 500$mm			100	150
3	Auger drill, motorized Luoyang shovel dry operation hole		−20	<1	70	150
4	Pile with hand-excavation	Cast-in-place concrete retaining wall	±50	<0.5	50	150
		Long steel casing retaining wall	±20	<1	100	200

b. The quality of the pile body shall be inspected. The sampling quantity shall not be less than 30% of the total number and not less than 20 piles.

c. The pile bearing capacity shall be tested, and the static load test method shall be adopted. The number of inspection piles shall not be less than 1% of the total number and not less than 3, and when the total number of inspection piles is less than 50, the number of inspection piles should not be less than 2.

②Quality requirements(acceptance).

a. The quality of the steel cage should meet the requirements of Table 3.9.

Table 3.9 Allowable Aviation of Cast-in-place Pile Construction

Item	Serial Number	Inspection Item	Allowable Deviation /mm	Inspection Method
Dominant Item	1	Main bar spacing	±10	Steel gauge
	2	Steel cage length	±100	Steel gauge
General Item	1	Reinforcement quality test	Design requirement	Sampling for test
	2	Stirrup spacing	±20	Steel gauge
	3	Steel cage diameter	±10	Steel gauge

b. The quality of cast-in-place pile should meet the inspection standards in the Table 3.10.

表 3.10 灌注桩成孔施工的允许偏差

项	序	项目	允许偏差	检查方法
主控项目	1	桩位	见表 3.8	开挖前量护筒,开挖后量桩中心
	2	孔深	+300 mm	重锤测或测钻杆、套管长、嵌岩桩应确保设计入岩深度
	3	桩体质量	按基桩检测技术规范	按基桩检测技术规范
	4	混凝土强度	设计要求	试件报告或钻芯取样送检
	5	承载力	按基桩检测技术规范	按基桩检测技术规范
一般项目	1	垂直度	见表 3.8	测套客或钻杆
	2	桩径	见表 3.8	井径仪或超声波检测
	3	泥浆比重(黏土或砂性土中)	1.15～1.20	用比重计测
	4	泥浆面标高(高于地下水位)	0.5～1.0	目测
	5	沉渣厚度	端承桩≤50 mm 摩擦桩≤150 mm	用沉渣仪或重锤测量
	6	混凝土坍落度	160～220 mm	坍落度仪
	7	钢筋笼安装深度	±100 mm	用钢尺量
	8	混凝土充盈系数	>1	检查每根桩的实际灌注量
	9	桩顶标高	±30 mm,−50 mm	水准仪,需扣除顶部浮浆层及劣质桩体

注:①桩径允许偏差的负值是指个别断面。
②采用复打、反插法施工的桩径允许偏差不受本表限制。
③H 为施工现场地面标高与桩顶设计标高的距离;d 为设计桩径。

Chapter 3 Subbase and Foundation Construction

Table 3.10 Allowable Deviation of Cast-in-place Pile Construction

Item	Serial Number	Item	Allowable Deviation	Inspection Method
Dominant Item	1	Pile position	see table 3.8	Measure the pipe casing before excavation, and measure the center of pile after excavation.
	2	Hole depth	+300mm	The weight measuring or measuring drill pipe, casing length and rock socketed-pile to ensure the designed depth into rock
	3	Pile body quality	According to technical specification of foundation pile testing	According to technical specification of foundation pile testing
	4	Concrete strength	Design requirement	Testing sample report or core sampling for inspection
	5	Bearing capacity	According to technical Specification of foundation pile testing	According to technical specification of foundation pile testing
General Item	1	Perpendicularity	See table 3.8	Test casing or drill pipe
	2	Pile diameter	See table 3.8	Caliper or ultrasonic testing
	3	Relative density of mud (in clay or sandy soil)	1.15—1.20	Test with hydrometer
	4	Mud level (above ground water level)	0.5—1.0	Visual inspection
	5	Sediment thickness	End bearing pile$\leqslant$50mm friction pile$\leqslant$150mm	Measure with dredger or hammer
	6	Concrete slump	160—220mm	Slump meter
	7	Installation depth of steel cage	±100mm	Measure with steel gauge
	8	Coefficient of concrete filling	>1	Check the actual filling capacity of each pile.
	9	Pile top elevation	±30mm, −50mm	Level, but the deduction of top slurry layer and poor quality pile is required.

Note: 1. The negative value of the allowable deviation of the pile diameter refers to individual section.
2. The allowable deviation of diameter of pile constructed by the re-driving and reverse inserting method is not limited by this table.
3. H is the distance between the ground elevation on the construction site and the design elevation on the pile top, and d is the designed pile diameter.

习 题

1. 地基处理方法一般有哪几种？各有什么特点？
2. 试述换土垫层法的材料要求及施工要点。
3. 什么是刚性基础？其构造要求中的主要参数是什么？针对不同的刚性基础有何要求？其施工要点是什么？
4. 筏形基础的特点有哪些？其适用范围是什么？其主要施工要点是什么？
5. 试解释端承桩和摩擦桩，它们的质量控制方法有何区别？
6. 吊桩时如何选择吊点？如何才能保证桩位准确、桩垂直？
7. 打桩顺序有哪几种？
8. 试述预制桩在施工中常易出现哪些问题？如何处理？
9. 灌注桩有何特点？灌注桩分哪几种？适用于什么情况？如何成孔？
10. 试述沉管灌注桩常见问题及处理方法。
11. 试述人工挖孔灌注桩中的优点和护壁方法。

Chapter 3 Subbase and Foundation Construction

Exercises

1. What are the general foundation treatment methods? What are their characteristics?

2. Try to describe the material requirements and construction points of soil replacement cushion method.

3. What is a rigid foundation? What are the main parameters of its structural requirements? What are the requirements for different rigid foundations? What are their construction points?

4. What are the characteristics of raft foundation? What is its scope of application? What are the main construction points?

5. Try to explain end bearing pile and friction pile and the difference of quality control methods between them?

6. How to select the lifting point when lifting the pile? How to ensure that the pile position is accurate and vertical?

7. What are the sequences of piling?

8. What problems often occur in the construction of precast piles and how to deal with them?

9. What are the characteristics of cast-in-place piles? What are the types of cast-in-place piles? What are they suitable for? How to form holes?

10. Discuss the common problems and treatment methods of cast-in-place pile.

11. Try to describe the advantages and wall protection methods of manual hole digging cast-in-place pile.

第4章　砌体工程施工

砌体工程是指采用规定配比拌制好的砂浆,将砌块按不同的组砌方法组合为符合设计要求的砌体。砖石结构具有就地取材方便、保温、隔热、隔声、耐火等良好性能,且可以节约钢材和水泥,不需大工型施工机械,施工组织简单等优点,但施工仍以手工操作为主,劳动强度大,生产效率低,目前采用新型墙体材料,改善砌体施工工艺已成为砌筑工程改革的重要发展方向。

砌体工程是一个综合的施工过程,包括材料运输、脚手架搭设和墙体砌筑等。

4.1　脚手架施工及施工机械

4.1.1　运输机械准备

1. 井架式升降机

井架式升降机是民用建筑工地常用的垂直方向的运输设备,如图 4.1 所示。它具有稳定性好、运输量大,除用型钢或钢管加工的定型井架之外,还可用脚手架材料搭设而成。

井架升降机的承载结构是井架,它是用角钢或钢管单节逐段安装的,为了保持井架的抗倾倒能力,必须用纤缆拉紧,纤缆数量根据安装高度而定。井架多为单孔井架,但也可构成两孔或多孔井架。井架侧边有槽钢或钢管构成的导轨,装料平台的滚轮沿导轨运行,起导向作用。装料平台由装在井架外的卷扬机通过钢丝绳滑轮组牵引而上升、下降。井架主股一侧铰装有吊臂,吊臂顶端有变幅滑轮组和起升滑轮组。吊臂用来吊钢筋,模板与轻型构件等。吊臂靠人工牵引,可作小于 180°的局部回转,覆盖面小,仅能靠近安装就位井架的重物。

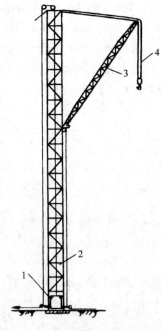

图 4.1　井架升降机
1—装料平台;2—井架;
3—吊臂;4—吊钩

Chapter 4　Masonry Works Construction

Masonry works refer to using the mortar mixed with the specified proportion to combine the blocks according to different masonry methods to meet the design requirements. This masonry structure has the advantages of convenient local materials, good thermal insulation, heat insulation, sound insulation, fire resistance, etc. , and can save steel and cement, without large-scale construction machinery, and it is simple in construction organization. But its construction is still mainly manual operation, labor intensity and low production efficiency. At present, the adoption of new wall materials and improvement of masonry construction technology has become an important development direction of masonry works reform.

Masonry works are a comprehensive construction process, which includes material transportation, scaffolding and wall masonry, etc.

4.1　Scaffold Construction and Construction Machinery

4.1.1　Preparation of Transportation Machinery

1. Derrick Elevator

Derrick elevator is common vertical transportation equipment used in civil construction sites, as shown in Fig. 4.1. It has advantages of good stability, large transportation volume and can be set up with scaffold materials in addition to the shaped derrick processed with section steel or steel pipe.

The load-bearing structure of derrick elevator is Derrick, which is installed with angle steel or steel pipe in single section by section. In order to maintain the anti-dumping ability of derrick, it must be tensioned with fiber cables, and the number of fiber cables depends on the installation height. Most derricks are single-hole derricks, but they can also form two-hole or multi-hole derricks. There are channel steel or steel pipe guide rails on the side of derrick. The rollers of the loading platform run along the guide rails, playing a guiding role. The loading platform is pulled up and down by the hoist outside the derrick through the wire rope pulley block. A boom is arranged on the side of main unit of the derrick, and a derricking pulley block and a lifting pulley block are arranged at the top of the boom. The boom is used for lifting steel bars, formwork and light components, etc. The boom can rotate less than 180° by manual traction, with small coverage, and can only be close to the weight installed in the derrick.

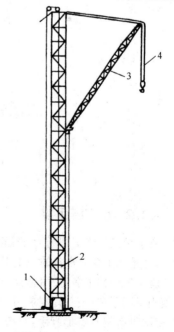

Fig. 4.1　Derrick Elevator
1—loading platform; 2—derrick;
3—boom; 4—lifting hook

井架式升降机一般只装备一台卷扬机作为提升机构，即可完成混凝土、砂浆、砌块、砖等物料垂直运输。其卷扬机可安装在距建筑物一定距离的地方，故操作安全方便，但技术性能差，工作效率也很低，并需要纤缆，占用场地大。

井架升降机的载重量一般是 5～20 kN，起升高度不超过 60 m，而小吊臂的井架升降机起重量一般只有 5～15 kN。

2. 龙门式升降机

龙门式升降机是建筑施工中常用的一种提升设备，构造简单、制作容易、用材少、装拆方便。它由两根三角形截面或矩形截面的立柱与天轮梁（横梁）组成的门架、起重平台、卷扬机及钢丝绳和滑轮组等组成，如图 4.2 所示。门架由若干节格构式金属标准节与横梁组成，底部安装在靠近建筑物专设的混凝土基础上，并分段与建筑物用拉杆锚固或拉设多根缆风绳，以保证门架工作时的稳定性。起重平台是用角钢或槽钢焊接而成的，两边设有滚轮可沿门架上轨道上、下运行。平台上铺有木板，两侧有围栏以保证安全。

龙门架宜进行材料、机具和小型预制构件的垂直运输，一般适用于中小型工程。

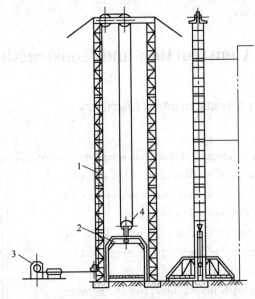

图 4.2 门式升降机
1—门架；2—起重平台；3—卷扬机；4—滑轮组

4.1.2 施工脚手架

脚手架是在施工现场为安全防护、工人操作和解决楼层水平运输而搭设的架体，系施工临时设施，也是施工企业常备的施工工具。

脚手架种类很多，按用途分有结构脚手架、装修脚手架和支撑脚手架等；按搭设位置分有外脚手架和里脚手架；按使用材料分有木、竹和金属脚手架；按设置形式分有单排、双排、多排满堂、满堂高、交圈脚手架及特形脚手架；按支固方式分有扣件式、碗扣式、门架式、悬挑式、吊式及附墙升降式、水平移动式等脚手架。下面介绍框架结构施工常用的几种脚手架。

The derrick elevator is usually equipped with only one hoist as the lifting mechanism, which can complete the vertical transportation of concrete, mortar, block, brick and other materials. Its hoist can be installed at a certain distance from the building, so it is safe and convenient to operate, but its technical performance is poor, its work efficiency is low, and it needs fiber cables, which occupies a large space.

The lifting capacity of derrick elevator is generally 5—20kN, and the lifting height cannot exceed 60m, while the lifting capacity of derrick elevator with small boom is generally only 5—15kN.

2. Gantry Elevator

Gantry elevator is common lifting equipment used in construction, simple structure, easy to make, less material, easy to assemble and disassemble. It consists of a portal frame, a lifting platform, a hoist machine, a steel wire rope and a pulley group which are composed of two vertical columns with triangular or rectangular cross-section and a skyscraper beam (cross-beam), etc., as shown in Fig. 4. 2. The portal frame is composed of several metal standard sections and beams with lattice structure. The bottom of the portal frame is installed on the concrete foundation close to the building and anchored or set guy ropes with tie rods between sections and buildings to ensure the stability of the portal frame when it is working. The lifting platform is welded with angle steel or channel steel and rollers are arranged on both sides to run up and down the rails of the portal frame. The platform is covered with wooden planks and fenced on both sides for safety.

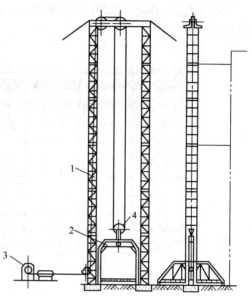

Fig. 4. 2 Gantry Elevator
1—portal frame; 2—lifting platform;
3—hoist; 4—pulley block

The gantry frame is suitable for materials, machines and small prefabricated components of the vertically transport, generally applicable to small and medium-sized projects.

4.1.2 Construction Scaffold

Scaffold is a kind of frame erected for safety protection, workers' operation and floor horizontal transportation at the construction site. It is a temporary construction facility and a common construction tool for construction enterprises.

There are many kinds of scaffolds, including structural scaffolds, decoration scaffolds and supporting scaffolds according to their uses; external scaffolds and internal scaffolds according to their erection positions; wood, bamboo and metal scaffolds according to their materials; single row, double row, multi row full hall, full height, cross circle scaffolds and special scaffolds according to their setting forms; According to the supporting methods, there are fastener type, bowl buckle type, gantry type, cantilever type, hanging type, wall attached lifting type, horizontal mobile scaffold, etc. The following introduces several kinds of scaffolds commonly used in frame structure construction.

1. 扣件式钢管脚手架

扣件式钢管脚手架装拆方便、搭设灵活,能适应建筑物平面及高度的变化;承载力大、搭设高度高、坚固耐用、周转次数多;加工简单、一次投资费用低、比较经济,故在建筑工程施工中使用最为广泛。它除用作搭设脚手架外,还可用以搭设井架、上料平台和栈桥等。但也存在着扣件(尤其中的螺杆、螺母)易丢易损、螺栓紧固程度差异较大、节点在力作用线之间有偏心或交汇距离等缺点。

(1)组成及其作用。扣件式钢管脚手架由钢管、扣件、底座、脚手板和安全网等组成,如图4.3所示。

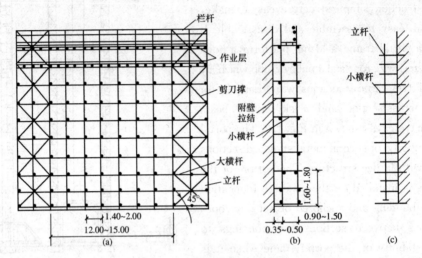

图 4.3 扣件式钢管外脚手架

①钢管。一般采用外径为48 mm、壁厚为3.5 mm的焊接钢管或无缝钢管,也可用外径为51 mm、壁厚为3 mm的焊接钢管。根据钢管在脚手架中的位置和作用不同,钢管则可分为立杆、纵向平杆(大横杆)、横向平杆(小横杆)、剪刀撑、斜杆和连墙件(抛撑)等,其作用如下:

a. 立杆。平行于建筑物并垂直于地面,是把脚手架荷载传递给基础的受力杆件。

Chapter 4 Masonry Works Construction

1. Steel Pipe Scaffold with Couplers

Steel pipe scaffold with couplers is easy to assemble and disassemble, flexible to set up, and can adapt to the changes of building plane and height. It has large bearing capacity, high erection height, strong durability, many turnover times, simple processing, low investment cost and relatively economical, so it is widely used in building construction projects. In addition to scaffolding, it can also be used to set up derrick, loading platform and trestle. However, there are also some shortcomings, such as the couplers(especially the screw and nut) easy to be lost and vulnerable, the difference of bolt tightening degree, the eccentricity or intersection distance between the force acting lines of the nodes, and so on.

(1) Composition and function. Steel pipe scaffold with couplers is composed of steel pipe, coupler, base, scaffold board and safety net, as shown in Fig. 4.3.

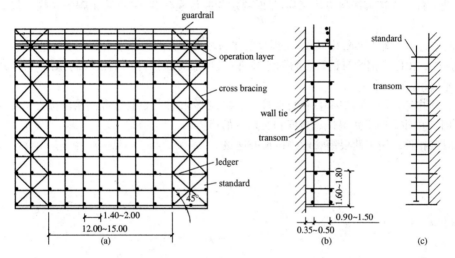

Fig. 4.3 Steel Pipe External Scaffold with Couplers

①Steel pipe. Generally, welded steel pipe or seamless steel pipe with outer diameter of 48mm and wall thickness of 3.5mm can be used, and welded steel pipe with outer diameter of 51mm and wall thickness of 3mm can also be used. According to the different position and function of steel pipe in scaffolding, steel pipe can be divided into vertical pole, longitudinal horizontal bar(large horizontal bar), transverse horizontal bar(small crass bar), cross bracing, diagonal bar and wall connecting member(throwing support), etc. Its functions are as follows:

a. Standard. The standard is parallel to the building and perpendicular to the ground, which is the load-bearing member to transfer the scaffold load to the foundation.

b. 纵向水平杆。平行于建筑物并在纵向水平连接各立杆,是承受并传递荷载给立杆的受力杆件。

c. 纵向水平扫地杆。连接立杆下端,是距底座下皮 200 mm 处的纵向水平杆,起约束立杆底端在纵向发生位移的作用。

d. 横向平杆。垂直于建筑物并在横向水平连接内、外排立杆,是承受并传递荷载给立杆的受力杆件。

e. 横向水平扫地杆。连接立杆下端,是位于纵向水平扫地杆上方处的横向水平杆,起约束立杆底端在横向发生位移的作用。

f. 剪刀撑。设在脚手架外侧面并与墙面平行的十字交叉斜杆,可增强脚手架的纵向刚度。

g. 水平斜拉杆。设在有连墙杆的脚手架内、外排立杆间的步架平面内的"之"字形斜杆,可增强脚手架的横向刚度。

h. 连墙件。连接脚手架与建筑物,是既要承受并传递荷载,又可防止脚手架横向失稳的受力杆件。

i. 抛撑。为了脚手架防止倾覆,在脚手架立面之外设置的斜撑。

②扣件。扣件是钢管与钢管之间的连接件,有可锻铸铁扣件和钢板轧制扣件两种,其基本形式有三种,如图 4.4 所示。

a. 直角扣件。用于两根垂直相交钢管的连接,依靠扣件与钢管表面间的摩擦力来传递载荷。

b. 回转扣件。用于两根任意角度相交钢管的连接。

c. 对接扣件。用于两根钢管对接接长的连接。

b. Ledger. The ledger is parallel to the building and connects the standard in the longitudinal horizontal direction. It is a load-bearing member which bears and transfers the load to the standard.

c. Base ledger. The base ledger is connected with the lower end of the vertical bar, which is a longitudinal horizontal bar 200mm away from the lower part of the base, which acts as a constraint on the vertical displacement of the bottom end of the standard.

d. Transom. The transom is perpendicular to the building and connects the internal and external standards horizontally. It is a stress-bearing member that bears and transfers the load to the standard.

e. Base transom. The base transom is connected with the lower end of the standard, which is located above the longitudinal horizontal bottom bar, and plays the role of restraining the horizontal displacement of the bottom end of the standard.

f. Cross bracing. The cross bracing, which is set on the outer side of the scaffold and parallel to the wall surface, can enhance the longitudinal stiffness of the scaffold.

g. Horizontal diagonal bar. The horizontal diagonal bar is a zigzag diagonal bar in the walking frame plane between the inner and outer rows of vertical bars of the scaffold with wall connecting bars, which can enhance the transverse rigidity of the scaffold.

h. Wall tie. Wall tie which connects scaffold with building is a kind of stress bar which can not only bear and transfer load, but also prevent scaffold from lateral instability.

i. Raking shore. Raking shore is a diagonal brace set outside the facade of the scaffold to prevent the scaffold from overturning.

②Coupler. Couplers are connectors between steel pipe and steel pipe. There are two kinds of couplers: malleable cast iron couplers and steel plate rolling couplers. There are three basic forms of couplers, as shown in Fig. 4.4.

a. Rigid coupler. Rigid coupler is used for the connection of two vertically intersecting steel pipes. The load is transmitted by the friction between the coupler and the steel pipe surface.

b. Swivel coupler. The swivel coupler is used for the connection of two steel pipes intersecting at any angle.

c. Butt coupler. The butt coupler is used for the connection of two steel pipes.

 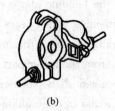 

图 4.4 扣件式形式

(a)直角扣件;(b)回转扣件;(c)对接扣件

③底座。设在立杆下端,是用于承受并传递立杆荷载给地基的配件。底座可用铸铁制成的标准底座(图 4.5),也可用钢管与钢板焊接而成(图 4.6)。

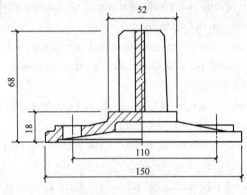 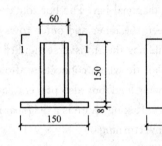

图 4.5 标准底座　　　　　　　图 4.6 焊接底座

④脚手板。脚手板是提供施工操作条件并承受和传递荷载给纵横水平杆的板件,当设于非操作层时起安全防护作用,可用竹、木、钢等材料制成。

⑤安全网。安全网是保证施工安全和减少灰尘、噪声、光污染的措施,包括立网和平网两部分。

(2)构造要点。扣件式钢管脚手架可用于搭设外脚手架、里脚手架、满堂脚手架、支撑架以及其他用途的架子。

①外脚手架。钢管外脚手架可分为双排和单排两种搭设方案[图 4.3(b)、(c)]。这里只介绍双排脚手架[图 4.3(b)]。

双排脚手架一般搭设高度 $H \leqslant 50$ m,如 $H > 50$ m 时则应分段搭设,或采用双立杆等构造措施,并需经过承载力校核计算。

Chapter 4 Masonry Works Construction

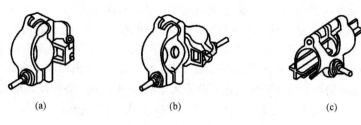

Fig 4.4 Types of Coupler
(a)rigid coupler;(b)swivel coupler;(c)butt coupler

③Base plate. Base plate is located at the lower end of the standard, which is used to bear and transfer the load of the pole to the foundation. The base plate can be a standard base made of cast iron(Fig. 4.5)or welded with steel pipe and steel plate(Fig. 4.6).

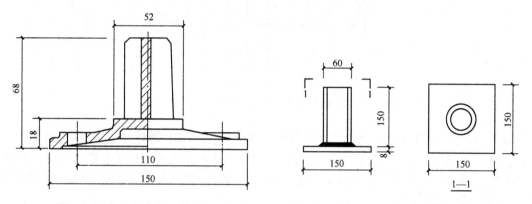

Fig. 4.5 Standard Base Plate

Fig. 4.6 Welded Base Plate

④Scaffold board. The scaffold board is a plate that provides construction operation conditions and bears and transfers loads to the horizontal and vertical bars. When it is set in the non operation layer, it plays a role of safety protection. It can be made of bamboo, wood, steel and other materials.

⑤Safety net. Safety net is a measure to ensure construction safety and reduce dust, noise and light pollution, including two parts of vertical network and horizontal network.

(2)Main points of construction. Fastener type steel pipe scaffold can be used to erect external scaffold, inner scaffold, full scaffold, support frame and other scaffolds.

①External scaffold

The external scaffold of steel pipe can be divided into two erection schemes: double row and single row[Fig. 4.3(b),(c)]. Only double row scaffold is introduced here[Fig. 4.3(b)].

Generally, the erection height of double row scaffold is less than or equal to 50m. If $H>50m$, it should be erected in sections, or double vertical poles and other structural measures should be adopted, and the bearing capacity should be checked and calculated.

a. 立杆。横距为 0.9~1.5 m(高层架子不大于 1.2 m);纵距为 1.4~2.0 m。立杆接长除顶层可采用搭接外,其余各层必须用扣件对接;两相邻立杆的接头位置不应设在同一步距内,且与相近的纵向水平杆距离不应大于等于 1/3 步距,如图 4.7 所示;立杆与纵向水平杆必须用直角扣件扣紧,不得隔步设置或遗漏;立杆顶端应高出女儿墙上皮 1.0 m,高出檐口上皮高度 1.50 m;每根立杆均应设置底座或垫块。立杆的垂直偏差应不大于架高的 1/300,并控制其绝对偏差值:架高≤20 m 时,不大于 50 mm;架高>20 m 而≤50 m 时,不大于 75 mm;架高>50 m 时应不大于 100 mm。

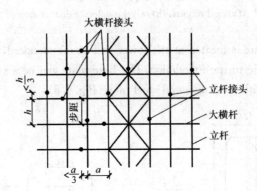

图 4.7 立杆、大横杆的接头位置

b. 大横杆。步距为 1.5~1.8 m。上下横杆的接长位置应错开布置在不同的立杆纵距中,与相近立杆的距离不大于纵距的三分之一,如图 4.7 所示。同一排大横杆的水平偏差不大于该片脚手架总长度的 1/250,且不大于 50 mm。相邻步架的大横杆应错开布置在立杆的里侧和外侧,以减少立杆偏心受力情况。

c. 小横杆。贴近立杆布置(对于双立杆,则设于双立杆之间),搭于大横杆之上并用直角扣件扣紧。在相邻立杆之间根据需要加设 1 根或 2 根。在任何情况下,均不得拆除作为基本构架结构杆件的小横杆。

a. Standard. The transverse distance is 0.9—1.5m(the high-rise frame is no more than 1.2m), and the longitudinal distance is 1.4—2.0m. In addition to the top layer that can be lapped, other layers must be butted with couplers. The joint position of two adjacent standards should not be within the same lift height, and the distance from the similar vertical horizontal poles should not be greater than or equal to 1/3 lift height, as shown in Fig. 4.7. The standard and the vertical horizontal pole must be fastened with rigid couplers, and should not be set or omitted at intervals. The top of the standard should be 1.0m higher than the top brick of the parapet wall and 1.50m higher than the top of the cornice; each standard should be provided with a base plate or cushion block. The vertical deviation of the standard shall not be greater than 1/300 of the frame height, and its absolute deviation value shall be controlled: when the scaffold height is $\leqslant$ 20m, it shall not be greater than 50mm; when the scaffold height is $>$ 20m but $\leqslant$ 50m, it shall not be greater than 75mm; when the scaffold height is $>$ 50m, it shall not be greater than 100mm.

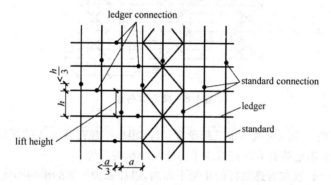

Fig. 4.7 Joint Position of Standard and Ledger

b. Ledger. The lift height is 1.5—1.8m. The lengthening positions of upper and lower ledgers shall be staggered in different vertical distances of standards, and the distance from adjacent standards shall not be greater than one third of the longitudinal distance, as shown in Fig. 4.7. The horizontal deviation of the same row of ledgers shall not be greater than 1/250 of the total length of the scaffold, and shall not be greater than 50mm. The ledgers of adjacent walking frames should be staggered in the inner and outer sides of the standards to reduce the eccentric stress of the standards.

c. Transom. Transom shall be arranged close to the standard(or between the two standards for double posts), and shall be erected on the ledger and fastened with rigid couplers. One or two additional standards shall be set between adjacent standards as required. Under no circumstances shall transoms which are members of the basic frame structure be removed.

d. 剪刀撑。35 m 以下脚手架除在两端设置剪刀撑外,中间应每隔 12～15 m 设一道。剪刀撑应连系 3～4 根立杆,斜杆与地面夹角为 45°～60°;35 m 以上脚手架,沿脚手架两端和转角处起,每 7～9 根立杆设一道,且每片架子不少于三道。剪刀撑应沿架高连续布置,在相邻两排剪刀撑之间,每隔 10～15 m 高加设一组长剪刀撑,如图 4.8 所示。

剪刀撑的斜杆除两端用旋转扣件与脚手架的立杆或大横杆扣紧外,在其中间应增加 2～4 个扣结点。

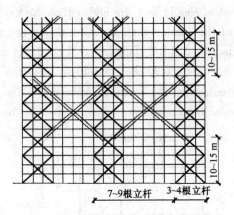

图 4.8 剪刀撑布置

e. 连墙件。可按二步三跨或三步三跨设置,其间距应不超过表 4.1 的规定,且连墙件一般应设置在框架梁或楼板附近等具有较好抗水平力作用的结构部位。

f. 水平斜拉杆。设置在有连墙杆的步架平面内,以加强脚手架的横向刚度。

g. 脚手板、护栏及挡脚板。脚手板一般应设置在 3 根横向水平杆上。当板长小于 2 m 时,允许设在 2 根横向水平杆上,但应将板两端可靠固定,以防倾翻;自顶层操作层往下计,宜每隔 12 m 满铺一层脚手板。作业层脚手板应铺满、铺稳,离墙 120～150 mm。在铺脚手板的操作层上必须设两道护栏和挡脚板。上栏杆高度≥1.1 m。挡脚板亦可用加设一道低栏杆(距脚手板面 0.2～0.3 m)代替。

表 4.1 连墙件的间距

脚手架类型	脚手架高度(m)	垂直间距(m)	水平间距(m)
双 排	≤50	≤6	≤6
	>50	≤4	≤6
单排	≤20	≤6	≤5

Chapter 4 Masonry Works Construction

d. Cross brace. In addition to setting cross brace at both ends of the scaffold below 35m, one cross brace shall be set every 12—15m in the middle. The cross brace shall be connected with 3—4 standards, and the angle between the cross braces and the ground is 45° to 60° and one for every 7—9 standards from both ends and corners of the scaffold for scaffolds with a height of more than 35m, and each scaffold shall be no less than three. The cross brace shall be arranged continuously along the frame height and a long length of cross brace shall be set every 10—15m between two adjacent rows of cross brace, as shown in Fig. 4. 8.

In addition to the two ends of the diagonal bar of the cross brace are fastened with the standards or ledgers of the scaffold with swivel couplers, 2—4 buckle nodes shall be added in the middle.

e. Wall tie. The wall tie can be set as two-step three spans or three-step three spans, and the spacing shall not exceed the provisions in Table 4. 1. Generally, the wall tie shall be set near the frame beam or floor and other structural parts with good horizontal force resistance.

f. Horizontal diagonal bracing. The horizontal diagonal bracing is set in the walking frame plane with wall tie to strengthen the transverse rigidity of the scaffold.

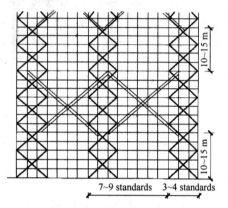

Fig. 4. 8 Cross Bracing Arrangement

g. Scaffold board, guardrail and toe board. Generally, the scaffold board should be set on three horizontal bars. When the length of the board is less than 2m, it is allowed to set it on two horizontal bars, but both ends of the board should be fixed reliably to prevent overturning; a layer of scaffold board should be laid every 12m from the top operation layer down. The scaffold board of the operation layer should be paved fully and stably, 120—150mm away from the wall. Two guardrails and toe boards must be set on the operation layer of the scaffold board. The height of upper rail is more than 1. 1m. The toe board can also be replaced by adding a low railing(0. 2—0. 3m away from the scaffold surface).

Table 4. 1 Spacing of Wall Tie

Types of Scaffold	Height of Scaffold(m)	Vertical Spacing(m)	Horizontal Spacing(m)
Double Row	≤50	≤6	≤6
	>50	≤4	≤6
Single Row	≤20	≤6	≤5

②里脚手架。里脚手架为室内作业架。

里脚手架常为"一"字形的分段脚手架,可采用双排或单排架。在无须搭设杆件式脚手架时,也可采用各种工具式脚手架(图4.9)。当作业层高≥2.0 m时,应按高处作业规定,在架子外侧设栏杆防护;用于高大厂房和厅堂的高度≥4.0 m的里脚手架应参照外脚手架的要求搭设。

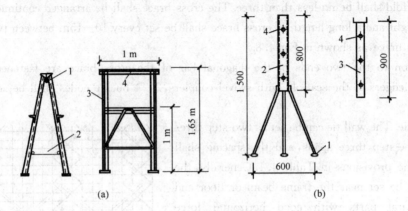

图 4.9　工具式里脚手架

(a)角钢折叠式；
1—铁铰链；2—挂钩；3—立柱；4—横楞
(b)套管式支柱
1—支脚；2—立管；3—插管；4—销孔

③满堂脚手架。满堂脚手架(图4.10)系指室内平面满设的,纵、横向各超过3排立杆的整块形落地式多立杆脚手架,可用于大面积楼板模板的支撑架,或用于天棚安装和装修作业以及其他大面积的高处作业。满堂脚手架也需设置一定数量的剪刀撑或斜杆,以确保在施工荷载偏于一边时,整个架子不会出现变形。

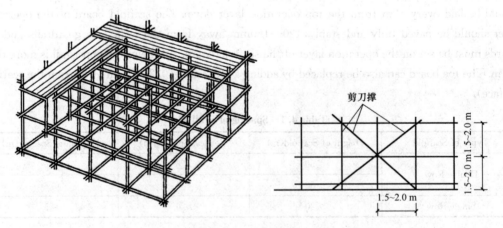

图 4.10　满堂脚手架

②Internal scaffold. The internal scaffold is an indoor working frame.

The internal scaffold is usually a "—" shaped sectional scaffold, which can be double row or single row. When there is no need to set up the rod scaffold, various tool scaffolds can also be used (Fig. 4. 9). When the height of the working floor is more than 2. 0m, railings shall be set outside the scaffold according to the regulations for high-altitude operation; the internal scaffold used for high-rise workshop and hall with the height ≥ 4. 0m shall be erected according to the requirements of external scaffold.

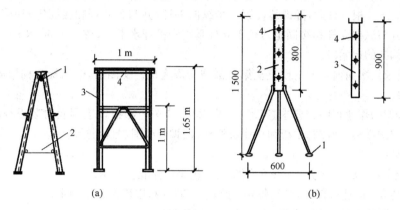

Fig. 4. 9 Tool Type Internal Scaffold
(a)angle folding; (b)tubular prop
1—Iron hinge; 2—hook; 3—standard; 4—transom;
1—stand bar; 2—vertical pipe; 3—intubation; 4—pin hole

③Full framing scaffold. The full framing scaffold (Fig. 4. 10) refers to the whole floor type multi pole scaffold with more than 3 rows of vertical and horizontal poles set in the indoor plane. It can be used as the support frame of large-area floor formwork, ceiling installation and decoration and other large-area high-altitude operations. A certain number of cross brace or diagonal bars should be set for full scaffold to ensure that the whole scaffold will not deform when the construction load is biased to one side.

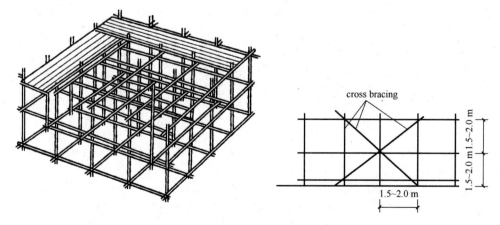

Fig. 4. 10 Full Framing Scaffold

(3)脚手架搭设与拆除。

①地基处理与底座安装。依据脚手架设计的计算结果进行搭设场地的平整、夯实、等地基处理,确保立杆有稳固的地基。然后按构架设计的立杆间距进行放线定位,铺设垫板和安放立杆底座,并确保位置准确、铺放平稳,不得悬空。使用双立杆时,应相应采用双底座,双底座放在一根槽钢底座板上(槽口朝上)。

②搭设作业。杆件搭设顺序:放置纵向水平扫地杆,自角部起依次向两边竖立杆,底端与纵向扫地杆用扣件连接固定后,装设横向扫地杆并与立杆固定(固定立杆底端前,应吊线确保立杆垂直),每边装起 3~4 根立杆后,随即装设第一步纵向平杆(并与各立杆扣接固定)和第一步横向平杆(靠近立杆并与纵向立杆扣接固定),校正立杆垂直和平杆水平后,按 40~60 N·m 力矩拧紧扣件螺栓,形成构架的起始段。

按上述要求依次向前延伸搭设直至第一部架交圈完成。交圈后,全面检查构架质量,设连墙件(或加抛撑)。

按第一步架的作业程序和要求搭设第二步、第三步……随搭设进程及时装设连墙件和剪刀撑,装设作业层间横杆,铺设脚手板和作业层栏杆、挡脚板或围护封闭措施。

③注意事项。

a. 严禁将外径 48 mm 与 51 mm 的钢管及其相应扣件混合使用;

b. 在设置第一排连墙件前,应约每隔 6 跨设一道抛撑,以确保架子稳定;

c. 连墙件和剪刀撑应及时设置,不得滞后超过 2 步;

d. 脚手架必须配合施工进度搭设,一次搭设高度不应超过相邻连墙杆以上两步;

e. 杆件端部伸出扣件之外的长度不得小于 100 mm;

f. 在顶排连墙件之上的架高(以纵向平杆计)不得多于 3 步,否则应每隔 6 跨架设一道撑拉措施;

g. 对接平板脚手板时,对接处的两侧必须设置间横杆。

Chapter 4 Masonry Works Construction

(3) Erection and removal of scaffold.

① Subbase treatment and base plate installation. According to the calculation results of scaffolding design, the subbase treatment such as leveling and compaction of the erection site is carried out to ensure that the vertical pole has a stable subbase. Then, the setting out and positioning shall be carried out according to the distance between the vertical poles designed by the framework, the base plate shall be laid and the base of the vertical pole shall be placed, and the position shall be accurate and stable, and shall be no hanging. When using double vertical poles, double bases should be used, and the double bases should be placed on a channel steel base plate (the notch facing up).

② Erection work. The erection sequence of the poles: Place the longitudinal horizontal bottom bar, erect the vertical poles from the corner to the two sides in turn, install the horizontal bottom bar and fix it with the vertical pipe after the bottom end is fixed with the fastener (before fixing the bottom end of the vertical rod, hang the wire to ensure the vertical of the vertical rod). After installing 3—4 vertical poles on each side, the first step of longitudinal horizontal bar (and buckle with each vertical pole) and the first step of horizontal bar shall be installed. After the vertical and horizontal vertical poles are corrected, the fastener bolts are tightened according to 40—60N·m torque to form the starting section of the frame.

According to the above requirements, the erection shall be extended forward until the completion of the first erection circle. After the completion of the circle, the quality of the frame shall be comprehensively checked, and the wall connecting parts (or bracing) shall be added.

According to the operation procedures and requirements of the first step, start the second step, and the third step… Along with the erection process, wall connecting parts and cross bracing shall be provided timely, cross bars between operation layers shall be installed, scaffold board and operation layer railing, toe board or enclosure measures shall be laid.

③ Matters needing attention.

a. It is forbidden to mix the steel pipe with outer diameter of 48mm and 51mm and its corresponding fasteners.

b. Before setting the first row of wall connecting parts, a bracing should be set every 6 spans to ensure the stability of the scaffold.

c. The wall connecting parts and cross bracing shall be set in time and shall not lag more than 2 steps.

d. The scaffold must be erected in accordance with the construction progress, and the erection height at one time shall not exceed two steps above the adjacent wall connecting rod.

e. The length of the rod end extending beyond the fastener shall not be less than 100mm.

f. The height of the top row of wall connecting parts (calculated by longitudinal horizontal bar) shall not be more than 3 steps, otherwise, a bracing and pulling measure shall be erected every 6 spans.

g. When butting the flat scaffold board, the cross bar must be set on both sides of the butt joint.

④脚手架拆除。

a. 拆架时应划出工作区标志和设置围栏,并派专人看守,严禁行人进入。拆除作业必须由上而下逐层进行,严禁上下同时作业。

b. 拆架时统一指挥,上下呼应,当解开与另一人有关的结扣时,应先行告知对方,以防坠落。

c. 连墙杆必须随脚手架逐层拆除,严禁先将连墙杆整层或数层拆除后再拆脚手架;分段拆除高差不应大于 2 步,如高差大于 2 步,应增设连墙杆加固;

d. 当脚手架采取分段、分立面拆除时,对不拆除的脚手架两端,应先按规范规定设置连墙杆和横向斜撑加固。

e. 各构配件严禁抛掷至地面。

f. 运至地面的构配件应及时检查、整修与保养,并按品种、规格随时码堆存放。

2. 碗扣式钢管脚手架

碗扣式脚手架是一种杆件承插锁固式钢管脚手架,采用带连接件的定型杆件,组装简便,具有比扣件式钢管脚手架较强的稳定承载能力,不仅可以组装各式脚手架,而且更适合构成各种支撑架。具有多功能、拼拆快速省力、安全可靠、易于加工、便于运输等特点,因此目前在建筑工程领域应用十分广泛(图 4.11)。

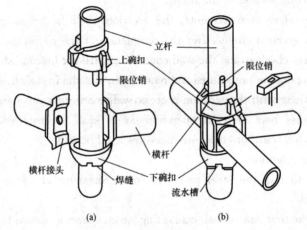

图 4.11 碗扣接头
(a)连接前;(b)连接后

Chapter 4 Masonry Works Construction

④Removal of scaffold.

a. When removing the scaffold, mark the working area and set up a fence, and assign special personnel to guard it. No pedestrians are allowed to enter. Demolition work must be carried out from top to bottom layer by layer, and simultaneous operation is strictly prohibited.

b. Unified command shall be given when removing the frame. When the knot related to another person is untied, the other party shall be informed in advance to prevent falling.

c. The wall connecting rod must be removed layer by layer along with the scaffold. It is strictly forbidden to dismantle the whole layer or several layers of the wall connecting rod before dismantling the scaffold. The height difference of sectional demolition shall not be greater than 2 steps. If the height difference is greater than 2 steps, wall connecting rod shall be added for reinforcement.

d. When the scaffold is demolished by sections and elevations, the two ends of the scaffold that are not to be demolished shall be reinforced with wall connecting rods and transverse diagonal braces according to the regulations.

e. It is forbidden to throw all components and fittings to the ground.

f. The components and parts transported to the ground shall be inspected, repaired and maintained in time, and stacked and stored at any time according to the varieties and specifications.

2. Cuplock Steel Pipe Scaffold

The cuplock scaffold is a kind of steel pipe scaffold with rod socket lock. It adopts the shaped rod with connecting pieces, which is easy to assemble and has stronger stable bearing capacity than the fastener type steel pipe scaffold. It can not only assemble various kinds of scaffolds, but also is more suitable for forming various supporting frames. It has the characteristics of multi-function, fast and labor-saving, safe and reliable, easy to process and easy to transport, so it is widely used in the field of building construction projects (Fig. 4. 11).

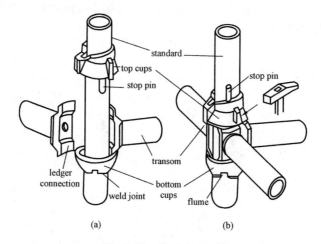

Fig. 4. 11 Cuplock Joint
(a) before connection; (b) after connection

碗扣式脚手架的杆配件及作用组成。

碗扣式钢管脚手架的杆配件有主构件、辅助构件、专用构件三类。

(1)主构件。主构件系用以构成脚手架主体的杆部件,包括立杆、顶杆、横杆、单排横杆、斜杆、立杆底座6类:

①立杆。立杆是脚手架的主要受力杆件,由一定长度的 $\phi 48\times 3.5$、Q235 钢管上每隔 0.60 m 安装一套碗扣接头,并在其顶端焊接立杆连接管制成。立杆有 3.0 m 和 1.8 m 两种规格。

②顶杆。顶杆即顶部立杆,其顶端没有立杆连接管,无法在顶端插入托撑或可调托撑等,有 2.10 m、1.50 m、0.90 m 三种规格。为了解决立杆、顶杆不通用,利用率低的问题,可将立杆的内销管改为下套管,实现立杆和顶杆的统一。改进后立杆规格有 1.2 m、1.8 m、2.4 m、3.0 m。两种立杆的基本结构如图 4.12 所示。

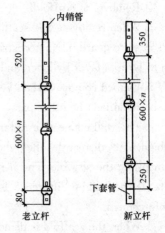

图 4.12 两种立杆的基本结构

③横杆。组成框架的横向连接杆件,由一定长度的 $\phi 48\times 3.5$、Q235 钢管两端焊接横杆接头制成,有 2.4 m、1.80 m、1.5 m、1.2 m、0.9 m、0.6 m、0.3 m 七种规格。

④单排横杆。主要用作单排脚手架的横向水平横杆,只在 $\phi 48\times 3.5$、Q235 钢管一端焊接横杆接头,有 1.4 m、1.8 m 两种规格。

⑤斜杆。斜杆是为增强脚手架稳定强度而设计的系列构件,在 $\phi 48\times 2.2$、Q235 钢管两端铆接斜杆接头制成,斜杆接头可转动,同横杆接头一样可装在下碗扣内,形成节点斜杆。有 1.69 m、2.163 m、2.343 m、2.546 m、3.00 m 等五种规格,分别适用于 1.20 m × 1.20 m、1.20 m × 1.80 m、1.50 m × 1.80 m、1.80 m × 1.80 m、1.80 m × 2.40 m 五种框架平面。

⑥底座。底座是安装在立杆根部,防止其下沉,并将上部荷载分散传递给地基基础的构件,有以下三种:

a. 垫座。由 150 mm×150 mm×8 mm 钢板和中心焊接连接杆制成,立杆可直接插在上面,高度不可调,只有一种规格(LDZ)。

Chapter 4　Masonry Works Construction

Composition of pole fittings and function of cuplock scaffold.

There are three types of pole fittings of cuplock steel pipe scaffold: main component, auxiliary component and special component.

(1) Main component. The main components are the rod components used to form the main body of the scaffold, including six types: standard, top bar, ledger, transom, bracing and base plate of standard.

①Standard. The standard is the main load-bearing member of the scaffold. It is made of a certain length of $\phi 48 \times 3.5$, Q235 steel pipe with a set of cuplock joints every 0.60m, and the top is welded with the connecting pipe. There are two types of standards, 3.0m and 1.8m.

②Top bar. The top bar is the top standard. There is no connecting pipe of the standard at the top, so it is impossible to insert supporting or adjustable supporting at the top. There are three specifications of 2.10m, 1.50m and 0.90m. In order to solve the problem that the standard and the top bar are not in common use and the utilization rate is low, the inner pin pipe of the standard can be changed into the casing pipe to realize the unification of the standard and the top bar. After improvement, the specifications of the standards are 1.2m, 1.8m, 2.4m and 3.0m. The basic structure of the two kinds of standard is shown in Fig. 4.12.

③Ledger. The ledger is a transverse connecting member of the frame, which is made of a certain length of $\phi 48 \times 3.5$, Q235 steel pipe welded ledger joints at both ends. There are seven specifications of 2.4m, 1.80m, 1.5m, 1.2m, 0.9m, 0.6m and 0.3m.

④Transom. Transom is mainly used as horizontal cross bar of single row scaffold. Transom joint is only welded at one end of $\phi 48 \times 3.5$ and Q235 steel pipe. There are two specifications of 1.4m and 1.8m.

⑤Bracing. The bracing is a series of components designed to enhance the stability and strength of scaffold. It is made of $\phi 48 \times 2.2$ and Q235 steel pipe ends by riveting the bracing joint. The bracing joint can be rotated and can be installed in the lower cuplock as the horizontal bar joint to form the joint bracing. There are five specifications of 1.69m, 2.163m, 2.343m, 2.546m and 3.00m, which are applicable to five frame planes of 1.20m × 1.20m, 1.20m × 1.80m, 1.50m × 1.80m, 1.80m × 1.80m and 1.80m × 2.40m.

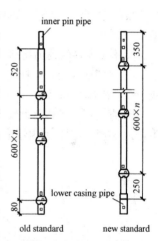

Fig. 4.12　Basic Structure of Two Kinds of Standard

⑥Base plate of standard. The base plate is installed at the root of the standard to prevent it from sinking and transmit the upper load to the subbase. There are three types as follows:

a. Sole Plate. It is made of 150mm × 150mm × 8mm steel plate and central welding connecting rod. The standard can be directly inserted on it, and the height is not adjustable. There is only one specification (LDZ).

b. 立杆可调座。由 150 mm×150 mm×8 mm 钢板和中心焊接螺杆并配手柄螺母制成,有 0.30 m 和 0.60 m 的两种规格。

c. 立杆粗细调座。基本上同立杆可调座,只是可调方式不同,由 150 mm×150 mm×8 mm 钢板、立杆管、螺管、手柄螺母等制成,只有 0.60 m 一种规格。

(2)辅助构件。辅助构件系用于作业面及附壁拉结等的杆部件,按其用途可分成 3 类:作业面辅助构件、连接的辅助构件、其他用途辅助构件。

①用于作业面的主要辅助构件。

a. 间横杆。为满足其他普通钢脚手板和木脚手板的需要而设计的构件,由 $\phi48\times3.5$、Q235 钢管两端焊接"∩"型钢板制成,可搭设于主架横杆之间的任意部位。有 1.2 m、(1.2+0.3)m 和 (1.2+0.6)m 三种规格。

b. 斜道板。用于搭设车辆及行人栈道,只有一种规格,坡度为 1∶3,由 2 mm 厚钢板制成,宽度为 540 mm,长度为 1 897 mm,上面焊有防滑条。

c. 挑梁。为扩展作业平台而设计的构件,有窄挑梁和宽挑梁。窄挑梁由一端焊有横杆接头的钢管制成,悬挑宽度为 0.30 m,可在需要位置与碗扣接头连接。宽挑梁由水平杆、斜杆、垂直杆组成,悬挑宽度为 0.60 m,也是用碗扣接头同脚手架连成一整体再接立杆。

d. 架梯。用于作业人员上下脚手架通道,由钢踏步板焊在槽钢上制成,两端有挂钩,可牢固地挂在横杆上,有 JT-255 一种规格。其长度为 2 546 mm,宽度为 540 mm,可在 1 800 mm×1 800 mm 框架内架设。

②用于连接的辅助构件。

a. 立杆连接销。立杆连接销是立杆之间连接的销定构件,为弹簧钢销扣结构,由 $\phi10$ 钢筋制成,有一种规格(LLX)。

b. 直角撑。为连接两交叉的脚手架而设计的构件,由 $\phi48\times3.5$、Q235 钢管一端焊接横杆接头,另一端焊接"∩"型卡制成,有一种规格(ZJC)。

b. Adjustable Plate of Standard. It is made of 150mm×150mm×8mm steel plate and central welding screw with handle nut. It has two specifications of 0.30m and 0.60m.

c. Adjustable Plate of Standard Thickness. It is basically the same as the adjustable plate of standard, but the adjustable mode is different. It is made of 150mm×150mm×8mm steel plate, standard, screw pipe, and handle nut, etc., with only one specification of 0.60m.

(2) Auxiliary components. Auxiliary components are rod components used for working scope and wall attachment, which can be divided into three categories according to their purposes: auxiliary components of working scope, auxiliary components of connection and auxiliary components for other purposes.

①Main auxiliary components for working scope.

a. Intermediate bearer. In order to meet the needs of other ordinary steel scaffold board and wooden scaffold board, the component is made of $\phi 48 \times 3.5$, Q235 steel pipe, welded with "∩" steel plate at both ends, and can be erected at any part between the cross bars of the main frame. There are three specifications: 1.2m, (1.2+0.3)m and (1.2+0.6)m.

b. Ramp slab. The ramp slab is used to set up the vehicle and pedestrian plank road. There is only one specification. The slope is 1:3. It is made of 2mm thick steel plate with 540mm width and 1897mm length. It is welded with anti-skid strip.

c. Cantilever beam. There are narrow cantilever beam and wide cantilever beam which are designed to expand the working platform. The narrow cantilever beam is made of a steel pipe welded with a cross bar joint at one end. The cantilever width is 0.30m, which can be connected with the bowl-type joint at the required position. The wide cantilever beam is composed of horizontal bar, inclined bar and vertical bar. The cantilever width is 0.60m. It is also connected with the scaffold by bowl-type joint and then connected with the vertical pole.

d. Erecting ladder. It is used for workers to get up and down the scaffold channel. It is made of steel step plate welded on channel steel, with hooks at both ends, which can be firmly hung on the cross bar, with JT-255 specification. Its length is 2546mm and width is 540mm. It can be erected in a frame of 1800mm×1800mm.

②Auxiliary components for connection.

a. Vertical pole connecting pin. The connecting pin of vertical pole is the pin fixing component connecting the vertical poles. It is a spring steel pin buckle structure, which is made of $\phi 10$ steel bar and has one specification(LLX).

b. Right angle brace. The component designed for connecting two crossed scaffolds is made of $\phi 48 \times 3.5$, Q235 steel pipe welded with cross bar joint at one end and "∩" type clamp at the other end, with one specification(ZJC).

c. 连墙撑。连墙撑是用于脚手架与墙体结构间的连接件,以加强脚手架抵御风荷载及其他水平荷载的能力,防止脚手架倒塌且增强稳定承载力的构件,碗扣式连墙撑和扣件式连墙撑两种型式。其中碗扣式连墙撑可直接用碗扣接头同脚手架连在一起,受力性能较好;

　　d. 高层卸荷拉结杆。高层卸荷拉结杆是高层脚手架卸荷专用构件,由预埋件、拉杆、索具螺旋扣、管卡等组成,其一端用预埋件固定在建筑物上,另一端用管卡同脚手架立杆连接,通过调节中间的索具螺旋扣,把脚手架吊在建筑物上,达到卸荷目的。

　　③其他用途辅助构件。

　　a. 立杆托撑。插入顶杆上端,用作支撑架顶托,以支撑横梁等承载物。由 U 形钢板焊接连接管制成,有一种规格(LTC)。当长度可调时为立杆可调托撑,有 0.60 m(KTC-60)长一种规格,可调范围为 0～600 mm。

　　b. 横托撑。用作重载支撑架横向限位,或墙模板的侧向支撑构件。由 $\phi 48 \times 3.5$、Q235 钢管焊接横杆接头,并装配托撑组成,可直接用碗扣接头同支撑架连在一起,有一种规格(HTC)。把横托撑中的托撑换成可调托撑可调范围为 0～300 mm,有一种规格(KHC-30)。

　　c. 安全网支架。安全网支架是固定于脚手架上,用以绑扎安全网的构件,由拉杆和撑杆组成,可直接用碗扣接头连接固定,有一种规格(AWJ)。

　　④专用构件。专用构件是用作专门用途的构件,共有 4 类。

　　由 0.3 m 长横杆和立杆、顶杆连接可组成支撑柱,作为承重构杆单独使用或组成支撑柱群。与其配套的有支撑柱垫座、支撑柱转角座和支撑柱可调座等专用构件。

　　a. 支撑柱垫座。支撑柱垫座是安装于支撑柱底部,均匀传递其荷载的垫座。由底板、筋板和焊于底板上的四个柱销制成,可同时插入支撑柱的四个立杆内,从而增强支撑柱的整体受力性能。支撑柱垫座有一种规格(ZDZ)。

c. Wall connecting brace. The wall connecting brace is used to connect the scaffold and the wall structure to strengthen the ability of the scaffold to resist the wind load and other horizontal loads, prevent the collapse of the scaffold and enhance the stable bearing capacity. There are two types of wall bracing, cuplock wall bracing and coupler wall bracing. Among them, the cuplock wall bracing can be directly connected with the scaffold by the cuplock joint, which has better mechanical performance.

d. High-rise unloading tie bar. The high-rise unloading tie rod is a special component for high-rise scaffold unloading, which is composed of embedded parts, pull rods, rigging screw buckles, pipe clamps, etc. one end of the tie rod is fixed on the building with embedded parts, and the other end is connected with the scaffold pole by pipe clamp. By adjusting the middle sling screw buckle, the scaffold is suspended on the building to achieve the purpose of unloading.

③Auxiliary components for other purposes.

a. Vertical pole support. Insert the upper end of the top rod to be used as the support frame to support the load such as the beam. It is made of U-shaped steel plate welded connection pipe, with one specification(LTC). When the length is adjustable, it is the adjustable support of vertical pole. It has a specification of 0.60m(KTC-60), and the adjustable range is 0—600mm.

b. Transverse brace. Transverse brace is used as lateral limit of heavy load support frame or lateral support member of wall formwork. It is composed of $\phi 48 \times 3.5$, Q235 steel pipe welding cross bar joint and assembling supporting brace, which can be directly connected with the support frame with bowl-type joint, and has one specification (HTC). Replace the supporting brace in the transverse brace with adjustable one, with the adjustable range of 0—300mm, and there is one specification(KHC-30).

c. Safety net bracket. The safety net bracket is a component fixed on the scaffold to bind the safety net. It is composed of a pull rod and a strut. It can be directly connected and fixed with a cuplock joint. There is one specification(AWJ).

④Special components. Special components are used for special purposes, and there are four types.

The supporting column can be composed of 0.3m long horizontal bar, vertical pole and top bar, which can be used as load-bearing member alone or form support column group. There are special components such as supporting column pedestal, supporting column corner seat and supporting column adjustable seat.

a. Supporting column pedestal. The supporting column pedestal is installed at the bottom of the supporting column to transmit its load evenly. It is made of base plate, rib plate and four column pins welded on the bottom plate, which can be inserted into the four vertical rods of the supporting column at the same time, so as to enhance the overall mechanical performance of the supporting column. There is one specification(ZDZ) for the support column pedestal.

b. 支撑柱转角座。作用同支撑柱垫座,但可以转动,使支撑柱不仅可用作垂直方向支撑,而且可以用作斜向支撑。其可调偏角为±10°,有一种规格(ZZZ)。

c. 支撑柱可调座。对支撑柱底部和顶部均适用,安装于底部作用同支撑柱垫座,但高度可调,可调范围为0~300 mm;安装于顶部即为可调托撑,同立杆可调托撑不同的是它作为一个构件需要同时插入支撑柱4根立杆内,使支撑柱成为一体。

d. 提升滑轮。提升滑轮是为提升小物料而设计的构件,与宽挑梁配套使用。由吊柱、吊架和滑轮等组成,其中吊柱可直接插入宽挑梁的垂直杆中固定,有THL一种规格。

⑤构造要求。用碗扣式钢管脚手架可方便地搭设双排外脚手架,拼拆快速省力,且特别适于搭设曲面脚手架和高层脚手架。目前,用其搭设"一杆到顶"(即脚手架全高均采用单立杆)的落地式脚手架的最大高度已达90.3 m。

a. 重型架。这种结构脚手架取较小的立杆纵距(0.90 m或1.20 m),用于重载作业或作为高层外脚手架的底部架。对于高层脚手架,为了提高其承载力和搭设高度,采取上、下分段,每段立杆纵距不等的组架方式,如图4.13所示。

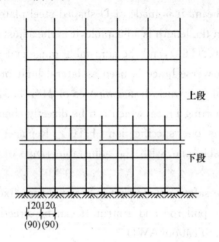

图4.13 分段组架布置

b. 普通架。普通架是最常用的一种,构造尺寸为1.50 m(立杆纵距)×1.20 m(立杆横距)×1.80 m(横杆步距)或1.80 m×1.20 m×1.80 m,可作为砌墙,模板工程等结构施工用脚手架。

c. 轻型架。主要用于装修、维护等作业荷载要求的脚手架,构架尺寸为2.40 m×1.20 m×1.80 m。

(3)搭设要点。

①组装顺序:立杆底座→立杆→横杆→斜杆→接头锁紧→脚手板→上层立杆→立杆连接销→横杆。

b. Support column corner seat. The corner seat of the supporting column acts the same as the pedestal of the supporting column, but it can be rotated, so that the supporting column can be used not only as vertical support, but also as oblique support. Its adjustable deflection angle is $\pm 10°$ and has one specification(ZZZ).

c. Supporting column adjustable seat. The adjustable seat of supporting column is applicable to both the bottom and the top of the supporting column. When installed at the bottom, the function is the same as that of the supporting column pedestal, but the height is adjustable, and the adjustable range is 0—300mm. The adjustable seat of supporting column installed on the top is the adjustable support. Different from the adjustable support of vertical pole, it needs to be inserted into four vertical poles of supporting column at the same time as a component, so that the supporting column can be integrated.

d. Lifting pulley. The lifting pulley is a component designed for lifting small materials, which is used together with wide cantilever beam. It is composed of davits, hangers and pulleys, in which the davits can be directly inserted into the vertical rods of wide cantilever beams for fixation. There is a specification of THL.

⑤Construction requirements. The cuplock steel pipe scaffold can be used to set up double row external scaffolds conveniently, which is fast and labor-saving. It is especially suitable to set up curved scaffolds and high-rise scaffolds. At present, the maximum height of the floor type scaffold with "one pole to the top" (that is, the whole height of the scaffold is single pole) has reached 90.3m.

a. Heavy scaffold. This kind of structural scaffold takes a small vertical distance (0.90m or 1.20m) for heavy load operation or as the bottom frame of high-rise external scaffold. For high-rise scaffold, in order to improve its bearing capacity and erection height, the upper and lower sections are segmented, and the vertical distance of each section is unequal, as shown in Fig. 4.13.

b. Common scaffold. The common scaffold is the most commonly used one, with the structural dimensions of 1.50m(vertical distance of vertical pole)×1.20m(horizontal distance of vertical pole)×1.80m(step distance of horizontal bar) or 1.80m×1.20m×1.80m. It can be used as scaffold for wall building, formwork and other structural construction.

c. Light scaffold. Light scaffold is mainly used for decoration, maintenance and other work load requirements. The scaffold size is 2.40m×1.20m×1.80m.

(3) Main points of erection.

①Assembly Sequence: base of vertical pole → vertical pole → horizontal pole → diagonal pole → joint locking → scaffold board → upper vertical pole → connecting pin of vertical pole → horizontal pole.

Fig. 4.13 Sectional Framing Arrangement

②注意事项。

a. 在已处理好的地基上按设计位置安放立杆垫座(或可调底座),其上再交错安装 3.0 m 和 1.8 m 长立杆,调整立杆可调座,使同一层立杆接头不在同一平面内。

b. 搭设中应注意调整架体的垂直度,最大偏差不得超过 10 mm。

c. 连墙杆应随脚手架的搭设而随时在设计位置设置,并尽量与脚手架和建筑物外表面垂直。

d. 脚手架应随建筑物升高而随时搭设,但不应超过建筑物 2 个步架。

3. 门(框组)式钢管脚手架

门式钢管脚手架是以门型、梯型及其他形式的钢管框架为基本构件,与连接杆件、辅件和各种配件组合而成的脚手架,它是国际上应用最为普遍的脚手架之一,已形成系列产品,各种配件多达 70 多种。可用来搭设各种用途的施工作业架子,如外脚手架、里脚手架、模板和其他承重支撑架、工作台等。

(1)主要组成部件。门式钢管脚手架由门式框架(门架)、交叉支撑(十字拉杆)和水平梁架(平行架)或脚手板构成基本单元(图 4.14)。将基本单元相互连接起来并增加梯子、栏杆等部件构成整片脚手架(图 4.15)。

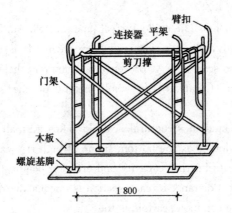

图 4.14 门式脚手架的基本组合单元

门式钢管脚手架的部件大致分为三类:基本单元部件、底座和托座、其他部件。

①基本单元部件。包括门架、交叉支撑和水平架等,如图 4.16 所示。

Chapter 4 Masonry Works Construction

②Matters needing attention.

a. Place the pedestal (or adjustable pedestal) of the vertical pole on the treated foundation according to the design position, and then install 3.0m and 1.8m long vertical poles alternately on it. Adjust the adjustable pedestal of the vertical pole, so that the joints of the vertical pole on the same floor are not in the same plane.

b. During erection, attention should be paid to the verticality of the frame body, and the maximum deviation should not exceed 10mm.

c. The wall connecting rod shall be set at the design position at any time with the erection of scaffold, and shall be perpendicular to the scaffold and the external surface of the building as far as possible.

d. The scaffold should be erected at any time with the rise of the building, but it should not exceed two lift height of the building.

3. Door-type(Frame Group)Steel Pipe Scaffold

Door-type steel pipe scaffold is a kind of scaffold which is composed of door type, ladder type and other forms of steel pipe frame as the basic components, connecting rods, auxiliary parts and various accessories. It is one of the most widely used scaffolds in the world, and has formed a series of products with more than 70 kinds of accessories. It can be used to set up all kinds of construction work scaffolds, such as external scaffold, internal scaffold, formwork, and other load-bearing support frame, working platform, etc.

(1) Main components. The door-type steel pipe scaffold is composed of door frame (vertical frame), cross support (cross tie rod), horizontal beam frame (walking frame) or scaffold board (Fig. 4.14). The basic units are connected with each other and the ladder, guardrail and other components are added to form the whole scaffold (Fig. 4.15).

The components of door-type steel pipe scaffold are roughly divided into three categories: basic unit components, baseplate and bracket, and other components.

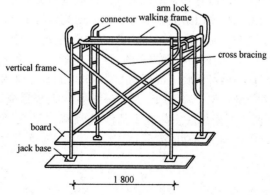

Fig. 4.14 Basic Combination Unit of Portal Scaffold

①Basic unit components. Basic unit components include portal frame, cross support and horizontal frame, etc., as shown in Fig. 4.16.

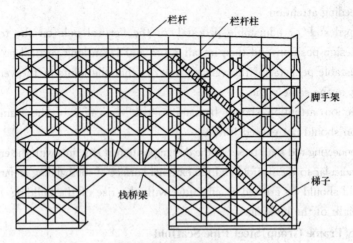

图 4.15 门式外脚手架

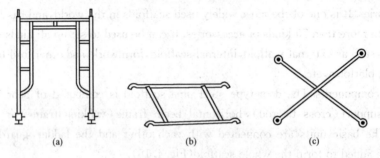

图 4.16 基本单元部件

门架是门式脚手架的主要部件,有多种不同形式。标准型是最基本的形式,主要用于构成脚手架的基本单元,一般常用的标准型门架的宽度为 1.219 m,高度有 1.9 m 和 1.7 m。

门架之间的连接,在垂直方向使用连接棒和锁臂,纵向使用交叉支撑,在架顶水平面使用水平架或脚手板。交叉支撑和水平架的规格根据门架的间距来选择,一般多采用 1.8 m。

②底座和托座。底座有三种:可调底座可调高 200~550 mm,主要用于支模架以适应不同支模高度的需要,脱模时可方便地将架子降下来。用于外脚手架时,能适应不平的地面,可用其将各门架顶部调节到同一水平面上。简易底座只起支承作用,无调高功能,使用它时要求地面平整。带脚轮底座多用于操作平台,以满足移动的需要。

Chapter 4 Masonry Works Construction

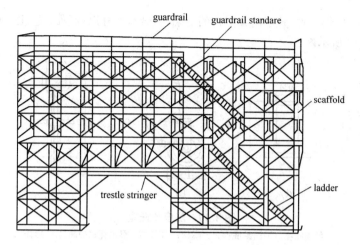

Fig. 4. 15 Door-type External Scaffold

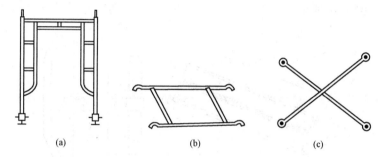

Fig. 4. 16 Basic Unit Components

Gantry is the main part of the portal scaffold, and there are many different forms. Standard type is the most basic form, which is mainly used to form the basic unit of scaffold. Generally, the width of standard type portal frame is 1.219m, and the height is 1.9m and 1.7m.

For the connection between the portal frames, the connecting rod and lock arm are used in the vertical direction, the cross support is used in the longitudinal direction, and the horizontal frame or scaffold board is used in the horizontal plane of the top of the frame. The specification of cross support and horizontal frame is selected according to the distance between the portal frames. Generally, 1.8m is used.

②Base plate and bracket. There are three kinds of base plates: adjustable base plate, which can be adjusted 200—550mm in height, is mainly used for supporting formwork to meet the needs of different formwork heights. It can be easily lowered when stripping formwork. When it is used for external scaffold, it can adapt to uneven ground, and can be used to adjust the top of each gantry to the same horizontal plane. The simple base plate only plays a supporting role and has no height adjustment function. The ground is required to be even when it is used. The caster base plate is mostly used for operation platform to meet the needs of mobile.

托座有平板和U形两种,置于门架竖杆的上端,多带有丝杠以调节高度,主要用于支模架。底座和托座如图4.17所示。

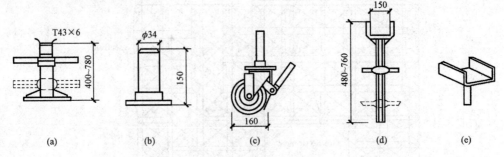

图4.17 底座和托座

(a)可调底座;(b)简易底座;(c)脚轮;(d)可调u形顶托;(e)简易U形托

③其他部件。有脚手板、连接棒、锁臂、钢梯、栏杆、连墙杆、脚手板托架等。钢脚手板、连接棒、锁臂的构造如图4.18所示。

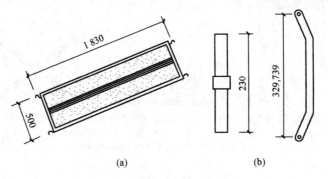

图4.18 其他部件

(a)钢脚手板;(b)连接棒和锁臂

(2)构造要点。

①脚手架部件之间的连接门式钢管脚手架部件之间的连接基本不用螺栓结构,而是采用方便可靠的自锚结构。

②确保脚手架的整体刚度。

a.门架之间必须满设交叉支撑。当架高≤45 m时,水平架应至少两步设一道;当架高>45 m时,水平架必须每步设置(水平架可用挂扣式脚手板和水平加固杆替代),其间连接应可靠。

b.整片脚手架必须适量设置水平加固杆,前三层宜隔层设置,三层以上则每隔3~5层设置一道。

Chapter 4 Masonry Works Construction

There are two kinds of brackets, flat plate and U-shaped, which are placed on the upper end of the vertical pole of the portal frame, with screw to adjust the height, which is mainly used for formwork support. The base and bracket are shown in Fig. 4. 17.

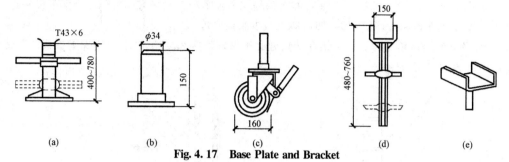

Fig. 4. 17 Base Plate and Bracket
(a)Adjustable base; (b)simple base; (c)caster; (d)adjustable U-shaped top bracket; (e)simple U-shaped support

C. Other components. Other components include scaffold board, connecting rod, lock arm, steel ladder, railing, wall connecting rod, scaffold board bracket, etc. The structural drawing of steel scaffold board, connecting rod and lock arm is shown in Fig. 4. 18.

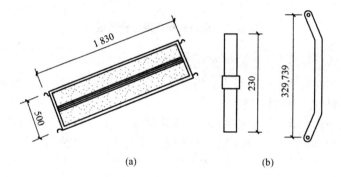

Fig. 4. 18 Other Components
(a)steel scaffold board; (b)connecting rod and flip lock

(2) Main points of construction.

①The connection between the components of the scaffold and the components of door-type steel pipe scaffold is not the bolt structure, but the convenient and reliable self anchored structure is used.

②Ensure the overall rigidity of scaffold.

a. There must be full cross supports between the portal frames. When the frame height is less than or equal to 45 m, the horizontal frame should be set at least in two steps; when the height of the frame is more than 45 m, the horizontal frame must be set every step (the horizontal frame can be replaced by the buckle type scaffold board and the horizontal reinforcement bar), and the connection between them should be reliable.

b. The whole scaffold must be provided with appropriate horizontal reinforcement bars. The first three layers should be set with separate layers, and above three layers should be set every 3—5 layers.

c. 在架子外侧面设置长剪刀撑($\phi 48$ 脚手钢管,长 6~8 m),其高度和宽度为 3~4 个步距(或架距),与地面夹角为 45°~60°,相邻长剪刀撑之间相隔 3~5 个架距。

d. 使用连墙管或连墙器将脚手架和建筑结构紧密连接,连墙点的最大间距,在垂直方向为 6 m,在水平方向为 8 m。一般情况下,在垂直方向每隔 3 个步距和在水平方向每隔 4 个架距设一点,高层脚手架应增加布设密度,低层脚手架可适当减少布设密度。

e. 做好脚手架转角处理。脚手架在转角处必须做好连接和与墙拉结,以确保脚手架的整体性(图 4.19)。处理方法为:

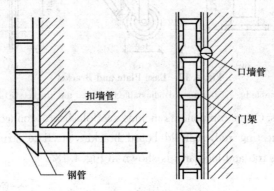

图 4.19 框组式脚手架的转角连接

Ⅰ利用回转扣直接把两片门架的竖管扣结起来;

Ⅱ利用钢管和扣件把处于角部两边的门架连接起来,连接杆可沿边长方向或斜向设置。另外,在转角处适当增加连墙点的布设密度。

③高层脚手架的构造措施。当脚手架的搭设高度超过 45 m 时,可分别采取以下构造措施:

a. 架高 20~30 m 内采用强力级梯架。以加强脚手架下部的整体刚度和承载能力;

b. 采用分段搭设或部分卸载措施(可参考扣件式钢管脚手架的作法),同时需在挑梁所在层及其上两层加设通长的大横杆。

(3)搭设技术与注意事项

①基底处理。应确保地基具有足够的承载力,在脚手架荷载作用下不发生塌陷和显著的不均匀沉降。

a. 基底必须严格夯实抄平。

b. 严格控制第一步门架顶面的标高,其水平误差不得大于 5 mm(超出时,应塞垫铁板予以调整)。

c. 在脚手架的下部加设通常的大横杆,并不少于 3 步,且内外侧均需设置。

②分段搭设与卸载构造的作法。当不能落地架设或搭设高度超过规定(45 m 或轻载的 60 m)时,可分别采取从楼板伸出支挑构造的分段搭设方式或支挑卸载方式。

Chapter 4 Masonry Works Construction

c. The long diagonal brace(ϕ48 scaffold steel pipe, 6—8m long) is set on the outer side of the scaffold, and its height and width are 3—4 steps(or frame distance), and the angle with the ground is 45° to 60° and the adjacent long diagonal brace is separated by 3—5 spacing.

d. The scaffold is closely connected with the building structure by wall connecting pipe or wall connecting device. The maximum distance between wall connecting points is 6m in vertical direction and 8m in horizontal direction. Generally, a point should be set every 3 steps in the vertical direction and 4 spacing in the horizontal direction. The layout density of high-rise scaffold should be increased, and the layout density of low-rise scaffold can be reduced appropriately.

e. Scaffold corner treatment. The scaffold must be connected and tied with the wall at the corner to ensure the integrity of the scaffold(Fig. 4.19). The treatment method is as follows:

Ⅰ. The vertical pipes of the two portal frames are directly connected with the rotary buckle.

Ⅱ. The steel pipe and fastener are used to connect the portal frame on both sides of the corner, and the connecting rod can be set along the side length direction or oblique direction. In addition, the layout density of connecting points should be increased at the corner.

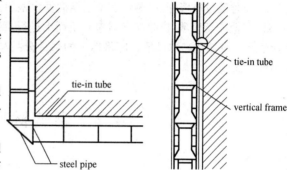

Fig. 4.19 Corner Connection of Frame Scaffold

Ⅲ. Construction measures of high-rise scaffold. When the erection height of scaffold exceeds 45m, the following structural measures can be taken respectively:

a. Within the height of 20—30m, the strong ladder scaffold shall be used to strengthen the overall stiffness and bearing capacity of the lower part of the scaffold.

b. Section erection or partial unloading measures shall be adopted(refer to the method of fastener type steel pipe scaffold). At the same time, large cross bars with full length shall be set up on the floor where the cantilever beam is located and the two floors above it.

(3)Erection technology and matters needing attention.

①Subbase treatment. It should be ensured that the subbase has sufficient bearing capacity, and no collapse and significant uneven settlement will occur under the load of scaffold.

a. The subbase must be compacted and leveled strictly.

b. The elevation of the top surface of the first step portal frame shall be strictly controlled, and the horizontal error shall not be greater than 5mm(when exceeding, iron pad shall be plugged for adjustment).

c. In the lower part of the scaffold, the usual large horizontal bar should be set up at least 3 steps, and the inside and outside sides should be set.

②Construction method of sectional erection and unloading structure. When it is unable to be erected on the ground or the erection height exceeds the specified value(45m or 60m for light load), the segmental erection method or cantilever unloading method can be adopted respectively.

③脚手架搭设程序:铺放垫木→拉线放底座→自一端起立门架,并随即装交叉支撑→装水平架(或脚手板)→装梯子→(视情况需要装设作加强用的大横杆)→装设连墙杆→插连接棒→安装上一步门架→装锁臂→照上述步骤,逐层向上安装→装加强整体刚度的长剪刀撑→装设顶部栏杆。

④注意事项。

a. 严格控制首层门型架的垂直度和水平度。使门架竖杆在两个方向的垂直偏差都控制在 2 mm 以内,门架顶部的水平偏差控制在 5 mm 以内。

b. 上下门架竖杆之间要对中,偏差不宜大于 3 mm。

c. 及时装设连墙杆,以避免在架子横向发生偏斜。

d. 其他注意事项:拆除架子时应自上而下进行,部件拆除的顺序与安装顺序相反。不允许将拆除的部件直接从高空掷下。应将拆下的部件分品种捆绑后,使用垂直吊运设备将其运至地面,集中堆放保管。

4.2　砌筑材料的准备

4.2.1　标准砖与砌筑前的准备工作

(1)标准砖是 240 mm×115 mm×53 mm 的立方体,这样砌筑 1 m³ 的墙大约需要 512 块砖;由于砍砖的损耗,实际砌筑时的用砖量要多。标准砖各个面的叫法是最大的面叫大面;长的一面叫条面;短的一面叫顶面(或者的叫丁面),当砌体条面朝外的称顺砖;丁面朝外的称丁砖,如图 4.20 所示。

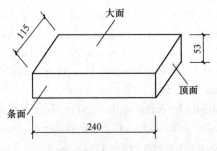

图 4.20　标准砖各个面

(2)由于烧结而成的砖表面有较多的粉尘且极易吸水,因此在使用前要提前浇水润湿,其润湿程度可在现场通过横断面润湿痕迹来判断,一般为 10~15 mm,但浇水湿润也不能使砖浸透,否则因不能吸收砂浆中的多余水分而影响与砂浆的粘结力,还会产生堕灰和砖滑动现象。夏季因水分挥发较快可在操作面上及时补水保持湿润。冬季则应提前润水并保证在使用前晾干表面水分。润水的作用是防止干砖过多吸收砂浆中的水分而降低了砂浆黏结力,同时也会因砂浆缺少生成强度所需的水分而影响其强度。

③Scaffold erection procedure: lay the wooden skid → pull the wire to place the base → erect the gantry from one end, and then install the cross support → install the horizontal frame (or scaffold board)→ install the ladder →(install the large cross bar for strengthening according to the situation)→ install the wall connecting rod → insert the connecting rod → install the upper door frame → install the lock arm → according to the above steps, install upward layer by layer → install the long cross brace to strengthen the overall stiffness → install the top rail.

④Matters needing attention.

a. The verticality and levelness of the first floor portal frame shall be strictly controlled. The vertical deviation of the vertical pole of the gantry in both directions is controlled within 2 mm, and the horizontal deviation of the top of the gantry is controlled within 5mm.

b. The vertical bars of the upper and lower gantry shall be aligned, and the deviation shall not be greater than 3mm.

c. Wall connecting rods shall be installed in time to avoid horizontal deflection of the scaffold.

d. Other matters needing attention: the removal of the scaffold should be carried out from top to bottom, and the order of component removal is opposite to that of installation. It is not allowed to throw the removed parts directly from high altitude. The removed parts shall be bundled by varieties and transported to the ground by vertical lifting equipment for centralized storage.

4.2　Preparation of Masonry Materials

4.2.1　Standard Brick and Preparation before Masonry

(1) The standard brick is a cube of 240 mm×115 mm×53mm, so about 512 bricks are needed to build a 1 m^3 wall; due to the loss of cutting bricks, the actual amount of bricks used is more. The name of each face of the standard brick is that the largest face is called the side; the long side is called the face; the short side is called the end (or called header). When the surface of masonry is outward, it is called stretcher; when the top surface is outward, it is called header, as shown in Fig. 4.20.

(2) Because there is much dust on the surface of the sintered brick and it is easy to absorb water, it should be watered in advance before use. The wetting degree can be judged by the cross-section wetting trace on site, which is generally 10—15mm. However, the brick cannot be soaked by watering, otherwise it will not absorb the excess water in the mortar and affect the bonding force with the mortar, and also cause the phenomenon of falling mortar and brick sliding. In summer, due to the rapid evaporation of water, water can be timely replenished on the operation surface to keep it moist. In winter, water should be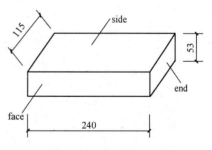

Fig 4.20　Each Face of the Standard Brick

moistened in advance and the surface moisture should be dried before use. The function of moistening is to prevent the dry brick from absorbing too much moisture in the mortar, which reduces the bonding force of the mortar. At the same time, the strength of the mortar will be affected due to the lack of water needed to generate strength.

（3）标准砖因砌筑中需进行模数组合,因此为避免砍砖带来的麻烦和不规则性,在采购中可专门定做 180 mm×115 mm×53 mm 规格的砖,以使组砌更方便且表面观感更规则。

（4）标准砖进场前应及时抽样复检。抽样工作应在现场监理工程师的监督下进行。每一生产厂家的砖进场后按烧结砖 15 万块、多孔砖 5 万块、灰砂砖及粉煤灰砖 10 万块各为一个验收批抽检一组。因抽检结果有滞后性,因而应提前进料并及时安排抽检以保证施工的正常开展。

房屋拆除后的砖剔除表面的砂浆可再利用,但应用在非承重部位。

4.2.2 砌筑砂浆

砂浆是单个的砖块、石块或砌块组合成砌体的胶结材料,同时又是填充块体之间缝隙的填充材料。砌筑砂浆应具备一定的强度、粘结力和工作度(或叫流动性、稠度)。

（1）砂浆的种类。砌筑砂浆是由骨料、胶结料、掺和料和外加剂组成。其强度等级有 M5～M15。砌筑砂浆一般分为水泥砂浆、混合砂浆、石灰砂浆三类。

①水泥砂浆。由水泥和砂子按一定比例混合搅拌而成,它可以配制强度较高的砂浆。水泥砂浆一般应用于基础、长期受水浸泡的地下室和承受较大外力的砌体。

②混合砂浆。由水泥、石灰膏、砂子拌合而成,一般用于地面以上的砌体施工。

③石灰砂浆。由石灰膏和砂子按一定比例搅拌而成的砂浆,完全靠石灰的气硬而获得强度。仅适用于干燥环境中的砌体施工或者临时性砌体施工。石灰砂浆有较好的和易性,有利于抹灰。

（2）砂浆搅拌完成后应在一定的使用时限内用完。对水泥砂浆的使用时限在冬季为 3 h,夏季为 2 h(30 ℃);混合砂浆冬季为 4 h,夏季为 3 h(30 ℃)。在使用时限内,当砂浆的和易性变差时,可以在灰盆内适当掺水拌和恢复其和易性后再使用,超过使用时限的砂浆不允许直接加水拌和使用,以保证砌筑质量。

（3）砂浆试块的制作十分重要,应在现场监理工程师的监督下,在搅拌机出料口随机取样严格按操作规程认真制作,避免因制作不当而不能如实反映砂浆的强度等级,给质量的评定带来麻烦。

(3) The modular combination is required for standard brick masonry. Therefore, in order to avoid the trouble and irregularity caused by brick cutting, 180mm×115mm×53mm bricks can be specially made in the procurement, so as to make the assembly more convenient and the surface appearance more regular.

(4) Standard bricks should be sampled and rechecked before entering the site. Sampling shall be carried out under the supervision of the on-site supervision engineer. After the bricks from each manufacturer enter the site, 150 000 fired bricks, 50 000 porous bricks, 100 000 lime sand bricks and fly ash bricks are selected and inspected as an acceptance batch. Due to the lag of sampling inspection results, it is necessary to send the bricks to the site in advance and arrange sampling inspection in time to ensure the normal development of construction.

Bricks from house breaking can be reused after the mortar on the brick surface is removed, but they can only be used in non load bearing parts.

4.2.2 Masonry Mortar

Mortar is a kind of binding material for masonry composed of single brick, stone or block, and also a filling material for filling gaps between blocks. Masonry mortar should have a certain strength, cohesion and workability(or liquidity, consistency).

(1) Types of mortar. Masonry mortar is composed of aggregate, binder, additive and admixture. Its strength grade is M5—M15. Masonry mortar is generally divided into cement mortar, mixed mortar and lime mortar.

①Cement mortar. Cement mortar is made by mixing cement and sand in a certain proportion. It can prepare mortar with high strength. Cement mortar is generally used in foundation, basement soaked in water for a long time and masonry with large external force.

②Mixed mortar. The mixed mortar is composed of cement, lime paste and sand, which is generally used for masonry construction above the ground.

③Lime mortar. Lime mortar is a kind of mortar made by mixing lime paste and sand in a certain proportion, and its strength is obtained completely by the air hardening of lime. It is only suitable for masonry construction in dry environment or temporary masonry construction. Lime mortar has good workability and is good for plastering.

(2) The mortar should be used up within a certain time limit after mixing. The service life of cement mortar is 3 h in winter and 2 h in summer(30 ℃), 4 h in winter and 3 h(30 ℃)in summer for mixed mortar. During the service life, when the workability of the mortar becomes poor, it can be mixed with water properly in the ash basin to restore its workability before use. The mortar beyond the service life is not allowed to be directly mixed with water to ensure the masonry quality.

(3) The production of mortar test block is very important. Under the supervision of the on-site supervision engineer, random sampling should be carried out at the outlet of the mixer, and the production should be carried out in strict accordance with the operation procedures, so as to avoid that the mortar strength grade cannot be reflected due to improper production, which will bring trouble to the quality evaluation.

(4)每一检验批且不超过 250 mm³ 砌体的各种类型及强度等级的砌筑砂浆,每台搅拌机应至少抽检一次。在搅拌机出料口随机取样制作砂浆试块(每盘砂浆只应制作一组试块)。

4.3 砖砌体施工

4.3.1 砖砌体的组砌形式

实心砖墙根据强度、保温隔热要求和砖的品种、材料的不同,可采用一顺一丁、梅花丁或三顺一丁的组砌形式,也可采用全顺、全丁、两平一侧等组砌形式,如图 4.21 所示。

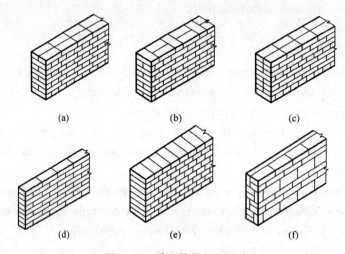

图 4.21 砖砌体的组砌形式
(a)一顺一丁;(b)梅花丁;(c)三顺一丁;(d)全顺;(e)全丁;(f)两平一侧

1. 一顺一丁砖墙的组砌法

这是一种最常见的组砌方法,一顺一丁砌法是由一皮顺砖与一皮丁砖互相间隔砌成,上下皮之间的竖向灰缝互相错开 1/4 砖长。这种砌法效率较高,操作较易掌握,墙面平整也容易控制。缺点是对砖的规格要求较高,如果规格不一致,竖向灰缝就难以整齐。另外在墙的转角、丁字接头和门窗洞口等处都要砍砖,在一定程度上影响了工效。它的墙面组砌形式有两种,一种是顺砖层上下对齐的称为十字缝,另一种顺砖层上下错开 1/2 砖的称为骑马缝。一顺一丁的两种砌法如图 4.22(a)、(b)所示。

(4) For each inspection batch of masonry mortar of various types and strength grades not exceeding 250 mm³, each mixer shall be randomly inspected at least once. At the outlet of the mixer, random samples are taken to make mortar test blocks(only one group of test blocks should be made for each batch of mortar).

4.3 Brick Masonry Construction

4.3.1 The Form of Brick Masonry

According to the strength, thermal insulation requirements and different types and materials of bricks, the solid brick wall can be constructed in the form of English bond, Flemish bond or Flemish garden wall bond, or can also be in the form of full stretcher bond, full header bond, and two stretchers and one soldier bond, as shown in Fig. 4-21.

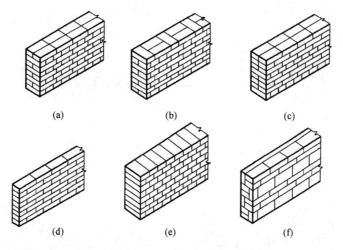

Fig. 4.21 the Form of Brick Masonry
(a) English bond; (b) Flemish bond; (c) Flemish garden wall bond; (d) full stretcher bond;
(e) full header bond; (f) two stretchers and one soldier bond

1. Construction Method of English Bond

This is one of the most common masonry methods. The English bond is composed of one stretcher and one header brick at intervals, and the vertical mortar joints between the upper and lower layers are staggered by 1/4 of the brick length. This method has high efficiency, easy operation and easy control of wall flatness. The disadvantage is that the specification of the brick is high. If the specification is not consistent, it is difficult for the vertical mortar joint to be neat. In addition, in the corner of the wall, T-joint, door and window openings should be cut, which affects the work efficiency to a certain extent. There are two forms of wall assembly: one is the cross joint which is aligned up and down along the brick layer, and the other is called interstice join when the upper and lower bricks are staggered along the brick layer. The two masonry methods of the English bond are shown in Fig. 4.22(a) and (b).

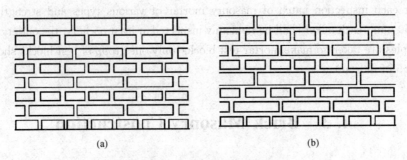

图 4.22　一顺一丁的两种砌法

用这种砌法时,调整砖缝的方法可以采用外七分头或内七分头,但一般都用外七分头,而且要求七分头跟顺砖走。采用内七分头的砌法是在大角上先放整砖,可以先把准线提起来,让同一条准线上操作的其他人员先开始砌筑,以便加快整体速度。但转角处有半砖长的"花槽"出现通天缝,一定程度上影响了墙体质量。一顺一丁墙的大角砌法如图 4.23~图 4.25 所示。

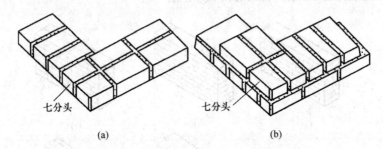

图 4.23　一顺一丁墙的大角砌法(一砖墙)

a)单层数；b)双层数

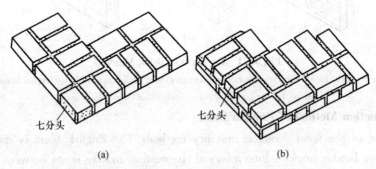

图 4.24　一顺一丁墙的大角砌法(一砖半墙)

(a)单层数；(b)双层数

Chapter 4 Masonry Works Construction

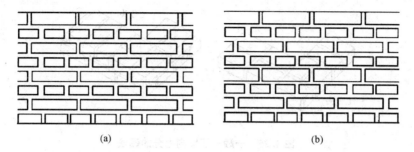

Fig. 4.22 Two Forms of the English Bond
(a) cross joint; (b) interstice joint

When using this kind of masonry method, the method of adjusting the brick joint can use the outer or the inner three-quarter closure brick, but generally use the outer three-quarter closure brick and the three-quarter closure brick is required to follow the stretcher brick. The method of internal three-quarter closure masonry is to put the whole brick on the large corner first. The guide line can be lifted first, and let other personnel on the same guide line start masonry first, so as to speed up the overall speed. But there is a half brick long vertical joint at the corner, which affects the quality of the wall to a certain extent. The large angle masonry method of the English bond is shown in Fig. 4.23—Fig. 4.25.

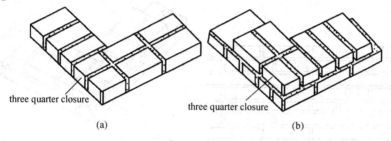

Fig 4.23 The Large Angle Masonry Method of the English Bond
(a) single course; (b) double-course

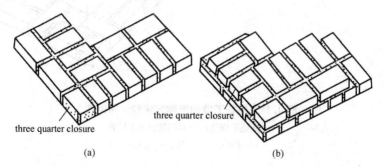

Fig. 4.24 The Large Angle Masonry Method of the English Bond
(a) single course; (b) double-course

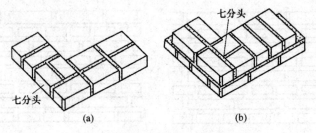

图 4.25　一顺一丁墙内七分的砌法
(a)单层数；(b)双层数

2. 梅花丁砌法

梅花丁砌法又称沙包式或十字式砌法，是在同一皮砖上采用两块顺砖夹一块丁砖的砌法。上皮丁砖坐中于下皮顺砖，上下两皮砖的竖向灰缝错开 1/4 砖长。梅花丁砌法的内外竖向灰缝每皮都能错开，竖向灰缝容易对齐，墙面平整度容易控制，特别是当砖的规格不一致时（一般砖的长度方向容易出现超长，而宽度方向容易出现缩小的现象），更显出其能控制竖向灰缝的优越性。这种砌法灰缝整齐，美观，尤其适宜于清水外墙。但由于顺砖与丁砖交替砌筑，影响操作速度，工效较低。梅花丁的组砌方法如图 4.26 所示。

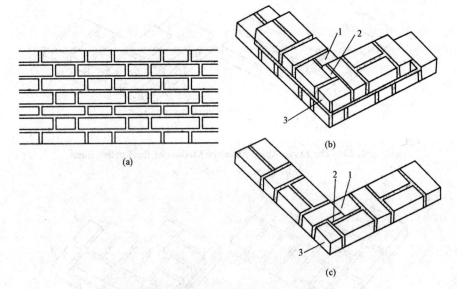

图 4.26　梅花丁墙的组砌方法和大角砌法
(a)梅花丁砌法；(b)双层数；(c)单层数
1—半砖；2—1/4 砖；3—七分头

Chapter 4 Masonry Works Construction

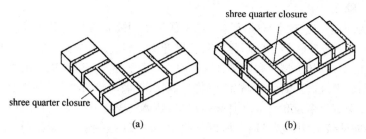

Fig. 4.25 The Internal Three-quarter Closure Masonry Method of the English Bond
(a) single course; (b) double-course

2. Flemish Bond

Flemish Bond masonry is also called sand bag or cross bond masonry, which is a method of two stretchers and one header on the same course. The vertical mortar joint of the upper and lower bricks is staggered by 1/4 of the length of the brick. The internal and external vertical mortar joints of the Flemish Bond masonry method can be staggered. The vertical mortar joints are easy to align, and the flatness of the wall surface is easy to control. Especially when the specifications of bricks are not consistent (the length direction of general bricks is easy to be super long, while the width direction is easy to shrink), the advantages of this method are more obvious. The mortar joints of this method are neat and beautiful, especially suitable for fair faced exterior walls. However, the operation speed is affected and the work efficiency is low due to the alternate laying of stretcher and header. The assembly method of the Flemish Bond is shown in Fig. 4.26.

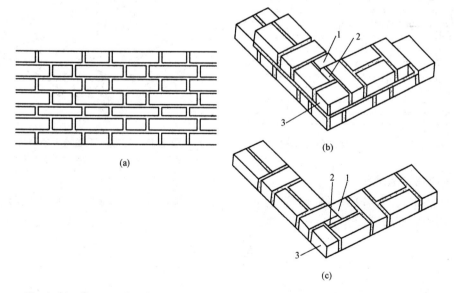

Fig. 4.26 Construction Method and Large Angle Masonry Method of Flemish Bond
(a) Flemish Bond; (b) double-course; (c) single course
1—half brick; 2—1/4 brick; 3—three-quarter brick

3. 三顺一丁砌法

三顺一丁砌法为采用三皮全部顺砖与一皮全部丁砖间隔砌成的组砌方法。上下皮顺砖间竖缝错开1/2砖长，上下皮顺砖与丁砖间竖向灰缝错开1/4砖长，同时要求山墙与檐墙(长墙)的丁砖层不在同一皮砖上，以利于错缝和搭接。这种砌法一般适用于一砖半以上的墙。这种砌法顺砖较多，砖的两个条面中挑选一面朝外，故墙面美观，同时在墙的转角处、丁字和十字接头处及门窗洞等处砍凿砖少，砌筑效率较高。缺点是顺砖层多，墙体的整体性较差。特别是砖比较潮湿或者砂浆较稀时容易向外挤出，出现"游墙"，所以与此相同缺点的五顺一丁砌法现在不用了。三顺一丁砌法一般以内七分头调整错缝和搭接。三顺一丁组砌形式如图 4.27 所示。三顺一丁砌法的大角做法如图 4.28 所示。

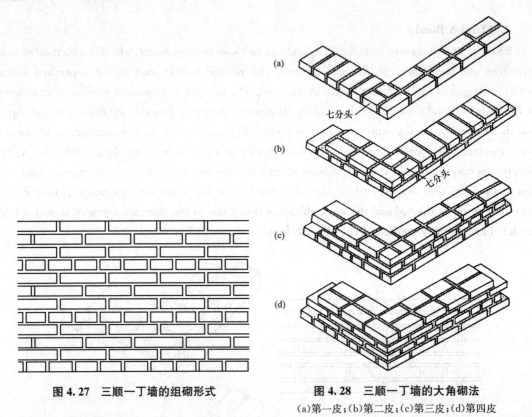

图 4.27 三顺一丁墙的组砌形式

图 4.28 三顺一丁墙的大角砌法
(a)第一皮；(b)第二皮；(c)第三皮；(d)第四皮

4. 其他几种组砌形式

(1)全顺砌法。全部采用顺砖砌筑，上下皮间竖向灰缝错开1/2砖长。这种砌法仅适用于砌半砖墙，如图 4.29 所示。

Chapter 4　Masonry Works Construction

3. Flemish Garden Wall Bond

The Flemish Garden Wall Bond is a group masonry method in which all the stretchers of three courses and all the headers of one course are laid at intervals. The vertical joints between the upper and lower stretcher courses are staggered by 1/2 of the brick length, and the vertical mortar joints between the upper and lower stretcher courses and the header courses are staggered by 1/4 of the brick length. At the same time, the header course of gable and eaves wall (long wall) is not on the same brick courses, so as to facilitate the staggered joint and overlap. This masonry method is generally applicable to more than one and a half brick walls. In this method, more stretchers are laid, and one of the two faces of the brick is selected to face outwards, so the wall is beautiful. At the same time, there are less bricks cut at the corner of the wall, T-shaped and cross joints, door and window openings, etc., and the masonry efficiency is high. The disadvantage is that there are more stretcher layers and the integrity of the wall is poor. In particular, when the brick is wet or the mortar is relatively thin, it is easy to extrude out, resulting in "floating wall". Therefore, the "five stretchers one header" method with the same shortcomings is not used now. Generally, the staggered joint and lap joint shall be adjusted within three-quarter brick for the Flemish Garden Wall Bond. The form of the Flemish Garden Wall Bond is shown in Fig. 4.27. The large angle construction method of the Flemish Garden Wall Bond is shown in Fig. 4.28.

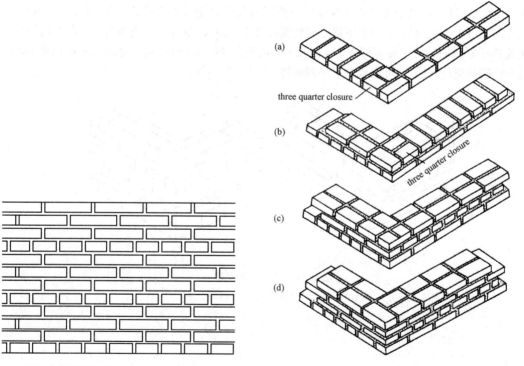

Fig. 4.27　Flemish Garden Wall Bond

Fig. 4.28　the Large Angle of Flemish Garden Wall Bond
(a) the first course; (b) the second course;
(c) the third course; (d) the fourth course

4. Other Forms of Masonry Bond

(1) Full stretcher bond. Full stretcher bond means that stretcher is laid all along and the vertical mortar joint between upper and lower layer is staggered by 1/2 brick length. This method is only suitable for half brick walls, as shown in Fig. 4.29.

· 249 ·

(2)全丁砌法。全部采用丁砖砌筑,上下皮间竖缝相互错开 1/4 砖长;这种砌法仅适用于砌圆弧形砌体,如烟囱、窨井等。一般采用外圆放宽竖缝,内圆缩小竖缝的办法形成圆弧。当烟囱或窨井的直径较小时,砖要砍成楔形砌筑,如图 4.30 所示。

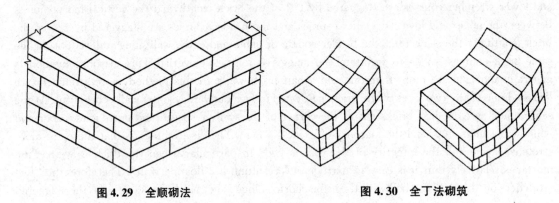

图 4.29　全顺砌法　　　　　　　图 4.30　全丁法砌筑

(3)二平一侧砌法。二平一侧砌法是采用二皮平砌与一皮侧砌的顺砖相隔砌成。这种砌法较费工,但节约用砖,仅适用于 180 mm 或 300 mm 厚的墙。连砌二皮顺砖(上下皮竖向灰缝相互错开 1/2 砖长),背后贴一侧砖(平砌层与侧砌层的竖向灰缝也错开 1/2 砖长)就组成了 180 mm 厚墙。连砌二皮丁砖或一顺一丁(上下皮之间竖缝错开 1/4 砖长)背后贴一侧砖(侧砖层与顺砖层之间竖缝错开 1/2 砖长,与丁砖层错开 1/4 砖长)就组成了 300 mm 厚的墙。每砌二皮砖以后,将平砌砖和侧砌砖里外互换,即可组成二平一侧墙体,如图 4.31 所示。

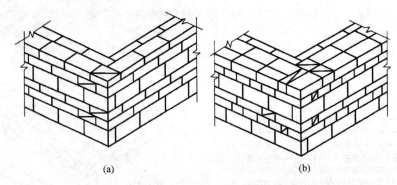

(a)　　　　　　　　　(b)

图 4.31　二平一侧砌法
(a)180 mm 厚墙体;(b)300 mm 厚墙体

(2) Full header bond. Full header bond means that header is laid all along and the vertical joints between upper and lower layer are staggered by 1/4 brick length. This method is only suitable for building circular arc masonry, such as chimney, inspection well, etc. Generally, circular arc is formed by relaxing the vertical joint on the outer circle and reducing the vertical joint in the inner circle. When the diameter of chimney or inspection well is small, the brick should be cut into wedge-shaped masonry, as shown in Fig. 4. 30.

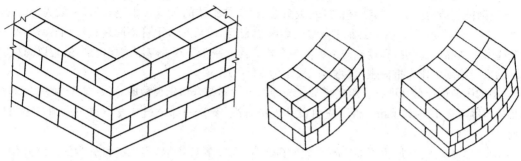

Fig. 4. 29　the Full Stretcher Bond　　　　　Fig. 4. 30　the Full Header Bond

(3) Two stretchers and one soldier bond. The two stretchers and one soldier bond is made of two paralleled bricks and one brick laid on its face with its side paralleled to the wall. This kind of masonry method is more labor-consuming, but saves bricks, and is only suitable for 180mm or 300mm thick walls. The 180mm thick wall is composed of two parallel bricks (the vertical mortar joints of the upper and lower courses are staggered by 1/2 of the brick length), and one side of the brick is pasted on the back (the vertical mortar joints of the paralleled bricks and side courses are also staggered by 1/2 brick length). The 300 mm thick wall is composed of two courses of headers or one stretcher one header (the vertical joints between the upper and lower layers are staggered by 1/4 of the brick length) and pasted with one side of the brick (the vertical joints between the side brick course and the stretcher course are staggered by 1/2 of the brick length, and staggered 1/4 of the brick length with the header layer). After each two courses of bricks are laid, the horizontal bricks and side bricks are exchanged inside and outside to form the wall of two stretchers and one soldier bond, as shown in Fig. 4. 31.

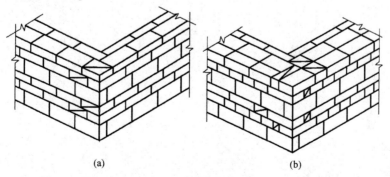

(a)　　　　　　　　　　　　　(b)

Fig. 4. 31　Two Stretchers and One Soldier Bond
(a) 180mm thick wall; (b) 300mm thick wall

上述各种砌法中,每层墙的最下一皮和最上一皮,在梁和梁垫的下面、墙的台阶水平面上均应用丁砖层砌筑。

5. 砌筑的基本要求及注意事项

(1)砌体上下层砖之间应错缝搭砌。搭砌长度为1/4砖长,方法如上述各图所示。为保证砌体的结构性,组砌时第一层和砌体顶部的一层砖为丁砌。

(2)砌体转角和内外墙应相互搭砌咬合,以保证有较好的结构整体性。

(3)砌体的结构性能与灰缝有直接的关系,因此要求砌体的灰缝大小应均匀,一般为10 mm,不大于12 mm,不小于8 mm,其水平灰缝的砂浆饱满度应≥80%,用百格网随进度抽查评定。竖向灰缝以不透亮,无瞎缝,有较好的饱满度为合格。为控制水平灰缝厚度的均匀度,通常以砌筑10匹砖的累计高度值作为指标进行检查,一般定为63~64 cm。

(4)砌体的设计尺寸应符合组砌模数。当建筑尺寸不符合组砌模数时,可在施工中进行调整,调整的范围应在10~20 mm。否则应由设计人员解决,调整的部位通常发生在门窗洞口和门垛等处。

(5)墙体纵横连接处,因不能或不易同时砌筑时,可留置施工缝,施工缝可分为踏步和外伸直槎式两种,施工缝高度不能超过一个楼层且≤4 m,有抗震设防要求或其他要求的,应按其规定处理。如加拉结钢筋,钢筋网片等,具体参见施工及验收规范。一般情况下宜留斜槎(图4.32)。

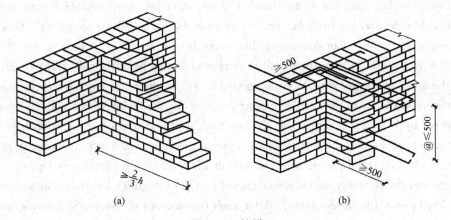

图 4.32 接槎
(a)斜槎;(b)直槎

In all the above mentioned masonry methods, the bottom and the top of each course of wall shall be constructed with header course under the beam and beam pad, and on the horizontal surface of the wall steps.

5. Basic Requirements of Masonry and Matters Needing Attention

(1) The upper and lower courses of masonry should be built by staggered joints. The length of staggered bond is 1/4 brick length, and the method is shown in the above drawings. In order to ensure the structure of masonry, the first course and the top course of masonry are header bond.

(2) Masonry corner and internal and external walls should be overlapped with each other to ensure better structural integrity.

(3) The structural performance of masonry is directly related to mortar joint, so the size of mortar joint of masonry shall be uniform, generally 10 mm, no more than 12 mm, no less than 8 mm, and the mortar plumpness of horizontal mortar joint shall be more than or equal to 80%, and bed joint mortar fullness measuring grid shall be used for random inspection and evaluation along with the progress. The vertical mortar joint is qualified if it is not transparent, has no blind joint and has good plumpness. In order to control the uniformity of the thickness of horizontal mortar joint, the accumulative height value of 10 masonry bricks is usually taken as the index for inspection, which is generally set as 63—64cm.

(4) The design size of masonry shall conform to the module of masonry. When the building size does not conform to the building module, it can be adjusted in the construction, and the adjustment range should be between 10 and 20mm. Otherwise, it should be solved by the designer, and the adjustment position usually occurs at the door and window opening and the door stack.

(5) When the vertical and horizontal joints of the wall cannot or are not easy to be built at the same time, the construction joints can be reserved, which are divided into two types: step type and extended toothing of brick wall type. The height of the construction joint shall not exceed one floor and is not more than 4m. If there are seismic fortification requirements or other requirements, they shall be treated according to the provisions, such as steel bar, steel mesh, etc. , refer to construction and acceptance specification for details. In general, stepped racking should be reserved(Fig. 4. 32).

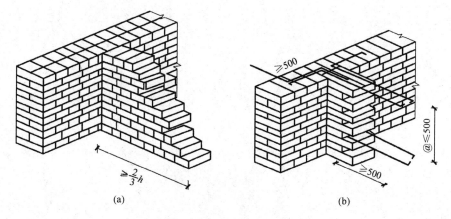

Fig. 4. 32 Racking Bond
(a) stepped racking; (b) serrated racking

(6)砌体的每班次或每日的砌筑高度应有一定的限制。防止因气候的变化或人为碰撞而发生变形和倾覆。同时砌筑高度的速度过快,砌体会因底部砂浆压缩变形而造成砌体的变形。具体要求可参照施工规定。如因工序需求且操作面特征允许,施工人员可根据施工经验来确定当日完成的高度。

(7)构造柱处的砌筑方法按构造要求应砌成大马牙槎(图4.33)。

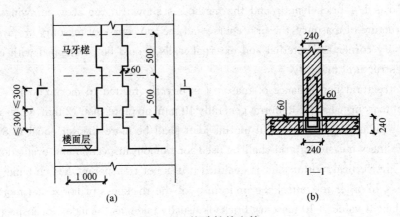

图 4.33 构造柱的砌筑
(a)立面图;(b)剖面图

标准砖每砌筑五层的高度为315~320 mm,为方便组砌,因此通常采用"五退五进"的砌筑方法,应先退砌后进砌,同时将拉结钢筋按设计要求砌入墙体中。

当两相邻的构造柱之间墙体净宽度≤365 mm时,造成施工困难,且不能保证此处墙体的稳定性;在安装模板中,极易损伤其结构性,因此应在图纸会审中提出建议,改为用混凝土与构造柱同时浇筑或构造柱分段浇筑的方法来解决,具体应由设计人员确定。

(8)砌筑过程中,脚手架支撑在墙上搭设时应注意以下几点。刚砌筑完成的墙上,门窗洞口边50 cm 范围内,空斗墙上,单砖墙上均不得用来支撑脚手架,单排脚手架钢管在墙上的支撑长度不应小于18 cm。

(6) The masonry height of each shift or each day should be limited to prevent deformation and overturning due to climate change or man-made collision. At the same time, if the masonry height is too fast, the masonry will deform due to the compression deformation of the bottom mortar. Specific requirements can refer to construction regulations. If the process needs and the characteristics of the operation surface permit, the construction personnel can determine the height of the day according to the construction experience.

(7) According to the structural requirements, the masonry method at the structural column should be built into big indented toothing(Fig. 4.33).

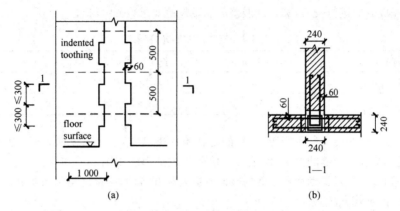

Fig. 4.33 **Masonry of the Structural Column**
(a) elevation; (b) section

The height of each five courses of standard brick is about 315—320mm. In order to facilitate the construction, the masonry method of "five back and five forward" is usually adopted. The back masonry should be carried out before the forward masonry. At the same time, the tie bar should be built into the wall according to the design requirements.

When the net width of the wall between two adjacent structural columns is less than or equal to 365mm, the construction is difficult, and the stability of the wall cannot be guaranteed. In the installation of formwork, the structure is easily damaged. Therefore, suggestions should be put forward in the joint review of drawings, which should be solved by pouring concrete and constructional column at the same time or pouring the structural column by sections, which shall be determined by the designer.

(8) In the process of masonry, the following points should be paid attention to when the scaffold support is erected on the wall. The newly built wall, within 50cm of the door and window opening, the empty bucket wall and the single brick wall shall not be used to support the scaffold, and the support length of single row scaffold steel pipe on the wall shall not be less than 18cm.

4.3.2 砖砌体的施工工艺

抄平放线(绑扎构造柱钢筋)→试摆砖→立匹数杆→组砌→清理。

(1)抄平放线。当垫层施工完成,应在四大角及平面几何特征变化处和立匹数杆的位置上,测出上述各确定点的实际高程,找出与设计标高的相对差值并标注在垫层上,同时记录在放线记录图上,以便确定和计算砌筑高度、灰缝厚度和组砌层数,保证砌体上口标高一致。放线则是利用控制桩找出基础中心线及交点,然后用墨斗弹出所需的线。通常是在垫层上弹出大放脚底宽边线;而轴线只需在基础大放脚底宽之外,引测弹出轴线线段即可,即方便基础砌筑也方便轴线控制与校核。放线完成后应进行校验,其允许误差应满足规范要求(表4-2)。

表4-2 放线尺寸的允许偏差

长度L、宽度B/m	允许偏差/mm	长度L、宽度B/m	允许偏差/mm
L(或B)≤30	±5	60<L(或B)≤90	±15
30<L(或B)≤60	±10	L(或B)>90	±20

(2)试摆砖。试摆砖又称干摆砖,其目的是在墙体砌筑前,沿墙的纵横方向,特别是在内外墙交接处,通过调节竖向灰缝的宽窄,保证每一层砖的组砌都能规则统一并符合模数。当砖的长宽尺寸有正负差时,要注意丁砌和顺砌砖的竖向灰缝要相互协调,尽量避免竖向灰缝大小不匀,否则应采用梅花丁的砌筑方法。

(3)立匹数杆。为了保证砌体在高度上层数统一,并控制灰缝大小和砌筑标高,墙体砌筑前应立匹数杆。匹数杆一般为木制,上面画有砖和灰缝的厚度、层数,门窗洞口及梁板构件标高位置等高度标识。用于基础施工的通常称小匹数杆。安装时,应根据垫层表面各点标高值,确定一个底平面标高值,以此确定值为依据,调整匹数杆标识并安放匹数杆,保证各点的组砌模数和上口标高一致。

4.3.2 Construction Process of Brick Masonry

Leveling and setting out (binding structural column reinforcement) → trial placing bricks → erecting story poles → assembling → cleaning.

(1) Leveling and setting out. When the cushion construction is completed, the actual elevation of the above-mentioned determination points shall be measured at the four corners, the change of plane geometric characteristics and the position of erecting story poles. The relative difference between the two points and the design elevation shall be found out and marked on the cushion. At the same time, it shall be recorded on the setting out record chart, so as to determine and calculate the masonry height, mortar joint thickness and the number of layers, so as to ensure the consistent elevation of the upper opening of masonry. Setting out is to use the control pile to find out the center line and intersection point of the foundation, and then pop up the required line with ink fountain. Generally, it is necessary to pop up the wide side line of stepped footing on the cushion; and the axis only needs to be outside the bottom width of stepped footing of the foundation, and the axis line can be drawn out by surveying, which is convenient for foundation masonry and axis control and check. Calibration shall be carried out after setting out, and the allowable deviation shall meet the requirements of specification (Table 4.2).

Table 4.2 Allowable Deviation of Setting out Dimension

Length L, width B/m	Allowable Deviation/mm	Length L, width B/m	Allowable Deviation/mm
L(or B)≤30	±5	60<L(or B)≤90	±15
30<L(or B)≤60	±10	L(or B)>90	±20

(2) Trial placing bricks. Trial laying brick is also known as dry laid brick. Its purpose is to ensure that the masonry of each layer of brick can be regular and uniform and conform to the modulus by adjusting the width of vertical mortar joint along the vertical and horizontal direction of the wall, especially at the junction of internal and external walls. When there is a positive and negative difference in the length and width of the brick, attention should be paid to the coordination of the vertical mortar joints between the joint and the brick laying, and try to avoid the uneven size of the vertical mortar joint, otherwise the Flemish bond should be adopted.

(3) Erecting story poles. In order to ensure that the number of upper layers of masonry is uniform, and to control the size of mortar joint and masonry elevation, the story poles should be erected before masonry. The story pole is generally made of wood, with height marks such as the thickness of brick and mortar joint, the number of layers, the elevation of door and window openings and beam and slab components. It is usually called small story pole for foundation construction. During installation, a bottom plane elevation value shall be determined according to the elevation value of each point on the cushion surface. Based on the determined value, the mark of the story pole shall be adjusted and the story pole shall be placed to ensure the consistency of the building modulus of each point and the upper opening elevation.

(4)大放脚砌筑。大放脚底层应采用丁砖砌筑并由低标高处开始,保证砌体搭砌合理和表面标高一致。砂浆应满铺,作到砂浆饱满,组砌正确,收阶对称,均匀一致,其每层的轴线偏差应不大于 10 毫米。不准用填芯码槽的方式砌筑。当退台收阶到 365 宽时,为防止偏差,应利用垫层上的轴线标志,采用线锤吊线校核其轴线,也可以利用控制桩采用经纬仪和拉通线的方法进行校核。

砌体施工时如需要留施工缝应以踏步槎为主,如需留直槎(凸槎),应按规定加拉结钢筋。

基础上如需预留穿墙管线的洞口,其位置和尺寸应计算准确,一般要求洞口上口标高高于管道上口标高 50 mm。基础上如设计有防潮层,则应按设计要求施工,一般采用 1∶2(内按水泥用量的 5‰加防水粉)水泥砂浆,在基础顶面按找平抹灰的方式施工。

(5)清理。基础完成后及时清理砖缝,墙面及落地砂浆等。

4.3.3 砌砖的技术要求

为保证墙面平整、垂直、灰缝均匀、层数一致,砌筑中可在四大角或相应的位置上立放匹数杆,匹数杆的立放依据是给定的标高点。最好是在下一层圈梁外侧,作出准确的标高点或画在构造柱钢筋上,依此点安放匹数杆,保证其竖向标高和尺寸的准确。与此同时用水平仪检查已作好的结构几大角或结构特征变化处的楼板面实际标高,如发现其误差值(与设计标高)较大,影响到砌体组砌层数的统一,则应用细石混凝土填平补齐后,再在上面试摆砖。特别应注意留窗洞处,此处可以少砌一层砖,但以后的各层砖与其他部位水平一直。试摆砖完成,匹数杆安好就可以依匹数杆拉线砌墙。拉线的长度一般不大于 10 m,否则应沿长度方向作挑线点,以保证线的平直。一般采用单股 0.6~0.8 mm 粗的尼龙线。为使墙体观感效果更好,可采用双面拉线的砌筑方法。为保证墙面的垂直度应按"三线一吊""五线一靠"的操作方法加以保证。所谓"三线一吊",是指砌筑完三层砖高时,就用吊线锤校验墙面垂直度,而"五线一靠"则是砌筑完五层砖高时,用靠线板(鱼尾板)垂直靠着墙的楞角检查校正垂直度。

(4) Stepped footing masonry. The bottom layer of stepped footing masonry shall be constructed with headers starting from the low elevation to ensure that the masonry construction is reasonable and the surface elevation is consistent. The mortar shall be fully paved until the mortar is full, the assembly is correct, the steps are symmetrical and uniform and the axis deviation of each layer shall not be greater than 10 mm. It is not allowed to use the method of filling core and laying slot for masonry. In order to prevent the deviation, the axis mark on the cushion should be used to check the axis with plumb hammer, or the control pile can be used to check by theodolite and pulling straight line method.

During masonry construction, if it is necessary to reserve construction joints, the stepped racking shall be the main method; if the serrated racking (convex racking) needs to be reserved, the reinforcement shall be added according to the regulations.

If it is necessary to reserve the hole through wall pipeline on the foundation, its position and size shall be calculated accurately. Generally, the elevation of the upper opening of the hole shall be 50mm higher than that of the pipe. If there is a damp proof layer designed on the foundation, the construction shall be carried out according to the design requirements. Generally, 1 : 2 cement mortars (5% of the cement amount plus waterproof powder) shall be used, and the construction method of leveling and plastering shall be adopted on the top surface of the foundation.

(5) Cleaning. After the foundation is completed, the brick joint, wall and floor mortar shall be cleaned in time.

4.3.3 Technical Requirements for Bricklaying

In order to ensure that the wall surface is flat and vertical, mortar joints are uniform and the number of layers is consistent, the story pole can be erected on the four corners or corresponding positions in the masonry, and the erection basis of the story pole is the given elevation point. It is better to make an accurate elevation point or draw it on the reinforcement of the structural column outside the ring beam of the next floor, and place the story pole according to this point to ensure the accuracy of its vertical elevation and size. At the same time, use the level to check the actual elevation of the finished structural corners or the floor slab surface where the structural characteristics change. If it is found that the error value (with the design elevation) is large, which affects the unity of masonry layers, fine aggregate concrete shall be used for filling and leveling, and then the bricks shall be placed on the upper floor. Special attention should be paid to the window opening, where a less layer of bricks can be laid, but the later layers of bricks are always level with other parts. After the trial placing of bricks is completed, the wall can be built according to the tie line of the story poles. The length of the tie line is generally no more than 10m. Otherwise, take-up points should be made along the length direction to ensure the straightness of the line. Generally, the single strand of 0.6—0.8mm thick nylon thread is used. In order to make the appearance of the wall better, the masonry method of double-sided tie line can be used. In order to ensure the perpendicularity of the wall, the operation method of "three layers and one hanging" and "five layers and one leaning" should be adopted. The so-called "three layers and one hanging" means that when the three layers of brick height are completed, the plumb is used to check the verticality of the wall surface, while the "five layers and one leaning" means that when the five layers of brick height are completed, the running plate (fishtail plate) is used to check and correct the verticality of the wall.

砌体的细部做法

(1)门窗洞口。门窗洞口的检查指标主要是高宽度误差。而标高位置则是必须得到保证的内容,特别是上口标高一定要准确,检查时通常按墙面正50线为参照点确定其高度方向的误差值,因而施工时一定要注意。门洞口的控制与楼板面标高和过梁标高均有关,楼面标高因施工误差出问题要多一些。

(2)管线槽的砌筑。在砌体施工中,墙内预埋水管、电线管是常见的现象。为保证砌体良好的结构性,应在承重较大的墙体部位采用合理的砌筑方法,以确保结构安全,通常采用下列处理方法。

①针对水平管道可在墙体中埋开槽的预制砌块,砌块长240 mm,同墙厚,高190 mm,管道后嵌固安装进去。

②单根PVCϕ20以内的管,可在墙中预埋,宜在墙厚1/2处埋设,以防止砌体偏心受压,此做法少砍砖,在顺砖中埋设较为理想,如为丁砖中埋设,则要求下一层砖为丁砖。管道周边应用砂浆填实。

③竖向暗装管道。竖向暗装管道单根的在墙厚1/2处,多根且因配电箱关系,需平墙面埋设,且在结构重要部位或管道较大较多处,应采用在墙中安放水平墙拉筋,并在竖向安模板浇灌混凝土的方法来补强。严禁在墙中开槽埋管,防止因墙体厚度削弱而影响结构安全。

(3)窗台的砌筑。窗台线在室内的高度应在抹灰及地面装饰完成后达到90 cm,因此砌筑时考虑到窗外流水坡面的关系,一般都预留40~50 mm作为抹灰找坡的空间,因此,一般应在楼层结构面上砌筑14层砖。安装前门窗供应商应在现场校核预留洞口尺寸,以确定合理的加工尺寸,保证窗框安装与装饰抹灰相互协调。

4.3.4 砖混结构中的混凝土构件施工

1. 构造柱

构造柱在墙体结构中与圈梁共同作用,提高结构的整体刚度,特别是竖向刚度,满足其抗震要求。

构造柱的布置主要在建筑物的四大角和内外墙交接处及其他按设计需要布置的位置。因此构造柱的模板安装形式和方法较多。

Details of Masonry

(1) Door and window openings. The inspection index of door and window openings is mainly height and width error. The elevation position is the content that must be guaranteed; especially the upper opening elevation must be accurate. When checking, the error value of the height direction is usually determined according to the height of 500mm of the wall as the reference point, so we must pay attention to it during construction. The control of the door opening is related to the floor surface elevation and lintel elevation, and the floor elevation has more problems due to construction error.

(2) Masonry of pipe duct. In masonry construction, it is a common phenomenon to embed water pipe and wire pipe in the wall. In order to ensure the good structure of masonry, reasonable masonry method should be adopted in the wall with large bearing capacity to ensure the structural safety. The following treatment methods are usually adopted.

① For horizontal pipeline, slotted precast block can be embedded in the wall. The block length is 240mm, the same thickness as the wall and the height is 190mm, and the pipe is embedded and installed afterwards.

② A single PVC pipe less than $\phi 20$ can be embedded in the wall, and it should be buried at 1/2 of the wall thickness to prevent the masonry from eccentric compression. In this way, less bricks are cut, and it is ideal to bury in the stretcher. If the pipe is embedded in the header, the next layer of brick is required to be header. The periphery of the pipeline shall be filled with mortar.

③ Vertical Concealed Pipeline. The vertical concealed pipe is located at 1/2 of the wall thickness. Multiple pipes due to the distribution box should be embedded in flat wall. In addition, horizontal wall tie bars should be placed in the wall and concrete should be poured in the vertical formwork for reinforcement in important parts of the structure or large number of pipelines. It is strictly forbidden to slot and bury pipes in the wall to prevent the structural safety from being affected by the weakening of wall thickness.

(3) Masonry of window sill. The indoor height of the window sill line should reach 90cm after plastering and floor decoration. Therefore, considering the relationship between the water slope outside the window and the slope surface, 40—50mm is generally reserved as the space for plastering and slope finding. Therefore, 14 layers of bricks should be laid on the structural surface of the floor. Before installation, the door and window supplier shall check the reserved hole size on site to determine the reasonable processing size and ensure the coordination of window frame installation and decoration plastering.

4.3.4 Construction of Concrete Components in Brick Concrete Structure

1. Structural Column

The structural column acts with the ring beam in the wall structure to improve the overall stiffness, especially the vertical stiffness, to meet the seismic requirements.

The structural column is mainly arranged at the four corners of the building and the junction of the internal and external walls and other positions according to the design requirements. Therefore, there are many forms and methods for formwork installation of constructional column.

构造柱的模板通常都采用木模板或定型组合钢模板。宜优先采用木模板,其接缝少,易加工安装,浇注出来的混凝土表面观感好。

为了保证模板安装质量,在墙体施工中,应解决好连接和固定模板的技术措施。通常采用下面的方法。

(1)在构造柱两侧墙体中留孔。在构造柱两侧墙体中留孔如图 4.34 所示。

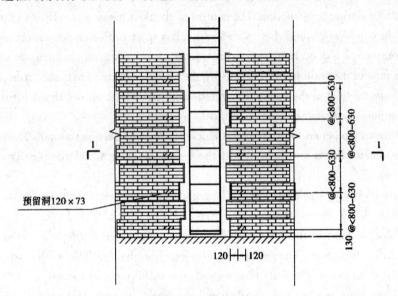

图 4.34　在构造柱两侧墙体中留孔示意图

(2)在墙体中预埋环形钢筋箍。在墙体中预埋环形钢筋箍如图 4.35 所示。

钢筋环箍长度 L＝墙厚＋模板厚＋夹固料表面尺寸＋安装间隙(30 mm)

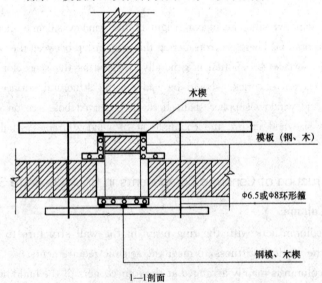

图 4.35　在墙体中预埋环形钢筋箍示意图

Chapter 4 Masonry Works Construction

The formwork of constructional column usually adopts wood formwork or set combined steel formwork. Wood formwork should be preferred, with few joints, easy processing and installation, and good appearance of concrete surface.

In order to ensure the quality of formwork installation, the technical measures of connecting and fixing formwork should be solved in the wall construction. The following method is usually used.

(1) Holes are reserved in the walls on both sides of the structural column. Holes are reserved in the walls on both sides of the constructional column as shown in Fig. 4.34.

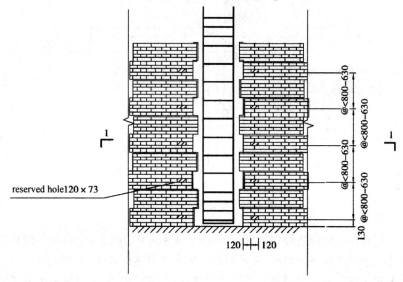

Fig. 4. 34 Schematic Diagram of Holes Reserved in Walls on Both Sides of Constructional Column

(2) Embedding ring reinforcement hoop in wall. The ring reinforcement hoop is embedded in the wall as shown in Fig. 4. 35.

Steel hoop length L = wall thickness + formwork thickness + surface dimension of clamping material + installation gap (30 mm)

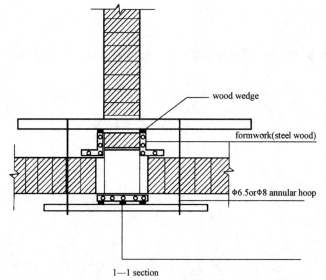

Fig. 4. 35 Schematic Diagram of Embedding Ring Reinforcement Hoop in Wall

(3)内外墙交接处构造柱做法。内外墙交接处构造柱做法如图4.36所示。

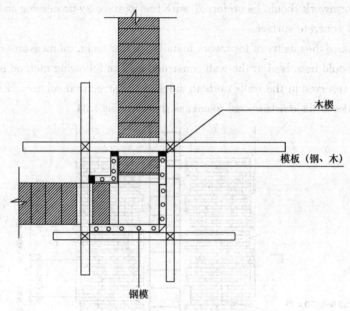

图4.36 转角处构造柱做法

环形箍直接埋在构造柱两边墙体的竖缝中,钢管或木方穿过环形箍后,应与模板紧密接触,可以在钢管或木方之间打木楔楔紧模板,在环形箍中加木楔楔紧钢管(图4.36)。

构造柱的混凝土浇筑应注意把握好单根浇筑速度,不宜过快,防止因混凝土自然固结沉陷,造成构造柱上部混凝土因受模板约束而产生表面裂纹。施工中可以采用二次捣固法防止此问题产生。同时应注意调整好钢筋露出部分的位置,防止偏移。

2. 圈梁

当圈梁上要安放预制楼板,其圈梁的施工方法同地圈梁,注意楼板上的夹具的安装间距。并在安放夹具的位置在模板内上口加内顶撑,保证夹具在紧固时模板的上口宽度满足设计要求。内顶撑随混凝土浇筑过程逐个取掉,周转使用。

3. 现浇楼板

现浇楼板与圈梁多为整浇。其模板安装形式一般可采用图4.37所示的方式。

(3) Construction method of structural column at the junction of internal and external walls. The method of structural column at the junction of internal and external walls is as shown in Fig. 4. 36.

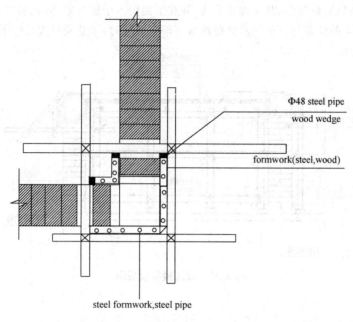

Fig. 4. 36 Construction Method of Structural Column at Corner

The ring hoop is directly embedded in the vertical joints of the walls on both sides of the structural column. After the steel pipe or wooden square passes through the ring hoop, it should be in close contact with the formwork. Wooden wedge can be used to wedge the formwork between the steel pipe or the wooden square, and the wooden wedge is added to the ring hoop to wedge the steel pipe(Fig. 4. 36).

Attention should be paid to the concrete pouring speed of single column, which should not be too fast, so as to prevent the surface cracks of the upper concrete of the structural column due to the constraint of the template due to the natural consolidation and settlement of the concrete. The second tamping method can be used to prevent this problem. At the same time, attention should be paid to adjust the position of exposed part of reinforcement to prevent deviation.

2. Ring Beam

When the precast floor slab is to be placed on the ring beam, the construction method of the ring beam is the same as that of the ground ring beam. Pay attention to the installation spacing of the clamps on the floor. And in the position where the fixture is placed, the inner top support is added at the upper opening of the formwork to ensure that the width of the upper opening of the formwork meets the design requirements when the fixture is tightened. The inner top support is removed one by one along with the process of concrete pouring and used in turn.

3. Cast-in-place Floor

The cast-in-place floor slab and ring beam are mostly integral casting. Generally, the installation form of formwork can be as shown in Fig. 4. 37.

模板的支撑体系应视楼板跨度大小及厚度,合理布置立杆纵横间距。一般为 800~1 000 mm,三道水平横杆,其中一道为扫地杆,距支承面约 200 mm,使支撑体系有较好的整体刚度和稳定性。

底模用层板或竹胶板制作,其接缝应平接,并在上面用不干胶封贴,防止漏浆。为防止模板移位,可用钉子在适当的位置与下面的木格栅连接。模板拆除时,注意及时拔出钉子,防止伤人。

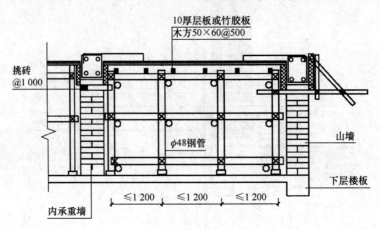

图 4.37 现浇楼板模板图

4.4 砌块施工

砌块代替黏土砖作为墙体材料,是墙体改革的一个重要途径。近几年来各地因地制宜,就地取材,以天然材料或工业废料为原材料,做成各种规格的砌块用于建筑物墙体结构,目前以小型混凝土砌块为主。砌块施工方法简易,改变了手工砌砖的落后面貌,减轻了劳动强度,提高了劳动生产率。

小型砌块施工,与传统的砖砌体砌筑工艺相似,也是手工砌筑,但在形状、构造上有一定的差异。本节只介绍小型空心砌块施工方法。

The support system of formwork shall be arranged reasonably according to the span size and thickness of floor slab, and the vertical and horizontal spacing of vertical poles shall be reasonably arranged. Generally, it is 800—1000mm. There are three horizontal cross bars, one of which is a bottom bar, which is about 200mm away from the supporting surface, so that the supporting system has good overall stiffness and stability.

The bottom formwork is made of laminated board or bamboo plywood, and its joints shall be flat connected and sealed with self-adhesive to prevent slurry leakage. In order to prevent the formwork from shifting, nails can be used to connect with the wood grid below in proper position. When the formwork is removed, the nails shall be pulled out in time to prevent injury.

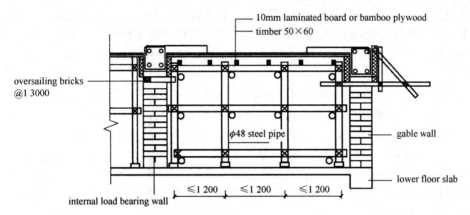

Fig. 4. 37 Cast-in-place Floor Formwork

4. 4 Block Construction

It is an important way to replace clay brick as wall material for wall reform. In recent years, local materials have been adapted to local conditions. Natural materials or industrial wastes have been used as raw materials to make blocks of various specifications for building wall structures. At present, small concrete blocks are mainly used. The construction method of block is simple, which changes the backwardness of manual bricklaying, reduces the labor intensity and improves the labor productivity.

Small-scale block construction, similar to the traditional brick masonry technology, is also manual masonry, but there are certain differences in shape and structure. This section only introduces the construction method of small-scale hollow block.

4.4.1 材料类型

小型砌块按构造形式分为空心和实心两种;按材料分有混凝土砌块、粉煤灰硅酸盐砌块、煤矸石硅酸盐砌块、加气泡沫混凝土砌块等。砌块高度为 380~940 mm 的称中型砌块,砌块高度小于 380 mm 的称小型砌块。

4.4.2 砌块施工工艺

砌块排列时,应根据砌块尺寸和垂直灰缝的宽度(8~12 mm)、水平缝的厚度(8~12 mm)计算砌块砌筑匹数和排数,尽量采用主规格。砌块一般采用全顺组砌,上下皮错缝 1/2 砌块长度,上下皮砌块应孔对孔、肋对肋,个别无法对孔砌筑时,可错孔砌筑,但其搭接长度不应小于 90 mm。如不能满足要求的搭接长度时,应在灰缝中设拉结钢筋。外墙转角处和纵横交接处,砌块应分皮咬槎,交错搭接。由于黏土砖与空心小型砌块的材料性能不同,对承重墙体不得采用砌块与黏土砖混合砌筑。

砌块应从外墙转交处或定位砌块处开始砌筑。砌块应底面朝上砌筑,称"反砌法"。生产小型空心砌块时,因抽芯脱模需要,孔洞模芯有一定的锥度,形成孔洞上口大底口小,因此利用底面朝上便于铺设砂浆,也便与对肋砌筑时砌块的摆放。若使用一端有凹槽的砌块时,应将有凹槽的一端接着平头的一端砌筑。砌块应逐块铺砌,全部灰缝应均匀填铺砂浆,水平灰缝宜用坐浆法铺浆。竖缝可先在砌块端头铺满砂浆(即将砌块铺浆的端面朝上,依次紧密排列、铺浆),然后将砌块上墙挤压至要求的尺寸;也可在砌筑好的砌块端头刮满砂浆,然后将砌块上墙。水平灰缝的砂浆饱满度不低于 90%,竖缝砂浆饱满度不得低于 60%。墙体临时间断处应设置在门窗洞口处,或砌成阶梯形斜槎(斜槎长度≥2/3 斜槎高度);如设置斜槎有困难时,也可砌成直槎,但必须采用拉结网片或采取其他构造措施。在圈梁底部或梁端支承处,一般可先用 C15 混凝土填实砌块孔洞后砌筑。

4.4.1 Types of Materials

Small-scale blocks are divided into two types, hollow and solid, according to their structural forms. They are concrete blocks, fly ash silicate blocks, gangue silicate blocks, and air entraining foam concrete blocks. Medium sized block with block height of 380—940mm and block height of less than 380mm are called small sized block.

4.4.2 Block Construction Technology

When the block is arranged, the number of blocks and rows shall be calculated according to the block size, the width of vertical mortar joint(8—12mm) and the thickness of horizontal joint(8—12mm), and the main specification shall be adopted as far as possible. Generally, the blocks are built in full stretcher, with staggered joints of 1/2 of the block length. The upper and lower layers should be hole to hole and rib to rib. If some blocks cannot be built to holes, staggered holes can be used, but the lap length should not be less than 90mm. If the required lap length cannot be met, tie bars shall be set in the mortar joint. At the corner of the external wall and the vertical and horizontal joints, the blocks shall be staggered and overlapped. Due to the different material properties of clay brick and hollow small sized block, the mixed masonry of block and clay brick is not allowed for bearing wall.

The blocks shall be laid from the transfer point of the external wall or the location block. The block should be built with the bottom face upward, which is called "reverse masonry method". In the production of small hollow block, due to the need of core pulling and demoulding, the hole core has a certain taper, which forms a hole with a large upper opening and a small bottom opening. Therefore, it is convenient to lay mortar by using the bottom face upward, which is also convenient for placing the block in the rib masonry. If a block with a groove at one end is used, the end with groove shall be laid next to the end with flat head. The blocks shall be paved one by one, all mortar joints shall be evenly filled with mortar, and the horizontal mortar joints shall be paved with mortar laying method. The vertical joint can be paved with mortar at the end of the block(i. e. , the end face of the block is upward, arranged closely and laid in sequence), and then the upper wall of the block is squeezed to the required size; or the mortar can be scraped on the end of the masonry block, and then the block is put on the wall. The mortar plumpness of horizontal mortar joint shall not be less than 90%, and that of vertical joint shall not be less than 60%. The temporary break of the wall should be set at the opening of doors and windows, or be built into stepped racking(length of stepped racking $\geqslant$ 2/3 height of stepped racking); if it is difficult to set stepped racking, it can also be built into serrated racking, but it must be connected with mesh or other structural measures. At the bottom of the ring beam or at the support of the beam end, the hole of the block can be filled with C15 concrete before masonry.

习 题

1. 砌筑工程中常用的建筑施工机械有哪些?
2. 简述砌筑用脚手架的作用及基本要求。
3. 脚手架的安全防护措施有哪些内容?
4. 砌筑用的砂浆有哪些种类?适用于什么场合?
5. 砖墙的组砌形式有哪些?
6. 砌筑时如何控制砌体的位置和标高?
7. 墙体接槎应如何处理?
8. 砖砌体工程质量要求有哪些?影响其质量的因素有哪些?

Chapter 4 Masonry Works Construction

Exercises

1. What kinds of construction machinery are commonly used in masonry works?
2. Briefly describe the function and basic requirements of masonry scaffold.
3. What are the safety protection measures of scaffold?
4. What kinds of mortar are used for masonry? What occasions are they suitable for?
5. What are the masonry forms of brick wall?
6. How to control the position and elevation of masonry?
7. How to deal with the wall joint?
8. What are the quality requirements of brick masonry works? What are the factors that affect its quality?

第 5 章　钢筋混凝土结构工程施工

5.1　模板工程施工

钢筋混凝土结构的模板工程,是混凝土构件成型的一个十分重要的组成部分。现浇混凝土结构使用的模板工程的造价约占钢筋混凝土工程总造价的 30%,总用工量的 50%。另外模板的选择也决定了混凝土工程的施工质量和模板的安装方法,考虑是否需要施工机械来配合才能完成模板的安装,因而选择模板不同也会造成施工方案不同。因此,采用先进的模板技术,对于提高工程质量、加快施工速度、提高劳动生产率、降低工程成本和实现文明施工,都具有十分重要的意义。

5.1.1　模板的种类

1. 按材料性质分类

模板是混凝土浇筑成型的模壳和支架。按材料的性质可分为木模板、钢模板、塑料模板和其他模板等。

(1)木模板。混凝土工程开始出现时,都是使用木材来做模板。一般多为松木和杉木,木材被加工成木板、木方,然后组合成构件所需的模板。

(2)组合钢模板。又称组合式定型小钢模,是目前使用较广泛的一种通用性组合模板。用它进行现浇钢筋混凝土结构施工,可事先按设计要求组拼成梁、柱、墙、楼板的大型模板,整体吊装就位,也可采用散装散拆方法,比较方便。组合钢模板的部件,主要由钢模板、连接件和支承件三部分组成。

①组合钢模板的类型及规格。钢模板主要包括平面模板和转角模板(图 5.1)两大类。

a. 平面模板。平面模板由面板和肋条组成,采用 Q235 钢板制成,面板厚 2.3 mm 或 2.5 mm,肋条上设有 U 形卡孔。

Chapter 5 Construction of Reinforced Concrete Structure

5.1 Formwork Construction

The formwork of reinforced concrete structure is a very important part of concrete component forming. The cost of formwork used in cast-in-place concrete structure accounts for about 30% of the total cost of reinforced concrete project and 50% of the total labor force. In addition, the selection of formwork also determines the construction quality of concrete engineering and the installation method of formwork. Considering whether the construction machinery is needed to complete the installation of formwork, the selection of formwork will also cause different construction schemes. Therefore, the use of advanced formwork technology is of great significance for improving the quality of the project, speeding up the construction speed, improving labor productivity, reducing project costs and realizing civilized construction.

5.1.1 Types of Formwork

1. Classification by Material Properties

The formwork is the shell and support of concrete casting. According to the properties of materials, it can be divided into wood formwork, steel formwork, plastic formwork and other forms.

(1) Wood formwork. When concrete works began to appear, wood was used as formwork. Generally, they are pine and fir wood. The wood is processed into boards and squares, and then combined into the formwork required by the components.

(2) Combined steel formwork. Combined steel formwork, also known as combined shaped small steel formwork, is a kind of universal combined formwork widely used at present. In the construction of cast-in-place reinforced concrete structure, large-scale formwork of beams, columns, walls and floors can be assembled in advance according to the design requirements, which can be hoisted in place as a whole, and can also be disassembled in bulk, which is more convenient. The components of composite steel formwork are mainly composed of steel formwork, connectors and supports.

①Types and specifications of combined steel formwork. Steel formwork mainly includes plane formwork and corner formwork(Fig. 5.1).

a. Plane formwork. The plane formwork is composed of panels and ribs, using Q235 steel plate with a thickness of 2.3mm or 2.5mm. The rib is provided with a U-shaped clamping hole.

平面模板的代号为 P，宽度以 100 mm 为基础，以 50 mm 为模数进级；长度以 450 mm 为基础，以 150 mm 为模数进级；肋板高为 55 mm，因此又称为 55 型组合钢模。平面模板利用 U 形卡和 L 形插销等，可以拼装成各种尺寸的大模板。U 形卡孔两边焊凸鼓，以增加 U 形卡的夹紧力；边肋倾角处有 0.3 mm 的凸棱，可增强模板的刚度并使接缝严密。平面模板的规格长度有：450 mm、600 mm、750 mm、900 mm、1 200 mm、1 500 mm；宽度有：100 mm、150 mm、200 mm、300 mm；高度 55mm。

b. 转角模板。转角模板(图 5.1)有阴角模板(代号为 E)、阳角模板(代号为 Y)和连接角模(代号为 J)三种，主要用于结构的转角部位。

转角模板的长度与平面模板相同，其中阴角模板的宽度有 150 mm×150 mm、100 mm×150 mm 两种；阳角模板的宽度有 100 mm×100 mm、50 mm×50 mm 两种；连接角模的宽度为 50 mm×50 mm。

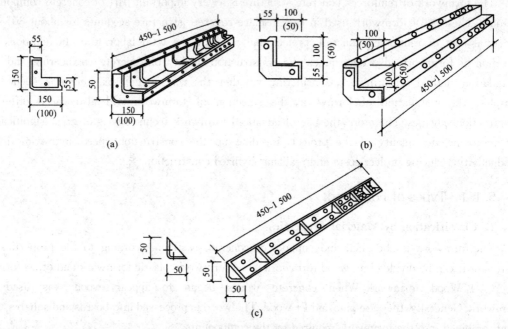

图 5.1　转角模板
(a)阴角模板；(b)阳角模板；(c)连接角模

Chapter 5 Construction of Reinforced Concrete Structure

The code of plane formwork is p, the width is based on 100mm, and the module is advanced by 50mm; the length is based on 450mm, and the module is upgraded by 150mm; the rib plate height is 55mm, so it is also called 55 type combined steel formwork. The plane formwork can be assembled into large formwork of various sizes by using U-shaped clamp and L-shaped bolt. The convex drum is welded on both sides of the U-shaped clamping hole to increase the clamping force of the U-shaped clamp; there is a 0.3mm convex edge at the inclination angle of the side rib, which can enhance the rigidity of the formwork and make the joint tight. The specification length of plane formwork is 450mm, 600mm, 750mm, 900mm, 1200mm, 1500mm; the width is 100mm, 150mm, 200mm, 300mm; the height is 55mm.

b. Corner borm work. There are three types of corner formwork(Fig. 5.1), i.e. internal corner formwork(code E), external corner formwork(Code: y) and connection angle formwork(Code: J), which are mainly used for the corner parts of the structure.

The length of corner formwork is the same as that of plane formwork, of which the width of internal corner formwork is 150mm×150mm and 100mm×150mm; the width of external angle formwork is 100mm×100mm and 50mm×50mm; the width of connecting angle formwork is 50mm×50mm.

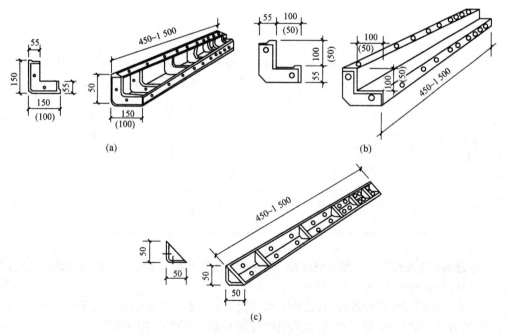

Fig. 5.1 Corner Formwork
(a)internal corner formwork; (b)external corner formwork; (c)connection angle formwork

钢模板规格编码如表 5.1。

表 5.1 钢模板规格编码

模板名称		模板长度/mm										
		450		600		750		900		1 200		1 500	
		代号	尺寸	代号	尺寸	代号	尺寸	代号	尺寸	代号	尺寸	代号	尺寸
平面模板代号 P	宽度/mm 代号												
	300	P3004	300×450	P3006	300×600	P3007	300×750	P3009	300×900	P3015	300×1 200	P3015	300×1 500
	250	P2504	250×450	P2506	250×600	P2507	250×750	P2509	250×900	P2512	250×1 200	P2515	250×1 500
	200	P2004	200×450	P2006	200×600	P2007	200×750	P2009	200×900	P2012	200×1 200	P2015	200×1 500
	150	P1504	150×450	P1506	150×600	P1507	150×750	P1509	150×900	P1512	150×1 200	P1515	150×1 500
	100	P1004	100×450	P1006	100×600	P1006	100×750	P1009	100×900	P1020	100×1 200	P1015	100×150
阴角模板（代号 E）		E1504	150×150×450	E1506	150×150×600	E1507	150×150×750	E1509	150×150×900	E1512	150×150×1 200	E1515	150×150×1 500
		E1004	100×150×450	E1006	100×150×600	E1007	100×150×750	E1009	100×150×900	E1012	100×150×1 200	E1012	100×150×1 500
阳角模板（代号 Y）		Y1004	100×100×450	Y1006	100×100×600	Y1007	100×100×750	Y1009	100×100×900	Y1012	100×100×1 200	Y1012	100×100×1 500
		Y0504	50×50×450	Y0506	50×50×600	Y0507	50×50×750	Y0509	50×50×900	Y0512	50×50×1 200	Y0512	50×50×1 500
连接角模（代号 J）		×J004	50×50×450	J0006	50×50×600	J0006	50×50×750	J0009	50×50×900	J0012	50×50×1 200	J0012	50×50×1 500

②组合钢模板连接配件。组合钢模板的连接配件主要包括 U 形卡；L 形插销、钩头螺栓、对拉螺栓、紧固螺栓和扣件等，如图 5.2 所示。

U 形卡用于相邻模板的连接，其安装距离一般不得大于 300 mm，即应每隔一个孔插一个 U 形卡，安装方向一顺一倒、相互交错，以抵消因打紧 U 形卡可能产生的位移。

L 形插销用于插入钢模板端部横肋的插销孔内，以加强两相邻模板接头处的刚度和保证接头处板面平整，使浇筑的混凝土表面无明显的痕迹。

钩头螺栓用于钢模板与内外钢楞的加固，安装间距一般不大于 600 mm，即应每隔两个孔插一个钩头螺栓，其长度应与采用的钢楞尺寸相适应。

Specifications and codes of steel templates are shown in Table 5.1.

Table 5.1 Specification Code of Steel Formwork

name of formwork			length of formwork/mm										
			450		600		750		900		1200		1500	
			code	size	code	size	code	size	code	size	code	size	code	size
plain formwork code P	width /mm	300	P3004	300×450	P3006	300×600	P3007	300×750	P3009	300×900	P3015	300×1200	P3015	300×1500
		250	P2504	250×450	P2506	250×600	P2507	250×750	P2509	250×900	P2512	250×1200	P2515	250×1500
		200	P2004	200×450	P2006	200×600	P2007	200×750	P2009	200×900	P2012	200×1200	P2015	200×1500
		150	P1504	150×450	P1506	150×600	P1507	150×750	P1509	150×900	P1512	150×1200	P1515	150×1500
		100	P1004	100×450	P1006	100×600	P1006	100×750	P1009	100×900	P1020	100×1200	P1015	100×150
Internal corner formwork (code E)			E1504	150×150×450	E1506	150×150×600	E1507	150×150×750	E1509	150×150×900	E1512	150×150×1200	E1515	150×150×1500
			E1004	100×150×450	E1006	100×150×600	E1007	100×150×750	E1009	100×150×900	E1012	100×150×1200	E1012	100×150×1500
External corner formwork (code Y)			Y1004	100×100×450	Y1006	100×100×600	Y1007	100×100×750	Y1009	100×100×900	Y1012	100×100×1200	Y1012	100×100×1500
			Y0504	50×50×450	Y0506	50×50×600	Y0507	50×50×750	Y0509	50×50×900	Y0512	50×50×1200	Y0512	50×50×1500
Connection angle formwork (code J)			J0004	50×50×450	J0006	50×50×600	J0006	50×50×750	J0009	50×50×900	J0012	50×50×1200	J0012	50×50×1500

②Connecting parts for combined steel formwork. The connecting parts of combined steel formwork mainly include U-shaped clamp, L-shaped bolt, hook bolt, split bolt, fastening bolt and fastener, etc., as shown in Fig. 5.2.

The U-shaped clamp is used for the connection of adjacent formworks, and the installation distance shall not be greater than 300mm. In other words, a U-shaped clamp shall be inserted every other hole, and the installation direction shall be in a clockwise and inverted manner and staggered with each other to offset the possible displacement caused by tightening the U-shaped clamp.

The L-shaped bolt is used to insert into the bolt hole of the transverse rib at the end of the steel formwork, so as to strengthen the rigidity of the joint of two adjacent formwork and ensure the flatness of the slab surface at the joint, so as to make the concrete surface without obvious trace.

The head bolt is used to reinforce the steel formwork and the inner and outer steel ridges, and the installation spacing is generally no more than 600mm, that is, a hook bolt should be inserted every two holes, and its length should be suitable for the size of the steel ridge used.

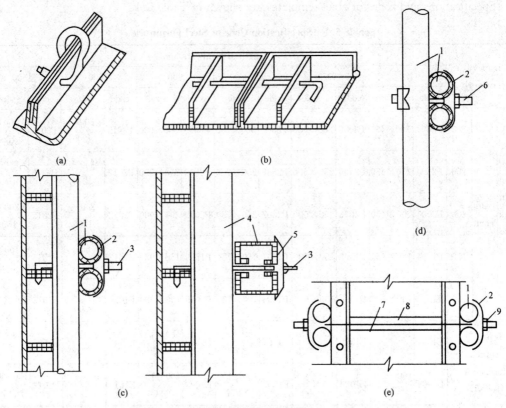

图 5.2 钢模板连接件
(a)U 形卡;(b)L 形插销;(c)钩头螺栓;(d)紧固螺栓;(e)对拉螺栓
1—圆钢管楞;2—"3"形扣件;3—钩头螺栓;4—内卷边槽钢楞;5—蝶形扣件;
6—紧固螺栓;7—对拉螺栓;8—塑料套管;9—螺母

紧固螺栓用于紧固内外钢楞,长度应与采用的钢楞尺寸相适应。

对拉螺栓用于连接墙壁两侧模板,以保持模板与模板之间的设计厚度,并承受混凝土的侧压力及水平荷载,并保证模板不产生变形。

扣件用于钢楞与钢楞或与钢模板之间的扣紧。按照钢楞的形状不同,可分别采用蝶形扣件和"3"形扣件。

③组合钢模板的支承件。组合钢模板的支承件,包括钢楞、柱箍、梁卡具、钢管架、扣件式钢管脚手架、平面可调桁架等。

钢楞又称龙骨,即支承模板的横档和竖档,分内钢楞与外钢楞。其内钢楞配置方向一般应与钢模板垂直,直接承受钢模板传来的荷载,其间距一般为 700~900 mm。外钢楞承受内钢楞传来的荷载,或用来加强模板结构的整体刚度和调整平直度。钢楞的材料有圆钢管、矩形钢管、内卷边槽钢、槽钢和角钢等。柱箍是用于直接支承和夹紧各类柱模的支承件,有扁钢、角钢、槽钢等多种形式,图 5.3 所示为柱箍。梁卡具又称梁托架,是一种将大梁、过梁等钢模板夹紧固定的装置,并承受混凝土的侧压力。梁卡具的种类较多,图 5.4 所示为扁钢与钢管组合的梁卡具。

Chapter 5 Construction of Reinforced Concrete Structure

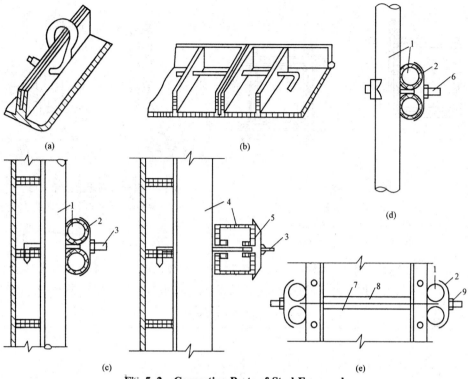

Fig. 5.2 **Connecting Parts of Steel Formwork**
(a)U-shaped clamp;(b)L-shaped bolt;(c)hooked bolt;(d)tightening bolt;(e)split bolt
1—round steel pipe ridge;2—"3"fastener;3—hooked bolt;4—inner crimping channel steel;
5—butterfly fastener;6— tightening bolt;7—split bolt;8—plastic sleeve;9—nut

The fastening bolts are used to fasten the inner and outer steel ridges, and the length shall be suitable for the size of the steel ribs used.

The split bolt is used to connect the formwork on both sides of the wall to maintain the design thickness between the formwork and the formwork, bear the lateral pressure and horizontal load of the concrete, and ensure that the formwork does not deform.

The fastener is used for fastening between steel ridge and steel ridge or steel formwork. Butterfly fastener and "3" fastener can be used according to the shape of steel rib.

③ Supporting parts of combined steel formwork. The supporting parts of combined steel formwork include steel ridge, column hoop, beam clamp, steel pipe frame, fastener type steel pipe scaffold, plane adjustable truss, etc.

Steel ridge is also known as keel, that is to support the horizontal and vertical formwork, and is divided into internal steel ridge and external steel ridge. In general, the direction of the inner steel ridge should be vertical to the steel formwork, and directly bear the load from the steel formwork, and the spacing is generally 700—900mm. The outer steel bar bears the load from the inner steel bar, or is used to strengthen the overall rigidity of the formwork structure and adjust the straightness. The materials of steel ridge include round steel pipe, rectangular steel pipe, inner crimping channel steel, channel steel and angle steel, etc. The column hoop is used to directly support and clamp various types of column formwork, including flat steel, angle steel, channel steel and other forms. Fig. 5.3 shows the column hoop. Beam clamp, also known as beam bracket, is a device to clamp and fix steel formwork such as girder and lintel, and bear the lateral pressure of concrete. There are many kinds of beam clamps. Fig. 5.4 shows the combination clamps of flat steel and steel pipe.

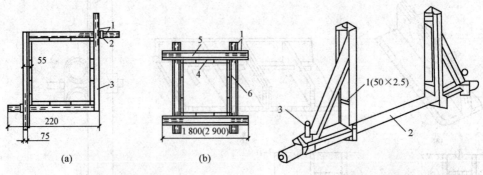

图 5.3 柱箍
(a)角钢型;(b)型钢型
1—插销;2—限位器;3—夹板;
4—模板;5—型钢 A;6—型钢 B

图 5.4 扁钢与钢管组合梁卡具
1—三脚架;2—底座;3—固定螺栓

钢管架又称钢支撑,用于大梁、楼板等水平模板的垂直支撑。钢管支柱由内外内两节钢管组成,可以伸缩以调节支柱的高度。钢管架的规格型式较多,目前常用的有 CH 和 YJ 两种(表 5.2)。

表 5.2 钢管架的规格型式

项目		CH-65	CH-75	CH-90	YJ-18	YJ-22	YJ-27
最小使用长度/mm		1 812	2 212	2 712	1 820	2 220	2 720
最大使用长度/mm		3 062	3 462	3 962	3 090	3 490	3 990
调节范围/mm		1 250	1 250	1 250	1 270	1 270	1 270
螺旋调节范围/mm		170	170	170	70	70	70
容许荷载	最小长度时/kN	20	20	20	20	20	20
	最大长度时/kN	15	15	12	15	15	12
质量/kg		12.4	13.2	14.8	13.87	14.99	16.39

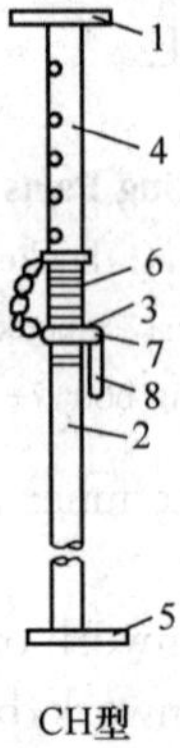

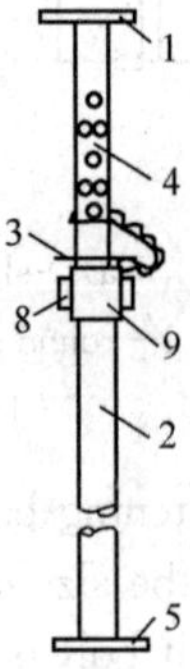

注:1. 1—顶板;2—套管;3—插销;4—插管;5—底座;6—转盘;7—螺管;8—手柄;9—螺旋套
2.CH 型相当于《组合钢模板技术规范》(GB/T 50214—2013)的 C-18 型、C-22 型和 C-27 型,其最大使用长度分别为 3 112 mm、3 512 mm、4 012 mm。

Chapter 5 Construction of Reinforced Concrete Structure

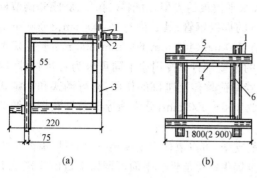

Fig. 5.3 Column Hoop
1—bolt; 2—stopper; 3—splint; 4—formwork;
5—rolled section steel A; 6—rolled section steel B
(a) angle steel type; (b) rolled section steel

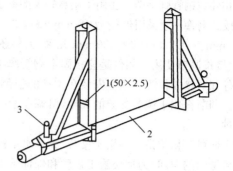

Fig. 5.4 Combination Clamps of
Flat Steel and Steel Pipe
1—tripod; 2—base; 3—fixed bolt

Steel pipe frame, also known as steel support, is used for vertical support of horizontal formwork such as girder and floor. The steel pipe column is composed of two sections of steel pipe inside and outside, which can be retracted to adjust the height of the column. There are many specifications and types of steel pipe frame, and CH and YJ are commonly used at present(Table 5.2).

Table 5.2 Specification and Type of Steel Pipe Frame

See drawing		CH型			YJ型		
Items		CH—65	CH—75	CH—90	YJ—18	YJ—22	YJ—27
minimum length of use/mm		1812	2212	2712	1820	2220	2720
maximum length of use/mm		3062	3462	3962	3090	3490	3990
adjustable range/mm		1250	1250	1250	1270	1270	1270
screw adjustable range/mm		170	170	170	70	70	70
allowable load	at minimum length/kN	20	20	20	20	20	20
	at maximum length/kN	15	15	12	15	15	12
weight/kg		12.4	13.2	14.8	13.87	14.99	16.39

Notes: 1. 1—top tray; 2—casing pipe; 3—bolt; 4—intubation; 5—base; 6—turntable; 7—screw pipe; 8—handle; 9—spiral sleeve

2. CH type is equivalent to C—18, C—22 and C—27 in technical specification for composite steel formwork (GB/T 50214—2013), and the maximum length of use is 3112mm, 3512mm and 4012mm respectively.

扣件式钢管脚手架主要用于层高较大的梁、板等水平模板的支架,由钢管、扣件、底座和调节杆等组成。钢管一般采用外径为 48 mm、壁厚 3.5 mm 的焊接钢管,其长度有 2 000 mm、3 000 mm、4 000 mm、5 000 mm 和 6 000 mm 几种,另外还配有 200 mm、400 mm、600 mm、800 mm 长的短钢管,供接长调节使用。扣件是连接固定钢管脚手架的重要部件,按用途不同可分为直角扣件、回转扣件和对接扣件等。底座安装在主杆的下部,起着将荷重传至基础的作用,有可调式和固定式两种。调节杆用于调节支架的高度,可调高度一般为 100～350 mm,分为螺栓调节杆和螺管调节杆两种。

平面可调桁架是一种用于楼板、梁等水平模板的支架,可以取代水平模板下的立柱,有效扩大施工楼层内的空间,为加快施工进度和保证施工安全提供了可能性。平面可调桁架目前在建筑工程中的类型较多,当施工荷载较小时,可用钢筋焊成;当跨度或施工荷载较大时,可用角钢或钢管制成,也可以制成两个半榀,在施工现场再拼装成整体。

图 5.5 所示为用角钢、扁钢和圆钢筋焊制而成的轻型桁架。两榀桁架组合后,其跨度可调范围为 2 100～3 500 mm。当梁的跨度大时,可以连续安装桁架,在桁架中间设置支柱。

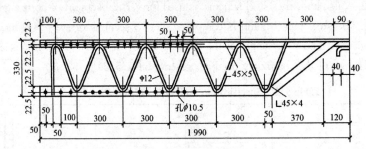

图 5.5 轻型桁架示意

Chapter 5 Construction of Reinforced Concrete Structure

Fastener type steel pipe scaffold is mainly used for the support of horizontal formwork such as beam and plate with large storey height, which is composed of steel pipe, fastener, base and adjusting rod. Generally, the welded steel pipe with an outer diameter of 48mm and a wall thickness of 3.5mm is used. Its length includes 2000mm, 3000mm, 4000mm, 5000mm and 6000mm. In addition, short steel pipes with length of 200mm, 400mm, 600mm and 800mm are also provided for lengthening adjustment. Fastener is an important part to connect and fix steel pipe scaffold. It can be divided into right angle fastener, rotary fastener and butt fastener according to different uses. The base is installed in the lower part of the main rod, which can transmit the load to the foundation, and can be adjusted or fixed. The adjusting rod is used to adjust the height of the bracket and the adjustable height is generally 100—350mm. The adjusting rod can be divided into bolt adjusting rod and screw tube adjusting rod.

The plane adjustable truss is a kind of support used for horizontal formwork such as floor slab and beam. It can replace the column under the horizontal formwork, effectively expand the space in the construction floor, and provide the possibility for speeding up the construction progress and ensuring the construction safety. At present, there are many types of plane adjustable truss in construction engineering. When the construction load is small, it can be welded with steel bars; when the span or construction load is large, it can be made of angle steel or steel pipe, or two half trusses, which can be assembled into a whole at the construction site.

Fig. 5.5 shows a light truss made by welding angle steel, flat steel and round steel bar. After the combination of the two trusses, the adjustable span is 2100—3500mm. When the span of the beam is large, the truss can be installed continuously, and the column can be set in the middle of the truss.

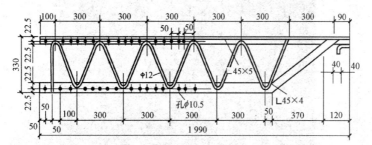

Fig. 5.5 Schematic Diagram of Light Truss

(3)全钢大模板。

①全钢大模板概述。全钢大模板是进行现浇剪力墙结构施工的一种工具式模板。它的单块模板面积较大,通常对于小开间而言,以一面现浇混凝土墙体为1块模板,对于大开间墙体而言,以几块模板拼装组成,一般为2块或3块。图5.6为其在工程上常用的种类之一。

图5.6 某工程应用的全钢大模板

全钢大模板是采用专业设计和按照墙体尺寸加工制作而成的一种工具式模板,一般与支架连为一体。由于它自重大,施工时需要配以相应的吊装和运输机械,用于浇筑现浇混凝土墙体。它具有安装和拆除简便、尺寸准确,另外由于选用钢板作为板面材料,钢板厚度一般为5~6 mm,因而避免了钢框木(竹)胶合板的刚度差、易变形、板面易损坏的弱点,钢材板面耐磨、耐久,一般可周转使用次数在200次以上。同时,钢材板面平整光洁且容易清理,有利于提高混凝土的质量。图5.7所示为某工程达到清水墙效果。

图5.7 某工程用全钢大模板浇筑混凝土后的清水墙效果

(3) All steel large formwork.

① Overview of all steel large formwork. All steel large formwork is a kind of tool type formwork for cast-in-place shear wall structure construction. Its single formwork area is large, and usually for small bay, one side of cast-in-place concrete wall is used as a formwork. For large bay wall, it is composed of several formworks, generally two or three. Fig. 5.6 shows one of the common types in project.

Fig. 5.6　All Steel Large Formwork Applied in a Project

All steel large formwork is a kind of tool type formwork which is made by professional design and processing according to the wall size, which is generally connected with the support. Due to its self weight, the construction needs to be equipped with the corresponding hoisting and transportation machinery, which is used for pouring the cast-in-place concrete wall. It has the advantages of simple installation and removal, and accurate size. In addition, due to the selection of steel plate as the board surface material, the thickness of steel plate is generally 5—6mm, thus avoiding the weakness of poor rigidity, easy deformation and easy damage of steel frame and wood(bamboo) plywood. The steel plate surface is wear-resistant and durable, and can be reused for more than 200 times. At the same time, the steel plate surface is smooth, bright and easy to clean, which is conducive to improve the quality of concrete. As shown in Fig. 5.7, a project achieves the effect of fair faced wall.

Fig. 5.7　Effect of Fair Faced Wall after Pouring
Concrete with All Steel Large Formwork

②全钢大模板的结构组成。全钢大模板由面板、加劲肋、竖楞、支撑桁架、稳定机构和操作平台、穿墙螺栓等组成,是一种现浇钢筋混凝土墙体的大型工具式模板。

a. 面板:面板是直接与混凝土接触的部分,通常采用钢面板(用 3~6 mm 厚的钢板制成)或胶合板面板(用 7~9 层胶合板)。

b. 加劲肋:加劲肋的作用是固定面板,可做成水平肋或垂直肋。加劲肋把混凝土传给面板的侧压力传给竖楞。加劲肋与金属面板焊接固定,与胶合板面板可用螺栓固定。加劲肋一般采用[65 或∟65 制作,肋的间距根据面板的大小、厚度及墙体厚度确定,一般为 300~500 mm。

c. 竖楞:竖楞的作用是加强大模板的整体刚度,承受模板传来的混凝土侧压力和垂直力并作为穿墙螺栓的支点。竖楞一般采用[65 或[80 槽钢制作,间距一般为 1.0~1.2 m。

d. 支撑桁架与稳定机构:支撑桁架用螺栓或焊接与竖楞连接在一起,其作用是承受风荷载等水平力,防止大模板倾覆。桁架上部可搭设操作平台。稳定机构为在大模板两端的桁架底部伸出支腿上设置的可调整螺旋千斤顶。在模板使用阶段,用以调整模板的垂直度,并把作用力传递到地面或楼板上;在模板堆放时,用来调整模板的倾斜度,以保证模板的稳定。

e. 操作平台:操作平台是施工人员操作场所,有两种做法:①将脚手板直接铺在水平弦杆上形成操作平台,外侧设栏杆。这种操作平台工作面虽小,但投资少,装拆方便;②在两道横墙之间的大模板的边框上用角钢连接成为搁栅,在其上满铺脚手板。优点是施工安全,但耗钢量大。

f. 穿墙螺栓:穿墙螺栓的作用是控制模板间距,承受新浇混凝土的侧压力,并能加强模板刚度。为了避免穿墙螺栓与混凝土粘结,在穿墙螺栓外边套一根硬塑料管或穿孔的混凝土垫块,其长度为墙体宽度。穿墙螺栓一般设置在大模板的上、中、下三个部位,上穿墙螺栓距模板顶部 250 mm 左右,下穿墙螺栓距模板底部 200 mm 左右。

Chapter 5　Construction of Reinforced Concrete Structure

②Structural composition of all steel large formwork. All steel large formwork is composed of panel, stiffening rib, vertical ridge, supporting truss, stabilizing mechanism and operating platform, through-wall bolt, etc. It is a large tool type formwork for cast-in-place reinforced concrete wall.

a. Panel. The panel is the part directly in contact with concrete, usually using steel panel(made of 3—6mm thick steel plate)or plywood panel(with 7—9 layers of plywood).

b. Stiffening rib. The role of stiffening rib is to fix the panel, which can be made into horizontal rib or vertical rib. The stiffening ribs transmit the lateral pressure from the concrete to the face plate and to the vertical ribs. The stiffening rib is welded with the metal panel and fixed with the plywood panel with bolts. The stiffening rib is generally made of [65 or ∟ 65, and the spacing of ribs is determined according to the size, thickness and wall thickness of the panel, which is generally 300—500mm.

c. Vertical ridge. The role of the vertical ridge is to strengthen the overall stiffness of the large formwork, bear the concrete lateral pressure and vertical force from the formwork, and act as the fulcrum of the through-wall bolt. The vertical ridge is generally made of [65 or [80 channel steel, and the spacing is generally 1.0—1.2m.

d. Supporting truss and stabilizing mechanism. The supporting truss is connected with the vertical ridge by bolts or welding, and its function is to bear horizontal force such as wind load to prevent the overturning of large formwork. An operation platform can be set up on the upper part of the truss. The stabilizing mechanism is an adjustable screw jack set on the outrigger at the bottom of the truss at both ends of the large formwork. In the stage of using the formwork, it is used to adjust the perpendicularity of the formwork and transfer the force to the ground or floor. When the formwork is stacked, it is used to adjust the inclination of the formwork to ensure the stability of the formwork.

e. Operation platform. The operation platform is the place for construction personnel to operate. There are two methods: ① directly lay the scaffold board on the horizontal chord to form the operation platform, and set the railing outside. Although the working space of this kind of operation platform is small, the investment is small, and the installation and disassembly are convenient; ② the frame of the large formwork between the two transverse walls is connected with angle steel to form joist, and the scaffold board is fully paved on it. The advantage is that the construction is safe, but the steel consumption is large.

f. Through-wall Bolt. The role of through-wall bolt is to control the spacing of formwork, bear the lateral pressure of fresh concrete, and strengthen the rigidity of formwork. In order to avoid the bonding between the through-wall bolt and the concrete, a hard plastic pipe or perforated concrete pad is set outside the wall bolt, and its length is the wall width. The through-wall bolts are generally set in the upper, middle and lower parts of the large formwork. The upper through wall bolt is about 250mm away from the top of the formwork, and the lower through wall bolt is about 200mm away from the bottom of the formwork.

g. 吊环：用螺栓与上部边框连接。材质为 Q235A，不准使用冷加工处理。

③全钢大模板的种类与构造形式。全钢大模板按照其构造和组拼方式的不同，用于浇筑墙体的全钢大模板主要有桁架式大模板和组合式大模板。

a. 桁架式大模板。桁架式大模板是我国最早采用的工业化模板。由板面、支撑桁架和操作平台组成，如图 5.8 所示。

板面由面板、横肋和竖肋组成。面板采用 4～5 mm 厚钢板，横肋用[8 槽钢，间距 300～330 mm，竖肋用[8 槽钢成组对焊接，与支撑桁架连为一体，间距 1000 mm 左右。桁架上方铺设脚手板作为操作平台，下方设置可调节模板高度和垂直度的地脚螺栓。

桁架式大模板通用性差。施工时，横墙和纵墙需分两次浇筑混凝土，同时还需要另配角模解决纵横墙间的接缝处理。适用于标准化设计的剪力墙施工，且目前很少采用。

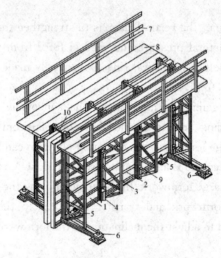

图 5.8　桁架式大模板构造示意
1—面板；2—水平肋；3—支撑桁架；4—竖肋；5—水平调整装置；
6—垂直调整装置；7—栏杆；8—脚手板；9—穿墙螺栓；10—固定卡具

b. 组合式大模板。组合式大模板是目前常用的一种模板形式。它通过固定于大模板上的角模，能把纵、横墙模板组装在一起，用以同时浇筑纵、横墙的混凝土，并可以利用模数条模板调整大模板的尺寸，以适应不同开间、进深尺寸的变化。

组合式大模板由板面、支撑系统、操作平台及连接件等部分组成，如图 5.9 所示。

g. Hoisting ring. It is connected with the upper frame with bolts. The material is Q235A. Cold working is not allowed.

③Types and structural forms of all steel large formwork. According to the different structures and assembly methods of all steel large formwork, truss type large formwork and combined large formwork are mainly used for pouring wall.

a. Truss type large formwork. Truss type large formwork is the first industrial formwork used in China. It is composed of panel, supporting truss and operation platform, as shown in Fig. 5. 8.

The panel is composed of face plate, transverse rib and vertical rib. The face plate is made of 4-5mm thick steel plate, the transverse rib is made of [8 channel steel, the spacing is 300—330 mm, and the vertical rib is welded in group with [8 channel steel, which is connected with the support truss as a whole, and the spacing is about 1000mm. The scaffold board is laid above the truss as the operation platform, and the anchor bolts that can adjust the height and perpendicularity of the formwork are set below.

Truss type large formwork has poor versatility. During the construction, the transverse wall and the longitudinal wall need to be poured twice, and at the same time, additional angle formwork is needed to solve the joint treatment between the vertical and horizontal walls. It is suitable for the construction of shear wall with standardized design, and it is rarely used at present.

b. Combined large formwork. Combined large formwork is a common formwork form at present. Through the angle formwork fixed on the large formwork, it can assemble the longitudinal and transverse wall formwork together to pour the concrete of the longitudinal and transverse walls at the same time, and can adjust the size of the large formwork by using the modulus strip formwork to adapt to the changes of the size of different openings and depth.

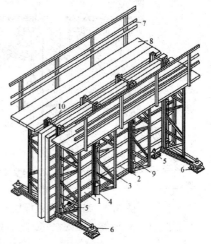

Fig. 5. 8 Composition Diagram of Combined Large Formwork

1—panel; 2—horizontal rib; 3—supporting truss;
4—vertical rib; 5—horizontal adjusting device;
6—vertical adjustment device; 7—railing;
8—scaffold board; 9— through-wall bolt;
10—fixed fixture

The combined large formwork is composed of panel, support system, operation platform and connecting parts, as shown in Fig. 5. 9.

(4)钢框胶合板模板。钢框胶合板模板拼装的大模板,由于钢框胶合板模板的钢框为热轧成型,并带有翼缘,刚度较好,组装大模板时可以省去竖向龙骨,直接将钢框胶合板和横向龙骨组装拼接。横向龙骨为两根[12槽钢,以一端采用螺栓,另一端为带孔的插板与板面相连,如图 5.10 所示。

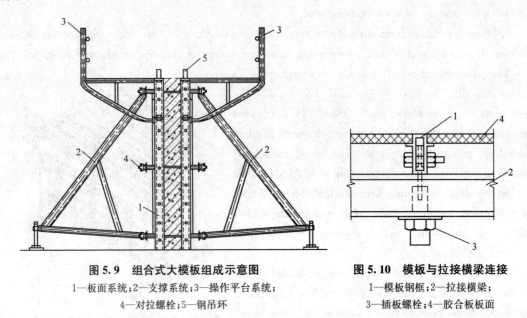

图 5.9　组合式大模板组成示意图
1—板面系统;2—支撑系统;3—操作平台系统;
4—对拉螺栓;5—钢吊环

图 5.10　模板与拉接横梁连接
1—模板钢框;2—拉接横梁;
3—插板螺栓;4—胶合板板面

大模板的上下端采用 L65×4 角钢和槽钢进行封顶和兜底,板面结构如图 5.11 所示。

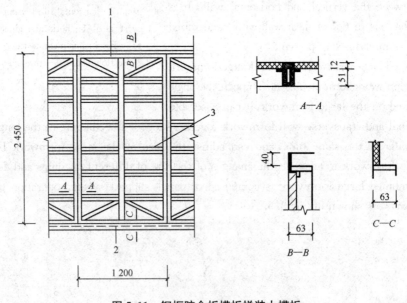

图 5.11　钢框胶合板模板拼装大模板
1—上拼角钢;2—下拼槽钢;3—钢框胶合板模板

(4) Steel frame plywood formwork. For the large formwork assembled by steel frame plywood formwork, the steel frame of steel frame plywood formwork is hot-rolled and has flange, so the rigidity is good. When assembling the large formwork, the vertical keel can be omitted, and the steel frame plywood and transverse keel can be directly assembled and spliced. The transverse keel is made of two [12 channel steel, one end of which is bolted, and the other end is a plug plate with holes, which is connected with the plate surface, as shown in Fig. 5.10.

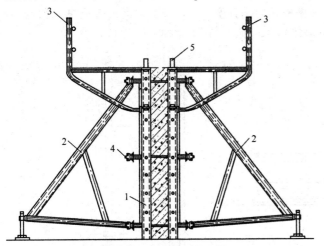

Fig. 5.9 Composition Diagram of
Combined Large Formwork

1—panel system; 2—support system; 3—operating platform system; 4—split bolt; 5—steel ring

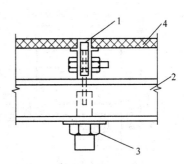

Fig. 5.10 Connection between
Formwork and Tie Beam

1—formwork steel frame; 2—tie beam;
3—plug plate bolt; 4—plywood board surface

The upper and lower ends of the large formwork are capped and bottomed with L65×4 and channel steel. The plate surface structure is shown in Fig. 5.11.

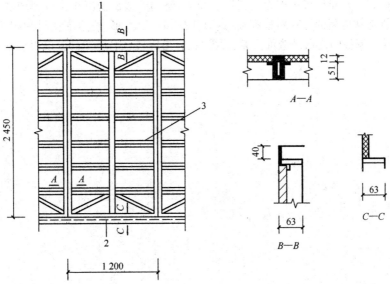

Fig. 5.11 Assembling Large Formwork with Steel Frame Plywood Formwork

1—upper angle steel; 2—lower channel steel; 3—steel frame plywood formwork

为了不在钢框胶合板板面上钻孔，而又能解决穿墙螺栓安装问题，同样设置一条 10 cm 宽的穿墙螺栓板带。该板带的四框与模板钢框的厚度相同，以使与模板能连为一体，板带的板面采用钢板。

角模用钢板制成，尺寸为 150 cm×150 cm，上下设四道加劲肋，与开间方向的大模板用螺栓连接固定在一起，另一侧与进深方向的大模板采用伸缩式搭接连接，如图 5.12 所示。

模板的支撑采用门形架。门架的前立柱为槽钢，用钩头螺栓与横向龙骨连接口其余部分用 48 钢管组成；后立柱下端设地脚螺栓，用以调整模板的垂直度。门形架上端铺设脚手板，形成操作平台。门形架上部可以接高，以适应不同墙体高度的施工。门形架构造如图 5.13 所示。

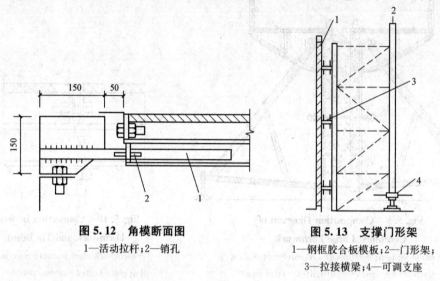

图 5.12　角模断面图
1—活动拉杆；2—销孔

图 5.13　支撑门形架
1—钢框胶合板模板；2—门形架；
3—拉接横梁；4—可调支座

用这种方法组装成的大模板，其造价可以降低 20%～30%，钢材用量节约 62%，质量可减轻 50%，因此，可以组装成大开间、大进深的大模板而不会使塔吊超载。

(5)塑料模板。塑料模板也称为塑壳定型模板，是随着钢筋混凝土预应力现浇密肋楼盖的出现，在实践中创造和应用的一种新型模板，其形状犹如一个方形大盆，支模时倒扣在支架上，口朝下、底朝上，这种模板的优点是造价低廉、拆模快、容易周转、成型美观，但仅用于钢筋混凝土结构的楼盖施工。

In order to avoid drilling holes on the plywood surface of steel frame and solve the installation problem of through wall bolts, a 10cm wide through wall bolt strip is also set up. The thickness of the four frames of the strip is the same as that of the steel frame of the formwork, so that it can be connected with the formwork. The plate surface of the strip is made of steel plate.

The angle formwork is made of steel plate with the size of 150cm×150cm. Four stiffening ribs are set up at the top and bottom, which are connected with the large formwork in the bay direction by bolts, and the large formwork in the depth direction on the other side is connected by telescopic lap connection, as shown in Fig. 5. 12.

The formwork is supported by portal frame. The front column of the portal frame is channel steel, which is connected with the transverse keel by hook head bolts, and the rest is made of <48 steel pipe; the lower end of the rear column is provided with anchor bolts to adjust the perpendicularity of the formwork. A scaffold board is laid on the upper end of the portal frame to form an operation platform. The upper part of the portal frame can be connected to accommodate the construction of different wall heights. See Fig. 5. 13 for portal structure.

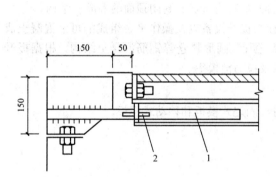
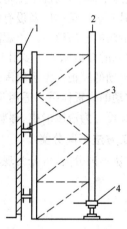

Fig. 5. 12 Section of Angle Formwork
1—movable rod; 2—pin hole

Fig. 5. 13 Supporting Portal Frame
1—steel frame plywood formwork; 2—portal frame;
3—tie beam; 4—adjustable support

The cost of the large formwork assembled by this method can be reduced by 20%—30%, the steel consumption can be saved by 62%, and the weight can be reduced by 50%. Therefore, the large formwork with large bay and large depth can be assembled without overloading the tower crane.

(5) Plastic formwork. Plastic formwork, also known as plastic shell shaped formwork, is a new type of formwork created and applied in practice with the emergence of reinforced concrete prestressed cast-in-place ribbed floor. Its shape is like a large square basin. When supporting the formwork, it is buckled on the support with the opening down and bottom up. The advantages of this kind of formwork are low cost, quick form removal, easy turnover and beautiful shape, but it is only used for the floor construction of reinforced concrete structure.

2. 按结构构件的类型分类

按结构构件的类型分类，有基础模板、柱模板、墙模板、楼板模板、梁模板、楼梯模板和其他各类构筑物模板等。

3. 按施工工艺条件分类

模板按施工工艺条件可分为现浇混凝土模板、预组装模板、大模板、跃升模板、水平滑动的隧道模板和垂直滑动的模板等。

（1）现浇混凝土模板。根据混凝土结构形状不同就地形成的模板，多用于基础、梁、板等现浇混凝土工程。模板支承体系多通过支于地面或基坑侧壁以及对拉的螺栓承受混凝土的竖向和侧向压力。这种模板适应性强，但周转较慢。

（2）预组装模板。由定型模板分段预组装成较大面积的模板及其支承体系，用起重设备吊运到混凝土浇筑位置。多用于大体积混凝土工程。

（3）大模板。由固定单元形成的固定标准系列的模板，多用于高层建筑的墙板体系。

（4）跃升模板。由两段以上固定形状的模板，通过埋设于混凝土中的固定件，形成模板支承条件承受混凝土施工荷载，当混凝土达到一定强度时，拆模上翻，形成新的模板体系。多用于变直径的双曲线冷却塔、水工结构以及设有滑升设备的高耸混凝土结构工程。

（5）水平滑动的隧道模板。由短段标准模板组成的整体模板，通过滑道或轨道支于地面、沿结构纵向平行移动的模板体系。多用于地下直行结构，如隧道、地沟、封闭顶面的混凝土结构。

（6）垂直滑动的模板。由小段固定形状的模板与提升设备以及操作平台组成的可沿混凝土成型方向平行移动的模板体系。适用于高耸的框架、烟囱、圆形料仓等钢筋混凝土结构。根据提升设备的不同，又可分为液压滑模、螺旋丝杠滑模以及拉力滑模等。

4. 按施工方法不同分类

按施工方法不同分类，可分为现场装拆式模板、固定式模板和移动式模板。

2. Classification by Types of Structural Members

According to the types of structural components, there are foundation formwork, column formwork, wall formwork, floor formwork, beam formwork, stair formwork and other types of structure formwork.

3. Classification according to Construction Process Conditions

According to the construction process conditions, the formwork can be divided into cast-in-place concrete formwork, pre-assembled formwork, large formwork, jump formwork, horizontal sliding tunnel formwork and vertical sliding formwork.

(1) Cast-in-place concrete formwork. According to the different shape of concrete structure, the formwork formed in situ is mostly used in cast-in-place concrete projects such as foundation, beam and slab. The formwork supporting system mostly bears the vertical and lateral pressure of concrete by supporting on the ground or the side wall of foundation pit as well as the bolts pulling against it. This kind of formwork has strong adaptability but slow turnover.

(2) Pre-assembled formwork. The larger formwork and its supporting system are assembled by sections of the fixed formwork, and lifted to the concrete pouring position by lifting equipment. It is mostly used in mass concrete project.

(3) Large formwork. The fixed standard series formwork formed by fixed unit is mostly used in the wallboard system of high-rise buildings.

(4) Jump formwork. The jumping formwork consists of more than two sections of fixed shape formwork. Through the fixed parts embedded in the concrete, the formwork supporting conditions are formed to bear the concrete construction load. When the concrete reaches a certain strength, the formwork is removed and turned up to form a new formwork system. It is mainly used in hyperbolic cooling tower with variable diameter, hydraulic structure and high-rise concrete structure with sliding equipment.

(5) Horizontal sliding tunnel formwork. The overall formwork is composed of short standard formwork, which is supported on the ground through slide or track and moves parallel along the longitudinal direction of the structure. It is mainly used for underground straight structure, such as tunnel, trench and concrete structure with closed top surface.

(6) Vertical sliding formwork. The formwork system is composed of small fixed shape formwork, lifting equipment and operation platform, which can move parallel to the concrete forming direction. It is suitable for high-rise frame, chimney, circular silo and other reinforced concrete structures. According to the different lifting equipment, it can be divided into hydraulic sliding mode, screw sliding mode and tension sliding mode.

4. Classification according to Different Construction Methods

According to the different construction methods, it can be divided into on-site assembling and disassembling formwork, fixed formwork and mobile formwork.

5.1.2 模板的作用与基本要求

1. 模板的作用

在钢筋混凝土结构施工中,刚从搅拌机中拌和出来的混凝土呈塑性流态,需要浇筑在与构件形状尺寸相同的模型内凝结硬化,才能形成所需的结构构件。模板系统是保证混凝土在浇筑过程中保持正确的形状和尺寸,在硬化过程中进行防护和养护的工具。

2. 对模板的基本要求

在现浇钢筋混凝土结构施工中,对模板系统的基本要求是:
(1)模板安装要保证结构和构件各部分的形状、尺寸及相互间位置的正确性。
(2)要有足够的强度、刚度和稳定性,以保证施工质量和施工安全。
(3)构造简单,装拆方便,能多次周转使用。
(4)接缝严密,不漏浆。
(5)用料节省,成本低。

5.1.3 模板的施工工艺

工地上最常用的模板为木(竹)胶合板模板、组合钢模板和全钢大模板,其施工工艺如下。

1. 木(竹)胶合板模板施工工艺

(1)墙模板安装。
①施工工艺流程:模板制作→安装门窗洞口模板→安装一侧墙模板→穿对拉螺栓套管并固定→安装另一侧墙模板→安装对拉螺栓及背楞→安装斜撑或地锚→校正加固→办理预检。
②施工要点。

a. 墙模板制作。依据模板设计制作图进行模板制作,先将方木背楞厚度方向上的上下两个面刨平、刨直,挑选厚度一致的背楞和胶合板。

根据单块模板制作大样图进行裁板,并对胶合板裁边处采用封口漆封边。

面板与背楞固定采用沉头螺钉或圆钉固定牢固。

同一面墙出现两块以上模板,其接缝处做法如图 5.14 所示。

5.1.2 The role and Basic Requirements of Formwork

1. The Role of Formwork

In the construction of reinforced concrete structure, the concrete just mixed from the mixer is in plastic flow state, which needs to be poured into the model with the same shape and size as the component to set and harden, so as to form the required structural components. The formwork system is a tool to ensure the correct shape and size of concrete in the pouring process, and to protect and maintain the concrete during the hardening process.

2. Basic requirements for formwork

In the construction of cast-in-place reinforced concrete structure, the basic requirements for the formwork system are as follows:

(1) The formwork installation shall ensure the correctness of the shape, size and mutual position of each part of the structure and components.

(2) Sufficient strength, rigidity and stability shall be provided to ensure construction quality and safety.

(3) The utility model has the advantages of simple structure, convenient assembly and disassembly, and can be used repeatedly.

(4) The joint is tight without leakage.

(5) Material saving and low cost.

5.1.3 Formwork Construction Technology

The most commonly used formwork on the construction site are wood (bamboo) plywood formwork, combined steel formwork and all steel large formwork, and the construction technology is as follows.

1. ConstructionTechnology of Wood(Bamboo)Plywood Formwork

(1) Wall formwork installation.

①Construction process: formwork production → installation of door and window opening formwork → installation of one side wall formwork → through split bolt sleeve and fixation → installation of opposite side wall formwork → installation of split bolt and back bolt → installation of diagonal brace or ground anchor → correction and reinforcement → pre inspection.

②Main points of construction.

a. Fabrication of wall formwork. The formwork is made according to the formwork design drawing. First, planed and straightened the upper and lower surfaces along the thickness direction of the square timber back, and then selected the back and plywood with the same thickness.

Cut the plate according to the detail drawing of single formwork, and sealing paint is used to seal the cutting edge of plywood.

The panel and back ridge are fixed firmly with countersunk screws or round nails.

If there are more than two formworks on the same wall, the method of joint is shown in Fig. 5.14.

单块模板制作完,按模板设计图纸对模板外形、尺寸、平整度、对角线进行检查,分规格平行叠放.其底层模板加垫木,距地面不小于 100 mm。

模板安装前,应均匀涂刷隔离剂。

b. 墙模板安装。按照位置线安装门窗洞口模板及预埋件或木砖,并与墙体钢筋固定。

分别对模板下口和模板之间接缝处粘贴海绵条。先安装角模,并与墙体钢筋临时固定,再安装一侧墙模,使模板稳定的坐落到基准面上,并设临时支撑。

按对拉螺栓的位置将套管固定,对有防水要求的混凝土墙应安装止水螺栓,清扫墙内杂物,随后将另一侧模板就位,依次将其余模板就位,并随时做好支撑。

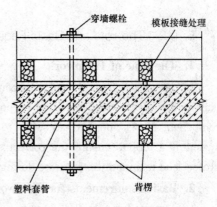

图 5.14 墙模板接缝处做法

背楞依据设计要求的规格、数量、间距自下而上安装,并随时拧紧对拉螺栓螺母固定。

斜撑位置应按模板设计要求进行,一般将斜撑固定在楼板中的预埋钢筋环上,每块模板立面宜采用两根以上斜撑,与地面角度宜为 45°～60°。

c. 墙模板校正加固。根据墙模板控制线校正墙模板位置,并利用方木、木楔与地锚将墙模下口固定。

模板上口吊线坠检查墙模板垂直度,通过斜撑校正墙身垂直度,调整合格后固定。

再次复查墙模板轴线位移、垂直度、截面尺寸、对角线偏差,调整无误后,锁定斜撑。

模板安装校正完毕后,将模板接缝及模板下口进行封严。

After the single formwork is manufactured, the shape, size, flatness and diagonal line of the formwork shall be checked according to the formwork design drawings, and the formwork shall be stacked in parallel according to the specifications. The bottom layer of formwork shall be padded, and the distance from the ground shall be no less than 100mm.

Before the formwork installation, the isolating agent shall be evenly applied.

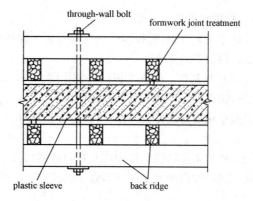

Fig. 5.14 Construction Method of Wall Formwork Joint

b. Wall formwork installation. According to the location line, install the door and window opening formwork and embedded parts or wood bricks, and fix them with the wall reinforcement.

Paste sponge strips on the bottom of the formwork and the joint between the formwork. First install the angle formwork, and fix it with the wall reinforcement temporarily, and then install the wall formwork on one side to make the formwork stably sit on the datum plane, and set up temporary support.

The casing pipe shall be fixed according to the position of split bolt. For concrete wall with waterproof requirements, water stop bolt shall be installed, and sundries inside the wall shall be cleaned. Then, the formwork on the other side shall be put in place, and the remaining formwork shall be put in place successively, and the support shall be made at any time.

The back ridge shall be installed from bottom to top according to the specification, quantity and spacing required by the design, and the split bolt and nut shall be tightened at any time for fixation.

The position of diagonal bracing shall be carried out according to the requirements of formwork design. Generally, the diagonal bracing shall be fixed on the embedded reinforcement ring in the floor slab. More than two diagonal braces should be used for the facade of each formwork, and the angle with the ground should be 45°—60°.

c. Correction and reinforcement of wall formwork. According to the control line of the wall formwork, the position of the wall formwork is corrected, and the lower opening of the wall formwork is fixed by square timber, wood wedge and ground anchor.

The verticality of the wall formwork shall be checked with a plumb at the upper end of the formwork. The verticality of the wall body shall be corrected by diagonal bracing, and fixed after the adjustment is qualified.

Recheck the axis displacement, perpendicularity, section size and diagonal deviation of the wall formwork again, and lock the diagonal brace after the adjustment is correct.

After the installation and correction of the formwork, the formwork joint and the lower end of the formwork shall be sealed.

(2)柱模板安装。

①施工工艺流程:模板制作→柱模安装→柱模板校正加固→办理预检及验收。

②柱模板安装要点。

a. 柱模板制作。依据模板设计制作图进行模板制作,先将木背楞厚度方向上的上下两个面刨平、刨直,挑选厚度一致的背楞和胶合板。

依据单块模板制作大样图,将胶合板按制作尺寸要求裁板,并对裁边处的胶合板采用封口漆封边。

胶合板与木背楞固定采用沉头螺钉或圆钉固定牢固,制作时应预留清扫口和振捣口。

柱模四角制作时宜采用企口做法,如图 5.15 所示。

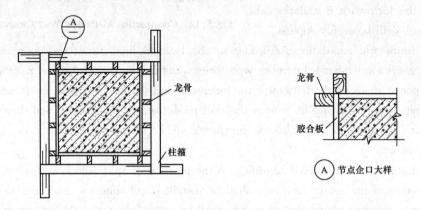

图 5.15 柱模节点示意图

单块模板制作完,按模板设计图纸对模板外形、尺寸、平整度、对角线进行检查,分规格平行叠放,其底层模板加垫木,距地面不小于 100 mm。

模板安装前,应均匀涂刷隔离剂。

b. 柱模板安装。每柱四片模板,四角相邻两块柱模板宜采用企口连接,模板安装时,沿柱模板边线外 2 mm 粘贴海绵条进行密封。

第一片模板就位后,设临时支撑或用柱主筋临时固定,然后依次将其余 3 片模板就位,并做好支撑。

依据设计要求的柱箍规格、数量、间距自下而上安装柱箍。

斜撑或拉杆的位置应按模板设计要求进行,一般将拉杆或斜撑固定在楼板中的预埋钢筋环上,柱模板立面均设两根拉锚或斜撑,与地面宜为 45°～60°。

(2) Column formwork installation.

①Construction process flow. formwork production → column formwork installation → column formwork correction and reinforcement → pre inspection and acceptance.

②Main points of column formwork installation.

a. Fabrication of column formwork. According to the formwork design and production drawing, the upper and lower sides in the thickness direction of the wood back edge shall be planed and straightened, and the back ridge and plywood with the same thickness shall be selected.

According to the detail drawing of single formwork, the plywood is cut according to the production size requirements, and the plywood edge at the cutting edge is sealed with sealing paint.

The plywood and wood back edge shall be fixed firmly with countersunk screw or round nail, and the cleaning and vibrating opening shall be reserved during the production.

The four corners of the column formwork should be made with tongue and groove, as shown in Fig. 5.15.

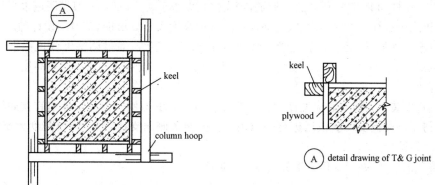

Fig. 5.15 Schematic Diagram of Column Formwork Joint

After the single formwork is manufactured, the shape, size, flatness and diagonal line of the formwork shall be inspected according to the formwork design drawings, and the formwork shall be stacked in parallel according to the specifications. The bottom layer formwork shall be padded and the distance from the ground shall not be less than 100mm.

Before the formwork installation, the isolating agent shall be evenly applied.

b. Column formwork installation. There are four formworks for each column, and the two adjacent column formworks at four corners should be connected with tongues and grooves. When installing the formwork, sponge strips should be pasted 2mm outside the edge line of the column formwork for sealing.

After the first formwork is in place, temporary support or temporary fixation with column main reinforcement shall be set, and then the remaining three formworks shall be placed in place and supported.

According to the specification, quantity and spacing of column hoop, the column hoop shall be installed from bottom to top.

The position of diagonal bracing or tie rod shall be carried out according to the requirements of formwork design. Generally, the tie rod or diagonal brace shall be fixed on the embedded reinforcement ring in the floor slab. Two anchor bolts or diagonal braces shall be set on the elevation of column formwork, and it shall be 45°—60° to the ground.

c. 柱模板校正加固。根据柱控制线校正柱模位置,并采用木楔与地锚将柱模下口固定。

将线坠分别吊于模板及相邻模板的上口,使线坠由模板上口延伸接近楼面检查模板的垂直度,并且通过拉锚或斜撑校正柱身垂直度和柱身扭向,调整偏差后固定。

(3)梁模板安装。

①施工工艺流程:梁模板制作→搭设梁模支撑→安装梁底模板→安装梁侧模→安装侧向支撑→校核梁截面尺寸并加固(→预检)。

②梁模板安装要点。按照梁模设计图,现场加工制作模板。先将木龙骨刨平、刨直,再挑选厚度一致的胶合板,按加工尺寸进行裁切,板边应涂刷封口漆,防止受潮变形,最后用铁钉将面板与龙骨钉牢。

搭设梁模板支撑。根据梁支撑体系平面布置图,安装梁支撑立杆,立杆下应铺设垫木。当立杆支设在钢筋混凝土楼板上时,垫木一般采用 50 mm×100×400 mm 木方。上下层立杆应对准,待顶板支撑搭设完毕后,应将梁、板支撑体系连成一体。每层立杆力求做到规格一致、横竖成排,合理设置水平拉杆和剪刀撑,留好施工通道,当支撑高度小于 4.5 m 时,一般设两道水平拉杆和剪刀撑,第一道水平拉杆(扫地杆)距楼、地面为 250 mm,高度大于 5 m 时,应专项设计确定。

在立杆上依次安放可调顶托、龙骨,铺设梁底模时拉十字线,通过调整可调顶托来校正梁底标高。当梁跨度大于等于 4 m 时,梁底应按设计要求起拱,如设计无要求,起拱高度宜为跨度的 1‰~3‰起拱。

梁侧模板安装。梁模安装应遵循"帮(侧模)包底(模)"的原则。侧模安装前宜在底模侧边粘贴海绵条,待梁侧模板安装就位后进行临时固定。安装梁侧向支撑,当梁侧模高度大于 600 mm 时,宜采用对拉螺栓。

校核梁截面尺寸,并用锁口方木和斜撑进行加固。

Chapter 5   Construction of Reinforced Concrete Structure

c. Correction and reinforcement of column formwork. According to the column control line, the position of column formwork is corrected, and the lower end of column formwork is fixed by wooden wedge and ground anchor.

Hang the plumb on the upper end of the formwork and adjacent formwork respectively to make the plumb extend from the upper end of the formwork to the floor, check the verticality of the formwork, correct the verticality and the torsion of the column body by pulling anchor or diagonal brace, and fix it after adjusting the deviation.

(3) Beam Formwork Installation.

①Construction process. Fabrication of beam formwork → erection of beam formwork support → installation of beam bottom formwork → installation of beam side formwork → installation of lateral support → check of beam section size and reinforcement(→ pre inspection).

② Main points of beam formwork installation. According to the design drawing of beam formwork, the formwork shall be processed and manufactured on site. First, the wood keel shall be planed and planed straight, then the plywood with the same thickness shall be selected and cut according to the processing size. The edge of the Board shall be painted with sealing paint to prevent moisture deformation. Finally, the panel and keel shall be nailed firmly with iron nails.

Erection of beam formwork support. According to the layout plan of the beam support system, the beam supporting pole shall be installed, and the cushion timber shall be laid under the vertical pole. When the vertical pole is supported on the reinforced concrete floor, the cushion timber is generally 50mm × 100 × 400mm. The upper and lower vertical poles should be aligned. After the erection of top floor support, the beam and slab support system should be connected as a whole. The vertical poles of each floor shall be consistent in specifications and arranged horizontally and vertically. The horizontal tie rod and cross bracing shall be set reasonably, and the construction channel shall be reserved. When the support height is less than 4.5m, two horizontal tie rods and cross bracing shall be generally set. The first horizontal tie rod (bottom horizontal pipe) shall be 250mm away from the floor and ground. When the height is greater than 5m, it shall be determined by special design.

The adjustable jacking and keel are placed on the vertical pole in turn. The cross line is drawn when laying the beam bottom formwork, and the elevation of the beam bottom is corrected by adjusting the adjustable jacking. When the span of the beam is more than or equal to 4m, the bottom of the beam should be arched according to the design requirements. If there is no design requirement, the arch height should be 1‰ to 3‰ of the span.

Beam side formwork installation: The installation of beam formwork should follow the principle of "side (side formwork) covering bottom (formwork)". Before the installation of side formwork, sponge strips should be pasted on the side of bottom formwork, and temporary fixation should be carried out after the beam side formwork is installed in place. When the height of beam side formwork is greater than 600mm, split bolts should be used.

Check the cross-section size of the beam and reinforce it with lock square timber and diagonal brace.

(4)顶板模板安装。

①施工工艺流程:搭设板模支撑→安装主、次龙骨→铺设顶板模板→调整模板标高→模板清理(→预检)。

②顶板模板安装要点。根据支撑体系设计搭设板模支撑,上下层支撑的立杆应对准并铺设垫木;支撑搭设完毕后,安装顶托,顶托的外露螺纹长度不得超过 300 mm。

支撑体系验收合格后,根据主、次龙骨平面布置图安装主、次龙骨;次龙骨应错开搭接,间距应均匀。

根据顶板模板排板图,采用硬拼法进行顶板模板的安装;相邻两块胶合板长边拼缝应在次龙骨上,并分别用铁钉钉牢。

拉十字线,调整可调顶托校正顶板标高。

当定板短向跨度≥4 m 时,模板应按短向跨度的 1‰~3‰起拱,起供的位置在板中,边缘部分要保持水平。

模板安装完后,用水准仪或拉线检查板面标高,并将模板表面清理干净,办理预检手续。

(5)模板拆除。木(竹)胶合板模板等现浇混凝土结构的模板及支架在拆除时,混凝土的强度必须达到设计要求。底模拆除时的混凝土强度要求见表 5.3。拆除承重模板的时间可参见表 5.4。

表 5.3 梁、板底模拆除时的混凝土强度要求

构件类型	构件跨度/m	达到设计的混凝土立方体抗压强度标准值的百分率/%
板	≤2	≥50
	>2,≤8	≥75
	>8	≥100
梁、拱、壳	≤8	≥75
	>8	≥100
悬臂构件	—	≥100

Chapter 5 Construction of Reinforced Concrete Structure

(4) Installation of roof formwork.

①Construction process: set up formwork support → install main and secondary keel → lay roof formwork → adjust formwork elevation → clear formwork(→ pre inspection).

②Main points of roof formwork installation. According to the design of the support system, the formwork support shall be set up, and the vertical poles of the upper and lower supports shall be aligned and the cushion timber shall be laid. After the support is erected, the jacking shall be installed, and the exposed thread length of the jacking shall not exceed 300mm.

After the support system is accepted, the main and secondary keels shall be installed according to the layout plan of main and secondary keels. The secondary keels shall be staggered and overlapped, and the spacing shall be uniform.

According to the layout of the roof formwork, the hard splicing method is adopted for the installation of the roof formwork. The long edge joint of the adjacent two pieces of plywood shall be on the secondary keel, and shall be nailed with iron nails respectively.

Draw cross line, adjust adjustable jacking and correct roof elevation.

When the short span of the roof is more than 4m, the formwork shall be arched according to 1‰ to 3‰ of the short span, and the position of the supply shall be in the slab, and the edge part shall be kept horizontal.

After the formwork is installed, the elevation of the board surface shall be checked by level gauge or guy wire, and the formwork surface shall be cleaned up and the pre inspection procedures shall be handled.

(5) Formwork removal. When the formwork and support of cast-in-place concrete structure such as wood(bamboo)plywood formwork are removed, the concrete strength must meet the design requirements. See Table 5.3 for concrete strength requirements during bottom formwork removal. See Table 5.4 for removal time of bearing formwork.

Table 5.3 Concrete Strength Requirements of Beam and Slab Bottom Formwork Removal

Component Type	Component Span/m)	Percentage Reaching the Standard Value of Concrete Cube Compressive Strength Designed/%
plate	≤2	≥50
	>2,≤8	≥75
	>8	≥100
beam, arch and shell	≤8	≥75
	>8	≥100
cantilever member	—	≥100

表 5.4 拆除承重模板的参考时间(单位:天)

水泥强度及品种	混凝土达到设计强度比/%	混凝土硬化时昼夜平均温度/℃					
		5	10	15	20	25	30
32.5普通水泥	50	12	8	6	4	3	3
	75	28	20	14	10	8	7
	100	55	45	35	28	21	18
42.5普通水泥	50	10	7	6	5	4	3
	75	20	14	11	8	7	6
	100	50	40	30	28	20	18
32.5矿渣水泥	50	18	12	10	8	7	6
	75	32	25	17	14	12	10
	100	60	50	40	28	24	20
42.5矿渣水泥	50	16	11	9	8	7	6
	75	30	20	15	13	12	10
	100	60	50	40	28	24	20

木(竹)胶合板模板拆除如下:
①墙、柱模板拆除。拆除墙、柱模板时,混凝土强度应以保证其表面及棱角不受损坏。

拆除模板顺序:按照模板设计要求的拆模顺序进行,先支的模板后拆,后支的先拆。

墙模板拆除时,应对模板进行临时固定,然后松开对拉螺栓螺母、斜撑,拆除模板下口木楔,拆除模板横楞,随后拆除对拉螺栓,使模板向后倾斜与墙体脱开;拆柱模板时,先拆拉锚、斜撑,后拆柱箍,如模板与混凝土面粘结时,可用撬棍轻轻撬动模板,但不得用大锤砸模板上口。门窗洞口模板应根据洞口宽度和混凝土强度掌握拆模时间。

模板吊出时,必须检查模板是否与墙体有钩挂的地方,且依次吊入模板插放架内。

模板应及时清理、修理,涂刷隔离剂,以便下次使用。

②梁、板模板拆除。拆除工艺流程:拆除梁侧模斜撑、连接件及侧模→下调顶托,拆除主、次龙骨→拆除顶板区域支撑体系→拆除顶板模板→拆除梁底支撑及模板→清理模板。

Chapter 5 Construction of Reinforced Concrete Structure

Table 5.4 Reference Time for Removal of Bearing Formwork (unit: days)

Strength and Variety of Cement	Concrete to Design Strength Ratio/%	Day and Night Average Temperature of Concrete Hardening/℃					
		5	10	15	20	25	30
32.5 portland cement	50	12	8	6	4	3	3
	75	28	20	14	10	8	7
	100	55	45	35	28	21	18
42.5 portland cement	50	10	7	6	5	4	3
	75	20	14	11	8	7	6
	100	50	40	30	28	20	18
32.5 slag cement	50	18	12	10	8	7	6
	75	32	25	17	14	12	10
	100	60	50	40	28	24	20
42.5 slag cement	50	16	11	9	8	7	6
	75	30	20	15	13	12	10
	100	60	50	40	28	24	20

The removal of wood(bamboo) plywood formwork is as follows:

①Removal of wall and column formwork. When removing the formwork of wall and column, the strength of concrete shall be such as to ensure that its surface and edges are not damaged.

Sequences of Removing Formwork

The formwork shall be removed according to the formwork design requirements. The formwork erected first shall be removed later, and the formwork erected later shall be removed first.

When the wall formwork is removed, the formwork shall be temporarily fixed, and then the split bolts, nuts and diagonal braces shall be loosened, the wooden wedge at the bottom of the formwork shall be removed, and the transverse edge of the formwork shall be removed, and then the split bolt shall be removed to make the formwork tilt backward and separate from the wall. When removing the column formwork, the anchor and diagonal brace shall be removed first, and then the column hoop shall be removed. If the formwork is bonded with the concrete surface, the formwork can be gently pried with a sled stick, but the upper opening of the formwork shall not be smashed with a sledge hammer. The removal time of formwork for door and window openings shall be controlled according to the width of opening and concrete strength.

When the formwork is lifted out, it is necessary to check whether the formwork is hooked with the wall, and lift it into the formwork insertion frame in turn.

The formwork shall be cleaned and repaired in time and coated with isolation agent for next use.

②Removal of beam and slab formwork: Removing process. Remove the diagonal bracing, connectors and side formwork of beam side formwork → lower the jacking, remove the main and secondary keel → dismantle the roof area support system → remove the roof formwork → remove the beam bottom support and formwork → clean the formwork.

梁、板模板拆除施工要点如下：

拆除梁、板模板要以同条件试块抗压强度试验报告为准，并应符合《混凝土结构工程施工质量验收规范》(GB 50204—2011)中的底模拆除时混凝土强度要求。填写拆模申请，经审批后方可拆模。

模板拆除宜逐跨依次拆除梁侧模板(帮包底)、顶板模板、梁底模板。

梁侧模板拆除。依次拆除梁侧模板的斜撑、锁口木方、连接件以及侧模。

顶板模板拆除。下调可调顶托，拆除主、次龙骨，再依次拆除支撑体系的斜杆、横杆及立杆，最后拆除顶板模板。

按拆除顶板模板的顺序拆除梁底支撑和底模。

后浇带模板的拆除应满足混凝土的强度要求，支顶应按施工技术方案执行。

拆下的模板应及时清理。钢管、龙骨、模板应分类码放，但不应集中放置，防止集中荷载过大，导致顶板出现裂缝。

(6)季节性施工要点。

①冬施模板安装后应按冬施要求对模板进行保温。

②冬施模板拆除时混凝土强度应达到受冻临界强度，并在拆模后加强混凝土保温覆盖。

2. 组合钢模板施工工艺

(1)柱模板安装施工。

①施工工艺流程：楼地面处理→焊模板定位筋→安装柱模板→安装柱箍→安装拉杆及斜撑并校正→模板预检。

②柱模板安装要点。当模板支设在回填土上时，应将回填土分层夯实，表面平整；当模板支在楼面上时，应沿柱边线外 2 mm 粘贴海绵条。若表面平整度偏差过大，应按照标高抹好水泥砂浆找平层，防止漏浆。

按照柱的位置线在预埋插筋上焊水平定位筋，每边不少于 2 点，从四面顶住模板，以固定模板位置，防止位移，如图 5.16 所示。

The main points of beam and slab formwork removal are as follows:

The removal of beam and slab formwork shall be subject to the compressive strength test report of test block under the same conditions, and shall meet the concrete strength requirements for bottom formwork removal in code for acceptance of construction quality of concrete structures (GB 50204—2011). Fill in the form removal application and remove the formwork after approval.

When removing the formwork, it is advisable to remove the beam side formwork (side covering bottom), roof formwork and beam bottom formwork in turn.

When removing beam side formwork, it is advisable to remove the diagonal brace, lock wood square, connector and side formwork of beam side formwork in turn.

When removing roof formwork, it is advisable to adjust the adjustable jacking, remove the main and secondary keels, then remove the inclined bar, cross bar and vertical pole of the support system in turn, and finally remove the roof formwork.

Remove the beam bottom support and bottom formwork according to the order of removing the roof formwork.

The removal of post cast strip formwork shall meet the strength requirements of concrete, and the top support shall be carried out according to the construction technical scheme.

The removed formwork shall be cleaned in time. The steel pipe, keel and formwork should be stacked by category, but should not be placed in a centralized way to prevent cracks in the roof caused by excessive concentrated load.

(6) Main points of seasonal construction.

①After the installation of winter construction formwork, the formwork shall be insulated according to the requirements of winter construction.

②When the formwork is removed in winter, the concrete strength should reach the freezing critical strength, and the concrete insulation cover should be strengthened after the formwork removal.

2. Construction Technology of Combined Steel Formwork

(1) Column formwork installation construction.

①Construction process: floor treatment → welding formwork positioning bar → installing column formwork → installing column hoop → installing tie rod and diagonal brace and correcting → formwork pre inspection.

②Main points of column formwork installation. When the formwork is supported on the backfill, the backfill shall be compacted in layers and the surface shall be flat; when the formwork is supported on the floor, sponge strips shall be pasted 2mm outside the column edge line. If the surface flatness deviation is too large, the cement mortar leveling layer shall be plastered according to the elevation to prevent slurry leakage.

According to the position line of the column, weld the horizontal positioning bar on the embedded joint bar, with no less than 2 points on each side. Support the formwork from four sides to fix the position of the formwork and prevent displacement, as shown in Fig. 5.16.

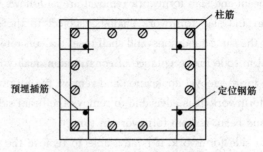

图 5.16 定位筋示意图

　　a. 柱模板安装。按柱子尺寸和位置线,将各块模板依次安装就位,用铅丝将模板与主筋临时绑扎固定后,用"U"形卡将相临模板连接卡紧;采用预组装法时,应预拼成一面一片(每面的边上带一个角模)或两面一片,依次吊装,安装完相临的两面再安装另外的两面。

　　b. 安装柱箍。按照模板设计的规格、间距,向下而上安装柱箍。柱截面较大时,应增设对拉螺栓;柱较高时,应按照模板设计将柱箍加密。

　　c. 支撑和拉杆的安装与校正。根据柱高、截面尺寸确定支撑、拉杆的数量,分别将其固定在预埋楼板内的钢筋环上。用经纬仪、线坠控制,用花篮螺栓、可调支撑调节,校正模板的垂直度。预埋的钢筋环距柱宜为 3/4 柱高,拉杆、支撑与地面夹角宜为 45°～60°。柱较高时,应根据需要增设支撑或拉杆。

　　安装群柱模板时,先安装两端柱模,校正、固定后,拉通线,再安装中间各柱。

　　(2)墙模板安装。墙模板的安装可采用就位安装法和预组装法。

　　①墙模板就位安装法。

　　a. 施工工艺流程:基层处理→焊模板定位筋→安装门窗洞口模板→安装第一步模板、穿对拉螺栓→安装内背楞调直并临时紧固→安装第二步模板、内背楞、对拉螺栓→外背楞安装并上紧对拉螺栓→加斜撑、拉杆、校正模板并紧固→模板预检。

　　b. 就位安装法施工要点如下:

　　当模板支设在回填土上时,应将回填土分层夯实,表面平整;当模板支在楼面上时,应沿柱边线外 2 mm 粘贴海绵条。若表面平整度偏差过大,应按照标高抹好水泥砂浆找平层,防止漏浆。

　　模板定位钢筋在墙两侧预埋插筋上点焊定位筋,间距依据支模方案确定,在墙对拉螺栓处 加焊定位钢筋,定位筋两端刷防锈漆。

a. Column formwork installation: According to the size and location line of the column, install each formwork in order, and use lead wire to temporarily bind and fix the formwork and the main reinforcement, and then use the "U" shaped clamp to connect and clamp the adjacent formwork. When the pre assembly method is adopted, it should be pre assembled into one piece on one side (with an angle mold on each side) or one piece on two sides, which should be hoisted in sequence. After the adjacent two sides are installed, the other two sides will be installed.

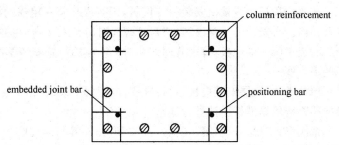

Fig. 5.16 Schematic Diagram of Positioning Bar

b. Install the column hoop: According to the specification and spacing of formwork design, column hoop shall be installed downward and upward. When the column planting surface is large, the split bolt should be added. When the column is high, the column hoop should be densified according to the formwork design.

c. Installation and correction of supports and tie rods: According to the column height and section size, the number of supports and tie rods are determined, and they are respectively fixed on the reinforcement ring in the embedded floor. The verticality of formwork shall be corrected by theodolite and plumb, adjusted by basket bolt and adjustable support. The distance between embedded reinforcement ring and column shall be 3/4 column height, and the angle between tie rod, support and ground shall be 45°—60°. When the column is high, support or tie rod shall be added as required.

When installing the column group formwork, first install the column formwork at both ends, correct and fix, pull through the line, and then install the middle columns.

(2) Wall formwork installation. The wall formwork can be installed in place and pre assembled.

①The method of wall formwork installation in place.

a. Construction process: base course treatment → welding formwork positioning bar → installing door and window opening formwork → installing the first step formwork and pulling bolt → installing the inner back bracket, straightening and temporarily fastening → installing the second step formwork, inner back ridge and split bolt → installing and tightening the split bolt on the outer back ridge → adding diagonal brace, pull rod, correcting formwork and fastening → formwork pre inspection.

b. The construction points of in place installation method are as follows:

When the formwork is supported on the backfill, the backfill shall be compacted in layers and the surface shall be flat. When the formwork is supported on the floor, sponge strips shall be pasted 2mm outside the column edge line. If the surface flatness deviation is too large, the cement mortar leveling layer shall be plastered according to the elevation to prevent slurry leakage.

The positioning reinforcement of formwork shall be spot welded on the embedded joint bars on both sides of the wall, and the spacing shall be determined according to the formwork erection scheme. The positioning reinforcement shall be welded at the wall opposite tie bolt, and both ends of the positioning bar shall be painted with antirust paint.

按照位置线安装门窗洞口模板,安放预埋件或木砖,并将模板与四周钢筋可靠固定。

按照墙模板边线依次安装一侧模板,用"U"形卡将相临的模板卡紧,同时穿入对拉螺栓和套管及顶帽。再安装另一侧模板,将对拉螺栓对应穿入孔内,并将模板做临时固定,防止倾覆。

安装完第一层模板,将内背楞按照对拉螺栓的位置安装就位,临时固定,同时初步调整该层模板的平直度。

照此方法,逐层安装模板、对拉螺栓、内背楞至顶,各相临层之间也用U形卡卡紧,墙体的对拉螺栓间距应依据模板设计进行加密。

依次将外背楞安装到位,同时,将"3"形扣件等与对拉螺栓连接,上紧对拉螺栓。随即用支撑和拉杆整体校正模板的平整顺直度和垂直度。模板安装完毕后,检查扣件、螺栓是否紧固,模板下口拼缝是否严密,并办理预检手续。

②墙模板预组装安装法。

a. 施工工艺流程:基层和定位处理→安装门窗洞口模板→安装一侧模板→安装支撑临时固定→对拉螺栓就位→清扫模内杂物→安装另一侧模板→安装支撑临时固定→安背楞、上紧对拉螺栓→加支撑、校正、固定→模板预检。

b. 预组装安装法施工要点如下:

楼地面处理、模板定位处理和门窗洞口模板安装与"就位安装法"相同。

安装一侧模板。将预组装模板吊运至工作面,按照模板位置线在地面上距模板面处通长粘贴海绵条,并在相临模板的一边侧面上距模板边沿 1.5 mm 处粘通长海绵条,然后,依据模板的位置线将一侧模板安装就位,用支撑做临时固定。

安装对拉螺栓。按照模板设计的间距将对拉螺栓、套管及顶帽穿入已安装的模板对拉螺栓孔内。

安装另一侧模板。清扫模板内杂物后,安装另一侧模板,同时将对拉螺栓对应穿入,做临时固定。

安装模板的背楞,注意相临板块接缝处水平背楞应拉通或可靠搭接,同时将对拉螺栓上紧,加支撑、调整模板并紧固。

Install the formwork of door and window openings according to the location line, place the embedded parts or wood bricks, and fix the formwork reliably with the surrounding reinforcement.

One side formwork shall be installed according to the side line of wall formwork, and the adjacent formwork shall be clamped tightly with "U" shaped clamp, and the split bolt, sleeve and top cap shall be penetrated at the same time. Then install the formwork on the other side, insert the split bolt into the hole, and fix the formwork temporarily to prevent overturning.

After the first layer of formwork is installed, the inner back is installed in place according to the position of split bolt, and is temporarily fixed. Meanwhile, the glancing flatness of this layer of formwork is preliminarily adjusted.

According to this method, the formwork, split bolt, inner back ridge to the top shall be installed layer by layer, and the U shaped clamp shall be used to clamp the adjacent layers. The spacing of bolts on the wall shall be densified according to the formwork design.

Install the outer back ridge in place in turn. At the same time, connect the "3" shaped fastener with the split bolt and tighten the split bolt. Then the flatness, straightness and perpendicularity of the formwork shall be corrected with the support and pull rod. After the formwork is installed, check whether the fasteners and bolts are tight, and whether the joint at the bottom of the formwork is tight, and handle the pre inspection procedures.

②Wall formwork pre assembly installation method.

a. Construction process: base course and positioning treatment → installation of door and window opening formwork → installation of one side formwork → installation of temporary support fixation → positioning of split bolt → cleaning of impurities in the formwork → installation of formwork on the other side → temporary fixation of support → installation of back edge and tightening of split bolt → support, correction and fixation → formwork pre inspection.

b. The construction points of pre assembly installation method are as follows:

Floor treatment, formwork positioning and formwork installation of door and window openings are the same as "in place installation method".

Install one side formwork. Lift the pre assembled formwork to the working face, paste sponge strips on the ground from the formwork surface according to the formwork location line, and stick long sponge strips at the position 1.5mm away from the edge of the formwork on one side of the adjacent formwork, and then install the formwork at one side according to the position line of the formwork, and fix it temporarily with supports.

Install the split bolt. According to the designed spacing of the formwork, the split bolt, sleeve and top cap shall be penetrated into the bolt hole of the installed formwork.

Install the other side formwork. After cleaning the sundries in the formwork, install the formwork on the other side, and at the same time, put the split bolt into the formwork for temporary fixation.

When installing the back ridge of formwork, pay attention to the horizontal back flange at the joint of adjacent plates, which should be pulled through or reliably overlapped. At the same time, the split bolt should be tightened, the support should be added, and the formwork should be adjusted and tightened.

(3)梁横板安装施工。

①施工工艺流程：弹线→搭设支架→安装梁底龙骨、模板→调整标高→安装梁钢筋→安装梁侧模板→侧模支撑并调整→模板预检。

②梁模板安装要点如下：

在墙、柱混凝土上弹出梁的轴线、标高控制线。

搭设梁支架。支架的间距根据模板设计，一般为 800～1 200 mm，支架搭设于回填土上时，应分层夯实，并将表面平整好，下垫通长木板；搭设于楼板上时，应加设垫木，并使上、下层立柱在同一竖向中心线上。梁支架的横杆宜与楼板支架拉通设置，若单独支设，立杆应加设双向剪刀撑和水平拉杆。

在支架立杆上安装顶托，顶托的外露螺纹长度不得超过 300 mm，在顶托上放置主、次龙骨，间距应均匀，然后安装梁底模板，并拉线找直，用可调顶托调整立杆的标高。当梁跨度≥4 m 时，应按照设计要求在梁中间起拱，如设计无要求时，按照全跨长度的 1‰～3‰起拱。

绑扎梁钢筋，经检查合格后办理隐检手续。

安装梁侧模板。模板接缝处距模板面处粘贴海绵条，用"U"形卡将梁侧模与梁底模通过连接角模连接，梁侧模板的支撑采用梁托架或三脚架、钢管、扣件与梁支架等连成整体，形成三角斜撑，间距宜为 700～800 mm；当梁侧模高度超过 600 mm 时，应加对拉螺栓。

安装完后，校正梁中心线、标高、起拱高度、断面尺寸。清理模板内杂物，经检查合格并办理预检手续。

(4)楼板模板安装施工。

①施工工艺流程：搭设支架→安装主、次龙骨→调整模板上皮标高并起拱→铺设模板块→非标板制作、安装→模板预检。

②楼板模板安装要点如下：

楼板模板的支架搭设前，地面及支脚的处理完毕，符合要求。支架搭设时，自边跨开始先完成一个格构的立柱、水平连接杆及斜撑安装，再逐排、逐跨向外扩展。支柱和水平杆的布置应考虑设置施工通道，立柱间距与龙骨间距应根据楼板厚度与模板规格、型号设计确定，支架搭设完成后应检查支架的标高和稳定性。

(3) Installation of Beam Transverse Plate

①Construction process: snapping line → erecting support → installing beam bottom keel and formwork → adjusting elevation → installing beam reinforcement → installing beam side formwork → supporting and adjusting side formwork → formwork pre inspection.

②The main points of beam formwork installation are as follows:

The axis and elevation control line of beam shall be popped on the wall and column concrete.

Set up beam support. The spacing of supports is generally 800—1200mm according to the design of formwork. When the support is erected on the backfill, it should be tamped in layers, and the surface should be leveled, with a full length of timber under it; when it is set on the floor, the upper and lower columns should be on the same vertical center line. The cross bar of the beam support should be connected with the floor support. If it is separately supported, the vertical pole should be added with two-way cross bracing and horizontal tie rod.

Install the jacking on the vertical pole of the support, and the exposed thread length of the jacking support shall not exceed 300mm. The main and secondary keels shall be placed on the top support, and the spacing shall be uniform. Then, the formwork at the bottom of the beam shall be installed, and the guy wire shall be used for straightening, and the elevation of the vertical pole shall be adjusted with adjustable jacking. When the span of the beam is more than or equal to 4m, the arch shall be arched in the middle of the beam according to the design requirements. If there is no design requirement, the arch shall be arched according to 1‰ to 3‰ of the whole span length.

Bind the beam reinforcement, and go through the hidden inspection procedures after passing the inspection.

Install the beam side formwork. Sponge strips shall be pasted at the joint of the formwork and the formwork at the bottom of the beam shall be connected with the side formwork of the beam with the U-shaped clamp. The support of the side formwork of the beam shall be connected with the beam bracket or tripod, steel pipe, fastener and beam support to form a triangular diagonal brace with the spacing of 700—800mm. When the height of the side formwork of the beam exceeds 600mm, the split bolt shall be added.

After installation, correct the beam center line, elevation, camber height and section size. Clean up the sundries in the formwork, pass the inspection and go through the pre inspection procedures.

(4) Installation and construction of floor formwork.

①Construction process: erection of support → installation of main and secondary keels → adjustment of formwork surface elevation and camber → laying of formwork plate → fabrication and installation of non-standard plate → formwork pre inspection.

②The main points of floor formwork installation are as follows:

Before the support of floor formwork is erected, the ground and sole timber are treated and meet the requirements. When erecting the support, the column, horizontal connecting rod and diagonal brace of a lattice structure shall be installed first from the side span, and then extended outward one by one and span by row. The layout of columns and horizontal bars should consider the setting of construction access. The spacing between columns and keels should be determined according to the thickness of floor slab, the specification and model of formwork. The elevation and stability of supports should be checked after the erection of supports.

在支架上安装顶托后,再安放主、次龙骨。

拉通线,调节可调托,调整龙骨的标高,当顶板跨度大于 4 m 时应按照设计要求在板的中部起拱,边缘部分要保持水平,当设计无要求时,应按照 1‰~3‰ 起拱。

铺设组合钢模板。先用阴角模与墙模或梁模连接,然后逐块向跨中铺设平模。相临两块模板用"U"形卡正反相间卡紧。

对于不合模数的窄条缝,应采用与模板同厚的木板,四面刨平,紧密嵌入缝中,保证接缝严密。

模板面安装完毕后,检查平整度与楼板板底标高、起拱高度,办理预检手续。

(5)组合钢模板拆除施工。组合钢模板拆除应遵循的原则:按照先支后拆,后支先拆;先拆非承重的模板,后拆承重部分的模板;自上而下;先拆侧向支撑,后拆竖向支撑。

①墙、柱模板拆除。墙体、柱模板拆除时,混凝土强度应能保证其表面及棱角不因拆除模板受损坏。

墙模板拆除。墙模板拆除应逐块拆除。先拆除斜拉杆或斜支撑,再拆除对拉螺栓及纵横龙骨或钢管卡,将 U 形卡或插销等附件拆下,然后用锤向外侧轻击模板上口,用撬棍轻轻撬动模板,使模板脱离墙体,将模板逐块传下码放;预组装模板应整体拆除、吊运。

②柱模板拆除。先拆除柱斜撑和拉杆,卸掉柱箍,再将连接每片柱模板的 U 形卡拆除,然后用锤向外侧轻击模板上口,使之松动,脱离混凝土。

③梁、板模板拆除。拆除时,应根据混凝土的强度填写拆模申请,经批准后,方可拆模。

梁、板模板拆除时,其混凝土同条件试块强度应满足表 5.3 的要求。

拆除梁与楼板模板的连接角模及梁侧模板,以使两相邻模板脱开。

After the top support is installed on the bracket, the main and secondary keels are placed.

Pull through the line, adjust the adjustable bracket, and adjust the elevation of the keel. When the span of the top plate is more than 4m, the middle part of the slab should be arched according to the design requirements, and the edge part should be kept horizontal. When there is no design requirement, the arch should be arched according to 1‰ to 3‰.

Lay composite steel formwork. First, the internal angle formwork is connected with the wall formwork or beam formwork, and then the flat formwork is laid to the middle of the span one by one. The two adjacent formworks shall be clamped with U shape clamp alternately.

For the narrow strip joints that do not conform to the modulus, the board with the same thickness as the formwork shall be used, planed on all sides, and closely embedded in the joint to ensure the tight joint.

After the formwork surface is installed, check the flatness, floor slab bottom elevation and arch camber height, and handle the pre inspection procedures.

(5) Removal construction of combined steel formwork. The principles to be followed in the removal of combined steel formwork are as follows: firstly support and then dismantle, then support first; remove the non bearing formwork first, then the bearing part formwork; from top to bottom; first remove the lateral support and then the vertical support.

①Removal of wall and column formwork. When the wall and column formwork is removed, the concrete strength shall be able to ensure that its surface and edges are not damaged due to the removal of formwork.

Removal of wall formwork. The wall formwork shall be removed one by one. First remove the diagonal tie rod or diagonal support, and then remove the split bolt, vertical and horizontal keel or steel pipe clamp. Remove the "U" shaped clamp or bolt and other accessories, and then gently tap the upper part of the formwork with a hammer, gently pry the formwork with a crowbar to separate the formwork from the wall, and transfer the formwork block by block. The pre assembled formwork shall be removed and lifted as a whole.

②Column formwork removal. First remove the diagonal bracing and tie rod of the column, remove the column hoop, and then remove the "U" shaped clamp connecting each column template, and then tap the upper part of the formwork with a hammer to make it loose and separate from the concrete.

③Removal of beam and slab formwork.

When demolishing, the formwork removal application shall be filled in according to the strength of concrete, and the formwork can be removed only after being approved.

When the beam and slab formwork is removed, the strength of concrete test block under the same conditions shall meet the requirements of Table 5.3.

Remove the connection angle formwork between beam and floor formwork, and beam side formwork to separate the two adjacent formworks.

下调支柱的可调托,先拆钩头螺栓,然后拆下"U"形卡和"L"形插销,再轻撬模板,或用锤子轻敲,拆下第一块,然后逐块、逐跨拆除模板。拆除的模板传递放到地面上,或搭设临时支架,托住下落的模板,严禁使模板自由落下。

拆除大跨度梁模时,应按照跨中至梁端的顺序松可调托、拆模板。

(6)季节性施工要点。

冬施期间应根据结构设计进行热工计算,必要时应在模板外采取相应的覆盖、保温措施。

冬期柱、墙模板拆除时混凝土强度应达到受冻临界强度,并在拆模后要用保温材料进行覆盖。

5.2 钢筋工程施工

5.2.1 钢筋的种类

钢筋混凝土结构中的钢筋,按生产工艺不同可分为:热轧钢筋、冷轧带肋钢筋、冷拉钢筋、冷拔钢丝、热处理钢筋、精轧螺纹钢筋、碳素钢丝、刻痕钢丝及钢绞线。

按化学成分不同可分为:碳素钢筋和普通低合金钢钢筋。

按钢筋直径大小可分为:钢丝(3~5 mm)、细钢筋(6~10 mm)、中粗钢筋(12~18 mm)、粗钢筋(>18 mm)。

热轧钢筋是建筑工程中用量最大的钢材品种之一,主要用于钢筋混凝土结构和预应力混凝土结构的配筋。按照《混凝土结构设计规范》(GB 50010—2010)的规定,混凝土结构的钢筋应按下列规定选用:纵向受力普通钢筋宜采用 HRB400、HRB500、HRBF400、HRBF500 钢筋,也可采用 HPB300、HRB335、HRBF335、RRB400 钢筋;梁、柱纵向受力普通钢筋应采用 HRB400、HRB500、HRBF400、HRBF500 钢筋;箍筋宜采用 HRB400、HRBF400、HPB300、HRB500、HRBF500 钢筋,也可采用 HRB335、HRBF335 钢筋;预应力筋宜采用预应力钢丝、钢绞线和预应力螺纹钢筋。

5.2.2 钢筋进场验收

钢筋进场验收包含以下内容:

(1)钢筋进场时应按下列规定检查性能及质量:

Adjust the adjustable bracket of the support. First remove the hook bolt, then remove the "U" shaped clamp and "L" bolt, and then pry the formwork, or tap with a hammer to remove the first piece, and then remove the formwork piece by piece and span by span. The removed formwork shall be transferred and placed on the ground, or temporary support shall be erected to support the falling formwork. It is strictly forbidden to make the formwork fall freely.

When removing the large-span beam formwork, the adjustable support and formwork should be loosened and removed according to the sequence from the middle span to the beam end.

(6) Main Points of Seasonal Construction

During winter construction, thermal calculation shall be carried out according to the structural design. If necessary, corresponding covering and thermal insulation measures shall be taken outside the formwork.

In winter, the concrete strength should reach the critical freezing strength when the formwork of column and wall is removed, and the insulation material should be used to cover the formwork.

5.2　Construction of Reinforcement Works

5.2.1　Types of Reinforcement

According to the different production processes, the steel bars in reinforced concrete structures can be divided into hot-rolled steel bars, cold-rolled ribbed bars, cold-drawn steel bars, cold-drawn steel wires, heat-treated steel bars, finished rolled ribbed bars, carbon steel wires, notched steel wires and steel hinge wires.

According to different chemical composition, it can be divided into carbon steel bar and ordinary low alloy steel bar.

According to the diameter of steel bar, it can be divided into steel wire(3—5mm), fine steel bar (6—10mm), medium thick steel bar(12—18mm) and coarse steel bar($>$18mm).

Hot rolled steel bar is one of the most widely used steel products in construction engineering. It is mainly used for reinforcement of reinforced concrete structure and prestressed concrete structure. According to the code for design of concrete structures (GB 50010—2010), the reinforcement of concrete structure shall be selected according to the following provisions: For longitudinal load-bearing ordinary reinforcement, HRB400, HRB500, HRBF400 and HRBF500 reinforcement should be used, and HPB300, HRB335, HRBF335 and RRB400 can also be used. For longitudinal stress of beam and column, HRB400, HRB500, HRBF400 and HRBF500 reinforcement shall be used. HRB400, HRBF400, HPB300, HRBF500 reinforcement shall be used for stirrup, and HRB335, HRBF335 reinforcement can also be used for stirrup. Prestressed steel wire, steel strand and prestressed ribbed bar should be used as prestressed reinforcement.

5.2.2　Site Acceptance of Reinforcement

The site acceptance of reinforcement includes the following contents:

(1) The performance and weight of reinforcement shall be checked according to the following provisions when entering the site.

①应检查生产企业的生产许可证证书及钢筋的质量证明书。
②应按国家现行有关标准的规定抽样检验屈服强度、抗拉强度、伸长率及单位长度质量偏差,屈服强度、抗拉强度、伸长率性能应符合如下的有关规定:

a. 钢筋单位长度质量偏差应符合表5.5的规定。

表5.5 钢筋单位长度质量偏差要求

公称直径/mm	实际质量与理论质量的偏差
≤12	±7%
14~20	±5%
≥22	±4%

b. 钢筋的规格和性能应符合国家现行有关标准的规定。常用钢筋的主要性能指标应符合《混凝土结构工程施工规范》(GB 50666—2011)附录B(常用钢筋的规格和力学性能)的规定,公称直径、公称截面面积、计算截面面积及理论质量应符合《混凝土结构工程施工规范》(GB 50666—2011)附录C的规定。

c. 对有抗震设防要求的结构,其纵向受力钢筋的性能应满足设计要求;当设计无具体要求时,对按一、二、三级抗震等级设计的框架和斜撑构件(含梯段)中的纵向受力钢筋应采用HRB335E、HRB400E、HRB500E、HRBF335E、HRBF400E或HRBF500E钢筋,其强度和最大力下总伸长率的实测值应符合下列规定:

钢筋的抗拉强度实测值与屈服强度实测值的比值不应小于1.25;
钢筋的屈服强度实测值与屈服强度标准值的比值不应大于1.30;
钢筋的最大力下总伸长率不应小于9%。

③经产品认证符合要求的钢筋,其检验批量可扩大一倍。在同一工程项目中,同一厂家、同一牌号、同一规格的钢筋连续三次进场检验均合格时,其后的检验批量可扩大一倍。

④钢筋的表面质量应符合国家现行有关标准的规定;钢筋应平直、无损伤、表面不得有裂纹、油污、颗粒状或片状老锈。

Chapter 5 Construction of Reinforced Concrete Structure

①The production license certificate of the manufacturer and the quality certificate of the steel bar should be checked.

②The yield strength, tensile strength, elongation and weight deviation per unit length shall be inspected by sampling according to the current national standards. The yield strength, tensile strength and elongation performance shall meet the following requirements:

①the weight deviation of reinforcement per unit length should meet the requirements of Table 5.5.

Table 5.5 Requirements for Weight Deviation of Reinforcement Per Unit Length

Nominal Diameter/mm	Deviation between Actual Weight and Theoretical Weigh
≤12	±7%
14~20	±5%
≥22	±4%

②the specification and performance of reinforcement should conform to the current national standards. The main performance indexes of commonly used steel bars shall comply with the provisions of Appendix B (specifications and mechanical properties of common reinforcement) of code for construction of concrete structures (GB 50666—2011). The nominal diameter, nominal section area, calculated section area and theoretical weight shall comply with Appendix C of code for construction of concrete structures (GB 50666—2011).

③for the structure with seismic fortification requirements, the performance of longitudinal load-bearing reinforcement should meet the design requirements. When there are no specific requirements in the design, the longitudinal load-bearing reinforcement in the frame and diagonal bracing members (including ladder) designed according to the seismic grade Ⅰ, Ⅱ and Ⅲ should be HRB335E, HRB400E, HRB500E, HRBF335E, HRBF400E or HRBF500E. The measured values of the strength and total extensibility under the maximum force of the reinforcement shall meet the following requirements:

The ratio of measured tensile strength to yield strength of reinforcement shall not be less than 1.25.

The ratio between the measured value of yield strength and the standard value of yield strength should not be greater than 1.30.

The total extensibility of reinforcement under the maximum force shall not be less than 9%.

③The inspection batch can be doubled for the steel bars that meet the requirements after product certification. In the same project, if the steel bars of the same manufacturer, same brand and same specification are qualified for three consecutive mobilization inspections, the subsequent inspection batch can be doubled.

④The surface quality of reinforcement shall comply with the provisions of current national standards. The reinforcement shall be straight, without damage, and the surface shall be free from cracks, oil stains, granular or flake old rust.

⑤当无法准确判断钢筋品种、牌号时,应增加化学成分、晶粒度等检验项目。

(2)成型钢筋进场时,应检查成型钢筋的质量证明书及成型钢筋所用材料的检验合格报告,并应抽样检验成型钢筋的屈服强度、抗拉强度、伸长率。检验批量可由合同约定,且同一工程、同一原材料来源、同一组生产设备生产的成型钢筋,检验批量不应大于 100 t。

(3)盘卷供货的钢筋调直后应抽样检验力学性能和单位长度质量偏差,其强度应符合国家现行有关产品标准的规定,断后伸长率、单位长度质量偏差应符合现行国家标准《混凝土结构工程施工质量验收规范》GB 50204 的有关规定。

(4)当发现钢筋脆断、焊接性能不良或力学性能显著不正常等现象时,应停止使用该批钢筋,并对该批钢筋进行化学成分检验或其他专项检验。

5.2.3 钢筋加工

钢筋加工是钢筋工程施工中的主导施工过程,主要包括钢筋调直、除锈、剪断、连接接长、弯曲成形等工作。每一道工序都关系到钢筋混凝土构件的施工质量,各个环节都应当严肃、认真地对待。

1. 钢筋的调直

钢筋调直可采用冷拉调直或调直切断机械进行。采用冷拉调直只是为了调直,而不是为了提高钢筋强度,可用调直冷拉率控制:HPB235、HPB300 光圆钢筋的冷拉率不宜大于 4%;HRB335、HRB400、HRB500、HRBF335、HRBF400、HRBF500 及 RRB400 带肋钢筋的冷拉率不宜大于 1%。

采用无延伸功能的机械设备调直的钢筋,可不进行本条规定的检验。

为提高施工机械化水平,在大中型钢筋混凝土工程中钢筋的调直宜采用钢筋调直切断机,它具有自动调直、定位切断、除锈、清垢等多种功能。钢筋调直切断机按调直原理,可分为孔模式和斜辊式;按传动原理,可分为液压式、机械式和数控式。

⑤When it is impossible to accurately judge the type and grade of steel bar, the inspection items such as chemical composition and grain size should be added.

(2) The quality certificate of the formed reinforcement and the inspection report of the materials used for the formed reinforcement shall be checked when the formed reinforcement enters the site, and the yield strength, tensile strength and extensibility of the formed reinforcement shall be inspected by sampling. The inspection batch can be agreed in the contract, and the inspection batch shall not be greater than 100 t for the formed reinforcement produced by the same project, the same raw material source and the same production equipment.

(3) After straightening, the mechanical properties and weight deviation per unit length of steel bars supplied by coils shall be inspected by sampling. The strength shall conform to the current national product standards. The elongation after fracture and weight deviation per unit length shall comply with the relevant provisions of the current national standard code for acceptance of construction quality of concrete structures(GB 50204).

(4) When brittle fracture, poor welding performance or abnormal mechanical properties are found, the use of this batch of reinforcement shall be stopped, and the chemical composition inspection or other special inspection shall be carried out for the batch of reinforcement.

5.2.3 Reinforcement Processing

Reinforcement processing is the leading construction process in the construction of steel bar works, mainly including steel bar straightening, derusting, cutting, connecting and lengthening, bending and forming, etc. Each process is related to the construction quality of reinforced concrete components, and each link should be taken seriously.

1. Straightening of Reinforcement

The steel bar can be straightened by cold drawing straightening or straightening cutting machine. The cold drawing straightening is only used to straighten, not to improve the strength of the steel bar. The cold drawing rate can be adjusted: The cold tensile rate of HPB235 and HPB300 plain round reinforcement should not be greater than 4%; the cold drawing rate of HRB335, HRB400, HRB500, HRBF335, HRBF400, HRBF500 and RRB400 ribbed bars should not be greater than 1%.

The steel bars straightened by mechanical equipment without extension function may not be subject to the inspection specified in this article.

In order to improve the level of construction mechanization, the steel bar straightening cutting machine should be used in the large and medium-sized reinforced concrete works. It has many functions such as automatic straightening, positioning cutting, rust removal and scale cleaning. According to the straightening principle, the steel bar straightening cutting machine can be divided into hole mode and inclined roller type. According to the transmission principle, it can be divided into hydraulic type, mechanical type and numerical control type.

2. 钢筋的切断

(1) 钢筋切断机的种类。钢筋下料时必须按钢筋下料长度切断。钢筋切断可采用以上所讲钢筋调直切断机,也可用钢筋切断机或手动切断器。手动切断器只是用于切断直径小于 16 mm 的钢筋;钢筋切断机可切断直径 40 mm 的钢筋。钢筋切断机按工作原理,可分为凸轮式和曲柄连杆式;按传动方式,可分为液压式和机械式。

(2) 钢筋切断机使用要求。

① 使用前应当认真检查刀片安装是否牢固,两刀片的间隙是否在 0.5～1 mm 范围内,电气设备是否正常,所有零件是否拧紧,经过空车试运转正常后方可使用。

② 在切断钢筋时,必须要握紧切断的钢筋,以防止钢筋末端摆或弹出伤人。在切断短钢筋时,不允许用手送料,应当用钳子夹住短筋送料。

③ 在进行切断钢筋时,必须是先调直后切断。在钢筋送料时,应在活动刀片退离固定刀片时进行,钢筋应放在刀刃的中部,并垂直于切断刀口。

④ 在切断机运转的过程中,不得对其进行修理和校正工作,不得随意取下防护罩,不得触及运转部位,不得将手放在刀刃切断位置,不得用手抹或嘴吹方式清理铁屑,上述一切工作均必须在停机后进行。

⑤ 严格执行钢筋切断机的操作规程,禁止切断规定范围以外的钢筋和其他材料,也不允许切断烧红的钢筋及超过刀刃硬度的材料。

⑥ 作业完毕后应清除刀具和刀具下边的杂物,清洁机体;认真检查各部螺栓的紧固程度及三角皮带的松紧度;调整活动刀片与固定刀片之间的空隙,更换磨钝的刀片。

⑦ 按有关规定定期保养,即对钢筋切断机需要润滑部位进行周期性维修保养;检查齿轮、轴承及偏心体的磨损程度,调整各部位的间隙。

3. 钢筋的弯曲

(1) 钢筋弯曲的种类。钢筋按计算的下料长度切断后,应按钢筋设计图纸、弯曲设备特点、钢筋直径、弯曲角度等进行画线,以便弯曲成设计的尺寸和形状。

(2) 钢筋弯曲机的使用要求。

① 根据钢筋弯曲加工的要求,钢筋弯曲直径是钢筋直径的不同倍数,不同直径的钢筋其弯曲直径也不能相同,弯曲时应根据钢筋直径来选用相应规格的心轴。

Chapter 5 Construction of Reinforced Concrete Structure

2. Steel Bar Cutting

(1) Types of steel bar cutter. When the steel bar is cut off, it must be cut off according to the length of the steel bar. The steel bar can be cut off by the steel bar straightening cutting machine mentioned above, as well as the steel bar cutter or manual cutting device. The manual cutter is only used to cut the steel bar with the diameter less than 16mm; the steel bar cutter can cut the steel bar with the diameter of 40mm. According to the working principle, the steel bar cutter can be divided into cam type and crank connecting rod type; according to the transmission mode, it can be divided into hydraulic type and mechanical type.

(2) Application requirements of steel bar cutter.

①Before using, it is necessary to carefully check whether the blade is firmly installed, whether the clearance between the two blades is within 0.5—1mm, whether the electrical equipment is normal, and whether all parts are tightened. It can be used only after the empty machine test run is normal.

②When cutting the steel bar, it is necessary to hold the cut steel bar tightly to prevent the end of the steel bar from swinging or ejecting to hurt people. When cutting the short reinforcement, it is not allowed to feed by hand, but should be clamped with pliers.

③When cutting the steel bar, it must be straightened first and then cut off. When the steel bar is fed, it should be carried out when the movable blade retreats from the fixed blade. The steel bar should be placed in the middle of the blade and perpendicular to the cutting edge.

④During the operation of the cutter, it is not allowed to repair and correct it, take down the protective cover at will, touch the running part, put the hand at the cutting position of the cutting edge, and clean the iron filings by hand wiping or mouth blowing. All the above work must be carried out after the machine is shut down.

⑤Strictly implement the operation procedures of the steel bar cutter. It is forbidden to cut the steel bars and other materials beyond the specified scope, or cut the burned hot steel bars and materials with the edge hardness exceeding the blade.

⑥After the operation, the tools and the sundries under the cutting tools should be removed, and the machine body should be cleaned; the tightening degree of bolts at each part and the tightness of the V-belt should be carefully checked; the gap between the movable blade and the fixed blade should be adjusted, and the blunt blade should be replaced.

⑦Regular maintenance shall be carried out according to relevant regulations, that is, periodic maintenance shall be carried out for the lubricating parts of steel bar cutter; the wear degree of gears, bearings and eccentric bodies shall be checked, and the clearance of each part shall be adjusted.

3. Bending of Steel Bars

(1) Types of steel bar bending. After the steel bar is cut according to the calculated cutting length, the line shall be drawn according to the reinforcement design drawing, bending equipment characteristics, reinforcement diameter, bending angle, etc., so as to bend to the designed size and shape.

(2) Application requirements of steel bar bending machine.

①According to the requirements of steel bar bending processing, the bending diameter of steel bar is different multiple of steel bar diameter, and the bending diameter of steel bar with different diameter cannot be the same. When bending, the mandrel with corresponding specification should be selected according to the diameter of steel bar.

②为了适应钢筋直径与心轴直径的变化,应在成型轴上加上一个偏心套,以调节心轴、钢筋和成型轴三者之间的间隙。

③在弯曲钢筋时,挡铁插座上安装的挡铁轴,应设置轴套,以便弯曲过程中消除钢筋与挡铁轴的摩擦。

④在进行弯曲钢筋时,应当将钢筋挡架上的挡板紧贴要弯曲的钢筋,以保证钢筋弯曲形状和尺寸的正确。

⑤挡铁轴的直径和强度,不应小于被弯曲钢筋的直径和强度;不经过调直的钢筋,禁止在弯曲机上进行弯曲;作业时应注意钢筋放入的位置、长度和工作盘旋转方向,以防止发生差错。

在钢筋弯曲的操作中,不允许更换弯曲附件,任何检修工作必须在停机后进行。

⑥每次钢筋弯曲结束后,应当及时清除铁锈和杂物等,检查机械的运转和部件磨损情况,并定期进行维修和保养。

5.2.4 钢筋连接

在进行钢筋混凝土结构工程施工中,常遇到钢筋长度不足而需连接的情况。钢筋连接的方式在建筑工程中主要有绑扎、焊接和机械连接三种。

1. 钢筋绑扎连接

钢筋绑扎连接是利用混凝土的黏结锚固作用,实现两根锚固钢筋的应力传递。为保证钢筋的应力能充分传递,必须满足施工规范规定的最小搭接长度的要求,且应将接头位置设在受力较小处。

(1)钢筋绑扎要求。

①钢筋接头宜设置在受力较小处。同一纵向受力钢筋不宜设置两个或两个以上接头。接头末端至钢筋弯起点的距离不应小于钢筋直径的10倍。

②钢筋绑扎搭接接头连接区段及接头面积百分率应符合要求。

③纵向受力钢筋绑扎搭接接头的最小搭接长度应符合相关规范规定。

(2)钢筋绑扎连接的控制要点。

①在钢筋搭接处,交叉点都应在中心和两端用钢丝扎牢。

②焊接骨架和焊接网采用绑扎连接时,应符合相关规范规定。

B. In order to adapt to the change of the diameter of steel bar and mandrel, an eccentric sleeve should be added to the forming shaft to adjust the gap among mandrel, reinforcement and forming shaft.

C. When bending the steel bar, the shaft sleeve should be set on the iron retaining shaft installed on the retaining iron socket, so as to eliminate the friction between the steel bar and the iron retaining shaft in the bending process.

D. When bending the steel bar, the baffle on the steel bar retaining frame should be close to the steel bar to be bent, so as to ensure the correct bending shape and size of the steel bar.

E. The diameter and strength of the iron retaining shaft shall not be less than the diameter and strength of the steel bar to be bent; the steel bar without straightening shall not be bent on the bending machine; during the operation, attention shall be paid to the position, length and rotation direction of the working disc of the steel bar to prevent errors.

During the operation of steel bar bending, it is not allowed to replace the bending accessories, and any maintenance work must be carried out after shutdown.

F. After each bending of steel bars, rust and sundries shall be removed in time, the operation of machinery and wear of parts shall be checked, and regular repair and maintenance shall be carried out.

5.2.4 Reinforcement Connection

In the construction of reinforced concrete structure, it is often encountered that the length of reinforcement is insufficient and needs to be connected. There are three main ways of reinforcement connection in construction works: binding, welding and mechanical connection.

1. Binding Connection of Reinforcement

The binding connection of reinforcement is to realize the stress transfer of two anchorage bars by using the bonding and anchoring effect of concrete. In order to ensure that the stress of reinforcement can be fully transferred, the minimum lap length specified in the construction specification must be met, and the joint position should be set at the place with less stress.

(1) Reinforcement binding requirements.

①The reinforcement joint should be set at the place with less stress. It is not suitable to set two or more joints for the same longitudinal load-bearing reinforcement. The distance from the end of the joint to the bending point of the reinforcement shall not be less than 10 times of the diameter of the reinforcement.

②The percentage of connection section and joint area of reinforcement binding lap joint shall meet the requirements.

③The minimum lap length of longitudinal stressed reinforcement binding lap joint shall conform to relevant specifications.

(2) Control points of reinforcement binding connection.

①At the overlapping of reinforcement, the intersection shall be firmly fastened with wire at the center and both ends.

②When binding connection is adopted for welding skeleton and welding net, it shall comply with relevant regulations.

2. 钢筋的焊接连接

钢筋采用焊接连接代替绑扎连接,不仅可以提高工作效率、降低工程成本,而且还可以改善结构受力性能、节省大量钢材。钢筋焊接常用的方法有闪光对焊、电弧焊、电渣压力焊、埋弧压力焊和气压焊等。

热轧钢筋的对接连接,应采用闪光对焊、电弧焊、电渣压力焊或气压焊等;钢筋骨架和钢筋网片的交叉焊接,宜采用电阻点焊;钢筋与钢板的 T 形连接,宜采用电弧焊或埋弧压力焊。电渣压力焊应用于柱、墙、烟囱等现浇混凝土结构中竖向受力钢筋的连接;不得用于梁、板等结构中水平钢筋的连接。

(1)闪光对焊。闪光对焊是利用对焊机使两段钢筋接触,通以低电压强电流,把电能转化为热能,使钢筋加热到接近熔点时施加轴向压力进行顶锻,使两根钢筋焊合在一起,形成对焊接头。

闪光对焊的质量检查主要包括外观检查和力学性能试验。其力学性能试验又包括抗拉强度和冷弯性能两个方面。

①闪光对焊接头的外观检查。闪光对焊接头表面应当无裂纹和明显烧伤,应有适当镦粗和均匀的毛刺;接头如有弯折,其角度不大于 4°,接头轴线的偏移不应大于 $0.1d$,亦不应大于 2 mm。外观检查不合格的接头,可将距接头左右各 15 mm 切除再重新焊接。

②光对焊接头力学性能试验。应按同一类型分批进行,每批切取 65%,但不得少于 6 个试件,其中 3 个做抗拉强度试验,3 个做冷弯性能试验。3 个接头试件抗拉强度实测值,均不应小于钢筋母材的抗拉强度规定值;试样应呈塑性断裂且破坏点至少有两个试件断于焊接接头以外。

(2)电弧焊。

①电弧焊的工作原理。电弧焊是利用弧焊机使焊条与焊件之间产生高温电弧,熔化焊条和高温电弧范围内的焊件金属,冷却凝固后形成焊接接头。电弧焊的工作原理如图 5.24 所示。弧焊机有直流和交流之分,常用的是交流弧焊机。焊条的种类较多,宜根据钢材级别和焊接接头形式选择焊条。焊条型号选用见表 5.21;焊条直径和焊接电流选用见表 5.22。

2. Welding Connection of Reinforcement

Using welding connection instead of binding connection can not only improve work efficiency and reduce engineering cost, but also improve structural mechanical performance and save a lot of steel. The common welding methods of steel bar are flash butt welding, arc welding, electroslag pressure welding, submerged arc pressure welding and gas pressure welding.

Butt connection of hot-rolled steel bars shall adopt flash butt welding, electric arc welding, electroslag pressure welding or gas pressure welding, etc. ; cross welding of reinforcement framework and reinforcement mesh shall adopt resistance spot welding; T-shaped connection between reinforcement and steel plate shall adopt arc welding or submerged arc pressure welding. Electroslag pressure welding should be used for the connection of vertical load-bearing steel bars in cast-in-place concrete structures such as columns, walls, chimneys, etc. , and should not be used for the connection of horizontal reinforcement in beams, slabs and other structures.

(1) Flash butt welding. Flash butt welding is to use butt welding machine to make two sections of steel bar contact, pass low voltage and strong current, convert electric energy into heat energy, make steel bar heated to close to melting point, exert axial pressure for upsetting, make two steel bars welded together to form butt welding joint.

The quality inspection of flash butt welding mainly includes appearance inspection and mechanical property test. The mechanical property test includes tensile strength and cold bending property.

①Appearance inspection of flash butt welded joint. The surface of flash butt welding joint shall be free of cracks and obvious burns, and shall be properly upset and evenly burred. If the joint is bent, its angle shall not be greater than 4°, and the offset of joint axis shall not be greater than 0. 1d nor more than 2mm. If the joint fails to pass the appearance inspection, the left and right 15mm away from the joint can be cut off and then welded again.

②Mechanical property test of butt welded joint. It shall be carried out in batches according to the same type, and 65% of each batch shall be cut, but no less than 6 specimens shall be tested, 3 of which shall be subjected to tensile strength test and 3 to cold bending property test. The measured values of tensile strength of the three joint specimens should not be less than the specified value of the tensile strength of the reinforcement base metal; the samples should be plastic fracture, and at least two specimens were broken outside the welded joint.

(2) Arc welding.

①Working principle of arc welding. Arc welding is the use of arc welding machine to produce high-temperature arc between the electrode and the weldment, which melts the metal within the range of the electrode and high-temperature arc, and forms a welding joint after cooling and solidification. The working principle of arc welding is shown in Fig. 5. 24. Arc welding machine can be divided into DC and AC, and AC arc welding machine is commonly used. There are many kinds of welding rod, so the welding rod should be selected according to the steel grade and welding joint form. See Table 5. 21 for electrode type selection and Table 5. 22 for electrode diameter and welding current selection.

②电弧焊接头质量检查。电弧焊接头的外观检查包括焊缝平顺,不得有裂纹,没有明显的咬边、凹陷、焊瘤、夹渣和气孔。用小锤敲击焊缝应发出与其金属同样的清脆声;焊缝尺寸与缺陷的偏差应符合规范。

坡口接头除应进行外观检查和超声波探伤外,还应分批切取1%的接头进行切片观察(指焊缝金属部分)。切片经磨平后,其内部没有裂缝大于规定的气孔和夹渣。经切片后的焊缝处,允许用相同的焊接工艺进行补焊。

(3)电渣压力焊。电渣压力焊在建筑工程施工中应用十分广泛,多用于现浇钢筋混凝土结构竖向钢筋的接长,但不适用于水平钢筋或倾斜钢筋(斜度小于4∶1)的连接,也不适用于可焊性较差的钢筋连接。

①电渣压力焊的工作原理是将两根钢筋安放成竖向对接形式,利用焊接电流通过两根钢筋端面间隙,在焊剂层下形成电弧和电渣过程,从而产生电弧热和电阻热,将两根钢筋端部熔化,然后施加压力使钢筋焊合。与电弧焊相比,该法具有工作条件好、工效高、成本低、易于掌握、节省能源和钢筋等优点。

②电渣压力焊的施工工艺。主要包括端部除锈、固定钢筋、通电引弧、快速顶压、焊后清理等工序,具体施工工艺过程如下:第一,钢筋在调直后,对两根钢筋端部120 mm范围内,进行认真除锈和清理杂质工作,以便于很好地焊接,确保焊接质量;第二,在钢筋电渣压焊机机头的上、下夹头,分别夹紧要焊接的上、下钢筋,钢筋应保持在同一条轴线上,一经夹紧不得出现晃动;第三,电渣引弧过程,即在焊接夹具夹紧上、下钢筋,钢筋端面处安放引弧钢丝球,将焊剂灌入焊剂盒,接通电源,引燃电弧;第四,造渣过程,由于电弧的高温作用,将钢筋端面周围的焊剂充分熔化,从而形成渣池;第五,电渣过程,钢筋端面处形成一定深度的渣池后,将上钢筋缓慢插入渣池中,此时电弧熄灭,渣池电流加大,由于渣池的电阻较大,温度迅速升至2 000 ℃以上,将钢筋端头熔化;第六,挤压过程,待钢筋端头熔化达一定程度后,施加一定挤压力,将熔化金属和熔渣从结合部挤出,同时切断电源;第七,接头焊完后,应停歇一定时间才能回收焊剂和卸下焊接夹具,并敲掉粘在钢筋上的渣壳;四周焊缝应均匀,凸出钢筋表面的高度应≥4 mm。

② Quality inspection of arc welding joint. The appearance inspection of arc welding joint includes smooth weld, no crack, no obvious undercut, depression, overlap, slag inclusion and air hole. When the weld is knocked with a small hammer, it shall emit the same clear and crisp sound as the metal; the deviation between the weld size and the defect shall be in accordance with the specification.

In addition to appearance inspection and ultrasonic flaw detection, 1% of the joint should be cut in batches for observation. After grinding, there is no crack in the chip, which is larger than the specified porosity and slag inclusion. The same welding process is allowed to be used for repair welding of the weld after slicing.

(3) Electroslag pressure welding. Electroslag pressure welding (ESW) is widely used in the construction of building projects. It is mostly used to lengthen the vertical reinforcement of cast-in-place reinforced concrete structure, but it is not suitable for the connection of horizontal reinforcement or inclined reinforcement (slope less than 4 : 1), and it is not suitable for the connection of reinforcement with poor weldability.

①The working principle of electroslag pressure welding is to place two steel bars in the form of vertical butt joint. The arc and electroslag process is formed under the flux layer by using the welding current through the end gap of the two reinforcement bars, thus generating arc heat and resistance heat, melting the ends of the two steel bars, and then applying pressure force to weld the steel bars. Compared with arc welding, this method has the advantages of good working conditions, high efficiency, low cost, easy to master, energy saving and steel bar saving.

②Construction technology of electroslag pressure welding. It mainly includes end derusting, reinforcement fixing, arc striking with electricity, rapid top pressing and post welding cleaning. The specific construction process is as follows: Firstly, after the steel bar is straightened, the rust and impurities at the ends of two steel bars within 120mm shall be carefully removed to facilitate good welding and ensure the welding quality. Secondly, the upper and lower clamps of the electroslag welding machine head are used to clamp the upper and lower steel bars to be welded respectively. The steel bars should be kept on the same axis, and once clamped, no shaking is allowed. Thirdly, the electroslag arc striking process is to clamp the upper and lower steel bars with the welding fixture, place arc striking wire ball at the end face of the steel bar, pour the flux into the flux box, connect the power supply and ignite the arc. Fourth, the slag making process: due to the high temperature of the arc, the flux around the end face of the steel bar is fully melted, thus forming a slag pool. Fifth, the electroslag process: after the slag pool with a certain depth is formed at the end of the steel bar, the upper steel bar is slowly inserted into the slag pool. At this time, the arc is extinguished and the slag pool current is increased. Due to the large resistance of the slag pool, the temperature rapidly rises to above 2000℃, and the steel bar end is melted. Sixth, the extrusion process: after the end of the steel bar is melted to a certain extent, a certain extrusion pressure is applied to extrude the molten metal and slag from the joint, and the power supply is cut off at the same time. Seventhly, after the joint is welded, the flux can be recovered and the welding fixture can be removed, and the slag shell adhered to the reinforcement can be knocked off. The welding joint around shall be uniform, and the height protruding from the surface of reinforcement shall be $\geqslant$ 4mm.

③电渣压力焊的质量检查。电渣压力焊的质量检查,主要包括外观检查和拉伸试验。

a. 外观检查。钢筋电渣压力焊的接头应逐个进行外观检查。接头的外观检查结果,应符合下列要求:四周焊包凸出钢筋表面的高度,应不得小于 4 mm;钢筋与电极的接触处,应无烧伤缺陷;接头处的弯折角不得大于 4°;接头处的轴线偏移不得大于钢筋直径的 0.1 倍,且不得大于 2 mm。

b. 拉伸试验。钢筋电渣压力焊的接头,应进行力学性能试验。在一构筑物中,应以 300 个同级别钢筋接头作为一批;在现浇钢筋混凝土多层结构中,应以每一楼层或施工区段中 300 个同级别钢筋接头作为一批,不足 300 个接头的仍应作为一批。

从每批钢筋接头中随机切取 3 个试件做拉伸试验,其试验结果:3 个试件的抗拉强度均不得小于该级别钢筋规定的抗拉强度。

当试验结果中有 1 个试件的抗拉强度低于规定值,应再切取 6 个试件进行复验;当复验中仍有 1 个试件的抗拉强度小于规定值,应确认该批接头为不合格品。

3. 钢筋的机械连接

钢筋的机械连接是最近几年迅速发展的一项钢筋连接技术。钢筋机械连接是通过机械手段将两根钢筋进行对接,其连接方法分类及适用范围见表 5.6。

表 5.6　钢筋机械连接方法分类及适用范围

机械连接方法		适用范围	
		钢筋级别	钢筋直径/mm
钢筋套筒挤压连接		HRB335,HRB400,RRB400	16～40
钢筋锥螺纹套筒连接		HRB335,HRB400,RRB400	16～40
钢筋全效粗直径直螺纹套筒连接		HRB400,RRB400	16～40
钢筋滚压直螺纹套筒连接	直接滚压	HRB400,RRB400	16～40
	挤肋滚压	HRB400,RRB400	16～40
	剥肋滚压		16～50

钢筋机械连接是通过连接件的机械咬合作用或钢筋端面的承压作用,使两根钢筋能够传递力的连接方法。钢筋机械连接接头质量可靠,现场操作简单,施工速度较快,无明火作业,不受气候影响,适应性很强,而且可用于可焊性较差的钢筋。

③ Quality inspection of electroslag pressure welding. The quality inspection of electroslag pressure welding mainly includes appearance inspection and tensile test.

a. Appearance inspection. The joints of steel bar electroslag pressure welding shall be inspected one by one. The appearance inspection results of the joint shall meet the following requirements: the height of the surrounding welding package protruding from the reinforcement surface shall not be less than 4mm; there shall be no burn defect at the contact between the reinforcement and the electrode; the bending angle at the joint shall not be greater than 4°; the axis deviation at the joint shall not be greater than 0.1 times of the reinforcement diameter and shall not be greater than 2mm.

b. Tensile test. Mechanical property test should be carried out for the joint of steel bar electroslag pressure welding. In a structure, 300 steel bar joints of the same grade shall be taken as a batch; in cast-in-place reinforced concrete multi-storey structure, 300 reinforcement joints of the same grade in each floor or construction section shall be taken as a batch and those less than 300 joints shall still be regarded as a batch.

Three specimens are randomly cut from each batch of reinforcement joints for tensile test. The test results show that the tensile strength of the three specimens shall not be less than the tensile strength specified for the grade of reinforcement.

When the tensile strength of one specimen in the test results is lower than the specified value, 6 more specimens shall be cut for re inspection; if the tensile strength of one specimen is still lower than the specified value, the batch of joints shall be confirmed as unqualified products.

3. Mechanical connection of reinforcement

Mechanical connection of reinforcement is a rapid development of reinforcement connection technology in recent years. Mechanical connection of reinforcement is to butt joint two steel bars by mechanical means. The classification and application scope of connection method are shown in Table 5.6.

Table 5.6 Classification and Application Scope of Mechanical Connection Method of Reinforcement

Method of mechanical connection		Scope of Application	
		Reinforcement Grade	Reinforcement diameter/mm
Steel sleeve extrusion connection		HRB335, HRB400, RRB400	16—40
Taper thread sleeve connection of reinforcement		HRB335, HRB400, RRB400	16—40
Full effect thick diameter straight thread sleeve connection of reinforcement		HRB400, RRB400	16—40
Steel bar rolling straight thread sleeve connection	Direct rolling	HRB400, RRB400	16—40
	Ribbed rolling		16—40
	Rib peeling and rolling		16—50

Mechanical connection of steel bar is a connection method that can transfer force between two steel bars through mechanical bite of connector or pressure bearing effect of steel bar end face. The mechanical connection joint of steel bar is reliable in quality, simple in site operation, fast in construction, no open fire operation, not affected by climate, and has strong adaptability, and can be used for reinforcement with poor weldability.

(1)钢筋套筒挤压连接。钢筋套筒挤压连接是将两根待连接钢筋插入特制的钢质连接套筒内,再采用专用挤压机在常温下对连接套筒进行加压,使钢质连接套筒产生塑性变形后与待连接钢筋端部形成机械咬合,从而形成可靠的钢筋连接接头。钢筋套筒挤压连接又分为径向挤压(图 5.17)和轴向挤压(图 5.18)两种作业方式。

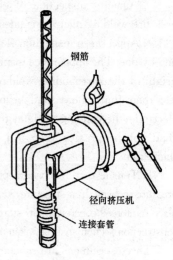

图 5.17 径向套筒挤压连接示意

钢筋套筒挤压连接的质量检查主要包括外观检查和拉力试验。

a. 外观检查。采用专用工具或游标卡尺进行检测。钢筋连接端的肋纹完好无损,连接处无油污、水泥等的污染。要检查接头挤压道数和压痕尺寸:钢筋端头离套筒中心不应超过 10 mm,压痕间距宜为 1~6 mm,挤压后的套筒接头长度为套筒原长度的 1.10~1.15 倍,挤压后套筒接头外径,用量规测量应能通过。量规不能从挤压套管接头外径通过的,可更换压模重新挤压一次,压痕处最小外径为套管原外径的 0.85~0.90 倍。挤压接头处不得有裂纹,接头弯折角度不得大于 4°。

b. 拉力试验。以词批号钢套筒且同一制作条件的 500 个接头为一个验收批,不足 500 个仍为一个验收批,从每验收批接头中随机抽取 3 个试件进行拉力试验;如试验结果中有 1 个试件不符合要求,应再抽取 6 个试件进行复验;如仍有 1 个试件不符合要求,则该验收批接头不合格。

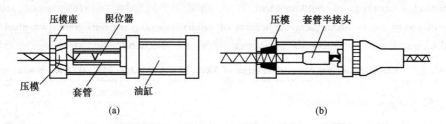

图 5.18 轴向套筒挤压连接示意

(1) Steel sleeve extrusion connection. Steel sleeve extrusion connection is to insert two steel bars to be connected into a special steel connecting sleeve, and then use a special extruder to pressurize the connecting sleeve at normal temperature, so that the steel connecting sleeve will form mechanical bite with the end of the steel bar to be connected after plastic deformation, so as to form a reliable steel bar connection joint. The steel sleeve extrusion connection is divided into radial extrusion (Fig. 5. 17) and axial extrusion (Fig. 5. 18).

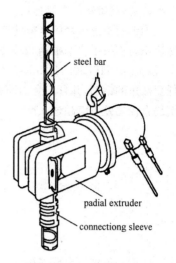

Fig. 5. 17　Extrusion Connection of Radial Sleeve

The quality inspection of steel sleeve extrusion connection mainly includes appearance inspection and tensile test.

a. Appearance inspection. Use special tools or vernier caliper for detection. The rib pattern at the connection end of the reinforcement is intact, and the joint is free from oil pollution, cement, etc. Check the number of extrusion passes and indentation size of the joint: the distance between the end of the reinforcement and the center of the sleeve shall not exceed 10mm, the spacing between the indentations shall be 1—6 mm, the length of the sleeve joint after extrusion shall be 1.10—1.15 times of the original length of the sleeve, and the outer diameter of the sleeve joint after extrusion shall be able to pass the gauge measurement. If the gauge cannot pass through the outer diameter of the casing joint, the press mould can be replaced and extruded again. The minimum outer diameter at the indentation is 0.85—0.90 times of the original outer diameter of the casing. There shall be no crack at the extruded joint, and the bending angle of the joint shall not be greater than 4°.

b. Tensile test. Take 500 joints of steel sleeve with the same manufacturing conditions as an acceptance batch, and less than 500 joints are still an acceptance batch. Three test pieces are randomly selected from each acceptance batch for tensile test; if one test piece fails to meet the requirements in the test results, another 6 test pieces shall be selected for re-inspection; if one test piece still fails to meet the requirements, the acceptance batch joint is unqualified.

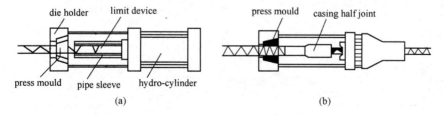

Fig. 5. 18　Extrusion Connection of Axial Sleeve

(2)钢筋锥螺纹套筒连接。

①钢筋锥螺纹套筒连接的连接原理是将两根待连接钢筋的端部和套筒预先加工成锥形螺纹,然后用力矩扳手将两根钢筋端部旋入套筒形成机械式钢筋接头(图5.19)。这种连接方式能在施工现场连接 HPB300、HRB335、HRB400 级直径为 16～40 mm 的同直径或异直径的竖向、水平和任意倾角的钢筋,并且不受钢筋有无螺纹及含碳量大小的限制。当连接不同直径钢筋时,所连接钢筋直径之差不应超过 9 mm。

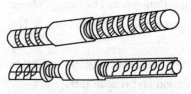

图 5.19 钢筋锥螺纹套筒连接示意

②钢筋锥螺纹套筒连接的特点及适用范围。加强锥螺纹套筒连接具有连接速度快、轴线偏差小、施工工艺简单、安全可靠、无明火作业、不污染环境、节约钢材、节省能源、可全天候施工、有利文明施工等特点,有明显的技术经济效益。适用于按一、二级抗震设防的一般工业与民用房屋及构筑物的现浇混凝土结构,尤其适用于梁、柱、板、墙、基础的钢筋连接施工。但不得用于预应力钢筋或经常承受反复动荷载及承受高应力疲劳荷载的结构。

③钢筋锥螺纹套筒连接的质量检查。钢筋锥螺纹套筒连接的抗拉强度必须大于钢筋的抗拉强度。锥形螺纹可用锥形螺纹旋切机加工;钢筋用套丝机进行套丝。钢筋接头拧紧力矩值见表5.7。

表 5.7 钢筋接头拧紧力矩值(mm)

钢筋直径	16	18	20	22	25	28	32	36	40
拧紧力矩值	118	145	177	216	275	275	314	343	343

钢筋锥螺纹套筒连接的接头质量,应符合以下要求。

a. 钢筋套丝牙型质量必须与牙型规格吻合,锥螺纹的完整牙数不得小于表5.8中的规定值;钢筋锥螺纹小端直径必须在卡规的允许误差范围内,连接套筒规格必须与钢筋规格一致。

(2)Taper thread sleeve connection of reinforcement.

①The connection principle of taper thread sleeve connection is that the ends and sleeves of two reinforcement bars to be connected are pre-processed into conical threads, and then the ends of two reinforcement bars are screwed into the sleeve with a torque wrench to form a mechanical reinforcement joint(Fig. 5.19). This connection method can connect HPB300, HRB335, HRB400 and the same diameter or different diameter of vertical, horizontal and arbitrary inclination reinforcement with diameter of 16—40 mm in construction site, and is not limited by the screw thread and carbon content of reinforcement. When connecting steel bars with different diameters, the diameter difference of the connecting bars shall not exceed 9mm.

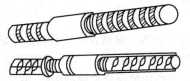

Fig. 5.19 Connection of Taper Thread Sleeve

②Characteristics and application scope of taper thread sleeve connection. The connection of reinforced taper thread sleeve has the characteristics of fast connection speed, small axis deviation, simple construction technology, safety and reliability, no open fire operation, no pollution to the environment, steel saving, energy saving, all-weather construction, and favorable for civilized construction. It is suitable for the cast-in-place concrete structure of general industrial and civil buildings and structures with seismic fortification of grade Ⅰ and Ⅱ, especially for the reinforcement connection construction of beams, columns, slabs, walls and foundations. However, it should not be used for prestressed steel bars or structures that often bear repeated dynamic loads and high stress fatigue loads.

③Quality inspection of taper thread sleeve connection. The tensile strength of taper thread sleeve connection must be greater than that of reinforcement. Taper thread can be processed by conical thread rotary cutting machine, and reinforcing steel bar can be threaded by thread threading machine. See Table 5.7 for tightening torque of reinforcement joint.

Table 5.7 Tightening Torque Values of Reinforcement Joints(mm)

Reinforcement diameter	16	18	20	22	25	28	32	36	40
Tightening torque value	118	145	177	216	275	275	314	343	343

The joint quality of taper thread sleeve connection shall meet the following requirements.

a. The profile quality of steel bar threading must be consistent with the profile specification, and the complete number of taper thread shall not be less than the specified value in Table 5.8; the diameter of small end of taper thread of reinforcement must be within the allowable error range of caliper gauge, and the specification of connecting sleeve must be consistent with the specification of reinforcement.

表 5.8　钢筋锥螺纹的完整牙数

钢筋直径/mm	16～18	20～22	25～28	32	36	40
最少完整牙数	5	7	8	10	11	12

b. 钢筋接头的拧紧力矩值检查。按每根梁、柱构件抽验 1 个接头；板、墙、基础底板构件每 100 个同规格接头作为一批，不足 100 个接头也作为一批。每批抽验 3 个接头，要求抽验的钢筋接头 100% 达到规定的力矩值。如发现 1 个接头不合格，必须加倍抽验，再发现 1 个接头达不到规定力矩值，则要求该构件的全部接头重新复拧到符合质量要求。若复检时仍发现不合格接头，则该接头必须采取贴角焊缝补强，将钢筋与连接套焊在一起，焊缝高不小于 5 mm。连接好的钢筋接头螺纹不准有一个完整的螺纹外露。

(3) 钢筋直螺纹套筒连接。

① 钢筋直螺纹套筒连接的连接原理。钢筋直螺纹套筒连接的连接原理是通过钢筋端头特制的直螺纹和直螺纹套管，将两根钢筋咬合在一起。与带螺纹钢筋套筒连接的技术原理相比，相同之处都是通过钢筋端头的螺纹与套筒内螺纹合成钢筋接头，主要区别在钢筋等强技术效应上。

② 钢筋直螺纹套筒连接的连接形式。钢筋直螺纹套筒连接有滚压直螺纹接头和镦头直螺纹接头两种。

a. 滚压直螺纹接头。滚压直螺纹接头也称为 GK 型锥螺纹钢筋连接，是在钢筋端头先采用对辊滚压，使钢筋端头应力增大，而后采用冷压螺纹（滚丝）工艺加工成钢筋直螺纹端头，套筒采用快速成孔切削成内螺纹钢套筒，这种对钢筋端部的预压冷硬化处理，使其强度比钢筋母材提高 10%～20%，因而使锥螺纹的强度也相应得到提高，弥补了因加工锥螺纹减小钢筋截面而造成接头承载力下降的缺陷，从而可提高锥螺纹接头的强度。

b. 镦头直螺纹接头。镦头直螺纹接头是在钢筋端头先采用设备顶压增径（镦头）使钢筋端头应力增大，而后采用套丝工艺加工成等直径螺纹端头，套筒采用快速成孔切削成内螺纹钢套筒，简称为镦头直螺纹接头或镦粗切削直螺纹接头。

Chapter 5 Construction of Reinforced Concrete Structure

Table 5.8 Complete Number of Taper Thread of Steel Bar

Reinforcement diameter/mm	16—18	20—22	25—28	32	36	40
Minimum complete number	5	7	8	10	11	12

b. Inspection of tightening torque value of reinforcement joint. One joint shall be inspected according to each beam and column member; every 100 joints of the same specification for slab, wall and foundation slab members shall be taken as one batch, and less than 100 joints shall also be taken as a batch. Three joints shall be inspected in each batch, and 100% of the steel joints shall meet the specified torque value. If one joint is found to be unqualified, it is necessary to double sampling and find that one joint fails to meet the specified torque value, and then all joints of the component shall be re-screwed to meet the quality requirements. If unqualified joint is still found during re-inspection, fillet weld shall be adopted for reinforcement of the joint, and the reinforcement shall be welded with the connection sleeve, and the weld height shall not be less than 5mm. The thread of the connected steel joint shall not be exposed with one complete thread.

(3) Reinforcement straight thread sleeve connection.

①Connection principle of straight thread sleeve connection. The connection principle of steel bar straight thread sleeve connection is that two steel bars are occluded together through the special straight thread and straight thread sleeve at the end of steel bar. Compared with the technical principle of sleeve connection of steel bar with screw thread, the same point is that the reinforcement joint is synthesized by the thread at the end of steel bar and the internal thread of sleeve, which is mainly different from the equal strength technical effect of steel bar.

②Connection form of straight thread sleeve connection. There are two kinds of connection forms of steel bar straight thread sleeve connection: rolling straight thread joint and upsetting head straight thread joint.

a. Rolling straight thread joint. Rolling straight thread joint is also known as GK type taper thread steel bar connection. The opposite roller rolling is used at the end of the steel bar to increase the stress at the end of the steel bar, and then the cold pressing thread (thread rolling) process is used to process the straight thread end of the steel bar. The sleeve is made of internal thread steel by fast drilling. This kind of precompression cold hardening treatment on the end of steel bar can increase its strength by 10%—20% compared with the base material of reinforcement, so that the strength of taper thread is also improved accordingly, which makes up for the defect that the bearing capacity of the joint decreases due to the reduction of the cross-section of the reinforcement due to the processing of the taper thread, so as to improve the strength of the taper thread joint.

b. Upsetting straight thread joint. The upsetting straight thread joint is used to increase the stress at the end of the steel bar by the top pressure of the equipment, and then the thread end of the same diameter is processed by the threading process. The sleeve is made of internal thread steel sleeve by rapid hole forming, which is referred to as upsetting straight thread joint or upsetting cutting straight thread joint.

③钢筋直螺纹套筒连接的特点。以上两种方法都能有效地增强钢筋端头材料的强度,使直螺纹接头与钢筋母材等强。这种接头形式使得在地震情况下结构的强度与延性的安全性具有更大的保证,钢筋混凝土截面对钢筋接头百分率放宽,大大方便了设计与施工;等强直螺纹接头施工采用普通扳手旋紧即可,对螺纹少旋入 1~2 丝不影响接头强度,省去了锥螺纹力矩扳手检测和疏密质量检测的繁杂程度,可提高施工工效;套筒丝距比锥螺纹套筒丝距少,可节省套筒钢材;此外,还有设备简单、经济合理、应用范围广等优点。

④等强直螺纹连接套筒的类型。主要有标准型(用于 HRB335 级、HRB400 级带肋钢筋)、扩口型(用于钢筋难于对接的施工)、变径型(用于钢筋变径时的施工)、正反螺纹型(用于钢筋不能转动时的施工)。套筒的抗拉设计强度不应低于钢筋抗拉设计强度的 1.20 倍。为确保接头强度大于现行国家标准中的 A 级标准,接头抗拉设计强度应取钢筋母材实测抗拉强度或取钢筋母材标准抗拉强度的 1.10 倍。

(4)机械连接接头的现场检验。钢筋机械连接接头的现场检验应按验收批进行。对于同一施工条件下采用同一批材料的同等级、同形式、同规格的钢筋接头,以 500 个接头为一个检验批,不足 500 个也作为一个检验批。每一个检验批,必须随机取 3 个试件做单向拉伸试验,按设计要求的接头性能 A、B、C 等级进行检验和评定。

5.2.5 钢筋配料与代换

1. 钢筋配料

钢筋配料是根据构件配筋图,绘出各种形状和规格的单根钢筋简图并加以编号,分别计算钢筋的下料长度、根数和重量,并绘制配料单,以作为钢筋加工的依据。

钢筋配料就是根据结构施工图,分别计算构件中各种钢筋的下料长度、根数及重量,并编制钢筋配料单。钢筋配料是确定钢筋材料计划、进行钢筋加工和结算的依据。承包商的施工管理人员必须认真对待这项工作。

(1)计算依据:

①外包尺寸。结构施工图中所标注的钢筋尺寸一律是外包尺寸,即钢筋外边缘至外边缘之间的长度,如图 5.20 所示。

③Characteristics of straight thread sleeve connection of reinforcement. The above two methods can effectively enhance the strength of steel bar end material, so that the straight thread joint is equal to the steel bar base material. This type of joint ensures the safety of structural strength and ductility under earthquake. The percentage of reinforced concrete cross-section joint is relaxed, which greatly facilitates the design and construction. The construction of equal strength straight thread joint can be tightened by ordinary wrench, and less screw in 1—2 thread does not affect the joint strength, thus eliminating the detection of cone thread torque wrench and density quality inspection, and the construction efficiency improved. Sleeve thread spacing is less than that of taper thread sleeve, and sleeve steel can be saved. In addition, it has the advantages of simple equipment, reasonable economy and wide application range.

④The types of equal strength straight thread connecting sleeves are mainly standard type(for HRB335, HRB400 reinforced bar with ribs), expansion type (for construction of reinforcement difficult to butt), reducer type (for construction of reinforcement diameter change), positive and negative thread type(for construction when the reinforcement cannot rotate). The design strength of sleeve shall not be less than 1.20 times of the design strength of the reinforcement. To ensure that the joint strength is greater than the standard of class A in the current national standard, the design strength of the joint tension shall be 1.10 times the measured tensile strength of the steel base metal or the standard tensile strength of the base metal.

(4) On-site inspection of mechanical connection joints. The on-site inspection of mechanical connection joints of steel bars shall be carried out according to the acceptance batch. For reinforcement joints of the same grade, form and specification using the same batch of materials under the same construction conditions, 500 joints shall be regarded as one inspection batch and less than 500 joints shall also be regarded as an inspection batch. For each inspection batch, 3 specimens must be randomly selected for unidirectional tensile test, and the joint performance A, B and C grade required by the design shall be inspected and evaluated.

5.2.5 Reinforcement Batching and Replacement

1. Reinforcement Batching

Reinforcement batching is to draw and number the single steel bar diagram of various shapes and specifications according to the component reinforcement drawing, calculate the cutting length, number and weight of reinforcement respectively, and draw the batching list as the basis for reinforcement processing.

Reinforcement batching is to calculate the cutting length, number and weight of various steel bars in the component according to the structural construction drawing, and prepare the steel bar batching list. Reinforcement batching is the basis for determining the reinforcement material plan, processing and settlement of reinforcement. The contractor's construction management personnel must take this work seriously.

(1)Calculation basis:

①Outer dimensions. The size of reinforcement marked in the structural construction drawing is the outer dimension, that is, the length from the outer edge of the reinforcement to the outer edge, as shown in Fig. 5.20.

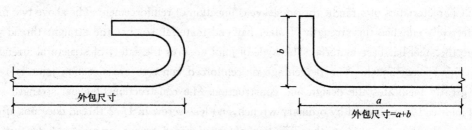

图 5.20　钢筋外包尺寸示意图

②量度差值。钢筋加工中需要进行弯曲。钢筋弯曲后,外边缘增长,内边缘缩短,但中心线长度不会发生变化。这样,钢筋的外包尺寸与钢筋中心线长度之间存在一个差值,这个差值称为量度差值。

计算钢筋下料长度时应扣除量度差值。否则,由于钢筋下料太长,一方面造成浪费;另一方面,可引起钢筋的保护层不够以及钢筋安装的不方便甚至影响钢筋的位置(特别是钢筋密集时)。

钢筋弯曲处的量度差值见表 5.9(推导略)。

表 5.9　钢筋弯曲量度差值表

弯曲角度	量度差值	弯曲角度	量度差值
45°	$0.5d_0$	60°	$0.85d_0$
90°	$2.0d_0$	135°	$2.5d_0$

③弯钩增加长度。规范规定,HPB300 级钢筋末端应做 180°弯钩,其弯弧内直径不应小于钢筋直径的 2.5 倍,弯钩的弯后平直部分长度不应小于钢筋直径的 3 倍。显然,此类钢筋下料长度要大于钢筋的外包尺寸。此时,计算中每个弯钩应增加一定的长度即弯钩增加长度。每个弯钩增加长度为 $6.25d$。(推导略)

④箍筋弯钩增加值。箍筋的弯钩形式如图 5.21 所示。有抗震或抗扭要求的结构应按图 5.21(a)形式加工箍筋,一般结构可按图 5.21(b)、(c)形式加工箍筋。箍筋弯后的平直部分长度,对一般结构,不宜小于箍筋直径的 5 倍,对有抗震要求的结构,不应小于箍筋直径的 10 倍。箍筋的下料长度应比其外包尺寸大,在计算中也要增加一定的长度即箍筋弯钩增加值。

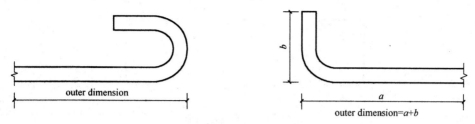

Fig. 5.20 Schematic Diagram of Steel Bar Outer Dimension

②Different value of measurement. Bending is required in the processing of steel bars. After bending, the outer edge increases and the inner edge shortens, but the length of the center line does not change. In this way, there is a difference between the outer dimension of reinforcement and the length of reinforcement center line, which is called different value of measurement.

The different value of measurement should be deducted when calculating the cutting length of reinforcement. Otherwise, because the steel bar is too long, on the one hand, it will cause waste; on the other hand, it may cause insufficient protective layer of steel bar and inconvenient installation of steel bar, and even affect the position of steel bar (especially when the reinforcement is dense).

See Table 5.9 for the different value of measurement at the bending point of reinforcement (the derivation is omitted).

Table 5.9 Different Value of Steel Bar Bending Measurement

Bending Angle	Different Value of Measurement	Bending Angle	Different Value of Measurement
45°	$0.5d_0$	60°	$0.85d_0$
90°	$2.0d_0$	135°	$2.5d_0$

③Increased length of hook. According to the specification, a 180° hook shall be made at the end of HPB300 steel bar, and the inner diameter of the bending arc shall not be less than 2.5 times of the diameter of the reinforcement, and the length of the straight part of the hook after bending shall not be less than 3 times of the diameter of the reinforcement. Obviously, the cutting length of this kind of reinforcement is larger than the outer dimension of reinforcement. At this time, each hook should be increased by a certain length, that is, the increased length of hook. The increased length of each hook is $6.25d$. (The derivation is omitted)

④Added value of stirrup hook. The hook form of stirrup is shown in Fig. 5.21. Stirrups shall be processed in the form of Fig. 5.21(a) for structures with seismic or torsional requirements, and stirrups can be processed in the form of Fig. 5.21(b) and (c) for general structures. The length of the straight part of stirrup after bending should not be less than 5 times of stirrup diameter for general structures and 10 times of stirrup diameter for structures with seismic requirements. The cutting length of stirrup should be larger than its outer dimension, and a certain length should be added in the calculation, that is, the added value of stirrup hook.

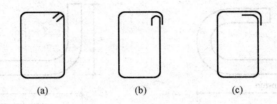

图 5.21 箍筋弯钩形式

箍筋弯钩增加值见表 5.10(推导略)。

表 5.10 箍筋弯钩增加值

箍筋形式	箍筋弯钩增加值	箍筋形式	箍筋弯钩增加值
135°/135°	$14d_0(24d_0)$	90°/180°	$14d_0(24d_0)$
90°/90°	$11d_0(21d_0)$		

注：表中括号内数据为有抗震要求时。

(2)计算公式：钢筋下料是根据需要将钢筋切断成一定长度的直线段。钢筋的下料长度就是钢筋的中心线长度。计算钢筋下料长度可按以下公式进行：

$$钢筋下料长度 = 外包尺寸 - 量度差值 + 弯钩增加长度（箍筋弯钩增加值）$$

例 1：试计算下列钢筋的下料长度，如图 5.22 所示。

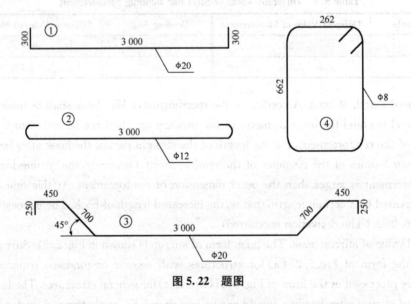

图 5.22 题图

解：钢筋下料长度 = 外包尺寸 - 量度差值 + 弯钩增加长度（箍筋弯钩增加值）

Chapter 5 Construction of Reinforced Concrete Structure

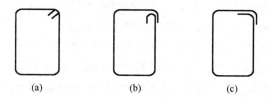

Fig. 5.21 Hook Form of Stirrup

The added value of stirrup hook is shown in Table 5.10(the derivation is omitted).

Table 5.10 Added Value of Stirrup Hook

Stirrup Form	Added Value of Stirrup Hook	Stirrup Form	Added Value of Stirrup Hook
135°/135°	$14d_0(24d_0)$	90°/180°	$14d_0(24d_0)$
90°/90°	$11d_0(21d_0)$		
Note: The data in brackets in the table refers to the seismic requirements.			

(2) Calculation formula: Reinforcement cutting is to cut the steel bar into a certain length of linear section according to the needs. The cutting length of the reinforcement is the length of the center line of the reinforcement. The cutting length of reinforcement can be calculated according to the following formula:

Cutting length of reinforcement=Outer dimension−measurement difference+Increased length of hook(added value of stirrup hook)

Example 1: Try to calculate the cutting length of the following reinforcement. The drawing is shown in Fig. 5.22.

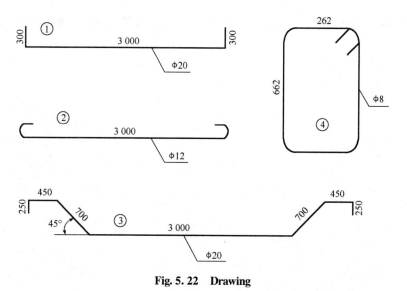

Fig. 5.22 Drawing

Solution: Cutting length of reinforcement=Outer dimension−measurement difference+increased length of hook(the added value of stirrup hook)

1号钢筋的下料长度为：$2\times300+3\ 000-2\times2\times20=3\ 520(\text{mm})$

2号钢筋的下料长度为：$3\ 000+2\times625\times12=3\ 150(\text{mm})$

3号钢筋的下料长度为：$2\times(250+450+700)+3\ 000-2\times2\times20-4\times0.5\times20=5\ 680(\text{mm})$

4号钢筋的下料长度为：$2\times(262+662)+14\times8-3\times2\times8=1\ 912(\text{mm})$

若有抗震要求，则下料长度为：$2\times(262+662)+24\times8-3\times2\times8=1\ 992(\text{mm})$

2. 钢筋代换

施工中如供应的钢筋品种和规格与设计图纸要求不符，可以进行代换。但代换时，必须充分了解设计意图和代换钢材的性能，严格遵守规范的各项规定。对抗裂性要求较高的构件，不宜用光面钢筋代换带肋钢筋；钢筋代换时不宜改变构件中的有效高度。

（1）当钢筋的品种、级别或规格需作变更时，应办理设计变更文件。当需要代换时，必须征得设计单位同意，并应符合下列要求：

①不同种类钢筋的代换，应按钢筋受拉承载力设计值相等的原则进行。代换后应满足混凝土结构设计规范中有关间距、锚固长度、最小钢筋直径、根数等要求。

②对有抗震要求的框架钢筋需代换时，应符合上条规定，不宜以强度等级较高的钢筋代替原设计中的钢筋；对重要受力结构，不宜用 HPB235 级钢筋代换带肋钢筋。

③当构件受抗裂、裂缝宽度或挠度控制时，钢筋代换时应重新进行验算；梁的纵向受力钢筋与弯起钢筋应分别进行代换。

代换后的钢筋用量不宜大于原设计用量的 5%，亦不低于 2%，且应满足规范规定的最小钢筋直径、根数、钢筋间距、锚固长度等要求。

（2）钢筋代换的方法有以下三种：

①当结构构件是按强度控制时，可按强度等同原则代换，称"等强代换"。如设计图中所用钢筋强度为 f_{y1}，钢筋总面积为 A_{s1}，代换后钢筋强度为 f_{y2}，钢筋总面积为 A_{s2}，则应使：

$$f_{y2}A_{s2} \geqslant f_{y1}A_{s1} \tag{5.1}$$

②当构件按最小配筋率控制时，可按钢筋面积相等的原则代换，称"等面积代换"，即：

$$A_{s2} \geqslant A_{s1} \tag{5.2}$$

式中　A_{s1}——原设计钢筋的计算面积；

A_{s2}——拟代换钢筋的计算面积。

Chapter 5 Construction of Reinforced Concrete Structure

The cutting length of No. 1 steel bar is: 2×300+3000−2×2×20=3520(mm)

The cutting length of No. 2 steel bar is 3000+2×625×12=3 150(mm)

The cutting length of No. 3 reinforcement is 2×(250+450+700)+3000−2×2×20−4×0.5×20=5680(mm)

The cutting length of No. 4 steel bar is 2×(262+662)+14×8−3×2×8=1912(mm)

If there are seismic requirements, the cutting length is 2(262+662)+24×8−3×2×8=1992(mm)

2. Replacement of Reinforcement

During construction, if the type and specification of steel bars supplied do not meet the requirements of design drawings, they can be replaced. But when replacing, it is necessary to fully understand the design intent and the steel performance of replacement and strictly abide by the specifications. It is not suitable to replace ribbed steel bars with smooth steel bars for components with high crack resistance requirements, and the effective height of components should not be changed when steel bars are replaced.

(1) When the variety, grade or specification of reinforcement need to be changed, the design change document shall be handled. When it is necessary to replace, it must be approved by the design unit and meet the following requirements:

①The replacement of different types of reinforcement should be carried out according to the principle that the design value of tensile bearing capacity of reinforcement is equal. After replacement, the requirements of spacing, anchorage length, minimum diameter and number of reinforcement in the code for design of concrete structures shall be met.

②When the frame reinforcement with seismic requirements needs to be replaced, it should comply with the above provisions, and it is not suitable to replace the reinforcement in the original design with the reinforcement with higher strength grade. For the important stressed structure, it is not suitable to replace the ribbed reinforcement with HPB235 reinforcement.

③When the member is controlled by crack resistance, crack width or deflection, the checking calculation shall be conducted again when the reinforcement is replaced. The longitudinal stressed reinforcement and bent reinforcement of beam shall be replaced separately.

The amount of reinforcement after replacement should not be more than 5% of the original design amount, or less than 2%, and should meet the requirements of the minimum reinforcement diameter, number, spacing, anchorage length, etc.

(2) There are three methods to replace reinforcement.

①When the structural members are controlled according to the strength, they can be replaced according to the principle of strength equivalence, which is called "equal strength substitution". If the strength of the reinforcement used in the design drawing is f_{y1}, the total area of the reinforcement is A_{s1}, the strength of the steel bar after replacement is f_{y2}, and the total area of the reinforcement is A_{s2}, then:

$$f_{y2}A_{s2} \geqslant f_{y1}A_{s1} \tag{5.1}$$

②When the member is controlled by the minimum reinforcement ratio, it can be replaced according to the principle of equal reinforcement area. It is called "equal area substitution", namely:

$$A_{s2} \geqslant A_{s1} \tag{5.2}$$

In the formula A_{s1}— the calculated area of the original design reinforcement;

A_{s2}— the calculated area of reinforcement to be replaced.

③当结构构件受裂缝宽度或挠度控制时,钢筋的代换需进行裂缝宽度或挠度验算。代换后,还应满足构造方面的要求(如钢筋间距、最小直径、最少根数、锚固长度、对称性等)及设计中提出的特殊要求(如冲击韧性、抗腐蚀性等)。

5.2.6 钢筋的安装验收

1. 现浇钢筋混凝土结构工程中钢筋安装施工

(1)施工工艺。

①柱钢筋安装。柱钢筋安装施工工艺流程:调整下层柱预留筋→套柱箍筋→绑扎竖向受力筋→画箍筋间距线→绑箍筋→检查验收。

柱钢筋安装施工要点如下:

第一,根据弹好的外皮尺寸线,检查预留钢筋的位置、数量、长度。绑扎前先整理调直预留筋,并将其上的水泥砂浆等清除干净。

第二,套柱箍筋。按图纸要求的间距,计算好每根柱箍筋的数量,将箍筋套在下层伸出的预留筋上。

第三,连接竖向受力筋。柱子主筋直径大于 16 mm 时宜采用焊接或机械连接,纵向受力钢筋机械连接及焊接接头连接区段的长度为 $35d$(d 为受力钢筋的较大直径),且不小于 500 mm,该区段内有接头钢筋面积占钢筋总面积百分率不宜超过 50%。

第四,画箍筋位置线。在立好的柱子竖向钢筋上,用粉笔画箍筋位置线,并加钢筋定距框。

第五,柱箍筋绑扎。按已画好的箍筋位置线,将已套好的箍筋往上移动,由上往下绑扎;箍筋转角处与柱主筋交点应采用兜扣绑扎,其余部位可采用八字扣绑扎;方柱箍筋的弯钩叠合处应沿柱子竖筋交错布置并绑扎牢固。圆柱宜采用螺旋箍筋。柱的第一道箍筋距地 50 mm,上下两端箍筋均按规定加密(柱净高 1/6 范围、柱长边宽度和不小于 500 mm,取三者中的最大尺寸);有抗震要求的地区,柱箍筋端头应弯成 135°,平直部分长度不小于 $10d$(d 为箍筋直径)。如设计要求为焊接箍筋,单面焊缝长度不小于 $10d$;柱筋保护层厚度应符合设计及规范要求,垫块应绑在柱主筋外皮上,间距宜为 1 000 mm(或用塑料卡卡在主筋上),以保证主筋保护层厚度的准确。

③When the structural members are controlled by the crack width or deflection, the replacement of reinforcement needs to check the crack width or deflection. After replacement, the structural requirements(such as spacing, minimum diameter, minimum number of bars, anchorage length, symmetry, etc.) and special requirements (such as impact toughness and corrosion resistance) proposed in the design shall also be met.

5.2.6 Installation and Acceptance of Reinforcement

1. Reinforcement Installation in Cast-in-place Reinforced Concrete Structure Project

(1) Construction technology.

① Column reinforcement installation. The construction process of column reinforcement installation: adjust the reserved reinforcement of lower column → sleeve the column stirrup → bind the vertical load-bearing reinforcement → draw the stirrup spacing line → bind the stirrup → check and accept.

The key points of column reinforcement installation are as follows:

First, check the position, quantity and length of the reserved reinforcement according to the snapped outer dimension line. Before binding, straighten the reserved reinforcement, and clean the cement mortar on it.

Second, set the stirrup of column. According to the spacing required by the drawing, the number of stirrups of each column shall be calculated, and the stirrup shall be set on the reserved reinforcement protruding from the lower layer.

Third, connect the vertical stress bar. When the diameter of the main reinforcement of the column is greater than 16mm, the welding or mechanical connection should be adopted. The length of the mechanical connection and welded joint section of the longitudinal load-bearing reinforcement is $35d$ (d is the larger diameter of the stressed reinforcement), and it is not less than 500mm. The percentage of the area with joint reinforcement in the total reinforcement area should not exceed 50%.

Fourth, draw the stirrup position line. On the vertical reinforcement of the column, the stirrup position line is drawn with chalk, and the reinforcement spacing frame is added.

Fifth, bind column stirrup. According to the drawn stirrup position line, move the stirrup which has been set up upward and bind it from top to bottom. The intersection point of stirrup corner and main reinforcement of column shall be bound with hoop buckle, and other parts can be bound with splay buckle. The overlapping position of hook of square column stirrup shall be staggered along the vertical reinforcement of column and bound firmly. Spiral stirrups should be used for columns. The first stirrup of the column is 50mm away from the ground, and the stirrups at the upper and lower ends are densified according to the regulations(The range of 1/6 clear height of the column and the width of the long side of the column shall not be less than 500mm, and the maximum size of the three shall be taken); in areas with seismic requirements, the end of column stirrup shall be bent to 135°, and the length of straight part shall not be less than $10d$ (d is stirrup diameter). If the design requirement is welding stirrup, the length of single side weld shall not be less than $10d$; the thickness of protective layer of column reinforcement shall meet the design and specification requirements; the cushion block shall be bound on the outside of main reinforcement of column, and the spacing shall be 1000mm(or clamped on the main reinforcement with plastic clamps), so as to ensure the accuracy of protective layer thickness of main reinforcement.

②墙体钢筋安装。墙体钢筋安装施工工艺流程：修整预留筋→绑竖向钢筋→绑水平钢筋→绑拉筋及定位筋→检查验收。

墙体钢筋安装施工要点如下：

第一，修整预留筋。将墙预留钢筋调整顺直，用钢丝刷将钢筋表面砂浆清理干净。

第二，钢筋绑扎。先立墙梯子筋，梯子筋间距不宜大于4 m，然后在梯子筋下部1.5 m处绑两根水平钢筋，并在水平钢筋上画好分格线，最后绑竖向钢筋及其余水平钢筋，梯子筋如图5.23所示；双排钢筋之间应设双形定位筋，定位筋间距不宜大于1.5 m。墙拉筋应按设计要求绑扎，间距一般不大于600 mm。墙拉筋应拉在竖向钢筋与水平钢筋的交叉点上，双形定位筋如图5.24所示；绑扎墙筋时一般用顺扣或八字扣，钢筋交叉点应全部绑扎；墙筋保护层厚度应符合设计及规范要求，垫块或塑料卡应绑在墙外排筋上，呈梅花形布置，间距不宜大于1 000 mm，以使钢筋的保护层厚度准确；墙体合模之后，对伸出的墙体钢筋进行修整，并绑一道水平梯子筋固定预留筋的间距。

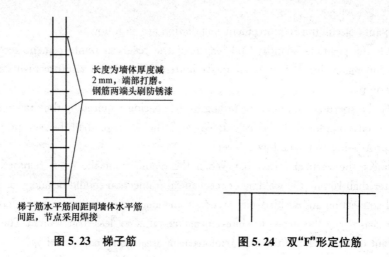

图5.23　梯子筋　　　　图5.24　双"F"形定位筋

第三，墙钢筋的连接。墙水平钢筋：墙水平钢筋一般采用搭接，接头位置应错开。接头的位置、搭接长度及接头错开的比例应符合规范要求。搭接长度末端与钢筋弯折处的距离不得小于$10d$，搭接处应在中心和两端绑扎牢固。墙竖向钢筋：直径大于或等于16 mm时，宜采用焊接（电渣压力焊）；小于16 mm时，宜采用绑扎搭接，搭接长度应符合设计及规范要求。

②Wall reinforcement installation. The construction process of wall reinforcement installation: trimming the reserved reinforcement → binding the vertical reinforcement → binding the horizontal reinforcement → binding the tie bar and positioning bar → inspection and acceptance.

The main points of wall reinforcement installation and construction are as follows:

First, trim the reserved reinforcement. Adjust and straighten the reserved reinforcement of the wall, and clean the surface of the reinforcement with steel wire brush.

Second, bind steel bar. First erect the ladder reinforcement of the wall, and the spacing of the ladder reinforcement should not be greater than 4m. Then, tie two horizontal steel bars at 1.5m lower part of the ladder reinforcement, and draw the grid division line on the horizontal reinforcement. Finally, tie the vertical reinforcement and other horizontal reinforcement. The ladder reinforcement is shown in Fig. 5.23. Double shape positioning bars should be set between double rows of steel bars, and the spacing of positioning bars should not be greater than 1.5m. The wall tie bars shall be bound according to the design requirements, and the spacing is generally not more than 600mm. The wall tie bar should be pulled at the intersection of vertical reinforcement and horizontal reinforcement, and the double shape positioning bar is shown in Fig. 5.24. When binding the wall reinforcement, it is generally to use straight clamp or splayed clamp, and all the crossing points of the reinforcement shall be bound. The thickness of the protective layer of the wall reinforcement shall meet the design and specification requirements. The cushion block or plastic clip shall be bound to the outer row of reinforcement bars in a quincunx shape, and the spacing shall not be greater than 1000mm, so as to make the thickness of the protective layer of the reinforcement accurate. After the wall is closed, the extended wall reinforcement shall be trimmed, and horizontal ladder reinforcement shall be bound to fix the spacing of the reserved reinforcement.

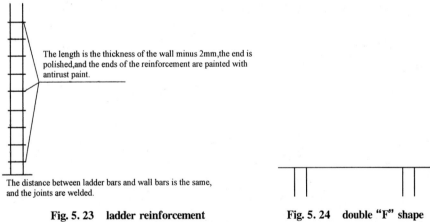

Fig. 5.23 ladder reinforcement   Fig. 5.24 double "F" shape positioning bar

Third, the connection of wall reinforcement. Horizontal reinforcement of wall: the horizontal reinforcement of wall is generally overlapped, and the joint position shall be staggered. The position, lap length and staggered ratio of joints shall meet the requirements of the specification. The distance between the end of lap length and the bending part of reinforcement shall not be less than $10d$, and the lap joint shall be firmly bound at the center and both ends. Vertical wall reinforcement: when the diameter is greater than or equal to 16mm, welding (electroslag pressure welding) should be adopted; when the diameter is less than 16mm, it is appropriate to use binding and overlapping, and the lap length should meet the design and specification requirements.

第四，剪力墙的暗柱和扶壁柱。剪力墙的端部、相交处、弯折处、连梁两侧、上下贯通的门窗洞口两侧一般设有暗柱或扶壁柱。暗柱或扶壁柱钢筋应先于墙筋绑扎施工，其施工方法与框架柱的施工方法相近。直径大于 16 mm 的暗柱或扶壁柱钢筋，应采用焊接（电渣压力焊）或机械连接（滚压直螺纹）。

第五，剪力墙连梁。连梁的第一道箍筋距墙（暗柱）50 mm，顶（末）层连梁箍筋应伸入墙（暗柱）内，并在连梁主筋锚固长度范围内满布。连梁的锚固长度、箍筋及拉筋的间距应符合设计及规范要求。

第六，剪力墙的洞口补强。当设计无要求时，应符合的规定有：矩形洞宽和洞高均不大于 800 mm 的洞口及直径不大于 300 mm 圆形洞口四边应各加 2 根加强筋；直径大于 300 mm 的圆形洞口应按六边形补强，每边各加 2 根加强筋；矩形洞宽和洞高大于 800 mm 的洞口四边应设暗柱和暗梁补强。另外，补强钢筋的直径、暗柱和暗梁设置应符合设计及规范要求。

③梁钢筋安装。梁钢筋安装施工工艺流程如下：

梁钢筋模内绑扎：画主梁箍筋间距→放主梁次梁箍筋→穿主梁底层纵筋及弯起筋→穿次梁底层纵筋并与箍筋固定→穿主梁上层纵向架立筋→按箍筋间距绑扎→穿次梁上层纵向钢筋→按箍筋间距绑扎。

梁钢筋模外绑扎（先在梁模板上口绑扎成形后再入模内）：画箍筋间距→在主次梁模板上口铺横杆数根→在横杆上面放箍筋→穿主梁下层纵筋→穿次梁下层钢筋→穿主梁上层钢筋→按 箍筋间距绑扎→穿次梁上层纵筋→按箍筋间距绑扎→抽出横杆落骨架于模板内。

梁钢筋安装施工要点如下：

第一，在梁侧模板上画出箍筋间距，摆放箍筋。

Fourth, the concealed column and buttress column of shear wall. Generally, concealed columns or buttress columns are set at the ends, intersections, bends, both sides of coupling beams, and both sides of door and window openings through which the upper and lower parts are connected. The reinforcement of concealed column or buttress column should be constructed before the binding of wall reinforcement, and the construction method is similar to that of frame column. Welding (electroslag pressure welding) or mechanical connection (rolling straight thread) shall be adopted for reinforcement of concealed column or buttress column with diameter greater than 16mm.

Fifth, shear wall coupling beam. The first stirrup of coupling beam is 50mm away from the wall (concealed column), and the top (end) coupling beam stirrup shall extend into the wall (concealed column) and be fully distributed within the anchorage length of main reinforcement of coupling beam. The anchorage length, stirrup and tie bar spacing of coupling beam shall meet the design and specification requirements.

Sixth, the opening reinforcement of shear wall. When there is no requirement in the design, the following requirements shall be met: two reinforcing bars shall be added to the four sides of the hole with the width and height of the rectangular hole not greater than 800mm and the circular hole with the diameter not greater than 300mm; the circular hole with the diameter greater than 300mm shall be reinforced by hexagon, with two reinforcing bars on each side; the four sides of the hole with the width and height of the rectangular hole greater than 800mm shall be provided with concealed columns and beams for reinforcement. In addition, the diameter of reinforcement bar, the setting of concealed column and concealed beam should meet the design and specification requirements.

③Beam reinforcement installation. The construction process of beam reinforcement installation is as follows:

Binding beam reinforcement inside the formwork: draw the stirrup spacing of the primary and secondary beams → place the stirrups of the main beam and the secondary beam → pass through the longitudinal reinforcement and bend the reinforcement at the bottom layer of the main beam and fix it with the stirrup → pass through the longitudinal frame of the main beam to erect the reinforcement according to the stirrup spacing → pass the longitudinal reinforcement of the upper layer of the secondary beam → bind according to the stirrup spacing.

Binding beam reinforcement outside the formwork (binding at the upper opening of the beam formwork before entering the formwork): draw the stirrup spacing → lay several cross bars on the upper edge of the primary and secondary beam formwork → place stirrups on the cross bar → pass through the longitudinal reinforcement of the lower layer of the main beam → pass the upper reinforcement of the secondary beam → pass the upper longitudinal reinforcement of the secondary beam → bind according to the stirrup spacing → draw out the cross bar and drop it into the formwork.

The main points of beam reinforcement installation are as follows:

First, draw the stirrup spacing on the beam side formwork and place the stirrup.

第二，先穿主梁的下部纵向受力钢筋及弯起钢筋，将箍筋按已画好的间距逐个分开；穿次梁的下部纵向受力钢筋及弯起钢筋，并套好箍筋；放主次梁的架立筋；隔一定间距将架立筋与箍筋绑扎牢固；调整箍筋间距使间距符合设计要求，绑架立筋，再绑主筋，主次梁同时配合进行。

第三，框架梁上部纵向钢筋应贯穿中间节点，梁下部纵向钢筋伸入中间节点，锚固长度及伸过中心线的长度要符合设计要求。框架梁纵向钢筋在端节点内的锚固长度也要符合设计要求。

第四，绑梁上部纵向筋的箍筋，宜用套扣法绑扎，如图 5.25 所示。箍筋的接头（弯钩叠合处）应交错布置在两根架立钢筋上，其余同柱。

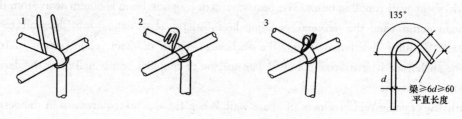

图 5.25　套扣绑扎示意图

第五，箍筋在叠合处的弯钩，在梁中应交错绑扎，箍筋弯钩为 $135°$，平直部分长度为 $10d$，如做成封闭箍时，单面焊缝长度为 $5d$。

第六，梁端第一个箍筋应设置在距离柱节点边缘 50 mm 处。梁端与柱交接处箍筋应加密，其间距与加密区长度均要符合设计要求。

第七，板、次梁与主梁交叉处，板的钢筋在上，次梁的钢筋居中，主梁的钢筋在下；当有圈梁或垫梁时，主梁的钢筋在上。在主、次梁受力筋下均应垫垫块（或塑料卡），保证保护层的厚度。纵向受力钢筋采用双层排列时，两排钢筋之间应垫以直径 25 mm 的短钢筋，以保持其设计距离。梁筋的搭接长度末端与钢筋弯折处的距离，不得小于钢筋直径的 10 倍。

Second, the longitudinal load-bearing steel bar and bending steel bar at the lower part of the main beam shall be penetrated first, and the stirrups shall be separated one by one according to the drawn spacing. The longitudinal load-bearing reinforcement and bending reinforcement at the lower part of the secondary beam shall be penetrated and set with stirrups. Place the supplementary reinforcement of primary and secondary beams. The supplementary reinforcement and stirrup shall be firmly bound at a certain interval. Adjust the spacing of stirrups to make the spacing meet the design requirements, bind the supplementary reinforcement, and then bind the main reinforcement, and the main and secondary beams shall be bound at the same time.

Third, the longitudinal reinforcement in the upper part of the frame beam should run through the middle node, and the longitudinal reinforcement at the lower part of the beam should extend into the middle node. The anchorage length and the length extending through the center line should meet the design requirements. The anchorage length of the longitudinal reinforcement of the frame beam in the end node should also meet the design requirements.

Fourth, the stirrup of longitudinal reinforcement on the upper part of the beam should be bound with sleeve buckle method, as shown in Fig. 5.25. The joint of stirrup (hook overlap) shall be staggered arranged on two supplementary reinforcement, and the rest shall be done in the same way as columns.

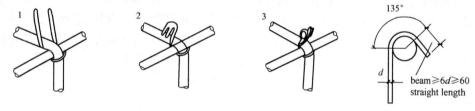

Fig. 5.25 Schematic Diagram of Buckle Binding

Fifth, the hook of stirrup at the overlapping position should be staggered bound in the beam. The hook of stirrup is 135°, and the length of straight part is $10d$. If closed hoop is made, the length of single side weld is $5d$.

Sixth, the first stirrup of beam end should be set at 50mm away from the edge of column joint. The stirrups at the joint of beam end and column shall be densified, and the spacing and length of densified area shall meet the design requirements.

Seventh, at the intersection of slab, secondary beam and main beam, the reinforcement of slab is in the upper part, the reinforcement of secondary beam isin the middle, and the reinforcement of main beam is in the lower part. When there is ring beam or pad beam, the reinforcement of main beam is in the upper part. Cushion block(or plastic clip) shall be padded under the bearing bars of main and secondary beams to ensure the thickness of protective layer. When the longitudinal load-bearing steel bars are arranged in two layers, the short steel bars with diameter of 25mm shall be padded between the two rows to keep the design distance. The distance between the end of the overlapping length of the beam reinforcement and the bending point of the reinforcement shall not be less than 10 times of the diameter of the reinforcement.

第八,框架节点处钢筋穿插十分稠密时,应特别注意梁顶面主筋间的净距要有 30 mm,以利浇筑混凝土。梁板钢筋绑扎时应防止水电管线将钢筋抬起或压下。

第九,梁钢筋的绑扎与模板安装之间的配合关系:梁的高度较小时,梁的钢筋架空在梁顶上绑扎,然后再落位;梁的高度较大($\geqslant 1.2$ m)时,梁的钢筋宜在梁底模上绑扎,其两侧模或一侧模后装。

④板钢筋安装。

a. 底板钢筋安装。底板钢筋安装施工工艺流程:弹出钢筋位置线→绑扎底板下铁钢筋→绑扎基础梁钢筋→绑扎底板上铁钢筋→绑扎墙、柱插筋→隐检验收。

底板钢筋安装施工要点如下:

第一,弹出钢筋位置线。根据设计图纸要求的钢筋间距弹出底板钢筋位置线和墙、柱、基础梁钢筋位置线。

第二,基础底板下铁钢筋绑扎。按底板钢筋受力情况,确定主受力筋方向(设计无指定时,一般为短跨方向)。施工时先铺主受力筋,再铺另一方向钢筋。底板钢筋绑扎可采用顺扣或八字扣,逐点绑扎,禁止跳扣。底板钢筋的连接:板的受力钢筋直径大于或等于 18 mm 时,宜采用机械连接;小于 18 mm 时,可采用绑扎连接。搭接长度及接头位置应符合设计及规范要求。当采用绑扎接头时,在规定搭接长度的任一区段内有接头的受力钢筋截面面积占受力钢筋总截面面积百分率,不宜大于 25%,可不考虑接头位置。当采用机械连接时,接头应错开,其错开间距不小于 $35d$(d 为受力钢筋的较大直径),且不小于 500 mm。任一区段内有接头的受力钢筋截面面积占受力钢筋总截面面积百分率,不宜大于 50%,接头位置下铁宜设在跨中 1/3 区域、上铁宜设在支座 1/3 区域;钢筋绑扎后应随即垫好垫块,间距不宜大于 1 000 mm,垫块厚度应确保主筋保护层厚度符合规范及设计要求。有防水要求的底板及外墙迎水面保护层厚度不应小于 50 mm。

Eighth, when the reinforcement at the frame joint is very dense, special attention should be paid to the clear distance between the main reinforcement of the beam top surface to be 30 mm, so as to facilitate the concrete pouring. When the beam and slab reinforcement is bound, the steel bar shall be prevented from being lifted or pressed down by the water and electricity pipeline.

Ninth, the coordination relationship between beam reinforcement binding and formwork installation: when the beam height is small, the beam reinforcement is overhead bound on the beam top, and then set down; when the beam height is large($\geqslant$ 1.2m), the beam reinforcement should be bound on the beam bottom formwork, and the two side formwork or one side formwork should be installed later.

④Slab reinforcement installation.

a. Installation of bottom slab reinforcement. The construction process of bottom slab reinforcement installation: set out the position line of reinforcement → bind bottom reinforcement of the slab → bind foundation beam reinforcement → bind top reinforcement of the slab → bind wall and column joint bars → hidden works inspection and acceptance.

The main points of installation and construction of bottom slab reinforcement are as follows:

First, set out the position line of reinforcement. According to the reinforcement spacing required by the design drawings, the position line of the bottom slab reinforcement and the position line of the wall, column and foundation beam reinforcement are marked.

Second, bind bottom reinforcement of the slab. According to the stress condition of the bottom slab reinforcement, the direction of the main load-bearing reinforcement is determined (if the design is not specified, it is generally the short span direction). During the construction, the main load-bearing reinforcement shall be laid first, and then the reinforcement in the other direction shall be laid. The reinforcement of the bottom slab can be tied with the straight or splayed clamp, binding point by point, and no jumping clamp is allowed. Connection of bottom slab reinforcement: mechanical connection shall be adopted when the diameter of stressed reinforcement of slab is greater than or equal to 18mm, and binding connection can be used when the diameter is less than 18mm. The lap length and joint position shall meet the design and specification requirements. When the binding joint is used, the percentage of the cross-sectional area of the stressed reinforcement with joint in the total cross-sectional area of the stressed reinforcement in any section of the specified lap length should not be greater than 25%, and the position of the joint may not be considered. When mechanical connection is adopted, the joints shall be staggered, and the staggering distance shall not be less than $35d$ (d is the larger diameter of the stressed reinforcement) and not less than 500mm. The percentage of the cross-sectional area of the stressed reinforcement with joint in the total cross-sectional area of the stressed reinforcement in any section should not be greater than 50%. The lower steel bar should be set in the middle 1/3 area of the span, and the upper steel bar should be set in the 1/3 area of the bearing. After the reinforcement is bound, the cushion block should be padded immediately, and the spacing should not be greater than 1000mm. The thickness of the cushion block should ensure that the thickness of the main reinforcement protection layer meets the specification and design requirements. The thickness of the protective layer of the bottom slab with waterproof requirements and the external wall on the water facing surface shall not be less than 50mm.

第三,基础梁钢筋绑扎。基础梁一般采用就地绑扎成形方式施工。基础梁高大于1 000 mm时,应搭设钢管绑扎架,将基础梁的架立筋两端放在绑扎架上,画出箍筋间距,套上箍筋,按已画好的位置与底板梁上层钢筋绑扎牢固。穿基础梁下层纵向钢筋,与箍筋绑牢。当纵向钢筋为双排时,可用短钢筋(直径不小于250 mm并且不小于梁主筋直径)垫在两层钢筋之间。抽出绑扎架,将已绑扎好的梁筋骨架落地。

第四,基础底板上铁钢筋绑扎。摆放钢筋马凳,间距不宜大于2 000mm,并与底板下铁钢筋绑牢。马凳架设在板下铁的上层筋上、上铁的下层筋下。马凳一般加工成"A"形或"工"字形,如图5.26、图5.27所示,并有足够的刚度。在马凳上绑扎上层定位钢筋,并在其上画出钢筋间距,然后绑扎纵、横方向钢筋。

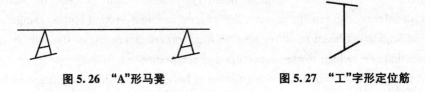

图5.26 "A"形马凳　　　　　　图5.27 "工"字形定位筋

第五。墙、柱插筋绑扎。根据弹好的墙、柱位置线,将墙、柱伸入基础底板的插筋绑扎牢固。插筋锚入基础深度应符合设计要求;插筋甩出长度应考虑接头位置,且不宜过长。其上部绑扎两道以上水平筋、箍筋及定位筋;其下部伸入基础底板部分也应绑扎两道以上水平或箍筋,以确保墙体插筋垂直,不发生位移。

第六,底板钢筋和墙、柱插筋绑扎完毕后,经检查验收并办理隐检手续,方可进行下道工序施工。

b.楼板钢筋安装。楼板钢筋安装施工工艺流程:放钢筋位置线→绑板下钢筋→绑板上钢筋及负弯矩钢筋→检查验收。

楼板钢筋安装施工要点如下:

Third, bind foundation beam reinforcement. The foundation beam is generally constructed by binding and forming on site. When the height of foundation beam is greater than 1000mm, steel pipe binding frame shall be set up. Place both ends of the erection reinforcement of the foundation beam on the binding frame, draw the stirrup spacing, cover the stirrup, and bind firmly with the upper reinforcement of the bottom beam according to the drawn position. The longitudinal steel bar passing through the lower layer of foundation beam shall be firmly bound with stirrup. When the longitudinal reinforcement is double row, the short reinforcement(diameter not less than 250mm and not less than the diameter of main reinforcement of beam) can be used to pad between the two layers of reinforcement. Pull out the binding frame and put the bound frame on the ground.

Fourth, bind top reinforcement of the slab. Place the steel bench, the spacing should not be greater than 2000mm, and it should be firmly bound with the bottom reinforcement on the base. The bench is erected on the top reinforcement of the bottom slab reinforcement and bottom reinforcement of the top slab reinforcement. The bench is generally processed into "A" shape or "I" shape, as shown in Fig. 5.26 and Fig. 5.27, with sufficient rigidity. The upper positioning reinforcement shall be bound on the bench, and the spacing of reinforcement shall be drawn on it, and then the longitudinal and transverse reinforcement shall be bound.

Fig. 5.26 "A" Shaped Bench

Fig. 5.27 "I" Shaped Positioning Bar

Fifth, bind wall and column joint bars. According to the snapped position line of the wall and column, the joint bars of the wall and column extending into the foundation slab shall be bound firmly. The anchorage depth of the joint bar into the foundation shall meet the design requirements; and the joint position shall be considered for the throw out length of the joint bar, and it shall not be too long. The upper part shall be bound with more than two horizontal bars, stirrups and positioning bars; the lower part extending into the foundation slab shall also be bound with more than two horizontal bars or stirrups to ensure that the wall joint bars are vertical and not displaced.

Sixth, after the binding of bottom slab reinforcement and wall and column joint bar, the next process can be carried out after inspection and acceptance and hidden inspection procedures.

b. Floor reinforcement installation. The construction process of floor reinforcement installation: setting out the position line of steel bar → binding bottom steel bar of the slab → binding top steel bar and the negative bending moment steel bar → inspection and acceptance.

The main points of floor reinforcement installation are as follows:

第一,在板面上画好主筋、分布筋的间距线。按画好的间距,先摆放下铁主受力筋,后放下铁分布筋,然后做水、电专业的管线预埋,最后摆放上铁主分布筋、上铁受力筋并绑扎。预埋件、预留洞等及时配合施工。

第二,绑扎板筋时一般用顺扣或八字扣,钢筋相交点全部绑扎。如板为双层钢筋时,两层钢筋间需加钢筋马凳,以确保上铁的位置。马凳架设在板下铁的上层筋上,上铁的下层钢筋下,马凳间距不宜大于 1 500 mm。负弯矩筋每个相交点均要绑扎。

第三,在钢筋的下面应垫好砂浆垫块或"H"形塑料垫块,间距宜为 1 000 mm。

第四,对于悬挑板,应在固定端 1/4 跨,且不大于 300 mm 的位置设"A"形通长马凳。

⑤楼梯钢筋安装。楼梯钢筋安装施工工艺流程:预留预埋件及检查→放位置线→绑板主筋→绑分布筋→检查验收。

楼梯钢筋安装施工要点如下:

第一,施工楼梯间墙体时,要做好预留预埋工作。休息平台板预埋钢筋于墙体内,做贴模钢筋。当设计为带肋钢筋时,钢筋应伸出墙外,模板应穿孔或做成分体形式。

第二,在楼梯段底模上用墨线分别弹出主筋和分布筋的位置。

第三,绑扎钢筋(先绑梁筋后绑板筋)。梁钢筋绑扎应按设计要求将主筋与箍筋分别绑扎;板筋绑扎时,应根据设计图纸主筋、分布筋的方向,先绑扎主筋后绑扎分布筋,每个点均应绑扎,一般采用八字扣,然后放马凳筋,绑上铁负弯矩钢筋及分布筋。马镫筋一般采用"Ⅱ"形,间距 1 000 mm。

第四,楼梯的中间休息平台钢筋应同楼梯段一起施工。

First, draw the spacing line of main reinforcement and distribution reinforcement on the base surface. According to the drawn spacing, the main steel bearing bars shall be placed and laid down first, and then the distribution bars shall be laid down, and then the pipelines of water and electricity shall be embedded, and finally, the main distribution bars and stress bars shall be placed and bound. The embedded parts and reserved holes shall be constructed in time.

Second, when binding the slab reinforcement, generally use the straight clamp or the splayed clamp, and all the intersection points of the reinforcement are bound. If the slab is double-layer reinforcement, a steel bench should be added between the two layers of reinforcement to ensure the position of the upper steel. The distance between steel benches should not be more than 1500mm when the steel benches are erected on the top reinforcement of the lower slab reinforcement and bottom reinforcement of the top slab reinforcement. Each intersection point of negative moment reinforcement shall be bound.

Third, mortar cushion block or "H" shaped plastic cushion block should be placed under the reinforcement, and the spacing should be 1000mm.

Fourth, for cantilever slab, A-shaped full-length bench should be set at the fixed end of 1/4 span and no more than 300mm.

⑤ Stair Reinforcement installation. Construction process of stair reinforcement installation: reserved embedded parts and inspection → setting out of position line → binding of main reinforcement of slab → binding of distribution reinforcement → inspection and acceptance.

The main points of stair reinforcement installation are as follows:

First, during the construction of staircase walls, it is necessary to make advance reserving work. The rest platform slab is embedded with reinforcement in the wall, which is used as formwork reinforcement. When it is designed as screw thread reinforcement, the reinforcement shall extend out of the wall, and the formwork shall be perforated or made into component body form.

Second, the position lines of main reinforcement and distribution reinforcement are respectively setting out with ink line on the bottom mold of stair section.

Third, binding steel bars(beam bars should be bound first and then plate bars should be bound). The main reinforcement and stirrup shall be bound separately according to the design requirements. When binding the plate reinforcement, according to the direction of the main reinforcement and distribution reinforcement in the design drawing, the main reinforcement shall be bound first and then the distributed reinforcement shall be bound, and each point shall be bound. Generally, the splayed clamp shall be used, and then the stirrup shall be placed, and the negative bending moment reinforcement and distribution reinforcement shall be bound. Bench reinforcement is generally in the shape of "Π", with a spacing of 1000mm.

Fourth, the middle rest platform reinforcement of stairs should be constructed together with the stair section.

5.3　混凝土工程施工

5.3.1　混凝土的配料

混凝土一般由水泥、骨料、水、外加剂以及各种矿物掺合料组成。将各种组分材料按已经确定的配合比进行拌制生产,首先要进行配料,确定混凝土配合比。

混凝土配合比是指混凝土各组成材料之间用量的比例关系。一般按重量计,以水泥质量为1,以水泥：砂：石子和水胶比来表示。

1. 混凝土配合比设计原则

应根据设计的混凝土强度等级以及混凝土施工和易性的要求确定,并应符合合理使用材料和经济的原则,对有防渗、抗冻等要求的混凝土,也应符合有关的专项规定。

2. 混凝土配合比的确定

混凝土配合比的确定是采用计算与试验相结合的方法。先根据经验,利用经验公式和图表(数据),综合考虑混凝土强度等级、耐久性要求、施工和易性要求及工地材料(水泥、粗细骨料)等情况,计算出"初步计算配合比"。再用施工所用材料进行试配、调整,得出基准配合比,最后通过强度检验(有抗冻、抗渗要求时还要作相应的检验),定出满足设计和施工要求、比较经济合理的混凝土配合比。

所谓混凝土施工配合比是指混凝土在施工过程中所采用的配合比。混凝土施工配合比一经确定就不能随意去改变,除非发生特殊情况。根据国家现行标准《普通混凝土配合比设计规程》(JGJ 55—2011)和《混凝土强度检验评定标准》(GB/T 50107—2010)的规定,混凝土施工配合比应由有相关资质的试验室提供(试验室配合比),也可以在试验室配合比的基础上根据施工现场砂石含水量情况进行调整。调整按以下步骤进行：

试验室配合比为：水泥：砂子：石子=1：X：Y,水灰比为w/c,测得现场砂子含水量为W_x,石子含水量为W_y,则施工配合比调整为：1：$X(1+W_x)$：$Y(1+W_y)$。

按试验室配合比每1 m³混凝土水泥用量为$C(kg)$计,水灰比不变,则换算后各种材料用量为：

5.3 Construction of Concrete Works

5.3.1 Concrete Batching

Concrete is generally composed of cement, aggregate, water, admixtures and various mineral admixtures. To mix and produce various component materials according to the determined mix proportion, the first step is to mix and determine the concrete mix proportion.

The concrete mix proportion refers to the proportion relationship between the amounts of each constituent material of concrete. According to the weight, the cement weight is 1, and it is expressed by cement : sand : gravel and water cement ratio.

1. Design Principle of Concrete Mix Proportion

It should be determined according to the designed concrete strength grade and the requirements of concrete construction workability, and should conform to the principle of rational use of materials and economy. For concrete with pit seepage and frost resistance requirements, it should also comply with relevant special provisions.

2. Determination of Concrete Mix Proportion

The determination of concrete mix proportion is based on the combination of calculation and test. First, according to the experience, using empirical formula and chart(data), considering the concrete strength grade, durability requirements, construction workability requirements and site materials(cement, coarse and fine aggregate), the "preliminary calculation mix proportion" is calculated. Then the construction materials are used for trial mixing and adjustment, and the reference mix proportion is obtained. Finally, through the strength test(corresponding inspection is required when there are freezing and impermeability requirements), the relatively economic and reasonable concrete mix proportion meeting the design and construction requirements is determined.

The so-called concrete construction mix ratio refers to the concrete in the construction process used in the mix proportion. Once the concrete construction mix ratio is determined, it cannot be changed at will unless special circumstances occur. According to the current national standards *Specification for mix proportion design of ordinary concrete*(JGJ 55—2011) and *Stand ard for evaluation of concrete compressive strength*(GB/T 50107—2010), the concrete construction mix proportion shall be provided by the laboratory with relevant qualification (laboratory mix proportion), and can also be adjusted according to the water content of sand and gravel at the construction site on the basis of laboratory mix proportion. The adjustment is carried out according to the following steps:

The laboratory mix proportion is: cement : sand : gravel $= 1 : X : Y$, water cement ratio is w/c, the water content of sand is W_x, the water content of stone is W_y, then the construction mix proportion is adjusted to $1 : X(1+W_x) : Y(1+W_y)$.

If the cement content per m³ of concrete in the laboratory is $C(\text{kg})$, and the water cement ratio remains unchanged, the amount of various materials after conversion is as follows:

水泥：$C'=C$
砂子：$G_砂=CX(1+W_x)$
石子：$G_石=CY(1+W_y)$
水：$W'=W-CXW_x-CYW_y$

例：设混凝土试验室配合比为：1：2.56：5.5，水胶比为0.64，每1 m³ 混凝土水泥用量为280 kg，测得砂子含水量为4%，石子含水量为2%，则施工配合比为：

1：2.56×(1+4%)：5.5×(1+2%) = 1：2.66：5.61

每1 m³ 混凝土材料用量为：

水泥：280 kg

砂子：280×2.66 = 744.8(kg)

石子：280×5.61 = 1 570.8(kg)

水：280×0.64−280×2.56×4%−280×5.5×2% = 119.7(kg)

(1)施工配料。施工现场所采用的搅拌机有一定的容量，因此还需要求出搅拌机每次搅拌需要多少原材料。如采用JZ350型搅拌机，其出料容量为0.35m³，则每次搅拌所需原材料为：

水泥：280×0.35 = 98.0(kg)（取用两袋水泥即100 kg）

砂子：744.8×100÷280 = 266.0(kg)

石子：1 570.8×100÷280 = 561.0(kg)

水：119.7×100÷280 = 42.8(kg)

为严格控制混凝土的配合比，原材料的计量必须准确。原材料每盘称量的允许偏差不得超过以下规定：水泥、掺合料为±2%；粗、细骨料为±3%；水、外加剂为±2%。

对称量用的各种计量设备要定期校验，对粗、细骨料含水量应经常测定，雨天施工时，应增加测定次数。

(2)外加剂、掺合料的使用。混凝土中掺用外加剂的质量及应用技术应符合现行国家标准《混凝土外加剂》(GB 8076/2008)、《混凝土外加剂应用技术规范》(GB 50119—2013)等和有关环境保护的规定。

混凝土中掺用矿物掺合料的质量应符合现行国家标准《用于水泥和混凝土中的粉煤灰》(GB 1596—2017)等的规定。矿物掺合料的掺量应通过试验确定。

5.3.2　混凝土的搅拌

混凝土搅拌是将水、水泥和粗细骨料进行均匀拌和的过程，且通过搅拌应使材料达到塑化、强化的作用，使不同细度、形状的散装物料搅拌成均匀、色泽一致、具有流动性的混凝土拌合物。混凝土的搅拌采用人工或机械进行搅拌。在工程中，一般采用机械搅拌，以保证拌和均匀和效率。除非没有条件的地方及拌和量极小时才采用人工搅拌。

Cement: $C' = C$
Sand: $G_{sand} = CX(1+W_x)$
Stone: $G_{stone} = CY(1+W_y)$
Water: $W' = W - CXW_x CYW_y$

For example: the concrete laboratory mix ratio is 1 : 2.56 : 5.5, the water cement ratio is 0.64, and the cement dosage per m³ of concrete is 280kg. The measured water content of sand is 4% and the water content of stone is 2%, then the construction mix ratio is:

1 : 2.56×(1+4%) : 5.5×(1+2%) = 1 : 2.66 : 5.61

The amount of concrete material per m³ is as follows:
Cement: 280kg
Sand: 280×2.66 = 744.8(kg)
Stone: 280×5.61 = 1570.8(kg)
Water: 280×0.64 − 280×2.56×4% − 280×5.5×2% = 119.7(kg)

(1) Construction Batching. The mixer used in the construction site has a certain capacity, so it is also necessary to find out how much raw material the mixer needs for each stirring. If JZ350 mixer is used, its discharge capacity is 0.35m³. The raw materials required for each mixing are as follows:

Cement: 280×0.35 = 98.0(kg)(take two bags of cement, i.e. 100kg)
Sand: 744.8×100÷280 = 266.0(kg)
Stone: 1 570.8×100÷280 = 561.0(kg)
Water: 119.7×100÷280 = 42.8(kg)

In order to strictly control the mix proportion of concrete, the measurement of raw materials must be accurate. The allowable deviation of weighing each stirring of raw materials shall not exceed the following provisions: cement and admixtures are ±2%; coarse and fine aggregates are ±3%; water and admixture are ±2%.

All kinds of measuring equipment used for symmetrical measurement should be checked regularly. The water content of coarse and fine aggregates should be measured frequently. When the construction is carried out in rainy days, the measurement times should be increased.

(2) Use of admixtures and blend. The quality and application technology of admixtures in concrete should conform to the current national standard *Concrete admixtures* (GB 8076—2008), *Code for utility of concrete admixtures* (GB 5119—2013) and other relevant environmental protection regulations.

The quality of mineral admixtures in concrete should conform to the current national standard *Fly ash used for cement and concrete* (GB 1596—2017). The content of mineral admixture should be determined by experiment.

5.3.2 Mixing of Concrete

Concrete mixing is the process of mixing water, cement and coarse and fine aggregate evenly. The material should be plasticized and strengthened by mixing, so that the bulk materials with different fineness and shape can be mixed into concrete mixture with uniform color and fluidity. The concrete shall be mixed manually or mechanically. In construction, mechanical mixing is generally used to ensure uniform mixing and efficiency. Unless there is no condition and the mixing volume is very small, manual mixing is adopted.

1. 人工拌制

人工拌制混凝土只适宜于野外、施工条件困难、混凝土用量不大而强度等级不高的混凝土。

人工拌制一般用铁板或包有薄钢板的木拌板，若用木制拌板应刨光、拼严、使不漏浆。人工拌制一般采用"三干三湿"法，即先将砂倒在拌板上，稍微摊平，再把水泥倒在砂上拌两遍，摊平加入石子再拌一遍，之后加适量的水后反复湿拌三遍，直至石子与水泥浆无分离现象为止。

为了搅拌均匀，人工搅拌混凝土必须有很大的劳动强度大和较大的坍落度。当水胶比不变时，人工搅拌要比机械搅拌多耗 10%～15% 的水泥用量。

2. 机械搅拌

(1) 搅拌机械。混凝土搅拌机的用途是将一定配合比的水泥、砂、石等拌和成均匀的混凝土拌合物。搅拌机按其原理分为自落式搅拌机和强制式搅拌机两类，如图 5.28、图 5.29 所示；按其出料方式分为倾翻式和非倾翻式两类；按搅拌筒的容量大小分为大型、中型和小型三种。

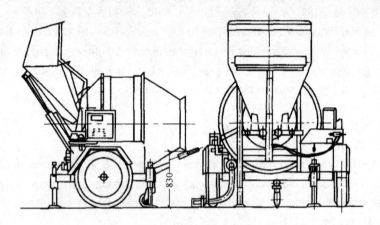

图 5.28 自落式混凝土搅拌机

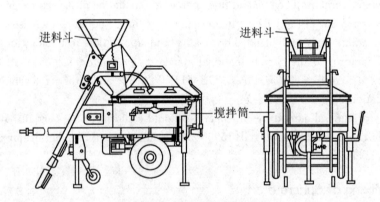

图 5.29 强制式搅拌机

Chapter 5 Construction of Reinforced Concrete Structure

1. Manual Mixing

Manual mixing concrete is only suitable for field, difficult construction conditions, small amount of concrete and low strength grade concrete.

Generally, iron plate or wood board wrapped with thin steel plate is used for manual mixing. If wood board is used, it should be planed and tightly assembled without leakage of slurry. Manual mixing generally adopts the "three dry and three wet" method, that is, first pour the sand on the mixing board, slightly level it, then pour the cement on the sand and mix it twice, spread it out and add stones, and then mix it again. After that, add appropriate amount of water and mix it wet for three times repeatedly until there is no separation between the stone and the cement slurry.

In order to mix evenly, the manual mixing concrete must have great labor intensity and a large slump. When the water cement ratio is constant, the manual mixing will consume 10%—15% more cement than the mechanical mixing.

2. Mechanical Mixing

(1) Mixing machine. The purpose of concrete mixer is to mix cement, sand and stone with certain mix proportion into uniform concrete mixture. According to its principle, the mixer can be divided into two types: self falling mixer and forced mixer, as shown in Fig. 5.28 and Fig. 5.29; according to its discharging mode, it can be divided into two types: tilting type and non-tipping type; according to the capacity of mixing drum, it can be divided into large, medium and small types.

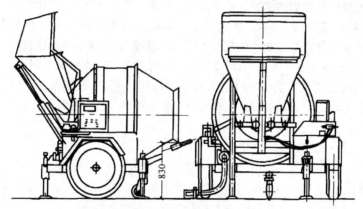

Fig. 5.28 Self Falling Concrete Mixers

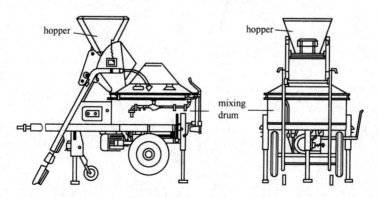

Fig. 5.29 Forced Mixers

①自落式搅拌机。自落式搅拌机适用于搅拌塑性和低流动性混凝土。其搅拌鼓筒是垂直放置的,利用自重作用,使鼓筒内的物料相互穿插、翻拌、混合以达到均匀搅拌的目的。目前应用较多的为锥形反转出料搅拌机(图5.28),它正转搅拌,反转出料,搅拌作用强烈具有构造简单、筒体和叶片磨损小、易于清理、操作维修和移动方便等特点。但动力消耗大、效率低,搅拌时间一般为90~120 s/盘。

由于自落式搅拌机对混凝土骨料有较大的磨损,因此正逐渐被强制式搅拌机所取代。

②强制式搅拌机。强制式搅拌机主要用于集中搅拌站、预制场等生产低流动性、干硬性和轻骨料混凝土。其分为水平轴强制式搅拌机和立轴强制式搅拌机(图5.29)两种。水平轴强制式搅拌机的搅拌机理是当水平轴带动叶片旋转时,筒内物料在做圆周运动的同时,也沿轴向来回运动,搅拌效果较好;立轴强制式搅拌机的搅拌机理是依据剪切原理设计的,它的搅拌筒是一个水平旋转的圆盘,在搅拌筒内有转动叶片,这些不同角度和位置的叶片转动时,克服了物料的摩擦力,产生环向、径向、竖向运动,而叶片通过后的空间,又由翻越的物料所填满。这样,由叶片强制产生的剪切位移,使物料搅拌得更均匀。

强制式搅拌机具有搅拌质量好、搅拌速度快、生产效率高、操作简便和安全可靠等优点。但机件磨损严重,一般需要用高强合金钢或其他耐磨材料做内层,如果底部的卸料口密封不好,水泥浆易漏掉,影响拌和质量。

混凝土搅拌机以其出料容量(m^3)×1 000标定规格。常用的有150 L、250 L、350 L等数种。

选择搅拌机型号,要根据工程量大小、混凝土坍落度和骨料尺寸等确定。既要满足技术上的要求,又要考虑经济效果和节约能源。

(2)搅拌(楼)站工艺布置。搅拌(楼)站是产生混凝土的主要基地,搅拌楼用于城市建设、水库大坝、道路桥梁预制构件厂等拌制各种塑性和干硬性混凝土;搅拌站用于建筑施工、道路、桥梁、混凝土构件厂及商品混凝土工厂等。搅拌(楼)站其布置是否合理,直接关系到产生效率和成本以及操作工人的劳动强度。搅拌(楼)站的工艺布置,应根据产生任务、现场条件、材料来源和机具设备等情况,尽量做到自动上料、自动称量、机动出料和集中控制,以利于逐步实现其机械化和自动化。

① Self falling mixer. Self falling mixer is suitable for mixing plastic and low fluidity concrete. The mixing drum is placed vertically, and the materials in the drum are interspersed, turned and mixed with each other to achieve the purpose of uniform mixing. At present, the cone reverse discharge mixer is widely used(Fig. 5. 28). It rotates forward and reverses the discharge, and the stirring effect is strong. It has the advantages of simple structure, small wear of cylinder and blade, easy cleaning, easy operation and maintenance, and easy to move. However, the power consumption is large and the efficiency is low. The stirring time is generally 90—120s/batch.

Because the self falling mixer has great wear on concrete aggregate, it is gradually replaced by forced mixer

②Forced mixer. Forced mixer is mainly used in centralized mixing plant and prefabrication yard to produce low fluidity, dry hard and lightweight aggregate concrete. It is divided into horizontal axis forced mixer and vertical shaft forced mixer(Fig. 5. 29). The mixing mechanism of the horizontal axis forced mixer is that when the horizontal axis drives the blades to rotate, the materials in the cylinder move in a circle and move back and forth along the axial direction at the same time, so the mixing effect is better; the mixing mechanism of the vertical shaft forced mixer is designed based on the shear principle, its mixing drum is a horizontal rotating disc, and there are rotating blades in the mixing cylinder, which have different angles and positions. When the blade rotates, it overcomes the friction force of the material and produces circumferential, radial and vertical movement. The space after the blade passes through is filled by the material. In this way, the shear displacement forced by the blade makes the material mix more evenly.

The forced mixer has the advantages of good mixing quality, fast mixing speed, high production efficiency, simple operation and safety. However, due to the serious wear of the machine parts, it is generally necessary to use high-strength alloy steel or other wear-resistant materials as the lining. If the discharge port at the bottom is not well sealed, the cement slurry is easy to leak out, which affects the mixing quality.

The concrete mixer is calibrated with its discharge capacity(m^3) × 1000. Commonly used are 150L, 250L, and 350L and so on.

The selection of mixer model should be determined according to the engineering quantity, concrete slump and aggregate size. It should not only meet technical requirements, but also consider economic effects and energy conservation.

(2) Process layout of mixing (plant) station. The mixing (plant) station is the main base for producing concrete. The mixing plant is used for mixing all kinds of plastic and dry hard concrete in urban construction, reservoir dam, road and bridge precast component plant. The mixing station is used for construction, road, bridge, concrete component plant and commercial concrete plant. Whether the layout of mixing(plant) station is reasonable or not is directly related to the production efficiency, cost and labor intensity of operators. The process layout of the mixing(plant) station should be based on the production task, site conditions, material sources, machines and equipment, etc., to achieve automatic feeding, automatic weighing, motorized discharging and centralized control as far as possible, so as to realize its mechanization and automation gradually.

搅拌楼主要由物料供给系统、称量系统、搅拌主机和控制系统四大部分组成。其生产流程一般是把砂、石、水泥等物料一次提升到楼顶料仓，各种物料按产生流程经称量、配料、搅拌，直到制成混凝土、出料装车。搅拌站和搅拌楼的组成一样，也由物料供给系统、称量系统、搅拌主机和控制系统四大部分组成。

(3)搅拌方法及要求。混凝土搅拌是为了将各组成材料拌成混合均匀、颜色一致的混凝土混合物。因此在搅拌过程中要确定正确的搅拌方法。

①原材料称量。混凝土各组成材料，水泥、砂、石、水、外加剂必须每次搅拌每次计量，除水和外加剂外，水泥、砂、石的计量应采用质量比。各组成材料称量应准确，其允许的偏差不得超过表5.11的规定。

表5.11 混凝土各组成材料称量的允许偏差

混凝土组成的材料名称	允许偏差
水泥和外掺混合材料	±2%
粗、细骨料	±3%
水、外加剂溶液	±2%

注：1. 各种衡器应定期校验，保持计量准确；
　　2. 经常测定骨料含水率，雨天施工时，应增加测定次数。

②混凝土搅拌的一般要求。

a. 搅拌混凝土前，应加水空转数分钟，然后将积水倒净，使搅拌筒内壁充分湿润。搅拌第一盘时，搅拌筒内壁上粘附的砂浆会造成砂浆损失，故石子用量应按配合比规定减半。

b. 搅拌好的混凝土要做到基本卸尽，在全部混凝土卸出之前不得再投入拌和物，更不能采取边出料边进料的方法来操作。

c. 严格控制水胶比和坍落度，未经试验人员同意不得随意加减用水量。

③投料顺序。投料顺序应从提高搅拌质量减少叶片、衬板的磨损，减少拌合物与搅拌筒的粘结，减少水泥飞扬，改善工作条件等方面综合考虑确定。常用的投料顺序有一次投料法、二次投料法和水泥裹砂法。

The mixing plant is mainly composed of four parts: material supply system, weighing system, mixing machine and control system. The production process is generally to lift sand, stone, cement and other materials to the roof silo at one time. All kinds of materials are weighed, batched and mixed according to the production process until the concrete is discharged and loaded. The composition of mixing stationis the same as that of mixing plant and is composed of four parts: material supply system, weighing system, mixing machine and control system.

(3) Mixing methods and requirements. The purpose of concrete mixing is to mix the constituent materials into concrete mixture with uniform mixing and uniform color. Therefore, the correct mixing method should be determined in the mixing process.

①Raw material weighing. Each constituent material of concrete, cement, sand, stone, water and admixtures must be mixed and measured every time. Except for water and admixture, the measurement of cement, sand and stone should adopt the mass ratio. Each component material shall be weighed accurately, and the allowable deviation shall not exceed the provisions in Table 5.11.

Table 5.11 Allowable Deviation of Weighing of Concrete Components

Material name of concrete composition	Allowable deviation
Cement and admixture	±2%
Coarse and fine aggregate	±3%
Water and admixture solution	±2%

Note: 1. All kinds of weighing instruments should be calibrated regularly to keep accurate measurement;
2. The moisture content of aggregate shall be measured frequently, and the frequency of determination shall be increased during construction in rainy days.

②General Requirements for Concrete Mixing

a. Before mixing concrete, water should be added and idled for several minutes, and then the accumulated water should be poured out to make the inner wall of the mixing drum fully wet. When mixing the first batch, the mortar adhered on the inner wall of the mixing drum will cause mortar loss, so the amount of stone should be reduced by half according to the mix proportion.

b. The mixed concrete should be basically discharged. Before all the concrete is discharged, no mixture should be put in again, nor can it be operated by the way of feeding while discharging.

c. The water cement ratio and slump shall be strictly controlled, and the water consumption shall not be increased or decreased at will without the consent of the test personnel.

③ Feeding sequence. The feeding sequence should be determined by improving the mixing quality, reducing the abrasion of blades and lining plates, the adhesion between the mixture and the mixing drum, the flying of cement and improving the working conditions. The commonly used feeding sequence includes one-time feeding method, two-time feeding method and cement wrapped sand method.

a. 一次投料法。一次投料法是在上料斗中先装石子(或砂),再加水泥和砂(或石子),然后一次投入搅拌机中。并同时向搅拌筒内装入拌和水(也可在筒内先装一半)。这样就将水泥加在砂、石之间,压住水泥,避免水泥飞扬;而且水泥和砂子又不至于粘结料斗底部,造成上料困难;水泥和砂先进入搅拌筒内形成水泥砂浆,可缩短包裹石子的时间。

b. 二次搅拌法。二次搅拌法的投料顺序有两种,一种是预拌水泥砂浆法,另一种是预拌水泥净浆。预拌水泥砂浆法是先将全部砂、水泥及 1/3 的水投入搅拌 20~30 s 后,再投入石子和 2/3 的水进行搅拌;预拌水泥净浆是将水泥和部分水进行净浆搅拌,然后再投入全部砂石和剩余的水进行搅拌。

二次投料法能使水泥较好地进行水化,水泥砂浆较易搅拌均匀,并较好包裹石子。因此二次投料法与一次投料法相比,可提高混凝土强度约 15%,节约水泥 15%~20%。

c. 水泥裹砂法。用这种方法拌制的混凝土称为造壳混凝土,是先在砂子表面形成一层水泥浆壳。其投料顺序见图 5.30。

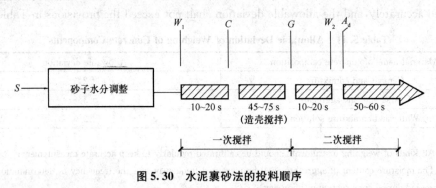

图 5.30 水泥裹砂法的投料顺序

④搅拌时间。为了使混凝土拌合物混合均匀和颜色一致,必须保证一定的搅拌时间。混凝土拌合物的搅拌时间与搅拌机的类型和混凝土混合料和易性有关,应根据混凝土拌和物要求的均匀性、混凝土强度增长的效果及生产效率等因素来确定适合的搅拌时间。

搅拌时间是指从全部材料投入搅拌筒起,到开始卸料为止所经历的时间。搅拌时间过短,混凝土的组成材料拌和不均匀,强度和和易性都会下降;搅拌时间过长,除生产效率降低外,还会使不坚硬的粗骨料,在大容量搅拌机中因脱角、破碎等影响混凝土的质量。混凝土的最短时间可按表 5.12 采用。

Chapter 5 Construction of Reinforced Concrete Structure

a. One-time feeding method. The one-time feeding method is to fill the hopper with stones (or sand), then add cement and sand (or stones), and then put them into the mixer once. At the same time, the mixing water is filled into the mixing drum (or half of the mixing water can be filled in the drum first). In this way, the cement is added between the sand and the stone to keep the cement from flying; moreover, the cement and sand will not stick to the bottom of the hopper, and will not result in feeding difficulties; The cement and sand first enter the mixing drum to form cement mortar, which can shorten the time of wrapping stones.

b. Two-time mixing method. There are two feeding sequence of two-time mixing method, one is premixed cement mortar method; the other is premixed cement paste. The method of premixed cement mortar is to mix all sand, cement and 1/3 water for 20—30s, and then add stones and 2/3 water for mixing; premixed cement paste is to mix the cement and part of water, and then put all sand and stone and remaining water for mixing.

The two-time feeding method can make the cement hydration better, the cement mortar is easy to mix evenly, and the stone can be wrapped well. Therefore, compared with the one-time feeding method, the two-time feeding method can improve the concrete strength by about 15% and save the cement by 15%—20%.

c. Cement wrapped sand method. The concrete mixed in this way is called shell making concrete, which first forms a layer of cement slurry shell on the surface of sand. The feeding sequence is shown in Fig. 5.30.

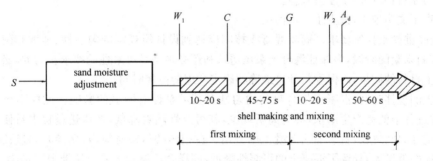

Fig. 5.30 Feeding Sequence of Cement Wrapped Sand Method

④Mixing time. In order to make the concrete mixture mix evenly and the color consistent, it is necessary to ensure a certain mixing time. The mixing time of concrete mixture is related to the type of mixer and the workability of concrete mixture. The suitable mixing time shall be determined according to the uniformity required by concrete mixture, the effect of concrete strength growth and production efficiency.

Mixing time refers to the time from the mixing of all materials to the beginning of discharging. If the mixing time is too short, the component materials of concrete will be mixed unevenly, and the strength and workability will be reduced; if the mixing time is too long, in addition to reducing the production efficiency, the non-hard coarse aggregate will also affect the quality of concrete due to angle separation and crushing in the large capacity mixer. The minimum time of concrete can be used according to Table 5.12.

表 5.12 混凝土搅拌的最短时间(s)

混凝土坍落度/mm	搅拌机类型	搅拌机出料量/L		
		<250	250~500	>500
≤30	强制式	60	90	120
	自落式	90	120	150
≥30	强制式	60	60	90
	自落式	90	90	120

注:1. 掺有外加剂时,搅拌时间应适当延长;
2. 轻混凝土可采用强制式搅拌机搅拌,特轻混凝土可采用自落式搅拌机搅拌,但搅拌时间应延长 60~90 s。

3. 商品混凝土的拌制

随着工程规模的逐步扩大、工程要求的不断提高和技术水平的不断发展,混凝土使用量越来越大,混凝土的强度等级、防水等级、耐久性要求也越来越高,分散落后的现场混凝土生产已难以满足现代工程建设技术进步的要求,在这种情况下,就出现了商品混凝土。

(1)商品混凝土介绍。商品混凝土是指由水泥、骨料、水以及根据需要掺加的外加剂、矿物掺和料等组分按一定比例,在搅拌(楼)站经计量、拌制后作为商品出售的,并采用专用运输车在规定时间内运至使用地点的混凝土拌合物。与现场拌制混凝土相比,具有以下优点:

①提高了混凝土质量。
②节约原材料,提高了经济效益。
③实现了文明施工,减少了环境污染。

(2)商品混凝土拌制要求。商品混凝土拌制应达到设计所规定的匀质性、强度和耐久性等要求外,在其尚未凝固阶段,为保证混凝土泵压送顺利还必须满足可泵性的要求。为此,必须严格地控制混凝土材料的质量,确保配料计量准确和确定合理的搅拌时间。

混凝土的搅拌时间:混凝土拌合物的均匀性对其可泵性有明显的影响。搅拌不充分的混凝土,在运输过程中就易产生泌水,甚至离析现象,其可泵性随着降低,在压送过程中易使输送管堵塞。如果搅拌时间过长,就会加速混凝土的凝结。在相同的运输时间内,搅拌时间过长的混凝土坍落度的损失就增大,导致了混凝土的可泵性降低,同样会给混凝土的压送带来困难,甚至发生堵塞现象。因此,应选择合理的搅拌时间,以保证混凝土拌合物的质量。合理搅拌时间应通过试验确定。

Chapter 5 Construction of Reinforced Concrete Structure

Table 5.12 Minimum Time of Concrete Mixing(s)

Concrete Slump/mm	Mixer Type	Discharge volume of mixer/L		
		<250	250~500	>500
≤30	Forcing type	60	90	120
	Self falling type	90	120	150
>30	Forcing type	60	60	90
	Self falling type	90	90	120

Note: 1. The mixing time should be extended appropriately when the admixture is added;
2. Light concrete can be mixed by forced mixer, and ultra light concrete can be mixed by self falling mixer, but the mixing time should be prolonged by 60—90s.

3. Mixing of Commercial Concrete

With the gradual expansion of engineering scale, the continuous improvement of engineering requirements and the continuous development of technical level, the amount of concrete used is increasing, and the requirements of concrete strength grade, waterproof grade and durability are also higher and higher. The scattered and backward on-site concrete production has been difficult to meet the requirements of modern engineering construction technology progress. In this case, commercial concrete appears.

(1) Introduction of Commercial Concrete. Commercial concrete refers to the concrete mixture which is made of cement, aggregate, water, admixtures, mineral admixtures, etc., which are sold as commodities after being measured and mixed at the mixing station according to a certain proportion, and transported to the place of use by special transport vehicles within the specified time. Compared with the on-site mixing concrete, it has the following advantages:

①The concrete quality is improved.

②Raw materials are saved and economic benefits are improved.

③Civilized construction is realized and environmental pollution is reduced.

(2) Requirements for commercial concrete mixing. In addition to the uniformity, strength and durability required by the design, the commercial concrete must meet the requirements of pumpability in order to ensure the smooth pumping of concrete at the stage of not setting. Therefore, it is necessary to strictly control the quality of concrete materials, ensure the accurate measurement of ingredients and determine the reasonable mixing time.

Mixing time of concrete: the uniformity of concrete mixture has obvious influence on its pumpability. If the concrete is not fully mixed, it is easy to produce bleeding and even segregation in the process of transportation. With the decrease of its pumpability, it is easy to block the delivery pipe in the process of pressure delivery. If the mixing time is too long, it will accelerate the setting of concrete. In the same transportation time, the loss of concrete slump increases when the mixing time is too long, which leads to the decrease of the pumpability of concrete, which also brings difficulties to the concrete compression, and even causes blockage. Therefore, reasonable mixing time should be selected to ensure the quality of concrete mixture. The reasonable mixing time should be determined by experiment.

5.3.3 混凝土的运输

由于混凝土自搅拌机中卸出以后应及时运送到浇筑地点,以确保在初凝前完成浇筑并满足施工时和易性的要求,因此选择混凝土运输方案时应综合考虑建筑结构特点及其施工方案、混凝土质量、运输距离、地形、道路、气候和现有设备条件等因素。对混凝土运输中可能产生的分层离析、水泥浆流失、坍落度变化、初凝等影响混凝土质量的现象,采取相应的技术措施以确保混凝土拌和物的质量。

1. 混凝土的运输要求

对混凝土拌合物运输的要求是:运输过程中,应保持混凝土的均匀性,避免产生分层离析现象,混凝土运至浇筑地点,应符合浇筑时所规定的坍落度;混凝土应以最少的中转次数,最短的时间,从搅拌地点运至浇筑地点,保证混凝土从搅拌机卸出后到浇筑完毕的延续时间不超过表 5.13 的规定。运输工作应保证混凝土的浇筑工作能连续进行。运送混凝土的容器应严密,其内壁应平整光洁,不吸水,不漏浆,粘附的混凝土残渣应经常清除。

表 5.13 混凝土从搅拌机中卸出后到浇筑完毕的延续时间(min)

混凝土强度等级	气温/℃	
	不高于 25	高于 25
C30 及 C30 以下	120	90
C30 以上	90	60

注:1. 掺用外加剂或采用快硬水泥拌制混凝土时,应按试验确定。
　　2. 轻骨料混凝土的运输、浇筑延续时间应适当缩短。

刚搅拌好的混凝土,由于内摩阻力、黏着力和重力等的作用,各种材料的位置处于固定位置,且分布均匀,此时混凝土拌和物处于相对平衡状态。在运输过程中,由于道路不平,运输工具的颠簸振动等影响,黏着力和内摩阻力将明显减少,特别是混凝土拌和物自高处下落时,失去平衡状态,在自重作用下向下沉落,质量越大,向下沉落的趋势越强。由于粗细骨料和水泥浆的质量不同,因而各自聚集在一定深度,形成分层离析现象。

分层离析现象对混凝土的质量是有害的,使混凝土的强度降低,容易形蜂窝或麻面,也增加捣实的困难。为此,对运输道路、运输时间和运输机具都有具体的要求。

5.3.3 Concrete Transportation

Since the concrete should be transported to the pouring site in time after it is discharged from the mixer, so as to ensure the completion of pouring before the initial setting and meet the requirements of workability during construction, the selection of concrete transportation scheme should comprehensively consider the characteristics of building structure and its construction scheme, concrete quality, transportation distance, terrain, road, climate and existing equipment conditions. In order to ensure the quality of concrete mixture, the corresponding technical measures should be taken to solve the problems that may affect the quality of concrete, such as layered segregation, loss of cement slurry, slump change, initial setting and so on.

1. Transportation Requirements of Concrete

The transportation requirements of concrete mixture are as follows: during the transportation, the uniformity of the concrete shall be maintained to avoid layered segregation. When the concrete is transported to the pouring site, the slump specified during pouring shall be met; The concrete shall be transported from the mixing place to the pouring place with the least transfer times and the shortest time, so as to ensure that the continuous time from the discharge of the mixer to the completion of the pouring shall not exceed the provisions of Table 5.13; the transportation work shall ensure the continuous pouring of concrete; the container for transporting concrete shall be tight, and its inner wall shall be smooth, bright and clean, without water absorption, slurry leakage and adhesion soil residue should be removed frequently.

Table 5.13 Duration from Discharge of Concrete from Mixer to Completion of Pouring(min)

Concrete strength grade	Temperature/℃	
	No more than 25	Above 25
C30 and below	120	90
Above C30	90	60

Note: 1. When mixing concrete with admixture or fast hardening cement, it should be determined according to the test.
2. The transportation and pouring duration of lightweight aggregate concrete should be appropriately shortened.

Due to the effect of internal friction, adhesion and gravity, the position of all kinds of materials is in a fixed position and evenly distributed. At this time, the concrete mixture is in a relatively balanced state. In the process of transportation, due to the influence of uneven road, bumpy and vibration of transportation tools, the adhesion and internal friction resistance will be significantly reduced, especially when the concrete mixture falls from a high place, it will lose its equilibrium state and sink downward under the action of self weight. The greater the mass, the stronger the tendency of downward settlement will be. Because of the different quality of coarse and fine aggregate and cement slurry, they gather in a certain depth and form layered segregation phenomenon.

The phenomenon of layered segregation is harmful to the quality of concrete, which reduces the strength of concrete, easily forms honeycomb or pitted surface, and increases the difficulty of tamping. For this reason, there are specific requirements for transportation time and transportation equipment.

2. 运输道路

为了防止混凝土运输中离析分层,要用最短的时间、最短的道路把混凝土运输到位,因此需要考虑道路的平坦和环形回路的布置,安排专人管理,避免拥挤堵塞。运输道的宽度要根据单行或双行及车辆的宽度而定。一般单轮手推车单行道宽度为 1.5~2.5 m,机动翻斗车为 2.5~3 m。

运输道路的布置应在施工方案时考虑,尤其是大体积的混凝土施工浇筑。应避免在已浇筑好的混凝土上行车,所以后退式的浇筑次序较好。特别注意运输道路、支凳的平坦和稳定,由于铺板缝子太大或不平,翻车现象时有发生。

3. 运输时间

混凝土应在初凝前完成浇筑,因此应尽可能减少运输时间,以增加施工浇筑时间。混凝土搅拌机卸料到浇筑完毕的延续时间,不宜超过表 5.13 的规定。

若运距较远可掺加缓凝剂,其延续凝结时间长短可由试验确定。使用快硬水泥或掺有促凝剂的混凝土,其运输时间应根据水泥性能及凝结条件确定。

4. 运输机具

混凝土运输主要分水平运输和垂直运输两方面,应根据施工方法、工程特点、运距的长短及现有的运输设备,选择可满足施工要求的运输工具。

水平运输常用的运输工具有:双轮手推车、架子车、自卸三轮汽车、轻便小型翻斗车、搅拌运输车等。

垂直运输常用的运输机械有各种升降机、卷扬机、塔吊、井架等,并配合采用吊斗等容器装运混凝土。

当混凝土的浇筑工程较集中和浇灌速度较稳定时,可采用带式运输机、混凝土泵等。下面简单介绍一些常用的机具:

(1)手推车。手推车是施工工地上普遍使用的水平运输工具,其种类有独轮、双轮和三轮等多种。手推车具有小巧、轻便等特点,不但适用于一般的地面水平运输,还能在脚手架、施工栈道上使用;也可与塔吊、井架等配合使用,解决垂直运输混凝土、砂浆等材料的需要(图 5.31)。

2. Transport Road

In order to prevent segregation and stratification in concrete transportation, the shortest time and shortest road should be used to transport the concrete in place. Therefore, it is necessary to consider the flatness of the road and the arrangement of circular circuit, and assign special personnel to manage it to avoid congestion. The width of the transport lane should be determined according to the width of one or two rows and vehicles. Generally, the width of one-way road with single wheel trolley is 1.5—2.5m, and that of motor dumper is 2.5—3m.

The layout of transportation road should be considered in the construction scheme, especially the construction and pouring of mass concrete. Driving on the poured concrete should be avoided, so the backward pouring sequence is better. Special attention should be paid to the smoothness and stability of the transportation road and stool. Due to the large or uneven seam of the pavement, rollover occurs from time to time.

3. Transport Time

The concrete should be finished pouring before the initial setting, so the transport time should be reduced as much as possible to increase the construction pouring time. The duration from the discharge of concrete mixer to the completion of pouring should not exceed the provisions in Table 5.13.

If the transport distance is far away, the setting time can be added with retarder, which can be determined by experiments. The transport time of fast setting cement or concrete mixed with accelerator shall be determined according to cement performance and setting conditions.

4. Transport Equipment

The concrete transportation is mainly divided into horizontal transportation and vertical transportation. According to the construction method, engineering characteristics, transportation distance and existing transportation equipment, the transport means that can meet the construction requirements should be selected.

The commonly used means of transport in horizontal transportation are wheelbarrows, rack trucks, self unloading three wheeled vehicles, light small dumper trucks, mixing trucks and so on.

Vertical transport commonly used transport machinery has all kinds of lifts, hoists, tower cranes, derrick, etc. , and with the use of tanks and other containers to transport concrete.

When the concrete pouring project is concentrated and the pouring speed is stable, belt conveyor, concrete pump and pneumatic transporter can be used. The following is a brief introduction to some commonly used machines and tools:

(1)Wheelbarrow. Wheelbarrow is a horizontal transportation tool widely used in construction site. Its types include single wheel, double wheel and three wheels. The trolley has the characteristics of small size and light weight. It is not only suitable for general horizontal transportation on the ground, but also can be used on scaffold and construction trestle; it can also be used with tower crane and derrick to meet the needs of vertical transportation of concrete, mortar and other materials(Fig. 5.31).

(2)机动翻斗车。系用柴油机装配而成的翻斗车,功率 735.5 W,最大行驶速度达 35 km/h。车前装有容量为 400 L、载重 1 000 kg 的翻斗。具有轻便灵活、结构简单、转弯半径小、速度快、能自动卸料、操作维护简便等特点。适用于短距离水平运输混凝土以及砂、石等散装材料,如图 5.32 所示。

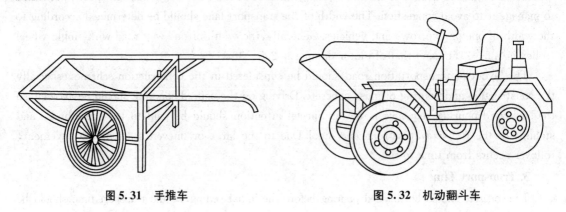

图 5.31 手推车　　　　　　　　　　图 5.32 机动翻斗车

(3)井架。主用于高层建筑混凝土灌注时的垂直运输机械,有井架、台灵拔杆、卷扬机、吊盘、自动倾卸吊斗钢丝缆、风绳等组成,具有一机多用、构造简单、装拆方便等优点。轻重高度一般为 25~40 m,如图 5.33 所示。

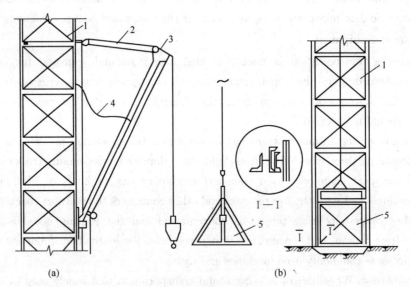

图 5.33 井架运输架
(a)井架台灵拔杆;(b)井架吊盘
1—塔臂;2—操作室;3—中心轴;4—涡轮箱;5—吊盘

(2) Mobile dumper. The dumper assembled with diesel engine has a power of 735.5W and a maximum speed of 35km/h. A tipping bucket with capacity of 400L and load 1000kg is installed in front of the vehicle. It has the characteristics of light and flexible, simple structure, small turning radius, fast speed, automatic unloading, easy operation and maintenance. It is suitable for short distance horizontal transportation of concrete, sand, stone and other bulk materials, as shown in Fig. 5.32.

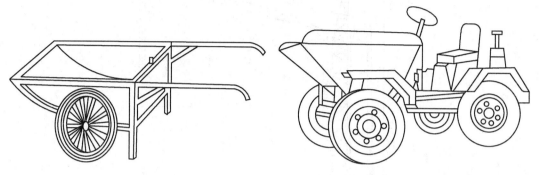

Fig. 5.31 Wheelbarrow 　　　　　　　Fig. 5.32 Mobile Dumper

(3) Derrick. It is mainly used for vertical transportation of concrete pouring in high-rise buildings. It is composed of derrick, platform pulling rod, hoist, hanging plate, automatic dumping bucket, and steel wire cable and wind rope. It has the advantages of multi-purpose, simple structure and convenient installation and disassembly. The height is generally 25—40m, as shown in Fig. 5.33.

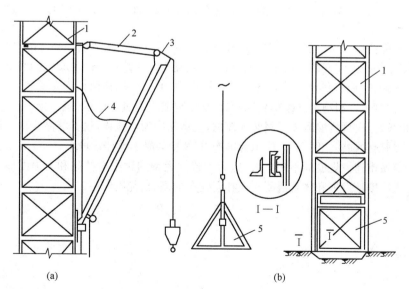

Fig. 5.33 derrick transport frame
(a) derrick flexible pulling rod; (b) derrick hanging plate
1—derrick; 2—steel wire rope; 3—flexible pulling rod;
4—turbine box; 5—hanging plate;

(4)塔式起重机。塔式起重机主要是用于大型建筑和高层建筑的垂直运输。利用塔式起重机与其他浇灌斗等机具相配合,可以很好地完成混凝土的垂直运输任务。塔式起重机通常有行走式、附着式、内爬式三种。工地多数选用行走式塔式起重机,它的工作幅度大,既能解决垂直运输,还能解决一定范围内的水平运输。图 5.34 所示为国产 TD-25 塔式起重机。

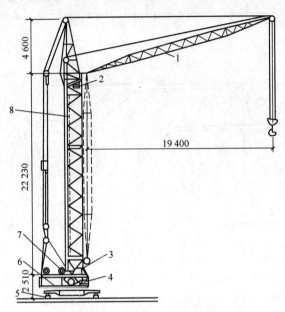

图 5.34　TD-25 塔式起重机
1—井架;2—钢丝绳;3—台灵拔杆;4—保险绳;
5—走轮;6—塔座;7—卷扬机;8—爬梯

(5)混凝土搅拌运输车。混凝土搅拌运输车是一种用于长距离运输混凝土的高效能机械,它是将运送混凝土的搅拌筒安装在汽车底盘上,而以混凝土搅拌站生产的混凝土拌和物冠状如搅拌筒内,直接运至施工现场,供浇灌作业需要。运输途中,混凝土搅拌筒始终在不停地作慢速转动,从而使筒内的混凝土拌和物可连续得到搅动,以保证混凝土通过长途运输后,仍不产生离析现象。在运输距离很长时,也可将混凝土干料装入筒内,在运输中加水搅拌,这样能减少由于长途运输而引起的混凝土坍落度损失。图 5.35 所示为国产 JC-2 型混凝土搅拌运输车。

目前,随着城市建设规模的不断扩大,应力求建立大型的集中搅拌站和大力发展商品混凝土的生产,以更大限度地发挥和体现混凝土搅拌运输车的高效、优质作用。

(4) Tower crane. Tower crane is mainly used for vertical transportation of large buildings and high-rise buildings. The vertical transportation of concrete can be well completed by using tower crane with other pouring bucket and other machines and tools. Tower crane usually has three types: walking type, attached type and internal climbing type. Most of the construction sites choose walking tower crane, which has a large working range. It can not only solve the vertical transportation, but also solve the horizontal transportation within a certain range. Fig. 5.34 shows the domestic TD-25 tower crane.

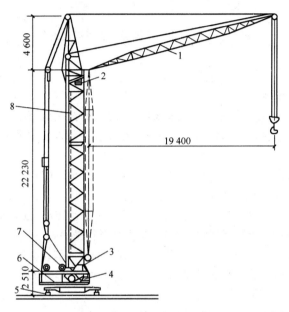

Fig. 5.34　TD-25 tower crane
1—Tower arm; 2—operation room; central shaft; 4—safety rope;
5—travelling wheel; 6—tower base; 7—winch; 8—ladder

(5) Concrete mixing truck. The concrete mixing truck is a kind of high-efficiency machinery for long-distance concrete transportation. It is to install the mixing drum for transporting concrete on the truck chassis, and the concrete mixture produced by the concrete mixing station, such as the mixing drum, is directly transported to the construction site for the needs of pouring operation. During the transportation, the concrete mixing drum always rotates slowly, so that the concrete mixture in the cylinder can be continuously stirred, so as to ensure that the concrete still does not produce segregation after long-distance transportation. When the transportation distance is very long, the dry material of concrete can also be loaded into the cylinder and mixed with water during transportation, which can reduce the loss of concrete slump caused by long-distance transportation. Fig. 5.35 shows the domestic JC-2 concrete mixer truck.

At present, with the continuous expansion of urban construction scale, large-scale centralized mixing plants should be striven to establish and the production of commercial concrete should be vigorously developed, so as to give full play to and reflect the high-efficiency and high-quality role of concrete mixing truck.

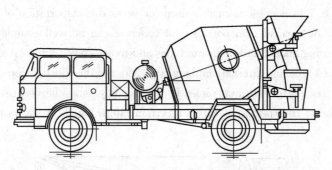

图 5.35　国产 JC—2 型混凝土搅拌运输车

5. 商品混凝土的运输

商品混凝土搅拌后应在 2 小时内运到现场，运输时间的控制应有必要的交通条件作保证，因此规划好运输路线，防止堵塞是极为重要（如有的城市只在夜间运输混凝土）。

(1)运输途中。

①混凝土搅拌运输车罐内严禁有积水，特别是涮罐或洗泵后应排放干净。另外，给压力水箱加水时，其后必须关闭的水管截止阀的。

②在装料和运输过程中，搅拌罐应保持 3～6 r/min 的慢速转动，以防止混凝土拌合物出现离析、分层等现象。装完料后，应高速搅拌，防止混凝土拌合物抛洒。

③混凝土拌合物应在初凝前卸料及施工。一般应根据外加剂缓凝时间长短、气温高低将时间控制在 4 小时内。对未初凝的剩料，应及时加入一定量的外加剂，使其恢复其坍落度。

④预拌混凝土的运输应保证施工的连续性。

⑤冬季施工时，应采取相应的保温措施，如保温套等。

(2)泵送(或自卸)。

①预拌混凝土的泵送应符合行业现行标准《混凝土泵送施工技术规程》(JGJ/T 10—2011)等的规定。

②确认混凝土泵和输送管无异物、无泄露后(泵水检查等)，应泵送水泥混合砂浆等。但须注意应分散开来，不得集中浇筑。

③检验混凝土拌合物的坍落度、可泵性等。不宜在运输和施工过程中往搅拌罐内任意加水，这样会改变混凝土的设计配合比，无法保证混凝土的强度等。当混凝土拌合物的坍落度不能满足要求时，可在符合混凝土设计配合比的前提下，适当加水，或加入一定量的减水剂，并高速搅拌均匀。

Chapter 5 Construction of Reinforced Concrete Structure

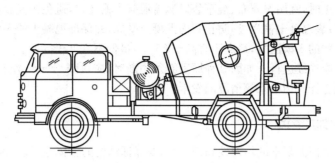

Fig. 5.35 Domestic JC-2 Concrete Mixer Truck

5. Transportation of Commercial Concrete

The commercial concrete should be transported to the site within 2 hours after mixing, and the control of transportation time should be guaranteed by necessary traffic conditions. Therefore, it is very important to plan the transportation route and prevent blockage(for example, concrete is transported only at night in some cities).

(1) In Transportation.

①No water is allowed in the tank of the concrete mixer truck, especially after washing the tank or pump, it should be discharged completely. In addition, when adding water to the pressure tank, the water pipe of the back stop valve must be closed.

②In the process of loading and transportation, the mixing tank should be kept at a slow speed of 3—6 r/min to prevent segregation and stratification of concrete mixture. After loading, the concrete should be mixed at high speed to prevent the concrete mixture from throwing.

③The concrete mixture shall be unloaded and constructed before the initial setting. Generally, it should be controlled within 4 hours according to the retarding time of admixture and the temperature. For the remaining material without initial setting, a certain amount of admixture should be added in time to restore its slump.

④The transportation of ready mixed concrete shall ensure the continuity of construction.

⑤In winter construction, corresponding insulation measures should be taken, such as insulation sleeve, etc.

(2) Pumping(or self unloading).

①The pumping of ready mixed concrete shall comply with the current industry standard *Technical specification for pumping construction of concrete*(JGJ/T10-2011).

②After confirming that the concrete pump and delivery pipe are free of foreign matters and leakage(pump water inspection, etc.), the cement mixed mortar shall be pumped. However, attention should be paid to dispersing and centralized pouring is not allowed.

③Check the slump and pumpability of concrete mixture. It is not suitable to add water arbitrarily to the mixing tank in the process of transportation and construction, which will change the design mix proportion of concrete and cannot guarantee the strength of concrete. When the slump of concrete mixture cannot meet the requirements, appropriate water can be added or a certain amount of water reducing agent can be added under the premise of meeting the design mix proportion of concrete, and the mixture can be evenly mixed at high speed.

④炎热季节施工时,宜用湿罩布、湿草袋等遮盖混凝土输送管,避免阳光直射。严寒季节施工时,宜用保温材料包裹混凝土输送管,防止管内混凝土受冻,并保证混凝土的入模温度。

⑤混凝土拌合物的入模温度,最高不宜高于35 ℃,最低不宜低于5 ℃。

⑥混凝土泵送应连续进行。如因某种原因必须中断时,应每隔15 min左右正反泵一次,防止出现输送管内混凝土拌合物堵塞等情况的发生。

5.3.4 混凝土的浇筑

混凝土浇筑要保证混凝土的均匀性和密实性,要保证结构的整体性、尺寸准确和钢筋、预埋件的位置正确,拆模后混凝土表面要平整、光洁。

为了防止浇筑混凝土出现露筋、裂缝、孔洞、蜂窝、麻面和影响混凝土结构的强度及整体性等情况,应根据所浇筑混凝土结构施工的特点,合理组织分段分层流水施工,并应根据总工程量、工期以及分层分段的具体情况,确定每工作班的工作量。再根据每班工程量和现有设备条件,选择混凝土搅拌机、运输及振捣设备的类型和数量进行施工,以确保混凝土工程的质量。

1. 混凝土浇筑的基本要求

(1)浇筑前。混凝土浇筑前应做好下列工作:

①对模板和支架、钢筋和预埋件进行检查记录;

②准备和检查材料、机具、运输道路;

③在模板上涂刷隔离剂,模板、工具(铁锹、小车等)应浇水润湿,模板内的垃圾、泥土以及钢筋上的油污等杂物应清洗干净,以防杂物混入混凝土中;模板的缝隙孔洞应堵严。

(2)浇筑。

①混凝土浇筑前不应发生初凝和离析现象,如已发生,可重新搅拌,使混凝土恢复流动性和粘聚性后再进行浇筑。

②为防止混凝土浇筑时产生离析,混凝土自由倾落高度不宜超过2 m,若超过2 m,应采用串筒、斜槽、溜管等下料。如图5.36(a)、(b)所示。当混凝土浇筑高度超过8 m时,则应节管振动串筒,即在串筒上每隔2、3节管安装一台振动器,如图5.36(c)所示。

④During construction in hot season, wet cloth and straw bag should be used to cover concrete delivery pipe to avoid direct sunlight. During the construction in severe cold season, it is advisable to wrap the concrete conveying pipe with thermal insulation material to prevent the concrete in the pipe from freezing and to ensure the temperature of concrete into the mold.

⑤The highest molding temperature of concrete mixture should not be higher than 35℃, and the lowest should not be lower than 5℃.

⑥Concrete pumping shall be carried out continuously. If it must be interrupted for some reason, the positive and negative pumps should be conducted every 15 minutes to prevent the occurrence of concrete mixture blockage in the delivery pipe.

5.3.4 Concrete Pouring

Concrete pouring should ensure the uniformity and compactness of concrete, the integrity and size of the structure, the correct position of reinforcement and embedded parts, and the concrete surface should be smooth, bright and clean after removing the formwork.

In order to prevent the occurrence of exposed reinforcement, cracks, holes, honeycomb, pitted surface and affecting the strength and integrity of the concrete structure, it is necessary to reasonably organize the sectional and layered flow construction according to the characteristics of the concrete structure construction, and determine the workload of each work team according to the total construction quantity, construction period and the specific situation of layered section. According to the quantities of each shift and the existing equipment conditions, the type and quantity of concrete mixer, transportation and vibration equipment are selected for construction, so as to ensure the quality of concreteworks.

1. Basic Requirements of Concrete Pouring

(1) Before pouring. The following work shall be done before concrete pouring:

①Check and record the formwork, support, reinforcement and embedded parts.

②Prepare and check materials, machines and tools, transportation roads.

③Brush the isolating agent on the formwork. Formwork, tools (shovels, wheelbarrows, etc.) should be watered and moistened. Waste materials, soil and oil stains on the steel bar should be cleaned to prevent debris from mixing into concrete. The gap and hole of the formwork should be blocked tightly.

(2) Pouring.

①The initial setting and segregation shall not occur before concrete pouring. If it has occurred, the concrete can be mixed again to restore the fluidity and cohesiveness of the concrete before pouring.

②In order to prevent segregation during concrete pouring, the free falling height of concrete should not be more than 2m. If it exceeds 2m, tumbling barrel, chute and elephant trunk should be used for discharging, as shown in Fig. 5.36(a)、(b). When the concrete pouring height is more than 8m, the section pipe should be vibrated, that is, one vibrator should be installed every 2—3 sections of pipe on the string pipe, as shown in Fig. 5.36(c).

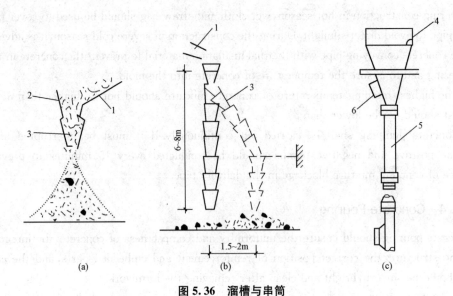

图 5.36 溜槽与串筒

(a)溜槽;(b)串筒;(c)节管振动串筒

1—溜槽;2—挡板;3—串筒;4—漏斗;5—节管;6—振动器

(3)分层浇筑。为了使混凝土各部位都振捣密实,混凝土必须分层浇捣,决不可一次下料过多。即每浇灌一适宜厚度时,应用振捣器或别的捣固方法进行捣固。混凝土每层浇灌的厚度应符合表 5.14 规定。

表 5.14 混凝土浇筑层的厚度

项次	捣实混凝土的方法		浇筑层的厚度/mm
1	插入式振捣		振捣器作用部分长度的 1.25 倍
2	表面振动		200
3	人工振动 (1)在基础无筋混凝土或配筋稀疏的结构中 (2)在梁、墙板、柱结构中 (3)在配筋密列的结构中		250 200 150
4	轻骨料混凝土	插入式振捣	300
		表面振捣(振动时需加载荷)	200

(4)连续作业。混凝土的浇筑工作,应尽可能连续作业。如必须间歇作业,其间歇时间应尽量缩短,并要在前层混凝土凝结(终凝)前,将次层混凝土浇筑完毕。间歇的最近时间应按所用水泥品种及混凝土凝结条件确定。即混凝土从搅拌机中卸出,经运输和浇筑完毕的延续时间不得超过表 5.13 的规定。

Chapter 5 Construction of Reinforced Concrete Structure

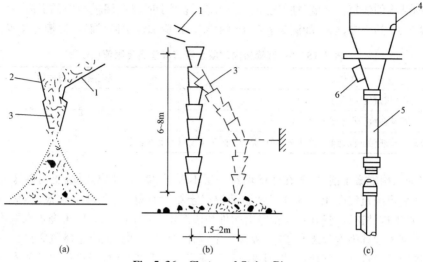

Fig. 5.36 Chute and String Pipe
(a) chute; (b) string pipe; (c) section pipe vibration
1—chute; 2—back plate; 3—string pipe; 4—hopper; 5—section pipe; 6—vibrator

(3) Layered pouring. In order to make all parts of the concrete vibrated and compacted, the concrete must be poured and vibrated in layers, and excessive discharging is not allowed at one time. That is, when pouring a suitable thickness, the vibrator or other tamping methods shall be used for tamping. The pouring thickness of each layer of concrete shall meet the requirements of Table 5.14.

Table 5.14 Thickness of Concrete Pouring Layer

item	Method of tamping concrete		Thickness of pouring layer/mm
1	Plug in vibration		1.25 times of the length of the action part of the vibrator
2	Surface vibration		200
3	Artificial vibration (1) In the foundation of reinforced concrete or sparsely reinforced structure (2) In the beam, wall slab, column structure (3) In the structure with dense reinforcement		250 200 150
4	Lightweight aggregate concrete	Plug in vibration	300
		Surface vibration (loading is required during vibration)	200

(4) Continuous operation. Concrete pouring shall be carried out continuously as far as possible. If intermittent operation is necessary, the interval time shall be shortened as far as possible. And the secondary layer of concrete shall be poured before the setting (final setting) of the front layer of concrete. The latest interval time shall be determined according to the type of cement used and concrete setting conditions. That is, the concrete is discharged from the mixer, and the duration of transportation and pouring shall not exceed the provisions of Table 5.13.

(5)竖向结构的浇筑。在墙、柱等竖向结构浇筑混凝土时,柱、墙模板内的混凝土浇筑倾落高度应符合表 5.15 的规定;当不能满足表 5.15 的要求时,应加设串筒、溜管、溜槽等装置。

表 5.15　柱、墙模板内混凝土浇筑倾落高度限值(m)

条件	浇筑倾落高度限值
粗骨料粒径大于 25 mm	≤3
粗骨料粒径小于等于 25 mm	≤6
注:当有可靠措施能保证混凝土不产生离析时,混凝土倾落高度可不受本表限制。	

浇筑竖向结构混凝土前,应先在底部填筑一层不大于 30 mm 厚与混凝土内砂浆成分相同的水泥砂浆,然后再浇筑混凝土。这样可使新旧混凝土结合良好,避免产生烂根、蜂窝、麻面现象。

竖向构件与水平构件连续浇筑时,应待竖向构件初步沉实后(为 1~1.5 h)再浇筑水平构件。

柱、墙混凝土设计强度等级高于梁、板混凝土设计强度等级时,混凝土浇筑应符合下列规定:

①柱、墙混凝土设计强度比梁、板混凝土设计强度高一个等级时,经设计单位同意,柱、墙混凝土可采用与梁、板混凝土设计强度等级相同的混凝土进行浇筑;

②柱、墙混凝土设计强度比梁、板混凝土设计强度高两个等级及以上时,应在交界区域采取分隔措施。分隔位置应在低强度等级的构件中,且距高强度等级构件边缘不应小于 500 mm;

③宜先浇筑高强度等级混凝土,后浇筑低强度等级混凝土。

(6)混凝土浇筑的布料点宜接近浇筑位置,应采取减少混凝土下料冲击的措施,并应符合下列规定:

①宜先浇筑竖向结构构件,后浇筑水平结构构件;

②浇筑区域结构平面有高差时,宜先浇筑低区部分再浇筑高区部分。

(7)浇筑混凝土时,应经常观察模板、支架、钢筋、预埋件和预留孔洞的情况,当发现有变形、移位时,应立即停止浇筑,并应在已浇筑的混凝土凝结前修整完好。

(5) Pouring of vertical structure. When pouring concrete for vertical structures such as walls and columns, the pouring height of concrete in column and wall formwork shall comply with the provisions in Table 5.15. When the requirements in Table 5.15 cannot be met, devices such as string pipe, chute and elephant trunk shall be added.

Table 5.15 Limit of Concrete Pouring Height in Column and Wall Formwork(m)

Condition	Limit of Pouring Height
Coarse aggregate size greater than 25mm	$\leqslant 3$
Coarse aggregate size less than or equal to 25mm	$\leqslant 6$

Note: when there are reliable measures to ensure that the concrete does not produce segregation, the concrete falling height cannot be limited by this table.

Before pouring the vertical structure concrete, a layer of cement mortar with the thickness of no more than 30mm and the same composition as the mortar in the concrete shall be filled at the bottom, and then the concrete shall be poured. In this way, the new and old concrete can be well combined and the rotten root, honeycomb and pockmarked surface can be avoided. When the vertical component and horizontal component are poured continuously, the horizontal component shall be poured again after the vertical component is initially settled(about 1—1.5h).

When the design strength grade of column and wall concrete is higher than that of beam and slab concrete, the concrete pouring shall meet the following requirements:

①When the design strength of column and wall concrete is one grade higher than that of beam and slab concrete, the concrete within the height range of beam and slab at column and wall position can be poured with concrete with the same design strength grade as beam and slab concrete with the approval of design unit.

②When the design strength of column and wall concrete is two or more grades higher than that of beam and slab concrete, separation measures shall be taken in the junction area. The separation position should be in the member of low strength grade, and the distance from the edge of high strength grade component should not be less than 500mm.

③High strength concrete should be poured first and then low strength concrete.

(6) The distribution point of concrete pouring should be close to the pouring position, and measures should be taken to reduce the impact of concrete blanking, which should meet the following requirements:

①The vertical structural members should be poured first, and then the horizontal structural members.

②When there is a height difference in the structural plane of the pouring area, the low area part should be poured first and then the high area part.

(7) When pouring concrete, the formwork, support, reinforcement, embedded parts and reserved holes shall be observed frequently. When deformation and displacement are found, pouring shall be stopped immediately and the poured concrete shall be repaired before setting.

2. 施工缝的设置

混凝土结构大多要求整体浇筑,如果由于技术上的原因或设备、人力上的限制、混凝土的浇筑不能连续进行,且停顿时间有可能超过混凝土的初凝时间,当中间的间歇时间预计将超过表 5.13 规定的时间时,则应留置施工缝。

设置施工缝应该严格按照规定,认真对待。由于该处新旧混凝土的结合力较差,是构件中薄弱环节,如果位置不当或处理不好,就会引起质量事故,轻则开裂、漏水,影响使用寿命;重则危及安全,不能使用。因此,应予以高度重视。

(1)施工缝的位置。由于混凝土的抗拉强度约为其抗压强度的 1/10,因而施工缝是结构中的薄弱环节,宜留在结构剪力较小的部位,同时要方便施工。柱应留水平缝,梁、板、墙应留垂直缝。留缝应符合下列规定:

①柱子留置在基础的顶面、梁或吊车梁牛腿的下面、吊车梁的上面、无梁楼板柱帽的下面,如图 5.37 所示。

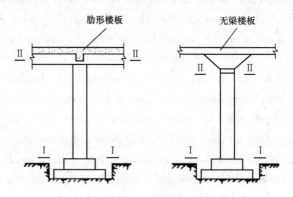

图 5.37 浇筑柱子的施工缝位置图

Ⅰ—Ⅰ、Ⅱ—Ⅱ标示施工缝的位置

②和板连成整体的大断面梁,留置在板底面以下 20～30 mm 处,当板下有梁托时,留在梁托下部。

③单向板,留置在平行于板的短边的任何位置。

④有主次梁的楼板,宜顺着次梁方向浇筑,施工缝应留置在次梁跨度中间 1/3 的范围内,如图 5.38 所示。

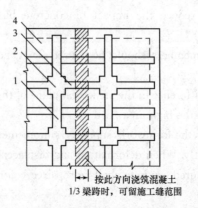

图 5.38 浇筑由主次梁楼板的施工缝位置图

1—柱;2—主梁;3—次梁;4—楼板

2. Setting of Construction Joints

Most concrete structures require integral pouring. If the concrete pouring cannot be carried out continuously due to technical reasons or equipment and manpower constraints, and the pause time may exceed the initial setting time of concrete, and the intermediate intermittent time is expected to exceed the time specified in Table 5.13, the construction joint shall be set.

The setting of construction joints should be strictly in accordance with the regulations and treated seriously. Because the binding force of the new and old concrete is poor, it is the weak link in the component. If the position is improper or the treatment is not good, it will cause quality accidents, such as cracking and water leakage, affecting the service life; if it is serious, it will endanger the safety and cannot be used. Therefore, great importance should be attached to it.

(1) Position of construction joints.

Since the tensile strength of concrete is about 1/10 of its compressive strength, the construction joint is the weak link in the structure, which should be left in the position with small shear force and convenient for construction. Horizontal joints shall be reserved for columns and vertical joints shall be reserved for beams, slabs and walls. The seam reservation shall meet the following requirements:

①The column is left on the top of the foundation, under the beam or crane beam bracket, above the crane beam, and under the column cap of the beamless floor, as shown in Fig. 5.37.

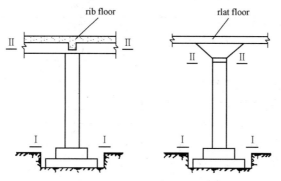

Fig. 5.37 Position of Construction Joint of Pouring Column

Ⅰ—Ⅱ and Ⅱ—Ⅱ indicate the position of construction joint;

②The large cross-section beam connected with the slab as a whole shall be left at 20—30mm below the bottom of the slab. When there is a beam bracket under the slab, it shall be left at the lower part of the beam bracket.

③ One-way slab, left in any position parallel to the short side of the slab.

④ The floor slab with primary and secondary beams should be poured along the secondary beam direction, and the construction joint should be left within 1/3 of the span of the secondary beam, as shown in Fig. 5.38.

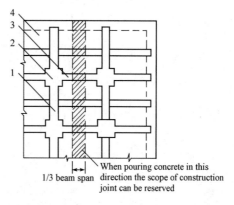

1/3 beam span — When pouring concrete in this direction the scope of construction joint can be reserved

Fig. 5.38 Position of Construction Joint Pouring from Main and Secondary Beam Floor

1—column; 2—main beam; 3—secondary beam; 4—floor

⑤墙,留置在门洞口过梁跨中 1/3 范围内,也可留在纵横墙的交接处。

⑥双向受力楼板、厚大结构、拱、薄壳、多层钢架及其他结构复杂的工程,施工缝的位置应按设计要求留置。

(2)施工缝的处理。在施工缝处继续浇筑混凝土时,应除掉水泥浮浆和松动石子,并用水冲洗干净,待已浇筑的混凝土的强度不低于 1.2 MPa 时才允许继续浇筑,混凝土达到这一强度的时间决定于水泥的标号、混凝土强度等级、气温等,可以根据试块试验确定,也可参照表 5.16 选用。同时,在结合面应先铺抹一层水泥浆或与混凝土砂浆成分相同的砂浆。

表 5.16 达到 1.2 MPa 强度所需龄期的试验结果

外界温度(℃)	水泥品种及强度等级	混凝土强度等级	期限/h
1~5 ℃	普通 42.5	C15	48
		C20	44
	矿渣 32.5	C15	60
		C20	50
5~10 ℃	普通 42.5	C15	32
		C20	28
	矿渣 32.5	C15	40
		C20	32
10~15℃	普通 42.5	C15	24
		C20	20
	矿渣 32.5	C15	32
		C20	24
15 ℃以上	普通 42.5	C15	20 以下
		C20	20 以下
	矿渣 32.5	C15	20
		C20	20

⑤The wall shall be left within 1/3 of the span of the lintel at the door opening, or at the junction of the vertical and horizontal walls.

⑥For two-way stressed floor slab, thick structure, arch, thin shell, multi-layer rigid frame and other complicated structures, the position of construction joint should be reserved according to the design requirements.

(2) Treatment of construction joint.

When pouring concrete at the construction joint, the cement laitance and loose stones should be removed and washed with water. The pouring can be continued only when the strength of the poured concrete is not less than 1.2MPa. The time when the concrete reaches this strength depends on the cement grade, concrete strength grade, air temperature, etc., which can be determined by the block test, or selected according to Table 5.16. At the same time, a layer of cement slurry or mortar with the same composition as concrete mortar shall be paved on the joint surface.

Table 5.16 Test Results of Age Required to Reach 1.2MPa Strength

Outside temperature(℃)	Cement type and strength grade	Concrete strength grade	Time limit/h
1—5 ℃	ordinary 42.5	C15	48
		C20	44
	slag 32.5	C15	60
		C20	50
5—10 ℃	ordinary 42.5	C15	32
		C20	28
	slag 32.5	C15	40
		C20	32
10—15 ℃	ordinary 42.5	C15	24
		C20	20
	slag 32.5	C15	32
		C20	24
Above 15 ℃	ordinary 42.5	C15	Below 20
		C20	Below 20
	slag 32.5	C15	20
		C20	20

①在已硬化的混凝土表面上继续浇筑混凝土之前，应清除垃圾、水泥薄膜、表面上松动沙石和软弱混凝土层，同时还应加以凿毛，用水冲洗干净并充分湿润，一般不少于 24 h，残留在混凝土表面的积水应予清除。

②注意施工缝位置附近回弯钢筋时，要做到钢筋周围的混凝土不受松动和损坏。钢筋上的油污、水泥砂浆及浮锈等杂物也应清除。

③在浇筑前，水平施工缝宜先铺上 10～15 mm 厚的水泥砂浆一层，其配合比与混凝土内的砂浆成分相同。

④从施工缝处开始继续浇筑时，要注意避免直接靠近缝边下料。机械振捣前，要向施工缝处推进，并距 80～100 cm 处停止振捣，但应加强对施工缝接缝的捣实工作，使其紧密结合。

5.3.5 混凝土的振捣

混凝土浇筑入模后，由于其内部骨料之间的摩擦力、水泥净浆的黏结力、拌合物与模板之间的摩擦力等因素会造成混凝土不能自动充满模板，且混凝土内部存在大量孔洞和空气，不能达到密实度的要求，这就会影响混凝土的强度、抗冻性、抗渗性和耐久性等。因此，必须在初凝前经过振捣，才能保证混凝土的密实度，制成符合要求的构件。

混凝土的振捣方法有机械振捣和人工振捣两种方法。现场施工主要采用机械振捣，只有在缺少振捣机械、工程量很小或者机械振捣不便的情况下才采用人工振捣的方法。

1. 机械振捣

振动机械的振动一般是由电动机、内燃机或压缩空气马达带动偏心块转动而产生的简谐振动。产生振动的机械将振动能量通过某种方式传递给混凝土拌合物使其受到强迫振动。在振动力作用下混凝土内部的粘着力和内摩擦力显著减少，使骨料犹如悬浮在液体中，在其自重作用下向新的位置沉落，紧密排列，水泥砂浆均匀分布填充空隙，气泡被排出，游离水被挤压上升，混凝土填满模板的各个角落并形成密实体积。机械振实混凝土可以大大减轻工人的劳动强度，减少蜂窝麻面的产生，提高混凝土的强度和密实性，加快模板周转，节约水泥 10%～15%。机械振捣器种类很多，建设工程上常用的为电动振捣器。按其传递振动方式又可分为内部振动器、表面振动器、外部振动器及振动台等四类，如图 5.39 所示。

①Before the concrete is poured on the hardened concrete surface, the garbage, cement film, loose sand and soft concrete layer on the surface shall be removed. At the same time, it shall be roughened, washed with water and fully wetted, generally no less than 24 h, and the residual water on the concrete surface shall be removed.

②Pay attention to the reinforcement bending near the construction joint to ensure that the concrete around the reinforcement is not loose and damaged. The oil stain, cement mortar and floating rust on the reinforcement should also be removed.

③Before pouring, a layer of 10—15 mm thick cement mortar should be laid on the horizontal construction joint, and its mix proportion is the same as that of the mortar in the concrete.

④When pouring from the construction joint, pay attention to avoid discharging materials directly to the joint. Before mechanical vibration, it is necessary to advance to the construction joints and stop the vibration from 80—100 cm, but the tamping of the joints should be strengthened so as to integrate them closely.

5.3.5 Concrete Vibration

After the concrete is poured into the mold, due to the friction between the internal aggregate, the bonding force of cement paste, the friction between the mixture and the formwork, the concrete cannot automatically fill the formwork, and there are a lot of holes and air in the concrete, which cannot meet the requirements of compactness, which will affect the strength, frost resistance, impermeability and durability of concrete. Therefore, it is necessary to vibrate before the initial setting to ensure the compactness of the concrete and to make components that meet the requirements.

There are two methods of concrete vibration: mechanical vibration and manual vibration. Mechanical vibration is mainly used in site construction, and manual vibration method is adopted only in case of lack of vibration machinery, small amount of work or inconvenient mechanical vibration.

1. Mechanical Vibration

The vibration by vibrating machinery is generally a simple harmonic vibration caused by eccentric block rotation driven by motor, internal combustion engine or compressed air motor. The vibration energy is transmitted to the concrete mixture by the vibration machine in a certain way, so that it is forced to vibrate. Under the action of vibration force, the internal adhesion force and internal friction force of concrete are significantly reduced, which makes the aggregate as if it is suspended in the liquid, sink to the new position under the action of its self weight, closely arranged, the cement mortar evenly distributes to fill the gaps, the bubbles are discharged, the free water is extruded and lifted, and the concrete fills all corners of the formwork and forms a dense volume. Mechanical vibration concrete can greatly reduce the labor intensity of workers, reduce the occurrence of honeycomb and pitting, improve the strength and compactness of concrete, accelerate the turnover of formwork, and save cement by 10%—15%. There are many kinds of mechanical vibrators, and electric vibrators are commonly used in construction projects. According to the vibration mode, it can be divided into internal vibrator, surface vibrator, external vibrator and shaking table, as shown in Fig. 5.39.

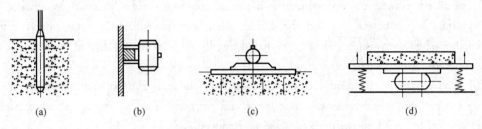

图 5.39 振动机械示意图
(a)内部振动器;(b)外部振动器;(c)表面振动器;(d)振动台

(1)振动机械的选择。

①内部振动器。又称插入式振动器,俗称振动棒。形式有硬管的、软管的。振动部分有锤式、棒式、片式等。振动频率有高有低。主要适用于大体积混凝土,基础、柱、梁、墙、厚度较大的板以及预制构件的捣实工作。当钢筋十分稠密或结构厚度很薄时,其使用就会受到一定的限制。插入式振动器结构如图 5.40 所示。

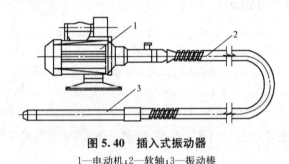

图 5.40 插入式振动器
1—电动机;2—软轴;3—振动棒

②表面振动器。又称平板式振动器。其工作部分是一个钢制或木制平板,板上装一个带偏心块的电动振动器。振动力通过平板传递给混凝土,由于其振动作用深度较小,仅适用于表面积大而平整的结构物,如平板、地面、屋面等构件。

③外部振动器。又称附着式振动器。这种振动器通常是利用螺栓或钳形夹具固定在模板外侧,不能与混凝土直接接触,借助模板或其他物体将振动力传递到混凝土。由于振动作用不能深远,仅适用于振捣钢筋较密、厚度较小以及不宜使用插入式振动器的结构构件。

④振动台。由上部框架和下部支架、支承弹簧、电动机、齿轮同步器、振动子等组成。上部框架是振动台的台面,上面可固定放置模板,通过螺旋弹簧支承在下部的支架上,振动台只能作上下方向的固定振动,适用于混凝土预制构件的振捣。

Chapter 5   Construction of Reinforced Concrete Structure

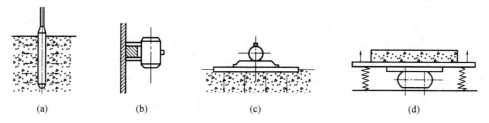

Fig. 5.39   schematic diagram of vibrating machinery
(a)internal vibrator; (b)external vibrator; (c)surface vibrator; (d)shaking table

(1)Selection of vibrating machinery.

①Internal vibrator. It is also known as plug-in vibrator, commonly known as vibrator. The forms are of hard tube and hose. The vibration parts are hammer type, rod type and plate type. The vibration frequency varies from high to low. It is mainly applicable to the tamping of mass concrete, foundation, column, beam, wall, thick slab and precast member. When the reinforcement is very dense or the thickness of the structure is very thin, its use will be limited. The structure of plug-in vibrator is shown in Fig. 5.40.

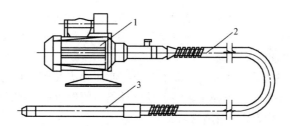

Fig. 5.40   Arrangement of Vibration Points
1—motor; 2—flexible shaft; 3—vibrator

②Surface vibrator, also known as plate vibrator. Its working part is a steel or wooden flat plate, which is equipped with an electric vibrator with eccentric block. The vibration force is transmitted to the concrete through the flat plate. Because of its small vibration depth, it is only applicable to large and flat structures, such as flat slab, ground, roof and other components.

③External vibrator, also known as attached vibrator. This kind of vibrator is usually fixed on the outside of the formwork by bolts or clamp fixture, which cannot directly contact with the concrete. With the help of formwork or other objects, the vibration force is transmitted to the concrete. Because the vibration effect cannot be far-reaching, it can only be applied to the structural members with dense reinforcement, small thickness and inappropriate use of plug-in vibrator.

④Vibration table. It is composed of upper frame and lower bracket, supporting spring, motor, gear synchronizer and vibrator. The upper frame is the table top of the shaking table, on which the formwork can be fixed and supported on the lower bracket by spiral spring. The shaking table can only make fixed vibration in the upper and lower directions, which is suitable for the vibration of precast concrete components.

(2)振动器作业要点：

①内部振动器（插入式振动器）的振捣方法有两种（图 5.41）：一种是垂直振捣，即振动棒与混凝土表面垂直，其特点是容易掌握插点距离、控制插入深度（不得超过振动棒长度的 1.25 倍）、不易产生漏振、不易触及钢筋和模板、混凝土受振后能自然沉实、均匀密实。另一种是斜向振捣，即振动棒与混凝土表面成一定角度，为 $40°\sim 45°$，其特点是操作省力、效率高、出浆快、易于排除空气、不会发生严重的离析现象、振动棒拔出时不会形成孔洞。

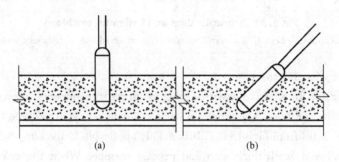

图 5.41　内部振动器振捣方法
(a)直插；(b)斜插

使用插入式振动器垂直操作时的要点是："直上和直下，快插与慢拔；插点要均匀，切勿漏插点；上下要插动，层层要扣搭；时间掌握好，密实质量佳；操作要小心，软管莫卷曲；不得碰模板，不得碰钢筋，用上 200 h，要润滑油振动 0.5 h，停歇 5 min"。

"快插慢拔"中快插是为了防止先将表面混凝土振实而无法振捣下部混凝土，与下面混凝土发生分层、离析现象；慢拔是为了使混凝土填满振动棒抽出时所形成的空隙。振动过程中，宜将振动棒上下略为抽动，以使上下混凝土振捣均匀。

振捣时插点排列要均匀，可采用"行列式"或"交错式"（图 5.42）的次序移动，且不得混用，以免漏振。每次移动间距应不大于振捣器作用半径的 1.5 倍，一般振动棒的作用半径为 30～40 cm。振动器与模板的距离不应大于振动器作用半径的 0.5 倍，并应避免碰撞模板、钢筋、芯管、吊环、预埋件或空心胶囊等。

(2) Main Points of vibrator operation:

①There are two vibration methods for internal vibrator(plug-in vibrator)(Fig. 5.41): one is vertical vibration, that is, the vibrator is perpendicular to the concrete surface, which is easy to master the distance of insertion point and control the insertion depth(no more than vibration 1.25 times of the length of the vibrator), it is not easy to produce missing vibration, and it is not easy to touch the reinforcement and formwork. After vibration, the concrete can be naturally and evenly compacted. The other is oblique vibration, that is, the vibrator and the concrete surface form a certain angle, about 40° to 45° and its characteristics are labor-saving, high efficiency, fast slurry, easy to remove air, no serious segregation phenomenon, and no hole will be formed when the vibrator is pulled out.

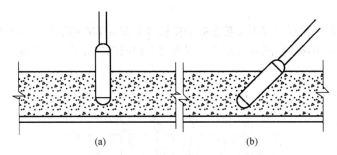

Fig. 5.41 Internal Vibrator Vibration Method
(a) vertical insertion; (b) oblique insertion

The main points of vertical operation with plug-in vibrator are as follows: "straight up and straight down, fast insertion and slow pull-out. The insertion points should be uniform, without missing the insertion points. The upper and lower should be inserted, and the layers should be buckled. The operation should be careful, and the hose should not be curled. Do not touch the formwork or steel bar. After using 200h, use lubricating oil to vibrate for 0.5h, and stop for 5min."

In the "fast insertion and slow pulling out" method, the fast insertion is to prevent the surface concrete from being vibrated but unable to vibrate the lower concrete, resulting in layering and segregation with the lower concrete. The purpose of the slow pulling out is to fill the gaps in the concrete. In the process of vibration, the vibrator should be slightly moved up and down to make the upper and lower concrete vibrate evenly.

During the vibration, the insertion points should be arranged evenly, and can be moved in the order of "determinant" or "interleaving" (Fig. 5.42), and should not be mixed to avoid missing vibration. The distance between each movement shall not be greater than 1.5 times of the action radius of the vibrator. Generally, the action radius of the vibrator is 30—40cm. The distance between the vibrator and the formwork shall not be greater than 0.5 times of the action radius of the vibrator, and collision with the formwork, reinforcement, core pipe, lifting ring, embedded parts or hollow capsules shall be avoided.

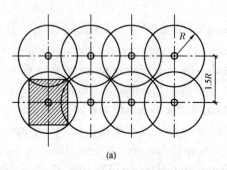 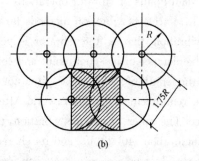

图 5.42 振动点的布置
(a)行列式;(b)交错式

分层振捣混凝土时,每层厚度不应超过振动棒长的 1.25 倍;在振捣上一层时,应插入下层 50 mm 左右,以消除两层之间的接缝,同时必须在下层混凝土初凝以前完成上层混凝土的浇筑,如图 5.43 所示。

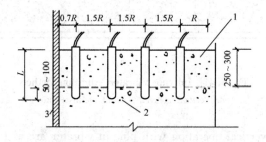

图 5.43 插入式振动器的插入深度
1—新浇筑的混凝土;2—下层已振捣尚未初凝的混凝土;3—模板
R—有效作用半径;L—振捣棒长度

振动时间要掌握恰当,过短混凝土不易被捣实,过长又可能使混凝土出现离析。一般每个插入点的振捣时间为 20~30 s,使用高频振动器时最短不应小于 10 s。而且以混凝土表面呈现浮浆,不再出现气泡,表面不再沉落为准。

②表面振动器(平板式振动器)。振动倾斜混凝土表面时,应由低处逐渐向高处移动,以保证混凝土振实。表面振动器在使用时,在每一位置应连续振动一定时间,一般为 25~40 s,以混凝土表面出现浆液,不再下沉为准;移动时成排依次振捣前进,前后位置和排与排间相互搭接应有 3~5 cm,防止漏振。

表面振动器的有效作用深度,在无筋或单筋平板中约为 200 mm,在双筋平板中约为 120 mm。

大面积混凝土地面,可采用两台振动器以同一方向安装在两条木杠上,通过木杠的振动使混凝土密实。

Chapter 5 Construction of Reinforced Concrete Structure

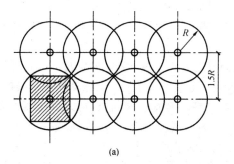
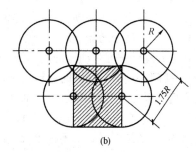

(a)   (b)

Fig. 5.42 Arrangement of Vibration Points
(a)determinant; (b)interleaving

When the concrete is vibrated in layers, the thickness of each layer shall not exceed 1.25 times of the length of the vibrator; when vibrating the upper layer, the lower layer shall be inserted for about 50 mm to eliminate the joint between the two layers. Meanwhile, the pouring of the upper layer concrete must be completed before the initial setting of the lower layer concrete, as shown in Fig. 5.43.

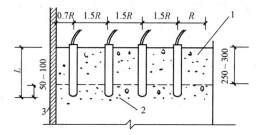

Fig. 5.43 Insertion Depth of Plug-in Vibrator
1—newly poured concrete; 2—vibrated but not initially set concrete in lower layer;
3—formwork
R—effective radius of action; L—length of vibrator

The vibration time should be controlled properly. If the vibration time is too short, the concrete is not easy to be tamped; and if it is too long, it may cause concrete segregation. Generally, the vibration time of each insertion point is 20—30 s, and the shortest time when using high-frequency vibrator should not be less than 10 s. Moreover, the surface of the concrete is free from bubbles and settlement.

②Surface vibrator(plate type vibrator). When the vibration inclines the concrete surface, it should move gradually from the low place to the high place to ensure the concrete vibration. When the surface vibrator is in use, it shall vibrate continuously at each position for a certain period of time; generally 25—40 s, subject to the presence of slurry on the concrete surface and no longer sinking; when moving, the vibrator shall be vibrated in rows in turn, and the front and back positions and rows shall overlap with each other for 3—5 cm to prevent missing vibration.

The effective action depth of the surface vibrator is about 200mm in the unreinforced or single rib plate and about 120mm in the double rib plate.

For large area concrete ground, two vibrators can be installed on two wooden bars in the same direction, and the concrete can be compacted by the vibration of wood bars.

在振动倾斜混凝土表面时,应由低处逐渐向高处移动,以保证混凝土振实。

③外部振动器(附着式振动器)。外部振动器的振动作用深度约为 25 cm。如构件尺寸较厚时,需在构件两侧安设振动器同时进行振捣。

待混凝土入模后方可开动振动器,混凝土浇筑高度要高于振动器安装部位。当构件尺寸较大时,需在构件两侧安设振动器同时进行振捣;一般是在混凝土入模后开动振动器进行振捣,混凝土浇筑高度须高于振动器安装部位,当钢筋较密或构件断面较深较窄时,也可采取边浇筑边振动的方法;外部振动器应与模板紧密连接,其设置间距应通过试验确定,一般为每隔 1~1.5 m 设置一个;振动时间的控制是以混凝土不再出现气泡,表面呈水平时为准。

④振动台。当混凝土构件厚度小于 20 cm 时,可将混凝土一次装满振捣,如厚度大于 20 cm,则需分层浇灌,每层厚度不大于 20 cm,或随浇随振。

振捣时间要根据混凝土构件的形状、大小及振动能力而定,一般以混凝土表面呈水平并出现均匀的水泥和不再冒气泡时,表示已振实,即可停止振捣。

(3)振捣器的故障和排除方法。振捣器结构比较简单,但因经常露天作业,经常移动,转速又高。很容易发生故障。表 5.17 列举了振捣器常见故障及其产生的原因和排除方法,供施工中参考。

表 5.17 振捣器的故障及其产生原因和排除方法

故障现象	故障原因	排除方法
电动机过热,机体温度过高(超过额定温度)	1. 工作时间太久 2. 定子受潮,绝缘程度降低 3. 负荷过大 4. 电源电压过大、过低、时常变动及三相不平衡 5. 导线绝缘不良,电流流入地中 6. 线路接头不紧	1. 停止作业,让其冷却 2. 应立即干燥 3. 检查原因、调整负荷 4. 用电压表测定,并进行调整 5. 用绝缘布缠好损坏处 6. 重新接紧线头
电动机有强烈的钝音,同时发生转速降低、振动力减小现象	1. 定子磁铁松动 2. 一相熔丝断开或内部断线	1. 应拆卸检修 2. 更换熔丝和修理断线处

Chapter 5　Construction of Reinforced Concrete Structure

When vibrating the inclined concrete surface, it should move gradually from low to high to ensure concrete vibration.

③External vibrator (attached vibrator). The vibration depth of external vibrator is about 25cm. If the size of the component is relatively thick, a vibrator should be installed on both sides of the component to vibrate at the same time.

The vibrator can be started after the concrete is put into the mold, and the concrete pouring height shall be higher than the installation position of the vibrator. When the size of the component is large, the vibrator should be installed on both sides of the component to vibrate at the same time; generally, the vibrator should be started after the concrete is put into formwork, and the concrete pouring height should be higher than the installation position of the vibrator. When the reinforcement is dense or the section of the component is deep and narrow, the method of vibration while pouring can be adopted. The external vibrator should be closely connected with the formwork, and the setting spacing should be determined by test, generally one should be set every 1— 1.5m. The control of vibration time is subject to the condition that there is no bubble in the concrete and the surface is horizontal.

④Shaking table. When the thickness of the concrete component is less than 20cm, the concrete can be filled and vibrated at one time. If the thickness is greater than 20cm, the concrete should be poured in layers, and the thickness of each layer should not be greater than 20cm, or the concrete should be vibrated along with the pouring.

The vibration time shall be determined according to the shape, size and vibration capacity of concrete components. Generally, when the surface of turbid concrete is horizontal and uniform cement appears and no bubbles are emitted, the vibration can be stopped.

(3) The breakdown of vibrator and its elimination method. The vibrator is relatively simple in structure, but due to frequent open-air operation, frequent movement and high speed, it is easy to break down. Table 5.17 lists the common faults of the vibrator and its causes and elimination methods for reference in construction.

Table 5.17　Breakdown of Vibrator and Its Causes and Elimination Methods

Breakdown Phenomenon	Cause of Breakdown	Elimination Methods
Motor overheating, body temperature too high (over rated temperature)	1. Working too long 2. The stator is damped and the insulation is reduced. 3. Overload 4. The power supply voltage is too large, too low, often changing and three-phase imbalance. 5. The insulation of the wire is poor and the current flows into the ground. 6. The line connector is not tight.	1. Stop work and let it cool down. 2. It should be dried immediately. 3. Check the cause and adjust the load. 4. Measure with voltmeter and adjust. 5. Wrap the damaged parts with insulating cloth. 6. Tighten the thread again.
The motor has a strong dull sound, at the same time, the speed and vibration force decrease	1. Loose stator magnet. 2. One phase fuse is broken or internal wire is broken.	1. It should be disassembled for maintenance. 2. Replace fuse and repair broken wire.

续表

故障现象	故障原因	排除方法
电动机线圈烧坏	1. 定子过热 2. 绝缘严重潮湿 3. 相间短路,内部混线或接线错误	必须部分或全部重新绕定子线圈
电动机或把手有电	1. 导线绝缘不良,漏电,尤其在开关盒接头处 2. 定子的一相绝缘破坏	1. 有绝缘胶布包好破裂处 2. 应检修线圈
开关冒火花,开关熔丝易断	1. 线间短路或漏电 2. 绝缘受潮、绝缘强度降低 3. 负荷过大	1. 检查修理 2. 进行干燥 3. 调整负荷
电动机滚动轴承损坏,转子、定子相互摩擦	1. 轴承缺油或油质不好 2. 轴承磨损而致损坏	更改滚动轴承
振动棒不振	1. 电动机转向反了 2. 单向合器部分机体损坏 3. 软轴和机体振动子之间的接头处没有接合好 4. 钢丝软轴扭断 5. 行星式振动子柔性铰链损坏或滚子与滚道间有油污	1. 需改变接线(交换任意两相) 2. 检查单向合器,必要时加以修理或更换零件 3. 将接头连接好 4. 重新用锡焊焊接或更换软轴 5. 检修柔性铰链和清除滚子与滚道间的油污,必要时更换橡胶油封
振动棒振动有困难	1. 电动机的电压现电源电压不符 2. 振动棒外壳磨损,漏入灰浆 3. 振动棒顶盖未拧紧或磨坏,漏入灰浆使滚动轴承损坏 4. 行星式振动子起振困难 5. 滚子现滚道间有油污 6. 软管衬簧和钢丝软轴之间摩擦太大	1. 调整电源电压 2. 更换振动棒外壳,清洗滚动轴承和加注润滑脂 3. 清洗或更换滚动轴承,更换或拧紧顶盖 4. 摇晃棒头尖对地面轻轻一碰 5. 清除油污,必要时更换油封 6. 修理钢丝软轴,并使软轴与软管衬簧的长短想适应

Chapter 5 Construction of Reinforced Concrete Structure

continued

Breakdown Phenomenon	Cause of Breakdown	Elimination Methods
The motor coil is burnt out.	1. Stator overheating 2. The insulation is seriously wet. 3. Phase to phase short circuit, internal mixed line or wiring error	The stator coil must be partially or completely rewound.
The motor or handle has power	1. Poor insulation of wire, leakage, especially at the switch box connector. 2. One phase insulation failure of stator.	1. Wrap the rupture with insulating tape 2. The coil should be repaired.
The switch sparks and the fuse is easy to break.	1. Interline short circuit or leakage 2. The insulation is damped and the insulation strength is reduced. 3. Overload	1. Inspection and repair 2. Dry 3. Adjust load
The rolling bearing of motor is damaged and the rotor and stator rub against each other.	1. The bearing is short of oil or the oil quality is poor. 2. The bearing is damaged due to wear.	Change rolling bearing
The vibrator doesn't vibrate.	1. The motor has turned in the opposite direction. 2. Part of the body of the one-way combiner is damaged. 3. The joint between the flexible axle and the body vibrator is not well connected. 4. Steel wire flexible axle twist. 5. The flexure hinge of planetary vibrator is damaged or there is oil stain between roller and raceway.	1. Need to change wiring (exchange any two phases) 2. Check the one-way combiner and repair or replace parts if necessary. 3. Connect the connector. 4. Solder or replace the flexible axle again. 5. Repair the flexure hinge and remove the grease between the roller and raceway, and replace the rubber oil seal if necessary.
It is difficult to vibrate the vibrator.	1. The voltage of the motor is inconsistent with that of the power supply. 2. The shell of the vibrator is worn and leaked into the mortar. 3. The top cover of vibrator is not tightened or worn, and the rolling bearing is damaged due to mortar leakage. 4. It is difficult to start vibration for planetary vibrator. 5. There is oil stain between roller raceways. 6. There is too much friction between the hose spring and the steel wire flexible axle.	1. Adjust the power supply voltage. 2. Replace the vibrator shell, clean the rolling bearing and add grease. 3. Clean or replace the rolling bearing, replace or tighten the top cover. 4. Shake the tip of the vibrator and gently touch the ground 5. Remove oil and replace oil seal if necessary. 6. Repair the steel wire flexible axle, and make the length of the flexible axle and hose lining spring adapt to each other.

续表

故障现象	故障原因	排除方法
胶皮套管破裂	1. 弯曲半径过小 2. 用力斜推振动棒或使用时间过多	割去一段重新连接或更新的胶皮套管
附着式振捣器机体内有金属撞击声	振动子锁紧螺栓松脱,振动子产生轴向位移	重新锁紧振动子,必要时更换锁紧螺栓
平板式振捣器的底板振动有困难	1. 振动子的滚动轴承损坏 2. V带松弛	1. 更换滚动轴承 2. 调整或更换电动机底座上的橡胶垫;调整或更换减振弹簧

2. 钢筋密集部位的振捣

钢筋密集部位多在框架梁柱节点处,这些位置受力复杂且有可能临近施工缝,在操作中因钢筋密集而造成混凝土入模困难,振捣也不方便,因此必须采取以下措施加以解决。

(1)从设计上考虑。设计必须保证施工的可能性,并提供浇筑条件。可从提高钢筋强度级别来减少钢筋直径得以增大钢筋间距,用焊接来代替绑扎、用对焊来代替帮条焊等措施来解决钢筋密集难浇筑的问题。

(2)控制石子粒径,调整配合比和坍落度,必要时用细石混凝土。

(3)振捣棒头加焊或套接刀片或钢钎。

(4)斜插振动棒振捣。

(5)人工振捣与机械振捣相结合,必要时可以人工为主。

5.3.6 混凝土的养护

养护是混凝土工艺中的一个重要环节。混凝土强度的增长过程,实质上是胶凝材料水泥凝结硬化的结果。水泥的凝结硬化与水泥的水化作用是分不开的。水化作用的正常进行,又与混凝土所处的环境条件紧密联系在一起。所谓环境条件,主要指的是温度和湿度两个方面。水化作用必须有适宜的温度和湿度。从理论上讲,水化作用停滞点为-15 ℃,但-3 ℃以下水化极慢,在自然条件下,混凝土必须在$+5$℃以上才能进行洒水养护。在蒸养条件下,也应该控制极限加热温度,并应控制升温、降温减速,防止温差过大,而出现温度裂缝。混凝土在常温下相对湿度是90%的条件下养护时,强度是标准强度100%,当时对湿度降至60%时,强度是标准强度55%~66%,失强约1/3,因此,混凝土必须在潮湿环境下或水中养护。

continued

Breakdown Phenomenon	Cause of Breakdown	Elimination Methods
Rupture of rubber sleeve	1. The bending radius is too small. 2. Push the vibrator with force or use it for too long.	Cut off a section and reconnect or replace with a new rubber sleeve.
There is metal impact sound in the body of attached vibrator.	The vibrator locking bolt is loose, and the vibrator produces axial displacement.	Re-lock the vibrator and replace the locking bolt if necessary.
It is difficult to vibrate the bottom plate of the plate vibrator.	1. The roller bearing of vibrator is damaged. 2. V-belt slack	1. Replace the rolling bearing. 2. Adjust or replace the rubber pad on the motor base; adjust or replace the damping spring.

2. Vibration of Dense Reinforcement Area

Most of the reinforcement concentrated parts are at the joints of frame beam and column. These positions are subject to complex stress and may be close to the construction joint. In operation, due to the dense reinforcement, it is difficult to cast concrete into the formwork and the vibration is not convenient. Therefore, the following measures must be taken to solve the problem.

(1) From the design point of view. The design must guarantee the possibility of construction and provide pouring conditions. In order to solve the problem of dense reinforcement and difficult pouring, the reinforcement diameter can be reduced by increasing the strength grade of reinforcement, and the spacing of reinforcement can be increased by welding instead of binding, and butt welding is used instead of rib welding.

(2) Control the stone particle size, adjust the mix proportion and slump, and use fine aggregate concrete if necessary.

(3) The vibrating rod head is welded or jacketed with blade or drilling steel.

(4) Vibration with inclined inserted vibrator.

(5) Manual vibration and mechanical vibration shall be combined, and if necessary, manual vibration can be given priority to.

5.3.6 Curing of Concrete

Curing is an important link in concrete technology. The growth process of concrete strength is essentially the result of setting and hardening of cementitious materials. The setting and hardening of cement is inseparable from the hydration of cement. The normal hydration is closely related to the environmental conditions of concrete. The so-called environmental conditions mainly refer to two aspects of temperature and humidity. Hydration must have appropriate temperature and humidity. Theoretically speaking, the stagnation point of hydration is 15℃, but the hydration is very slow below 3℃. Under natural conditions, the concrete must be above + 5℃ for watering curing. Under the condition of steam curing, the limit heating temperature should also be controlled, and the temperature rise, cooling and deceleration should be controlled to prevent temperature cracks due to excessive temperature difference. When the relative humidity of concrete is 90% at room temperature, the strength is 100% of the standard strength. When the humidity drops to 60%, the strength is 55%—66% and the strength loss is about 1/3. Therefore, the concrete must be cured in humid environment or water.

混凝土养护的目的,就是要创造各种条件,使水泥充分水化,加速混凝土硬化,防止在成型后因暴晒、风干、干燥、寒冷等自然因素的影响,造成混凝土表面脱皮、起砂、出现干缩裂缝,严重的甚至使混凝土内部结构疏松,降低混凝土的强度。因此混凝土浇筑后,必须根据水泥品种、气候条件和工期要求加强养护。

养护的方法很多,目前有自然养护、蒸汽养护、干燥养护、太阳能养护、养护剂养护等,要因时因地制宜,选择较好的养护方法。

1. 自然养护

混凝土自然养护,是指混凝土浇筑完毕后,在大气温度不低于 5 ℃和表面覆盖浇水润湿的条件下,按照要求的养护期所进行的养护工艺,分浇水养护和喷洒塑料薄膜养护两种。无论采取哪一种工艺养护,首先都应遵守下列规定:

第一,在浇筑完成 12 h 以内进行覆盖养护。

第二,混凝土强度达到 1.2 MPa 以后,才允许操作人员行走、安装模板和支架;但不得作冲击性或类似劈打木材的操作。

第三,不允许用悬挑构件作为交通运输的通道或作工具、材料的停放场。

(1)浇水养护。在自然气温高于 5 ℃的条件下,用草袋、麻袋、锯末等覆盖混凝土,并在上面经常浇水,普通混凝土浇筑完毕后,应在 12 h 内加以覆盖和浇水,浇水次数以能保持足够的湿润状态为宜。在一般气候条件下(气温为 15 ℃以上),在浇筑后最初 3 d,白天每隔 2 h 浇水一次,夜间至少浇水 2 次。在以后的养护中,每昼夜至少浇水 4 次。在干燥的气候条件下,浇水次数应适当增加,浇水养护时间一般以达到标准强度的 60%左右为宜。

在一般情况下,硅酸盐水泥、普通硅酸盐水泥及矿渣硅酸盐水泥拌制的混凝土,其养护天数不应少于 7 d;火山灰质硅酸盐水泥及矿渣硅酸盐水泥拌制的混凝土,其养护天数不应少于 14 d;矾土水泥拌制的混凝土,其养护天数不应少于 3 d;掺用缓凝型外加剂,或有根据该水泥的技术性能加以确定。

①在外界气温低于 5 ℃时,不允许浇水。

②养护用水与拌制用水相同。

③厚大体积混凝土的养护,在炎热气候条件下,应采用降温措施。

The purpose of concrete curing is to create various conditions to make the cement fully hydrated, accelerate the concrete hardening and prevent the concrete surface from peeling, sanding and drying shrinkage cracks due to natural factors such as exposure to the sun, air drying, drying and cold after molding, resulting in concrete surface peeling, sanding, drying cracks, and even make the internal structure of concrete loose, reduce the strength of concrete. Therefore, after concrete pouring, it is necessary to strengthen curing according to the cement varieties, climatic conditions and construction period requirements.

There are many curing methods. At present, there are natural curing, steam curing, drying curing, solar energy curing, maintenance agent curing and so on. It is necessary to choose a better curing method according to the time and local conditions.

1. Natural Curing

The natural curing of concrete refers to the curing process carried out according to the required curing period under the condition that the atmospheric temperature is not less than 5℃ and the surface is covered with water and wetted after the concrete pouring, which is divided into two kinds: watering curing and spraying plastic film curing. No matter which process is adopted, the following regulations shall be followed first:

First, cover curing should be carried out within 12h after pouring.

Second, only when the concrete strength reaches 1.2MPa, can the operator walk and install the formwork and support; however, it is not allowed to carry out impact or similar wood splitting operation.

Third, cantilever members are not allowed to be used as transportation channels or parking places for tools and materials.

(1) Watering curing. Under the condition that the natural temperature is higher than 5℃, cover the concrete with straw bags, gunny bags, sawdust, etc., and water them frequently. After the ordinary concrete is poured, it should be covered and watered within 12h, and the watering times should be enough to keep it moist. Under general climatic conditions (the temperature is above 15℃), concrete shall be watered every 2 hours in the day and at least twice at night in the first three days after pouring. In the future curing, concrete shall be watered at least 4 times every day and night. In dry climate conditions, the watering times should be increased appropriately, and generally, the watering curing time should reach about 60% of the standard strength.

In general, the curing days of concrete mixed with Portland cement, ordinary Portland cement and slag Portland cement should not be less than 7 days; The curing days of concrete mixed with pozzolanic Portland cement and slag Portland cement shall not be less than 14 days. For the concrete mixed with bauxite cement, the curing time shall not be less than 3 days. The concrete mixed with retarding admixture is determined according to the technical performance of the cement.

①When the outside temperature is lower than 5℃, watering is not allowed.

②The water for curing is the same as that for mixing.

③For the curing of thick mass concrete, cooling measures should be taken in hot climate.

(2)喷膜养护。喷膜养护是在混凝土表面喷洒一至两层塑料薄膜,它是将塑料溶液喷洒在混凝土表面上,溶液挥发后,塑料与混凝土表面结合成一层薄膜,使混凝土表面空气隔绝,封闭混凝土中的水分不再被蒸发,而完成水化作用。这种养护方法一般使用于表面积大的混凝土施工和缺水地区。

①常用的养护剂。

树脂型养护剂。以树脂为基料,以高挥发性液体做溶剂配制成溶液。喷涂于混凝土表面,当溶剂挥发后,有10%～50%的固体物质残留在混凝土表面,形成一层薄膜,使水分封闭在混凝土表面,形成一层薄膜,使水封闭在混凝土内,达到养护的目的。

油乳型养护剂。以石蜡和热亚麻仁油作基料,以水作乳化液,硬脂酸和三乙醇胺作稳定剂。配方为:石蜡12%、熟亚麻仁油20%、硬脂酸4%、三乙醇胺3%、水61%。

煤焦油养护剂。将煤焦油用溶剂稀释至适合喷涂的稠度即成。

沥青和地沥青养护剂。用水作乳化液制成。

②喷膜养护的操作要点。

喷洒压力以0.2～0.3 MPa为宜,喷出来的塑溶液呈较好的雾状为最佳。压力小,不易形成雾状;压力大,会破坏混凝土表面,喷洒时应离混凝土表面50 cm。

喷洒时间,应掌握混凝土水分蒸发情况,在不见浮水,混凝土表面以手指按无指印时,即可进行喷洒。过早会影响塑料薄膜与混凝土表面结合,过迟则影响混凝土强度。

溶液喷洒厚度以溶液的耗用量衡量,通常以每平方米耗用养护剂2.5 kg为宜,喷洒厚度要求均匀一致。

通常要喷洒两次,第一次成膜后再喷洒第二次,喷洒要求有规律,固定一个方向,前后两次的走向应互相垂直。

溶液喷洒后很快形成塑料薄膜,为达到养护目的,必须保护薄膜的养护性,要求不得有撕坏破裂,不得在薄膜上行走、拖拉工具,如发现损坏应及时补喷,如气温较低,应设法保温。

2. 蒸汽养护

蒸汽养护是缩短养护时间的有效方法之一。所谓蒸汽养护,是将就浇捣成型的混凝土构件置于固定的养护坑(或窑)内,然后通过蒸汽,使混凝土在较高的温度和湿度条件下,迅速凝结硬化,达到要求的强度。

一般蒸汽养护可分为四个阶段:

(2) Spray film curing. Spray film curing is to spray one or two layers of plastic film on the concrete surface. It is to spray the plastic solution on the concrete surface. After the solution volatilizes, the plastic and the concrete surface are combined to form a layer of film, so that the air on the concrete surface is isolated, and the water in the closed concrete is no longer evaporated, and the hydration is completed. This curing method is generally used in concrete construction with large surface area and water shortage area.

①Commonly used curing agents.

Resin curing agent: Resin was used as base material and high volatile liquid as solvent to prepare solution, spraying on the concrete surface. When the solvent volatilizes, 10%—50% of the solid matter remains on the concrete surface, forming a layer of film to seal the water on the concrete surface and form a film to seal the water in the concrete to achieve the purpose of curing.

Oil emulsion curing agent: Paraffin and hot linseed oil were used as base materials, water as emulsion, stearic acid and triethanolamine as stabilizers. The formula is: paraffin 12%, hot linseed oil 20%, stearic acid 4%, triethanolamine 3%, and water 61%.

Coal tar curing agent: Dilute the coal tar with solvent to the consistency suitable for spraying.

Asphalt and ground pitch curing agent: It is made with water as emulsion.

②Operation points of spray film curing.

The best spraying pressure is 0.2—0.3MPa, and the best spray is the plastic solution in good fog shape. Small pressure is not easy to form fog and high pressure will damage the concrete surface. Spraying should be 50cm away from the concrete surface.

Spraying time: It is necessary to master the water evaporation of concrete. When there is no floating water and there is no fingerprint on the concrete surface, spraying can be carried out. Too early will affect the combination of plastic film and concrete surface, too late will affect the strength of concrete.

The spraying thickness of the solution is measured by the consumption of the solution. Generally, the consumption of curing agent is 2.5kg per square meter, and the spraying thickness is required to be uniform.

Usually spray twice, the first time after the film is formed, and then spray the second time. The spraying should be regular and fixed in one direction, and the direction of the two times should be perpendicular to each other.

The solution will be formed into plastic film quickly after spraying. In order to achieve the curing purpose, the curing property of the film must be protected. It is required that there should be no tearing and cracking, walking on the film or dragging tools. If any damage is found, spray it in time. If the temperature is low, try to keep warm.

2. Steam curing

Steam curing is one of the effective methods to shorten the curing time. The so-called steam curing is to place the concrete components formed by pouring and tamping in the fixed curing pit (or kiln), and then through the steam, the concrete can quickly set and harden under the conditions of high temperature and humidity, so as to achieve the required strength.

General steam curing can be divided into four stages:

(1)静停阶段:就是指混凝土浇筑完毕至升温前在室温下先放置一段时间。这主要是为了增强混凝土对升温阶段结构破坏作用的抵抗能力。一般为 2~6 h(干硬性混凝土为 1 h)。

(2)升温阶段:就是混凝土原始温度上升到恒温阶段。温度急速上升,会使混凝土表面因体积膨胀太快而产生裂缝。因而必须控制升温速度,一般为 10~25 ℃/h(干硬性混凝土为 35~40 ℃/h)。

(3)恒温阶段:是混凝土强度增长最快的阶段.恒温的温度随水泥品种不同而异,普通水泥的养护温度不得超过 80,矿渣水泥、火山灰水泥可提高到 90 ℃~95 ℃。一般恒温时间为 5~8 h,恒温加热阶段应保持 90%~100% 的相对湿度。

(4)降温阶段:在降温阶段内,混凝土已经硬化,如降温过快,混凝土会产生表面裂缝,因此降温速度应控制。一般情况下,构件厚度在 10 cm 左右时,降温速度不大于 20~30 ℃/h。

为了避免由于蒸汽温度骤然升降而引起混凝土构件产生裂缝变形,必须严格控制升温和降温的速度。出槽的构件温度与室外温度相差不得大于 40 ℃;当室外为 0 ℃以下时,相差不得大于 20 ℃。

此外,对抗渗、抗冻、高标号的混凝土,在蒸汽养护后,应继续浇水湿润,进行自然养护,为避免暴晒和风干。

蒸汽养护通常用于预制混凝土构件生产线和冬期施工,其养护方法很多,如坑槽蒸养、力窑蒸养、隧道窑蒸养等。其主要设备是用蒸汽养护坑。

普通蒸汽养护坑是混凝土构件厂最常见的蒸汽设备构件叠放在坑内,尽可能提高填充系数,以节约蒸汽。蒸汽管每米钻 4~6 个喷气孔,称为花管。安装在坑壁的下部,使蒸汽上升。排水沟外设有自动流水器以排除冷凝水。坑盖、坑壁除保证结构强度外,中部应填充保温材料,以减少热能损耗。蒸汽养护坑的构造如图 5.44(a)所示。

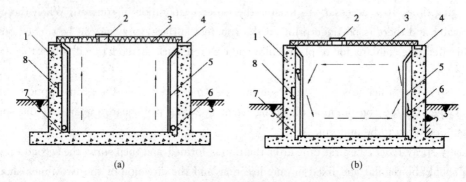

图 5.44 蒸汽养护坑
(a)普通蒸汽养护坑;(b)热介质循环蒸汽养护坑
1—坑壁;2—观察及降温口;3—盖板;4—水封槽;
5—护壁槽钢及梯;6—蒸汽管;7—水沟;8—测温装置

Chapter 5 Construction of Reinforced Concrete Structure

(1) Static stop stage: It refers to that the concrete is placed at room temperature for a period of time from the completion of concrete pouring to the temperature rise. This is mainly to enhance the resistance of concrete to the damage of the structure in the heating stage. It is generally 2—6h(1h for dry hard concrete).

(2) Heating up stage: The original temperature of concrete rises to the constant temperature stage. The rapid rise of temperature will cause cracks on the surface of concrete due to rapid volume expansion. Therefore, it is necessary to control the heating rate, which is generally 10—25℃/h(35—40℃/h for dry hard concrete).

(3) Constant temperature stage: It is the fastest growth stage of concrete strength. The temperature of constant temperature varies with different cement types. The curing temperature of ordinary cement shall not exceed 80, and slag cement and pozzolanic cement can be increased to 90℃—95℃. Generally, the constant temperature time is 5—8h, and the relative humidity of 90%—100% should be maintained in the constant temperature heating stage.

(4) Cooling stage: In the cooling stage, the concrete has hardened, if the temperature is cooling too fast, the concrete will produce surface cracks. Therefore, the cooling speed should be controlled. In general, when the thickness of the component is about 10cm, the cooling rate is not more than 20—30℃/h.

In order to avoid cracks and deformation of concrete components caused by sudden rise and fall of steam temperature, the speed of temperature rise and temperature fall must be strictly controlled. The temperature difference between the components out of the tank and the outdoor temperature shall not be greater than 40℃; when the outdoor temperature is below 0℃, the difference shall not be greater than 20℃.

In addition, the impermeable, frost resistant and high-grade concrete should continue to be watered and moistened after steam curing, so as to avoid exposure to the sun and air drying.

Steam curing is usually used in the production line of precast concrete components and in winter construction. There are many curing methods, such as pit and trough steam curing, power kiln steam curing, tunnel kiln steam curing and so on. The main equipment is steam curing pit.

The common steam curing pit is the most common steam equipment in the concrete component factory. The components are stacked in the pit to increase the filling coefficient as much as possible to save steam. About 4—6 air-jet holes are drilled in the steam pipe per meter, which is called flower tube. It is installed in the lower part of the pit wall to raise the steam. The drainage ditch is equipped with an automatic water flow device to drain the condensed water. In addition to ensuring the structural strength, the middle part of the pit cover and wall shall be filled with insulation materials to reduce the heat loss. The structure of steam curing pit is shown in Fig. 5.44(a).

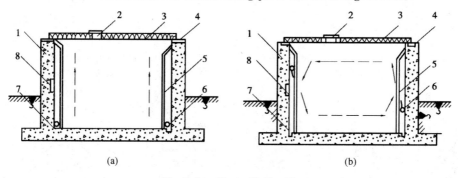

Fig. 5.44 Steam Curing Pit

(a) Common steam curing pit; (b) Hot medium circulating steam curing pit

1—pit wall; 2—observation and cooling port; 3—cover plate; 4—water sealing groove;
5—wall protection channel steel and ladder; 6—steam pipe; 7—ditch; 8—temperature measuring device

热介质循环蒸汽养护坑的功能同普通蒸汽养护坑,但普通蒸汽养护坑的缺点是坑内冷空气与蒸汽温度不易调匀。热介质循环蒸汽养护坑是上、下均安装蒸汽管,并将喷气孔改为蒸汽嘴,使蒸汽按一定方向循环流动,加强了热交换,并可进行自动调节,现已普遍使用。

施工现场由于条件限制,现浇预制构件一般可采用临时性地面或地下的养护坑,上盖养护罩或用简单的帆布、油布覆盖。

5.4 混凝土工程质量要求

5.4.1 混凝土工程质量缺陷

1. 现浇结构的外观质量缺陷

现浇结构的外观质量缺陷,应由监理(建设)单位、施工单位等各方根据其对结构性能和使用功能影响的严重程度,按表5.18确定。

表5.18 现浇结构外观质量缺陷

名称	现象	严重缺陷	一般缺陷
露筋	构件内钢筋未被砼包裹而外露	纵向受力钢筋有露筋	其他钢筋有少量露筋
蜂窝	混凝土表面缺少水泥浆而形成石子外露	构件主要受力部位有蜂窝	其他部位有少量蜂窝
孔洞	混凝土中孔穴深度和长度均超过保护层厚度	构件主要受力部位有孔洞	其他部位有少量孔洞
夹渣	混凝土中夹有杂物且深度超过保护层厚度	构件主要受力部位有夹渣	其他部位有少量夹渣
疏松	混凝土中局部不密实	构件主要受力部位有疏松	其他部位有少量疏松

The function of hot medium circulation steam curing pit is the same as that of ordinary steam curing pit, but the disadvantage of ordinary steam curing pit is that the temperature of cold air and steam in the pit is not easy to be adjusted evenly.

The steam maintenance pit of hot medium circulation is that steam pipes are installed at both the upper and lower parts, and the air jet hole is changed into a steam nozzle, so that the steam circulates in a certain direction, enhances the heat exchange, and can be automatically adjusted. It is now widely used.

Due to the limited conditions in the construction site, the cast-in-place precast components can generally adopt temporary ground or underground curing pit, covered with protective cover or covered with simple canvas and oilcloth.

5.4 Quality Requirements of Concrete Works

5.4.1 Quality Defects of Concrete Works

1. Appearance Quality Defects of Cast-in-place Structure

The appearance quality defects of cast-in-place structure shall be determined by the supervision (construction) unit, construction unit and other parties according to the severity of the impact on the structure performance and use function as shown in Table 5.18.

Table 5.18 Appearance Quality Defects of Cast-in-place Structure

Name	Appearance	Serious Defects	General Defects
Exposed reinforcement	The reinforcement in the component is not wrapped by concrete and exposed.	There is exposed reinforcement in longitudinal stressed reinforcement.	Other reinforcement has a small amount of exposed reinforcement.
Bug holes	The concrete surface is lack of cement slurry and the stone is exposed.	There are bug holes in the main stress part of the component.	There are a few bug holes in other parts.
Holes	The depth and length of the hole in the concrete exceed the thickness of the protective layer.	There are holes in the main stressed parts of the component	There are a few holes in other parts.
Slag inclusion	There are impurities in the concrete and the depth exceeds the thickness of the protective layer	There are slag inclusions in the main stress parts of the components.	There is a small amount of slag in other parts.
Loose	The concrete is not dense locally.	The main stress part of the component is loose.	There is a little loose in other parts.

续表

名称	现象	严重缺陷	一般缺陷
裂缝	缝隙从砼表面延伸至砼内部	构件主要受力部位有影响结构性能或使用功能的裂缝	其他部位有少量不影响结构性能或使用功能的裂缝
连接部位缺陷	构件连接处砼缺陷及连接钢筋、连接铁件松动	连接部位有影响结构传力性能的缺陷	连接部位有基本不影响结构传力性能的缺陷
外形缺陷	缺棱掉角、棱角不直、翘曲不平、飞出凸肋等	清水砼构件内有影响使用功能或装饰效果的外形缺陷	其他砼构件有不影响使用功能的外形缺陷
外表缺陷	构件表面麻面、掉皮、起砂、玷污等	具有重要装饰效果的清水砼构件有外表缺陷	其他砼构件有不影响使用功能的外表缺陷

注:1. 现浇结构的外观质量不应有严重缺陷。对已经出现的严重缺陷,应由施工单位提出技术处理方案,并经监理(建设)单位认可后进行处理,对经处理的部位,应重新检查验收。
2. 现浇结构的外观质量不宜有一般缺陷。

对已经出现的一般缺陷,应由施工单位按技术处理方案进行处理,并重新检查验收。

2. 现浇结构尺寸允许偏差

(1)现浇结构不应有影响结构性能和使用功能的尺寸偏差。混凝土设备基础不应有影响结构性能和设备安装的尺寸偏差。

对超过尺寸允许偏差且影响结构性能和安装、使用功能的部位,应由施工单位提出技术处理方案,并经监理(建设)单位认可后进行处理,对经处理的部位,应重新检查验收。

检查数量:全数检查。

检验方法:量测,检查技术处理方案。

(2)现浇结构和混凝土设备基础拆模后的尺寸偏差应符合表5.19、表5.20的规定。

Chapter 5 Construction of Reinforced Concrete Structure

continued

Name	Appearance	Serious Defects	General Defects
Crack	The gap extends from the concrete surface to the concrete interior.	There are cracks that affect the structural performance or service function in the main stressed parts of the component.	There are a few cracks in other parts that do not affect the structural performance or service function.
Defects in connection	Concrete defects at the joint of components and looseness of connecting steel bars and iron pieces	There are some defects in the connection which affect the force transfer performance of the structure.	There are some defects in the connecting parts which do not affect the force transfer performance of the structure.
Outside defect	Lack of edges and corners, irregular edges and corners, warping and uneven, flying out and convex ribs, etc.	There are outside defects in fair faced concrete components that affect the use function or decoration effect.	Other concrete components have outside defects which do not affect the use function.
Appearance defect	Pitting, peeling, sanding and contamination on the surface of components	Fair faced concrete members with important decorative effect have appearance defects.	Other concrete components have appearance defects that do not affect the use function.

Note: 1. The appearance quality of cast-in-place structure should not have serious defects. For the serious defects that have appeared, the construction unit shall put forward the technical treatment scheme, which shall be treated after being approved by the supervision(construction)unit, and the treated parts shall be inspected and accepted again.
2. The appearance quality of cast-in-place structure should not have general defects.

The general defects that have already appeared should be handled by the construction units according to the technical treatment plan and checked and accepted again.

2. Allowable Size Deviation of Cast-in-place Structure

(1) The cast-in-place structure should not have the size deviation which affects the structure performance and use function. The concrete equipment foundation shall not have the dimensional deviation which affects the structural performance and equipment installation.

For the parts that exceed the allowable size deviation and affect the structural performance, installation and use functions, the construction unit shall propose a technical treatment scheme, which shall be treated after being approved by the supervision(construction) unit, and the treated parts shall be re-inspected and accepted.

Inspection quantity: full inspection.

Test method: measure and check the technical treatment plan.

(2) The dimensional deviation of cast-in-place structure and concrete equipment foundation after formwork removal shall comply with the provisions in Table 5.19 and Table 5.20.

检查数量:按楼层、结构缝或施工段划分检验批。在同一检验批内,对梁、柱和独立基础,应抽查构件数量的10%,且不少于3件;对墙和板,应按有代表性的自然间抽查10%,且不少于3间;对大空间结构,墙可按相邻轴线间高度5m左右划分检查面,板可按纵、横轴线划分检查面,抽查10%,且均不少于3面;对电梯井应全数检查;对设备基础应全数检查。

检验方法:量测检查。

表 5.19 现浇结构尺寸允许偏差和检验方法

项目			允许偏差/mm	检验方法
轴线位置	基础		15	钢尺检查
	独立基础		10	
	墙、柱、梁		8	
	剪力墙		5	
垂直度	层高	≤5 m	8	经纬仪或吊线、钢尺检查
		>5 m	10	经纬仪或吊线、钢尺检查
	全高(H)		$H/1\,000$ 且≤30	经纬仪、钢尺检查
标高	层高		±10	水准仪或拉线、钢尺检查
	全高		±30	
截面尺寸			+8,−5	钢尺检查
电梯井	井筒长、宽对定位中心线		+25,0	钢尺检查
	井筒全高(H)垂直度		$H/1\,000$ 且≤30	经纬仪、钢尺检查
表面平整度			8	2 m靠尺和塞尺检查
预埋设施中心线位置	预埋件		10	钢尺检查
	预埋螺栓		5	
	预埋管		5	
预埋洞中心线位置			15	钢尺检查

注:检查轴线、中心线位置时,应沿纵、横两个方向量测,并取其中的较大值。

Inspection quantity: divide the inspection lot by floor, structural joint or construction section. In the same inspection lot, 10% of the number of components shall be sampled for beams, columns and independent foundations, and no less than 3 components shall be sampled; for walls and slabs, 10% of the representative natural rooms should be selected, and no less than 3 rooms should be selected; for the large space structure, the wall can be divided into inspection surfaces according to the height of about 5m between adjacent axes, and the inspection surface of slab can be divided according to the longitudinal and transverse axis, and the random inspection is 10% and no less than 3 sides; the elevator shaft should be fully inspected; the equipment foundation should be fully inspected.

Test method: measurement and inspection

Table 5.19 Allowable Deviation and Inspection Method of Cast-in-place Structure Size

Item			Allowable Deviation/mm	Test Method
Axis position	Foundation		15	Steel tape inspection
	Pad foundation		10	
	Wall, column and beam		8	
	Shear wall		5	
Verticality	Storey height	≤5 m	8	Inspection by theodolite, plumb line and steel tape
		>5 m	10	Inspection by theodolite, or plumb line and steel tape
	Full height(H)		H/1 000 and ≤ 30	Inspection by theodolite, and steeltape
Elevation	Storey height		±10	Inspection by level, line and steel tape
	Full height		±30	
Section size			+8,−5	Steel tape inspection
Elevator shaft	The length and width of the shaft to the positioning center line		+25,0	Steel tape inspection
	Verticality of full height(H) of shaft		H/1 000 and ≤ 30	Theodolite and steel tape inspection
Surface flatness			8	Inspection with 2m guiding rule and feeler gauge
Location of center line of embedded facilities	Embedded parts		10	Steel tape inspection
	Embedded bolt		5	
	Embedded pipe		5	
Center line position of embedded hole			15	Steel tape inspection
Note: when checking the position of the axis and center line, the measurement should be conducted along the longitudinal and horizontal directions, and the larger value should be taken.				

表 5.20 混凝土设备基础尺寸允许偏差和检验方法

项目		允许偏差/mm	检验方法
坐标位置		20	钢尺检查
不同平面的标高		0,-20	水准仪或拉线、钢尺检查
平面外形尺寸		±20	钢尺检查
凸台上平面外形尺寸		0,-20	钢尺检查
凹穴尺寸		+20,0	钢尺检查
平面水平度	每米	5	水平尺、塞尺检查
	全长	10	水准仪或拉线、钢尺检查
垂直度	每米	5	经纬仪或吊线、钢尺检查
	全高	10	
预埋地脚螺栓	标高(顶部)	+20,0	水准仪或拉线、钢尺检查
	中心距	±2	钢尺检查
预埋地脚螺栓孔	中心线位置	10	钢尺检查
	深度	+20,0	钢尺检查
	孔垂直度	10	吊线、钢尺检查
预埋活动地脚螺栓锚板	标高	+20,0	水准仪或拉线、钢尺检查
	中心线位置	5	钢尺检查
	带槽锚板平整度	5	钢尺、塞尺检查
	带螺纹孔锚板平整度	2	钢尺、塞尺检查

注:检查坐标、中心线位置时,应沿纵、横两个方向量测,并取其中的较大值。

5.4.2 常见混凝土的质量问题的处理

1. 混凝土脱模后常见的质量问题及产生的主要原因如下

(1)麻面:麻面是结构构件表面呈现无数的缺浆小凹坑而钢筋无外露。这类缺陷主要是由于模板表面粗糙或清理不干净;木模板在浇筑混凝土前湿润不够;钢模板脱模剂涂刷不均匀;混凝土振捣不足,气泡未排出等。

Chapter 5 Construction of Reinforced Concrete Structure

Table 5.20 Allowable Deviation and Inspection Method of Concrete Equipment Foundation Size

Item		Allowable deviation/mm	Test method
Coordinate position		20	Steel tape inspection
Elevation of different planes		0, −20	Inspection by level, line and steel tape
Plane dimension		±20	Steel tape inspection
Plane dimension on boss		0, −20	Steel tape inspection
Cavity size		+20, 0	Steel tape inspection
Plane levelness	Per meter	5	Level and feeler gauge inspection
	Full length	10	Inspection by level, line and steel tape
Verticality	Per meter	5	Inspection by theodolite, or plumb line and steel tape
	Full height	10	
Embedded anchor bolt	Elevation(top)	+20, 0	Inspection by theodolite, or plumb line and steel tape
	Center distance	±2	Steel tape inspection
Embedded anchor bolt hole	Centerline location	10	Steel tape inspection
	Depth	+20, 0	Steel tape inspection
	Verticality of holes	10	Inspection by plumb line and steel tape
Anchor plate of embedded movable anchor bolt	Elevation	+20, 0	Inspection by theodolite, or plumb line and steel tape
	Centerline location	5	Steel tape inspection
	Flatness of slotted anchor plate	5	Inspection by steel tape and feeler gauge
	Flatness of anchor plate with threaded hole	2	Inspection by steel tape and feeler gauge

Note: when checking the coordinates and the center line location, the measurement should be conducted along the longitudinal and horizontal directions, and the larger value should be taken.

5.4.2 Treatment of Common Concrete Quality Problems

1. The Common Quality Problems After Concrete Demoulding and the Main Causes are As Follows

(1) Pitted surface: pitted surface is the surface of structural components showing numerous small pits of lack of slurry and no exposed reinforcement. This kind of defect is mainly due to the rough surface of formwork or unclean cleaning; the wood formwork is not wet enough before pouring concrete; the coating of steel formwork release agent is uneven; the concrete vibration is insufficient and the bubbles are not discharged.

(2) 露筋：露筋是钢筋暴露在混凝土外面。产生的原因主要是浇筑时垫块过少，垫块位移钢筋紧贴模板；石子粒径过大，钢筋过密，水泥砂浆不能充满钢筋周围空间；混凝土振捣不密实，拆模方法不当，以致缺棱掉角等。

(3) 蜂窝：蜂窝是结构构件表面混凝土由于砂浆少、石子多，石子间出现空隙，形成蜂窝状的孔洞。其原因是材料配合比不准确（浆少、石子多）；搅拌不均匀造成砂浆与石子分离；振捣不足或过振；模板严重漏浆等。

(4) 孔洞：孔洞是指混凝土结构内部存在空隙，局部或全部没有混凝土。这种现象主要是由于混凝土严重离析，石子成堆，砂浆分离；混凝土捣空；泥块、杂物掺入等造成。

(5) 缝隙及夹层：缝隙和夹层是将结构分隔成几个不相连接的部分。产生的原因主要是施工缝、温度缝和收缩缝处理不当；混凝土内有杂物等。

(6) 裂缝：结构构件产生的裂缝的原因比较复杂，有外荷载引起的裂缝，由变形引起的裂缝和由施工操作不当引起的裂缝等。

(7) 混凝土强度不足：造成混凝土强度不足的原因是多方面的，主要是由混凝土配合比设计、搅拌、现场浇捣和养护等方面的原因造成。

2. 混凝土质量缺陷的防治与处理的方法

(1) 对数量不多的小蜂窝、麻面、露筋的混凝土表面，可用 1∶2～1∶2.5 水泥砂浆抹面补修。在抹砂浆前，须用钢丝刷和压力水清洗润湿．补抹砂浆初凝后要加强养护。

(2) 当蜂窝比较严重或露筋较深时，应凿去蜂窝、露筋周边松动、薄弱的混凝土和个别突出的集料颗粒，然后洗刷干净，充分润湿，再用比原混凝土强度等级高一级的细石混凝土填补，仔细捣实，加强养护。

(3) 对于影响构件安全使用的空洞和大蜂窝，应会同有关单位研究处理，有时应进行必要的结构检验。补救方法一般可在彻底清除软弱部分及清洗后用高压喷枪或压力灌浆法修补。

(2) Exposed reinforcement: exposed reinforcement means that the reinforcement is exposed outside the concrete. The main reasons are that there are too few cushion blocks during pouring, the cushion blocks are displaced, and the reinforcement is close to the formwork; the stone particle size is too large, the reinforcement is too dense, and the cement mortar can not fill the space around the reinforcement; the concrete vibration is not dense, and the formwork removal method is improper, resulting in the lack of edges and corners.

(3) Bug holes: bug holes are honeycomb shaped holes formed on the surface of structural members due to the lack of mortar and more stones. The reasons are as follows: the material mix proportion is not accurate (less slurry, more stones); the mortar is separated from the stone due to uneven mixing; insufficient or excessive vibration of vibration; serious mortar leakage of formwork.

(4) Hole: hole refers to the existence of voids in the concrete structure, and there is no concrete locally or completely. This phenomenon is mainly due to the serious segregation of concrete, piles of stones, separation of mortar, concrete hollowing, and mixing of mud and debris.

(5) Gap and interlayer: the gap and interlayer are the parts that separate the structure into several unconnected parts. The main causes are improper treatment of construction joints, temperature joints and contraction joints, and impurities in concrete.

(6) Cracks: the causes of cracks in structural members are relatively complex, including cracks caused by external load, cracks caused by deformation and cracks caused by improper operation.

(7) Insufficient concrete strength: there are many reasons for the insufficient concrete strength, which are mainly caused by the concrete mix proportion design, mixing, on-site pouring and curing.

2. The Prevention and Treatment Methods of Concrete Quality Defects Are As Follows:

(1) For a small number of small bug holes, pitted surface and exposed reinforcement concrete surface, 1 : 2—1 : 2.5 cement mortar can be used for surface repair. Before plastering the mortar, the steel wire brush and pressure water should be used to clean and moisten the mortar. After the initial setting of the supplementary plastering mortar, curing should be strengthened.

(2) When the bug holes are serious or the exposed reinforcement is deep, the loose and weak concrete and individual protruding aggregate particles around the bug holes and exposed reinforcement shall be chiseled out, then cleaned and fully wetted, and then filled with fine aggregate concrete one grade higher than the original concrete strength grade, carefully tamped and strengthened for curing.

(3) The cavity and large bug holes which affect the safe use of components should be studied and treated together with relevant units, and sometimes necessary structural inspection should be carried out. Generally, the remedy can be repaired by high-pressure spray gun or pressure grouting method after the soft parts are completely removed and cleaned.

(4)对于宽度大于 0.5 mm 的裂缝,宜采用水泥灌浆;对于宽度小于 0.5 mm 的裂缝,宜采用化学灌浆。在灌浆前,对裂缝的数量、宽度、连通情况及漏水情况等作全面观测,以便作出切合实际情况的补强方案。作为补强用的灌浆材料,常用的有环氧树脂浆液(能补缝宽 0.2 mm 以上的干燥裂缝)和甲凝(能补修 0.05 mm 以上的干燥裂缝)等。作为防渗堵漏用的灌浆材料,常用的有丙凝(能灌入 0.01 mm 以上的裂缝)和聚氨酯树脂(能灌入 0.015 mm 以上的裂缝)等。

5.3 混凝土工程施工安全技术

混凝土工程施工安全技术的规定如下:

(1)模板施工安全技术。在施工现场应戴安全帽,所用工具应装在工具包内;垂直运输模板或其他材料时,应有统一指挥,统一信号;拆模时应专人负责安全监督和设立警戒标志;高空作业人员应经过体格检查,不合格者不得进行高空作业;高空作业应穿防滑鞋,系好安全带。模板在安全系统未钉牢固之前,不得上下;未安装好的梁底板或挑檐等模板的安装与拆除,必须有可靠的技术措施,确保安全。非拆模人员不准在拆模区域内通行。拆下的模板应将铁钉尖头朝下,并及时运至指定堆放地点,拔除钉子,分类堆放整齐。

(2)钢筋施工安全技术。在高空绑扎和安装钢筋时,必须注意不要将钢筋集中堆放在模板或脚手架的某一部位,以保安全;特别是悬臂构件,还要检查支撑是否牢固。在脚手架上不要随便放置工具、箍筋或短钢筋,避免放置不稳滑下伤人。焊接或绑扎竖向放置的钢筋骨架时,不得站在已绑扎或焊接好的箍筋上工作。搬运钢筋的工人须戴帆布垫角、围裙及手套;除锈工人应戴口罩及风镜;电焊工人应带防护镜并穿工作服。300~500 mm 的钢筋短头禁止用机器切割。在有电线通过的地方安装钢筋时,必须小心谨慎,勿使钢筋碰着电线。

(4) For the cracks with width greater than 0.5mm, cement grouting should be used; for cracks with width less than 0.5mm, chemical grouting should be used. Before grouting, the number, width, connection and leakage of cracks shall be observed comprehensively, so as to make reinforcement scheme suitable for the actual situation. As the grouting materials for reinforcement, the commonly used ones are epoxy resin grout(which can repair the dry cracks with a width of more than 0.2mm) and Metronidazole(MMA)(which can repair the dry cracks above 0.05mm). As grouting materials for anti-seepage and plugging, the commonly used grouting materials are acrylic cement(which can be poured into the cracks above 0.01mm) and polyurethane resin(which can be poured into the cracks above 0.015mm), etc.

5.3 Construction Safety Technology of Concrete Works

The safety regulations for concrete construction are as follows:

(1) Safety technology of formwork construction. Safety helmet should be worn on the construction site, and the tools used should be installed in the tool kit; when the formwork or other materials are transported vertically, there should be unified command and signal; Special personnel shall be responsible for safety supervision and set up warning signs during formwork removal; personnel working at height shall undergo physical examination, and those who fail to pass the examination shall not be allowed to work at height; anti slip shoes and safety belt shall be worn for high-altitude operation. The formwork shall not go up or down before the safety system is nailed firmly; reliable technical measures must be taken for the installation and removal of the formwork such as beam bottom plate or overhanging eaves that have not been installed to ensure safety. Non demoulding personnel are not allowed to pass through the formwork removal area. The removed formwork shall face down the iron nails and timely transport them to the designated stacking place, pulling out the nails and stacking in order.

(2) Safety technology of reinforcement construction. When binding and installing steel bars at high altitude, it is necessary to pay attention not to stack the reinforcement in a certain part of the formwork or scaffold to ensure safety, especially for cantilever components, check whether the support is firm. Do not place tools, stirrups or short steel bars on the scaffold to avoid unstable sliding and hurting people. When welding or binding the vertically placed reinforcement framework, it is not allowed to work on the bound or welded stirrups. Workers carrying steel bars shall wear canvas corner pads, aprons and gloves; derusting workers shall wear masks and goggles; welding workers shall wear protective glasses and work clothes. It is forbidden to cut the short end of 300—500mm steel bar by machine. When installing steel bars in places where electric wires pass, care must be taken not to make the steel bars touch the wires.

(3)混凝土施工安全技术。在进行混凝土施工前,应仔细检查脚手架、工作台和马道是否绑扎牢固,如有空头板应及时搭好,脚手架应设保护栏杆。运输马道宽度:单行道应比手推车的宽度大400 mm以上;双行道应比两车宽度大700 mm以上。搅拌机、卷扬机、皮带运输机和振动器等接电要安全可靠,绝缘接地装置良好,并应进行试运转。搅拌台上操作人员应戴口罩;搬运水泥工人应戴口罩和手套,有风时戴好防风眼镜。搅拌机应由专人操作,中途发生故障时,应立即切断电源进行修理;运转时不得将铁锹伸入搅拌机筒内卸料;其机械转动外露装置应加保护罩。采用井字架和拔杆运输时,应设专人指挥;井字架上卸料人员不能将头或脚伸入井字架内,起吊时禁止在拔杆下站人。振动器操作人员 必须穿胶鞋;振动器必须设专门防护性接地导线,避免火线漏电发生危险,如发生故障应立即切断电源修理。夜间施工时应设足够的照明灯;深坑和潮湿地点施工时,应使用36 V以下低压安全照明。

习 题

1. 试述模板的作用和要求。
2. 试述梁模板和楼板模板的安装。
3. 跨度≥4 m的梁模板需要起拱吗?起拱多少?柱、墙模板需要起拱吗?
4. 试述组合钢模板的组成及各自的作用。
5. 何时需进行模板设计?模板及支架设计时应考虑哪些荷载?
6. 拆模的顺序如何?应注意哪些事项?
7. 试述振动器作业要点。
8. 混凝土养护的目的是什么?混凝土养护有哪些方法?

(3) Safety technology of concrete construction. Before concrete construction, the scaffold, worktable and catwalk shall be carefully checked to see if they are firmly bound. If there is any empty end plate, it shall be set up in time, and the scaffold shall be provided with protective railings. The width of the catwalk for transportation: the width of one-way road should be more than 400mm larger than that of wheelbarrow; the width of double lane should be more than 700mm larger than that of two vehicles. The mixer, winch, belt conveyor and vibrator should be connected safely and reliably, and the insulation grounding device should be good, and the test run should be carried out. The mixer, winch, belt conveyor and vibrator should be connected safely and reliably, and the insulation grounding device should be good, and the test run should be carried out. Operators on mixing platform should wear masks; cement handling workers should wear masks and gloves, and wear windproof glasses when there is wind. The mixer should be operated by a special person. When the fault occurs halfway, the power supply should be cut off immediately for repair. The shovel should not be discharged in the mixer barrel when running and the protective cover should be added to the mechanical rotation device. Special personnel shall be assigned to direct the transportation of the derrick and the derrick mast; the unloading personnel on the derrick shall not extend their head or feet into the derrick, and it is forbidden to stand under the derrick during lifting. The vibrator operator must wear rubber shoes; the vibrator must be equipped with special protective grounding wire to avoid the danger of leakage of live wire. In case of failure, the power supply shall be cut off immediately for repair. Sufficient lighting shall be provided during night construction; low voltage safety lighting below 36V shall be used for construction in deep pits and humid places.

Exercises

1. Try to describe the role and requirements of the formwork.
2. Try to describe the installation of beam formwork and floor formwork.
3. Does the beam formwork with span $\geqslant 4m$ need to be arched? How much is arched? Does the column and wall formwork need to be arched?
4. Try to describe the composition of combined steel formwork and their respective functions.
5. When will formwork design be required? What loads should be considered when designing formwork and scaffolds?
6. What is the order of form removal and what should be paid attention to?
7. Try to describe the key points of vibrator operation.
8. What is the purpose of concrete curing? What are the concrete curing methods?

第6章 预应力混凝土结构工程施工

6.1 预应力混凝土概述

6.1.1 预应力混凝土的基本概念

由于混凝土的抗拉性能很差,使钢筋混凝土存在两个无法解决的问题:一是在使用荷载作用下构件通常是带裂缝工作的;二是从保证结构耐久性出发,必须限制裂缝宽度。为了要满足变形和裂缝控制的要求,则需增大构件的截面尺寸和钢筋用量,这将导致自重过大,使钢筋混凝土结构用于大跨度或承受动力荷载的结构成为不可能或很不经济。

理论上讲,提高材料的强度可以提高构件的承载力,从而达到节省材料和减轻构件自重的目的。但在普通钢筋混凝土构件中,提高钢筋强度却很难收到预期的效果。这是因为对配置高强度钢筋的钢筋混凝土构件而言,承载力可能已不是控制条件,起控制作用的因素可能是裂缝宽度或构件的挠度。当钢筋应力达到 $500 \sim 1\,000 \text{ N/mm}^2$ 时,裂缝宽度将很大,无法满足使用要求。因而,钢筋混凝土结构中采用高强度钢筋是不能充分发挥其作用的,而提高混凝土强度等级对提高构件的抗裂性能和控制裂缝宽度的作用也极其有限。

混凝土抗拉强度及极限拉应变值都很低。其抗拉强度只有抗压强度的 $1/10 \sim 1/18$,极限拉应变仅为 $0.000\,1 \sim 0.000\,15$,即每米只能拉长 $0.1 \sim 0.15$ mm,超过后就会出现裂缝。而钢筋达到屈服强度时的应变却要大得多,为 $0.000\,5 \sim 0.001\,5$,如 HPB300 级钢筋就达 1×10^{-3}。对使用上不允许开裂的构件,受拉钢筋的应力只能用到 $20 \sim 30 \text{ N/mm}^2$,不能充分利用其强度。对于允许开裂的构件,当受拉钢筋应力达到 250 N/mm^2 时,裂缝宽度已达 $0.2 \sim 0.3$ mm。

Chapter 6 Prestressed Concrete Structure Construction

6.1 General Introduction to Prestressed Concrete

6.1.1 Basic Concepts of Prestressed Concrete

Concrete is strong in compression, but weak in tension; thus, there are two problems that cannot be solved in simple reinforced concrete: first, the members usually work with cracks under load; second, in order to ensure the durability of the structure, the crack width must be limited. In order to meet the requirements of deformation and crack control, it is necessary to increase the section size of the members and the number of reinforcing bars, which will result in excessive self-weight and will make it impossible or uneconomical for reinforced concrete structures to be used for large span or dynamic load-bearing constructions.

Theoretically, increasing the strength of materials can improve the bearing capacity of members and can achieve the purpose of saving materials and reducing the self-weight of the members. However, in simple reinforced concrete members, it is difficult to achieve the expected objectives by increasing the strength of the reinforcing bars. This is because the bearing capacity may not be a controlling factor for reinforced concrete members with high-strength reinforcing bars, and the controlling factor may be the width of the crack or the deflection of the members. When the stress of the reinforcing bar reaches 500—1000N/mm^2, the crack width will be large and it can't meet the requirements for use. Therefore, high-strength reinforcing bars in reinforced concrete members can't fully play their role and increasing the strength grade of concrete plays limited role in improving the crack resistance of the members and controlling the crack width.

The tensile strength and the variation of the ultimate tensile stress of concrete are low. The tensile strength is only 1/18 to 1/10 of the compressive strength and the ultimate tensile stress variation is only 0.0001 to 0.0015, which means one meter concrete can only be elongated 0.1—0.15mm, and if it exceeds the ultimate tensile stress variation, cracks will appear. However, the stress variation is much greater when the reinforcing bar reaches its yield strength, which is 0.005 to 0.0015. For example, the stress variation of reinforcing bar HPB 235 reaches 1×10^{-3}. For members which are not allowed to crack, the stress of the tensile reinforcing bar can only be up to 20—30N/mm^2, and in this situation, the strength of reinforcing bars cannot be efficiently utilized. For members which are allowed to crack, when the stress of the tensile reinforcing bar is up to 250N/mm^2, the crack width will be 0.2—0.3mm.

为了避免钢筋混凝土结构的裂缝过早出现,充分利用高强度钢筋及高强度混凝土,可以设法在结构构件承受使用荷载前,预先对受拉区的混凝土施加压力,使它产生预压应力来减小或抵消荷载所引起的混凝土拉应力,将结构构件的拉应力控制在较小范围,甚至处于受压状态,以推迟混凝土裂缝的出现和开展,从而提高构件的抗裂性能和刚度,这种预先施加预压应力的钢筋混凝土就叫作预应力混凝土。

6.1.2 预应力混凝土的类型

(1)预应力混凝土按预应力大小可分为:
①全预应力混凝土;
②部分预应力混凝土。
(2)预应力混凝土按施工方式不同可分为:
①预制预应力混凝土;
②现浇预应力混凝土;
③叠合预应力混凝土。
(3)预应力混凝土按预加应力的方法不同可分为:
①先张法预应力混凝土;
②后张法预应力混凝土。
(4)预应力混凝土按预应力筋与混凝土是否粘结又可分为:
①无粘结预应力混凝土;
②有粘结预应力混凝土。

6.1.3 预应力混凝土的特点

与普通钢筋混凝土相比,预应力混凝土具有以下特点:
①构件的抗裂性能较好。
②构件的刚度较大。由于预应力混凝土能延迟裂缝的出现和开展,并且受弯构件要产生反拱,因而可以减小受弯构件在荷载作用下的挠度。
③构件的耐久性较好。由于预应力混凝土能使构件不出现裂缝或减小裂缝宽度,因而可以减少大气或侵蚀性介质对钢筋的侵蚀,从而延长构件的使用期限。
④施工要求高。工序较多,施工较复杂,且需要张拉设备和锚具等设施。

Chapter 6 Prestressed Concrete Structure Construction

In order to avoid the early appearance of cracks in reinforced concrete structure and to make full use of high-strength of reinforcing bars and high-strength of concrete, some means can be taken to apply stress to reinforcing bars in the tensile zone before applying service loads to the members. In this way, the prestress is produced to reduce or counteract the tensile stress caused by imposed loads, and this can confine the tensile stress of members within specific limits, or put the members into compression, result in delaying the appearance and restraining the developing of cracks in concrete; thereby the crack resistance and stiffness of structural members can be improved. This kind of concrete with prestress is called prestressed concrete.

6.1.2 Different Types of Prestressed Concrete

(1) Classification by magnitude of prestress:
① Fully prestressed concrete;
② Partially prestressed concrete.
(2) Classification by Construction Methods:
① Precast prestressed concrete;
② Cast-in-place prestressed concrete;
③ Filigree concrete.
(3) Classification by prestressing methods:
① Pre-tensioned concrete;
② Post-tensioned concrete.
(4) Classification by bond performance of reinforcing bars:
① Unbonded prestressed concrete;
② Bonding prestressed concrete.

6.1.3 Features of Prestressed Concrete

Compared with ordinary reinforced concrete, the following are the features of prestressed concrete:

① The crack resistance is good.

② The stiffness is strong. Because prestressed concrete can delay the appearance and expansion of cracks and the flexural members exert an upward force, it can reduce the bending deflection under load.

③ The durability is better. Prestress concrete can prevent the appearance of cracks or reduce crack width and can prevent the erosion of reinforcing bars from atmosphere or corrosive materials, so the service life of members can be extended.

④ The requirements for construction are high. The construction procedure and techniques are complicated and stretching equipment and anchorage and other devices are required.

注:预应力混凝土不能提高构件的承载能力,也就是说,当截面和材料相同时,预应力混凝土与普通钢筋混凝土受弯构件的承载能力相同,与受拉区钢筋是否施加预应力无关。

6.2 先张法施工

6.2.1 先张法施工原理

先张法是在浇筑混凝土构件之前对预应力筋进行张拉,并将预应力筋临时锚固在台座或钢模上,然后浇筑混凝土构件,待混凝土达到一定强度(一般不低于混凝土强度标准值的75%),并使预应力筋与混凝土之间有足够粘结力时,放松预应力,预应力筋弹性回缩,借助于混凝土与预应力筋间的粘结,对混凝土产生压应力的一种施工方法。先张法也是为了提高钢筋混凝土构件的抗裂性能以及避免钢筋混凝土构件过早出现裂缝的一种方法,多用于预制构件厂生产定型的中小构件。

先张法生产有台座法和台模法两种。用台座法生产时,预应力筋的张拉、锚固、混凝土的浇筑和养护以及预应力筋的放松等工序都在台座上进行,预应力筋的张拉力由台座承受。台模法为机组流水、传送带生产方法,此时预应力筋的张拉力由钢模承受。目前,先张法生产构件主要采用长线台座法,一般台座长度在 100~150 m 之间,本节主要介绍台座法生产预应力混凝土构件的预应力施工方法。

6.2.2 先张法施工用机具设备

1. 台座

台座是先张法施工张拉和临时固定预应力筋的支撑结构,它承受预应力筋的全部张拉力,因此要求台座具有足够的强度、刚度和稳定性,以免因台座的变形、倾覆和滑移而引起预应力的损失,从而确保先张法生产的构件的质量。

台座按构造型式分为墩式台座和槽式台座两种。选用时,根据构件种类、张拉吨位和施工条件而定。

(1)墩式台座。以混凝土墩作为承力结构的台座称为墩式台座。墩式台座是由台墩、台面与横梁组成,如图 6.1 所示。

Chapter 6　Prestressed Concrete Structure Construction

Notes: Prestressed concrete cannot improve the bearing capacity of members; that is to say, when the section size and materials are the same, the bearing capacity of flexural members are the same in both prestressed concrete and ordinary reinforced concrete, so the bearing capacity of member has nothing to do with the prestressing of reinforcing bars in the tensile zone.

6.2　Pre-tensioning Construction

6.2.1　Pre-tensioning Construction Principles

Pre-tensioning is a construction method by which the members get compressive stresses. In this method, the tendons are stretched before casting the members. First, the tendons are anchored to the pre-tensioning bed or steel formwork temporarily, and then the concrete is cast; when the concrete gets sufficient strength(usually at least 75% of the its design strength) and the bond between the tendons and concrete reaches certain point, the tendons are cut and they rebound and the compressive stress is produced with the bond between tendon and concrete. The pre-tensioning is a method to improve the crack resistance and to avoid early appearance of cracks in reinforced concrete members. It is most commonly used in precast members factories to produce small and medium-sized shaped members.

There are two ways of providing platforms forming pre-tensioning concrete elements: stressing abutment and moulds. In the method of stressing abutment, the stretching and anchoring of tendons, the casting and curing of concrete and the releasing of tendons are all operated on stressing abutment, and the tension of the tendons is borne by the abutment. Stressing mould is a kind of operation in unit flow on conveyor, and the tension of the tendons is borne by steel formwork. At present, the pre-tensioning for members mainly uses the longline pre-tensioning mould with the length of 100—150m. This part mainly focuses on construction method of prestressed members in stressing abutment method.

6.2.2　Machinery and Equipment for Pre-tensioning Construction

1. Stressing Abutment

The stressing abutment is a support structure for stretching and temporary anchoring in pre-tensioning, and it bears the full tension of the tendon; therefore, the stressing abutment must have sufficient strength, stiffness and stability to avoid the loss of prestress caused by deformation, overturning, and creep of the abutment to ensure the quality of the members.

The abutment can be classified into pier abutment and groove abutment. The selection of the stressing abutment depends on the type of members, the tension force and the construction conditions.

(1) Pier abutment. Concrete pier is the bearing structure in pier abutment and it is composed of pier, platform surface and beam(Fig. 6.1).

(2)槽式台座。槽式台座由钢筋混凝土压杆、上下横梁及台面组成。由于其能承受较大张拉力和倾覆力矩,故常用于生产中小型吊车梁、屋架等大型构件,如图6.2所示。

2. 张拉夹具和机具

先张法的夹具分两类:一类是锚固夹具,其作用是将预应力筋固定在台座上;一类是张拉夹具,其作用是张拉时夹持预应力筋。预应力筋类型不同,采用的夹具形式也不同。锚固夹具与张拉夹具都是可以重复使用的工具。

先张法常采用的预应力筋有钢丝和钢筋两类,夹具也分为钢筋夹具和钢丝夹具。

(1)预应力钢丝的夹具和张拉机具。

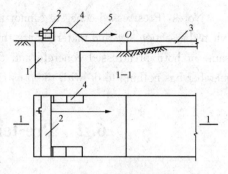

图 6.1 墩式台座

1—钢筋混凝土墩式台座;2—横梁;
3—混凝土台面;4—牛腿;5—预应力筋

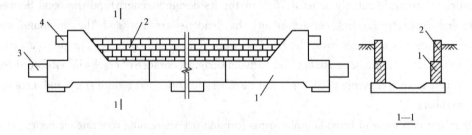

图 6.2 槽式台座

1—压杆;2—砖墙;3—下横梁;4—上横梁

①预应力钢丝夹具。钢丝的锚固一般可采用以下几种夹具或锚具:

a. 预应力钢丝锚固夹具。常用的钢丝锚固夹具有圆锥形夹具、楔形夹具两种形式,两者均属于锥销式体系。锚固时将锥塞或楔块击入套筒,借助摩擦阻力将钢筋锚固。如图6.3所示。

圆锥形夹具适用于锚固直径3～5 mm的冷拔低碳钢丝,也适用于锚固直径5 mm的碳素(刻痕)钢丝,如图6.3(a)、(b)所示。楔形夹具由锚板与楔块两部分组成,楔块的坡度为1/20～1/15,两侧面刻倒齿,如图6.5(c)所示。每个楔块可锚1～2根钢丝,适用于锚固直径为3～5 mm的冷拔低碳钢丝及碳素钢丝。

(2) Groove abutment. Groove abutment is composed of reinforcement concrete pressure bar, top beam, bottom beam and platform surface. Because it can bear large tension and overturning force moment, it is mainly used in manufacturing large members such as small and medium-sized crane beams, trusses and so on (Fig. 6.2).

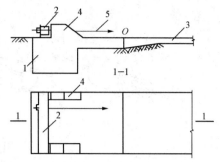

Fig. 6.1 Pier Abutment
1—reinforced concrete pier; 2—beam;
3—concrete platform surface;
4—fixing bracket; 5—tendon

2. Stretching Clamps and Machines

There are two kinds of stretching clamps in pre-tensioning construction. One is anchoring clamp, whose function is to anchor the tendons to the pre-tensioning bed, and the other is stretching clamps, whose function is to grip the tendon when stressing. The selection of clamps is based on the types of tendons. The anchoring clamps and the stretching clamps are reusable.

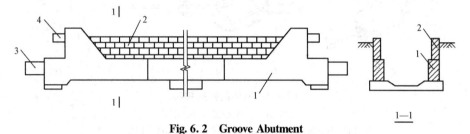

Fig. 6.2 Groove Abutment
1—pressure bar; 2—brick walls; 3—bottom beam; 4—top beam

The commonly used tendons in pre-tensioning construction are steel wires and tendons, so the stretching clamps are classified into steel wire clamps and tendon clamps.

(1) The clamps and equipment used for stressing steel wires.

① The clamps for stressing wires.

The following are the commonly used clamps:

a. Anchoring clamps.

The commonly used anchoring clamps are cone-shaped clamps and wedge-shaped clamps and they both belong to the tapered pin connection system. When anchoring, the chuck or the wedge block is hit into the sleeve to anchor the steel wire with the frictional resistance they generated (Fig. 6.3).

The cone-shaped clamps are suitable for anchoring the hard-drawn plain mild steel wire with the diameter of 3—5mm or carbon steel wire(indented), as shown in Fig. 6.3(a), (b). Wedge-shaped gripper is composed of anchor slab and wedge block. The slope of the wedge block is 1/20—1/15, and it has inverted teeth on two sides (left and right), as shown in Fig. 6.3(c). Every wedge block can anchor one or two steel wires. The steel wires can be hard-drawn plain mild steel wire with the diameter of 3—5mm and plain-carbon steel wire.

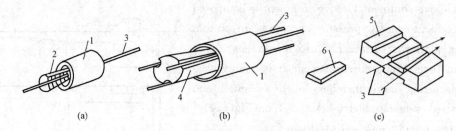

图 6.3　钢丝锚固夹具
(a)圆锥齿板式；(b)圆锥槽式；(c)楔形
1—套筒；2—齿板；3—钢丝；4—锥塞；5—锚板；6—楔块

b. 预应力钢丝张拉夹具。预应力钢丝常用的张拉夹具主要分为钳式、偏心式、楔形夹具三种，如图 6.4 所示。

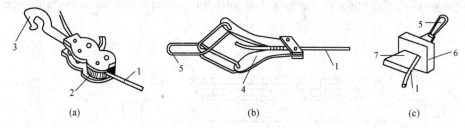

图 6.4　钢丝的张拉夹具
(a)钳式；(b)偏心式；(c)楔形
1—钢丝；2—钳齿；3—拉钩；4—偏心齿条；5—拉环；6—锚板；7—楔块

c. 预应力钢丝墩头锚具。钢丝的锚固除可采用锚固夹具外，还可以采用镦头锚具，墩头锚具采用冷墩形成。

②预应力钢丝的张拉机具。预应力钢丝的张拉机具主要有：卷扬机、电动螺杆张拉机、四横梁油压千斤顶等。选择张拉机具时，为了保证设备、人身安全和张拉力准确，张拉机具的张拉力应不小于预应力筋张拉力的 1.5 倍，张拉机具的行程应不小于预应力筋张拉伸长值的 1.1～1.3 倍。在使用张拉机具施工时，要求机具能准确控制钢丝的拉力，稳速增大拉力。

Chapter 6 Prestressed Concrete Structure Construction

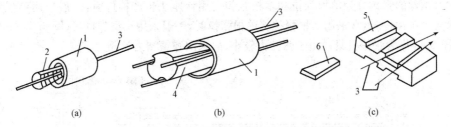

Fig. 6.3 Anchoring Clamps for Stressing Steel Wire

(a) cone-shaped gripper with indented plate; (b) cone-shaped gripper with grooved chuck;
(c) wedge-shaped gripper with wedge block

1—sleeve; 2—indented plate; 3—steel wire; 4—conical chuck;
5—anchor slab; 6—wedge block;

b. Stretching clamps for wires. The commonly used stretching clamps for wires are pincer type, eccentric type and wedge-shaped type, as shown in Fig. 6.4.

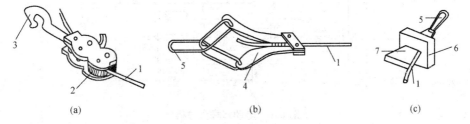

Fig. 6.4 Stretching Clamps for Wires

(a) pincer type; (b) eccentric type; (c) wedge-shaped type

1—steel wire; 2—clamp teeth; 3—hook; 4—eccentric rack;
5—pull ring; 6—anchor slab; 7—wedge block

c. The button-head anchorage for stressing steel wires. The steel wires can be anchored by anchoring clamps and button-head anchorages that are cast in cold header.

②Stretching equipment used for stressing steel wires.

The main stretching equipment for stressing wires are: crab winch, electrical stretching machine with twisted rod and hydraulic stretching jack with four beams. In order to ensure the safety of equipment and person and the accuracy of tension, the selection of the stretching equipment should consider the following factors: the tension of the stretching machine should be no less than 1.5 times of the tension of stressing steel wire and the working journey of the stressing machine should be no less than 1.1—1.3 times of the elongation of the prestressed steel wire; when using the stressing machine, make sure to control the tension exactly and increase the tension steadily.

· 439 ·

预应力钢丝的张拉分为单根张拉和多根张拉。用台座法生产构件时,一般采用单根张拉;用机组流水法生产构件时,常采用多根张拉。单根张拉时,一般采用小型卷扬机(图6.5)或电动螺杆张拉机(图6.6)作为张拉机具。由于张拉力较小,故采用弹簧测力计测力。

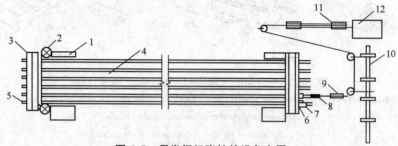

图 6.5 用卷扬机张拉的设备布置

1—台座;2—放松装置;3—横梁;4—钢筋;5—镦头;6—垫块;7—穿心式夹具;
8—张拉机具;9—弹簧测力计;10—固定梁;11—滑轮组;12—卷扬机

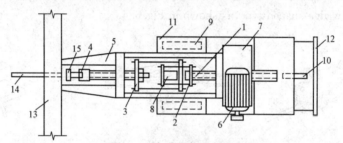

图 6.6 电动螺杆张拉机

1—螺杆;2、3—拉力架;4—张拉夹具;5—顶杆;6—电动机;7—齿轮减速器;
8—测力计;9.10—车轮;11—底盘;12—手把;13—横梁;14—钢筋;15—锚固夹具

多根张拉时,一般采用拉杆式千斤顶张拉(图6.7)。

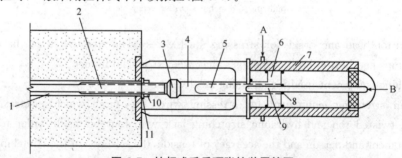

图 6.7 拉杆式千斤顶张拉装置简图

1—预留孔道;2—预应力筋;3—连接器;4—拉杆;5—副缸;
6—主缸活塞;7—主缸;8—油封;9—副缸活塞;10—锚固螺母;11—垫片

Chapter 6　Prestressed Concrete Structure Construction

The stretching of steel wires can be classified into individual wire stretching and group wire stretching. When members are cast on abutment, individual wire stretching is preferred, and when they are cast in unit flow, group wire stretching is preferred. When stretching the individual wire, we often use small sized stretching winch(Fig. 6. 5)or electrical stretching machine with twisted rod(Fig. 6. 6). Because the tension is small, the spring dynamometer can be adopted in measuring the tension.

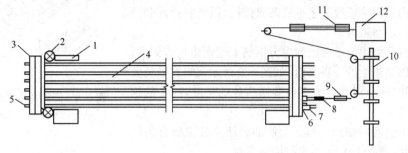

Fig. 6. 5　The Distributed Positions of Stretching with Crab Winch
1—abutment; 2—release equipment; 3—beam; 4—wire; 5—button-head anchorage;
6—washer; 7—pierced clamp; 8—stretching equipment; 9—spring dynamometer;
10—fixing beam; 11—pulley; 12—crab winch

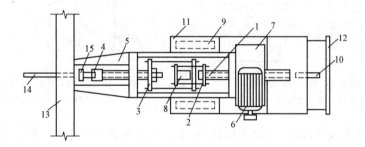

Fig. 6. 6　Electrical Stretching Machine with Twisted Rod
1—twisted rod; 2, 3—pulling rack; 4—stretching clamp; 5—crown bar; 6—motor;
7—reduction gear; 8—dynamometer; 9, 10—wheel; 11—chassis;
12—handle; 13—beam; 14—tendon; 15—anchoring clamp

When group wires are stretched at the same time, the drawbar jack is frequently used(Fig. 6. 7)

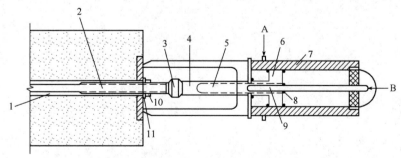

Fig. 6. 7　The Distributed Positions of Stretching with Drawbar Jack
1—preformed hole; 2—steel wire; 3—connector; 4—pulling rack; 5—secondary cylinder;
6—main cylinder piston; 7—main cylinder; 8—oil seal; 9—secondary cylinder piston;
10—anchoring nut; 11—washer

(2)预应力钢筋的夹具和张拉机具。

①预应力钢筋的锚固夹具。预应力钢筋锚固常用销片式夹具、镦头式夹具和螺丝端杆夹具等。

销片式夹具由套筒和锥形销片组成,如图6.8所示。销片可采用两片或三片式。套筒内壁锥角要与锥片的锥角吻合。销片的凹槽内采用热模锻工艺直接锻出齿纹,以增强销片和预应力钢筋之间的摩阻力。

镦头夹具是指用冷镦机把预应力钢筋末端镦粗加以固定,镦头卡在锚固垫板上的一种夹具,如图6.9所示。当直径较大时需压模加热,锻打成型。为了检验镦头处的强度,镦头的钢筋必须经过冷拉。

②预应力钢筋的张拉夹具。钢筋的张拉夹具主要分为压销式张拉夹具(图6.10)、穿心式张拉夹具等。

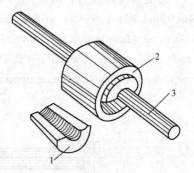

图6.8 圆锥两片式销片夹具
1—销片;2—套筒;3—预应力钢筋

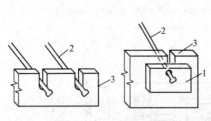

图6.9 镦头式锚具
1—垫片;2—镦头钢丝;3—承力板

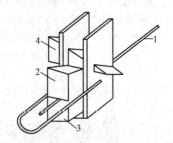

图6.10 压销式张拉夹具
1—钢筋;2—销片(楔);3—销片;4—压销

③预应力钢筋的张拉机具。预应力钢筋的张拉设备分单根张拉设备或多根成组张拉设备。

单根张拉时,一般采用小型卷扬机或电动螺杆张拉机作为张拉机具(图6.5、图6.6)。其原理和张拉钢丝相同,但张拉力可达300～600 kN。当单根钢筋长度不大时,也可采用拉杆式千斤顶或穿心式千斤顶张拉。如图6.7所示。

成组张拉需要较大张拉力的张拉设备,一般采用油压千斤顶进行张拉,如图6.11所示,这种装置由于千斤顶行程小,需多次回油,工效较低。

Chapter 6 Prestressed Concrete Structure Construction

(2) Clamps and stretching equipment for stressing tendons.

① Anchoring clamps for stressing tendons.

The commonly used anchoring clamps for tendons are dowel clamp, button-head clamp, screwed head clamp and so on. The dowel clamp is composed of wedge and sleeve, as shown in Fig. 6.8. There are two or three wedges and the bevel angle in the sleeve should comply with the bevel angel on the clip. Serration is forged in the groove with hot mould to increase the frictional resistance between the wedge and the tendon.

Upset clamp is to fix the set head of the tendon which is upset by cold header by the way of wedging the set head to anchoring plate, as shown in Fig. 6.9. When the diameter of the tendon is large, the mould should be heated to forge the set head. To ensure the strength of the set head, upset tendon should be drawn in cold.

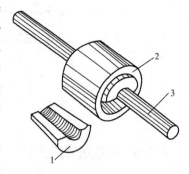

Fig. 6.8 Conical Dowel with Two Wedges
1—wedge; 2—sleeve; 3—tendon

② Stretching clamps for stressing tendons.

The stretching clamps for tendons are mainly compressive pin stretching clamps(Fig. 6.10) and pierced stretching clamps.

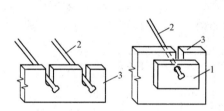

Fig. 6.9 Button-head Anchorage
1—attached plate; 2—button-head tendon;
3—stressing plate

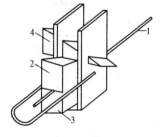

Fig. 6.10 Compressive Pin Stretching Clamp
1—tendon; 2—wedge;
3—wedge; 4—pin

③ Stretching machines for stressing tendons.

The stretching equipment for tendons are classified into individual stretching and group stretching. When stretching tendons individually, small sized crab winch or electrical machine with twisted rods are normally utilized, (Fig. 6.5, Fig. 6.6). The working principle of stretching is the same as stretching steel wires, but the tension can be up to 300—600kN. When the tendon is not too long, drawbar jack or pierced jack can be utilized in stretching, as shown in Fig. 6.7.

The group stretching of tendons need stretching machines with great tension, so hydraulic jack is usually utilized in stretching, as shown in Fig. 6.11. Because the short working journey and the repeated suction of oil, the work efficiency of hydraulic jack is low.

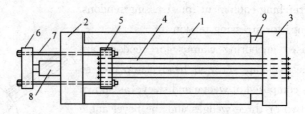

图 6.11 四横梁油压千斤顶张拉装置
1—台座;2—前横梁;3—后横梁;4—预应力筋;5、6—拉力架横梁

6.2.3 先张法施工工艺

先张法施工工艺流程如图 6.12 所示。

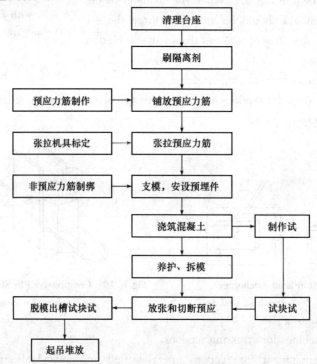

图 6.12 先张法施工工艺流程图

先张法施工过程示意图如图 6.13 所示。

Chapter 6 Prestressed Concrete Structure Construction

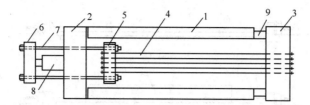

Fig. 6.11 Stretching Equipment of Hydraulic Jack with Four Beams
1—pre-tensioning bed; 2—front beam; 3—back beam; 4—tendon; 5,6—beam of pulling rack;
7—screwed rod; 8—hydraulic jack; 9—releasing equipment

6.2.3 Pre-tensioning Construction Process

The process flow of pre-tensioning construction is shown in Fig. 6.12:

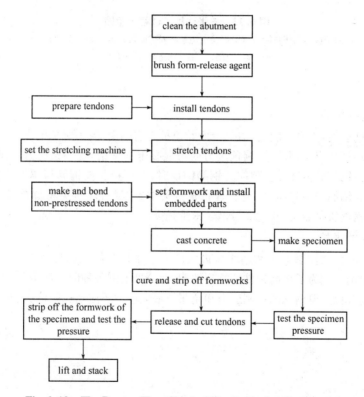

Fig. 6.12 The Process Flow Chart of Pre-tensioning Construction

The schematic diagram of the pre-tensioning construction procedure is shown in Fig. 6.13.

· 445 ·

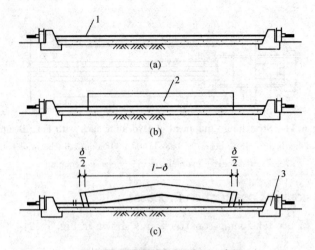

图 6.13 先张法施工过程示意图
(a)预应力筋张拉阶段;(b)混凝土浇筑和养护阶段;(c)预应力筋放张阶段
1—预应力筋;2—混凝土构件;3—台座

1. 预应力筋的铺设

在预应力筋铺设前,为便于脱模,需对台面及模板先刷隔离剂;为避免铺设预应力筋时因其自重下垂使隔离剂玷污预应力钢筋,影响预应力筋与混凝土的粘结,应在预应力筋设计位置下面先放置好垫块或定位钢筋后再铺设。预应力钢筋铺设时,钢筋接长或钢筋与螺杆的连接,可采用套筒双拼式连接器。钢筋采用焊接时,应合理布置接头位置,尽可能避免将焊接接头拉入构件中。钢丝接长可借助钢丝拼接器用 20~22 号钢丝密排扎牢。

2. 预应力筋的张拉

(1)预应力筋的张拉控制应力和超张拉最大应力。张拉时的控制应力直接影响预应力的效果,需按设计规定选用。为了提高构件的抗裂性能,部分抵消因各种因素产生的预应力损失,施工时,一般要进行超张拉。但钢筋的控制应力和超张拉最大应力不应超过限值(表 6.1)。

Chapter 6 Prestressed Concrete Structure Construction

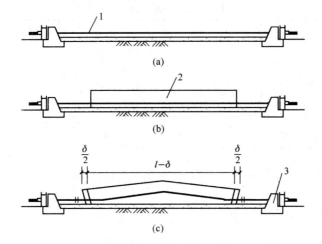

Fig. 6.13 The Schematic Diagram of the Pre-tensioning Construction
(a) the stretching of tendons; (b) the casting and curing of concrete;
(c) the releasing of tendons
1—tendon; 2—concrete member; 3—abutment

1. Installing the Tendons

Before installing the tendons, it's necessary to brush form-release agent onto the casting beds and the internal surfaces of formworks to prevent adhesion of the concrete to casting beds and allow the formworks easy to be stripped off. In order to avoid staining the tendons with form-release agent caused by drooping with self-weight and affect the bond between tendons and concrete, it's necessary to place block spacers under the designed location of the tendons before installing the tendons. When installing the tendons, sleeve double joint connector can be used in the connection of tendons or the connection of tendon and rod. When welding the tendons, the joints should be reasonably arranged to avoid pulling the welded head into the member. The connectors can be used to tie up the steel wire (grade 20—22) closely in connecting them together.

2. Stretching the Tendons

(1) Control stress and ultimate stress of tendons. When stretching, the control stress directly affects the effect of prestress, so the control stress should comply with the design requirement. To improve the crack resistance of the members and partly counteract the loss of prestress, tendons are generally subject to ultra stretching in construction. The control stress and ultimate stress in ultra stretching should not exceed the extreme value(Table 6.1).

表 6.1　张拉控制应力 σ_{con} 的允许值

项次	预应力筋种类	张拉方法	
		先张法	后张法
1	消除应力钢丝、钢绞线	$0.75f_{ptk}$	$0.75f_{ptk}$
2	冷轧带肋钢筋	$0.70f_{ptk}$	
3	精轧螺纹钢筋		$0.85f_{pyk}$

注：f_{ptk} 为预应力筋极限抗拉强度标准值；f_{pyk} 为预应力筋屈服强度标准值。

(2)张拉程序(表 6.2)。预应力筋张拉程序应按设计规定进行，若设计无规定时，可采取下列程序之一：

0→105%σ_{con} 持荷 2 min→σ_{con}(锚固)。

0→103%σ_{con}(锚固)。

σ_{con} 为预应力筋张拉控制应力。

①第一种张拉程序中，超张拉 5% 并持荷 2 min，其目的是在高应力状态下加速预应力松弛早期发展，以减少应力松弛引起的预应力损失。采用第一种张拉程序时，千斤顶回油至稍低于 σ_{con}，再进油至 σ_{con}，以建立准确的预应力值。

②第二种张拉程序，超张拉 3% 是为了弥补应力松弛引起的损失，根据"常温下钢筋松弛性能的试验研究"一次张拉 0→σ_{con}，比超张拉持荷再回到控制应力 0→1.05σ_{con}→σ_{con}，(持荷 2 min)应力松弛大 2%～3%，因此，一次张拉到 1.03σ_{con} 后锚固，是同样可以达到减少松弛效果的。这种张拉程序施工简单，一般常被采用。

以上两种张拉程序是等效的，可根据构件类型、预应力筋与锚具种类、张拉方法、施工速度选用。

表 6.2　先张法预应力筋张拉程序

预应力筋种类	张拉程序
钢筋	0→初应力→1.05σ_{con}(持荷 2 min)→0.9σ_{con}→σ_{con}(锚固)
钢丝、钢绞线	对于夹片式等具有自锚性能的锚具： 普通松弛力筋 0→初应力→1.03σ_{con}(锚固) 低松弛力筋 0→初应力→σ_{con}(持荷 2min 锚固)

Chapter 6 Prestressed Concrete Structure Construction

Table 6.1 The Allowable Value of Control Stress σ_{con}

Item	Types of Tendons	Stretching Methods	
		Pre-tensioning	Post-tensioning
1	stress-relieved wires, strands	$0.75 f_{ptk}$	$0.75 f_{ptk}$
2	cold-rolled ribbed bars	$0.70 f_{ptk}$	
3	prestressed screwed bars		$0.85 f_{pyk}$

Note: f_{ptk} is the standard value of ultimate tensile strength of tendon; f_{pyk} is the standard value of yield strength of tendon

(2) Stretching process (Table 6.2).

The stretching process for tendons should comply with the design requirement, and if there is no specific requirement in design, we can take either one from the following processes.

$0 \rightarrow 105\% \sigma_{con}$ kept at this level for 2 minutes $\rightarrow \sigma_{con}$ (anchoring).

$0 \rightarrow 103\% \sigma_{con}$ (anchoring).

σ_{con} is the control stress for stretching tendons.

① In the first process, the tension should be 5% more than the control stress and kept at that level for two minutes, so as to accelerate prestress slackness under high stress in the early stage and reduce the relaxation loss caused by stress slackness. In this process, let hydraulic jack low pressure oil return to allow the stress in tendons a bit lower than σ_{con}, followed by high pressure oil inflow up to σ_{con} for establishing the precise prestress in tendons.

② In the second process, the tension should be 3% more than the control stress to counteract the loss of prestress caused by relaxation. According to "Experimental study on the slackness properties of tendons at room temperature", the prestress relaxation loss in stretching from 0 to σ_{con} one time is 2%—3% larger than the relaxation loss in the process of over stretching and then returning to control stress as $0 \rightarrow 1.05\sigma_{con} \rightarrow \sigma_{con}$ (kept tension at that level for 2 minutes). Therefore, when tension is up to $1.03\sigma_{con}$ one time, anchor the tendons, and this will reduce the relaxation loss. The construction of this stretching process is easy, so it is frequently adopted.

③ The effects of the above two processes are almost the same, so the selection of the process is based on the types of members, the types of tendons and anchorages, the methods of stretching and the speed of construction.

Table 6.2 The Stretching Process of Tendons in Pre-tensioning

The Types of Tendons	Stretching Process
Tendon	$0 \rightarrow$ initial stress $\rightarrow 1.05\sigma_{con}$ (keep tension for 2 min) $\rightarrow 0.9\sigma_{con} \rightarrow \sigma_{con}$ (anchor it)
Steel wire, strands	For the clipped anchors and other anchors with self anchorage: Common relaxation tendons: $0 \rightarrow$ initial stress $\rightarrow 1.03\sigma_{con}$ (anchor it) Low relaxation tendons: $0 \rightarrow$ initial stress $\rightarrow \sigma_{con}$ (keep tension for 2 min and then anchor it)

(3)预应力值校核。预应力钢筋的张拉力,一般采用油压表控制、伸长值校核。预应力筋张拉锚固后实际建立的预应力值与工程设计规定检验值的相对允许偏差为±5%。预应力钢丝的张拉力,张拉锚固后采用钢丝应力测力仪检查钢丝的预应力值,伸长值只用作张拉操作中参考,其相对允许偏差为±5%。

3. 混凝土浇筑与养护阶段

预应力筋张拉完毕后即应浇筑混凝土。混凝土的浇筑应一次完成,不允许留设施工缝。混凝土的用水量和水泥用量必须严格控制,以减少混凝土由于收缩和徐变而引起的预应力损失。预应力混凝土构件浇筑时必须振捣密实(特别是在构件的端部),以保证预应力筋和混凝土之间的粘结力。预应力混凝土构件混凝土的强度等级一般不低于 C30;当采用高强钢丝、钢绞线、热处理钢筋做预应力筋时,混凝土的强度等级不宜低于 C40。采用平卧迭浇法制作预应力混凝土构件时,其下层构件混凝土的强度需达到 5 MPa 后,方可浇筑上层构件混凝土并应有隔离措施。

混凝土可采用自然养护或蒸汽养护。在台座上用蒸汽养护时,温度升高后,预应力筋膨胀而台座的长度并无变化,因而引起预应力筋应力减小,这就是温差引起的预应力损失。为了减少这种温差应力损失,应保证混凝土在达到一定强度之前,温差不能太大(一般不超过 20 ℃),故在台座上采用蒸汽养护时,其最高允许温度应根据设计要求的允许温差(张拉钢筋时的温度与台座温度的差)经计算确定。当混凝土强度养护至 7.5 MPa(粗钢筋)或 10 MPa(钢丝、钢绞线配筋)以上时,则可不受设计要求的温差限制,按一般构件的蒸汽养护规定进行。这种养护方法又称为二次升温养护法。在采用机组流水法用钢模制作、蒸汽养护时,由于钢模和预应力筋同样伸缩,所以不存在因温差而引起的预应力损失,可以采用一般加热养护制度。

4. 预应力筋的放张

当混凝土达到设计规定的放松强度之后,可在台座上放松受拉预应力筋(称为"放张"),对预制构件施加预应力。预应力筋放张过程是预应力的传递过程,是先张法构件能否获得良好质量的一个重要生产过程。应根据放张要求,确定合理的放张顺序、放张方法及相应的技术措施。

(3) Checking prestress value. When stretching the tendons, the oil pressure gauge is generally adopted to control the tension and check the elongation value. The allowable deviation between the actual prestress value after stretching and anchoring the tendon and the test value of tendon in design requirement is ±5%. That is to say, the tension of the tendon, the prestress value checked by dynamometer after stretching and anchoring and the elongation value are just for reference in stretching, and there is an allowable deviation of ±5%.

3. Casting and Curing of Concrete

Concrete should be cast after the tendons are stretched. The casting of concrete should be completed one time, and construction joints are not allowed. The water consumption in concrete and in cement must be strictly controlled to reduce the prestress loss caused by shrinkage and creep of concrete. When the members are cast, the concrete must be vibrated and compacted strongly (especially at the ends) to ensure the bond between the tendons and the concrete. Concrete strength grade for prestressed concrete members should be no lower than C30. When high-strength steel wire, steel strand, heat-treated tendons are used as tendons, the strength grade of concrete should be no lower than C40. When prestressed concrete members are constructed in forms of filigree concrete, the topping concrete shall be applied after the concrete strength of filigree decking has gained 5MPa while the isolation measures shall be adopted.

The curing of concrete can adopt natural curing or steam curing. When curing with steam on abutment, with the rise of temperature, the tendons expand but the length of the abutment does not change. Therefore, the prestress of the tendons is reduced, which is called prestress loss due to temperature difference. In order to reduce the prestress loss caused by temperature difference, the temperature difference should not be too large (generally not exceeding 20℃) before the concrete reaching certain strength. When curing with steam on the abutment, the ultimate allowable temperature shall be determined by calculation based on the allowable temperature difference (the difference of temperature in stretching the tendons and on the abutment) in the design requirements. When the concrete strength is up to above 7.5MPa (large-diameter tendons) or above 10MPa (steel wire and strand placement), it's no need to consider the temperature difference in the design requirements, and the curing can be carried out according to the steam curing specifications of general members. This method is also called the second heating curing method. When the concrete is cast on the steel formwork in the unit flow method and cured with steam, the steel formwork and the tendon expand or shrink in the same ratio, so there is no stress loss due to temperature difference, so general heating curing method can be adopted.

4. Releasing Prestressed Tendons

When the concrete reaches the design releasing strength, the tendons can be released on the abutment to apply prestress to the fabricated members. The process of releasing tendons is a transfer of prestress and it is a critical process for producing prestressed members with good quality. The releasing should comply with the design requirements, follow reasonable releasing orders and releasing methods and take corresponding technical measures.

(1)放张要求。放张预应力筋时,混凝土强度必须符合设计要求。

当设计无要求时,不得低于设计的混凝土强度标准值的75%。对于重叠生产的构件,要求最上一层构件的混凝土强度不低于设计强度标准值的75%时方可进行预应力筋的放张。

过早放张会引起较大的预应力损失或产生预应力筋滑动。预应力混凝土构件在预应力筋放张前要对混凝土试块进行试压,以确定混凝土的实际强度。

(2)放张顺序。

①对承受轴心预压的构件,所有预应力筋应同时放张。

②承受偏心预压的构件,应先放张预压力较小区域的预应力筋,再同时放张压力较大区域的预应力筋。

③当不能按上述规定放张时,应分阶段、对称、交错地放张。

(3)放张方法。对于预应力钢丝混凝土构件,分两种情况放张。配筋不多的预应力钢丝放张采用剪切、割断和熔断的方法自中间向两侧逐根进行,以减少回弹量,利于脱模。配筋较多的预应力钢丝放张应采用同时放张的方法,以防止最后的预应力钢丝因应力突然增大而断裂或使构件端部开裂。

对于预应力钢筋混凝土构件,放张应缓慢进行。预应力筋放张前,应拆除侧模,使构件自由收缩。配筋不多的预应力钢筋,可采用逐根加热熔断或借预先设置在钢筋锚固端的楔块等单根放张。配筋较多的预应力钢筋,所有钢筋应同时放张,放张可采用楔块或砂箱等装置进行缓慢放张。

现将几种常见的放张方法介绍如下。

①砂箱放张法。放张的装置在预应力筋张拉前放置在非张拉端。张拉前将砂箱活塞全部拉出,箱内装满干砂,让其顶着横梁。张拉时箱内砂被压实,承受横梁反力。放张预应力筋时,打开出砂口,让砂慢慢流出,活塞缩回,逐渐放松预应力筋,如图6.14所示。

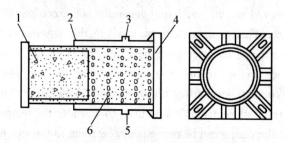

图6.14 砂箱装置构造图
1—活塞;2—钢套箱;3—进砂口;
4—钢套箱底板;5—出砂口;6—砂子

Chapter 6 Prestressed Concrete Structure Construction

(1) Releasing requirements.

When releasing the tendons, the strength of the concrete must meet the design requirements. If there is no design requirement for releasing, the concrete strength should not be less than 75% of the standard strength value. For the members in flapped casting, the tendons should not release until the concrete strength on the upper layer reaches at least 75% of the standard strength value.

Premature releasing of tendons will result in more prestress loss or slippage of tendons. Before releasing, it is necessary to make a compressive test to check the strength of the concrete.

(2) Releasing order.

① For the members bearing concentric prestress, all the tendons should be released at the same time.

② For the members bearing eccentric prestress, the tendons in the areas with less prestress should be released first and then tendons in the areas with more prestress can be released simultaneously.

③ When it is impossible to follow the above releasing principle, the releasing should be staged, symmetric and staggered.

(3) Releasing method.

For prestressed steel wire concrete members, there are two releasing methods. The prestressed steel wires with small amount of reinforcement are released by the method of shearing, cutting and fusing from the middle to the sides one by one to reduce the amount of rebound, and this is good for strip off the formworks. Prestressed steel wires with more reinforcement should be released at the same time to prevent the breaking of the last steel wire due to a sudden increase in stress or the cracking at the ends of the member.

For prestressed tendon concrete members, the releasing should carry out slowly. Before releasing the tendons, the side formwork should be removed to make the members shrink freely. The concrete members with a small number of tendons are released individually by fusing or by the wedge block preserved in the anchorage. Members with more tendons should be released at the same time, and the releasing can be carried out slowly with wedge block, sand box or other equipment.

Now, we will introduce several commonly used releasing methods:

①Sand box releasing. The releasing device is placed on the non-stretching end before the tendons are stretched. Before stretching, pull out the sandbox piston and fill the sandbox with dry sand to support the beam. When stretching, the sand inside the box is compacted and subjected to the beam reaction force. When releasing the tendons, open the sand outlet and let the sand flow out slowly, and the piston retracts, so the tendon released slowly, as shown in Fig. 6.14.

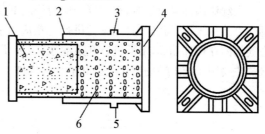

Fig. 6.14 Structural Diagram of Sand Box
1—piston; 2—steel box; 3—sand inlet
4—steel box bottom plate; 5—sand outlet; 6—sand

②千斤顶放张法。在台座固定端的承力架与横梁之间张拉前就安放两个千斤顶,待混凝土达到规定放松强度后,即可让两千斤顶同步回程,使拉紧的预应力筋慢慢回缩,将预应力筋放松。

③楔块放张法。楔块装置在台座与横梁之间,放张预应力筋时,旋转螺母使螺杆向上运动,带动楔块向上移动,钢块间距变小,横梁向台座方向移动,便可同时放张预应力筋(图6.15),楔块放张一般用于张拉力不大于300 kN的情况。

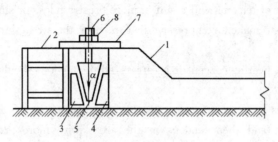

图 6.15　楔块放张
1—台座;2—横梁;3、4—钢块;5—钢楔块;
6—螺杆;7—承力板;8—螺母

④氧割法。直接用氧炔焰沿构件端部将锚固在台座上的预应力筋切断,这种放松预应力筋的方法对预应力冲击很大,易产生裂缝和造成大批预应力损失。氧割操作人员只准沿横向站立,严禁站在预应力筋上进行操作。

⑤手工法。即采用各种手工机具将预应力筋沿构件端部锯断或剪断,此法费工费时。

预应力筋全部放松后,可用"乙炔—氧气"烧割或用电弧切割外露钢筋,切割时要防止烧伤端部混凝土,切割后的外露端头,应用砂浆封闭或涂刷防蚀材料,防止生锈。

6.3　后张法施工

6.3.1　后张法施工原理

在制作构件或块体时,在放置预应力筋的部位留设孔道,待混凝土达到设计规定的强度后,将预应力筋穿入预留孔道内,用张拉机具将预应力筋张拉到规定的控制应力,然后借助锚具把预应力筋锚固在构件端部,最后进行孔道灌浆(也有不灌浆的),这种预加应力的方法称为后张法。后张法主要用于施工现场制作大型和重型的构件。预应力后张法构件生产示意图如图6.16所示。

②Jack releasing. Before stretching, place two jacks between the bearing frame and the beam at the anchoring end of the abutment. After the concrete reaches the specified releasing strength, the two jacks can be released simultaneously, so the stretched prestressed tendon retract lowly and is released finally.

③Wedge releasing. The wedge is placed between the abutment and the beam. When releasing the tendon, twist the nut and the bolt move upward to drive the wedge upward, so the distance between the steel blocks is narrowed and the beam slips to the abutment. This can release the tendons simultaneously(Fig. 6.15). Wedge releasing is generally adopted when the tension is less than 300 kN.

④Oxy-acetylene cutting. Directly cut the tendons anchored to the abutment along the end of the member with an oxygen lance to release the tendons. This method has a large impact on the prestress, so it is easy to cause cracks and great prestress loss. The operator of the oxygen lance must stand sideways and mustn't stand on the tendons.

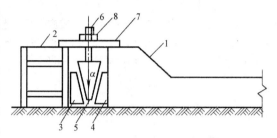

Fig. 6.15 Wedge Releasing
1—abutment; 2—beam; 3,4—steel block;
5—steel wedge; 6—screw bolt;
7—stressing plate; 8—screw nut

⑤ Manual releasing. In this method, the tendons are sawn or cut by different types of manual equipment along the end of the members, so it takes more time and labor.

After releasing all the tendons, the exposed tendons should be cut with "oxy-acetylene" or with electrical arc. When cutting, it is necessary to prevent the external concrete on the end from being burnt, and the exposed ends of the tendons should be sealed with mortar or brushed with anti-corrosion material to prevent rusting.

6.3 Post-tensioning Construction

6.3.1 Post-tensioning Construction Principles

When making member or block, the holes are preformed in the places where the tendons will be placed. After the concrete reaches the design strength, the tendons are inserted into the preformed holes and are pulled to the specified control stress with stretching equipment, and then the tendons are anchored to the end of the member with anchorages. Finally, the preformed holes are grouted (sometimes no grouting). This method of prestressing is called posttensioning method. It is mainly used to make large and heavy members on the construction site. The production procedure diagram of prestressed concrete in post-tensioning method is shown in Fig. 6.16.

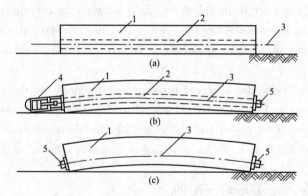

图 6.16 预应力混凝土后张法生产示意图
(a)制作混凝土构件;(b)后放钢筋;(c)锚固和孔道灌浆
1—混凝土构件;2—预留孔道;3—预应力筋;4—千斤顶;5—锚具

6.3.2 后张法施工用机具设备

1. 锚具

锚具是后张法结构构件中为保持预应力筋拉力并将其传递到混凝土上的永久性锚固装置。锚具按其锚固钢筋或钢丝数量分为单根粗钢筋锚具、钢筋束锚具和钢绞线束以及钢丝束锚具。

(1)单根粗钢筋预应力筋锚具。单根粗预应力钢筋在后张法施工时,根据构件长度和张拉工艺要求有一端张拉和两端同时张拉两种张拉方式。锚具与预应力钢筋的基本配套组合有三种:即两端张拉时,预应力筋两端均采用螺丝端杆锚具[图 6.17(a)];一端张拉一端固定时,张拉端采用螺丝端杆锚具,固定端采用帮条锚具[图 6.17(b)];一端张拉一端固定时,张拉端采用螺丝端杆锚具,固定端采用镦头锚具[图 6.17(c)]。

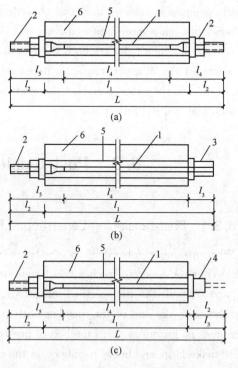

图 6.17 预应力筋与锚具连接图
1—预应力筋;2—螺丝端杆锚具;3—帮条锚具;
4—镦头锚具;5—孔道;6—混凝土构件

Chapter 6 Prestressed Concrete Structure Construction

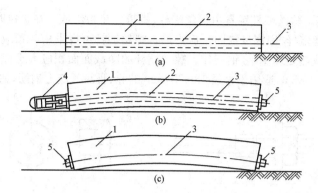

Fig. 6.16 The Schematic Diagram of the Post-tensioning Construction
(a) casting concrete member; (b) placing tendon; (c) anchoring and grouting
1—concrete member; 2—preformed hole; 3—tendon; 4—jack; 5—anchorage

6.3.2 Machinery and Equipment for Post-tensioning Construction

1. Anchorage

Anchorage is a permanent anchoring device in post-tensioning structural member to maintain the tension of the tendon and transfer it to the concrete. The anchorages can be classified by their hold numbers of wires or bar tendons into single large-diameter bar anchorage, multi-bar anchorage, multi-strand anchorage and wire-bundled anchorage.

(1) Single large-diameter bar anchorage. In post-tensioning construction, the single large-diameter bar can be stretched at one end or at both ends based on the length of members or the construction technology. There are three ways for the connection of anchorage and bar as follows: when both ends stretched, the ends of bar are anchored by screwed head anchorage[Fig. 6.17(a)]; when one end anchored and one end stretched, the stretched end is anchored by screwed head anchorage and the anchored end is anchored by buddle welded rods anchorage[Fig. 6.17(b)]; when one end anchored and the other end stretched, the stretched end is anchored by screwed head anchorage and the anchored end is anchored by button-head anchorage[Fig. 6.17(c)].

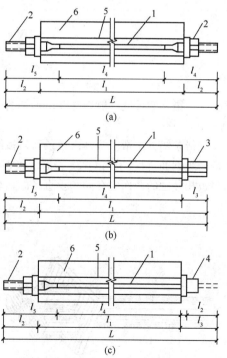

Fig. 6.17 Connection Between Tendon and Anchorage
1—tendon; 2—screwed head anchorage; 3—buddle welded rods anchorage; 4—button-head anchorage; 5—preformed hole; 6—concrete member

①螺丝端杆锚具。螺丝端杆锚具由螺丝端杆、螺母和垫板组成。在张拉时,将螺丝端杆与预应力筋对焊,张拉螺丝端杆,用螺母锚固预应力筋。螺丝端杆可以采用与预应力筋同级冷拉钢筋制作,也可采用冷拉或热处理45号钢制作。螺丝端杆的净截面面积应大于或等于预应力筋截面面积,型号有LM18～LM36,适用于直径18～36 mm的Ⅱ、Ⅲ级预应力钢筋,如图6.18所示。

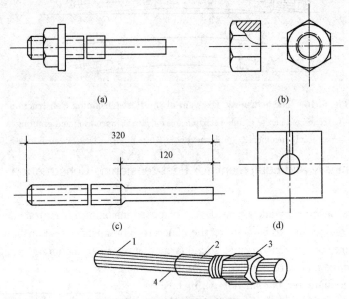

图 6.18 螺丝端杆锚具
a)螺丝端杆锚具;b)螺丝端杆;c)螺母;d)垫板
1—钢筋;2—螺丝端杆;3—螺母;4—焊接接头

锚具的长度一般为320 mm,当为一端张拉或预应力筋的长度较长时,螺杆的长度应增加30～50 mm。

②帮条锚具。帮条锚具由帮条和衬板组成,其构造如图6.19、图6.20所示。帮条采用与预应力筋同级钢筋,衬板采用Q235普通低碳钢板。帮条焊接应在冷拉前进行,三根帮条应互成120°,与衬板相接触的截面应在同一垂直平面,以免受力扭曲。

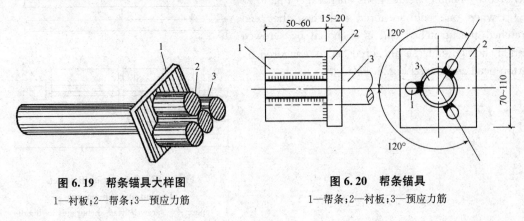

图 6.19 帮条锚具大样图
1—衬板;2—帮条;3—预应力筋

图 6.20 帮条锚具
1—帮条;2—衬板;3—预应力筋

①Screwed head anchorage.

Screwed head anchorage is composed of end bolt, nut and attached plate. When stretching, weld the end bolt and the tendon together, and then stretch the end bolt and anchor the tendon by the nut. The end bolt can be made of cold-drawn wire as the same grade of the tendon or be made of tendon grade 45 treated by cold or heat drawing. The section size of end bolt should be the same as or bigger than the section size of the tendon.

The types of screwed head anchorages are LM18 to LM36, which are suitable for anchoring the tendons graded I, II with diameter of 18 to 36mm, as shown in Fig. 6. 18.

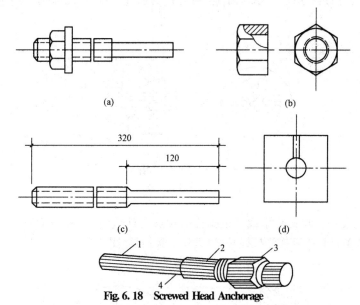

Fig. 6. 18 Screwed Head Anchorage
a)screwed head anchorage; b)end bolt; c)nut; d)attached plate
1—tendon; 2—screwed head bolt; 3—nut; 4—welding connector

The screwed head anchorage is normally 320 mm long. When the tendon is stretched on one end or when the tendon is too long, the screwed rod should be elongated by 30—50 mm.

②Bundle-welded rods anchorage. The bundle-welded rods anchorage is composed of several rods and a steel base plate, as shown in Fig. 6. 19, Fig. 6. 20. The rods are made of bars as the same grade of the tendon, and the steel plate is made of low carbon steel plate grated Q235. The rods should be welded together at pitch angle of 120° before stretching. The sections of the rods connected to the steel plate should be in the same vertical plane to avoid deflection under tension.

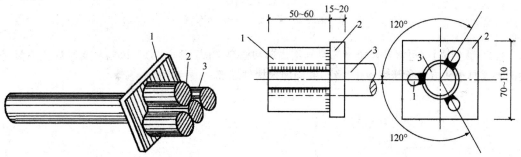

Fig. 6. 19 Detail Drawing of
Bundle-welded Rods Anchorage
1—steel base plate; 2—rod; 3—tendon

Fig. 6. 20 Bundle-welded Rods Anchorage
1—rod; 2—steel base plate; 3—tendon

③镦头锚具。当一端张拉时,采用镦头锚具可降低成本。镦头是直接在预应力筋端部热镦、冷镦或锻打成型。

(2)预应力钢筋束、钢绞线束及钢丝束锚具。预应力钢筋束、钢绞线束及钢丝束常用的锚具有 JM 型、XM 型、QM 型、KT-Z 型以及固定端用的镦头锚具等。

JM 型锚具由锚环和夹片组成。根据夹片数量和锚固钢筋类型、根数,有光 JM12-3~6、螺 JM12-3~6 和绞 JM12-5~6 等几种,如图 6.21 所示。JM 锚具的夹片属分体组合型,锚环为单孔,有方形和圆形两种。JM 型锚具利用楔块原理锚固多根预应力筋,既可作张拉端锚具,又可作固定端锚具和工具锚具。

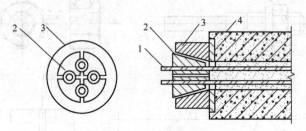

图 6.21 JM-12 锚具

1—预应力筋;2—夹片;3—锚环;4—垫板

XM 型锚具由孔锚板和夹片组成。根据锚固预应力筋数量,可分为单根 XM 型锚具和多根 XM 型锚具。XM 型锚具既可作张拉锚具,也可作工具锚具(图 6.22)。

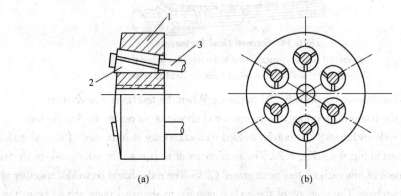

图 6.22 XM 型锚具

(a)装配;(b)锚板

1—锚板;2—夹片(三片);3—钢绞线

KT-Z 型锚具(可锻铸铁锥型锚具),由锚环和锚塞组成(图 6.23)。该锚具属半埋式锚具,使用时将锚具小头嵌入承压钢板中焊牢,共同埋入构件端部。

③ Button-head anchorage. When the tendon is stretched at one end, it is economical to use button-head anchorage. The button head is forged at the end of the tendon by cold upsetting, hot upsetting or forging.

(2) Multi-bar anchorage, multi-strand anchorage and wire-bundled anchorage.

The commonly used anchorages for multi-bar anchorage, multi-strand anchorage and wire-bundled anchorage are the types of JM, XM, QM, KT-Z and button-head anchorages on the anchored end, etc.

JM anchorage is composed of wedge plate and wedges. Based on the number of wedges and the type and number of tendons it anchors, it can be classified into Plain JM12-3—6, Screwed JM12-3—6 and Twisted JM12-3—6, as shown in Fig. 6.21. The wedges of the JM belong to spilt connection system, and the wedge plate is in the shape of square or circle. JM anchorage is to anchor a number of tendons with the principle of wedging. It can be used as stretching anchorage, fixing anchorage and tool anchorage.

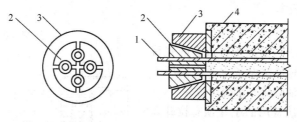

Fig. 6.21 JM 12 Anchorage
1—tendon; 2—wedge; 3—wedge plate; 4—attached plate

XM anchorage is composed of anchor plate and wedges, as shown in Fig. 6.22. Based on the number of tendons it anchors, it can be classified into individual tendon XM anchorage and group tendons XM anchorage. XM anchorage can be used as stretching or fixing anchorage.

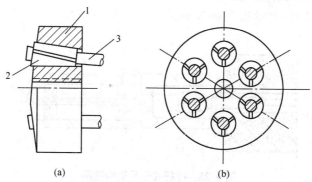

Fig. 6.22 XM Anchorage
(a) anchor set; (b) anchor plate
1—anchor plate; 2—wedge(three pieces); 3—steel strands

KT-Z anchorage, a tapered anchorage made of malleable cast iron, is composed of anchor ring and anchor male cone[Fig. 6.23]. This is half embedded anchorage. When used, the small end of it is inserted into the stressing steel plate and is welded to it permanently and then they are embedded in the end of the member.

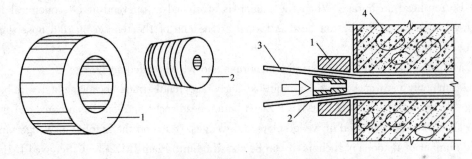

图 6.23 钢质锥形锚具安装示意图
1—锚环;2—锚塞;3—钢丝束;4—构件

固定端用的镦头锚具(图 6.24),由锚固板和带镦头的预应力筋组成。一般用以替代 KT-Z 锚具和 JM 型锚具,降低成本。

2. 张拉设备

预应力筋的张拉工作,必须配置成套的张拉机具设备。后张法用的张拉设备主要由液压千斤顶、高压油泵和外接油管等三部分组成。

(1)千斤顶。后张法常用的千斤顶有拉杆式千斤顶,也称拉伸机(代号 YL)、锥锚式千斤顶(代号 YZ)和穿心式千斤顶(代号 YC)三种。

拉杆式千斤顶如图 6.25 所示,主要用于张拉采用螺丝端杆锚具的粗钢筋、锥型螺杆锚具钢丝束和镦头锚具的钢丝束。常用的是 YL-60 型,其最大张拉力为 600 kN,张拉行程为 150 mm,活塞面积为 16 200 mm²,最大工作油压为 40 N/mm²。

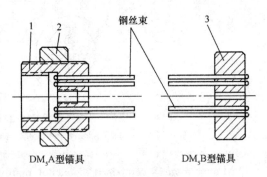

图 6.24 镦头锚具
1—锚环;2—螺母;3—锚板

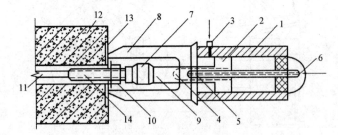

图 6.25 拉杆式千斤顶构造图
1—主油缸;2—主缸活塞;3—进油孔;4—回油缸;5—回油活塞;6—回油孔;7—连接器;8—传力架;9—拉杆;10—螺母;11—预应力筋;12—混凝土构件;13—预埋铁板;14—螺丝端杆

Chapter 6 Prestressed Concrete Structure Construction

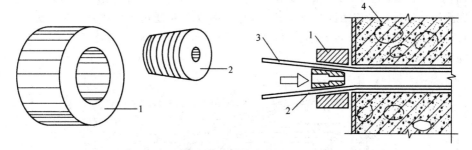

Fig. 6.23 Installation Diagram of Steel Tapered Anchorage
1—anchor ring; 2—anchor male cone; 3—steel strands; 4—member

The anchorage on the anchored end is composed of anchor plate and button-head tendon, as shown in Fig. 6.24. It is used to replace the KZ-T and JM anchorages to reduce cost.

2. Stretching Equipment

The stretching of tendons should be equipped with a set of stretching machines. The stretching equipment in post-tensioning are hydraulic jack, fuel injection pump and extension tubing.

(1) Jack.

The commonly used jacks in post-tensioning are drawbar jacks, which are classified into drawbar jack (code name YL), corn archor jack (code name YZ) and pierced jack (code name YC).

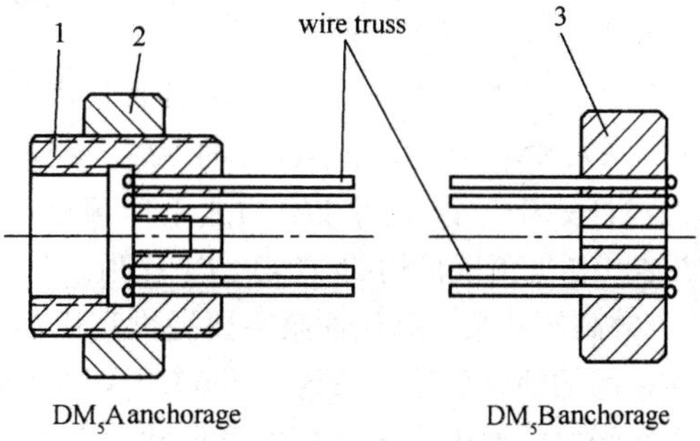

Fig. 6.24 Button-head Anchorage
1—anchor ring; 2—nut; 3—anchor plate

Drawbar jacks (Fig. 6.25) are mainly used for stretching large-diameter tendons anchored by screwed head anchorage, grouped steel wire anchored by conical bolt anchorage or button-head anchorage. The commonly used drawbar jack is the type of YL-60. The ultimate tension is 600kN; the jack stoke is 150mm; the section size of the piston is 16200mm², and the ultimate working oil pressure is 40N/mm².

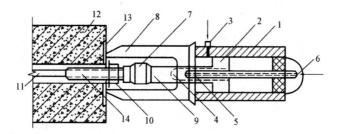

Fig. 6.25 Structure Diagram of Drawbar Jack
1—main cylinder; 2—main cylinder piston; 3—inlet hole; 4—oil returning cylinder;
5—oil returning piston; 6—oil returning hole; 7—connector; 8—load transfer frame; 9—pulling rack;
10—nut; 11—tendon; 12—concrete member; 13—preserved steel plate; 14—screwed head rod

锥锚式千斤顶如图 6.26 所示,主要用于张拉以 KT-Z 型锚具为张拉锚具的钢筋束和钢绞线束以及以钢质锥型锚具为张拉锚具的钢丝束。常用的有 YZ-36 型和 YZ-60 型。前者的最大张拉力为 360 KN,张拉行程为 300 mm,最大工作油压为 25.4 N/mm²。后者的最大张拉力为 600 kN,张拉行程为 150～300 mm,最大工作油压为 30 N/mm²。

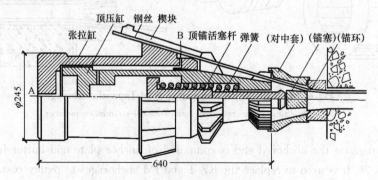

图 6.26 锥锚式千斤顶构造图

穿心式千斤顶是我国目前常用的张拉千斤顶,如图 6.27 所示,主要用于张拉 JM－12 型(图 6.28)、XM 型和 QM 型锚具的预应力钢丝束、钢筋束和钢绞线束。穿心式千斤顶加以改装,可作为拉杆式千斤顶使用和锥锚式千斤顶使用。YC 型千斤顶常用的有 YC60、YC20D、YCD120、YCD200 和无顶压机构的 YCQ 型千斤顶。

(2)高压油泵。高压油泵主要提供高压油,与千斤顶配套使用,是千斤顶的动力和操纵部分。目前常用的油泵型号有:ZB0.8/500、ZB0.6/630、ZB4/500 和 ZB10/500 等。ZB4/500 型油泵是预应力筋张拉的通用油泵。其外形尺寸为 745 mm×494 mm×1 052 mm,采用 10 号或 20 号机械油,油箱容量为 42 升,有 2 个出油嘴,每个出油嘴的额定排量为 2 升/秒。

Chapter 6   Prestressed Concrete Structure Construction

Cone anchor jacks (Fig. 6.26) are mainly used for stretching grouped bars and grouped strand anchored by KT-Z anchorage and steel wire bundle anchored by steel conical anchorage. The commonly used types of cone anchor jacks are YZ-36 and YZ-60. The ultimate tension of YZ-36 is 360kN; the jack stroke is 300 mm, and the ultimate working oil pressure is 25.4N/mm². The ultimate tension of YZ-60 is 600kN; the jack stroke is 150—300mm, and the ultimate working oil pressure is 30N/mm².

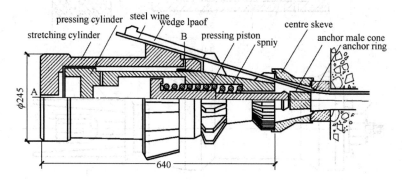

Fig. 6.26   Structure Diagram of Cone Anchor Jack

Pierced jack is a kind of most commonly used jack in China, as shown in Fig. 6.27. It is mainly used for stretching grouped bars, grouped strands and grouped steel wire anchored by the anchorage types JM-12(Fig. 6.28), XM and QM. After modified, the pierced jack can be used as drawbar jack and conical wedge jack. The YC style jack includes YC 60, YC 20D, YCD 120, YCD 200 and YCQ.

(2) Fuel injection pump. Fuel injection pump mainly provides high pressure oil, and it is used with jack as the power and operation part of it. The commonly used pump types are ZB0.8/500, ZB0.6/630, ZB4/500 and ZB10/500, etc. ZB4/500 is the universal pump for tendon stretching with the dimension of 745mm×494mm×1052mm. It consumes the hydraulic oil 10# or 20#, and the oil tank capacity is 42L with 2 nozzles. The nominal flow rate of the nozzle is 2L/second.

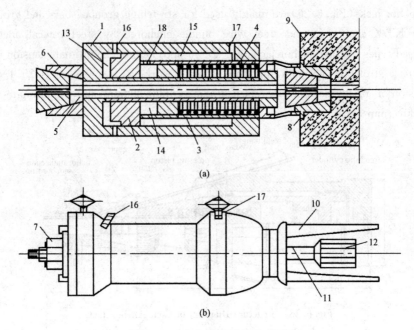

图 6.27 穿心式千斤顶构造图
(a)构造与工作原理;(b)加撑脚后的外貌
1—张拉油缸;2—顶压油缸;3—顶压活塞;4—弹簧;5—预应力筋;6—工具锚;
7—螺帽;8—锚环;9—构件;10—撑脚;11—张拉杆;12—连接器;13—张拉工作油室;
14—顶压工作油室;15—张拉回程油室;16—张拉罐油嘴;17—顶压罐油嘴;18—油孔

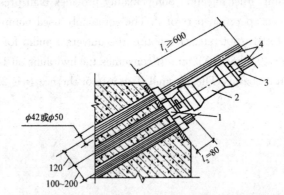

图 6.28 JM12 型锚具和 YC60 型千斤顶的安装示意图
1—工作锚;2—YC60 型千斤顶;3—工具锚;4—预应力筋束

Chapter 6 Prestressed Concrete Structure Construction

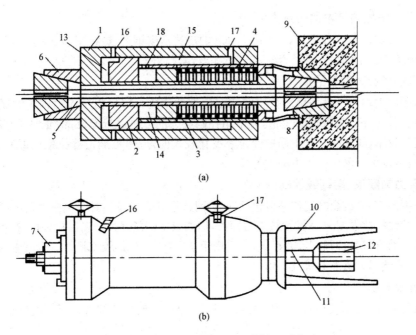

Fig. 6.27 Structure Diagram of Pierced Jack

1—stretching cylinder; 2—pressing cylinder; 3—pressing piston; 4—spring;
5—tendon; 6—anchor; 7—nut; 8—anchor ring; 9—member; 10—support brace;
11—stretching rod; 12—connector; 13—stretching oil cavity; 14—pressing oil cavity;
15—stretching retuning oil cavity; 16—nozzle of stretching cylinder;
17—nozzle of pressing cylinder; 18—oil-hole

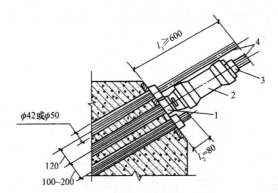

Fig. 6.28 Installation Diagram of JM12 Anchorage and YC 60 Jack

1—fixed anchor; 2—YC60 jack; 3—tool anchor; 4—grouped bars

6.3.3 后张拉施工中的预应力筋

后张法中常用的预应力筋有直径 3～5 mm 的冷拔低碳钢丝、冷拉钢筋、碳素钢丝、钢绞线、直径 6～10 mm 的热处理钢筋、直径 25～32 mm 精轧螺纹钢筋。

1. 单根粗预应力钢筋的制作

单根粗预应力钢筋一般采用冷拉钢筋或精轧螺纹钢筋。单根粗预应力钢筋的制作,包括配料、对焊、冷拉等工序。为保证其质量,宜采用控制应力的方法进行冷拉;钢筋配料时应根据钢筋的品种测定冷拉率,如果在一批钢筋中冷拉率变化较大时,应尽可能把冷拉率相近的钢筋对焊在一起进行冷拉,以保证钢筋冷拉力的均匀性。

2. 预应力钢筋束、钢绞线的制作

预应力钢筋束、钢绞线的制作一般包括开盘冷拉、下料和编束。对于预应力钢绞线束,在张拉前应采用钢绞线抗拉强度85%的预拉应力预拉,但如果出厂前经过低温回火处理,则可不必预拉。钢绞线在切断前,在切口两侧各 50 mm 处,应用铅丝绑扎,以免钢绞线松散。编束时,把钢筋或钢绞线理顺,用铅丝每 1 m 左右绑扎一道。

3. 钢丝束的制作

钢丝束的制作,随锚具形式不同制作方式也有差异,一般包括调直、下料、编束和安装锚具等工序。

用钢质锥形具锚固的钢丝束,其制作和下料长度计算基本上和钢筋束相同。

用镦头锚具锚固的钢丝束,其下料长度应力求精确,对直的或一般曲率的钢丝束,下料长度的相对误差要控制在 $L/5\,000$ 以内,并且不大于 5 mm。为此,要求钢丝在应力状态下切断下料,下料的控制应力为 $300\ \text{N/mm}^2$。钢丝下料长度,取决于锚具的类型及张拉的方式。

用锥形螺杆锚固的钢丝束,经过矫直的钢丝可以在非应力状态下料。

为防止钢丝扭结,必须进行编束。在平整场地上先把钢丝理顺平放,然后在其全长中每隔 1 m 左右用 22 号铅丝编成帘子状(图 6.29),再每隔 1 m 放一个按端杆直径制成的螺丝衬圈,并将编好的钢丝帘绕衬圈围成束绑扎牢固。

6.3.3 Tendons in Post-tensioning Construction

The commonly used tendons in post-tensioning construction are made from cold drawing low carbon steel wire, cold drawing tendon, carbon steel wire, steel strands, heat-treated tendon with the diameter of 6—10mm, refined-rolled helically ribbed bar with the diameters ranging from 25—32mm.

1. Preparation of single large-diameter PC bars

The single large-diameter PC bars are mainly made from cold drawing low carbon steel wire or refined-rolled helically ribbed bars. The procedure of producing single large-diameter PC bars includes material selection, welding, cold drawing and so on. To control the quality of single large-diameter PC bar, the cold drawing is subject to stress control, and in selecting the materials, it is necessary to test the cold drawing ratio based on the steel grade. If the cold drawing ratio varies dramatically in a group of bars, it is better to weld the bars with the similar cold drawing ratio to ensure the uniformity of the cold drawing force.

2. Preparation of Grouped Bars and Strand

The procedures of preparing grouped PC bars and stands are unfolding and cold drawing, cutting and braiding. For grouped stands, they should be prestressed with the tension that is equal to 85% of their tensile strength before stretching. If they are treated by low temperature tempering in factory, it is no need to take prestressing. Before cutting the stands, it is necessary to bind up the strands at 50mm on each side of the incision with lead wire to avoid loosing. In braiding, straighten the PC bars or steel strands, and bind up them with lead wire about every 1m.

3. Preparation for Grouped Wires

The production methods of grouped wires vary with the different types of anchorages. The procedures of producing grouped wires are straightening, cutting, braiding, anchoring and so on.

For grouped steel wires anchored by conical steel anchorage, the producing method and the calculation of the length of the steel wire is almost the same as that of grouped PC bars.

For grouped steel wire anchored by button-head anchorage, the calculation of the length of the steel wires should be as accurate as possible. For straight grouped wires or grouped wires with general curvature, the relative error tolerance of the length should be less than $L/5000$ and not more than 5mm. Therefore, the steel wire is required to cut off in the state of stress, and the control stress of the blanking is $300N/mm^2$. The length of the wires depends on the type of anchorage and the way of stretching.

For grouped steel wires anchored by conical screwed rod, the straightened steel wire can be cut in the non-stressed state.

To prevent the steel wires from kinking, they must be braided. First, straighten and lay them horizontally on the plane ground. Second, braid the lead wire type No. 22 into chain shape every 1m all along the steel wires, as shown in Fig. 6.29. Thirdly, install a core tube with the same diameter as the screw rod every 1m of the steel wire, and then wrap the steel chains around the steel liner into a bundle and tie them up tightly.

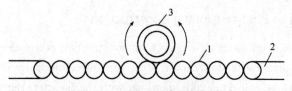

图 6.29 钢丝束编束示意图
1—钢丝；2—铅丝；3—衬圈

6.3.4 后张法施工工艺

后张法构件制作的施工工艺流程如图 6.30 所示。下面主要介绍孔道的留设、预应力筋的张拉和孔道灌浆三部分内容。

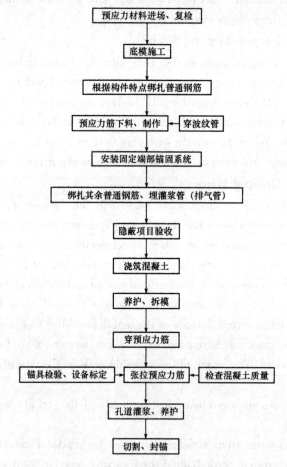

图 6.30 后张法生产工艺流程图

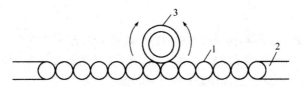

Fig. 6.29 Diagram of Braiding Grouped Wire
1—steel wire; 2—lead wire; 3—core tube

6.3.4 Post-tensioning Construction Procedure

The process flow chart of posttensioning is shown in Fig. 6.30. The following part is mainly about the three procedures: preforming holes, stretching tendons and grouting.

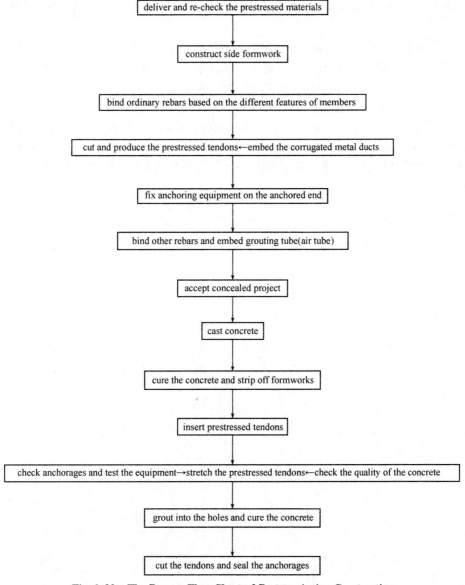

Fig. 6.30 The Process Flow Chart of Post-tensioning Construction

1. 孔道的留设

孔道形状有直线、曲线、折线，由设计方根据构件的受力性能，并参考张拉锚固体系来决定。孔道直径对于粗钢筋来说比预应力筋直径大 10~15 mm；对于钢丝束或钢绞线束比其大 5~10 mm。孔道间距不小于 50 mm；孔道至边缘净距不小于 40 mm。

孔道留设是后张法构件制作中的关键工作，孔道留设常用的方法有钢管抽芯法、胶管抽芯法和预埋波纹管法。前两者所用的钢管和胶管可重复使用，造价低廉但施工较烦琐；后者为一次性埋入铁皮管或波纹管，虽施工简单但造价较高。施工时，依据实际情况选用恰当的孔道留设方法。

(1) 钢管抽芯法。钢管抽芯法用于直线孔道的留设。构件的模板和非预应力钢筋安装完成后，把钢管预埋在需要留设孔道的部位。一般采用钢筋井字架固定钢管，接头处用铁皮套管连接，如图 6.31 所示。在混凝土浇筑和养护期间，每隔一段时间要慢慢转动钢管一次，防止钢管与混凝土粘结，待混凝土初凝后、终凝前抽出钢管在构件中形成孔道。为了保证预留孔道的质量，施工时应注意以下几点：

①钢管应平直、光滑，预埋前应除锈、刷油，安放位置要准确。钢管不直，则在转动和抽管时易将混凝土管壁挤裂。钢管位置的固定，一般采用钢筋井字架，钢筋井字架间距一般在 1~2 m，浇筑混凝土时，应防止振动器直接接触钢管，以免产生变形和位移。

②每根钢管长度一般不超过 15 m，以便于旋转和抽管，钢管两端应各伸出构件外 500 mm 左右。较长的构件留孔可采用两根钢管，中间用套管连接，如图 6.31 所示。白铁皮套管直径不宜太大，长度不宜太短。直径太大则在混凝土浇筑时，水泥砂浆容易流进套管中，使转管和抽管困难；套管太短则在钢管旋转时，钢管接头容易脱出套管，严重者可能导致水泥砂浆堵塞孔道。

Chapter 6 Prestressed Concrete Structure Construction

1. Making Preformed Holes

Based on the load bearing properties of the members and the stretching and anchoring characteristics of the tendons, the holes are designed to the shape of straight line, curve line or bending line. The diameter of the holes should be 10—15mm larger than the diameter of the large-diameter rebar and should be 5—10mm larger than that of grouped steel wire or steel strand. The distance between holes should be no less than 50mm and the distance between the edge of the hole and the edge of the member should be no less than 40mm.

The preforming of holes is the critical work in the fabrication of members in post-tensioning method. The commonly used methods for preforming holes are withdrawing embedded steel tubes, withdrawing embedded rubber hoses and embedding corrugated metal ducts. The steel tubes and the rubber hoses can be reused, so it is economical, but the construction process is quite complex. The embedded metal tubes or corrugated metal ducts can only be used once, so it is expensive, but the construction process is simple. The selection of methods in making preformed holes in construction is based on the actual situation.

(1) Holes preformed by withdrawing embedded steel tubes.

Holes preformed by withdrawing embedded steel tubes is suitable for preforming straight holes. After the installation of the member's formwork and the unstressed reinforcement, embed the steel tube in the place reserved for the hole. Generally, the steel tube is fixed by two-way steel frame and the joints are connected by galvanized iron casing, as shown in Fig. 6. 31. During the period of concrete casting and curing, the steel tube should be rotated slowly once in a while to prevent the steel tube from bonding with the concrete. After the initial setting and before the final setting of the concrete, the steel tube should be withdrawn to form a hole in the member. In order to ensure the quality of the preformed hole, pay attention to the following points in construction:

① The steel tube should be straight and smooth. Before it is embedded, rust removal and oil brushing should be carried out. The embedded position should be accurate. If the steel tube is not straight, it is easy to squeeze the concrete wall around the hole and cause crack in it. We mainly use two-way steel frame to fix the steel tube and the distance between the frames should be 1—2m. When casting the concrete, keep the vibrator away from the steel tube to avoid causing deformation and displacement.

② In order to rotate and withdraw the tube easily, generally the length of the steel tube is no more than 15m. Preformed holes for long members can take two steel tubes connected by iron casting, as shown in 6. 31. The iron casting should be neither too large nor too short. When it is large, the cement paste can flow into the casting easily, which makes it hard to rotate and withdraw the tube. When the casting is short, the connector of the steel tube is easy to detach from casting, which may even lead to the blocking of the air tube by the cement paste.

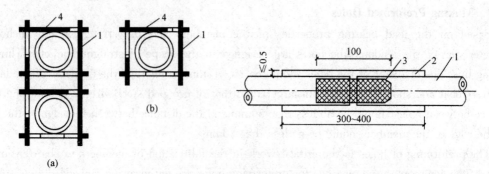

图 6.31 钢管连接方法
1—钢管；2—白铁皮套管；3—硬木塞；4—支架

③恰当掌握抽管时间。抽管过早，会造成塌孔；抽管太晚，混凝土与钢管粘结牢固，摩阻力增大，抽管困难，严重时可能抽不出钢管。具体的抽管时间与混凝土的性质、气温和养护条件有关。一般是掌握在混凝土初凝以后终凝以前，手指按压混凝土表面不粘浆又无明显手指印痕时，即可抽管。在常温下，抽管时间在混凝土浇筑后 3~6 h。

为了保证顺利抽管，混凝土浇筑顺序应合理安排，对预应力屋架来讲，若在气温较高的季节施工时，混凝土的浇筑应从上弦开始，然后自屋架两端方向中间一起浇筑下弦杆混凝土，保证在整榀屋架浇完混凝土后不太长的时间内抽管；反之，在气温较低的季节施工时，混凝土浇筑完成后需较长的时间才能抽管，则其浇筑顺序应从下弦开始，在上弦中间汇合，待下弦混凝土有较长时间养护后抽管。

④抽管顺序和方法。抽管顺序宜先上后下进行。抽管时，必须速度均匀，边抽边转，并与孔道保持在一条直线上。抽管后，应及时检查孔道，并做好孔道的清理工作，以免孔道中有水泥浆等从而增加以后穿筋的困难。

由于孔道灌浆的需要，每个构件在与孔道垂直的方向，应留设若干个灌浆孔和排气孔，孔距一般不大于 12 m，孔径 20 mm。留设灌浆孔或排气孔时，可用木塞或铁皮管成孔。

Chapter 6 Prestressed Concrete Structure Construction

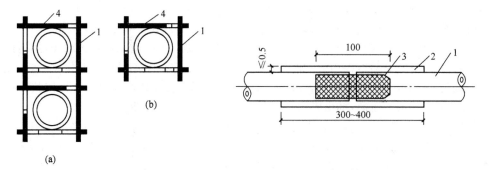

Fig. 6.31 The Connection of Steel Tubes
(a) two-way frame for double tubes; (b) two-way frame for single tube
1—steel tube; 2—galvanized iron casting; 3—wooden plug; 4—frame

③ Master the proper time to withdraw the steel tube. If the steel tube is drawn too early, the hole will collapse. When it is drawn too late, the bond between concrete and the steel tube is firm and the friction resistance increases, so it is hard to draw the steel tube and even worse, the steel tube may not be drawn. The proper time to withdraw the steel tube is related to the nature of the concrete, room temperature and curing condition. The proper time to withdraw the steel tube can be determined by pressing the concrete surface by your fingers after the initial setting and before the final setting. If your fingers don't get sticky slurry and the concrete surface doesn't get obvious finger marks, the steel tube can be drawn. After the initial setting and before the final setting of the concrete, the steel tube can be withdrawn. At room temperature, the proper time to withdraw the steel tube is 3—6 hours after concrete casting.

In order to guarantee successful withdrawing of the steel tube, the concrete casting sequence should be reasonably arranged. For example, in the case of prestressed roof truss, if it is constructed in the season with high room temperature, the concrete casting should start from the upper chord, and then cast the concrete of the bottom chord from both ends to the middle of the roof truss, so as to ensure that the steel tube can be withdrawn within a short time after the casting of the whole roof truss; on the contrary, if it is constructed in the season with low room temperature, it takes a long time for the steel tube to be withdrawn after the completion of concrete casting, so the concrete casting sequence should start from the bottom chord and converge in the middle of the upper chord, and the steel tube can be drawn out after the bottom chord concrete has been cured for a long time.

④ Master the withdrawing sequence and method. The withdrawing of the steel tube should be carried out from top to bottom. The withdrawing of the steel tube must be carried out in even speed with rotating and should be kept in a straight line with the hole. After withdrawing the steel tube, the hole should be checked in time and cleaned carefully, so as to avoid the difficulty of inserting tendons due to the presence of cement paste in the hole.

Due to the need of grouting, several grouting holes and air holes should be set up in the direction perpendicular to the preformed holes in every member. The distance between the holes is generally less than 12m, with a diameter of 20mm. When the grouting hole or air hole is reserved, wooden plug or iron tube can be used to form holes.

(2)胶管抽芯法。胶管抽芯法用于留设直线、曲线和折线孔道。胶管一般用 5～7 层夹布胶管或者预应力混凝土专用的钢丝网胶皮管。后者与钢管的使用方法相同,只不过混凝土浇筑后无须转动。前者在使用前,必须充水或充气。将胶管一端外表面削去 1～3 层胶皮或帆布,然后插入带有粗丝扣的一端密闭的钢管,再用铅丝把胶管和钢管连接处密缠牢固。胶管的另一端接上充水或充气用的阀门,采用同样的方法密封,如图 6.36 所示。抽管前,先放水或放气降压使胶管孔径变小,从而使胶管与混凝土脱离,抽出成孔。

胶管用钢管井字架固定,直线孔道每隔 400～500 mm 一道,曲线孔道应适当加密。对于充气或充水的胶管,在浇筑混凝土前,胶管中应充入压力为 0.6～0.8 MPa 的压缩空气或压力水,此时胶管直径可增大约 3 mm,当抽管时,放出压缩空气或压力水,胶管孔径缩水,与混凝土脱开,随即抽出胶管,形成孔道。胶管抽芯留孔与钢管抽芯相比,它弹性好,便于弯曲,因此,它不仅可留设直线孔道,也能留设曲线孔道。

用胶管留孔时,构件长度在 20～30 m 以内可用整接头。对于充气或充水胶管,其接头处应做好密封,防止漏气或漏水,接头形式如图 6.32 所示。胶管的抽管顺序,应先上后下,先曲后直。

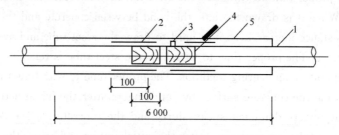

图 6.32 胶管连接方法
1—胶管;2—白铁皮套管;3—钉子;4—厚 1 mm 的钢管;5—硬木塞

Chapter 6 Prestressed Concrete Structure Construction

(2) Holes preformed by withdrawing embedded rubber hoses. Holes preformed by withdrawing embedded rubber hoses is suitable for preforming holes in the shape of straight line, curved line and bending line. The rubber hose is made from fabric hose with 5—7 canvas layers or made from specific fabric wire hose for prestressed concrete. The usage method of the fabric wire hose is the same as that of the steel tube except that it is no need to rotate the hose after concrete casting. For the usage of fabric hose with canvas layer, it should be filled with water or air before using. On one end of the hose, cut off 1—3 layers of rubber or canvas skins on the outer surface of it, then insert a short, one end plugged and the open one coarsely threaded steel pipe, and then use low carbon bright wire to tightly wrap the lapped joint between the hose and pipe. The other end of the hose is connected to valve filling water or air, and then seal it with the same method as the former one, as shown in Fig. 6.32. Before withdrawing the rubber hoses, discharge the air or water in the hose to decrease the pressure to make the hose smaller, so the rubber hose is separated from the concrete, and it can be withdrawn to form the hole.

The hose is practically positioned by utilizing the bars of grid pattern. For the straight hole, the distance between holes should be 400—500mm. For the curved holes, the distance between holes should be shorter. For the rubber hoses filled with water or air, before casting concrete, the compressive water or air with the pressure of 0.6—0.8MPa should be filled into the rubber hoses. The diameter of the rubber hose may expand about 3mm after filled with water or air. When it is time to withdraw the hose, discharge the compressive water or air to make the rubber hoses smaller and it is separated from the concrete, and then withdraw the hose to form the hole. Compared with withdrawing embedded steel tubes, the withdrawing of rubber hoses has the advantage of better elasticity, so it is easy to bend. Therefore, it is not only used in the preforming of straight holes, but also used in curved holes.

When the holes are preformed by withdrawing rubber hoses and the length of the member is between 20—30m, we can use integrated connector for the tubes. For the rubber hoses filled with water or air, the connection should be sealed to avoid leaking of water or air. The way of connection can be shown in Fig. 6.32.

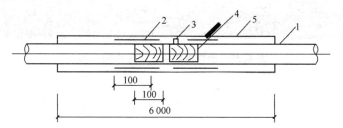

Fig. 6.32 The Connection Method of the Rubber Hoses
1—rubber hoses; 2—galvanized iron casting; 3—nail;
4—1 mm thick steel tube; 5—wooden plug

(3)预埋波纹管成孔。预埋管法是利用与孔道直径相同的波纹管埋在构件中,无须抽出。根据材质可分为金属波纹管和塑料波纹管。目前应用最广的是金属螺纹管。金属波旋管是用冷轧钢带或镀锌钢管在卷管机上压波后螺旋咬合而成。一般每根长度为4~6 m,当长度不足时,采用大一号的同型螺旋管连接,接头管长度应大于200 mm,用密封胶带或塑料热塑管封口。金属波纹管有镀锌和非镀锌之分,波纹有单波和双波之分,形状有圆形和扁形之分,如图6.33、图6.34所示。金属波纹管要求咬口牢固,波纹高度符合要求,弯曲的波纹管不渗水,环向刚度足够。目前在桥梁的长孔道中应用的比例在增加。(图6.35)

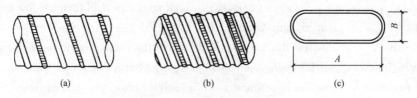

图6.33 金属波纹管示意图

(a)圆形单波纹;(b)圆形双波纹;(c)扁形

图6.34 金属波纹管

图6.35 金属波纹管在工程中的应用

(3) Holes preformed by embedded corrugated ducts.

Embedded corrugated ducts is a method to preform holes by embedding corrugated metal ducts with the same diameter as the duct. In this method, it's no need to withdraw the corrugated ducts. According to the difference of the materials, the corrugated ducts can be classified into corrugated metal ducts and corrugated plastic ducts. The most widely used is the corrugated metal ducts and it is formed by spirally bonding the corrugated cold-rolled steel belt or galvanized steel tube on a coiler. Generally, the length of the corrugated metal ducts is 4 to 6 m. When it is not long enough, it should be connected to other ducts with same type of spiral joint with larger diameter. The length of the duct joint should be more than 200 mm, and the duct joints should be sealed with sealing tape or plastic thermoplastic pipe. The corrugated metal ducts can be classified into galvanized and non-galvanized ones, and the corrugations can be classified into single wave and double wave. The shape of the corrugated metal ducts can be cylinder or ellipsoid, as shown in Fig. 6.33, Fig. 6.34. According to the requirement for the corrugated metal ducts, the connection should be firm; the height of the corrugations should meet the requirement; the water will not seep through the curved corrugated metal ducts; the hoop stiffness should be sufficient. Now, it is increasingly applied in the construction of long tunnels in bridges(Fig. 6.35).

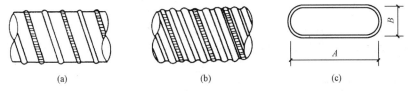

Fig. 6.33　Schematic Diagram of Corrugated Metal Ducts

(a) cylindrical tube corrugated duct with single wave;

(b) cylindrical tube corrugated duct with double wave;

(c) ellipsoidal tube

Fig. 6.34　Corrugated Metal Ducts

Fig. 6.35　The Application of Corrugated Metal Ducts in Engineering

塑料波纹管其刚度比金属波纹管大,如图 6.36 所示,其孔壁摩擦系数小,不易被混凝土振捣棒振瘪,价格比金属波纹管高。塑料波纹管的材料有高密度聚乙烯和聚丙烯。塑料波纹管的接管可采用热接,即用电加热板加热管口至热塑状态后,加压使管口对接在一起。

图 6.36　塑料波纹管在工程中的应用

预留孔道质量要求:
①预留孔道的规格、数量、位置和形状应符合设计要求;
②预留孔道的定位应牢固,浇筑混凝土时不应出现移位和变形;
③孔道应平顺,端部的预埋锚垫板应垂直于孔道中心线;
④成孔用管道应密封良好,接头应严密且不得漏浆;
⑤在曲线孔道的波峰部位应设置泌水管,灌浆孔与泌水管的孔径应能保证浆液畅通。排气孔不得遗漏或堵塞;
⑥曲线孔道控制点的竖向位置偏差应符合表 6.3 的规定。

表 6.3　曲线孔道控制点的竖向位置允许偏差(mm)

截面高(厚)度	$h \leqslant 300$	$300 \leqslant h \leqslant 1\ 500$	$h > 1\ 500$
允许偏差	±5	±10	±15

2. 穿预应力筋

穿入前,应检查预应力钢筋或预应力钢筋束的规格、总长是否符合要求。同时要考虑钢筋接头位置是否符合规范的规定。

The stiffness of the corrugated plastic ducts is larger than the corrugated metal ducts, as shown in Fig. 6.36 and the friction coefficient is small, so it is hard to be dented by the concrete vibrating rod, but the price is higher than corrugated metal ducts. The materials for corrugated plastic ducts are high density polyethylene and polypropylene. The connection of the corrugated plastic ducts can use thermal joint, which means heating the plastic duct end to thermoplastic state with the electric heating plate and connecting the ends with pressurizing them together.

Fig. 6.36 The Application of Corrugated Plastic Ducts in Engineering

Quality requirements for preforming holes:

①The specification, quantity, position and shape of the preformed holes shall meet the design requirements;

②The positioning of the preformed holes should be firm, and there should be no displacement and deformation when casting concrete;

③The inner surface of the preformed hole should be smooth and regular, and the embedded anchor plate at the end of the hole should be perpendicular to the center line of the hole;

④Tubes used for the reservation of the holes should be sealed well, and joints should be tight without leaking;

⑤Vents should be set at the peak of the curved hole. The grouting hole and the vent should be large enough to ensure the flowing of the cement paste, and the air hole should not be omitted or blocked.

⑥ The vertical position deviation of the control point in curved hole should meet the specification in Table 6.3.

Table 6.3 Allowable Vertical Position Deviation of Control Point in Curved Hole unit: mm

height(thickness) of section	$h \leqslant 300$	$300 \leqslant h \leqslant 1500$	$h > 1500$
allowable deviation	±5	±10	±15

2. Inserting Prestressed Tendons

Before inserting tendons, check whether the specifications and total length of the prestressed tendons or the grouped PC bars meet the design requirements. At the same time, consider whether the position of the joint meets the requirements of the design specification.

穿预应力筋时，带有端杆螺丝的预应力筋，应将丝扣保护好，以免损坏。钢筋束或钢丝束应将钢筋或钢丝顺序编号，并套上穿束器。先把钢筋或穿束器的引线由一端穿入孔道，由另一端穿出，然后逐渐将预应力钢筋或预应力钢丝束拉出到另一端（图 6.37）。穿入后应保护好预应力筋的螺纹。

图 6.37　某预制预应力桥梁构件穿筋后的现场图片

3. 预应力筋的张拉

预应力筋张拉前，应提供构件混凝土的强度试验报告。当混凝土的立方体强度满足设计要求时，方可施加预应力。如设计无要求，则不应低于强度等级的 75%。

(1)预应力筋的张拉方式（图 6.38、图 6.39）。对于预应力筋张拉应符合设计要求，当设计无具体要求时，应符合下列规定：当孔道为抽芯成型时，对曲线预应力筋和长度大于 24 m 的直线预应力筋，应在两端张拉，对于长度不大于 24 m 的直线预应力筋，可在一端张拉；当孔道为预埋波纹管时，对曲线预应力筋和长度大于 30 m 的直线预应力筋，宜在两端张拉，对于长度不大于 30 m 的直线预应力筋可在一端张拉。

两端张拉有下列两种方式：两端同时张拉、一端先张拉并锚固后再张拉另一端。前一种适用于锚具变形损失不大、且设备充足时；后一种主要在只有一台张拉设备或为了减少锚具变形损失时应用。

When inserting the tendons with threaded rod, it is important to protect the thread from destroying. The tendon or grouped wires shall be numbered sequentially and fitted in a barrel-gripped pull through device. First, penetrate the lead wire of the tendon or the pull through device into the hole from one end and draw it out from the other end, and then gradually draw the prestressed tendon or grouped PC bars to the other end(Fig. 6.37). The threads of the prestressed tendons should be protected well in the process.

Fig. 6.37 Inserting Tendons in Precast Prestressed Member

3. The Stretching of Prestressed Tendons

Before the prestressed tendons stretched, the strength test report of the concrete member shall be provided. Prestressing can only be applied when the cube strength of the concrete meets the design requirements. If there is no specific design requirement, the concrete strength should not be less than 75% of the standard strength value.

(1) The stretching method of prestressed tendons(Fig. 6.38, Fig. 6.39). The stretching of prestressed tendons should meet the design requirements. When there is no specific requirement in the design, the following requirements shall be met: when the hole is preformed by withdrawing tubes, the curved prestressed tendons and the straight prestressed tendons longer than 24m shall be stretched at both ends; for the straight prestressed tendon not longer than 24m, it can be stretched at one end. When the hole is preformed by embedding corrugated ducts, the curved prestressed tendon and the straight prestressed tendon longer than 30m should be stretched at both ends; for straight prestressed tendon not longer than 30m, it can be stretched at one end.

There are two methods for stretching at both ends: one is stretching both ends simultaneously and the other is stretching and anchoring one end, and then stretching the other end. The former method is suitable for the anchorage with less deformation loss and the construction with sufficient equipment. The latter is mainly used when there is only one stretching device or when it needs to reduce the deformation loss of the anchorage.

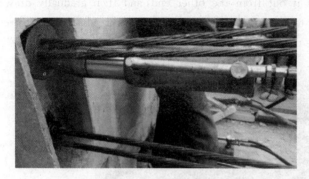 

图 6.38　后张法预应力筋张拉现场图片　　图 6.39　后张法预应力筋锚固现场图片

(2)预应力筋之间张拉顺序。当预应力筋数量多于设备数量时,不能做到同时张拉,需要分批张拉。对于同一批次内,应同时张拉,但张拉端要对称。对于不同批次预应力筋的张拉顺序,应遵守以下原则:对称张拉(即构件不产生扭转与侧弯);尽量减少张拉设备移动次数。

①当构件同一截面有多根预应力筋须分批张拉时,则应考虑混凝土弹性压缩对预应力筋的有效预应力值的影响。所以,先一批张拉的预应力筋,其张拉力应加上由于后几批预应力筋张拉时所产生的混凝土弹性压缩所造成的预应力损失值,使分批张拉完成后,每根预应力筋的张拉基本相等。

②平卧重叠浇筑的构件,宜先上后下逐层进行张拉。为了减少上下层之间因摩阻引起的预应力损失,可逐层加大张拉力。但底层张拉力不宜比顶层张拉力大 5%(钢丝、钢绞线、热处理钢筋)或 9%(冷拉 HRB335、HRB400、HRB500 级钢筋),且最大张拉应力:冷拉 HRB335、HRB400、HRB500 级钢不得超过屈服强度的 90%,钢丝、钢绞线不得超过屈服强度的 75%,热处理钢筋不得超过标准强度的 70%。张拉后的实际预应力值的偏差不得超过规定值的 5%。

预应力筋的张拉一般采用应力控制方法,但应校核预应力筋的伸长值。预应力筋的实际伸长值,宜在初应力约为 10% 时开始量测,如实际伸长值比计算伸长值大 10% 或小于 5%,应暂停张拉,在采取措施予以调整后,方可继续张拉。预应力筋在锚固过程中,应检查张拉端预应力筋的内缩量,内缩量的数值不得大于表 6.4 中的规定。

Chapter 6 Prestressed Concrete Structure Construction

Fig. 6.38 Stretching Prestressed Tendons in Post-tensioning

Fig. 6.39 Anchoring Prestressed Tendons in Post-tensioning

(2) Stretching sequence for the prestressed tendons. When the number of prestressed tendons is more than the number of equipment, the prestressed tendons cannot be stretched simultaneously, so they need to be stretched in batches. For the same batch, they shall be stretched at same time, but the stretching end shall be symmetrical. For the stretching sequence of prestressed tendons in different batches, the following principles should be obeyed: stretch them symmetrically (no torsion and lateral bending in the member); minimize the moving times of the stretching device.

①When there are multiple prestressed tendons in the same section to be stretched in batches, it's necessary to consider the effect of elastic compression of concrete on the effective prestressing value of the prestressed tendons. Therefore, the tensile force of the first batch of prestressed tendons should be added to the prestress loss caused by the elastic compression of the concrete produced by stretching of the subsequent batches of prestressed tendons, so that after the prestressed tendons are stretched in different batches, the tensile force of each prestressed tendon is almost the same.

②For member that is laid horizontally and cast in filigree, the stretching of tendons should carry out layer by layer from top to bottom. In order to reduce the prestress loss caused by the friction between the upper and lower layers, the tensile force can be increased layer by layer. However, the tensile strength of the bottom layer should not be 5% (for steel wire, steel strand, heat treated tendon) or 9% (for cold drawing tendon graded HRB335、HRB400、HRB500) more than the tensile strength of the top layer, and the maximum tensile stress shall not exceed 90% of the yield strength (for cold drawing tendon graded HRB335、HRB400、HRB500) or shall not exceed 75% of the yield strength (for steel wire and steel strand), or shall not exceed 70% of the standard strength (for heat treated tendon). The deviation of the actual prestress value after stretching shall not exceed 5% of the specified value.

The stretching is generally carried out with stress control, but the elongation of the prestressed tendons should be checked. The actual elongation value of the prestressed tendon should be measured when the initial stress is about 10% of the total stress. If the actual elongation value is 10% more or 5% less than the calculated elongation value, the stretching should be suspended. Only after taking effective measures to adjust the equipment, the stretching can be continued. During the anchoring process of the tendons, the internal shrinkage of the prestressed tendon at the stretching end shall be checked. The internal shrinkage value shall not be greater than that specified in Table 6.4.

表 6.4　锚固阶段张拉端预应力筋的内缩量允许值

锚具类别	内缩量允许值/mm
支承式锚具(镦头锚具、带有螺丝端杆的锚具等)	1
锥塞式锚具	5
夹片式锚具	5
每块后加的锚具垫板	1

注:1. 内缩量是指预应力筋锚固过程中,由于锚具零件之间和锚具与预应力筋之间的相对移动和局部塑性变形造成的回缩量;
　　2. 当设计对内缩允许值有专门规定时,可按设计规定确定。

4. 孔道灌浆

预应力筋锚固张拉后,应及时进行孔道灌浆(图 6.40、图 6.41),其主要作用是保护预应力筋,防止其锈蚀,增加结构的整体性和耐久性,提高结构的抗裂性能。并使预应力筋与结构混凝土形成整体。因此,孔道灌浆宜在预应力筋张拉锚固后尽早进行。

孔道灌浆用的砂浆,除应满足强度和粘结力要求外,应具有较大的流动性和较小的干缩性、泌水性。因此,孔道灌浆应采用强度等级不低于 42.5 级普通硅酸盐水泥配置的水泥浆;对空隙大的孔道,可采用水泥砂浆灌浆,水泥浆及砂浆强度均不应小于 20 MPa。水泥浆的水灰比宜在 0.4 左右,搅拌后 3 h 泌水率宜控制在 2% 左右,最大不得超过 3%,当需要增加孔道灌浆的密实性时,水泥浆中可掺入对预应力筋无腐蚀作用的外加剂。

灌浆前混凝土孔道应用压力水冲刷,确保孔道混凝土湿润和洁净。孔道灌浆可采用电动灰浆泵。水泥浆倒入灰浆泵时,必须过筛,以免水泥块或其他杂物进入泵体或孔道,影响灰浆泵正常运动或堵塞孔道。在孔道灌浆过程中,灰浆泵内应始终保持有一定的灰浆量,以免空气进入孔道而形成空腔。

灌浆应缓慢均匀地进行,不得中断,并排气通顺,在灌满孔道并封闭排气孔后,宜再继续加至 0.5~0.6 MPa,并稳定一定时间,再封闭灌浆孔,以确保孔道灌浆的密实性。对于用不加外加剂的水泥浆灌浆时,必要时可掌握时机,可采用二次灌浆法。

Chapter 6 Prestressed Concrete Structure Construction

Table 6.4 Allowable Internal Shrinkage Values of Prestressed Tendon at Stretching End in Anchoring

Anchorage Category	Allowable Internal Shrinkage Value/mm
button-head canchor, screwed head anchor etc.)	1
tapered plug anchorage	5
clipped anchor	5
additional anchor plate	1

Note: ①internal shrinkage value refers to the shrinkage value caused by relative movement or partial plastic deformation among anchoring devices, or between anchorages and prestressed tendons in anchoring the tendons.
②When there is specific requirement for the allowable internal shrinkage value in the design, we should follow the design requirement.

4. Grouting

After the tendons are anchored and stretched, the grouting should be carried out in time (Fig. 6.40, Fig. 6.41). The main function of grouting is to protect the prestressed tendon from rusting, so as to increase the integrity and durability of the structure, to improve the crack resistance of the structure, and to integrate the prestressed tendons and the concrete together. Therefore, the grouting should be carried out as soon as possible after the prestressed tendons are anchored and stretched.

Mortar used in grouting should meet the strength and adhesion requirements and also should have greater fluidity and less shrinkage and water loss. Therefore, cement paste for grouting should be made from ordinary portland cement graded 42.5 or higher; for large-diameter holes, cement mortar grouting can be adopted, and the strength grade of the cement paste and mortar should not be less than 20MPa. The ratio of the water and cement in cement paste should be about 0.4, and the water loss rate should be about 2% and should not exceed 3% in 3 hours after stirring. When it is necessary to increase the compactness of the grouting, the cement paste can be mixed with admixtures that have no corrosive effect on prestressed tendons.

Before grouting, the holes should be washed by pressure water to ensure that the hole is wet and clean. The electric mortar pump can be used for the grouting. When the cement paste is poured into the mortar pump, it must be screened to prevent the cement block or other debris from entering the pump or the hole, because they may influence the work of the mortar pump or they may cause blocking in the hole. During the grouting procedure, the mortar pump should always keep a certain amount of mortar to prevent air from entering the hole to form cavity.

The grouting should be carried out slowly and evenly without interruption, and the discharging of air should be smooth. After fully filling the preformed hole and sealing the grout vent tubes, give 0.5—0.6MPa additional pressure and keep it stable for an appropriate intervals, and then seal the grouting hole, so as to ensure the compactness of grouting in holes. For grouting with cement paste without adding admixture, if it is necessary, secondary grouting is adopted in the proper time.

灌浆顺序应先下后上，以避免上层孔道灌浆，而把下层孔道堵塞。曲线孔道灌浆，宜由低点压入水泥浆。至最高点排气孔中排出空气及溢出浓浆为止。为确保曲线孔道最高处或锚具端部灌浆密实，宜在曲线孔道最高处设立泌水竖管，使水泥浆下沉，泌水上升到泌水管内排出，并利用压入竖管内的水泥浆回流，以保证曲线孔道最高处和锚固区的灌浆密实。

图 6.40　后张法预应力预制构件灌浆

图 6.41　后张法预制预应力构件封端处理

The grouting should be carried out from bottom to top to avoid the blocking of the bottom hole when grouting the top hole. For grouting the curved hole, the cement paste is pressed into the holes from the low point and the grouting will not stop until all the air is discharged and cement paste flows out of the air hole at the highest point. In order to ensure that the highest point of the curved hole or the end of the anchor is grouted compactly, it is advisable to set up overflow pipe at the highest point of the curved hole to make the cement paste sink, to make bleeding water rise into and then discharged from the overflow pipe, and to take advantage of the backflow of the cement paste pressed into the stand tube, so that it can ensure the compactness of the grouting in the highest point and the anchoring zone.

Fig. 6.40 Grouting for Precast Member in Post-tensioning

Fig. 6.41 Sealing the Ends of Precast Member in Post-tensioning

6.4 无粘结预应力混凝土施工

6.4.1 无粘结预应力混凝土施工原理

后张法预应力混凝土中,预应力筋分为有粘结与无粘结两种。凡是预应力筋张拉后,通过灌浆或其他措施使预应力筋与混凝土产生粘结力,在使用荷载作用下,构件的预应力筋与混凝土不产生相对滑动的预应力筋称为有粘结,反之为无粘结。无粘结预应力施工方法是:在预应力筋表面刷涂料并包裹塑料布后,如同普通钢筋一样,先铺设在安装好的模板内,浇混凝土,待混凝土达设计要求强度后,进行张拉锚固。

无粘结预应力混凝土具有施工简单,不需预留孔道和灌浆,张拉阻力小,易于弯成曲线形状等优点。但预应力筋强度一般会降低10%~20%,对锚具要求高。

6.4.2 无粘结预应力混凝土施工用机具设备

1. 锚具

无粘结预应力筋的锚固体系宜采用夹片式锚具和镦头式锚具。

(1)张拉端采用夹片式锚具时,可采用下列做法:当锚具凸出混凝土表面时,其构造由锚环、夹片、承压板、螺旋筋组成,如图6.42(a)所示;当锚具凹进混凝土表面时,其构造由锚环、夹片、承压板、塑料塞、螺旋筋、钩螺丝和螺母组成,如图6.42(b)所示。

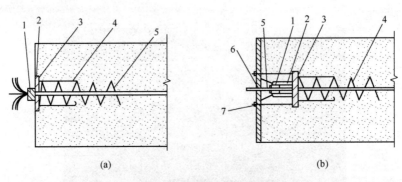

图6.42 夹片锚具系统张拉端构造
(a)夹片锚具凸出混凝土表面;(b)夹片锚具凹进混凝土表面
1—夹片;2—锚环;3—承压板;4—螺旋筋;
5—无粘结预应力筋;6—塑料塞;7—钩螺丝和螺母

Chapter 6 Prestressed Concrete Structure Construction

6.4 Unbonded Prestressed Concrete Construction

6.4.1 Unbonded Prestressed Concrete Construction Principles

For prestressed concrete in post-tensioning construction, the prestressed tendons can be classified into bonded ones and unbonded ones. After the tendons stretched, the bond is generated between prestressed tendons and concrete by grouting or other measures. When the member is under load, if there is no relative sliding between the tendons and the concrete, the tendons are bonded tendons; otherwise, they are called unbonded tendon. The following is the construction method for unbonded tendons: first, coat the prestressed tendon and wrap it with plastic sheet; second, place it in the pre-set formwork and cast the concrete, just as ordinary tendon; third, when the concrete reaches the design strength, stretch it and anchor it. The construction of unbonded prestressed concrete has many advantages. For example, the construction procedure is simple; it is no need to preform holes or carry out grouting; the tensile resistance is small; it is easy to bend to curved one, etc. However, the strength of unbonded prestressed tendons is generally reduced by 10% to 20% and the construction of unbonded prestressed tendons has high requirement for anchorages.

6.4.2 Equipment in Unbonded Prestressed Concrete

1. Anchorages

The preferably anchoring systems of unbonded prestressed tendons are clipped anchorages and button-head anchorages.

(1) When the stretching end is anchored by clipped anchorage, pay attention to the following requirements: when the anchorage protrudes from the concrete surface, it is mainly composed of anchor ring, wedges, stressing plate, spiral ribs, as shown in Fig. 6.42(a); when the anchorage is recessed into the concrete surface, it is mainly composed of anchor ring, wedges, stressing plate, plastic plug, spiral ribs, hook screw and a nut, as shown in Fig. 6.42(b).

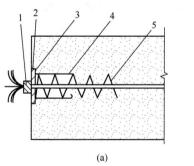 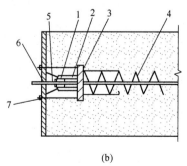

(a)　　　　　　　　　　　　　　　(b)

Fig. 6.42 Structure of the Stretching End in Clipped Anchorage

(a) clipped anchorage protruding from the concrete surface;
(b) clipped anchorage recessed into the concrete surface;
1—wedges; 2—anchor ring; 3—stressing plate; 4—spiral ribs;
5—unbonded prestressed tendon; 6—plastic plug; 7—hook screw and nut

(2)夹片式锚具系统的固定端必须埋设在板或梁的混凝土中,可采用下列做法:

挤压锚具的构造由挤压锚具、承压板和螺旋筋组成,如图6.43(a)所示。挤压锚具应将套筒等组装在钢绞线端部经专用设备挤压而成。

焊板夹片锚具的构造由夹片锚具、锚板与螺旋筋组成,如图6.43(b)所示。该锚具应预先用开口式双缸千斤顶以预应力筋张拉力的0.75倍预紧力将夹片锚具组装在预应力筋的端部。

压花锚具的构造由压花端及螺旋筋组成,如图6.43(c)所示。

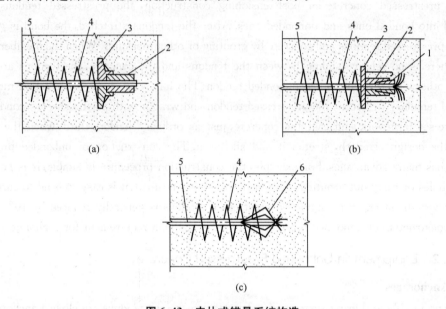

图6.43　夹片式锚具系统构造
(a)挤压锚具;(b)焊板夹片锚具;(c)压花锚具
1—夹片;2—锚环;3—承压板;4—螺旋筋;5—无粘结预应力筋;6—压花端

(3)镦头锚具系统的张拉端和固定端可采用下列做法:

张拉端的构造由锚环、螺母、承压板、塑料保护套和螺旋筋组成,如图6.44(a)所示。

固定端的构造由镦头锚板和螺旋筋组成组成,如图6.44(b)所示。

锚具的规格、质量应符合设计及应用技术规程的要求。其性能应符合现行国家标准《预应力筋用锚具、夹具和连接器》(GB/T 14370—2007)中相关规定。

(2) The anchored end of the clipped anchorage must be embedded into the concrete of the slab or beam with the following method: the extruded anchor is composed of extruded anchorage, stressing plate and the spiral ribs, as shown in Fig. 6.43(a). The extruded sleeve end anchor is produced by inserting a strand into a steel coupler or the like and extruding the both using a specified equipment.

Welded plate clipped anchorage is composed of clipped anchorage, anchoring plate and spiral ribs, as shown in Fig. 6.43(b). In this anchorage, the clipped anchorage is fixed to the end of the prestressed tendon with 75% of the tension stress of stretching the tendon by open double cylinder jack.

The embossed anchor is composed of the embossed end and spiral ribs, as shown in Fig. 6.43(c).

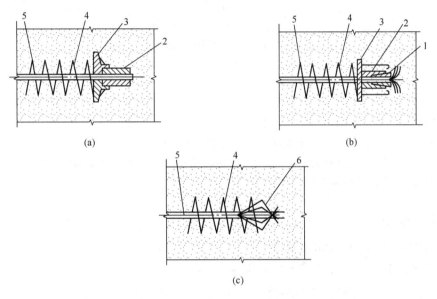

Fig. 6.43 The Structure of the Clipped Anchorage System
(a) extruded anchorage; (b) welded plate clipped anchorage;
(c) embossed anchorage
1—clip; 2—anchor ring; 3—stressing plate; 4—spiral rib;
5—unbonded prestressed tendon; 6—embossed end

(3) The structure of the stretching end and the anchoring end of the button-head anchorages are as follows:

The stretching end is composed of anchor ring, nut, stressing plate, plastic protective cover and spiral ribs, as shown in Fig. 6.44(a).

The anchoring end is composed of button-head anchoring plate and spiral ribs, as shown in Fig. 6.44(b).

The specification and quality of the anchorage shall meet the requirements of the design and application technical specification. Their properties should meet the current related specification in China, which is *Standards for Anchorages, Clamps and Connectors for Prestressed Tendons* (GB/T 14370—2007).

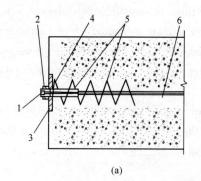 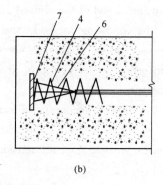

图 6.44 镦头锚具系统构造
(a)张拉端；(b)固定端
1—锚环；2—螺母；3—承压板；4—螺旋筋；
5—塑料保护套；6—无粘结预应力筋；7—镦头锚板

2. 张拉设备

无粘结预应力筋张拉时可选用 YC 系列的油压千斤顶,可在狭小场地及高空进行张拉。在梁板顶面或墙壁侧面的斜槽内张拉无黏结预应力筋时,宜采用变角张拉装置。变角张拉装置的关键部件是变角块,安装时注意块间的槽口搭接,保证变角轴线向结构外侧弯曲。张拉设备及仪表由专人使用和管理,定期维护和校验,校验期限不宜超过半年。

6.4.3 无粘结预应力混凝土施工用预应力筋

1. 无粘结预应力筋的组成及要求

无粘结预应力筋由单根钢绞线涂抹建筑油脂外包塑料套管组成(图 6.45)。无粘结预应力束的钢材,一般可选用 7 根直径 5 mm 的碳素钢丝束,7 根直径 5 mm 或 4 mm 的钢丝铰合而成的钢绞线(图 6.46)。表面涂料是为了长期保护预应力束不受腐蚀要求,一般选用 1 号或 2 号建筑油脂作为无粘结的表面涂层,要求表面涂层材料应满足下列要求:

(1)较好的化学稳定性,在 $-20\ ℃\sim +70\ ℃$ 内不裂,不变脆,不流淌。
(2)与周围混凝土、钢材等材料等不起化学作用。
(3)不会被腐蚀,具有不透水性,不吸湿性,良好润滑性,一般选用建筑专用油脂。

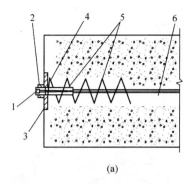 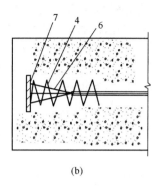

Fig. 6.44 The Structure of the Button-head Anchorage
1—anchor ring; 2—nut; 3—stressing plate; 4—spiral rib;
5—plastic protective cover; 6—unbonded prestressed tendon;
7—upset anchoring plate

2. Stretching Equipment

YC series hydraulic jacks can be utilized in stretching of unbonded prestressed tendons and they can be operated in narrow place and in high altitude. When the unbonded PT tendons are stretched in the skewed stressing pan (pocket) formed on the top of the beam or the side of the wall, it is preferred to be done by the adjustable angle aid device. The key part of the adjustable angle aid is the variable angle block. When the block is installed, pay attention to the overlapping of the notches of the blocks to ensure that the variable axis bends to the outside of the structure. The stretching equipment and meters are used and managed by designated employee(s) and should be regularly maintained and verified. The verification period should not exceed half a year.

6.4.3 Prestressed Tendons in Unbonded Prestressed Concrete

1. Composition and Requirements of Unbonded Prestressed Tendons

Unbonded prestressed tendon is composed of single steel strand coated with building grease and wrapped by outer plastic sleeve (Fig. 6.45). The unbonded prestressed grouped steel wires are mainly made of stranding seven carbon steel wire bundles each with the diameter of 5mm and seven steel wires(Fig. 6.46) with the diameter of 5mm or 4mm. The surface coating is used to protect the prestressed tendons from corrosion for a long time. Generally, the building grease #1 or #2 is used in the coating of unbonded tendons. The coating should meet the following requirements:

(1) It should have good chemical stability; it will not crack or Brittle or flow in the temperature ranging from -20 to $+70$ Celsius centigrade;

(2) It should not have chemical reaction with concrete, steel and other materials nearby;

(3) It will not be corroded; it should have the property of impermeability, humidity resistance and good lubricity; specified building grease is generally used in coating.

无粘结预应力筋护套材料应有足够韧性、耐磨、耐冲击性,对周围材料无侵蚀作用,在规定温度范围内,低温不脆化,高温化学性能稳定,防惨性、延伸度好。宜选用高密度聚乙烯,有可靠经验也可采用聚丙烯,但不宜采用聚氯乙烯。无粘结预应力筋护套材料应满足以下要求:

(1)在 $-20\ ℃\sim +70\ ℃$ 温度范围内,低温不脆化,高温化学性能稳定。
(2)有足够韧性、抗破损性。
(3)对周围材料无侵蚀作用。
(4)防水性、延伸度好。

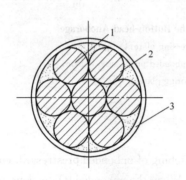

图 6.45 无粘结预应力筋截面图
1—钢绞线或钢丝束;2—油脂;3—塑料护套

图 6.46 无粘结预应力用钢绞线

2. 无黏结预应力筋的制作

无粘结预应力束制作通常采用缠纸工艺、挤压涂层工艺两种方法。缠纸工艺是在缠纸机上连续作业,完成编束、涂油、镦头、缠塑料布和切断等工序。挤压涂层工艺主要是钢丝通过涂油装置涂油,涂油后的钢丝束通过塑料挤压机成型塑料薄膜,再经冷却筒槽成型塑料套管。这种工艺效率高,质量好,设备性能稳定。

Chapter 6 Prestressed Concrete Structure Construction

The materials for the protective sleeves of the unbonded prestressed tendons should have the properties of good ductility, wearing resistance and impact resistance; it should not corrode the materials nearby; it should not brittle in low temperature and should have stable chemical stability in high temperature within specified temperature range; it should have good properties of impermeability and extension. High-density polyethylene is preferred. If you have reliable experience, you can also use polypropylene, but it is not suitable to use polyvinyl chloride. The materials for the protective sheath of the unbonded prestressed tendons should meet the following requirements:

(1) It should not brittle in low temperature and should have stable chemical stability in high temperature ranging from -20 to $+70$ Celsius.

(2) It should have the property of good ductility and wearing resistance.

(3) It should not corrode the materials nearby.

(4) It should have good property of good impermeability and extension.

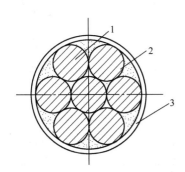

Fig. 6.45 Sectional Pan of Unbonded Prestressed Tendon
1—multistrand or wire bundle;
2—grease; 3—protective plastic sheath

Fig. 6.46 Unbonded Prestressed Steel Strand

2. The Production of Unbonded Prestressed Tendons

The unbonded prestressed tendons can be produced in two methods: the wrapping method and the extrusion coating method. The wrapping method is a continuous operation on the wrapping machine to complete the processes of bundling, oiling, upsetting, wrapping the plastic sheet and cutting. The extrusion coating method is mainly the application of oil to the steel wire through the oiling device. The oiled wire bundle is dressed with plastic film when it passes a plastic extruder, and then it goes through cooling cylinder trench to form plastic sleeve. This process has advantages of high efficiency, good quality and stable performance.

6.4.4 无粘结预应力混凝土施工工艺

无粘结预应力混凝土施工工艺流程如图 6.47 所示。

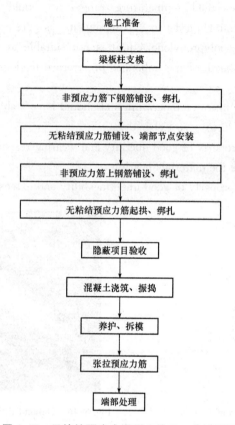

图 6.47 无粘结预应力混凝土施工工艺流程图

1. 无粘结预应力筋的铺放与定位

无粘结预应力筋在平板结构中一般为双向曲线配置。铺设前应根据双向钢丝束交点的标高差,绘制铺设顺序图,避免相互穿插。铺设前先根据设计图计算出各点标高、位置、反弯点位置、波峰位置,然后将马凳就位(马凳间距 1~2 m),再铺设钢丝束,一般是钢丝束波峰低的底层钢丝束先铺设,然后依次铺设波峰高的上层钢丝束。对波峰高度及水平位置进行调整,检查无误后用铅丝绑牢(图 6.48)。无粘结预应力曲线筋或折线筋的末端的切线应与承压板相垂直,曲线段的起始点至张拉锚固点应有不少于 300 mm 的直线段。

6.4.4 The Construction Process of Unbonded Prestressed Concrete

The construction process of unbonded prestressed concrete is shown as Fig. 6.47.

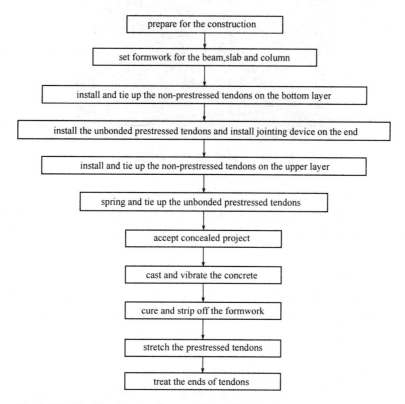

Fig. 6.47 **The Construction Process Chart of Unbonded Prestressed Concrete**

1. The Installation and Positioning of Unbonded Prestressed Tendons

The unbonded prestressed tendons in slab structure are generally installed in curved lines in two ways. Before installing, it is necessary to draw the installation process plan based on the elevation variation of the joints of the steel wire bundles in two ways to avoid crossing. Calculate the elevation and position of each joint, the position of the contraflexure point and the position of the peak based on the plan; put the support chair in place (every 1 to 2m) and install the steel wire bundle. The steel wire bundles with low peak height on the bottom lay are installed first, and then install the steel wire bundles with higher peak height on the upper layer. After installing, adjust the height and horizontal position of the peak, and tie up them tightly after checking (Fig. 6.48). The tangent of the end of the straight unbonded prestressed rib or the curved unbonded prestressed rib should be perpendicular to the bearing plate. The starting point of the curved segment to the anchoring point of stretching should have a straight line of not less than 300mm.

图 6.48　无粘结预应力铺设现场图片

张拉端固定:张拉端的承压板应用钉子固定在端模板上或用点焊固定在钢筋上。当张拉端采用凹入式作法时,可用塑料穴模或泡沫塑料、木块等形成凹口。混凝土浇筑时,严禁踏压撞碰预应力筋、支撑钢筋及端部预埋件,张拉端固定端混凝土必须振实,混凝土强度等级不低于 C30,梁不低于 C40。

2. 预应力筋的张拉

张拉前清理承压板面,检查承压板后面的混凝土质量,有缺陷应先修补处理。张拉时先张拉板,后张拉梁。板中的无粘结预应力筋,可依次张拉;梁中的无粘结预应力筋采用对称张拉。无粘结曲线预应力筋的长度超过 35 m 时,宜采取两端张拉。当无粘结预应力筋长度超过 70 m 时,宜采取分段张拉。

无粘结预应力的张拉(图 6.49)与普通后张法带有螺丝端杆锚具的有粘结预应力钢丝束张拉方法相似。张拉程序一般采用 $0 \rightarrow 103\% \sigma_{con}$ 进行锚固。由于无粘结预应力一般为曲线配筋,故应采用两端同时张拉。无粘结预应力束的张拉顺序,应根据其铺设顺序,先铺设的先张拉,后铺设的后张拉。无粘结预应力束一般长度大,有时又呈曲线形布置,如何减少其摩阻损失值是一个重要的问题。摩阻损失值,可用标准测力计或传感器等测力装置进行测定。施工时,为降低摩阻损失值,宜采用多次重复张拉工艺。

Chapter 6 Prestressed Concrete Structure Construction

Fig. 6.48 Installing the Unbonded Prestressed Tendon

The anchoring of the stretching end: The stressing plate on the stretching end should be anchored to the end formwork by nails or anchored to the tendons by tack welding. When the stretching end is anchored with recessing method, the recessing opening is formed by plastic cavity formwork, cellular plastic, wooden blocks and so on. When the concrete is cast, it is forbidden to stamp, press and collide the prestressed tendons, bearing bars and embedded end members. The concrete on the anchoring of the stretching end must be vibrated firmly, and the concrete strength grade should not be lower than C30, and the concrete strength grade for beams should not be lower than C40.

2. The Stretching of Prestressed Tendons

Before stretching, clean the surface of the stressing plate and check the quality of the concrete under the stressing plate, and if there is defect, make sure to repair it first. When stretching, stretch the prestressed tendons in the slab first and then stretch the prestressed tendons in the beam. The unbonded prestressed tendons in the slab should be stretched in sequence; the unbonded prestressed tendons in the beam should be stretched symmetrically. When the length of the unbonded curved prestressed tendon exceeds 35m, it should be stretched at both ends. When the length of the unbonded prestressed tendon exceeds 70m, segmental stretching should be adopted.

The stretching method of the unbonded prestressed tendons(Fig. 6.49) is similar with that of bonded prestressed wire strand with screwed head rod in posttensioning. In the stretching process, when the control stress increases from $0 \to 103\%$ σcon, the unbonded prestressed tendons are anchored. Since the unbonded prestressed tendon is generally equipped with curved tendons, it should be stretched at both ends. The stretching sequence of the unbonded prestressed tendons is based on the installing sequence of the tendons. That means, stretch the tendons installed first, and then stretch the tendons installed latter. The unbonded prestressed steel wire bundles are generally long and sometimes they are arranged in a curve, so how to reduce the value of friction loss is an important issue. The friction loss value can be measured by a force measuring device such as a standard dynamometer or a sensor. In order to reduce the friction loss value during construction, it is advisable to use multiple repeated stretching processes.

无粘结预应力筋张拉伸长值校核与有粘结预应力筋相同;对超长无粘结筋由于张拉初期的阻力大,初拉力以下的伸长值比常规推算伸长值小,应通过试验修正。

张拉时,无粘结筋的实际伸长值宜在初应力为张拉控制应力10%左右时开始测量,量测得到的伸长值,必须加上初应力以下的推算伸长值,并扣除混凝土构件在张拉过程中的弹性压缩值。张拉中,严防钢丝被拉断,要控制同一截面的断裂不得超过2%,最多只允许1根。

图 6.49　无粘结预应力张拉

3. 端部锚头处理

张拉后,应对锚固区进行保护,必须有严格密封措施,防止水汽进入腐蚀预应力筋。无粘结预应力筋锚固后的外露长度不小于 30 mm,外露多余的预应力筋宜用手提砂轮锯切割,不得用电弧焊切割。

端部处理方法有以下两种:

在孔道中注入油脂并加以封闭[图 6.50(a)]。

在孔道中注入环氧水泥砂浆,其抗压强度不低于 35 MPa[图 6.50(b)]。

Chapter 6 Prestressed Concrete Structure Construction

The method of checking the elongation value of unbonded prestressed tendons is the same as that of the bonded prestressed tendons; For the long unbonded prestressed tendons, due to the large resistance in the initial stage of the stretching, if the elongation value under the initial tensile force is smaller than the conventional estimated elongation value, it should be corrected by experiment.

When the unbonded prestressed tendon is stretched, the actual elongation value of it should be measured when the initial stress is about 10% of the tensile control stress. The measured elongation value must be added to the estimated elongation value below the initial stress, and must be deduced by the elastic compression value of the concrete member in the process of stretching. In stretching, the steel wire is strictly prevented from being broken. The breaking of the steel wire in the same section should not exceed 2%, and only one steel wire can be allowed to break at most.

Fig. 6.49 The Stretching of Unbonded Prestressed Tendon

3. The Treatment of the End Anchor

After stretching, the anchoring zone should be protected, so there must be secure sealing treatment to prevent moisture from entering and corroding the prestressed tendons. The exposed length of the unbonded prestressed tendon after anchoring is not less than 30mm. The protruding tails of tendons are cut off using a hand-held abrasive disk saw after anchored and arc cutting should not be used.

The following are the two methods for the treatment of the end anchor:

Injecting grease into the hole and sealing it, as shown in Fig. 6.50(a).

Injecting epoxy cement mortar into the hole and the compressive strength of the mortar is not less than 35MPa, [Fig. 6.50(b)].

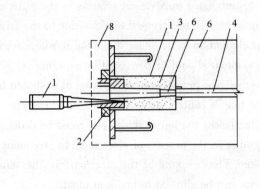 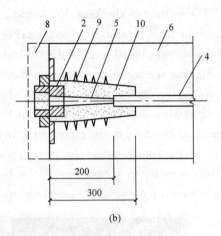

(a)   (b)

图 6.50 锚头端部处理方法

1—油枪；2—锚具；3—端部孔道；4—有涂层的无粘结预应力束；5—无涂层的端部钢丝；
6—构件；7—注入孔道的油脂；8—混凝土封闭；9—端部加固螺旋钢筋；10—环氧树脂水泥砂浆

对凹入式锚固区，锚具表面经上述处理后，再用微胀混凝土或低收缩防水砂浆密封。对凸出式锚固区，可采用外包钢筋混凝土圈梁封闭。锚固区混凝土或砂浆净保护层最小厚度：梁为 25 mm，板为 20 mm。

习 题

1. 什么是预应力混凝土？预应力混凝土的特点有哪些？
2. 先张法和后张法中常用的张拉机具有哪些？
3. 简述先张法的施工工艺。
4. 简述后张法的施工工艺。
5. 简述后张法施工中孔道的留设方式及施工方法。
6. 简述无粘结预应力混凝土的特点及施工工艺。

Chapter 6 Prestressed Concrete Structure Construction

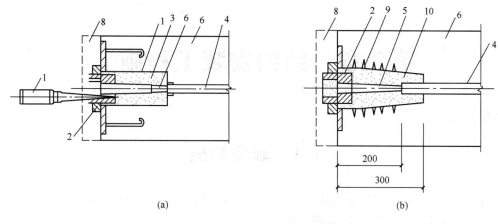

Fig. 6.50 The Treatment of the End Anchor
(a) injecting grease into the hole; (b) injecting epoxy cement mortar
1—oil gun; 2—anchorage; 3—the end of the hole;
4—unbonded prestressed steel truss with coating;
5—end steel wire without coating; 6—member;
7—grease injected into the hole; 8—concrete closure;
9—end reinforced spiral reinforcement;
10—epoxy resin cement mortar

For recessed anchoring zone, after the treatment of the anchor surface with the above method, it should be sealed with micro-expansion concrete or low-shrinkage waterproof mortar. For the protruding anchoring zone, it can be sealed or enclosed by RC tie beams. The minimum thickness of the net protection layer by concrete or mortar in anchoring zone should be 25mm for beam and 20 mm for slab.

Exercises

1. What is prestressed concrete? What are the characteristics of prestressed concrete?
2. What are the mainly used stretching equipment in pre-tensioning and post-tensioning construction?
3. Briefly describe the construction process of pre-tensioning.
4. Briefly describe the construction process of post-tensioning.
5. Briefly describe the methods in preforming holes and the construction process of these methods in post-tensioning construction.
6. Briefly describe the characteristics of unbonded prestressed concrete and its construction process.

第7章 结构安装工程施工

7.1 起重机械

结构安装工程中常用的起重机械,主要有桅杆式起重机(图7.1)、自行杆式起重机(履带式起重、汽车式起重机、轮胎式起重机)及塔式起重机。

7.1.1 自行杆式起重机

自行杆式起重机可分为:履带式起重机、轮胎式起重机、汽车起重机三种。

自行杆式起重机的优点是灵活性大,移动方便,能为整个建筑工地服务。起重机是一个独立的整体,一到现场即可投入使用,无须进行拼接等工作,施工起来更方便,只是稳定性稍差。

1. 履带式起重机

履带式起重机主要由动力装置、传动机构、行走机构(履带)、工作机构(起重杆、滑轮组、卷扬机)和平衡重等组成。如图7.2所示,是一种360°全回转的起重机,其操作灵活,行走方便,能负载行驶。缺点是稳定性较差。行走时对地面破坏较大,行走速度慢,在城市中和长距离转移时,需要拖车进行运输。目前它是结构吊装工程常用的机械之一。

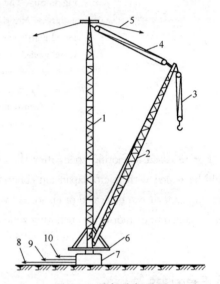

图7.1 桅杆式起重机

1—拔杆;2—起重臂;3—起重滑轮组;4—变幅滑轮组;
5—缆风绳;6—回转盘;7—底座;8—回转索;
9—起重索;10—变幅索

Chapter 7 Installation of Prefabricated Structural Members Construction

7.1 Lifting Machinery

The main lifting machinery frequently used in installation of prefabricated structural members are mast crane (Fig. 7.1), self-propelled boom crane (track-mounted crane, truck-mounted crane and tyre crane) and tower crane.

7.1.1 Self-propelled Boom Crane

The self-propelled boom crane can be classified into track-mounted crane, truck-mounted crane and tyre crane. They are flexible and they can move freely, so they can work for the whole construction site. The crane is an independent part, so it can be put into use as soon as it arrives at the site without splicing or other treatment. It is convenient in operation, but its stability is slightly poor.

1. The Track-mounted Crane

Track-mounted crane is mainly composed of the mobile power units, the transmission mechanism, the mobile mechanism (crawler), the lifting mechanism (boom, pulley and winch) and counter weight. As shown in Fig. 7.2, it is a full-rotating 360° crane. It is convenient in operation and it can walk freely on the site. Furthermore, it can move with load. However, its stability is poor and it damages the road surface. Because its speed is low, when it needs to be transported in a long distance in the city, it should be transported on the trailer. It is one of the most frequently used machinery in lifting installation construction.

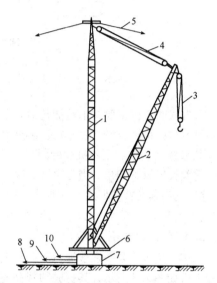

Fig. 7.1 Mast Crane
1—withdrawing pole; 2—boom;
3—lifting pulley; 4—variable pulley;
5—guy cable; 6—rotating platform;
7—base frame; 8—rotating rig;
9—lifting rig; 10—variable rig

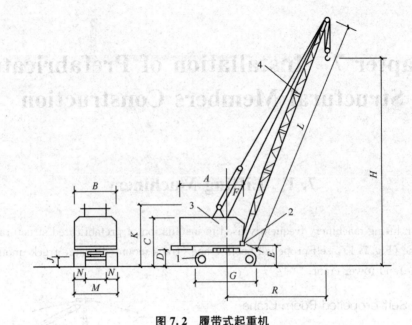

图 7.2　履带式起重机

1—行走装置；2—回转装置；3 机身；4—起重臂

A、B、…、N—外形尺寸符号；L—起重臂长度；H—起重高度；R—起重半径

　　履带式起重机主要技术性能包括 3 个主要参数：起重量 Q 起重半径 R 和起重高度 H。起重量一般不包括吊钩、滑轮组的重量，起重半径 R 是指起重机回转中心至吊钩的水平距离，起重高度 H 是指起重吊钩中心至停机面的距离。三个工作参数间存在着互相制约的关系，其取值大小取决于起重臂长度及其仰角。当起重臂长度一定时，随着仰角的增大，起重量和起重高度增加，而起重半径减小；当起重臂的仰角不变时，随着起重臂长度的增加，起重半径和起重高度增加，而起重量减小。

2. 汽车式起重机

　　汽车式起重机是装在普通汽车底盘上或特制汽车底盘上的一种起重机，也是一种自行式全回转起重机如图 7.3 所示。其行驶的驾驶室与起重操作室是分开的，其具有行驶速度高、机动性能好的特点。

　　但吊重时需要打支腿，因此不能负载行驶，也不适合在泥泞或松软的地面上工作。

Chapter 7   Installation of Prefabricated Structural Members Construction

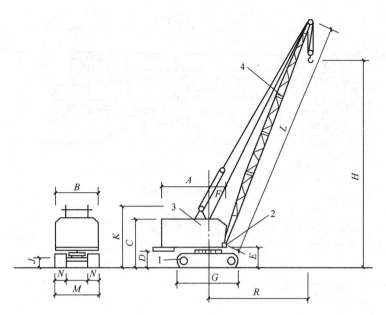

Fig. 7. 2   Track-mounted Crane

1—mobile mechanism; 2—rotating mechanism; 3—crane body; 4—boom

A, B,⋯,N—representing the size of different parts; L—the length of boom;

H—lifting height; R—lifting radius

The main technical performance of track-mounted crane includes three main parameters: lifting weight Q, lifting radius R and lifting height H. Lifting weight generally does not include the weight of hooks and pulley. Lifting radius R refers to the horizontal distance from the revolving center to the hook, and lifting height H refers to the distance from the center of the lifting hook to its standing level. There is a mutual restrictive relationship among the three working parameters of lifting weight, lifting radius and lifting height. The value depends on the length of boom and its boom angle. When the length of the boom is fixed, with the increase of boom angle, the lifting weight and height increase, but the lifting radius decreases; when the boom angle remains the same, with the increase of the boom length, the lifting radius and height increase, the lifting weight decreases.

2. Truck-mounted Crane

Truck-mounted crane is a kind of crane mounted on the chassis of ordinary automobile or special automobile. It is also a self-propelled full-rotating crane, as shown in Fig. 7. 3. The driving cab is separated from the lifting operation room. This kind of crane has the characteristics of high speed and good maneuverability. But when lifting very heavy members, it should be supported by brackets, so it can't move with load, and it's not suitable for working on muddy or soft ground.

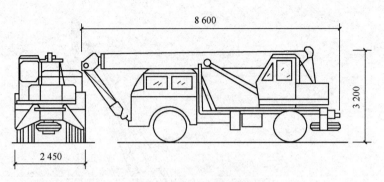

图 7.3 汽车式起重机

在使用汽车式起重机时不准负载行驶或不放下支腿就起重,在起重工作之前要平整场地,以保证机身基本水平(一般不超过3°),支腿下要垫硬木块。支腿伸出应在吊臂起升之前完成,支腿的收入应在吊臂放下搁稳之后进行。

3. 轮胎式起重机

轮胎式起重机是把起重机安装在加重型轮胎和轮轴组成的特制底盘上的一种自行式全回转起重机,如图 7.4 所示。随着起重量的大小不同,底盘下装有若干根轮轴,配备有 4～10 个或更多个轮胎。吊装时一般用四个支腿支撑以保证机身的稳定性,必要时,支腿下可加垫木,以扩大支撑面;构件重力在不用支腿允许荷载范围内也可不放支腿起吊。轮胎式起重机与汽车式起重机的优缺点基本相似,其行驶均采用轮胎,故可以在城市的路面上行走不会损伤路面。轮胎式起重机可用于装卸和一般工业厂房的安装和低层混合结构预制板的安装工作。

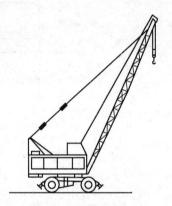

图 7.4 轮胎式起重机

7.2 钢筋混凝土结构单层厂房结构安装

单层工业厂房由于构件类型少,数量多,除基础在施工现场就地浇筑外,其他构件一般均为预制构件。其主要构件有柱、吊车梁、屋架、薄腹梁、天窗架、屋面板、连系梁、地基梁、各种支撑等,如图 7.5 所示。尺寸大、重量重的大型构件(柱、屋架等)一般在施工现场就地制作;中小型构件则集中在构件厂制作,然后运送到施工现场安装。

Chapter 7 Installation of Prefabricated Structural Members Construction

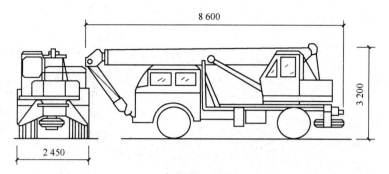

Fig. 7.3 Truck-mounted Crane

When truck-mounted crane working, it is forbidden to move with load or to lift without brackets. Before lifting, the site should be leveled to ensure the horizontal level of the crane (generally no more than 3 degrees), and hard wood blocks should be padded under the brackets. The brackets should be set before the rising of the boom and be withdrawn after the boom is put down.

3. Tyre Crane

Tyre crane is a kind of crane mounted on the special chassis composed by heavy-duty tyre and axle. It is also a self-propelled full-rotating (360°) crane, as shown in Fig. 7.4. There are several axles and 4 to 10 heavy-duty tyres or more under the chassis. The number of the axle and tyre is determined by the lifting weight. When lifting, four brackets are usually used to support the crane to ensure its stability. When necessary, the hard wood blocks should be padded under the brackets. When the lifting weight is in the load capacity range of the crane without brackets, it is no need to set the brackets. The advantages and disadvantages of tyre crane and truck-mounted crane are similar. They are all driven by tyres, so they can walk on the urban road without damaging the road surface. Tyre crane can be used for loading and unloading, installation of general industrial building and installation of prefabricated panels in low-rise masonry-concrete composite structures.

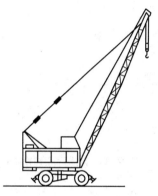

Fig. 7.4 Tyre Crane

7.2 Installation of Single-storey PRC Industrial Buildings

In the installation of single-storey PRC industrial buildings, it will use a large number of members but the types of members are limited, so the members are generally prefabricated, except that the foundation is cast in situ at the construction site. Its main members are columns, crane girders, roof trusses, thin web girder, skylights truss, roof slab, connecting beams, foundation beams and other struts, as shown in Fig. 7.5. Large-scale members (columns, roof trusses, etc.) with large size and weight are usually manufactured in situ, while small and medium-sized members are manufactured in the component factory and then transported to the construction site for installation.

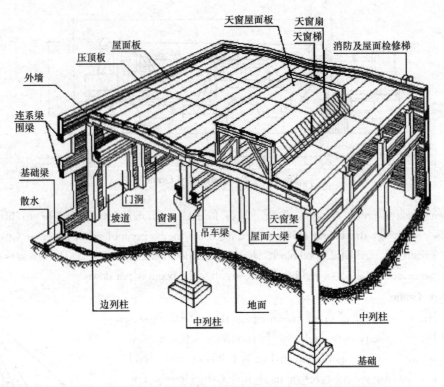

图 7.5 装配式钢筋混凝土排架结构的单层厂房构建组成

7.2.1 吊装前的准备工作

由于工业厂房吊装的构件种类、数量较多,为了进行合理而有序的安装工程,构件吊装前要做好各项准备工作,其内容有:场地清理和道路修筑;基础的准备;各种构件运输、就位和堆放;构件的强度、型号、数量和外观等质量检查;构件的拼装与加固;构件的弹线、编号以及吊具准备等。

1. 场地清理与道路铺设

起重机进场之前,安装现场平面布置图,标出起重机的开行路线,清理道路上的杂物,进行平整压实。回填土或松软土地基上,要用枕木或厚钢板铺垫。正确敷设水电管线。雨季施工,要做好排水工作,准备一定数量的抽水机械,以便及时排水。

2. 基础准备

杯型基础的准备工作主要是在柱吊装前对杯底抄平和在杯口顶面弹线。

Chapter 7 Installation of Prefabricated Structural Members Construction

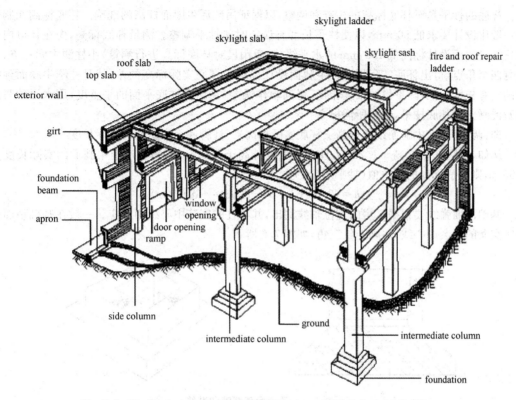

Fig. 7.5 Prefabricated Members of Single-storey PRC Industrial Building

7.2.1 Preparations Before Lifting

Because there are so many kinds and large quantities of members in the installation of single-storey PRC industrial building, in order to carry out installation projects in reasonable order, various preparations should be made before lifting, including site clearance and temporary access road construction, the works related to the members such as preparation of stockpiling foundation, transportation and stacking, checking of strength, type, quantity and appearance of members, splicing and strutting of members, marking with ink line and giving ID numbers to members and the preparation of sling devices and so on.

1. Site Clearance and Temporary Access Road Construction

Before the crane enters the construction site, install the layout of the site plan, mark the traveling route of the crane, clean up the debris on the road and level up and compact the road of the crane. When the travelling route is on the backfill or soft soil foundation, cross sleeper or thick steel plate should be used to pave the road. It is necessary to lay hydropower pipelines correctly. The drainage work should be done well in rainy season, so a certain amount of pumping machinery should be prepared to drain water in time.

2. Preparation in the Foundation

The preparations for pocket foundation are checking and adjusting the elevation of the bottom of the pocket and marking lines with chalk on the top of the pocket foundation.

杯底的抄平是对杯底标高的检查和调整,以保证吊装后牛腿面标高的准确。杯底标高在制作时一般比设计要求低 50 mm,以便柱子长度有误差时能抄平调整。测量杯底标高,先在杯口内弹出比杯口顶面设计标高低 100 mm 的水平线,随后用尺对杯底标高进行测量,小柱测中间一点,大柱测四个角点,得出杯底实际标高。牛腿面设计标高与杯底实际标高的差,就是柱子牛腿面到柱底的应有长度,与实际量得的长度相比,得到制作误差,在结合柱底平面的平整度,用水泥砂浆或细石混凝土将杯底抹平,垫至所需标高。

即:调整值=(牛腿面设计标高－杯底实际标高)－柱脚底至牛腿面的实际长度

例如,实际杯底标高－1.20 m,柱牛腿面设计高＋7.80 m,量得柱底至牛腿面的实际长度为 8.95 m,则杯底标高的调整值(抄平厚度)为

$$\Delta h=(7.80+1.20)-8.95=+0.05(\text{m})$$

基础顶面弹线要根据厂房的定位轴线测出,并与柱的安装中心线相对应。一般在基础顶面弹十字交叉的安装中心线,并画上红三角,如图 7.6 所示。

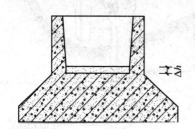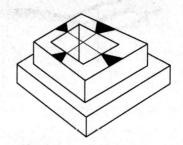

图 7.6 基础标高调整和准线

最后,将找平好的杯型基础的杯口部分加以覆盖,防止杂物落入其内。当检查发现杯口的定位轴线与中心线的偏差超过±10 mm,或杯口上下部分的尺寸与杯基中心线相差超过规范允许值时,杯口应进行修整,以保证柱子的正确安装。

3. 构件的运输和堆放

(1)构件的运输应符合下列规定:

a. 构件运输应严格执行所制定的运输技术措施。

b. 运输道路应平整,有足够的承载力、宽度和转弯半径。

Chapter 7 Installation of Prefabricated Structural Members Construction

The purpose of checking and adjusting the elevation of the bottom of the pocket is to ensure the accuracy of the elevation of the corbel surface after lifting. The bottom elevation of the pocket is generally 50mm lower than the design requirement, so that the column length can be adjusted when there are errors. To measure the bottom elevation of the pocket, first mark the horizontal line on the wall inside the pocket, 100mm lower than the designed top elevation of the pocket, then measure the bottom elevation of the pocket with a ruler to get the actual bottom elevation of the pocket; when the column is small, measure the middle point of it; while the column is big, measure the four corners. The difference between the designed top elevation of the corbel and the actual elevation of the bottom of the pocket shall be the desired length between the corbel top to the bottom of the column. Compared with the actual length measured, the fabrication error is obtained. Then it is time to smooth and raise the bottom of the foundation with cement mortar or fine aggregate stone concrete to the required elevation based on the flatness of the surface. To illustrate in detail:

The adjustment value = (The designed elevation of the corbel surface − the actual elevation of the pocket bottom) − the actual length from the column bottom to the surface of the corbel.

For example, if the actual elevation of the bottom of the pocket is −1.20m, the designed elevation of the corbel surface is +7.80m, and the actual length from the column bottom to the corbel surface is 8.95 m, and then the adjustment value of the pocket bottom elevation is:

$$\Delta h = (7.80 + 1.20) - 8.95 = +0.05(\text{m})$$

The place to mark with ink lines on the top of the pocket foundation should be located on base of alignment axis and in accordance with the installation central line of the column. Generally, we set a cross-sectional installation central line with ink on the top of the pocket foundation and draw red triangles on the ends of the ink lines, as shown in Fig. 7.6.

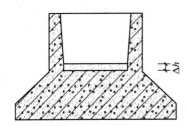 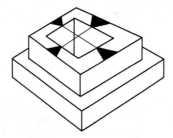

Fig. 7.6 The Adjustment of the Foundation Elevation and the Setting of Installation Base Line

Finally, cover screeded pocket mouth of the preformed foundation to avoid debris falling in. When the error between the alignment axis and the installation central line is larger than 10 mm or the error between the size of the pocket mouth and the central line of pocket bottom is larger than the allowable range, the pocket mouth should be adjusted to ensure the correct installation of the column.

3. The Transportation and Stacking of Members

(1) The transportation of the members should follow the rules:

a. Transportation of members shall strictly follow the formulated technical requirements.

b. The roads for transporting members should be flat, with sufficient bearing capacity, width and turning radius.

c. 高宽比比较大的构件运输,应采用支承框架、固定架、支撑或用捯链等予以固定,不得悬吊或堆放运输。支承架应进行设计计算,应稳定、可靠和装卸方便。

d. 当大型构件采用半拖或平板车运输时,构件支承处应设转向装置,防止构件侧向扭转断,并避免构件在运输时滑动、变形或相互碰撞。

e. 运输时,各构件应拴牢于车厢上。

总之,构件运输既要合理组织,提高运输效率,又要保证构件不损坏、不变形、不倾倒,确保质量和安全。构件运输时的混凝土强度,一定要符合设计规定,如设计无要求时应遵守《混凝土结构工程施工质量验收规范》(GB 50204—2015)的规定。否则运输中振动较大,构件容易损坏。构件的垫点和装卸车时的吊点,不论上车运输或卸车堆放都应按设计进行。"Γ"形等形状的构件都属特型构件,叠放在车上或堆放在现场上的构件,构件之间的垫木应在一条垂直线,且厚度相等。经核算需要加固的必须加固。对于重心较高、支承面较窄的构件,应采用支架固定,严防在运输途中倾倒。大型构件因其不易调头,必须根据其安装方向确定装车方向,并在支承处设转向装置。

(2)构件的堆放应符合下列规定:

a. 构件堆放场地应压实平整,周围应设排水沟,严防因地面下沉而使构件倾斜。

b. 构件应严格按平面布置图堆放,并满足吊装方法和吊装方向的要求,同时还应按类型和吊装顺序做到配套堆放,目的是避免二次倒运。

c. 构件应按设计支承位置堆放平稳,底部应设置垫木。对不规则的柱、梁、板,应专门分析确定支承和加垫方法。

d. 垫点应接近设计支承位置,异形平面垫点应由计算确定,等截面构件垫点位置亦可设在离端部 $0.207L$(L 为构件长度)处。柱子则应避免柱裂缝,一般易将垫点设在距牛腿 300~400 mm 处。同时构件应堆放平稳,底部垫点处应设垫木,应避免搁空而引起翘棱。

e. 屋架、薄腹梁等重心较高的构件,应直立放置,除设支承垫木外,应在其两侧设置支撑使其稳定,支撑不得少于两道。

Chapter 7　Installation of Prefabricated Structural Members Construction

c. When transporting members with large depth-width ratio, they should be fixed by supporting frames, fixtures, hand chain hoist and so on. They should not be suspended or stacked in transporting. The supporting frames should be designed and calculated carefully to make it stable, reliable and convenient for loading and unloading.

d. When large members are transported by semi-trailer or flat car, steering devices should be installed at the supporting zone to prevent lateral torsion of members and avoid sliding deformation or collision of members during transportation.

e. When transporting, the members should be fastened firmly to the carriage.

In a word, members transportation should be organized reasonably to improve transportation efficiency, and in transportation, we should ensure that members are not damaged, deformed and dumped and we should guarantee the safety and quality. When transporting members, the strength of concrete must conform to the design requirements. If there is no specific design requirement, it should comply with the "*Code for Acceptance of Construction Quality of Concrete Structural Engineering*" (GB 50204—2015). Otherwise, the members are easy to be damaged due to the vibration in transportation. The support points of the members and the pick points in loading, unloading or stacking should be located based on the design. no matter when they are loaded, transported or stacked. When transporting or stacking the special-shaped members like the shape of "Γ", the sole timber between the members should be in a vertical line with the same thickness. If it is necessary, it should be strutted. For the members with high gravity center and narrow supporting surface, they should be fixed by brackets to prevent dumping during transportation. Because it is difficult for large members to turn their heads, loading direction is determined by their installation direction, and steering device should be installed at the supporting zone.

(2) The stacking of the members should follow the rules:

a. The site for stacking members should be compacted and leveled, and drainage ditches should be set around the site to prevent the inclination caused by ground subsidence.

b. Members should be stacked strictly according to the plane layout, and meet the requirements of lifting method and direction. At the same time, they should be stacked in accordance with lifting type and sequence to avoid secondary transportation.

c. Members should be stacked smoothly according to the designed supporting position, and sole timbers should be installed at the bottom. For special shaped columns, beams and plates, methods for supporting and cushioning should be specially analyzed and determined.

d. The standing position should be close to the designed supporting zone. For special shaped members, the standing position should be calculated, and the standing position for members with constant section can also be located at $0.207L$ (L is the length of the component) away from the end. Pillars should avoid cracks, so the standing positions for pillars should be located at 300—400mm away from the corbel. At the same time, the members should be stacked smoothly, and sole timbers should be set at the cushion point of the bottom to avoid buckling.

e. For members with high gravity center like roof trusses, thin web girder, etc., they should be placed upright. In addition to supporting sole timbers, brackets should be set on both sides to make them stable, and brackets should be no less than two layers

f. 重叠堆放的构件应采用垫木隔开,上下垫木应在同一垂线上。堆放高度梁、柱不宜超过2层;大型屋面板不宜超过6层。堆垛间应留置2 m宽的通道。

g. 装配式大板应采用插放法或背靠法堆放,堆放架应经设计计算确定。

4. 构件的检查与清理

预制构件在生产过程中,可能会出现外形尺寸方面的误差及在构件表面产生一些缺陷等问题。因此对预制构件必须进行检查和清理,以保证构件吊装的质量。

构件安装前应对所有构件进行全面检查。

(1) 数量:各类构件的数量是否与设计的件数相符。

(2) 强度:安装时混凝土的强度应不低于设计强度等级的70%。对于一些大跨度或重要构件,如屋架,则应达到100%的设计强度等级。对于预应力混凝土屋架,孔道灌浆强度应不低于15 N/mm²。

(3) 外形尺寸:构件的外形尺寸,预埋件的位置和尺寸,吊环的位置和规格,接头的钢筋长度等是否符合设计要求,具体检查内容:

a. 柱子。检查总长度,柱脚底面平整度,柱脚到牛腿面的长度,截面尺寸,预埋件的位置和尺寸。

b. 屋架。检查总长度及跨度,是否与轴线尺寸相吻合。屋架侧向弯曲,连接屋面板、天窗架等构件用的预埋件的位置等。

c. 吊车梁。检查总长度、高度、侧向弯曲、预埋件位置等。

d. 外表面。检查构件外表有无损伤、缺陷、变形;预埋件上有无粘砂浆等污物;吊坏有无损伤、变形,能否穿卡环或钢丝绳等。预埋件上若粘有砂浆等污物,均应清除,以免影响拼装与焊接。

总之,构件检查应做记录,对不合格的构件,应会同有关单位研究,并采取适当措施,才可进行安装。

5. 构件的弹线和编号

构件经检查合格后,即可在构件表面上弹出中心线,以作为构件安装、对位、校正的依据。对形状复杂的构件,还要标出它的重心和绑扎点的位置。具体要求是:

Chapter 7 Installation of Prefabricated Structural Members Construction

f. For members stacked overlappingly, they should be separated by sole timbers, and the sole timbers from the top to the bottom should be on the same vertical line. The stacking of beams and columns should not exceed 2 stories, and the stacking of large roof slabs should not exceed 6 stories. Passages with the width of no less than 2m should be reserved between the stacking sites.

g. Assembled boards should be stacked by insertion method or backrest method. The stacking racks should be set by proper design and calculation.

4. The Inspection and Cleaning of Members

In the production process of prefabricated members, there may be errors in shape and size and some defects on the surface of members. Therefore, the prefabricated members must be inspected and cleaned to ensure the quality of component lifting.

All the members should be inspected thoroughly before lifting.

(1) Quantity inspection: Check whether the number of the members match the number in design.

(2) Strength inspection: When installing, the strength of concrete should not be less than 70% of the designed strength grade. For some large-span or important members like roof truss, the strength of the concrete should be the same as the designed strength grade. For the prestressed concrete roof truss, the grouting strength should not be less than $15N/mm^2$.

(3) Physical dimensions inspection: Check whether the physical dimensions of members, the positions and sizes of embedded members, the positions and specifications of suspension rings and the length of the joints for reinforcing bars meet the design requirements, including the following aspects:

a. Column. Inspect the total length, the flatness of the base of the column, the length from the base of the column to the surface of the corbel, the size of the section, the position and size of the embedded parts.

b. Roof truss. Check whether the total length and span are consistent with the axis size and inspect the lateral bending of roof truss and the location of embedded parts for connecting roof panels, skylight truss, etc.

c. Crane girder. Inspect the total length, height, lateral bending, position of embedded parts, etc.

d. Surface. Inspect the appearance of members to make sure that there is no damage, defect and deformation, to make sure that there is no mortar and or other dirt on the embedded parts, to make sure there is no damage or deformation on the rings and whether snap rings or wire-ropes can punch through. If there is any dirt on the embedded parts, such as mortar, it should be removed to avoid affecting splicing and welding.

In a word, component inspection should be recorded. Unqualified members should be studied in cooperation with relevant units and appropriate measures should be adopted before installation.

5. Marking Reference Line on the Members with Black Chalk and Giving Them ID Numbers

After the members are inspected, the central lines can be marked with black chalk on the surface of the members to serve as the basis for component installation, alignment and correction. For members with complex shapes, the center of gravity and the binding points should also be marked. There are some specific requirements as follows:

(1)柱子:在柱身三面弹出安装中心线,所弹中心线的位置与柱基杯口面上的安装中心线相吻合。对矩形截面的柱子,可按几何中线弹出;对工字形截面的柱子为便于观测和避免视差,则应靠柱边弹出控制准线。此外,在柱顶与牛腿面上还要弹出安装屋架及吊车梁的定位线,如图7.7所示。对工字形截面的柱子为便于观测和避免视差,则应靠柱边弹出控制准线。

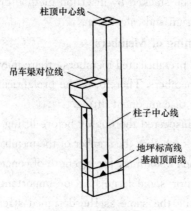

图 7.7 柱子弹线

(2)屋架:屋架上弦顶面应弹出几何中心线,并从跨中向两端分别弹出天窗架、屋面板或檩条的安装定位线;在屋架两端弹出安装中心线。

(3)梁:在梁的两端及顶面弹出安装中心线。在弹线的同时,应按图纸对构件进行编号,号码要写在明显部位。不易辨别上下左右的构件,应在构件上标明记号,以免安装时将方向搞错。

(4)编号:应按图纸将构件进行编号。

7.2.2 构件的吊装

构件吊装过程主要有绑扎、吊升、就位、临时固定,校正、最后固定等工序。

Chapter 7 Installation of Prefabricated Structural Members Construction

(1) Column: mark the installation central lines with chalk on three sides of the column and make sure that the position of the central lines on the column match the installation central lines on the top of the pocket base. For columns with rectangular cross section, the installation central lines can be marked according to their geometric central line; for I- shaped columns, to avoid parallax and facilitate observation, the installation lines should be marked on the edge of the column. In addition, the positioning lines for installing the roof truss and the crane girder should also be set on the top of the column and the surface of the corbel, as shown in Fig. 7. 7.

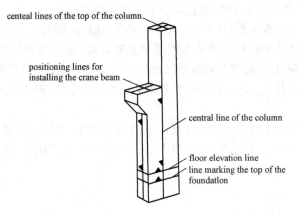

Fig. 7. 7 Marking Lines with Chalk on the Members

(2) Roof truss: the geometric central line should be marked on the top chord of the roof truss, and the installation and positioning line of skylight truss, roof slab or purl should also be marked with chalk from the mid-span to both ends; the installation central line should be marked at both ends of the roof truss.

(3) Beam: the installation central lines should be marked at both ends and top surface of the beam. When marking the lines with chalk, the members should be marked with ID number according to the drawings, and the ID numbers should be written on the parts that can be seen easily. For members with difficulty in identifying directions, clear marks should be marked on the members to avoid confusing in installation.

(4) Giving ID number: mark ID numbers on the members according to drawings.

7.2.2 The Installation of Members

The installation process of members mainly includes binding, lifting, alignment, temporary fixing, correction and permanent fixing.

1. 柱子的吊装

(1)柱子的绑扎。绑扎就是使用吊装索具、吊具绑扎构件,并做好吊升准备的操作。

柱的绑扎方法、绑扎位置和绑扎点数,应根据柱的形状、长度、截面、配筋、起吊方法和起重机性能等因素确定。由于柱起吊时吊离地面的瞬间由自重产生的弯矩最大,其最合理的绑扎点位置,应按柱产生的正负弯矩绝对值相等的原则来确定。一般中小型柱(自重13 t以下)大多数绑扎一点;重型柱或配筋少而细长的柱(如抗风柱),为防止起吊过程中柱的断裂,常用绑扎两点甚至三点。对于有牛腿的柱,其绑扎点应选在牛腿以下200 mm处;工字型断面和双肢柱,应选在矩形断面处,否则应在绑扎位置用方木加固翼缘,防止翼缘在起吊时损坏。

根据柱起吊后柱身是否垂直,分为斜吊法和直吊法,相应的绑扎方法如下:

①斜吊绑扎法。当柱平卧起吊的抗弯刚度满足要求时,可采用斜吊绑扎法,如图7.8所示。此法特点是柱不需要翻身,吊起后呈倾斜状态,由于吊索歪在柱的一边,起重钩低于柱顶,因此,起重臂可以短些。当柱身较长,起重机臂长不够时,此方法较方便,但因柱身倾斜,就位对中比较困难。

②直吊绑扎法。当柱平放起吊的抗弯强度不足,需将柱由平放转为侧立然后起吊时,可采用直吊绑扎法,如图7.9所示。起吊后,铁扁担高过柱顶,柱身呈直立状态,柱子垂直插入杯口。

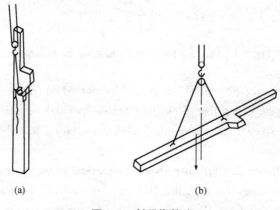

(a)   (b)

图7.8 斜吊绑扎法

(a)—一点斜吊法;(b)两点斜吊法

Chapter 7 Installation of Prefabricated Structural Members Construction

1. The Installation of Columns

(1) Binding the columns.

Binding is the operation of using rigging and sling devices to bind components to prepare for lifting. The binding method, binding location and the number of the binding points should be determined by the shape, length, cross section, reinforcement notes, lifting method, crane performance, etc. When the column is lifted off the ground, the bending moment generated by its own weight is the largest, so the most reasonable binding point should be the point that can generate equal positive and negative bending moment. Generally speaking, for small and medium-sized columns (less than 13t), there is only one binding point. For heavy columns or slender columns with few reinforcements (e.g. wind columns), in order to prevent the column from breaking during lifting, there are usually two or three binding points. For columns with corbels, the binding point should be 200mm below the corbel. For columns with the cross section "I" shaped or with two bearing rods, the binding point should be at the rectangular cross section; otherwise, the edge on the binding point should be protected by square timber to prevent damaging in lifting.

In lifting, the column may stay vertical or oblique, so the lifting methods can be classified into vertical lifting and oblique lifting. The binding requirements in different lifting methods are as follows:

①Oblique binding method: when the flexural rigidity meets the requirements for lifting the column in horizontal position, the oblique binding method can be used, as shown in Fig. 7.8. The characteristic of this method is that the column does not need to be turned over and it stays oblique in lifting. Because the suspension cable is crooked on one side of the column and the crane hook is lower than the top of the column, the boom can be shorter. When the column is very long and the boom is not long enough, this method is more convenient. However, because the column is oblique in lifting, it is difficult to make the positioning.

②Vertical binding method: when the flexural rigidity can't meet the requirements for lifting the column in horizontal position, the vertical binding method can be used, as shown in Fig. 7.9. Before lifting, the column should be turned over from horizontal position to lateral position, as shown in Fig. 7.9. In lifting, the horizontal lifting beam is higher than the top of the column, and the column stays vertically, so the column can be inserted perpendicularly into the pocket mouth.

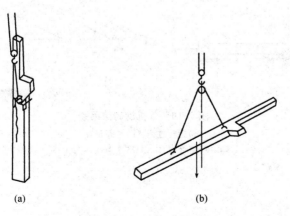

(a)   (b)

Fig. 7.8 Binding Methods in Oblique Lifting

(a) one binding point in oblique lifting; (b) two binding points in oblique lifting

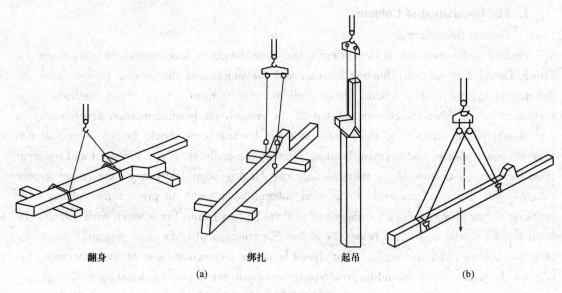

图 7.9 直吊绑扎法
(a)一点直吊法;(b)两点直吊法

(2)柱子的吊升。柱子的吊升方法,根据柱子质量、长度、起重机性能和现场施工条件而定。一般可分为旋转法和滑行法两种。

①旋转法。柱子吊升时,起重机边升钩,边回转起重杆,使柱子绕柱脚旋转而吊起之后插入杯口。为了便于操作和起重机吊升时起重臂不变幅,柱子在预制和堆放时,应使柱子的绑扎点、柱脚中心、杯口中心三点均位于起重机的同一起重半径的圆弧上。该圆弧的圆心为起重机的回转中心,半径为圆心到绑扎点的距离。柱子堆放时,应尽量使柱脚靠近杯口,以提高吊装速度,如图 7.10 所示。

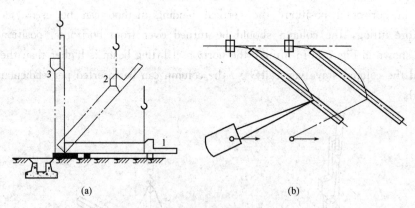

图 7.10 单机旋转法吊装
(a)旋转过程;(b)平面布置
1—柱平放时;2—起吊途中;3—直立

Chapter 7 Installation of Prefabricated Structural Members Construction

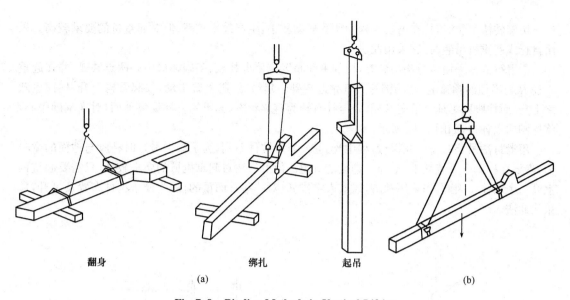

翻身　　　　　　　　　绑扎　　　起吊
(a)　　　　　　　　　　　　　　　　　　　(b)

Fig. 7.9 Binding Methods in Vertical Lifting

(a)one binding point in vertical lifting; (b)two binding points in vertical lifting

(2)The lifting of columns.

The lifting methods of the columns depend on their weight, length, crane performance and site construction conditions. Generally, they can be classified into two kinds: rotation method and sliding method.

①Rotation lifting: when the column is being hoisted, the crane lifts the hook while turning the lifting rod so that the column rotates around the foot and then it is inserted into the pocket mouth. In order to facilitate operation and keep the length of the boom unchanged, the binding point, the center of the column and the center of the pocket mouth should be located on the arc of the same lifting radius of the crane when the column is fabricated and stacked. The center of the arc is the rotating center of the crane, and the radius is the distance from the center to the binding point. When stacking the column, the foot of the column should be as close as possible to the pocket mouth to improve the lifting speed, as shown in Fig. 7.10.

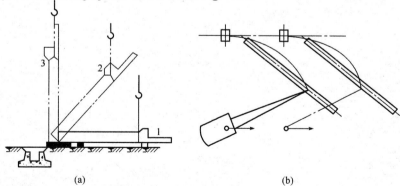

(a)　　　　　　　　　　　(b)

Fig. 7.10 Single Crane Rotation Lifting Method

(a)rotating process; (b)plane layout

1—horizontal position; 2—position in lifting; 3—vertical position

用旋转法吊装时,柱在吊装过程中所受振动较小,生产效率较高,但对起重机的要求较高。采用自行式起重机吊装时,宜采用此法。

②滑行法。采用滑行法吊装时,柱的平面布置应使绑扎点、基础杯口中心两点共弧,并在起重半径 R 为半径的圆弧上,柱的绑扎点宜靠近基础。起吊时,起重臂不动,仅重钩上升,柱顶也随之上升,而柱脚则沿地面滑向基础,直至柱身转为直立状态,起重钩将柱提离地面,对准基础中心,将柱脚插入杯口,如图 7.11 所示。

用滑行法吊装时,柱在滑行过程中受到振动,对构件不利,为了减少滑行时柱脚与地面的摩阻力,需要在柱脚下设置托木、滚筒并铺设滑行道。但滑行法对起重机械的要求较低,只需要起重钩上升一个动作。因此,当采用独脚拔杆、人字拔杆、对一些长而重的柱,为便于构件布置及吊升,常采用此法。

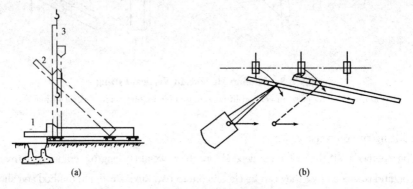

图 7.11 单机滑行法吊装
(a)滑行过程;(b)平面布置
1—柱平放时;2—起吊途中;3—直立

③双机抬吊旋转法。双机抬吊旋转法,是用一台起重机抬柱的上吊点,另一台抬柱的下吊点,柱的布置应使两个吊点与基础中心分别处于起重半径的圆弧上,起吊绑扎点尽量靠近杯口。主机起吊上柱,副机起吊柱脚。随着主机起吊,副机进行跑吊和回转,将柱脚递送至杯口上方,主机单独将柱子就位,如图 7.12 所示。

Chapter 7 Installation of Prefabricated Structural Members Construction

When the column is hoisted by rotating method, the vibration is small and the production efficiency is high, but the requirement for the crane is high. when the self-propelled crane is used in lifting, this method should be adopted.

②Sliding lifting: when the column is hoisted by sliding method, the binding point and the center of the pocket mouth should be located on the same arc with the lifting radius R and the binding point of the column should be close to the foundation. When lifting, the boom does not move, but the lifting hook rises to lift the top of the column and the column foot slides to the foundation. When the column becomes vertical, the crane hook hoists the column to the center of the pocket mouth and insert it into the pocket mouth, as shown in Fig. 7.11.

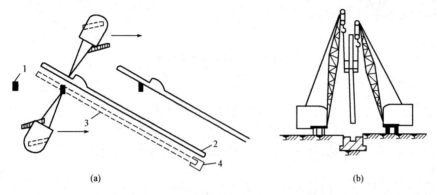

Fig. 7.11 Single Crane Sliding Lifting Method
(a)sliding process; (b)plane layout
1—horizontal position; 2—position in lifting; 3— vertical position

In sliding method, the column bears great vibration, which is harmful to the member. In order to reduce the friction between the column foot and the ground in sliding, it is necessary to set up supporting timber and roller at the column foot and lay a sliding path. The requirement for lifting machinery is low because the crane only needs to lift the hook. Therefore, this method is often used in single pole and herringbone pole or for some long and heavy columns to facilitate members layout and lifting.

③Double cranes rotation lifting method: in this method, one crane is used to hoist the upper pick point and the other crane is used to hoist the lower pick point. The two pick points and the center of the pocket mouth should be located on the same arc of lifting radius, and the binding points should be as close as possible to the pocket mouth. The main crane hoists the upper column and the auxiliary crane hoists the foot of the column. When the main crane hoists the column, the auxiliary crane moves and rotates to deliver the column foot to the top of the pocket mouth, and then the main crane inserts the column into the pocket mouth, as shown in Fig. 7.12.

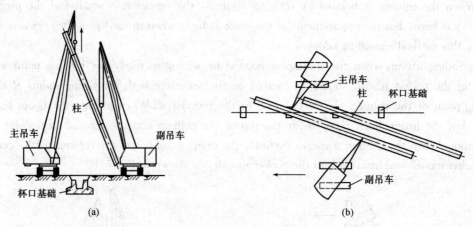

图 7.12 双机抬吊旋转法
(a)递送过程；(b)平面布置

④双机抬吊滑行法。当柱的重量较大，使用一台起重机无法吊装时，可以采用双机抬吊。柱应斜向布置，起吊绑扎点尽量靠近基础杯口。两台起重机停放位置相对而方，其吊钩均应位于基础上方。起吊时，两台起重机以相同的升钩、降钩、旋转速度工作，故宜选择型号相同的起重机，如图 7.13 所示。

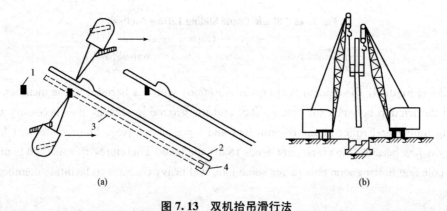

图 7.13 双机抬吊滑行法
(a)俯视图；(b)立面图
1—基础；2—柱预制位置；3—柱翻身后位置；4—滚动支座

吊装步骤：柱翻身就位→柱脚下设置托板、滚筒，铺好滑道→两机相对而立、同时起钩将柱吊离地面→同时落钩、将柱插入基础杯口。

Chapter 7 Installation of Prefabricated Structural Members Construction

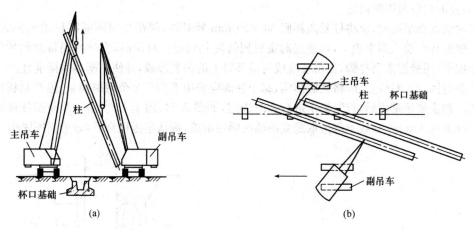

Fig. 7.12 Double Cranes Rotation Lifting Method
(a)lifting process;(b)plane layout

④Double cranes sliding lifting method: when the column is very heavy and it can't be hoisted by one crane, this method can be adopted. In this method, the column should be laid obliquely and the binding point should be as close as possible to the pocket mouth of the foundation. The two cranes are parked in opposite positions, and their hooks should be located above the foundation. When lifting, the two cranes work at the same speed in lifting hook, descending hook and rotating, so it is better to choose the same type of cranes, as shown in Fig. 7.13.

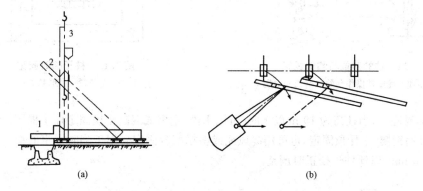

Fig. 7.13 Double Cranes Sliding Lifting Method
(a)plan view;(b)elevation
1—pocket foundation;2—column prefabrication position;
3—the position after turning over;4—rolling bracket

The lifting steps: turn over the column →set up supporting timber and roller at the column foot and lay a sliding path→ park the two cranes in opposite position and hoist the column with hooks at the same time → descend the hooks and insert the column into the pocket mouth.

(3)柱的对位与临时固定。

①对位。直吊法时,应将柱悬离杯底 30～50 mm 处对位,斜吊法时则需将柱送至杯底,在吊索的一侧的杯口插入两个楔子,再通过起重机回转使其对位。对位时,在柱四周每侧向杯口内放入 2 个楔子,用撬棍拨动柱脚,使吊装准线对准杯口上的吊装准线,并使柱基本保持垂直。

②临时固定。对位后,应将柱底落实,每个柱面应采用不少于 2 个钢楔楔紧,但严禁将楔子重叠放置。初步校正垂直后,打紧楔子进行临时固定,如图 7.14、图 7.15 所示。对重型柱或细长柱以及多风或风大地区,在柱上部采取稳妥的临时固定措施,确认牢固可靠后,方可指挥脱钩。

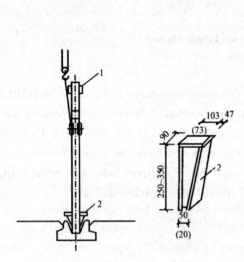

图 7.14 柱的对位与临时固定
1—安装缆风绳或挂操作台的夹箍;2—钢楔

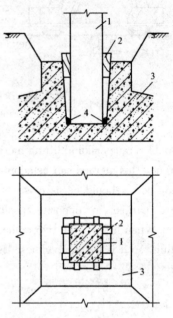

图 7.15 柱子临时固定
1—柱子;2—楔子;3—杯形基础;4—石子

柱子临时固定,当柱高为 10 m 以下时,可用木楔、钢楔或混凝土楔固定柱子根部;当柱高大于 10 m 时,可用钢楔、千斤顶固定,也可用缆风绳或斜撑配合固定。用于临时固定的楔子,宜露出杯口 100～150 mm,以便柱子校正时调整。

Chapter 7 Installation of Prefabricated Structural Members Construction

(3) The alignment and temporary fixing of column.

①The alignment: in vertical lifting, the alignment can be carried out when the column is 30 to 50mm above the pocket mouth; and in oblique lifting, first, insert the bottom of the column into the pocket mouth, insert two chocks into the pocket mouth on the side of the suspension cable and rotate the crane to make the alignment. In this process, insert two chocks into each side of the column, pry the foot of the column with the craw bar to make the center of the lifting line on the column align with the central lifting line on the pocket mouth and keep the column basically vertical.

②Temporary fixing: After positioning, make sure that the bottom of the column lay steadily on the foundation and the column is chocked up by steel chocks. There are at least two steel chocks on each side of the column and they mustn't be overlapped. After initial positioning, drive the chocks to make temporary fixing, as shown in Fig. 7.14 and Fig. 7.15. For some heavy columns and thin and long columns, or in some areas with frequent or strong wind, the top of the column should be fixed firmly and then release the hook after making sure that the column is fixed firmly.

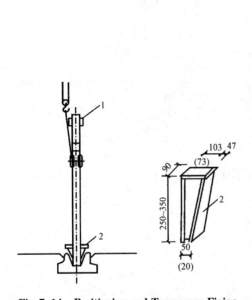

Fig. 7.14 Positioning and Temporary Fixing
1—clip for guy cable or control panel; 2—steel chock

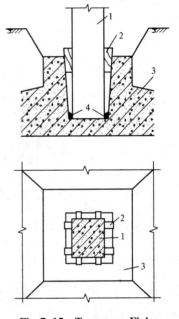

Fig. 7.15 Temporary Fixing
1—column; 2—chock;
3—pocket foundation; 4—gravel

When the height of the column is less than 10m, the root of the column can be fixed by wood chocks, steel chocks or concrete chocks; when the height of the column is more than 10m, it can be fixed by steel chock and jack, or it can be fixed by guy cable or inclined strut. The chocks used for temporary fixing should expose the pocket mouth 100—150mm so that the column can be adjusted with it in correction.

(4)柱的校正与最后固定。

①校正。柱的校正是一项重要工作,如果柱的吊装对位不够准确,就会影响与柱相连接的吊车梁、屋架等构件吊装的准确性。柱的校正包括平面位置校正、标高和垂直度的校正。标高校正在吊装前通过调整杯底标高已经校正;平面位置校正通过对位在临时固定前已经校正。

柱的校正主要是垂直度的校正,用两台经纬仪从柱的两个垂直方向同时观测柱的正面和侧面的中心线进行校正,如图 7.16 所示。校正柱时,严禁将楔子拔出,在校正好一个方向后,应稍打紧两面相对的 4 个楔子,方可校正另外一个方向。待完全校正好后,除将所有楔子按规定打紧外,还应采用石块将柱底脚与杯底四周全部楔紧。采用缆风绳或斜撑校正柱时,应在杯口第二次浇筑的混凝土强度达到设计强度的 75% 时,方可拆除缆风或斜撑。

校正方法:有敲打楔块法、千斤顶校正法、钢管撑杆斜顶法及缆风绳校正法等,如图 7.17 所示。

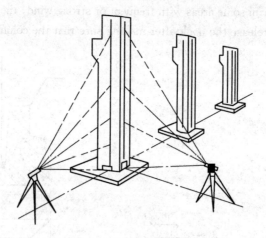

图 7.16 柱子的垂直度校正

Chapter 7 Installation of Prefabricated Structural Members Construction

(4) Correction and permanent fixing of columns.

①Correction: If the alignment of column is not accurate, it will affect the accuracy of installing crane girders, roof trusses and other members connected with columns, so the column should be corrected accurately after alignment. Column correction includes plane position correction, elevation correction and perpendicular correction. The elevation has been corrected by adjusting the elevation of the bottom of the pocket foundation before lifting, and the plane position has been corrected by alignment before temporary fixing. Therefore, the correction of the column mainly involves perpendicular correction. Two theodolites are used to observe the central line of the front and side of the column simultaneously from the two vertical directions of the column, as shown in Fig. 7.16. When correcting the column, it is forbidden to pull out the chock. After correcting one direction, drive firm the two chocks on this side and the other two chocks on the opposite side, and then correct the column from the other vertical direction. After finishing correction, drive all the chocks home and chock the foot of the column firmly with grovels from every side. When the guy cable or inclined strut is used to correct the column, the guy cable or the inclined strut can be removed only when the second pouring of the concrete at the pocket mouth reaches 75% of the designed strength.

The correction methods mainly include driving chocks, jacking, inclined supporting by steel pipe strut and guy cables, as shown in Fig. 7.17.

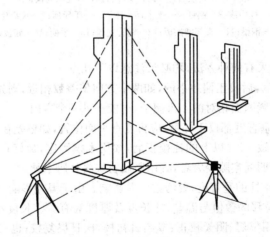

Fig. 7.16 The Perpendicular Correction of Column

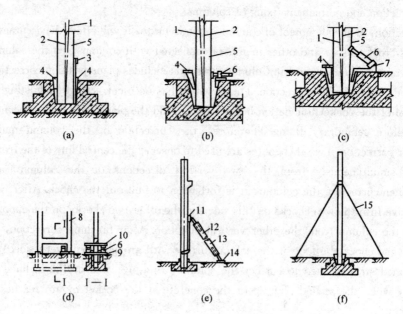

图 7.17 柱子校正方法
(a)钢钎法;(b)千斤顶平顶法;(c)千斤顶斜顶法;
(d)千斤顶立顶法;(e)钢管支撑斜顶;(f)缆风绳校正法
1—铅垂线;2—柱中线;3—钢钎;4—楔子;5—柱子;6—千斤顶;7—铁簸箕;8—双肢柱;
9—垫木;10—钢梁;11—头部摩擦板;12—钢管支撑;13—手柄;14—底板;15—缆风绳

在实际施工中,无论采有何种方法,均必须注意以下几点:

a. 应先校正偏差大的,后校正偏差小的,如两个方向偏差数相近,则先校正小面,后校正大面。校正好一个方向后,稍打紧两面相对的4个楔子,再校正另一个方向。

b. 柱在两个方向的垂直度都校正好后,应再复查平面位置,如偏差在5 mm以内,则打紧8个楔子,并使其松紧基本一致。8 t以上的柱校正后,如用木楔固定,最好在杯口另用大石块或混凝土块塞紧,柱底脚与杯底四周空隙较大者,宜用坚硬石块将柱脚卡死。

c. 在阳光照射下校正柱的垂直度,要考虑温差影响。由于温差影响,柱将向阴面弯曲,使柱顶有一个水平位移。水平位移的数值与温差、柱长度及厚度等有关。长度小于10 m的柱可不考虑温差影响。细长柱可利用早晨、阴天校正;或当日初校,次日晨复校;也可采取预留偏差的办法来解决。

Chapter 7 Installation of Prefabricated Structural Members Construction

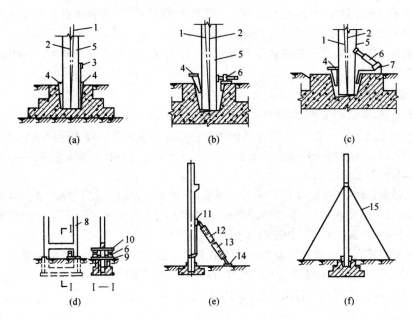

Fig. 7.17 Correction Methods of Columns
(a) correction by steel pin; (b) correction by horizontal jacking; (c) correction by inclined jacking;
(d) correction by vertical jacking; (e) correction by inclined steel pipe strut; (f) correction by guy cables
1—lead line; 2—central line of the column; 3—steel pin; 4—chock; 5—column; 6—jack; 7—steel sheathing;
8—double struts; 9—timber; 10—steel girder; 11—friction plate; 12—steel pipe strut;
13—handle; 14—sole plate; 15—guy cable

No matter which correction method is adopted, the following rules should be followed:

a. Correct the direction with large deviation first and then correct the direction with small derivation. If the deviations of two directions are similar, correct the direction of the small side first and then correct the big side. After correcting one direction, slightly drive the four chocks on the opposite sides of this direction, and then correct the other direction.

b. After the perpendicular correction of the column from both directions, the plane position should be re-checked. If the deviation is less than 5mm, eight chocks should be tightened with similar condition. If the column is more than 8t and chocked with wooden chock after correction, it is better to use big stone or concrete block to chock the column firmly. If the gap between the foot of the column and the bottom of the pocket is large, the foot of the column should be chocked with hard stone.

c. The effect of temperature difference should be taken into account when correcting the verticality of column under sunlight. Because of the influence of temperature difference, the column will bend to the night side, so the top of the column has a horizontal displacement. The horizontal displacement is related to temperature difference, column length and thickness. For columns less than 10m in length, it is no need to consider the effect of temperature difference. For columns more than 10m in length, they can be corrected in the morning or in cloudy day, or they can be corrected on the first day and checked in the morning of the second day, and they can be corrected with the reserving deviation method.

②最后固定。柱子经临时固定后,必须经过平面位置校正、标高和垂直度的校正后方可做最后固定。对校正完毕的柱子经有关部门检查合格后,应及时进行最后固定。

当使用混凝土楔子时,可一次浇筑至基础顶面。混凝土强度应作试块检验,冬期施工时,应采取冬期施工措施。

2. 吊车梁的吊装

吊车梁的类型,通常有T形、鱼腹型和组合型等。其长度一般有6 m、12 m,质量一般为3~5 t。吊车梁的安装为了稳定的需要,应在柱永久固定和柱间支撑安装完毕后进行。吊车梁安装必须在柱子杯口第二次浇筑的混凝土强度标准值达到50%以上时方可进行。其安装程序为:绑扎、起吊、就位、临时固定、校正和最后固定。

(1)绑扎、吊升、就位与临时固定。吊车梁吊装时,应两点绑扎,对称起吊,吊钩应对称梁的重心,以便使梁起吊后保持水平,吊索收紧后与梁的水平夹角不得小于45°,吊车梁的两端用溜绳控制,以免在吊升过程中碰撞柱子,如图7.18所示。

吊车梁对位时不宜用撬棍在纵轴方向撬动吊车梁,以防使柱身受挤动产生偏差。

吊车梁对位后,由于梁本身稳定性好较好,仅用垫铁垫平即可,不需采取临时固定措施,但当梁的高宽比大于4时,宜用铁丝将吊车梁临时绑在柱上。

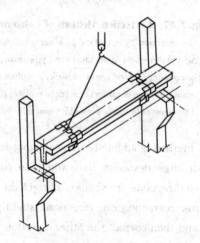

图7.18 吊车梁吊装

(2)校正和最后固定。吊车梁的校正主要是平面位置和垂直度的校正。吊车梁的标高取决于柱牛腿标高,在柱吊装前已经调整。如仍存在误差,可待安装吊车轨道时进行调整。

吊车梁的校正工作一般在屋面构件安装校正并最后固定后进行。因为在安装屋架、支撑等构件时,可能引起柱子偏差影响吊车梁的准确位置。但对重量大的吊车梁,脱钩后撬动比较困难,应采取边吊边校正的方法。

吊车梁垂直度校正一般采用吊线锤的方法检查,如存在偏差,在梁的支座处垫上薄钢板调整。

吊车梁的平面位置的校正通常用通线法和平移轴线法。

Chapter 7 Installation of Prefabricated Structural Members Construction

②The permanent fixing of column: the permanent fixing can be carried out after the plane position correction, elevation correction and perpendicular correction. After the corrected columns have passed the inspection by the relevant departments, they should be fixed permanently in time. Fine aggregate stone concrete with higher strength should be poured into the pocket mouth to make permanent fixing.

When wooden or steel chocks are used in temporary fixing, the concrete should be poured twice. In the first pouring, it should stop when the concrete reaches the bottom of the chock and when the concrete strength reaches no less than 30% of its design strength, the chocks can be pulled out and then pour the concrete again to the top of the pocket foundation; when concrete chocks are used in temporary fixing, the concrete can be poured to the top of the pocket foundation once. Concrete strength should be tested by test cubes, and in winter, some special construction measures should be taken.

2. The Installation of Crane Girder

The types of crane girders usually include T type, fish-belled type and combination type. Its length is generally 6m or 12m, and its weight is generally 3—5t. For the sake of stability, the installation of crane girders should be carried out after the column is permanently fixed and the strut between columns is installed. The installation of crane girders can only be carried out when the concrete strength poured into the pocket mouth at the second time reaches more than 50% of its standard strength. Its installation steps are: binding, lifting, alignment, temporary fixing, correction and permanent fixing.

(1) Binding, lifting, alignment and temporary fixing.

When installing the crane girder, it should be bound at two points and hoisted symmetrically, and the hook should align with the gravity center of the beam so as to keep the beam horizontal in lifting. The horizontal angle between the sling and the beam should be no less than 45 degrees. The two ends of the crane girder should be controlled by a tagline so as to avoid collision with the column during lifting, as shown in Fig. 7.18.

(2) Correction and permanent fixing

The correction of crane girder is mainly the correction of plane position and verticality. The elevation of crane girder depends on the elevation of column corbel, which has been adjusted before the installation. If there are still errors, adjustments can be made when crane tracks are installed. The correction of crane girders is usually carried out after the roof members are installed and fixed. When installing roof truss, struts and other members, they it may cause column deviation and affect the exact location of crane girder. But for heavy crane girders, it is difficult to pry after releasing the hook, so the correction can be carried out in lifting.

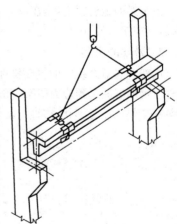

Fig. 7.18 The Installation of Crane Girder

The verticality of crane girder is usually checked by lifting wire hammer and if there is deviation, it can be corrected by padding thin steel plate on the bracket of the beam. The horizontal position of the crane girder is usually rectified by the through-line method and the translational axis method.

①通线法。根据柱子轴线用经纬仪和钢尺，准确地校核厂房两端一跨内的四根吊车梁位置，对吊车梁的纵轴线和轨距校正好之后，再依据校正好的端部吊车梁，沿其轴线拉上钢丝通线，逐根拨正，如图 7.19 所示。

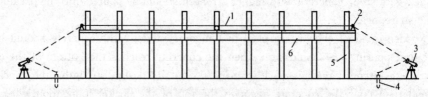

图 7.19　通线法校正吊车梁的平面位置
1—钢丝；2—支架；3—经纬仪；4—木桩；5—柱；6 吊车梁

②平移轴线法。在柱列边设置经纬仪，逐根将杯口中柱的吊装准线投影在吊车梁顶面处的柱身上，并做好标志。若按照准线到柱定位轴线的距离为 a，则标志距吊车梁定位轴线应为 $\lambda-a$（一般 $\lambda=750$ mm），据此逐根拨正吊车梁的安装中心线如图 7.20 所示。

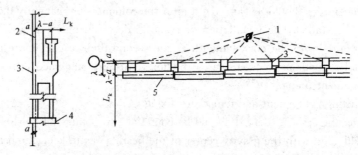

图 7.20　平移轴法校正吊车梁的平面位置
1—经纬仪；2—标志；3—柱；4—柱基础；5—吊车梁

吊车梁的最后固定，是在吊车梁校正完毕后，用连接钢板与柱侧面、吊车梁顶端的预埋铁件相焊接，并在接头处、吊车梁与柱的空隙处支模，浇筑细石混凝土。

3. 屋架的吊装

屋架是屋盖系统中的主要构件，除屋架之外，还有屋面板、天窗架、支撑、天窗侧板及天沟板等构件。钢筋混凝土预应力屋架一般在施工现场平卧叠浇生产，吊装前应将屋架扶直、就位。屋架吊装的主要工序有绑扎、扶直与就位、吊升、对位、校正、最后固定等。

Chapter 7　Installation of Prefabricated Structural Members Construction

① Through-line method: Check the positions of the four crane girders accurately with theodolite and steel ruler in one span of the single-story PRC industrial building based on the column axis. After correcting the longitudinal axis and track gauge of both end spans of crane girders, a steel wire is drawn along its axis and then the other three beams are corrected one by one based on the position of the steel wire, as shown in Fig. 7. 19.

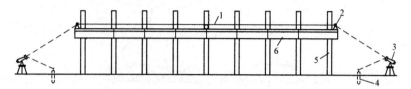

Fig. 7. 19　Correcting the Plane Position of the Crane Girder with Through-line Method
1—steel wire; 2—bracket; 3—theodolite; 4—wood pile; 5—column; 6—crane girder

② The translational axis method: Set a theodolite at one side of the columns, and project the lifting alignment of the columns on the top of the crane girder one by one, mark the projection on the top of the crane girder. If the distance from the alignment line to the column alignment axis is a, the distance from the mark to the alignment axis of the crane girder should be $\lambda - a$ (generally $\lambda = 750$ mm). Based on this number, correct the installation central line of the crane girders one by one, as shown in Fig. 7. 20.

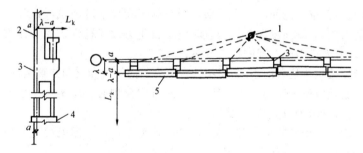

Fig. 7. 20　Correcting the Plane Position of the Crane Girder with Translational Axis Method
1—theodolite; 2—projection mark; 3—column; 4—column foundation; 5—crane girder

The permanent fixing of the crane girder is carried out after the crane girder has been corrected. The crane girder is permanently fixed by welding the steel plate and the steel members embedded in the column and the top of the crane girder and by pouring fine stone concrete with mold into the conjunction and gap between the crane girder and the column.

3. The Installation of Roof Truss

Roof truss is the main component of roof system. In addition to roof truss, there are roof panel, skylight truss, strut, skylight side panel and ceiling panel. Prestressed reinforce concrete roof truss is usually produced by horizontal filigree casting at construction site. The roof truss should be straight and in place before lifting. The main procedures of roof truss installation include binding, erecting and positioning, lifting, alignment, correction, and permanent fixing.

(1)屋架的绑扎。屋架绑扎点应设在上弦节点处,左右对称。吊点的数目及位置一般由设计确定,设计无规定时应经吊装验算确定。当屋架跨度小于18 m时采用两点绑扎;屋架跨度为18～24 m时采用四点绑扎;跨度大于30 m时采用9 m横吊梁、四点绑扎。以降低吊装高度和减小吊索对屋架上弦的轴向压力。翻身扶直屋架时,吊索与水平线的夹角α不宜小于60°,吊装时不宜小于45°。屋架的绑扎方法如图7.21所示。

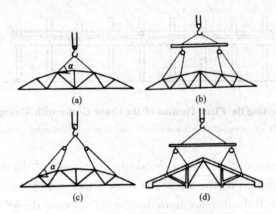

图7.21　屋架的绑扎方法
(a)屋架跨度≤18 m时;(b)屋架跨度18～24 m时;(c)屋架跨度≥30m时;
(d)三角形组合屋架

(2)屋架的扶直与就位。钢筋混凝土屋架一般在施工现场平卧浇筑,吊装前应将屋架扶直就位。屋架是平面受力构件,侧向刚度差。扶直时由于自重会改变杆件的受力性质,容易造成屋架损伤,所以必须采取有效措施或合理的扶直方法。

按照起重机与屋架相对位置的不同,屋架扶直分为正向扶直和反向扶直两种方法。

①正向扶直。起重机位于屋架下弦一侧,吊钩对准屋架中心。屋架绑扎起吊过程中,应使屋架以下弦为轴心,缓慢旋转为直立状态,如图7.22(a)所示。

②反向扶直。起重机位于屋架上弦一侧,吊钩对准屋架中心。屋架绑扎起吊过程中,使屋架以下弦为轴心,缓慢旋转为直立状态,如图7.22(b)所示。

正向扶直和反向扶直的最大不同点是:起重机在起吊过程中,对于正向扶直时要升钩并升臂;而在反向扶直时要升钩并降臂。一般将构件在操作中升臂比降臂较安全,故应尽量采用正向扶直。

Chapter 7 Installation of Prefabricated Structural Members Construction

(1) The binding of roof truss. The binding point of roof truss should be located symmetrically at the top nodes. The number and location of pick points are generally determined by the design. When the design is not specified, the number and location of pick points should be determined by the checking calculation. When the span of the roof truss is less than 18m, two-point binding method is adopted; when the span of the roof truss is 18—24m, four-point binding method is adopted; when the span is more than 30m, nine-meter cross beam and four-point binding method is used to reduce the lifting height and the axial pressure of the suspension cable on the upper chord of the roof truss. When turning over to upright roof truss, the angle α between suspension cable and horizontal line should not be less than 60 degrees, and it should not be less than 45° in lifting. The binding method of the roof truss is shown in Fig. 7.21.

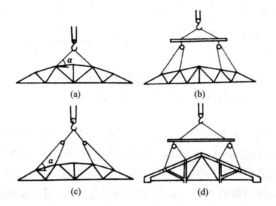

Fig. 7.21 The Binding Methods of the Roof Truss
(a) the roof truss span $\leqslant 18$ m; (b) the roof truss span between 18—24m
(c) the roof truss span $\geqslant 30$ m; (d) assembled triangle roof truss

(2) Erecting and positioning of roof trusses. Reinforced concrete roof truss is usually produced by horizontal filigree casting at construction site and it needs erecting and positioning before lifting. The roof truss is a plane force member with poor lateral stiffness. In erecting, the self-weight will change the force nature of the roof truss and easily cause its damage, so effective measures or reasonable erecting methods must be taken.

According to the different relative positions between crane and roof truss, roof truss erecting methods can be classified into two methods: forward erecting and backward erecting.

①Forward erecting: The crane is located on the side of the lower chord of the roof truss, and the hook is aligned with the center of the roof truss. In the process of binding and lifting, the roof truss should rotate around the lower chord slowly to be straight, as shown in Fig. 7.22 (a).

②Backward erecting: The crane is located on the side of the upper chord of the roof truss, and the hook is aligned with the center of the roof truss. In the process of binding and lifting, the roof truss should rotate around the lower chord slowly to be straight, as shown in Fig. 7.22 (b).

In the process of forward erecting, the crane lifts the hook and lifts the boom, but in backward erecting, the crane lifts the hook and descends the boom. Generally speaking, lifting boom is safer than descending the boom, so the forward erecting method is preferred.

屋架扶直后，应立即进行就位。就位指移放在吊装前最近的便于操作的位置。屋架就位位置应在事先加以考虑，它与屋架的安装方法、起重机械的性能有关，还应考虑到屋架的安装顺序，两端朝向，尽量少占场地，便利吊装。就位位置一般靠柱边斜放或以 3~5 榀为一组平行于柱边。屋架就位后，应用 8 号铁丝、支撑等与已安装的柱或其他固定体相互拉结，以保持稳定。当屋架就为位置与屋架的预制位置在起重机开行路线同一侧时，称同侧就位[图 7.22(a)]。当屋架就为位置与屋架的预制位置在起重机开行路线各一侧时，称异侧就位[图 7.22(b)]。

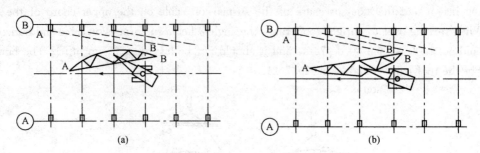

图 7.22 屋架的扶直
(a)正向扶直，同侧就位；(b)反向扶直，异侧就位

(3)屋架的吊升、对位与临时固定。在屋架吊离地面约 300 mm 时，将屋架引至吊装位置下方，然后再将屋架吊升超过柱顶 300 mm，然后缓慢下落在柱顶上，进行屋架与柱顶的对位。

屋架对位应以建筑物的定位轴线为准。因此在屋架吊装前，应用经纬仪或其他工具在柱顶放出建筑物的定位轴线。如柱顶截面中线与定位轴线偏差过大时，可逐渐调整纠正。

屋架对位后，立即进行临时固定。临时固定稳妥后，起重机才可脱钩离去。

第一榀屋架的临时固定方法，应在其上弦杆拴缆风绳作临时固定。缆风绳应采用两侧布置，每边不得少于 2 根。当跨度大于 18 m 时，宜增加缆风绳数，间距不得大于 6 m，如图 7.23 所示。也可将屋架与抗风柱连接作为临时固定。

Chapter 7 Installation of Prefabricated Structural Members Construction

After erecting the roof truss, the positioning should be carried out immediately. Positioning means move and locate the roof truss in the place that is nearest to the lifting and is convenient for operation. The positioning of roof truss should plan beforehand. It should consider the installation method and sequence, the direction of roof truss ends and the performance of cranes. The positioning of the roof truss should take less space and be convenient for lifting. Generally speaking, the roof truss can be placed obliquely along the column or placed parallelly to the column with three to five pieces together. After the positioning of the roof truss, it should be bound to the installed columns or other fixed objects with the No. 8 iron wire, struts or other things to maintain stability. When the positioning and the roof truss prefabrication position is on the same side of the crane route, it is called same side positioning [Fig. 7. 22 (a)]. When the positioning and the roof truss prefabrication position is on the different sides of the crane route, it is called the opposite side positioning [Fig. 7. 22 (b)].

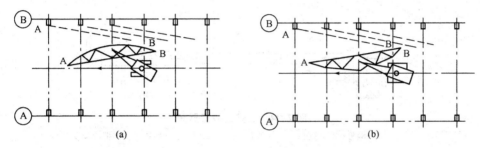

Fig. 7. 22 The Erecting of Roof Truss
(a) forward erecting method, same side positioning;
(b) backward erecting method, opposite side positioning

(3) The lifting, alignment and temporary fixing of roof truss. When the roof truss is lifted about 300mm above the ground, move it to the lifting point, and lift it about 300mm above the column, and then fall slowly on the top of the column to align the roof truss with the column.

The alignment of roof truss should be based on the positioning axis of the building. Therefore, before the roof truss is lifted, theodolite or other tools are used to find the positioning axis of the building and mark it with ink on the top of the column. If the deviation between the central line of the column top section and the positioning axis of the building is too large, it should be corrected gradually.

After the alignment of the roof truss, temporary fixing should be carried out immediately. After making sure the temporary fixing is firm and stable, the crane can release the hook and leave.

In temporary fixing of the first piece of roof truss, guy cable should be used to fasten the upper chord. The cables should be set at the left and right sides of the roof truss and there must be at least two cables on each side. When the span of the roof truss is more than 18m, it is advisable to increase the number of cables, and the space between the cables should not be greater than 6 m, as shown in Fig. 7. 23. The roof truss can also be connected with the wind beam as temporary fixing.

第二榀屋架的临时固定,是以第一榀屋架为支承点,用两根屋架工具式校正器(图 7.24)撑牢在前一榀屋架上,以后各榀屋架的临时固定也都是用两根屋架工具式校正器撑牢在前一榀屋架上,如图 7.25 所示。

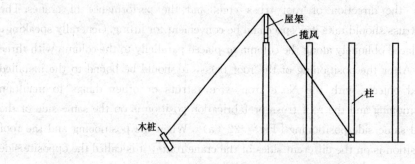

图 7.23　第一榀屋架临时固定

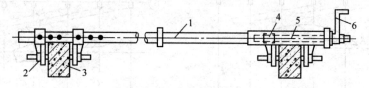

图 7.24　工具式校正器图

1—钢管；2—撑脚；3—屋架上弦；4—螺母；5—螺杆；6—摇把

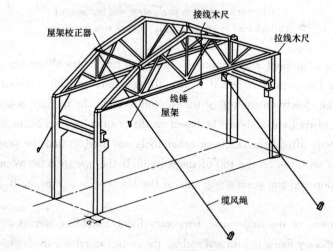

图 7.25　屋架的临时固定

Chapter 7 Installation of Prefabricated Structural Members Construction

The temporary fixing of the second roof truss is supported by the first roof truss, and it is strutted firmly to the first roof truss with two correcting appliances (Fig. 7.24). The temporary fixing of the other roof trusses all take the method of strutting the former roof truss with correcting appliance, as shown in Fig. 7.25.

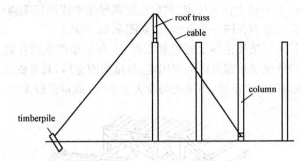

Fig. 7.23 The Temporary Fixing of the First Piece of Roof Truss

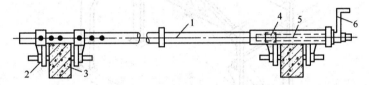

Fig. 7.24 Correcting Appliance

1—steel pipe; 2—strut anchor; 3—upper chord;
4—screw nut; 5—screw rod; 6—handle

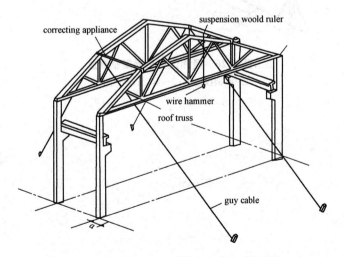

Fig. 7.25 The Temporary Fixing of the Roof Truss

(4)屋架的校正与最后固定。屋架经对位、临时固定后,主要校正垂直度偏差。施工规范规定:屋架上弦(在跨中)对通过两支座中心垂直面的偏差不得大于 $h/250$(h 为屋架高度)。屋架的垂直度偏差可用锤球或经纬仪检查。用经纬仪检查方法是在屋架上安装三个卡尺,卡尺与屋架的平面垂直,一个安装在屋架上弦中点附近,另两个安装在屋架两端。自屋架几何中心向外量出一定距离(一般为 500 mm)在卡尺上作出标志,然后在距离屋架中线同样距离(500 mm)处安置经纬仪,观察三个卡尺上的标志是否在同一垂直面上,如图 7.26 所示。

有偏差时,转动工具式屋架校正器上螺栓加以校正,在屋架两端的柱底上嵌入斜垫铁。屋架校正完毕后,应立即按设计规定用螺母或电焊固定,待屋架固定后,起重机方可松卸吊钩。

中、小型屋架,一般均用单机吊装,当屋架跨度大于 24 m 或重量较大时,应采用双机抬吊。

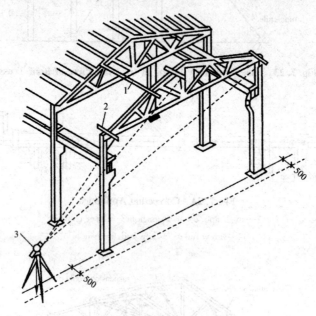

图 7.26 经纬仪校正屋架垂直度
1—屋架校正器;2—卡尺;3—经纬仪

Chapter 7 Installation of Prefabricated Structural Members Construction

(4) The correction and permanent fixing of the roof truss.

After the alignment and temporary fixing of the roof truss, the correction of its verticality deviation should be carried out. According to the construction specification, the deviation between the upper chord of the roof truss (in the span) and the vertical plane between two pedestals should not larger than $h/250$ (h is the height of the roof truss). The verticality deviation of the roof truss can be checked by plummet or theodolite. In checking with theodolite, install three calipers on the roof truss; one is installed at the center of the upper chord of the roof truss and the other two are installed at the two ends of the roof truss. The direction of the caliper should be perpendicular to the plane of the roof truss. Measure a certain distance (usually 500mm) from the geometric center of the roof truss and mark it with ink on the caliper, and then place a theodolite at the same distance (500mm) from the central line of the roof truss, and check whether the signs on the three calipers are on the same vertical plane, as shown in Fig. 7. 26.

If there is derivation, adjust the screw bolt on the correcting appliance to correct the position of the roof truss and put oblique steel cushion on the top of the column under the ends of the roof truss. After the correction of the roof truss, it should be fixed permanently with screw nut or welding based on the design specification. After the permanent fixing of the roof truss, the cane can release the hook.

For middle or small sized roof truss, it can be installed by one crane. When the span of the roof truss is longer than 24m or roof truss is very heavy, two cranes shall be used in installation.

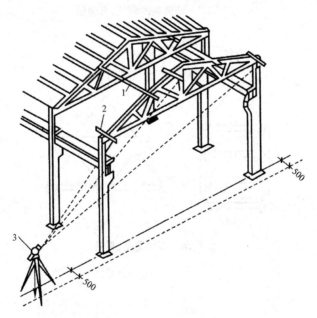

Fig. 7. 26 Verticality Correction of the Roof Truss by Theodolite
1—correcting appliance; 2—caliper; 3—theodolite

4. 天窗架及屋面板的吊装

(1)天窗架吊装。天窗架可与屋架拼装成整体同时吊装,亦可单独吊装,视起重机的起重能力和起吊高度而定。前者高空作业少,但对起重机要求较高,后者为常用方式,安装时需待天窗架两侧屋面板安装后进行。钢筋混凝土天窗架一般采用两点或四点绑扎,如图 7.27 所示。其校正、临时固定亦可用缆风绳、木撑或工具式夹具进行。

(2)屋面板吊装。屋面板均预埋有吊环,为充分发挥起重机效率,屋面板多采用一钩多块叠吊或平吊法,如图 7.28 所示。板的安装应自两边檐口左右对称地逐块安向屋脊或两边左右对称地逐块吊向中央,以免支承结构不对称受荷。因此,有利于下部结构的稳定,屋面板对位、校正后应立即焊接牢固,每块板的焊接角点不应少于 3 个。

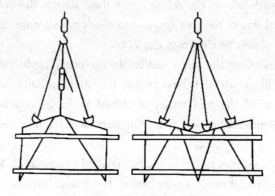

图 7.27 天窗架的绑扎、吊装图

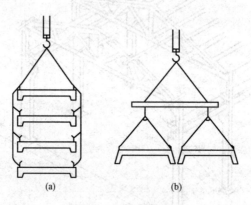

图 7.28 多块叠吊、平吊

Chapter 7 Installation of Prefabricated Structural Members Construction

4. The Installation of Skylight Shelf and Roof Slab

(1) The installation of skylight shelf. Based on the crane performance and the lifting height, the skylight shelf can be assembled and installed with the roof truss or it can be installed individually. When it is installed with the roof truss, it needs less operation high above the ground, but it requires high performance of the crane. When it is installed individually, it is a common way and it should be carried out after the installation of the roof slab on the left and right. Reinforced concrete skylight shelf should be bound at two points or four points, as shown in Fig. 7.27. The correction and temporary fixing can be carried out with guy cable, wood strut and clippers.

(2) The installation of roof slab. Suspension rings are embedded in the roof slab, so the roof slabs can be installed several pieces once a time to increase the efficiency. When one hook hoists several overlapping pieces of roof slabs, it is called overlapping lifting; when one hook hoists several pieces in horizontal position, it is called horizontal lifting, as shown in Fig. 7.28. In order to avoid asymmetric loads on the support structure, the installation of the roof slab should begin from the eaves at the two sides and go on symmetrically to the ridge or begin from the two sides symmetrically to the center. For the sake of substructure stability, after the alignment and correction of the roof slab, it should be welded immediately and there should be at least three welding points.

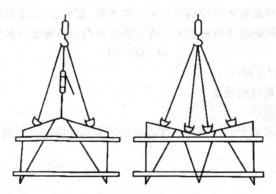

Fig. 7.27 The Binding and Lifting of Skylight Shelf

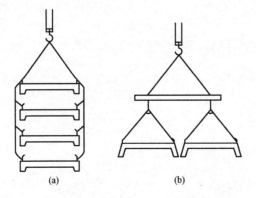

(a)   (b)

Fig. 7.28 Overlapping Lifting and Horizontal Lifting

7.2.3 结构吊装方案

单层工业厂房结构的特点是：平面尺寸大、承重结构的跨度与柱距大、构件类型少、重量大，厂房内还有各种设备基础(特别是重型厂房)等。因此，在拟定结构安装方案时，应着重解决起重机选择、结构安装方法、起重机械开行路线三个方面问题。

1. 起重机的选择

起重机是结构安装工程的主导设备，它的选择直接影响结构安装的方法，起重机的开行路线以及构件的平面布置。

(1)起重机类型选择。起重机的选择，应根据厂房外形尺寸，构件尺寸和质量，以及安装位置和施工现场条件等因素综合考虑。

①对于一般中小型工业厂房，由于外形平面尺寸较大，构件的质量与安装高度却不大，因此选用履带式起重机最为适宜。

②对于结构高度和长度较大的工业厂房，则应选塔式起重机吊装。

③在缺乏自行式起重机的地方，可采用桅杆式起重机安装。

④大跨度重型工业厂房，应结合设备安装来选择起重机类型。

⑤当一台起重机无法吊装时，可选择两台起重机抬吊。

(2)起重机型号和起重臂长度的选择。起重机的类型确定之后，还需要进一步选择起重机的型号及起重臂的长度。所选起重机应满足三个工作参数：起重量、起重高度、工作幅度的要求。

①起重量 Q。起重机的起重量必须大于所安装最重构件的质量与索具质量之和。

$$Q \geqslant Q_1 + Q_2 \tag{7.1}$$

式中　Q——起重机的起重量(t)；

　　　Q_1——所吊最重构件的质量(t)；

　　　Q_2——索具的质量(t)。

②起重高度 H。起重机的起重高度必须满足所吊装构件的高度要求，如图 7.29 所示。

7.2.3 Structural Installation Scheme

The structural characteristics of single-storey industrial building are: large plane size, large span and column spacing, a small number of component types, large weight, with various equipment foundations (especially in heavy industrial buildings), etc. Therefore, when making structural installation scheme, we should focus on the following three aspects: the selection of crane, the structure installation method and the cranes' route.

1. The Selection of Crane

Crane is the dominant equipment in structural installation engineering. Its selection directly affects the method of structural installation, the traveling route and the plane layout of members.

(1) The selection of crane type.

The selection of crane type should consider the dimension of the industrial building, the size and weight of the members, the installation location and construction site conditions.

①For small and medium-sized industrial building, because of its large plane size and small component weight and installation height, track-mouted crane is the most suitable choice.

②For the high and long industrial building, tower crane is the most suitable choice.

③If there is no self-propelled crane, mast crane can be used in installation.

④For large span and heavy duty industrial buildings, the selection of crane type should be based on the types of members and other equipment.

⑤When one crane can't meet the installation requirement, two cranes can be used at the same time.

(2) The selection of crane model and boom length.

Once the type of the crane is selected, the model of the crane and its boom length should be considered. The crane should meet requirements of the three main parameters: lifting weight Q, lifting height H and lifting radius R.

①Lifting weight Q.

The lifting weight of the crane must be greater than the sum of the weight of the heaviest component and the rigging.

$$Q \geqslant Q_1 + Q_2 \tag{7.1}$$

Q—the lifting weight(t);

Q_1—the weight of the heaviest component(t);

Q_2—the weight of the rigging(t);

②Lifting height H.

The lifting height of the crane must meet the height requirements of the members, as shown in Fig. 7.29.

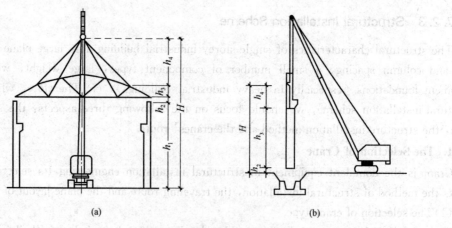

图 7.29 起重高度计算图
(a)安装屋架;(b)安装柱子

$$H \geqslant h_1 + h_2 + h_3 + h_4 \tag{7.2}$$

式中　H——起重机的起重高度(m);
　　　h_1——安装点的支座表面高度,从停机地面算起(m);
　　　h_2——安装对位时的空隙高度,不小于 0.3 m;
　　　h_3——绑扎点至构件吊起时底面的距离(m);
　　　h_4——绑扎点至吊钩中心的索具高度(m)。

③起重半径 R。起重半径的确定,可以按三种情况考虑。

a. 当起重机可以开到构件附近去吊装时,对起重半径没有什么要求,只要计算出起重量和起重高度后,便可以查阅起重机资料来选择起重机的型号及起重臂长度,并可查得在一定起重量 Q 及起重高度 H 下的起重半径 R;还可为确定起重机的开行路线以及停机位置做参考。

b. 当起重机不能够开到构件附近去吊装时,应根据实际所要求的起重半径 R、起重量 Q 和起重高度 H 这三个参数,查阅起重机起重性能表或曲线来选择起重机的型号及起重臂的长度。

c. 当起重臂需跨过已安装好的构件(屋架或天窗架)进行吊装时,应计算起重臂与已安装好的构件不相碰的最小伸臂长度。计算方法有数解法和图解法如图 7.30 所示。

Chapter 7 Installation of Prefabricated Structural Members Construction

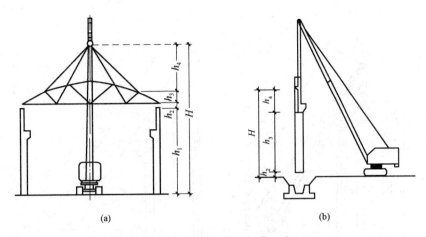

Fig. 7. 29 The Lifting Height Requirement
(a) the installation of roof truss; (b) the installation of column

$$H \geqslant h_1 + h_2 + h_3 + h_4 \qquad (7.2)$$

H—the lifting height of the crane(m);
h_1—the height of the surface of pedestal at the installation point(m);
h_2—the height of the gap between the component and the pedestal, usually no less than 0.3m;
h_3—the distance between the binding point to the bottom of the component(m);
h_4—the distance between the binding point to the rigging at the center of the hook(m).

③Lifting radius R.

The determination of the lifting radius can be considered in the following three cases:

a. When the crane can get access to the component to hoist it, there is no special requirement for the lifting radius. When the lifting height and the lifting weight is determined, the crane model, the boom length and the lifting radius can be determined based on some reference data. The determination of these factors will provide useful information in making the crane route and location.

b. When the crane can't get access to the component to hoist it, the selection of the crane model and the boom length should be based on the requirements for the lifting radius R, lifting weight Q and lifting height H and the data of the crane performance chart or curve.

c. If the boom has to span over some installed components (roof truss or skylight shelf), the minimum length of the boom that can span over the components without bumping into them should be calculated. There are numerical and graphic methods for calculation, as shown in Fig. 7.30.

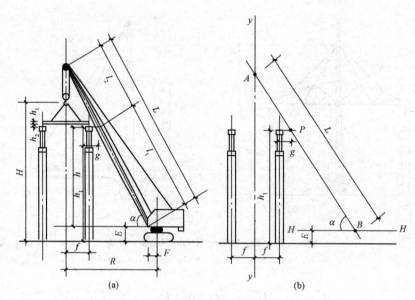

图 7.30 安装屋面板时,起重臂最小长度计算简图
(a)数解法;(b)图解法

④最小起重臂长度的确定。当起重机的起重臂须跨过已吊好的构件上空去吊装构件时,为了不与已吊好的构件相碰,臂长必须满足最小值。最小起重臂长度 L 可按下列方法计算。

a. 数解法。

要求的最小起重臂长,可按下式计算:

$$L \geqslant L_1 + L_2 = h/\sin\alpha + (f+g)/\cos\alpha \tag{7.3}$$

式中 L——起重臂最小长度(m);

h——起重臂下铰点至屋面板吊装支座的垂直高度(m),$h = h_1 - E$;

h_1——停机地面至屋面板吊装支座的高度(m);

f——起重吊钩需跨过已安装好结构的水平距离(m);

g——起重臂轴线与已安装好结构之间在已安构件顶面标高的水平距离,至少取 1 m。

为了使起重臂长度最小,可把上式进行一次微分,并令 $dL/d\alpha = 0$。

在 α 的可能区间$(0, \pi/2)$仅有

$$\alpha = \arctan \sqrt[3]{\frac{h}{f+g}} \tag{7.4}$$

又有 $d^2L/d\alpha^2 > 0$ 知,L 有最小值。

把 α 值代入式(7.3),即可求出最小起重臂的长度。当计算起重半径时:

Chapter 7 Installation of Prefabricated Structural Members Construction

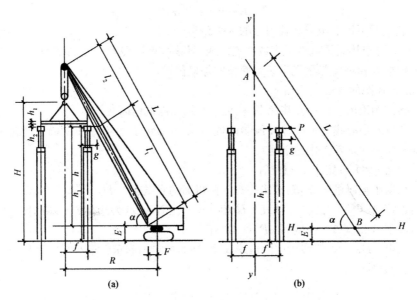

Fig. 7.30 The Calculation of the Minimum Length of Boom in Installing the Roof Slab
(a) numerical method; (b) graphic method

④ The calculation of the minimum length of the boom. If the boom has to span over some installed members (roof truss or skylight shelf), the minimum length of the boom should meet the requirement that it can span over the members without bumping into them. Suppose the minimum boom length is L, it can be calculated with the following methods:

a. Numerical method: L can be calculated in the following formula:

$$L \geqslant L_1 + L_2 = h/\sin\alpha + (f+g)/\cos\alpha \tag{7.3}$$

In the formula L—the minimum boom length

h—the vertical distance between the base hinge of the boom and the pick point of the roof slab in installation. $h = h_1 - E(m)$;

h_1—the distance between the ground and the top of the pedestal of the roof slab(m);

f—the horizontal distance between the hook and the installed component that need to be passed through(m);

g—the horizontal distance between the axis of the boom and the elevation of the top of the installed component, at least 1m.

In order to minimize the length of the boom, the above formula can be differentiated once and made $dL/d\alpha = 0$.

Only in the possible interval of $(0, \pi/2)$

$$\alpha = \arctan\sqrt[3]{\frac{h}{f+g}} \tag{7.4}$$

Because $d^2L/d\alpha^2 > 0$, L has the minimum value.

The minimum boom length can be calculated by replacing the α with the actual value in formula (7.3). When calculating the lifting radius:

$$R = F + L\cos\alpha \tag{7.5}$$

式中 F——起重臂下铰点至回转轴中心的水平距离(m)。

根据 R 和 L 可查用起重机性能表或性能曲线,复核起重量 Q 及起重高度 H,如能满足构件吊装要求,即可根据 R 值确定起重机吊装屋面板的停机位置。

b. 图解法。作图的方法步骤如下:

第一步:按比例绘出构件的安装标高,柱距中心线和停机地面线。

第二步:在柱距中心线上定出臂杆顶端位置 A(d 为吊钩中心到臂杆顶端定滑轮中心的最小距离,是保证滑轮组和正常工作的空间)。

第三步:根据 $g=1$ m 定出 P 点位置。

第四步:根据起重起机的 E 值,绘出平行于停机面的直线 $H-H$。

第五步:连接 A、P 并延长使之与 $H-H$ 相交于一点 B(此点为起重臂下端的铰点中心)。

第六步:高于 A 得到 A_1,连接 A_1、P,得 B_1 等。

第七步:量出 AB、A_1B_1 等线段中的最小长度,即为所求的起重臂最小长度近似值。

① 起重机台数的确定

$$N = \frac{1}{TCK}\sum \frac{Q_i}{P_i} \tag{7.6}$$

式中 N——起重机台数;

T——工期(d);

C——每天工作班数;

K——时间利用系数,一般取 0.8~0.9;

Q_i——每种构件的安装工程量(件或 t);

P_i——每种构件的安装工程量(件或 t)。

注意:决定起重机台数时,应考虑构件装卸、拼装和就位的需要。

2. 结构安装方法

单层工业厂房的结构吊装,通常有两种方法:分件吊装法和综合吊装法。

(1)分件吊装法。分件吊装法就是起重机每开行一次只安装一类或一、二种构件。通常分三次开行即可吊完全部构件,如图 7.31 所示。这种吊装法的一般顺序是:起重机第一次开行,安装柱子;第二次开行,吊装吊车梁、连系梁及柱向支撑;第三次开行,吊装屋架、天窗架、屋面板及屋面支撑等。

分件吊装法的主要优点是:

Chapter 7 Installation of Prefabricated Structural Members Construction

$$R = F + L\cos\alpha \tag{7.5}$$

F—Horizontal distance from base hinge point of the boom to the center of the rotating axis(m).

According to the calculated value of R and L, check the lifting weight Q and lifting height H in the crane performance chart or curve; if they can meet the requirement of component installation, the location of crane in lifting the roof truss can be determined according to R.

b. Graphic method: the steps are as follows:

Step 1: Draw the installation elevation of the members, the central line of column distance and the ground line of the crane according to some scale.

Step 2: Locate A (the top of the boom) on the central line of the column distance (d is the minimum distance from the center of the hook to the top of the center of the pulley on the top of the boom, which is the distance to ensure normal working space of the pulley group).

Step 3: Determine the position of P based on $g = 1m$.

Step 4: Draw a straight-line $H-H$ parallel to the ground line of the crane based on the value of E.

Step 5: Connect A and P and extend them to intersect with $H-H$ at point B (this point is the hinge center of the lower boom).

Step 6: Get A_1 above A, connect A_1, P, get B_1, etc.

Step 7: Measure the length of the AB, $A_1 B_1$ and other line segments, and the minimum value of them is approximate minimum length of the boom.

⑤The calculation of the required number of the cranes

$$N = \frac{1}{TCK} \sum \frac{Q_i}{P_i} \tag{7.6}$$

N—the required number of the cranes;

T—construction period(d);

C—the daily working shifts;

K—The time utilization coefficient, and it is generally 0.8—0.9;

Q_i—Installation quantity of each component(piece or t);

P_i—Installation quantity of each component(piece or t);

Note: When calculating the number of cranes, consider the requirements of loading, unloading, splicing and positioning of members.

2. Installation Method

The mainly structural installation methods for single-storey industrial building are classified members sequent installation and comprehensive installation.

(1) Classified members sequent installation.

In classified members sequent installation, the crane only installs one or two kinds of members at a time. Usually, the members of the industrial building can be installed in three times, as shown in Fig. 7.31. The general sequence of this installation method is: first, install the columns; second, install the crane girders, connecting beams and column bracing; third, install the roof trusses, skylight shelf, roof slab and roof support, etc.

The advantages of classified members sequent installation are:

①构件便于校正；
②构件可以分批进场，供应单一，吊装现场不拥挤；
③吊具变换次数较少，且操作易熟练，吊装速度快；
④可以根据不同构件选用不同性能的起重机械，有利于发挥机械效率，减少施工费用。

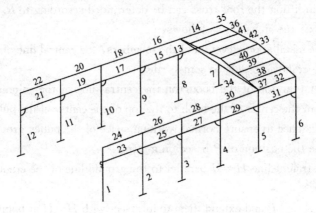

图 7.31　分件安装时的构件吊装顺序
图中数字表示构件吊装顺序，其中
1～12—柱；13～32—单数是吊车梁，双数是连系梁；33、34—屋架；35～42—屋面板

分件吊装法的主要缺点：
①不能为后续工程及早提供工作面；
②起重机开行路线长。

（2）综合吊装法。综合吊装法，又称节间吊装法，即一台起重机每移动一次，就吊装完一个节间内的全部构件，如图 7.32 所示。其顺序是：先吊装完这一节间柱子，柱子固定后立即吊装这个节间的吊车梁、屋架和屋面板等构件；完成这一节间吊装后，起重机移至下一个节间进行吊装，直至厂房结构构件吊装完毕。

综合吊装法的主要优点是：
①由于是以节间为单位进行吊装，因此其他后续工种可以进入已吊装完的节间内进行工作，有利于加速整个工程的进度。
②起重机开行路线短。

综合吊装法的主要缺点是：
由于同时吊装多种类型构件，机械不能发挥最大效率；构件供应现场拥挤，校正困难。故目前较少采用此法。

①It is easy to correct the members;
②Members can be transported to the site in batches, so the site will not be crowded;
③The change of sling devices will be less and the operation can be more familiar, so the installation will be fast.
④Different types of crane will be selected according to their performance, so it can improve mechanical efficiency and reduce construction costs.

The disadvantages of partial installation method are:
①It can't provide workplace to the following work as soon as possible;
②The route of the crane is long.

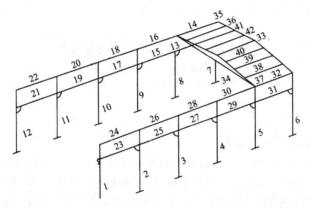

Fig. 7. 31 The Sequence of Installing Members in Classified Members Sequent Installation Method
The numbers represent the installation sequence
1—12—columns;13—32—the odds are the crane girders, the evens are connecting beams;
33, 34—roof trusses;35—42—roof slabs

(2)Comprehensive installation method

The comprehensive installation method, also known as the multi-section installation method, means that every time the crane moves, it installs all the members in one section, as shown in Fig. 7. 32. The installation sequence is as follows: first, install all the columns in one section and fix them, and then install the crane girders, roof truss and roof slabs, etc. ;after the members of this section are all installed, the crane moves to the next section for installation and the installation sequence will continue to the end.

The advantages of comprehensive installation method are:
①After the installation in one section, the following work can be carried on in this section, so the work efficiency can be improved.
②The root of the crane is short.

The disadvantages of comprehensive installation method are:
Different types of members are installed at the same time, the work efficiency of the crane will be low;different members are transported to the site at the same time, so it is crowded on the site and it is difficult to correct them. So, this installation method is rarely adopted.

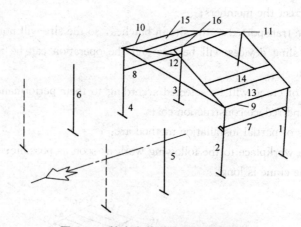

图 7.32 综合安装时的构件吊装顺序
图中数字表示构件吊装顺序,其中
1~6—柱;7、8—吊车梁;9、10—连系梁;11、12—屋架;13~16—屋面板

3. 起重机开行路线及停车位置

起重机的开行路线及停机位置,与起重机的性能、构件的尺寸、重量、构件的平面位置、构件的供应方式以及吊装方法等问题有关。当吊装屋架、屋面板等屋面构件时,起重机大多是沿着跨中开行的。当吊装柱子时,根据厂房跨度大小、柱子尺寸和重量,以及起重机性能,可以沿着跨中开行,也可以沿着跨边开行。

如果用 L 表示厂房跨度,用 b 表示柱的开间距离,用 a 表示起重机开行路线到跨边的距离,那么,起重机除了满足起重量、起重高度要求以外,起重半径 R 还应满足一定条件,如:

(1)吊装柱子时:

a. 当起重半径 $R \geqslant L/2$ 时,起重机沿跨中开行,每个停机位可吊两根柱子,如图 7.33(a)所示;

b. 当 $R \geqslant \sqrt{\left(\dfrac{L}{2}\right)^2 + \left(\dfrac{b}{2}\right)^2}$ 则可吊装四根柱,如图 7.33(b)示;

c. 当 $R < L/2$ 时,起重机需沿跨边开行,每个停机位置吊装 1~2 根柱,如图 7.33(c)所示;

d. $R \geqslant \sqrt{(a)^2 + \left(\dfrac{b}{2}\right)^2}$ 时,起重机在跨内靠边开行,每个停机点则可吊二根柱子,如图 7.33(d)所示。

Chapter 7 Installation of Prefabricated Structural Members Construction

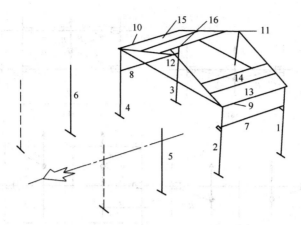

Fig. 7.32 The Sequence of Installing the Members in Comprehensive Installation Method
The numbers represent the installation sequence
1—12—columns; 7, 8—crane girders; 9, 10—connecting beams;
11, 12—roof shelf; 13—16—roof slabs

3. The Crane Route and Its Standing Positions in Installation

The route and the location of crane are related with the performance of crane, the size and weight of members, the plane position of members, the supply mode of members, the installation method, etc.

In the installation of roof truss and roof slab, the crane moves along the center of the span.

In the installation of columns, the crane can move along the center of the span or the border of the span based on the length of the span, the size and weight of the column and the performance of the crane.

We can use L to represent the span of the industrial building, b represent distance between the columns, and a represent distance between the crane route and the border of the span. The crane should meet the requirements of lifting weight and lifting height, and the lifting radius R should also meet the requirement in the following situations:

①In the installation of columns:

a. When the lifting radius $R \geqslant L/2$, the crane moves along the center of the span. At every standing position, the crane can install two columns, as shown in Fig. 7.33 (a);

b. When $R \geqslant \sqrt{\left(\frac{L}{2}\right)^2 + \left(\frac{b}{2}\right)^2}$, the crane can install four columns at one standing position, as shown in Fig. 7.33 (b);

c. When $R < L/2$, the crane should move beside the border of the span, and it can install 1—2 columns at every standing position, as shown in Fig. 7.33 (c);

d. When $R \geqslant \sqrt{(a)^2 + \left(\frac{b}{2}\right)^2}$, the crane should move along the border of the span, and it can install two columns at one standing position, as shown in Fig. 7.33(d).

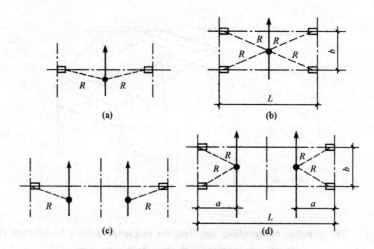

图 7.33 起重机的开行路线及停机位置举例

(2)屋架扶直就位及屋盖系统吊装时,起重机大多数在跨中开行。

图 7.34 所示是单跨厂房采用分件吊装法时起重机开行路线及停车位置图。起重机从Ⓐ轴线进场,沿跨外开行吊装Ⓐ列柱,再沿Ⓑ轴线跨内开行吊装Ⓑ列柱,然后转到Ⓐ轴线扶直屋架并将其就位,再转到Ⓑ轴线吊装Ⓑ列吊车梁、连系梁,随后转到Ⓐ轴线吊装Ⓐ列吊车梁、连系梁。最后转到跨中吊装屋盖系统。

屋架扶直就位及屋盖系统吊装时,起重机大多数在跨中开行。

当单层厂房面积大或具有多跨结构时,为加快工程进度,可将建筑物划分为若干段,选用多台起重机同时施工。每台起重机可单独作业,完成一个区段的全部吊装工作也可以选用不同性能的起重机协同作业,有的专门吊柱,有的专门吊屋盖系统结构,是一种大流水施工方法。

Chapter 7 Installation of Prefabricated Structural Members Construction

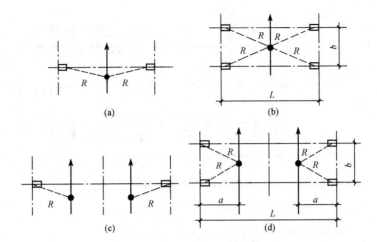

Fig. 7.33 The Route and the Standing Position of the Crane

(2) During erecting and positioning of the roof truss and the installation the roof covering systems, the crane typically moves along the center of the span.

The crane route and the crane standing position of installing single span industrial building in classified members sequent installation method is shown in Fig. 7.34. First, the crane enters the site along the axis Ⓐ and installs the columns in line A when it moves beside the border of the span. Second, the crane installs the columns in line B when it moves along the axis Ⓑ in the span. Third, it moves to axis Ⓐ to carry out the erecting and alignment. Fourth, it comes to axis Ⓑ to install crane girders, connecting beams. Fifth, it moves to axis Ⓐ again to install the crane girders and connecting beams. Sixth, it moves to the center of the span to install the roof system.

When the roof truss is in place and the roof system if hasted, most of the cranes operate in the middle of the span.

When the single-storey industrial building has large area or has several spans, the construction can be divided into several sections and several cranes can work at the same time to improve the work efficiency. One crane can install all the members in one section or several cranes can work together to install different kinds of members in one section. For example, some cranes only install the columns and some cranes install the roof system, and this is a way of stream line construction method.

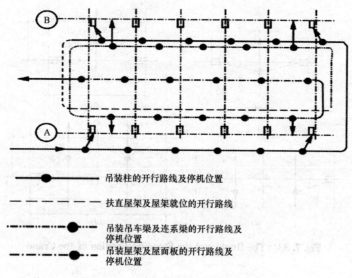

图 7.34 起重机开行路线及停机点位置

7.2.4 构件平面位置的布置

单层工业厂房构件的平面布置,是吊装工程中一件很重要的工作,如果构件布置得合理,可以免除构件在场内的二次搬运,充分发挥机械效益,提高劳动生产率。

关于构件的平面布置,它与吊装方法、起重机性能、构件制作方法等有关。所以应该在确定了吊装方法和起重机后,根据施工现场的实际情况,进行制定平面构件的合理布置方案。

1. 构件的平面布置原则

(1)每跨的构件宜布置在本跨内,如有困难时,也可布置在跨外便于安装的地方。

(2)构件的布置,应便于支模及浇筑混凝土;若为预应力混凝土构件,要留出抽管,穿筋的操作场地。

(3)构件的布置,要满足安装工艺的要求,尽可能布置在起重机的工作幅度内,尽量减少起重机负荷行驶的距离及起伏起重臂的次数。

(4)构件的布置,力求占地最少,保证起重机械、运输车辆的道路畅通。起重机回转时,机身不得与构件相碰。

(5)构件布置时,要注意安装朝向,避免在安装时空中调头,影响安装进度和安全。

(6)构件均应在坚实的地基上浇筑,新填土要加以夯实,以防下沉。

构件的平面布置,分为预制阶段的平面布置和吊装阶段的平面布置两种。

Chapter 7 Installation of Prefabricated Structural Members Construction

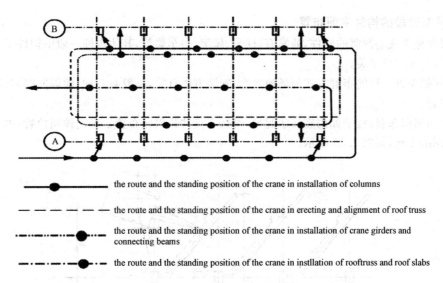

———●——— the route and the standing position of the crane in installation of columns

— — — — — the route and the standing position of the crane in erecting and alignment of roof truss

—··—··—●—··— the route and the standing position of the crane in installation of crane girders and connecting beams

—·—·—●—·— the route and the standing position of the crane in instllation of rooftruss and roof slabs

Fig. 7.34 The Route and the Standing Positions of the Cranes

7.2.4 The Layout of the Plane Position of the Members

The layout of the plane position of the members is an important work in the installation of single-storey industrial building. If the members are reasonably laid on the construction site, it will avoid being moved again on the site and it will improve the work efficiency.

The plane layout of the members is related with the installation methods, the crane performance, the fabrication methods, etc., so the plane position of the members should be based on the practical situation of the site and the selection of the installation method and the type of crane.

1. The Layout Principle of the Members

(1) The members for every span should be laid in the span. If it is difficult to do so, lay them outside the span but be convenient for installation.

(2) The layout of the members should be convenient for setting formwork and casting concrete. If the members are made of precast concrete, leave enough space for withdrawing steel tube and inserting tendons.

(3) The layout of the members should meet the requirements of installation and it is better to be laid within the work span of the crane and be suitable for reducing the travelling distance of the crane under load and reducing the times of moving boom.

(4) The layout of the members should take least space to leave enough space for the moving of crane and other vehicles. When the crane rotates, it mustn't bump into the members.

(5) When determining the layout of the members, pay attention to the installation direction. The direction of the members should meet the installation requirement, so there will be no need for the members to turn around in the air to influence the installation speed and safety.

(6) The members should be cast on solid ground, and the newly filled soil should be tamped to avoid settlement.

The layout of the members can be classified into the layout in the prefabrication stage and the layout in the installation stage.

2. 预制阶段的构件平面布置

需要在施工现场预制的构件,通常有:柱子、屋架、吊车梁等,其他构件一般由构件工厂或现场以外制作,运来进行吊装。

(1)柱的布置。柱的布置方式与场地大小、安装方法有关,一般有三种:即斜向布置、纵向布置及横向布置。

①柱的斜向布置:柱子如用旋转法起吊,可按三点共弧斜向布置。确定预制位置,可采用作图法,其作图的步骤,如图7.35所示。

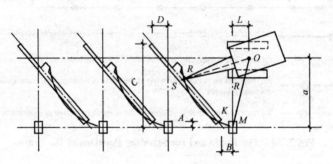

图7.35 柱子的斜向布置

a. 确定起重机开行路线到柱基中线的距离 a 和起重机吊装柱子时与起重机相应的工作幅度 R,起重机的最小工作幅度 R_{min} 有关,要求:

$$R_{min} < a \leqslant R \tag{7.7}$$

同时,开行路线不要通过回填土地段,不要靠近构件,防止起重机回转时碰撞构件。

b. 确定起重机的停机点。安装柱子时,起重机位于所吊柱子的横轴线稍后的范围内比较合适;这样,司机可看到柱子的吊装情况便于安装对位。停机点确定的方法是,以要安装的基础杯口中心 M 为圆心,所选的工作幅度 R 为半径,画弧相交开行路线于 O 点,O 点即为安装那根柱子的停机点。

c. 确定柱的预制位置。以停机点 O 为圆心,OM 为半径画弧,在靠近柱基的弧上任选一点 K 作为预制时柱脚中心。K 点选定后,以 K 为圆心,柱脚到吊点的长度为半径画弧,与 OM 半径所画的弧相交于 S,连 KS 线,得出柱中心线,即可画出柱子的模板位置图。量出柱顶、柱脚中心点到柱列纵横轴线的距离 A、B、C、D,作为支模时的参考。

Chapter 7 Installation of Prefabricated Structural Members Construction

2. The Plane Layout of the Members in Prefabrication Stage

The members that need to be prefabricated on the construction site are mainly the following: columns, roof truss, crane girders, etc. Other members are mainly made in the factory or outside the construction site and they are transported to the site.

(1) The layout of the columns. The layout of the columns is related with the size of the site and the installation method. Generally, there are three methods: oblique layout, vertical layout and horizontal layout.

① The oblique layout of the columns. If the column is hoisted by rotating method, it can be laid obliquely based on the requirement of three points (center of the pocket mouth, center of the end of the column and the binding point) on the same arc. The prefabrication position can be determined by the graphic method. The steps for graphic method are shown in Fig. 7.35.

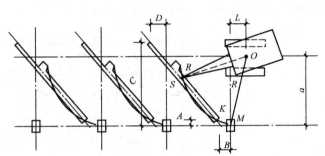

Fig. 7.35 The Oblique Layout of the Column

a. Determine the distance (a) from the route of the crane to the center of the foundation and the working range (R) of the crane in lifting columns. The minimum working range (R_{min}) of the crane should meet the following requirement:

$$R_{min} < a \leqslant R \tag{7.7}$$

Notes: the route of the crane should avoid the backfill ground and keep a distance from the members to avoid bump into them.

b. Determine the standing position of the crane. When installing the column, the crane is suitably located at the place slightly backward to the transverse axis of the column. In this way, the hoister can see the lifting of the column easily and this is good for installation and alignment. The method to determine the standing position is to take the center of the pocket mouth M as the peripheral core, the working range R as the radius, and draw an arc. The arc and the route of the crane will intersect at the point O, and O will be the standing position in the installation of the column for the pocket base.

c. Determine the prefabrication position of the column. Take the standing position O as the peripheral core and OM as the radius, draw an arc. Select one point (K) that is near the foundation as the center of the end of the column, take K as the peripheral core and the distance between the end of the column to the bind point as the radius, draw an arc. The arc will intersect with the previous drawn arc (radius OM) at the point of S. Connect K and S and the line KS is the central line of the column. Up to now, the location of the formwork for prefabricating the column can be drawn. Measure the distance from the center of the end of the column to the horizontal axis as A; measure the distance from the center of the end of the column to the vertical axis as B; measure the distance from center of the top of the column to the horizontal axis as C; measure the distance from the center of the top of the column to the vertical axis as D. These values will provide help in setting framework for the column.

布置柱时，要注意柱牛腿的朝向，避免安装时在空中调头。当柱布置在跨内时，牛腿应面向起重机；布置在跨外时，牛腿应背向起重机。

布置柱时，有时由于场地限制或柱身过长，无法做到三点（杯口、柱脚、吊点）共弧，可根据不同情况，布置成两点共弧。两点共弧的布置方法有两种：一是将杯口、柱脚共弧，吊点放在工作幅度 R 之外，如图 7.36(a)所示。安装时，先用较大的工作幅度 R' 吊起柱子，并升起重臂，当工作幅度变为 R 后，停止升臂，随之用旋转法安装柱子。另一种方法是：将吊点、杯口共弧，安装时采用滑行法，即起重机在吊点上空升钩，柱脚向前滑行，直到柱子成直立状态，起重臂稍加回转，即可将柱子插入杯口，如图 7.36(b)所示。

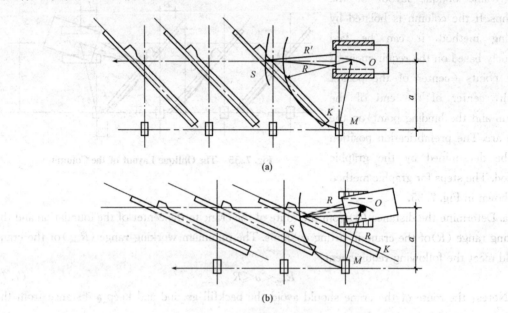

图 7.36 两点共弧布置法
(a)杯口与柱脚两点共弧；(b)吊点与杯口两点共弧

Chapter 7 Installation of Prefabricated Structural Members Construction

Pay attention to the direction of the corbel in the column layout. The direction of the corbel should be convenient for installation and avoid turning around in the air. When the column is laid in the span, the corbel should face the crane and when the column is laid outside the span, the corbel should back to the crane.

When the three points (the center of the pocket mouth, the center of the column end and the binding point) can't be set on the same arc because of the limitation of the construction site or the length of the column, two points of the three can be set on the same arc. The following are the two methods for setting two points on the same arc: First, set the center of the pocket mouth and the center of the end of the column on the same arc, and the binding point can be outside the working range R, as shown in Fig. 7.36 (a).

The distance from the binding point to the standing point is R'. Hoist the column with the working range R' and lift the boom. When the working range of crane becomes R, stop the boom and install the column with rotating method.

Second, set the binding point and the center of the pocket mouth on the same arc and install the column with sliding method. The crane lifts the hook over the binding point and the end of the column slides forward. When the column becomes vertical, rotate the crane and insert the column into the pocket mouth, as shown in Fig. 7.36 (b).

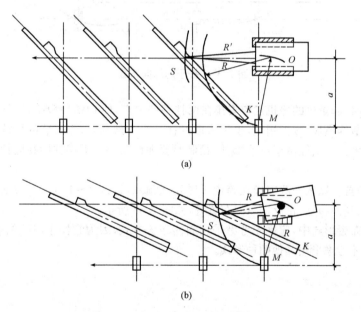

Fig. 7.36 Layout of Two Points on the Same Arc
(a) The center of the pocket mouth and the end of the column on the same arc;
(b) The binding point and the center of the pocket mouth on the same arc

②柱的纵向布置：对一些较轻松的柱起重机能力有富余，考虑到节约场地，方便构件制作，可顺柱列纵向布置，如图7.37所示。柱子纵向布置，绑扎点与杯口中心两点共弧。

若柱子长度大于12 m，柱子纵向布置宜排成两行，如图7.37(a)所示；

若柱子长度小于12 m，则可叠浇排成一行，如图7.37(b)所示。

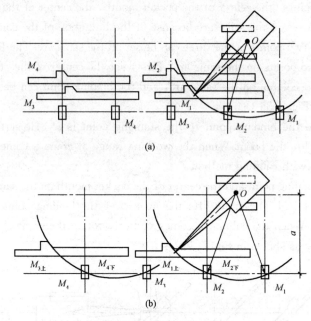

图7.37 柱子的纵向布置

柱纵向布置时，起重机的停机点应安排在两柱基的中点，使$OM_1 = OM_2$，这样，每一停机点可吊两根柱。为了节约模板，减少用地，也可采取两柱叠浇。预制时，先安装的柱放在上层，两柱之间要做好隔离措施。上层柱由于不能绑扎，预制时要埋设吊环。下层柱则在底模预留砂孔，便于起吊时穿钢丝绳。

(2)屋架的布置。屋架一般安排在跨内平卧迭浇预制，每迭3～4榀。布置的方式有三种：正面斜向布置、正反斜向布置、正反向纵向布置等，如图7.38所示。

在上述三种布置形式中，应优先考虑采用斜向布置方式，因为它便于屋架的扶直就位。只有在场地受限制时才考虑采用其他两种形式。

Chapter 7 Installation of Prefabricated Structural Members Construction

②The vertical layout of the columns. For small or light columns, the crane can hoist them easily, so in order to save space and be convenient for prefabrication, the columns can be laid vertically, as shown in Fig. 7. 37. The binding point and the center of the pocket mouth should be on the same arc.

When the column is longer than 12m, the columns should be laid in two rows, as shown in Fig. 7. 37 (a);

When the column is less than 12m, they can be laid in one row, as shown in Fig. 7. 37 (b).

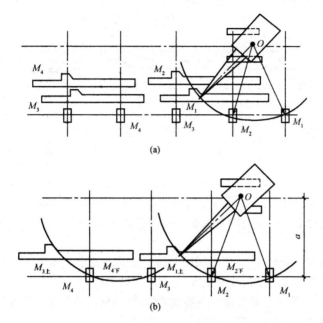

Fig. 7. 37 The Vertical Layout of the Columns

When the columns laid vertically, the standing position of the crane should be in the middle of the two columns, so that $OM_1 = OM_2$, so the crane can install two columns on one standing position. Two columns can be cast overlappingly to save formwork and space. The column installed first should be laid on the top and some protection measures should be taken between the two columns. It is impossible to bind the column on the top, so embed suspension ring in casting. When setting the formwork for the column on the bottom, create a sand filled slot in the bottom formwork to facilitate wire rope lings through-in to lift the columns.

(2) The layout of the roof trusses. The roof trusses are usually cast in the span with overlapping casting method. There are 3 or 4 pieces of trusses overlapped together. The following three are the methods for laying roof trusses: face-face oblique layout, face-back oblique layout and face-back layout, as shown in Fig. 7. 38.

Among the above three layout methods, the face-face oblique layout method is preferred because it is convenient for erecting and positioning. When the site is too limited for this method, the other two methods can be adopted.

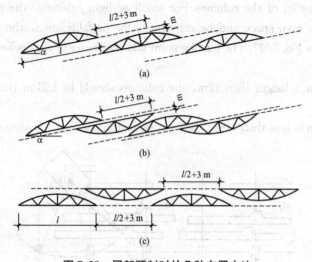

图 7.38 屋架预制时的几种布置方法
(a)正面斜向布置;(b)正反斜向布置;(c)正反向纵向布置

屋架正面斜向布置时,下弦与厂房纵轴线的夹角 $\alpha=10°\sim20°$;预应力屋架的两端应留出($l/2$)$+3$ m 的距离(l 为屋架的跨度)。如用胶皮管预留孔道时,距离可适当缩短。

屋架之间的间隙可取 1 m 左右以上支模及浇筑混凝土。屋架之间相互搭接的长度视场地大小及需要而定。

在布置屋架的预制位置时,还应考虑到屋架的扶直排放要求及屋架扶直的先后次序,先扶直的放在上层。对屋架两端朝向及预埋件位置,也要注意做出标记。

(3)吊车梁的布置。当吊车梁安排在现场预制时,可靠近柱基顺纵轴线或略做倾斜布置,也可插在柱子的空当中预制,或在场外集中预制等。

3. 安装阶段构件的就位布置及运输堆放

由于柱子在预制时,即已按吊装阶段的堆放要求进行了布置,所以柱子在两个阶段的布置是一致的。一般先吊柱子,以便腾出场地堆放其他构件。所以吊装阶段构件的堆放,主要是指屋架、吊车梁、屋面板等构件。

(1)屋架的吊装阶段布置。为了适应吊装阶段吊装屋架的工艺要求,首先用起重机把屋架由平卧转为直立,这叫屋架的扶直或翻身起扳。屋架扶直以后,用起重机把屋架吊起并移到吊装前的堆放位置,叫就位。堆放方式一般有两种:即斜向就位和纵向就位。

Chapter 7 Installation of Prefabricated Structural Members Construction

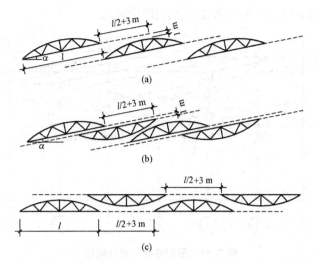

Fig. 7.38 The Different Layout Methods in Prefabricating Roof Trusses
(a) face-face oblique layout; (b) face-back oblique layout; (c) face-back vertical layout

When taking the face-face oblique layout, the angle between the lower chord and the longitudinal axis of the industrial building is α and $\alpha = 10° — 20°$. Suppose l is the length of the span, for prestressed roof truss, leave $l/2+3m$ for both the ends of the truss. If rubber hoses are used for preforming holes for placing pre-stressing tendons, the spaces between the trusses can be decreased appropriately.

The intervals between the roof trusses should be kept at least about 1 m for giving space of erecting formwork and pouring concrete. The length of the overlapped part of the trusses is up to the practical needs and condition on the site.

When making the layout of the roof trusses, we should consider the needs of erecting and the erecting sequence. The roof trusses being erected first should be laid on the top. Mark the direction of the roof truss and the position for embedding members.

(3) The layout of crane girders. When the crane girders are planned to be prefabricated on site, they can be arranged along the horizontal axis of the foundation or slightly oblique to it or they can be cast on the open spaces between the columns or prefabricated outside the construction site.

3. The Plane Layout, Transportation and Stacking of the Members in Installation Stage

The columns are laid reasonably when cast in situ, so the layout of the columns is the same with the prefabrication stage. The columns will be installed first to make room for other members, so the layout of the members in installation stage mainly involves the stacking of roof truss, crane girders, roof slabs, etc.

(1) The layout of the roof trusses in installation stage.

In order to meet the technological requirements of lifting roof trusses, the roof truss should be transformed from horizontal to vertical by crane, which is called erecting or turning over of roof truss. After the roof truss is straightened, it is lifted by a crane and moved to the stacking position, which is called positioning. Oblique and vertical positioning are the two general ways.

①屋架斜向就位如图7.39所示,可按下述作图法确定。

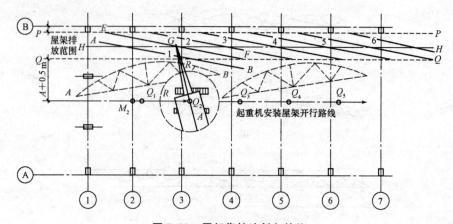

图7.39 屋架靠柱边斜向就位
(虚线表示屋架预制时的位置)

a. 确定起重机吊装屋架时的开行路线及停机位置:吊装屋架时,起重机一般沿跨中开行。需要在跨中标出开行路线(在图上画出开行路线)。

停机位置的确定,以要吊装屋架的设计位置轴线中心为圆心,以所选择的起重半径R为半径画弧线交于开行路线于O点,该点即为吊装该屋架时的停机点。

b. 确定屋架的就位范围:屋架宜靠柱边就位,即可利用柱子作为屋架就位后的临时支撑。所以要求屋架离开柱边不小于0.2 m。

外边线:场地受限制时,屋架端头可以伸出跨外一些。这样,我们首先可以定出屋架就位的外边线$P-P$。

内边线:起重机在吊装时要回转,若起重机尾部至回转中心距离为A,那么在距离起重机开行路线$A+0.5$ m范围内不宜有构件堆放。所以,由此可定出内边线$Q-Q$;在$P-P$和$Q-Q$两线间,即为屋架的就位范围。

c. 确定屋架的就位位置:屋架就位范围确定之后,画出$P-P$与$Q-Q$的中心线$H-H$,那么就位后屋架的中心点均应在$H-H$线上。

屋架斜向就位位置确定方法是:以停机点O_2为圆心,起重半径R为半径,画弧线交于$H-H$线上于G点,G点即为②轴线就位后屋架的中点。再以G点为圆心,以屋架跨度的1/2为半径,画线交于$P-P$、$Q-Q$两线于E和F点,连接EF,即为②轴线屋架就位的位置。其他屋架就位位置均应平行此屋架。

① The oblique positioning is shown in Fig. 7.39 and it can be determined by the following graphic method.

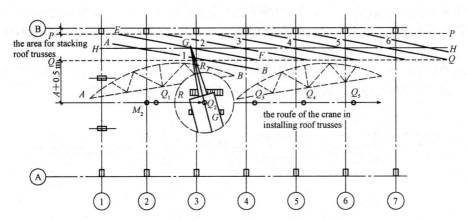

Fig. 7.39   Oblique Positioning of Roof Trusses
(the dotted lines represent the prefabrication position of roof trusses)

a. Determine the route and the standing position of the crane in installing roof trusses: in installing the roof trusses, the crane moves along the central line of the span. Draw the central line of the span and mark it as the route of the crane.

Take axis center of designed position of the roof truss as the peripheral core, the lifting radius R as the radius, and draw an arc. The arc intersects with the route of the crane at a point O, and O is the standing position of the crane in installing the roof truss.

b. Determine the positioning range of the roof trusses: The positioning of the roof trusses should be next to the columns, so that the columns can be contemporary supporting in positioning. The distance between the column and the roof truss should be no less than 0.2m.

The exterior border line: when the construction site is limited, the end of the roof truss can extent the span so it is necessary to determine the exterior border line P—P.

The interior border line: The crane will rotate in installing the roof trusses. If the distance from the end of the crane to the rotating center is A, the range $A+0.5$m besides the route of the crane is not suitable for stacking roof trusses. Determine the interior border Q—Q, and the range between Q—Q and P—P is the place for roof positioning.

c. Determine the point for roof truss positioning: After determining the range of the positioning, draw the central line H—H of the range between P—P and Q—Q, and the center of the roof trusses after positioning should be on the line H—H.

The oblique positioning method is: Take the standing position O_2 as peripheral core, the lifting radius as the radius and draw an arc. The arc will intersect with H—H at a point G, so G is the center of the roof truss in axis ② after positioning. Take G as the peripheral core, 1/2 of the span as the radius, and draw an arc. The arc will intersect with P—P and Q—Q at the point E and F, connect E and F. The line EF is the positioning place of the roof truss on axis ②. The positioning place of other roof trusses should be parallel to line EF.

对于①轴线的屋架而言,当安装好抗风柱时,需要退到②轴线屋架附近就位。

②屋架纵向就位如图 7.40 所示。

屋架纵向就位,一般以 4~5 榀为一组靠近边柱顺轴线纵向排列。屋架与柱之间,屋架与屋架之间的净距不小于 0.2 m,相互之间用铅丝绑扎牢靠。每组之间应留出 3 m 左右的间距,作为横向通道。

每组屋架就位中心线,应安排在该组屋架倒数第二榀安装轴线之后 2 m 外。这样可以避免在已安装好的屋架下绑扎和起吊屋架;起吊以后也不会和已安装好的屋架相碰。

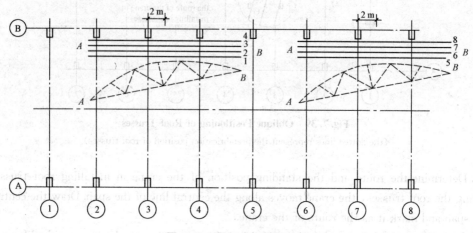

图 7.40 屋架成组纵向就位

(2)吊车梁、连系梁、天窗架和屋面板的运输堆放。

单层工业厂房的吊车梁、连系梁、天窗架和屋面板等,一般在预制厂集中生产,然后运至工地安装。

构件运至现场后,应根据施工组织设计规定的位置,按编号及吊装顺序进行堆放。

吊车梁、连系梁、天窗架的就位位置,一般在吊装位置的柱列附近,不论跨内跨外均可,条件允许时也可随运输随吊装。

屋面板则由起重机吊装时的起重半径决定。当在跨内布置时,后退 3~4 个节间沿柱边堆放;在跨外布置时,应后退 1~2 个节间靠柱边堆放,以屋架吊装停机点附近、起重半径内旋转路程短为标准。每 6~8 块为一叠堆放。

Chapter 7 Installation of Prefabricated Structural Members Construction

For the roof trusses on axis ①, after the installation of the anti-wind columns, the positioning place should be near axis ②.

②The standing position of roof truss is shown in Fig. 7. 40.

A group of 4 to 5 pieces of standing roof trusses is arranged near the outer columns and lines up with their longitudinal axis. The distance between the roof truss and column and the distance between the roof trusses should be no less than 0. 2m. The adjacent members should be firmly tied with lead wire. The distance between different groups of roof trusses should be about 3m for the horizontal passage.

The central line of the roof trusses after positioning shall be arranged 2m away from the axis of the penultimate roof truss in this group. This will avoid the operation of binding and lifting the roof truss under the installed roof trusses and it will avoid bumping into the installed roof trusses.

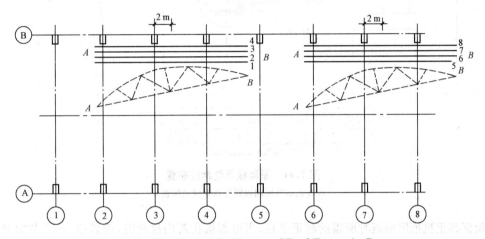

Fig. 7. 40 The Vertical Positioning of Roof Trusses in Group

(2) The transportation and stacking of crane girders, connecting beams, skylight shelves and roof slabs

The crane girders, connecting beams, skylight shelves and roof slabs for single-storey industrial building are prefabricated in prefabrication industrial building and then they are transported to and installed at the construction site. When they are transported to the construction site, they should be stacked at the position stipulated in the construction organization design and the stacking should follow the numbers marked on the members and the installation sequence.

The positioning place of the crane girders, connecting beams and skylight shelves should be near the columns. They can be stacked in or out of the span. If it is necessary, it can be installed directly after being transported.

The positioning place of the roof slabs is determined by the lifting radius of the crane. When they are stacked in the span, the stacking place should be near the columns and back 3 or 4 sections from the installation place. When they are stacked out of the span, they should back 1 or 2 sections from the installation place and should be near the standing position of the crane and shorten the rotating distance. 6—8 pieces can be stacked together as a group.

吊车梁、连系梁的就位位置，一般在其吊装位置的柱列附近，跨内跨外均可。也可以从运输车上直接吊装，不需在现场排放。屋面板的就位位置，跨内跨外均可，如图7.41所示。

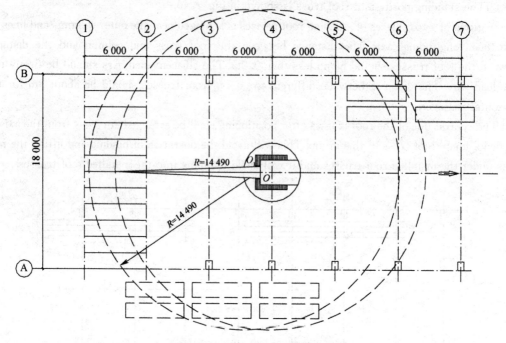

图7.41 屋面板吊装就位布置
（虚线表示当屋面板跨外布置时之位置）

根据起重机吊屋面板时所需的起重半径，当屋面板在跨内排放时，应后退3～4节间开始排放；若在跨外排放，应后退1～2节间开始排放。

实际工程中，构件的平面布置会受很多因素影响，制定时要密切联系现场实际情况，确定出切实可行的构件平面布置图。排放构件时，可将纸板按一定的比例剪切出各类构件形状的小模型，在同样比例的平面图上进行布置和调整。经研究可行后，给出构件平面布置图。

7.3 装配式施工

装配式混凝土结构是由预制混凝土构件通过可靠的连接方式装配而成的混凝土结构，包括装配整体式混凝土结构、全装配混凝土结构等。在建筑工程中，简称装配式建筑；在结构工程中，简称装配式结构。

The stacking place of the crane girders and the connecting beams should be near the columns for the installation of them. They can be stacked in or out of the span. They can be also installed directly when they are transported. The stacking place of roof slabs can be in or out of the span, as shown in Fig. 7. 41.

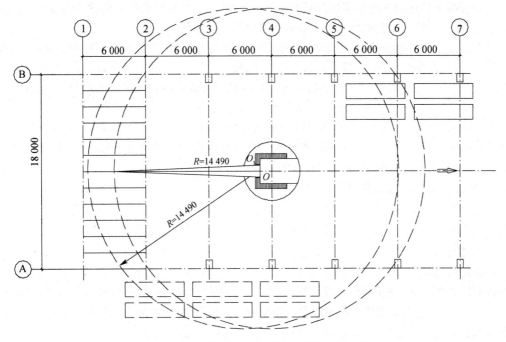

Fig. 7. 41　The Stacking Place of Roof Slabs
(The dotted lines represent the stacking place out of the span)

According to the lifting radius of the crane, when the roof slabs are arranged in the span, they should drive back 3 or 4 sections from the installation place; when they are arranged out of the span, they should drive back 1 or 2 sections from the installation place.

In practice, the plane layout of the members will be influenced by many factors, so the plan of the layout should be based on the practical condition of the site. In designing, we can cut cardboard to the shape of the members in certain scale and simulate the layout on the plan and adjust the position to make it reasonable. After that, the plane layout plan can be drawn.

7. 3　Construction of Prefabricated Concrete Structure

Prefabricated concrete structure is a concrete structure composed of precast concrete members through reliable connection and it includes hybrid concrete construction structure, total precast concrete construction structure, etc. In construction engineering, it is called prefabricated construction, and in structural engineering, it is called prefabricated structure.

装配整体式混凝土结构是国内外建筑工业化最重要的生产方式之一,具有提高建筑质量、缩短工期、节约能源、减少消耗、清洁生产等诸多优点。

7.3.1 预制混凝土构件制作与运输

预制混凝土构件制作前应审核预制构件深化设计图纸,并根据构件深化设计图纸进行模具设计,其中影响构件性能的变更修改应由原施工图设计单位确认。

预制混凝土构件制作前,应根据构件特点制定生产方案,明确各阶段质量控制要点,具体内容包括:生产计划及生产工艺、模具计划及模具方案、技术质量控制措施、成品存放、保护及运输方案等内容。必要时应进行预制构件脱模、吊运、存放、翻转及运输等相关内容的承载力、裂缝和变形验算。

预制混凝土构件按照产品种类有预制外墙板、内墙板、叠合板、楼梯板、阳台板、梁和柱等。无论哪种形式的预制构件生产主流程基本相同,包括:模具清扫与组装、钢筋加工安装及预埋件安装、混凝土浇筑及表面处理、养护、脱模、存储、标识、运输。

1. 装配式结构的基本构件

装配式结构的基本构件主要包括柱、梁、剪力墙、楼(屋)面板、楼梯、阳台、空调板、女儿墙等,这些主要受力构件通常在工厂预制加工完成,待强度符合规定要求后进行现场装配施工。

2. 围护构件

围护构件是指围合、构成建筑空间,抵御环境不利影响的构件,本章中只展开讲解外围护墙和预制内隔墙相关内容。外围护墙用以抵御风雨、温度变化、太阳辐射等,应具有保温、隔热、隔声、防水、防潮、耐火、耐久等性能。内隔墙起分隔室内空间作用,应具有隔声、隔视线的性能以及某些特殊要求。

3. 运输

预制构件的运输应制定运输计划及方案,包括运输时间、次序、堆放场地、运输线路、固定要求、堆放支垫及成品保护措施等内容。对于超高、超宽、形状特殊的大型构件的运输和堆放应采取专门质量安全保证措施。

Chapter 7 Installation of Prefabricated Structural Members Construction

Prefabricated hybrid concrete construction structure is one of the most important production modes in construction industrialization in China and overseas. It has many advantages, such as improving construction quality, shortening construction period, saving energy, reducing consumption and pollution.

7.3.1 The Production and Transportation of Precast Concrete Members

Before fabricating concrete members, we should check the design drawing and design the formwork for the members according to the drawing. If we have to make some changes that will influence the performance of the members, the changes have to be confirmed by design unit.

Before fabricating the concrete members, the production plan has to be made based on the characteristics of the members and the quality controlling methods have to be clear. The following are the specific aspects: production plan and production method, formwork plan and method, quality controlling measures, the plan of stacking, protection and transportation, etc. If it is necessary, make bearing strength test, crack and deformation test in deforming, lifting, stacking, turning over and transportation of the members.

The prefabricated members are classified into prefabricated exterior slab, interior slab, overlapped slab, staircase slab, balcony slab, beams and columns, etc. The production procedures are almost the same and they may include the following steps: the clearance and assembling of formwork, the installation of reinforcing bars and embedded members, the casting of concrete and the treatment of the concrete face, curing, stripping off formworks, stacking, labelling and transportation.

1. The Basic Members of Prefabricated Structure

The basic members of prefabricated structure are columns, beams, shear walls, slabs, staircase, balcony, air conditioning slab, parapet wall, etc. These members are prefabricated in prefabrication industrial building and when their strength reaches the design specification, they can be transported to the site and installed on the site.

2. Enclosing Members

Enclosing members are used to enclose and form building space to resist the influence of the environment. In this part, we only introduce the content related with exterior enclosing wall and prefabricated interior wall. The exterior enclosing walls are to resist the wind and rain, the temperature changes, solar radiation, etc. , so they should have the performance of thermal insulation, heat insulation, sound insulation, waterproof, moisture-proof, fire resistance, durability and so on. Interior walls are to separate the interior space, so they should have the performance of sound insulation, sight insulation and some special requirements.

3. Transportation

We should make transportation plan for prefabricated members and the plan should include the following aspects: the transportation time, transportation sequence, the stacking place, the transportation route, the fixing requirement, the stacking brackets and the protection measures in transportation, etc. Special protection measures should be taken for the transportation and stacking of super-high, super-wide and special-shaped members.

(1)制混凝土构件运输宜选用低平板车,并采用专用托架,构件与托架绑扎牢固。

(2)预制混凝土梁、楼板、阳台板宜采用平放运输;外墙板宜采用竖直立放运输;柱可采用平放运输,当采用立放运输时应防止倾覆。

(3)预制混凝土梁、柱构件运输时平放不宜超过2层。

(4)搬运托架、车厢板和预制混凝土构件间应放入柔性材料,构件应用钢丝绳或夹具与托架绑扎,构件边角或锁链接触部位的混凝土应采用柔性垫衬材料保护。

7.3.2 预制混凝土构件的连接

装配式结构中,构件与接缝处的纵向钢筋应根据接头受力、施工工艺等情况的不同,选用钢筋套筒灌浆连接、焊接连接、浆锚搭接连接、机械连接、螺栓连接、栓焊混合连接、绑扎连接、混凝土连接等连接方式。这里介绍钢筋套筒灌浆连接。

钢筋套筒灌浆连接是在预制混凝土构件内预埋的金属套筒中插入钢筋,并在套筒中灌浆。钢筋套筒灌浆连接是一种因工程实践的需要和技术发展而产生的新型的连接方式。该连接方式弥补了传统连接方式(焊接、机械连接、螺栓连接等)的不足,得到了迅速的发展和应用。

1. 分类

按照钢筋与套筒的连接方式不同,该接头分为全灌浆接头、半灌浆接头两种,见图7.42。全灌浆接头是传统的灌浆连接接头形式,套筒两端的钢筋均采用灌浆连接,两端钢筋均是带肋钢筋。半灌浆接头是一端钢筋用灌浆连接,另一端采用非灌浆方法(例如螺纹连接)连接的接头。

2. 应用

钢筋套筒灌浆连接主要适用于装配整体式混凝土结构的预制剪力墙、预制柱等预制构件的纵向钢筋连接,也可用于叠合梁等后浇部位的纵向钢筋连接。

3. 对接头性能、套筒、灌浆料的要求

钢筋套筒灌浆连接接头在同截面布置时,接头性能应达到钢筋机械连接接头的最高性能等级,国内建筑工程的接头应满足国家行业标准《钢筋机械连接技术规程》(JGJ 107—2016)中的Ⅰ级性能指标。套筒的各项指标应符合《钢筋连接用灌浆套筒》(JG/T 398—2019)的要求。灌浆料的各项指标应符合《钢筋连接用套筒灌浆料》(JG/T 408—2019)的要求。

Chapter 7　Installation of Prefabricated Structural Members Construction

(1) The precast concrete members should better be transported on low flat truck and facilitated with specialized brackets, and the brackets and members should be bound together.

(2) The prefabricated concrete beams, slabs, balcony slabs should be laid horizontally in transportation; the exterior slabs should be laid vertically in transportation; the columns should better be horizontal in transportation and if they have to be laid vertical, some measures should be taken to prevent falling down.

(3) In transporting prefabricated beams and columns, if they are laid horizontally and overlapped, they should be overlapped at most two layers.

(4) In transporting prefabricated members, the gaps between the members should be filled with flexible material. The members should be bound with steel wires or grippers with the brackets. Some flexible material should be used to cover the edges and binding points of the members to protect them.

7.3.2　The Splicing of Precast Concrete Members

In precast concrete structures, the way of splicing longitudinal rebar within joints of the members are mainly influenced by the bar stress conditions in the junctions, the workmanship and so on. The following are the generally adopted connection methods: grouting sleeve splicing, welded splicing, grout encased lap splicing, mechanical splicing, bolting, bolting-welding, lap splicing, casting, etc. In this part, we mainly focus on the method of bar joint with grouting sleeve splicing.

In this method, insert reinforcing bars into the sleeves pre-embedded in the members and grout mortar into the sleeves. This is a new technology based on practical needs and technology development. This method makes up the shortage of traditional splicing methods, such as welding, mechanical splicing and bolting, so it is developed rapidly.

1. Classification

Based on the splicing method of the bar and the grouting sleeve, it can be classified into fully grouting sleeve joint and half grouting sleeve joint, as shown in Fig. 7.42. Fulling grouting sleeve joint is a traditional splicing method, in which both ends of ribbed reinforcing bar are connected with sleeves in grouting. Half grouting sleeve joint is a splicing method in which one end of the reinforcing bar is connected with sleeve in grouting, and the other end is connected with sleeve in other methods, such as bolting.

2. Application

The splicing method of bar joint with sleeve grouting is mainly suitable for vertical joints in prefabricated shear wall, prefabricated columns and other members in prefabricated hybrid concrete construction structure and it is also suitable for the splices of longitudinal rebar in composite beams and other precast members where post-pour concrete to be applied.

3. The Requirements for Joints, Sleeve and Grouting Material

The performance of the sleeve joints should meet the highest grade in the ganging performance of steel bars. In domestic construction engineering, the joints should meet the grade I performance index of the national industry standard *Technical Specification for Mechanical* Splicing of Steel Reinforcing Bars (JGJ 107—2016). The indicators of sleeve should meet *The of Grouting Sleeve for Rebars Splicing* (JG/T 398—2019). The indexes of grouting material should meet Cementitious Grout for Sleeve of Rebar Splicing (JG/T 408—2019).

建筑施工技术(中英文对照版)

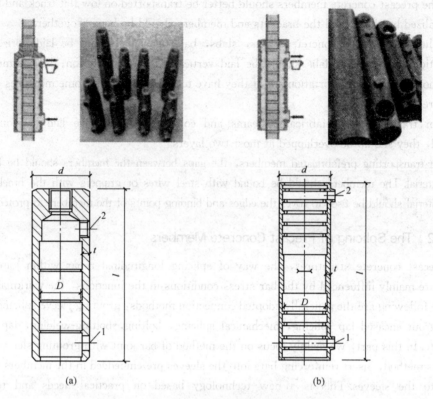

图 7.42　灌浆套筒剖面图
(a)半灌浆接头；(b)全灌浆接头
1—灌浆孔；l—套筒总长；2—排浆孔；d—套筒外径；D—套筒锚固段环形突起部分的内径

7.3.3　预制施工工艺流程

装配整体式混凝土框架结构是全部或部分框架梁、柱采用预制构件构建成的装配整体式混凝土结构。简称装配整体式框架结构。

装配整体式混凝土框架结构是以预制柱(或现浇柱)、叠合板、叠合梁为主要预制构件，并通过叠合板的现浇以及节点部位的后浇混凝土而形成的混凝土结构，其承载力和变形满足现行国家规范的应用要求(图 7.43)。

Chapter 7 Installation of Prefabricated Structural Members Construction

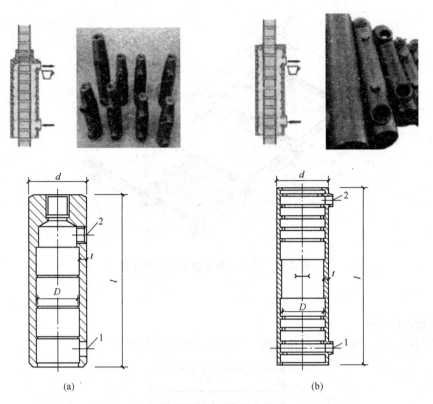

Fig. 7.42 Sectional Drawing of Grouting Sleeves

(a) semi-grout joint; (b) full filling paddle joint

1—grouting hole; l—the total length of sleeve; 2—an outlet for discharging mortar;
d—outer diameter of the sleeve; D—inter diameter of the sleeve in anchoring part

7.3.3 The Construction Process of Prefabrication Construction

The prefabricated hybrid concrete frame structure is a structure that is fully or partly composed of frame beams and columns spliced by prefabricated members.

The prefabricated hybrid concrete frame structure is a concrete structure mainly composed of prefabricated columns (or cast-in-situ columns), prefabricated slabs and prefabricated beams and the overlapped members are cast in situ and the joints are cast in post pouring. Its bearing capacity and deformation should meet the current application requirements in China (Fig. 7.43).

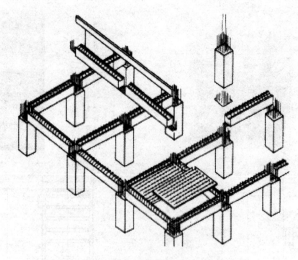

图 7.43　装配式整体框架结构示意图

1. 预制构件安装

(1)预制柱安装。预制框架柱吊装施工工艺流程如图 7.44 所示。

图 7.44　预制框架柱吊装施工工艺流程图

施工要点如下：

①根据预制柱平面各轴的控制线和柱框线校核预埋套管位置的偏移情况，并做好记录。若预制柱有小距离的偏移，需借助协助就位设备进行调整。

Chapter 7 Installation of Prefabricated Structural Members Construction

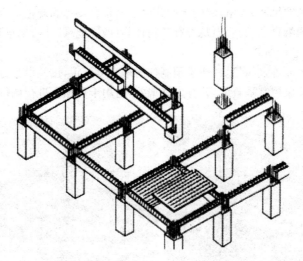

Fig. 7.43 Schematic Diagram of Prefabricated Hybrid Concrete Frame Structure

1. The Installation of Prefabricated Members

(1) The installation of prefabricated columns

The construction process chart of installing prefabricated frame columns is shown in Fig. 7.44.

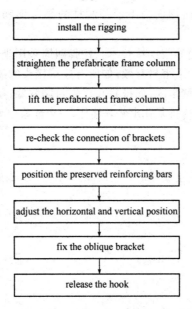

Fig. 7.44 The Construction Process Chart of Installing
Prefabricated Frame Columns

The key points in construction are as follows:

①Check the deviation of the embedded grout sleeves according to the reference line to each axis of the prefabricated column and its cross-section profile marking lines, and record it. If the prefabricated column deviates a little bit, it needs to be adjusted with the positioning equipment.

②检查预制柱进场的尺寸、规格，混凝土的强度是否符合设计和规范要求，检查柱上预留套管及预留钢筋是否满足图纸要求，套管内是否有杂物；同时做好记录，并与现场预留套管的检查记录进行核对，无问题方可进行吊装。

③吊装前在柱四角放置金属垫块，便于预制柱的垂直度校正，按照设计标高，结合柱子长度对偏差进行确认。用经纬仪控制垂直度，若有少许偏差，运用千斤顶等进行调整（图7.45）。

图 7.45　预制框架柱吊装

④柱初步就位时应将预制柱钢筋与下层预制柱的预留钢筋初步试对，无问题后准备进行固定。

⑤预制柱接头连接采用套筒灌浆连接技术。柱脚四周采用坐浆材料封边，形成密闭灌浆腔，保证在最大灌浆压力（约 1 MPa）下密封有效；如所有连接接头的灌浆口都未被封堵，当灌浆口漏出浆液时，应立即用胶塞进行封堵牢固；如排浆孔事先封堵胶塞，摘除其上的封堵胶塞，直至所有灌浆孔都流出浆液并封堵后，等待排浆孔出浆并堵住；一个灌浆单元只能从一个灌浆口注入，不得同时从多个灌浆口注浆。

2. 套筒灌浆

套筒灌浆是装配式施工中的重要环节。套筒灌浆的施工步骤如图7.46 所示。

Chapter 7 Installation of Prefabricated Structural Members Construction

②Check the size and specifications of prefabricated columns; check whether the strength of the concrete meet the requirement of the design and specification; check whether the preserved sleeve and reinforcing bars meet the requirement of the drawings; check whether there is debris in the sleeve. Record the data and compare it with the record in preserving sleeves. If there is no discrepancy, the columns can be installed.

③Place metal pads in the four corners of the column before lifting to facilitate the perpendicular correction of the prefabricated column, and check the deviation according to the designed elevation and the length of the column. Use theodolite to control the perpendicular, and if there is small deviation, use jacks to adjust it(Fig. 7. 45).

Fig. 7. 45 The Installation of Prefabricated Frame Columns

④In positioning, try to align the reinforcing bars in the prefabricated columns with the reinforcing bars on the lower layer, and if they can be aligned, the columns can be fixed.

⑤ Technology of slicing with sleeve grouting is usually adopted to column-to-column connection. The perimeter joints under the lower end of the upper column sections are sealed with grouting materials to form a closed grouting cavity to ensure effective sealing under the maximum grouting pressure (approx. 1MPa); If all the inlet ports interconnected with the cavity are not blocked, when the injected grout overflowing from the inlet grouting ports, these ports shall be immediately sealed firmly by using rubber plugs; then if the outlet ports are plugged previously, remove the rubber plugs, until all inlet ports flow out of grout and then block them and wait for grout flow out of the outlet ports and then block them; for one grouting connection unit, grout can be only injected from one grouting port, and the method that multiple inlet ports are used to inject grout at the same time is not allowed.

2. Sleeve Grouting

Sleeve grouting is an important work in the construction of prefabricated structure. The construction process of sleeve grouting is shown in Fig. 7. 46.

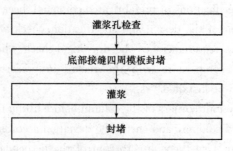

图 7.46　套筒灌浆的施工工序

套筒灌浆施工是确保竖向结构可靠连接的过程，施工品质决定了建筑物的结构安全。因此，施工时必须特别重视。应符合《钢筋套筒灌浆连接应用技术规程》(JGJ 355—2015)的规定。

(1)灌浆孔检查。检查灌浆孔的目的是确保灌浆套筒内畅通，没有异物。套筒内不畅通会导致灌浆料不能填充满套筒，造成钢筋连接不符合要求。检查方法如下，使用细钢丝从上部灌浆孔伸入套筒，如从底部可伸出，并且从下部灌浆孔可看见细钢丝，即畅通。如果钢丝无法从底部伸出，说明里面有异物，需要清除异物直到畅通为止。

(2)接缝四周用模板进行封堵。为了提高封堵效率，采用预埋定位螺栓加模板封堵的方法。假设预制剪力墙板采用墙板和保温的两层构造，其中接缝部位保温板断开，方便接缝封堵，接缝采用四周模板封堵方法，外侧模板通过预埋在上下两层墙板内的螺栓进行固定，内侧模板通过预埋在混凝土楼板内的螺栓和木方、木楔进行调整和固定，内墙板封堵均采用木楔和木方模板。

(3)灌浆。灌浆前应首先测定灌浆料的流动度，使用专用搅拌设备搅拌砂浆，之后倒入圆截锥试模，进行振动排出气体，提起圆截锥试模，待砂浆流动扩散停止，测量两方向扩展度，取平均值，要求初始流动度大于等于 300 mm，30 min 流动度大于等于 260 mm。

灌浆时需要制作灌浆料抗压强度相同的试块两组，试件尺寸采用 40 mm×40 mm×160 mm 的棱柱体。灌浆工程采用灌浆泵进行，灌浆料由下端靠中部灌浆孔注入，当其余套筒出浆孔有均匀浆液流出，视为灌浆完成。

Chapter 7 Installation of Prefabricated Structural Members Construction

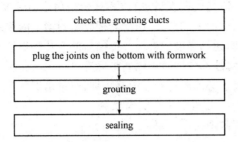

Fig. 7.46 **The Construction Process of Sleeve Grouting**

Sleeve grouting is a process to ensure the reliable connection of vertical structures, and its quality ensures the safety of the whole structure. Therefore, we should pay special attention to the quality of sleeve grouting and the construction process should meet Technical Specification for Grout Sleeve Splicing of Rebars (JGJ 355—2015). (1)Check the grouting ducts. The aim of checking the grouting duct is to ensure there is no debris in the duct and the duct is not blocked. If there is debris in the duct, the mortar can't fill in the sleeve and the connection of reinforcing bars will be influenced. The checking methods are as follows: insert a steel wire from the outlet port on the top, and if it can pass through the outlet port on the bottom and the steel wire can be seen from the outlet port on the bottom, that means the duct is unblocked; if the steel wire cannot pass through the outlet port on the bottom, it indicates that there is debris in the duct, and the debris must be moved to make the duct clear.

(2) Plugging joints with formwork. To improve the plugging efficiency, the method of embedding positioning bolt and plugging formwork is usually adopted. Suppose that the prefabricated shear wall is constructed by wallboard and thermal insulation board, in which the insulation board at the joint is disconnected to facilitate joint plugging, the joint is closed with formwork from transverse perimeter. The outer formwork is fixed by bolts embedded in the upper and lower wallboards. The inner formwork is adjusted and fixed by bolts embedded in the concrete floor slab and wooden cubes and wooden wedges. The inner wallboard is plugged by wooden wedges and wooden formwork.

(3) Grouting. Before grouting, the fluidity of grouting material should be measured first. In measuring, stir the mortar with special mixing equipment, and then pour the mortar into the conical mortar testing die and vibrate it to discharge the gas. Raise the testing die and wait till the mortar stop flowing, and then measure the expansion length from two directions. The average expansion length is the fluidity. The initial fluidity should be $\geqslant$ 300mm and the 30min fluidity should be $\geqslant$260mm.

When grouting, two groups of test cubes with the same compressive strength as grouting material are needed. The shape of cube is prismatic and the size of it is 40mm×40mm×160mm. The grouting project is carried out by grouting pump. The grouting material is grouted from the middle outlet at the bottom of the duct. When the other outlets leak mortar evenly, the grouting is finished.

(4)封堵。注浆完成后,通知监理人员进行检查,合格后进行注浆孔的封堵,封堵要求与原墙面平整,并及时清理墙面上、地面上的余浆,用配套橡胶塞封堵。灌浆结束后24 h内不得对墙板施加振动冲击等影响,待灌浆料达到强度后拆模。拆模后灌浆孔和螺栓孔用砂浆进行填堵。外墙板灌浆完成后,外保温板水平缝隙部位按防火隔离带标准进行二次施工。

套筒灌浆施工人员灌浆前必须经过专业灌浆培训,经过培训并且考试合格后方可进行灌浆作业工作。套筒灌浆前灌浆人员必须填写套筒灌浆施工报告书,如表7.1套筒灌浆施工报告书所示。灌浆作业的全过程要求监理人员必须进行现场旁站。

表 7.1 套筒灌浆施工报告书

套筒灌浆施工报告书					
项目名称:		施工日期:		施工部位(构件编号):	
灌浆开始时间: 灌浆结束时间:		灌浆责任人:		监理责任人:	
砂浆注入管理记录	室外温度: ℃		水量: ℃	砂浆批号:	
	水温: ℃		流动值: mm	备注:	
	灌浆时浆体温度: ℃				

7.3.4 水、电、暖等预留与预埋

1. 水暖安装洞口预留

(1)当水暖系统中的一些穿楼板(墙)套管不易安装时,可采用直接预埋套管的方法,埋设于楼(屋)面、空调板、阳台板上,包括地漏、雨水斗等,需要顶先埋设套管。有预埋管道附件的预制构件在工厂加工时,应做好保洁工作,避免附件被混凝土等材料污染、堵塞。

(2)由于预制混凝土构件是在工厂生产现场组装,和主体结构间靠金属件或现浇处理进行连接的。因此,所有预埋件的定位除了要满足距墙面、穿越楼板和穿梁的结构要求外,还应给金属件和墙体留有安装空间,一般距两侧构件边缘不小于40 mm。

Chapter 7 Installation of Prefabricated Structural Members Construction

(4) Sealing.

After the grouting is finished, the supervisor shall be notified to check whether the grouting meet the requirement. If the grouting is qualified, we can seal the outlet ports. In sealing, the surface of the outlet ports should be smooth and the mortar on the wall or on the floor should be cleaned. The outlet port is plugged with the prepared matching rubber plug. The wallboard shall not be subjected to vibration and other impact within 24 hours after grouting. Strip off the formwork after the grouting material reaches its strength. The grouting holes and bolt holes should be filled with mortar after removing the formwork. After the grouting of the external wallboard is finished, the horizontal joints of the external thermal insulation board should be constructed again according to the standard of fire-proof isolation zones.

The workers in charge of sleeve grouting should undergo professional grouting training. Only after they pass the training examination, they can do the job. Before grouting, the worker must fill in the sleeve grouting construction report, as shown in Table 7.1. The supervisor should supervise the whole process of grouting operation.

Table 7.1 Sleeve Grouting Construction Report

Sleeve Grouting Construction Report			
Project name:		Construction date:	The grouting part (component number):
Start time: End time:		Person responsible for grouting:	Supervisor:
The grouting record	Outdoor temperature: ℃	Quantity of water: ℃	Serial number of the mortar:
	Water temperature: ℃	Flow value: mm	Notes:
	Mortar temperature: ℃		

7.3.4 Preforming Holes and Embedding Members for Water, Electricity and Heating System

1. Preforming Holes for Water and Heating System

(1) When it is hard to install some sleeves that pass through the floor slabs or walls, they can be embedded in them. For example, the sleeves can be embedded in the floor slabs, air conditioning slabs, balcony slabs, floor drain, rainwater drain and so on. The prefabricated members with embedded sleeves should be clean in fabrication, so that the sleeve can avoid being polluted or blocked by concrete or other materials.

(2) Because the precast concrete members are fabricated in the precast concrete factories and are spliced on the construction site, they are connected to the main structure by metal parts or by cast-in-place processing. Therefore, the positioning of the all embedded members should not only meet the structural requirements of the wall, floor slab and beams, but also it must leave space for the purpose of applying the metal connection parts and splicing of wall, generally no less than 40 mm from the edges of the members on both sides.

(3)装配式建筑宜采用同层排水。当采用同层排水时,下部楼板应严格按照建筑、结构、给水排水专业的图纸,预留足够的施工安装距离。并且应严格按照给水排水专业的图纸,预留好排水管道的预留孔洞。

2. 电气安装预留预埋

(1)预留孔洞。预制构件一般不得再进行打孔、开洞,特别是预制墙应按设计要求标高预留好过墙的孔洞,重点注意预留的位置、尺寸、数量等应符合设计要求。

(2)预埋管线及预埋件。电气人员对预制墙构件进行检查,检查需要预埋的箱盒、线管、套管、大型支架埋件等是否漏设,规格、数量、位置等是否符合要求。

预制墙构件中主要埋设:配电箱、等电位联结箱、开关盒、插座盒、弱电系统接线盒(消防显示器、控制器、按钮、电话、电视、对讲等)及其管线。

预埋管线应畅通,金属管线内外壁应按规定做除锈和防腐处理,清除管口毛刺。埋入楼板及墙内管线的保护层不小于 15 mm,消防管路保护层不小于 30 mm。

(3)防雷、等电位联结点的预埋。装配式建筑的预制柱是在工厂加制作的,两段柱体对接时,较多采用的是套筒连接方式:一段柱体端部为套筒,另一段为钢筋,钢筋插入套筒后注浆。如用柱结构钢筋作为防雷引下线,就要将两段柱体钢筋用等截面钢筋焊接起来,达到电气贯通的目的。选择柱体内的两根钢筋作为引下线和设置预埋件时,应尽量选择预制墙、柱的内侧,以便于后期焊接操作。

预制构件生产时应注意避雷引下线的预留预埋,在柱子的两个端部均需要焊接与柱筋同截面的扁钢作为引下线埋件。应在设有引下线的柱子室外地面是 500 mm 处,设置接地电阻测试盒,测试盒内测试端子与引下线焊接。此处应在工厂加工预制柱时做好预留,预制构件进场时由现场管理人员进行检查验收。

Chapter 7 Installation of Prefabricated Structural Members Construction

(3) In prefabricated structure, same floor drainage system is usually adopted. In same floor drainage system, the lower floor slabs should reserve enough space for installation construction based on the drawings of construction, structure and water supply and drainage and should reserve ducts for drainage pipelines according to the water supply and drainage drawings.

2. Preforming Holes and Embedding Members for Electricity

(1) Preforming holes. Prefabricated members are generally not allowed to perforate or open holes. Especially for prefabricated walls, they should preform holes passing through walls according to the design requirements. Pay special attention to the location, size and quantity of the holes, and make sure that they meet the design requirements.

(2) Embedding pipelines and embedded parts. The electrician inspects the prefabricated members for walls and checks whether the boxes, pipelines, sleeves and large supporting embedded parts are missing and whether their specifications, quantities and locations meet the requirements.

In prefabricated members, the mainly embedded parts are: distribution board, equipotential bonding wire converging boxes, switch boxes, socket boxes, junction boxes, the junction boxes for weak current circuit system (fire display, controller, buttons, telephone, TV, intercom, etc.) and its conduits.

The embedded pipelines should not be blocked. The inner and outer walls of the metal pipelines should be treated with rust removal and anti-corrosion and the burrs at the mouths of the pipelines should be removed. The protective layer of the pipelines embedded in the floor slab or wall should be no less than 15mm and the protective layer of the firefighting pipelines is no less than 30mm.

(3) Embedding lightning protection and equipotential connector.

The columns for prefabricated construction are fabricated in concrete industrial factory. When two columns are connected together, sleeve grouting is usually adopted. In this method, the end of one column is sleeve, and the end of the other column is reinforcing bar; the reinforcing bar insert into the column and they are grouted together. If we use the reinforcing bars in the columns as the downward conductor in lightning protection, the reinforcing bars in two columns should be welded together by reinforcing bar with same section to achieve the purpose of electrical penetration. When choosing the reinforcing bars in the columns as the down conductor in lightning protection and considering the place for embedding them, the reinforcing bars on the inner side of the prefabricated wall or column are preferred because it is easy to weld them.

When fabricating the columns, the down conductor in lightning protection should be reserved or embedded. At the ends of the columns, flat steel with the same section as the reinforcing bars should be welded as the embedded parts of down conductor. The grounding resistance test box should be set 500mm away from the room with down conductor column, and the test terminal should be welded with down conductor. The place for down conductor should be reserved when fabricating the columns in the industrial building, and the site management personnel should make acceptance inspection when the columns enter the site.

预制构件应在金属管道入户处做等电位联结,卫生间内的金属构件应进行等电位联结,应在预制构件中预留好等电位联结点。整体卫浴内的金属构件应在构件内完成等电位联结,并标明和外部联结的接口位置。

为防止侧击雷,应按照设计图纸的要求,将建筑物内的各种竖向金属管道与钢筋连接,部分外墙上的栏杆、金属门窗等较大金属物要与防雷装置相连,结构内的钢筋连成闭合回路作为防侧击雷接闪带。均压环及防侧击雷接闪带均须与引下线做可靠连接,预制构件处需要按照具体设计图纸要求预埋连接点。

3. 整体卫浴安装预留预埋

(1)施工测量卫生间截面进深、开间、净高、管道井尺寸、窗高、地漏、排水管口的尺寸、预留的冷热水接头、电气线盒、管线、开关、插座的位置等,此外应提前确认楼梯间、电梯的通行高度、宽度以及进户门的高度、宽度等,以便于整体卫浴部件的运输。

(2)卫生间地面找平,给水排水预留管口检查,确认排水管道及地漏是否畅通无堵塞现象,检查洗脸面盆排水孔是否可以正常排水,给水预留管口进行打压检查,确认管道无渗漏水问题。

(3)按照整体卫浴说明书进行防水底盘加强筋的布置,加强筋布置时应考虑底盘的排水方向,同时应根据图纸设计要求在防水底盘上安装地漏等附件。

习 题

1. 简述自行杆式起重机的不同类型及其特点。
2. 单层工业厂房构件吊装前的准备工作包括哪些?
3. 简述柱子吊装的基本流程。
4. 简述屋架吊装的基本流程。
5. 简述起重机开行路线及停机位置的设计原则。
6. 简述装配整体式混凝土框架结构的施工工艺流程。

For the prefabricated members, the equipotential bonding shall be made to the metal service piping at the penetrations and so as the metal items in the bathroom. The equipotential bonding points in the prefabricated elements shall be made ready for connection. The equipotential bonding in the integral bathroom shall also be provided to its metal members, and position for its external bonding connection shall be marked.

In order to prevent lateral lightning strikes, vertical metal pipelines in the building should be connected with reinforcing bars according to the requirements of design drawings. Some large metal objects such as railings on external walls and metal doors and windows should be connected with lightning protection devices. The reinforcing bars in the building should be connected into a closed circuit as lightning protection band. The pressure equalizing ring and lightning protection band should be connected with the down conductor, and connectors should be embedded in the prefabricated members according to the requirements of specific design drawings.

3. Preforming and Embedding in Integral Bathroom Installation

(1) Measure the depth, width and clear height of the bathroom; measure the size of conduit shaft, the height of the window, the size of floor drain and drainage pipeline; make sure the position of the joints of hot and cold water supply pipelines, electrical installation box, pipelines, switches, sockets and so on. In addition, the height and width of stair case and elevator, as well as the height and width of entrance door should be confirmed in advance so as to facilitate the transportation of integral bathroom members.

(2) Make floor leveling in the bathroom; check the mouths of the reserved ducts for water drainage pipelines to make sure they are unblocked, and check whether the washbasin drainage hole is blocked. Make pressure inspection in the mouth of the reserved ducts for water supply pipelines to ensure that there is no water creeping and leaking.

(3) Install additional reinforcement under the base of integral bathroom according to the manufactures' instruction. In the construction, the drainage direction of the bottom should be considered. At the same time, install floor drain and other accessories on the base of the waterproof layer according to the requirements of design drawings.

Exercises

1. Describe the different types and their characteristics of self-propelled boom crane.
2. What are the preparations before lifting prefabricated members of single-storey PRC industrial building?
3. Briefly describe the operation process of lifting columns.
4. Briefly describe the operation process of lifting roof truss.
5. Describe the principles in designing the route and the standing positions of the cranes.
6. Briefly describe the construction process of the prefabricated hybrid concrete frame structure.

第8章 防水工程施工

8.1 屋面防水

屋面工程应根据建筑物的性质、重要程度、使用功能要求以及防水层合理使用年限，按不同等级进行设防，并应符合表8.1的要求。

表8.1 屋面防水等级和设防要求

项目	屋面防水等级			
	Ⅰ级	Ⅱ级	Ⅲ级	Ⅳ级
建筑物类别	特别重要或对防水有特殊要求的建筑	重要的建筑和高层建筑	一般的建筑	非永久性的建筑
防水合理使用年限	25年	15年	10年	5年
设防要求	三道或三道以上防水设防	二道防水设防	一道防水设防	一道防水设防
防水层选用材料	宜选用合成高分子防水卷材、高聚物改性沥青防水卷材、金属板材、合成高分子防水涂料、细石防水混凝土等材料	宜选用高聚物改性沥青防水卷材、合成高分子防水卷材、金属板材、合成高分子防水涂料、高聚物改性沥青防水涂料、细石防水混凝土、平瓦、油毡瓦等材料	宜选用高聚物改性沥青防水卷材、合成高分子防水卷材、三毡四油沥青防水卷材、金属板材、高聚物改性沥青防水卷材、合成高分子防水涂料、细石防水混凝土、平瓦、油毡瓦等材料	可选用二毡三油沥青防水卷材、高聚物改性沥青防水涂料等材料

Chapter 8 Waterproofing Construction

8.1 Roof Waterproofing

Roof waterproofing can be classified into different grades according to the properties, importance and function of the building and the service life of the waterproof layer. It should meet the requirements as shown in Table 8.1.

Table 8.1 Roof Waterproofing Grade and Requirements

Item	Roof Waterproofing Grade			
	Grade I	Grade II	Grade III	Grade IV
building types	particularly important buildings or buildings with special requirements for waterproofing	important buildings or high-rise buildings	general buildings	non-permanent buildings
reasonable service life of waterproof layer	25 years	15 years	10 years	5 years
layers of waterproof membrane	at least three	two	one	one
the suitable waterproofing materials	synthetic polymer waterproofing sheet membrane, polymer modified bituminous waterproofing sheet membrane, metal sheet, synthetic polymer waterproof coating, fine aggregate stone waterproof concrete and so on	polymer modified bituminous waterproofing sheet membrane, synthetic polymer waterproofing sheet membrane, metal sheet, synthetic polymer waterproof coating, polymer modified bituminous waterproof coating, fine aggregate stone waterproof concrete, flat tile, linoleum tile and so on	polymer modified bituminous waterproofing sheet membrane, synthetic polymer waterproofing sheet membrane, waterproofing sheet membrane with three layers of nonwoven polyester felt and four layers of liquid bituminous, metal sheet, polymer modified bituminous waterproof coating and synthetic polymer waterproof coating, fine aggregate stone waterproof concrete, flat tile, linoleum tile and so on	waterproofing sheet membrane with two layers of nonwoven polyester felt and three layers of liquid bituminous, polymer modified bituminous waterproof coating and so on

屋面防水根据防水材料的使用，主要有刚性防水屋面、卷材防水屋面、涂膜防水屋面或几种复合屋面等常见的类型。

8.1.1 高聚物改性沥青防水卷材屋面

1. 卷材防水层细部构造

(1)变形缝(图8.1)。
(2)出屋面管道与排气管(图8.2)。
(3)水落口(图8.3)。
(4)屋面出入口(图8.4)。
(5)泛水(图8.5)。
(6)檐沟和自由落水檐口(图8.6)。

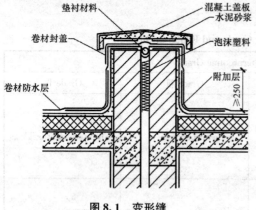

图8.1 变形缝

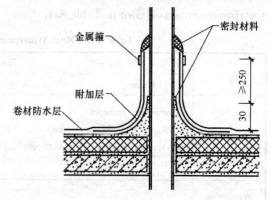

图8.2 出屋面管道与排气管

Chapter 8 Waterproofing Construction

The waterproofing of roofs can be classified into different types according to the waterproof materials. The following are the main types of roofs in waterproofing: rigid waterproofing roof, sheet membrane waterproofing roof, coating waterproofing roof or combined waterproofing roof.

8.1.1 Roof Waterproofing with Polymer Modified Bituminous Waterproofing Sheet Membrane

1. The Structure of Waterproof Layer with Sheet Membranes in Details
(1) Deformation joint (Fig. 8.1).
(2) Outlet pipe and discharging pipe (Fig. 8.2).
(3) Scupper (Fig. 8.3).
(4) Roof hatch (Fig. 8.4).
(5) Flashing (Fig. 8.5).
(6) Gutter and eave allowing free discharge (Fig. 8.6).

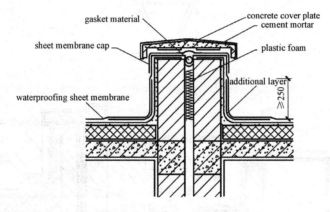

Fig. 8.1 Deformation Joint

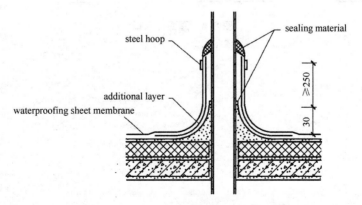

Fig. 8.2 Outlet Pipe and Discharging Pipe

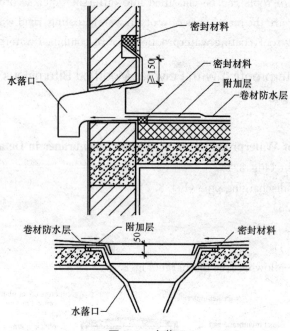

图 8.3 水落口

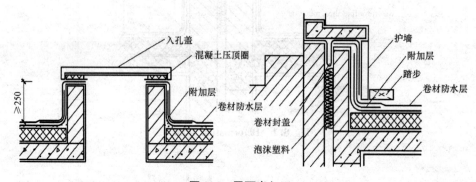

图 8.4 屋面出入口

Chapter 8 Waterproofing Construction

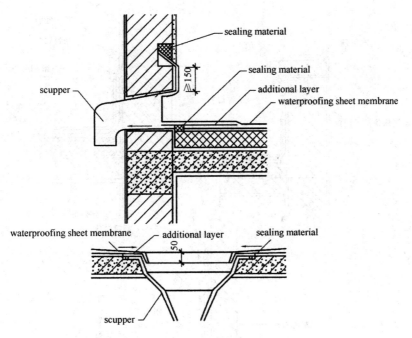

Fig. 8.3 Scupper

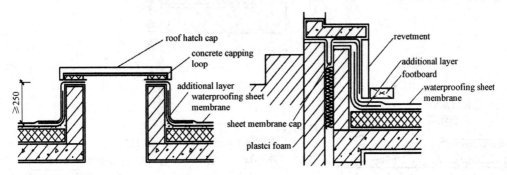

Fig. 8.4 Roof Hatch

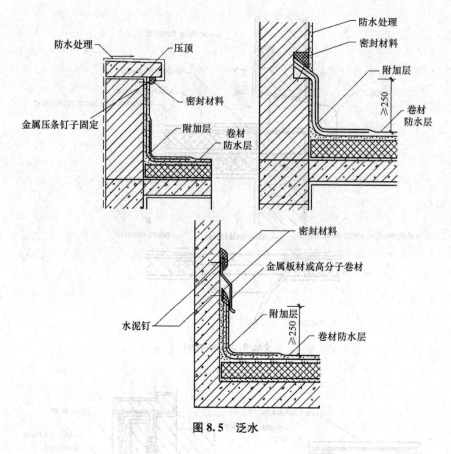

图 8.5 泛水

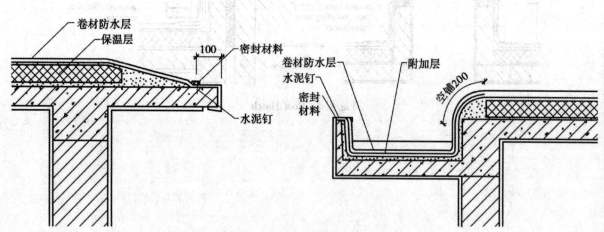

图 8.6 檐沟和自由落水檐口

2. 找平层施工

(1)卷材、涂膜防水层的基层应设找平层,常用的找平层分为:水泥砂浆、细石混凝土、沥青砂浆找平层。

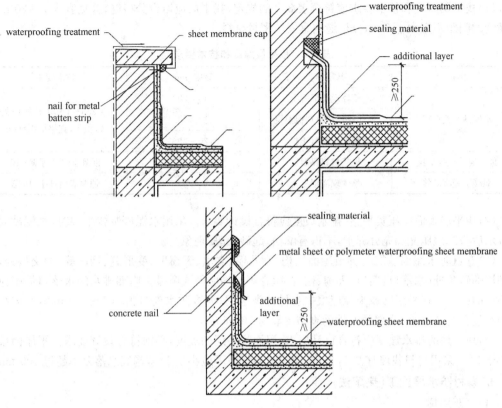

Fig. 8.5 Flashing

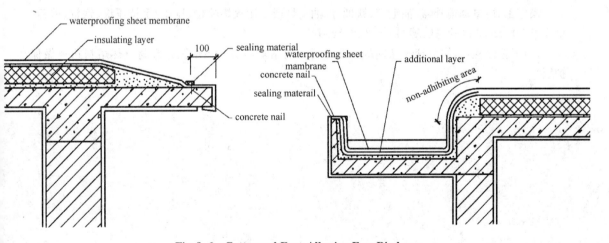

Fig. 8.6 Gutter and Eave Allowing Free Discharge

2. Levelling of the Substrate

(1) The substrate for sheet and fluid-applied waterproofing materials should be leveled. The commonly used levelling materials are cement mortar, fine aggregate concrete and cement asphalt mortar.

(1)找平层厚度和技术要求应符合表 8.2 的规定;找平层应留设分格缝,缝宽宜为 5～20 mm,纵横缝的间距不宜大于 6 m,分格缝内宜嵌填密封材料。

表 8.2　找平层厚度和技术要求

类别	基层种类	厚度/mm	技术要求
水泥砂浆找平层	整体现浇混凝土	15～20	1:2.5～1:3(水泥:砂)体积比,宜掺抗裂纤维
水泥砂浆找平层	整体或板状材料保温层	20～25	1:2.5～1:3(水泥:砂)体积比,宜掺抗裂纤维
水泥砂浆找平层	装配式混凝土板	20～30	1:2.5～1:3(水泥:砂)体积比,宜掺抗裂纤维
细石混凝土找平层	板状材料保温层	30～35	混凝土强度等级 C20
混凝土随浇随抹	整体现浇混凝土	—	原浆表面抹平、压光

(2)找平层表面应压实平整,排水坡度应符合设计要求。采用水泥砂浆找平层时,水泥砂浆抹平收水后应二次压光和充分养护,不得有酥松、起砂、起皮现象。

(3)卷材防水屋面基层与突出屋面结构(女儿墙、立墙、天窗壁、变形缝、烟囱等)的交接处,以及基层的转角处(水落口、檐口、天沟、檐沟、屋脊等),均应做成圆弧。内部排水的水落口周围应做成略低的凹坑。找平层圆弧半径应根据卷材种类选用,沥青防水卷材为 100～150 mm,合成高分子防水卷材为 20 mm,高聚物改性沥青防水卷材 50 mm。

(4)找平层的排水坡度应符合设计要求,屋面找平层的坡度,必须符合设计要求。平屋面坡度不小于 3%,采用材料找坡宜 2%,天沟、檐沟纵向找坡不应小于 1%,沟底水落差不超过 200 mm。

3. 卷材防水层施工(热熔法)

(1)主要用具。

清理用具:高压吹风机、小平铲、笤帚;

操作工具:电动搅拌器、油毛刷、铁桶、汽油喷灯或专用火焰喷枪、压子、手持压滚、铁辊、剪刀、量尺、1 500 mmϕ30 管(铁、塑料)、划(放)线用品。

(2)工艺流程:基层清理→涂刷基层处理剂→铺贴附加层→热熔铺贴卷材→热熔封边→做保护层。

①The thickness and technique requirements of levelling should comply with the specifications shown in Table 8.2. Expansion joint should be made in levelling and the width of the expansion joint should be 5 to 20mm. The distance between adjacent longitudinal and transversal expansion joints should not be larger than 6m. The sealant should be filled in the expansion joints.

Table 8.2 The Levelling thickness and the Technical Requirements

Types of Levelling	Types of Substrate	Thickness/mm	Technical Requirements
sand/cement/mortar	monolithic cast-in-situ concrete	15—20	volume ratio of cement and sand: 1 : 2.5—1 : 3. It's better to add some anti-cracking fiber
	monolithic insulation or insulation board	20—25	
	prefabricated concrete slabs	20—30	
fine aggregate stone concrete	insulation board	30—35	the strength level of the concrete: C20
concrete casting followed by instant levelling	monolithic cast-in-situ concrete	—	the mortar surface should be clean and smooth

②The levelling should be compacted with smooth surfaces, and the runoff slope should meet the technical specifications. When levelling with cement mortar, the mortar should be smoothed and cured again after it reaches to initial set while before final set. There should be no scale, laitance or sand dust on the surface of the mortar levelling.

③The intersection of the waterproofing sheet substrate and the projecting parts (such as parapet wall, stud wall, skylight wall, deformation joint and chimney) and the corner of the substrate (such as the scupper, eaves, inner drain, gutter and ridge) should be in the shape of arc. The slightly lower depression shall be formed where scupper located. The arc radius of the levelling should be based on the types of sheets. For bituminous sheet, the leveling arc radius should be 100—150mm; for polymer waterproofing sheet membrane, it should be 20mm; for polymer modified bituminous waterproofing sheet membrane, it should be 50mm.

④The runoff slope of levelling should meet the design requirements. The slope for flat roof should be not less than 3% and the other filler materials are laid to form slope of at least 2%. Slopes for roof inner drain and gutter should be not less than 1% and the water drop between bottoms at the opposite end should not exceed 200mm.

3. The Laying of Waterproofing Sheet (Hot-melt Application Method)

(1) The main tools.

Cleaning tools: high pressure blower, small flat shovel, broom.

Operating tools: motor agitator, paint brush, metal bucket, gas torch or spray gun, presser, hand press roller, iron roller, scissors, measuring scale, 1500mm ϕ 30 pipe (iron or plastic) and tools for setting out and drawing construction lines.

(2) The process flow: clean the substrate surface →brush substrate treatment agent →lay the additional layer →lay the waterproofing sheet membrane with hot-melt application method →seal the edge with hot-melt application method →make productive layer.

(3)操作工艺：

①基层清理：施工前将验收合格的基层清理干净。

②涂刷基层处理剂：在基层表面满刷一道用汽油稀释的氯丁橡胶沥青胶粘剂，涂刷应均匀，不透底。

③铺贴附加层：管根、阴阳角部位加铺一层卷材，按规范及设计要求将卷材裁成相应的形状进行铺贴。

④铺贴卷材：将改性沥青防水卷材按铺贴长度进行裁剪并卷好备用，操作时将已卷好的卷材，用 $\phi 30$ 的管穿入卷心，卷材端头比齐开始铺的起点，点燃汽油喷灯或专用火焰喷枪，加热基层与卷材交接处，喷枪距加热面保持 300 mm 左右的距离，往返喷烤，当卷材的沥青开始熔化时，手扶管心两端向前缓缓滚动铺设，要求用力均匀、不窝气，铺设压边宽度应掌握好，满贴法搭接宽度为 80 mm，条粘法搭接宽度为 100 mm。

⑤热熔封边：卷材搭接缝处用喷枪加热，压合至边缘挤出沥青粘牢，卷材末端收头用橡胶沥青嵌缝膏嵌固填实。

⑥保护层施工：平面做水泥砂浆或细石混凝土保护层；立面防水层施工完，应及时稀撒石碴并抹水泥砂浆形成保护层。

(4)质量标准。

主控项目：卷材防水层所用卷材及其配套材料，必须符合设计要求；卷材防水层不得有渗漏和积水现象。

一般项目：卷材防水层的搭接缝应粘结牢固，密封严密，不得有皱折，翘边和鼓泡等缺陷，防水层的收头应与基层粘结并固定牢固，缝口封严，不得翘边；卷材的铺贴方向应正确，卷材的搭接宽度的允许偏差为 ±10 mm。检验方法：观察和尺量检查。

(5)成品保护。施工人员必须穿软底鞋操作，并避免在施工完的涂层上走动，以免鞋钉及尖硬物将卷材划破。防水卷材施工完后，应及时做细石砼保护层，减少不必要的返修。严禁在已施工好的防水层上堆放物品，特别是钢筋等。

(3) The construction process:

①Clean the substrate surface: Clean the qualified substrate before laying.

②Brush substrate treatment agent: Brush neoprene bituminous adhesive diluted with gasoline on the substrate surface. The brushing should be even and complete, so all the surface is covered.

③Lay the additional layer: Lay another layer of sheet on the foot of the pipe and the intersection of the vertical and horizontal surfaces. Cut the sheet into the required size and shape and lay them.

④Lay sheet: Cut the modified bituminous sheet into the required length and roll them up ready for use. Then insert a $\phi 30$ pipe into the roll center and align the sheet end with the starting line of the substrate. Light the gasoline blow torch or other special welding torch to heat the seam between two sheets and the substrate. The distance between torch and the heat-melt surface of the sheet should be about 300mm. Heat polyethylene film of the sheet and substrate at the interface back and forth to make it nearly melted (not flowing), then lay the sheet with rolling it slowly with hands holding the two ends of the pipe. The hands on the two ends should move with same speed and force. For the sheets fully stuck to substrate, the overlap margin should be 80mm and 100mm for partly bonded sheets.

⑤Seal the edges with hot-melt application method. The joints between the sheets should be heated with gas torch and be pressed until the bitumen flows out to bond them together. The ends of the sheets should be sealed firmly to the substrate with rubber asphalt sealant.

⑥Lay the protective covering: for horizontal face, sand/cement mortar or fine aggregate stone cement can be used for protective covering. For vertical face, after waterproofing operation, stone ballast should be sprinkled on it and the cement/mortar should be cast on it to form the protective covering.

(4) Quality standard.

For key quality control assurance: the sheets and the accessory materials used in waterproof layers should comply with the design requirement. There should be no leakage and ponding occurring after the waterproof layers have been done.

For general quality control assurance: the overlap margin between the sheets should be bonded together and sealed firmly. There should be no defects like wrinkling, warping and bubbling. The ends of the waterproof layer should be bonded to the substrate and the joints should be sealed firmly. The laying direction of the sheet should be correct and the allowable variation of the overlap margin width is ± 10mm. The checking method: observation and measuring with scale.

(5) Finished product protection.

The roofers must wear soft-soled shoes and they should not walk on the finished sheet membrane to avoid scratching the sheet with hobnails or hard objects. After the laying of waterproofing sheet membranes, fine aggregate stone concrete protective covering should be laid in time. It is strictly prohibited to pile up items especially steel bars on the finished waterproof layer.

(6)安全措施。防水材料均为易燃品,储存时应放在干燥和远离火源的地方,施工现场严禁烟火,并应随时配备有灭火器。

施工操作人员应戴防护手套。进入现场戴好安全帽、防毒口罩及其他安全防毒用具,扎紧袖口、裤脚,手不得直接与沥青、油性材料接触。

严格按照施工方案的要求作业。材料堆放必须远离火源,有毒性物品必须封存,并且配置消防器材;运送材料时,当心坠落伤人,施工过程中严禁靠近烟火,同时,必须随身携带灭火器材。

操作人员操作时,如有头痛、恶心现象,应停止作业。患有材料刺激过敏人员,不得参加相应的工作。所用须调配材料,应先在地面调好后,再送至使用地区,材料用后要及时封存好剩余材料。

在外墙等高处作业时,必须拴好安全带。并搭设脚手架,铺设好操作平台,保证其稳固性。点火时,汽油喷灯不准对人,正在燃烧的汽油喷灯不得放在工件或地面上。作业结束后,应将汽油喷灯关好,检查作业地点,确认无着火危险,方可离开。

8.1.2 合成高分子防水卷材屋面

合成高分子防水卷材与高聚物改性沥青防水卷材铺贴方法不同,可以采用冷粘法,即在常温下采用胶结剂等材料进行卷材与基层、卷材与卷材间黏结的施工方法;自粘法,即采用自粘胶的防水卷材进行搭接粘合的施工方法。

1. 主要用具

基层处理用具:高压吹风机、平铲、钢丝刷、扫帚;材料容器:大小铁桶;弹线用具:量尺、小线、色粉袋;裁剪卷材用具:剪刀;涂刷用具:滚刷、油刷、压辊、刮板。

2. 工艺流程及主要操作工艺(冷粘法)

材料准备→基层清理→基层处理剂涂刷→复杂部位增强处理→卷材表面涂胶粘剂→基层表面涂胶粘剂→铺贴卷材→卷材压实排气→卷材接头、收头黏结、密封→蓄水试验→保护层

(6) Safety measures.

The waterproofing materials are inflammable, so they should be stored in the dry places and be kept away from fire. There should be no smoking on the construction site. Fire extinguisher should be available anytime.

The roofers should wear protective gloves. They should wear safety cap, anti-poison mask and other safety gas-proof tools on site. They should tighten the cuffs and trousers and do not touch the bituminous and oily materials directly.

The construction must follow the requirements of the construction plan. The store of materials must be away from fire and the toxic items must be sealed; fire extinguisher must be equipped on the site. When transporting materials, avoid falling down and injuring people. No fires and smoking on the construction site and roofers should carry extinguisher all the time.

If roofers feel headache or nausea, they must stop the work. The roofers suffering from material irritation allergies are not allowed to take part in the work.

For the combined materials, they should be collocated well and then transferred to the working area, and the remaining materials should be sealed and stored in time.

The roofers must fasten seat belts when working in high rise places, such as external wall. The scaffold shall be set up and the operation platform shall be laid to ensure the stability. The gas torch is not allowed to face people in lighting and the gas torch shall not be placed on the workpiece or the ground. After the work is finished, turn off the gas torch and check the operation site. Only after making sure that there is no fire, they can leave the site.

8.1.2 Synthetic Polymer Waterproofing Sheet Applied Roof

The laying method of the synthetic polymer waterproofing sheet membrane is different from the laying method of polymer modified bituminous waterproofing sheet membrane. We can adopt cold adhesion method. That is a method in which the sheet can bond with the substrate and with other sheets in the help of adhesive or other binding materials at room temperature. We can also adopt self-adhesive method. That is the construction method in which the waterproofing sheet membrane can bond with the help of self-adhesive agent. Hot-air welding method is also adoptable. In this construction method, the sheet can bond at the overlap margin with the help of hot-air welding gun.

1. The Main Tools

Tools for substrate treatment: high pressure hairdryer, shovel, steel wire brush, broom; material container: large and small iron barrel; tools for tapping ink line: measuring scale, string cable, toner bag; cutting tools: scissors; painting tools: rotary broom, paint brush, compression roller, and scraper.

2. The Process Flow and Main Construction Process (Cold Adhesion Method)

Prepare the materials → clean the substrate surface → brush substrate treatment agent → further treatment for some parts with complex structure → brush adhesive on the surface of the sheet → brush adhesive on the surface of the substrate → lay the sheet → compact the sheet to squeeze out the air → bind and seal the joints and ends → carry out leakage test → make protective covering.

(1)基层清理。应做好基层作业条件的检查和验收工作,对不合格部位应进行修补,防水层施工前应对基层进行清理。

(2)涂布基层处理剂。基层处理剂按卷材配套采用,一般为聚氨酯—煤焦油的二甲苯溶液或氯丁橡胶乳液。采用聚氨酯防水涂料时,可按甲料∶乙料∶甲苯=1∶1.5∶(2~3)或甲料∶乙料=1∶3的比例配制,搅拌均匀后使用。常用卷材与基层处理剂配套使用参见表8.3。

表 8.3 基层处理剂

主体防水材料名称	基层处理剂量名称
三元乙丙—丁基橡胶卷材	聚氨酯底胶甲∶乙∶二甲苯=1∶1.5∶1.5~3
氯化聚乙烯—橡胶共混卷材	氯丁胶乳,BX-12胶粘剂
氯磺化聚乙烯	氯丁胶沥青胶乳

涂布方法可使用油漆刷和长柄滚刷,蘸满药剂后在基层上均匀涂刷。应先用油漆刷在各复杂部位涂刷,再用长柄滚刷进行大面积涂刷。注意涂刷薄厚均匀,不得漏底,涂布量0.15~0.2 kg/m²,涂后需干燥4 h以上,才能进行下一工序施工。采用氯丁橡胶乳液做基层处理剂时,可采用喷涂的方法。喷涂后需干燥12 h左右,才可进行下一工序施工。

(3)屋面各复杂部位的防水增强处理。对屋面上各阴阳角、排水口等易渗漏部位,应按设计要求进行防水增强处理。采用合成高分子涂膜附加层时,应选用与卷材材性相配套的品种,通常用聚氨酯防水涂料进行处理。将甲料与乙料按1∶2的比例混合并搅拌均匀,在各增强部位涂刷,涂刷宽度应距中心250 mm以上,厚度1.5 mm以上,多遍涂刷,每遍干燥后涂刷下一遍,涂膜应固化24 h后,才能进行卷材铺贴施工。

采用卷材、夹铺胎体涂膜或胶粘带(常温用硫化丁基橡胶粘带)进行增强处理时,应按基层的形式预先裁剪好卷材、胎体材料(聚酯无纺布、化纤无纺布、玻璃纤维布)或胶粘带。

(1) Clean the substrate surface. The inspection and the following acceptance of operation conditions of the substrate should be done. Amendment shall be made in time if any part is unqualified. The substrate surface must be cleaned before laying the waterproof layer.

(2) Brush substrate treatment agent.

The substrate treatment agent should match with the types of sheets. We generally use the xylene fluid or neoprene latex in polyurethane-coal tar. When using polyurethane waterproof membrane, its proportion can follow the ratio: Part A : Part B : Methyl benzene =1 : 1.5 : (2—3) or Part A : Part B =1 : 3. The materials should be fully mixed and stir well before using. The commonly used sheets and the matching substrate treatment agent can be shown in Table 8.3.

Table 8.3 Substrate Treatment Agent

Waterproofing Materials	Substrate Treatment Agent
ethylene propylene-butyl rubber	polyurethane rubber A : B : xylene=1 : 1.5 : 1.5—3
chlorinated polyethylene-rubber blended	neoprene latex, BX-12 adhesive
chlorosulfonated polyethylene	neoprene bituminous rubber

Paint brush and long handle rotary broom can be used to brush the agent evenly on the substrate. For some parts with complex structure, brush agent with paint brush. For the large area, use long handle rotary broom to brush the agent. The agent should be brushed evenly and cover all the substrate. The amount of the agent for brushing is 0.15—0.2kg/m^2. After brushing, it takes at least 4 hours to dry before the next step. When the treatment agent is neoprene latex, the spray coating method can be adopted. It takes about 12 hours to dry before the next process.

(3) Waterproofing enhancement treatment for complex parts of the roof.

For some parts easy to leak water, like the intersection of the vertical and horizontal surfaces and the outlet, waterproofing enhancement treatment should be taken according to the design requirements. When the additional membrane is made of synthetic polymer sheet, the treatment agent should match the types of the sheets. We generally use polyurethane waterproof coating as the substrate treatment agent. Mix A material and B material with the ratio of 1 : 2 and stir them well. Brush them on the parts that need enhancement treatment with a width of at least 250mm around the center of the parts. The thickness of the coating should be at least 1.5mm and the coating should be brushed several times. The coating should be dry before brushing next coating and it takes 24 hours to dry before laying the sheet.

When the sheet, additional fiber (polyester non-woven fabric, non-woven man-made fabric, glass fiber cloth) or adhesive tape (usually butyl sulfide rubber tape) are used as the enhancement treatment agent, they should be cut into the required shape before laying.

（4）胶粘剂的涂制方法有两种，一种只在基层上涂刷，另一种要求在基层表面和卷材粘结面均要涂刷。无论一面涂刷或两面涂刷，均要求涂布均匀，不露底，不能堆积。如果采用空铺法、条粘法或点粘法铺贴时，应按预先确定的位置和尺寸涂刷，涂刷这些部位时，仍按上述原则进行。具体操作：将胶粘剂容器打开，用电动搅拌器或手动搅拌均匀，然后开始涂刷。

①卷材表面涂刷。展开待铺卷材于平坦干净的基层上，用长柄滚刷蘸满胶粘剂，均匀地涂刷于卷材粘结表面。应注意留出两边搭接宽度（满粘法不小于 80 mm，空铺法、条粘法、点粘法不小于 100 mm）不刷。要求厚薄均匀，不得漏涂。然后晾干，待基本不粘结手后（为 10~20 min），用纸筒芯卷起待铺。

②基层表面涂刷。用长柄滚刷蘸满胶粘剂，在基层上均匀涂刷。涂刷应一次均匀涂布，不得在一处反复涂刷，以防止将基层处理剂（底胶）带起。涂刷顺序应按卷材铺贴顺序，先铺先涂。涂刷过后，需晾干（10~20 min），手触不粘时，即可开铺卷材。

（5）卷材铺贴。一般要根据屋面坡度和有无震动确定卷材铺贴方向，但合成高分子卷材铺贴时，多采用平行于屋脊的方向铺贴。平行屋脊铺贴施工方便；卷材主要搭接缝（长边接缝）应顺水流方向，渗漏隐患少，施工速度快，接头少，省材料。

铺贴顺序按先高跨后低跨、先远跨后近跨、先复杂部位后大面积、从低标高向高标高铺贴的原则进行。

根据屋面形状和尺寸，将卷材进行配置，并根据预先提出卷材配置方案图。再按照卷材铺贴图，弹出准线。将已涂胶粘剂的卷材置于铺贴起点，迎着主导风向进行铺贴。

铺贴时在卷材筒芯插入铁管（$\phi 30 \times 1\,500$ mm），由两人各执铁管一端将卷材展开。铺贴卷材应松紧适度，既不过力拉伸，又不松弛出皱折、鼓泡。

对转角和阴阳角处要仔细，防止出现空鼓现象。沿准线铺贴时，应隔一定距离粘结固定一下，使卷材稳定在基层预定位置。

(4) The brushing method of the adhesive. There are two brushing method of the adhesive. One is brushing the adhesive on the substrate only and the other is brushing them on substrate and the side of the sheet adhesive to the substrate. The brushing should be even and flat, and the adhesive should cover all the surface of the substrate. If the adhesive between the sheet and the substrate is non-bonded, tape bonded or point bonded, the place and size of brushing the adhesive should be determined beforehand. The operation process is as follows: first, open the adhesive container; second, stir it fully with electric or hand mixer, and then brush it on the substrate or sheet.

①Brush adhesive on the surface of the sheets: Unroll the sheet on the flat and clean substrate and brush the adhesive on the side of the sheet adhesive to the substrate evenly. Long handle rotary broom can be used in this brushing. Do not brush adhesive on the overlap margin of the sheet. For fully bonded sheet, the overlap margin is at least 80mm; for non-bonded, tape bonded or point bonded sheet, the overlap margin is at least 100mm. The adhesive should be brushed evenly and fully. Then let it dry and after the hands touch on the applied adhesive and feel basically no sticking (about 10—20min), roll up the sheet with its paper core and wait for laying.

②Brush adhesive on the surface of the substrate. Brush the adhesive evenly on the substrate with the long handle rotary broom. Brush once on one point and brush on the part that lays first. After brushing adhesive, it needs 1 to 20 minutes to dry and after the hands touch on the applied adhesive and feel basically no sticking (about 10—20 min), the laying of the sheet can begin.

(5) Laying of the sheets.

According to the slope of roof and whether there is vibration, we can determine the direction of laying the sheet membranes. For synthetic polymer sheet, the laying direction is typically parallel to the ridge lines. Sheet laying parallel to ridges is convenient. The long side overlapping shall be in the direction of water flow, as a result, the laying becomes time-saving, while producing less overlaps and potential leakage, and materials can be saved.

The laying sequence should follow the rules: Lay sheet on the span with higher altitude and then the span with lower altitude; lay sheet on the span that is far and then on the span that is near; lay sheet on the parts with complex structure and then on the open and flat area; lay the sheet on the span from the lower altitude to the higher altitude.

Make layout plan for the sheets based on the shape and size of the roof and prepare the sheets ready to place according to the plan. Mark the alignment on the substrate according to the plan of the sheets. Locate the sheet with applied adhesive at the starting point, unfold and bond it following the direction contrary to the prevailing wind.

Insert a steel tube ($\phi 30 \times 1500$mm) into the center of roll and two roofers hold two ends of the tube separately to unroll the sheet. The sheet should be moderately tight, so that it is not stretched hard, nor slack with wrinkles or blisters during the laying.

Operations should be careful at the edges, inner and outer corners to avoid hollows present in the layers. When laying along the alignment, it should be bonded and fixed at intervals to retain the sheet at the predetermined position on substrate surfaces.

建筑施工技术(中英文对照版)

一卷铺毕,应立即自一端开始用长柄干净滚刷沿卷材宽度方向推压前进,使窝于卷材下面的空气排出。继而用外包胶皮重 30 kg 左右、长 300 mm 的铁辊顺序滚压一遍,使卷材与大面基层粘牢。对立面卷材及大面上不牢之处,应使用手持压辊将其压牢压实。

卷材铺好并与基层粘合后,应即将相邻的卷材搭接缝粘结牢靠。接缝粘结时,应先将接缝处清理干净,采用与卷材配套的接缝专用粘结剂,在接缝处均匀涂刷,做到厚薄均匀,不漏刷,不堆积。要根据专用粘结剂的性能,控制好涂刷与粘合的间隔时间;以及根据每平方米粘结剂用量决定其厚度;双组份的粘结剂要根据配合比进行制备。粘合应按顺序进行,要注意排除空气,辊压牢靠,不皱折、不鼓泡。

3. 热熔焊接法主要操作工艺

(1)当找平层涂刷基层处理剂干燥后,首先粘贴加强层。

(2)铺贴大面积卷材时,先打开卷材的一端对准弹好的标准线,然后将卷材头倒退卷回 1 m 左右,一人扶卷材,另一人手持火焰喷枪(宜采用两把或多把喷枪同时分段加热),点燃后调好火焰,使火焰成蓝色,将喷枪对准卷材与基层交界面,使喷枪与卷材保持最佳距离,从卷材一侧向另一侧缓缓移动,使基层与卷材同时加热,当卷材底面的热熔胶熔化并发黑色光泽时,负责卷材铺贴的人员就可以缓缓滚压粘贴,摊滚操作应紧密配合加热熔化速度进行。

(3)待端部粘贴好后,摊滚操作人员站向卷材对面,火炬喷枪移向反面,继续进行粘贴。摊滚粘贴时,操作人员必须注意卷材沿所弹标准线铺贴,滚铺时应排除卷材下面的空气,卷材边缘应有热熔胶溢出,并趁热用刮板将熔胶刮至接缝处封严。

(4)摊铺滚贴 1~2 m 后,另外一人用压辊趁热滚压严实,使之平展,不得有皱折。

(5)熔化热熔胶时,应特别注意卷材边缘的热熔胶要充分热熔,确保搭接质量。铺贴复杂部位及表面不平整处,应扩大烘热卷材面,使整片卷材处于柔软状态,便于与基层粘贴平整、严实。

(6)用条粘法时,每幅卷材的每边粘贴宽度不应小于 150 mm。

(7)施工时应严格控制摊滚速度和火焰烘烤距离,摊滚过快、烘烤距离太远、热溶胶未达到熔化温度,会造成卷材与基层粘结不牢;摊滚过慢、烘烤距离太近,火焰容易将热熔胶烧流、烧焦或烧穿卷材,施工人员必须熟练掌握这一操作。

After laying one roll of sheet, use the long handle rotary broom to press and roll compactly on it from one end to squeeze out the air, and then press and roll the sheet with steel roller wrapped in rubber, with the weight of 30kg and the width of 300mm to make the sheet bond with the substrate firmly. For some parts that is not bonded well, use hand roller to press and roll them again.

After the laying and bonding of the sheet, bond the overlap joints of the sheets immediately. Clean the joints first and then bond them with the specific adhesive. In brushing the adhesive, the thickness should be even and the adhesive should cover all the joints and there should be no wrinkling. It is necessary to control the interval between brushing and bonding according to the performance of the specific adhesive and to determine the thickness of the adhesive according to the amount of adhesive per square meter; for adhesive composed of A and B materials, mix them fully together according to the prescribed ratio. The bonding of the sheet should go on step by step and the rolling should be compacted to squeeze out the air. There should be no wrinkling and bubbling in the bonding.

3. The Main Construction Process of Hot-melt Application Method

(1) After brushing the treatment agent, lay the reinforced layer.

(2) When lay the sheet with large size, align one end of the sheet with the marked alignment and roll the sheet backward for about 1m. One roofer holds the sheet and the other roofer hold the gas torch and it's better to have several pairs working at different parts at the same time. Light the gas torch and adjust the flame to make the flame become blue, and then use the flame to heat the intersection of the substrate and the sheet. Keep the gas torch at an optimal distance from the sheet and slowly move it from one side of the sheet to the other, so that it can heat the substrate and the sheet at the same time. When the adhesive on the sheet begins to melt and glosses black, the roofer holding the sheet can unroll the sheet slowly and the speed of unrolling should match the speed of heating.

(3) After the bonding of the end, the roofer holding the gas torch can move to the opposite side of the end and begin to heat the left part of the sheet with the same method. The roofer holding the sheet will also move to the other side and begin to unroll the sheet. In unrolling, the sheet should align with the alignment and there should be no air under the sheet. At the edges of the sheet, the adhesive will melt and split after heating, and at that time, scrape the adhesive to the joint to seal the joint.

(4) After unrolling and laying the sheet for 1—2m, another roofer should roll it firmly with compression roller to make it flat and firm, and there should be no wrinkling.

(5) In the melting of the adhesive, make sure that the adhesive on the edges melt fully to ensure the quality of the lapping joints. When the sheet lays on the parts with complex structure or the substrate is not flat, the sheet should be heated at large area to make the sheet soft and easy to make it flat and firm.

(6) In tape bonding method, the bonding width for every edge of the sheet should be no less than 150mm.

(7) In laying, control the unrolling speed and the heating area. If it is too fast to unroll the sheet or the gas torch is too far from the sheet or the adhesive is not heated enough, the sheet will not adhere to the substrate fully. If it is too slow to unroll the sheet or the gas torch is too near to the sheet, the sheet will easily be burnt. Therefore, the roofers must be skillful in the operation.

4. 自粘法铺贴合成高分子防水卷材的操作要点

(1)基层处理剂干燥后,即可铺贴加强层,铺贴时应将自粘胶底面的隔离纸完全撕净,宜采用热风焊枪加热,加热后随即粘贴牢固,溢出的自粘胶随即刮平封口。

(2)铺贴大面积卷材时,应先仔细剥开卷材一端背面隔离纸约 500 mm,将卷材头对准标准线轻轻摆铺,位置准确后再压实。

(3)端头粘牢后即可将卷材反向放在已铺好的卷材上,从纸芯中穿进一根 500 mm 长钢管,由两人各持一端徐徐往前沿标准线摊铺,摊铺时不能收紧,但也不能有皱折和扭曲。

(4)在摊铺卷材过程中,另一人手拉隔离纸缓缓掀剥,必须将自粘胶底面的隔离纸完全撕净。

(5)铺完一层卷材,即用长把压辊从卷材中间向两边顺次来回滚压,彻底排除卷材下面空气,为粘结牢固,可以使用大压辊再一次压实。

(6)搭接缝处,为提高可靠性,可采用热风焊枪加热,加热后随即粘贴牢固,溢出的自粘胶随即刮平封口,最后将接缝口用密封材料封严,宽度不小于 10 mm。

(7)铺贴立面、大坡面卷材时,应用热风焊枪加热后粘贴牢固。

5. 保护层施工

(1)防水层铺贴完毕,清扫干净,经淋(蓄)水检验,检查验收合格后,方可进行保护层的施工。

(2)云母或蛭石保护层不得有粉料,撒铺应均匀,不得露底,多余的云母或蛭石应清除。

(3)水泥砂浆保护层的表面应抹平压光,并设表面分格缝,分格面积宜为 $1\ m^2$。

(4)块体材料保护层应留设分格缝,分格面积不宜大于 $100\ m^2$,分格缝宽度不宜小于 20 mm。

(5)细石混凝土保护层,混凝土应密实,表面抹平压光,并留设分格缝,分格面积不大于 $36\ m^2$。

(6)浅色涂料保护层应与卷材粘结牢固,厚薄均匀,不得漏涂。

(7)水泥砂浆、块材或细石混凝土保护层与防水层之间应设置隔离层。

4. Main Operating Skills of Synthetic Polymer Waterproofing Sheet Membrane in Self-adhesive Method

(1) After the substrate treatment agent is dry, the reinforced layer can be laid. Peel back the protective paper attached to the bottom of self-adhesive agent completely, and it is better to heat the adhesive agent with hot air welding gun. After heating the surface of self-adhesive sheet, paste it firmly on the substrate and scrape the spilled adhesive agent to seal the joints.

(2) When lay the sheet with large size, peel back the protective paper attached to the bottom of self-adhesive sheet for about 500mm from one end of it and align the end with the alignment and lay it loosely. After the aligning, it can be pressed firmly.

(3) After the end section is firmly bonded, the large portion of the remaining roll-type sheet can be rolled back standing by on it and insert a steel pipe with the length of 500mm. Two roofers hold the two ends separately and unroll the sheet slowly along the alignment. The unrolling should not be tightened, and there should be no wrinkling and distortion.

(4) In laying, one roofer should peel back the protective paper slowly, and the protective paper should be peeled back completely.

(5) After laying one layer of sheet, compress it firmly from the middle to the left and right sides with long handle compression roller. This operation will repeat several times until the air under the sheet is completely squeezed out. To bond it firmly, a bigger compression roller can be used to compress it again.

(6) To improve the sealing quality on the overlap joints, it is better to heat the adhesive with hot air welding gun. After being heated, paste it firmly on the substrate and scrape the spilled adhesive agent to seal the joints. At last, the sealing materials will be used to seal the joints and the width of it should be no less than 10mm.

(7) In laying sheets on the vertical face or pitched roof, hot air welding gun is used to heat the sheets and so the sheets can be bonded to the substrate firmly.

5. The Making of Protective Covering

(1) After laying the waterproof layer and clean it, carry out a leakage test. After the inspection and acceptance of the waterproof layer, the protective covering can be laid.

(2) If the protective covering is made of mica or vermiculite, the mica or vermiculite should be scattered evenly and cover all the base. The excessive mica or vermiculite should be removed.

(3) If the protective covering is made of mortar, the surface of the mortar should be smooth and compact, and set expansion joints to make the mortar into $1m^2$ pieces.

(4) If the protective covering is made of block material, set expansion joints to make the block into pieces no larger than $100m^2$, and the width of expansion joint should be not less than 20mm.

(5) If the protective covering is made of fine aggregate stone concrete, the concrete should be compact and the surface should be smooth. Set expansion joints to make the concrete into pieces no larger than $36m^2$.

(6) If the protective covering is made of light color coating, it should bond to the sheet firmly. The width of the coating should be uniform and should cover all the base.

(7) Set seperation layer between the protective covering of mortar, block material or fine aggregate stone concrete and the waterproof layer.

(8)刚性保护层与女儿墙、山墙之间应预留宽度为 30 mm 的缝隙,并用密封材料嵌填严密。

(9)高低跨的屋面,如为无组织排水时,低屋面受水冲滴的部位应加铺一层整幅的卷材,再设 300～500 mm 宽的板材加强保护;如为有组织排水时,水落管下应加设钢筋混凝土水簸箕。

6. 质量标准

(1)主控项目:所用卷材及其配套材料,必须符合设计要求;卷材防水层不得有渗漏或积水现象;卷材防水层在天沟、檐沟、檐口、水落口、泛水、变形缝和伸出屋面管道的防水构造,必须符合设计要求。

(2)一般项目:卷材防水层的搭接缝应粘(焊)结牢固,密封严密,不得有皱折、翘边和鼓泡等缺陷;防水层的收头应与基层粘结并固定牢固,封口严密,不得翘边。卷材防水层上的撒布材料和浅色涂料保护层应铺撒或涂刷均匀,粘结牢固;水泥砂浆、块材或细石混凝土保护层与卷材防水层间应设置隔离层;刚性保护层的分格缝留置应符合设计要求。排汽屋面的排汽道应纵横贯通,不得堵塞。排气管应安装牢固,位置正确,封闭严密。卷材的铺贴方向应正确,卷材搭接宽度的允许偏差为±10 mm。

7. 成品保护

(1)施工人员应认真保护已经做好的防水层,严防施工机具等把防水层戳破;施工人员不允许穿带钉子的鞋在卷材防水层上走动。

(2)穿过屋面的管道,应在防水层施工以前进行,卷材施工后不应在屋面上进行其他工种的作业。如果必须上人操作时,应采取有效措施,防止卷材受损。

(3)屋面工程完工后,应将屋面上所有剩余材料和建筑垃圾等清理干净,防止堵塞水落口或造成天沟、屋面积水。

(4)施工时必须严格避免基层处理剂、各种胶粘剂和着色剂等材料污染已经做好饰面的墙壁、檐口等部位。

(5)水落口处应认真清理,保持排水畅通,以免天沟积水。

8. 安全环保措施

(1)防水工程施工前,应编制安全技术措施,使用书面形式向全体操作人员进行安全技术交底工作,并办理签字手续备查。

(8) Reserve a gap of 30mm in width between the rigid protective covering and the parapet wall or the gable and the gap should be filled compactly with sealing material.

(9) If the roof has different altitudes and there is no organized drainage on the roof, the area undergoing water flow from higher altitude on the lower altitude roof should be added with a layer of complete sheet, and then set a sheet with the width of 300—500mm to strengthen the protective covering; if there is organized drainage system, the reinforced concrete water dustpan should be laid under the rain conductor.

6. Quality Control

(1) For key quality control assurance: the sheets and the accessory materials should meet the design requirement. There should be no leakage and ponding on the waterproof layer. The waterproofing sheet membrane should meet the design requirements in the parts with complex structure, such as the gutter, the inner drain, eaves, scupper, flashing, deformation joints and the pipes penetrating the roof.

(2) For general quality control assurance: The overlap joints for the waterproofing sheet membrane should be bonded firmly and sealed compactly. There should be no defects like wrinkling, warping and bubbling. The ends of the sheet should bond with the substrate firmly and be sealed firmly. The scattered material and the light color coating should be even and firm. There should be separation layer between the protective covering of mortar, block material or fine aggregate stone concrete and the waterproof layer. The reservation of the expansion joints in rigid protective covering should comply with the design requirements. The pipes for discharging steam should be connected vertically and horizontally, and should not be blocked. The steam discharging pipes should be installed firmly at the right position and the gaps around the pipes should be sealed. The laying direction of the sheet should be correct and the allowable variation of the overlap width of the sheet is ± 10mm.

7. Finished Product Protection

(1) The roofers should carefully protect the laid waterproof layer, and they must protect the waterproof layer from being destroyed by the machinery and tools. The roofers wearing shoes with hobnails are forbidden to walk on the laid sheet.

(2) The pipes penetrating the roof shall be laid before the laying of waterproof layer, and it is not allowed to carry out other work after laying waterproof layer. If roofers must stand on the sheet in laying, effective measures should be taken to prevent the damage to the sheet.

(3) After the roof works are completed, all the remaining materials and construction waste on the roof shall be cleaned up to prevent them blocking the scuppers and causing water leakage and ponding.

(4) During laying, prevent the substrate treatment agent, adhesive agent and coloring paste from polluting the finished walls, eaves, etc.

(5) Clean the areas beside the scupper to prevent blocking and causing ponding on the roof.

8. Measures for Safety and Environment Protection

(1) Before the construction of the waterproofing project, safety technical measures should be compiled, and all the roofers should be provided with the written form of safety technical measures and they should sign name on it, and the signature documents should be filed for reference.

(2)施工过程中,应有专人负责督查,严格按照安全规程进行各项操作,合理使用劳动防护用品,操作人员不得赤脚或穿短袖衣服进行作业,防止胶粘液溅泼和污染,应将袖口和裤脚扎紧,应戴手套,不得直接接触油溶型胶泥油膏。接触有毒材料应戴口罩并加强通风。施工时禁止穿带高跟鞋、带钉鞋、光滑底面的塑料鞋和拖鞋,以确保上下屋面、在屋面上行走及上下脚手架的安全。

(3)患有皮肤病、支气管炎、结核病、眼病以及对胶泥油膏有过敏的人员,不得参加操作。

(4)操作时应注意风向,防止下风操作以免人员中毒、受伤。在较恶劣条件下,操作人员应戴防毒面具。

(5)运输线路要畅通,各项运输设施应牢固可靠,屋面孔洞及檐口应有安全防护措施。

(6)为确保施工安全,对有电器设备的屋面工程,在防水层施工时,应将电源临时切断或采取安全措施,对施工照明用电,应使用36 V安全电压,对其他施工电源也应安装触电保护器,以防发生触电事故。

(7)操作现场禁止吸烟。严禁在卷材或胶泥油膏防水层的上方进行电、气焊工作,以防引起火灾和损伤防水层。

(8)必须切实做好防火工作,备有必要且充足的消防器材,一旦发生火灾,严禁用水灭火。

(9)施工现场及作业面的周围不得存放易燃易爆物品。

8.1.3 涂膜防水屋面

涂膜防水屋面是在屋面基层上涂刷防水涂料,经固化后形成一层有一定厚度和弹性的整体涂膜,从而达到防水目的的一种防水屋面形式。

1. 涂膜防水屋面的典型构造层次如图8.7所示。
2. 材料要求

所采用的防水涂料、胎体增强材料、密封材料等应有产品合格证书和性能检测报告,材料的品种、规格、性能等技术指标应符合现行国家产品标准和设计要求。

(1)高聚物改性防水涂料质量要求(表8.4)。

Chapter 8 Waterproofing Construction

(2) In the construction, there should be specific person who is in charge of inspection to make sure that all the works follow the safety requirements and the labor protection articles are used reasonably. The roofers should not work on the site with bear feet or with T-shirt. They should tighten the cuffs and trousers and wear gloves and do not touch the adhesive agent directly. They should wear mask and ventilation should be strengthened when there are toxic materials. It is forbidden to wear shoes with high heels and hobnails or plastic shoes with smooth soles or slippers because it is not safe when walking on the roof and scaffold.

(3) People suffering from skin diseases, bronchitis, tuberculosis, pinkeye or allergy to binding unction are not allowed to take part in the operation.

(4) Observe the wind direction and roofers should not operate in the downwind to avoid being injured or poisoned. If the toxic gas is strong, roofers should wear the respirators.

(5) The routes for transporting materials should be smooth and the transport facilities should be solid and reliable, and protective measures should be taken around the gutters and eaves.

(6) When the roof has electrical equipment, cut off the power or set protective measures in the construction of waterproof layer. For construction lighting, 36 voltage should be used and electric shock protectors should be laid for other construction power sources to prevent electric shock accidents.

(7) Smoking is prohibited on the construction site. It is strictly forbidden to take electric or gas welding works on the waterproof layer of sheet or binding unction, in case of causing fire or damaging the waterproof layer.

(8) Safety measures should be taken to prevent fire, and necessary and sufficient fire fighting equipment should be provided. If there is fire on the site, it is strictly prohibited to use water to extinguish fire.

(9) Flammable and explosive materials shall not be stored on the construction site and around the working area.

8.1.3 Roof with Waterproof Coating

In the construction of waterproof coating, brush waterproof materials on the substrate of the roof and when the material solidifies, it will form an integrated coating film with certain thickness and elasticity to achieve the goal of waterproofing.

1. The Typical Structure of Waterproof Coating is Shown in Fig. 8.7

2. The Requirements for the Materials

The waterproof coating materials, matrix reinforcing materials, sealing materials and other materials shall have product qualification certificates and performance test reports, and the technical indexes, such as the type of the materials, the specifications and the performance shall meet current national product standards and design requirements.

(1) The quality requirements for polymer modified waterproof coating (Table 8.4).

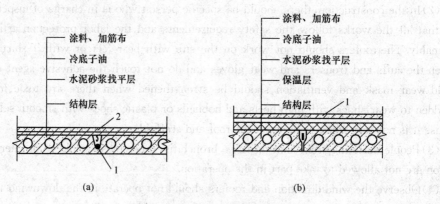

图 8.7 涂膜防水屋面构造
(a)无保温层涂膜屋面；(b)有保温层涂膜屋面
1—细石混凝土；2—油膏嵌缝

表 8.4 高聚物改性防水涂料质量要求

项目		质量要求
固体含量/%		≥43
耐热度(80 ℃,5 h)		无流淌、起泡和滑动
柔性(-10 ℃)		3 mm 厚,绕 ϕ20 mm 圆棒,无裂纹、无断裂
不透水性	压力/MPa	≥0.1
	保持时间/min	≥30 min 不渗透
延伸(20±2 ℃)/mm		≥4.5

(2)合成高分子防水涂料的质量要求(表 8.5)。

表 8.5 合成高分子防水涂料质量要求

项目		质量要求		
		反应固化型	挥发固化型	聚合物水泥涂料
固体含量/%		≥94	≥65	≥65
拉伸强度/MPa		≥1.65	≥1.5	≥1.2
断裂延伸率/%		≥350	≥300	≥200
柔性/℃		-30,弯折无裂纹	-20,弯折无裂纹	-10,绕 ϕ10 mm 棒无裂纹
不透水性	压力/MPa	≥0.3		
	保持时间/min	≥30		

由于合成高分子材料本身的优异性能，以此为原料制成的合成高分子防水涂料有较高的强度和延伸率，优良的柔韧性、耐高低温性能、耐久性和防水能力。

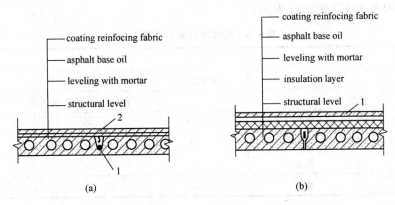

Fig. 8.7 The Structure of Waterproof Coating
(a) roof with waterproof coating without insulation Layer; (b) roof with waterproof coating with insulation layer
1—fine aggregate stone concrete; 2—sealing with unction

Table 8.4 The Quality Requirements for Polymer Modified Waterproof Coating

Items		Quality Requirements
solid content/%		≥43
heat resistance(80℃,5h)		no flowing, bubbling and sliding
flexibility(−10℃)		with the thickness of 3mm and no crack or breaking when wrap the round bar of ϕ20mm
watertightness	pressure/MPa	≥0.1
	retension time/min	≥30min no leakage
extension(20±2℃)/mm		≥4.5

(2) The quality requirements for synthetic polymer waterproof coating (Table 8.5).

Table 8.5 The Quality Requirements for Synthetic Polymer Waterproof Coating

Items		Quality Requirements		
		Reaction-solidified Type	Volatile-solidified Type	Polymer Cement Coating
solid content/%		≥94	≥65	≥65
tensile strength/MPa		≥1.65	≥1.5	≥1.2
fracture elongation/%		≥350	≥300	≥200
flexibility/℃		−30, no crack at bending	−20, no crack at bending	−10, no crack when wrapping the round bar of ϕ10mm
Watertightness	pressure/MPa	≥0.3		
	Retention time/min	≥30		

Because of the excellent performance of the synthetic polymer material, the synthetic polymer waterproof coating made from this material has the good performance of high strength and elongation, excellent flexibility, high and low temperature resistance, durability and waterproof ability.

(3)胎体增强材料质量要求(表 8.6)。

表 8.6 胎体增强材料质量要求

项目		质量要求		
		聚酯无纺布	化纤无纺布	玻纤布
外观		均匀、无团状、平整、无折皱		
拉力(不小于,N/50 mm)	纵向	150	45	90
	横向	100	35	50
延伸率(不小于,%)	纵向	10	20	3
	横向	20	25	3

3. 技术关键要求

(1)所有节点防水施工时,均应先填密封材料。

(2)应在先涂布的涂层干燥或固化成膜(不粘脚)后,方可涂布下一遍涂料。

(3)涂膜防水层的施工顺序应按"先高后低,先远后近"的原则进行,同一屋面上先涂布阴阳角和排水较集中的水落口、天沟、檐口、天窗下等节点部位,再进行大面积涂布。

(4)各遍涂层之间的涂布方向应相互垂直。涂层间每遍涂布的退槎和接槎应控制在 50~100 mm。

(5)管道等根部直径 500 mm 范围内,找平层应抹出高度不小于 30 mm 的圆台,其根部四周应铺贴胎体增强材料,宽度和高度不应小于 300 mm;管道上的涂膜收头处应用防水涂料多道涂刷,并应用密封材料封严。

(6)采用两层胎体增强材料时,上下层不得相互垂直铺贴,搭接缝应错开,其间距不应小于幅度的 1/3。

(7)胎体增强材料应加铺在涂层中间,下面涂层厚度不小于 1 mm;上层的涂层厚度不小于 0.5 mm。

(3) The quality requirements for matrix reinforcing materials(Table 8.6).

Table 8.6 The Quality Requirements for Matrix Reinforcing Materials

Items		Quality Requirements		
		Polyester Non-woven Fabric	Chemical Fiber Non-woven Fabric	Glass Fabric
surface		even, flat, no clustering, no wrinkling		
Tension(no less than, N/50 mm)	vertical	150	45	90
	horizontal	100	35	50
elongation rate (no less than, %)	vertical	10	20	3
	horizontal	20	25	3

3. Special Technical Requirements

(1) When taking waterproofing construction on the joints, they should be filled with sealing materials.

(2) When brushing the coating for several times, leave enough time for the prior coating to dry or solidify (until it doesn't stick to feet) before brushing it again.

(3) The sequence of applying coating waterproof layer should follow the principle of " higher locations first and then the lower, farther places first and then the near". On the same roof, the joints with complex structure should be coated first, for example, the inner and outer corners, the scupper, the gutter, eaves, skylight walls, etc. After that, coat the large and open areas.

(4) The direction of brushing each coating should be perpendicular to each other. The width of the end and side overlap of each following coating layer shall be controlled in the range of 50—100mm.

(5) The levelling of the scope around the root of the pipe with the diameter of 500mm should be a circular table 30mm higher than adjacent layers. The matrix reinforced material should be laid around the root of the pipe, and the height and width of it should be no less than 300mm. The ends of the coating on the pipes should be brushed with waterproof materials for several times and be sealed with sealing material.

(6) When two layers of matrix reinforcing materials are used, they should not be laid perpendicular to each other and the overlap of them should be staggered with the spacing not less than 1/3 of the its width.

(7) The matrix reinforcing materials should be laid between the coating materials. The thickness of the coating material on the lower side should be no less than 1mm, and the thickness of the coating material on the upper side should be no less than 0.5mm.

4. 施工工艺

涂膜应根据防水涂料的品种分层分遍涂布，不得一次涂成；应待先涂的涂层干燥成膜后，方可涂后一遍涂料。需铺设胎体增强材料时，屋面坡度小于15%时，可平行屋脊铺设；屋面坡度大于15%时，应垂直于屋脊铺设。胎体长边搭接宽度不应小于50 mm，短边搭接宽度不应小于70 mm。采用二层胎体增强材料，上下层不得相互垂直铺设，搭接缝应错开，其间距不应小于幅宽的1/3。

应按照不同屋面防水等级，选定相应的防水涂料及其涂膜厚度。高聚物改性沥青防水涂料，在屋面防水等级为Ⅱ级时不应小于3 mm；合成高分子防水涂料，在屋面防水等级为Ⅲ级时不应小于1.5 mm。

施工要点：防水涂膜应分层分遍涂布，第一层一般不需要刷冷底子油。待先涂的涂层干燥成膜后，方可涂布后一遍涂料。在板端、板缝、檐口与屋面板交接处，先干铺一层宽度为150～300 mm塑料薄膜缓冲层。铺贴玻璃丝布或毡片应采用搭接法，长边搭接宽度不小于70 mm，短边搭接宽度不小于100 mm，上下两层及相邻两幅的搭接缝应错开1/3幅宽，但上下两层不得互相垂直铺贴。铺加衬布前，应先浇胶料并刮刷均匀，然后立即铺加衬布，再在上面浇胶料刮刷均匀，纤维不露白，用辊子滚压实，排尽布下空气。

必须待上道涂层干燥后方可进行后道涂料施工，干燥时间视当地温度和湿度而定，一般为4～24 h。

涂膜防水屋面应设涂层保护层。

5. 质量控制标准

(1) 主控项目：

①防水涂料、胎体增强材料、密封材料和其他材料必须符合质量标准和设计要求。施工现场应按规定对进场的材料进行抽样复验。

②涂膜防水屋面施工完后，应经雨后或持续淋水24 h的检验。若具备作蓄水检验的屋面，应做蓄水检验，蓄水时间不小于24 h。必须做到无渗漏、不积水。

4. Construction Process

The coating should be taken for several times and in several layers based on the types of the waterproof materials. When brushing the coating, leave enough time for the prior coating to dry or solidify to form membrane before brushing it again. In laying the matrix reinforcing material, it can be laid in the direction parallel to the ridge when the slope of the roof is less than 15%; when the slope of the roof is larger than 15%, the matrix reinforcing materials should be laid in the direction perpendicular to the ridge. The overlap joints of the matrix on the longer side should be no less than 50mm and the overlap joints of the matrix on the shorter side should be no less than 70mm. When two layers of matrix reinforcing materials are used, they should not be perpendicular to each other and the joints of them should be staggered with no less than 1/3 of the its width.

The type of the waterproof materials and the thickness of the coating should be based on waterproofing grade of the roof. For polymer modified bituminous waterproof material, when the waterproofing grade is II, the coating thickness should be no less than 3mm. For synthesized polymer waterproof material, when the waterproofing grade is III, the coating thickness should be no less than 1.5mm.

The main operating skills: the coating should be taken for several times and in several layers, and it's no need to brush adhesive bitumen primer. When brushing the coating, leave enough time for the prior coating to dry or solidify to form membrane before brushing it again. In the joints of the board, eaves and slabs, plastic film with the width of 150mm to 300mm should be laid first as buffer layer. When laying textile glass fabric or mat, overlap joints method should be adopted. The overlap joints on the longer side should be no less than 70mm and the overlap joints on the shorter side should be no less than 100mm. The overlap joints for the adjacent layer and pieces should be staggered about 1/3 of its width and the upper layer and lower layer should not be laid perpendicular to each other. Before laying the backing fabric, bonding adhesive should be spread and scraped evenly on the substrates, and then the backing fabric is applied immediately. Again the successive bonding adhesive are provided onto the newly laid fabric by the same operations using appropriate tools to cover all the fabric. Finally apply pressure onto the fabric with the roller to squeeze out the air under it.

Leave enough time for the prior coating to dry and then the next coating can be applied. The time for drying depends on the temperature and humidity of the construction site and it is normally 4 to 24 hours.

Protective covering should be laid for roof with waterproof coating layer.

5. Quality Standard

(1) For key quality control assurance:

①The waterproof coating materials, matrix reinforcing materials, sealing materials and other materials should comply with the quality standard and design requirement. The incoming waterproof materials must be sampled and retested on site in accordance with the relative requirements.

②After the construction of roof with waterproof coating, it should be tested by rainwater or spray water for 24 hours. If the condition is permitted, water storage test can be carried out for 24 hours. There must be no leakage and ponding on the roof.

③天沟、檐沟必须保证纵向找坡符合设计要求。
④细部防水构造(如:天沟、檐沟、檐口、水落口、泛水、变形缝和伸出屋面的管道)必须严格按照设计要求施工,必须做到全部无渗漏。
(2)一般项目:
①涂膜防水层应表面平整、涂布均匀,不得有流淌、皱折、鼓泡、裸露胎体增强材料和翘边等质量缺陷,发现问题,及时修复。涂膜防水层与基层应粘结牢固 涂膜防水层的平均厚度应符合规定和设计要求,涂膜最小厚度不应小于设计厚度的80%。
②涂膜防水层上采用细砂等粒料做保护层时,应在涂布最后一遍涂料时,边涂布边均匀铺撒,使相互间粘结牢固,覆盖均匀严密,不露底。

6. 成品保护

涂膜防水层施工进行中或施工完后,均应对已做好的涂膜防水层加以保护和养护,养护期一般不得少于7d,养护期间不得上人行走,更不得进行任何作业或堆放物料。

7. 安全环保措施

(1)溶剂型防水涂料易燃有毒,应存放于阴凉、通风、无强烈日光直晒、无火源的库房内,并备有消防器材。
(2)使用溶剂型防水涂料时,施工现场周围严禁烟火,应备有消防器材。施工人员应着工作服、工作鞋、戴手套。操作时若皮肤上沾上涂料,应及时用沾有相应溶剂的棉纱擦除,再用肥皂和清水洗净。

8.2 地下防水

8.2.1 地下工程防水等级

现行规范规定地下工程防水等级及其相应的适用范围见表8.7;地下工程防水设防要求见表8.8及表8.9。

③ The vertical slope of the gutter and the inner drain should meet the design requirement.

④ The **waterproofing** for details in structure should strictly follow the design requirement, such as on the **gutter**, the inner drain, eaves, scupper, flashing, deformation joints and the pipes penetrating the roof, and there must be no leakage on those parts.

(2) For general quality control assurance:

① The surface of the coating waterproof layer should be smooth and the coating should be even. There must be no defects like flowing, wrinkling, bubbling, warping and exposing of matrix reinforcing material. If there is any problem, some measures should be taken immediately to solve it. The coating waterproof layer should be bonded closely with the substrate. The thickness of the coating waterproof layer should meet the requirements and design standard and it should be at least 80% of the designed thickness.

② When the protective covering for the coating waterproof layer is made by fine sand, the fine sand should be sprinkled evenly when brushing the last coating. It should be carried out at the same time of brushing so that they can bond together closely and cover the surface of the substrate fully.

6. Finished Product Protection

In and after the construction of coating waterproof layer, protection and curing should be carried out. The curing should be no less than 7 days and it is forbidden to walk, work or stack materials on it.

7. Measures for Safety and Environment Protection

① Solvent waterproof coating is flammable and toxic. It should be stored in a cool, ventilated, sun-free, fire-free store house with firefighting equipment.

② When using solvent waterproof coating, fire should be strictly prohibited around the construction site and firefighting equipment should be provided. Roofers should wear work clothes, shoes and gloves. If the coating sticks to skin, it should be removed immediately by cotton yarn moistened by relevant dissolvent and then washed with soap and clear water.

8.2 Underground Works Waterproofing

8.2.1 Waterproof Grade of Underground Works

The current specifications and codes for waterproof grade of underground works and their application scope can be shown in Table 8.7; the requirements for waterproofing of underground works can be shown in Table 8.8 and Table 8.9.

表8.7 地下工程防水等级及其适用范围

防水等级	标准	适用范围
一级	不允许渗水,结构表面无湿渍	人员长期停留的场所;因有少量湿渍会使物品变质、失效的贮物场所及严重影响设备正常运转和危及工程安全运营的部位;极重要的战备工程
二级	不允许漏水,结构表面可有少量湿渍 工业与民用建筑:总湿渍面积不应大于总防水面积(包括顶板、墙面、地面)的1/1 000;任意100 m² 防水面积上的湿渍不超过1处,单个湿渍的最大面积不大于0.1 m² 其他地下工程:总湿渍面积不应大于总防水面积的6/1 000;任意100 m² 防水面积上的湿渍不超过4处,单个湿渍的最大面积不大于0.2 m²,隧道工程还要求平均渗透量不大于0.15 L/(m²·d)	人员经常活动的场所;在有少量湿渍的情况下不会使物品变质、失效的贮物场所及基本不影响设备正常运转和工程安全运营的部位;重要的战备工程
三级	有少量漏水点,不得有线流和漏泥砂 任意100 m² 防水面积上的漏水点数不超过7处,单个漏水点的最大漏水量不大于2.5 L/d,单个湿渍的最大面积不大于0.3 m²	人员临时活动的场所;一般战备工程
四级	有漏水点,不得有线流和漏泥砂 整个工程平均漏水量不大于2 L/(m²·d);任意100 m² 防水面积的平均漏水量不大于4 L/(m²·d)	对渗漏水无严格要求的工程

地下防水工程的主要形式有:防水混凝土结构防水、卷材防水和涂膜防水等。

Chapter 8 Waterproofing Construction

Table 8.7 Waterproof Grade of Underground Works and Its Application Scope

Waterproof Grade	Standards	Application Scope
I	no leakage on the works and no wet stain on the surface of the works	places where people stay permanently, storage places where a little wet stain will cause deterioration and expiration of the stored things, the parts that determine the operation of equipment and affect the safety of the building, very important defense preparation engineering
II	no leakage on the works and a little wet stain on the surface of the works For underground works of industrial and civil engineering: the area of the wet stain should not be larger than 1/1000 of the whole waterproof area (including the ceiling, wall and floor); for a waterproof area of 100m^2, there is at most one scope of wet stain and its area is no more than 0.1m^2. For other underground works: the area of the wet stain should not be larger than 6/1000 of the whole waterproof area; for a waterproof area of 100m^2, there are at most four scopes of wet stain and the area of any one of them is no more than 0.2m^2; for tunnel works, the average infiltration capacity should not be more than 0.15L/($m^2 \cdot d$).	places where people often stays, storage places where a little wet stain will not cause deterioration and expiration of the stored things, the parts that will not determine the operation of equipment and affect the safety of the building, important defense preparation engineering
III	There are some wet spots, but there are no wet streaks and bleeding soil or sand and no leakage of sediment; for waterproof area of 100m^2, there are at most seven scopes of wet stain and the area of any one of them is no more than 0.3m^2, and the most leakage capacity is no more than 2.5L/d.	places where people stay contemporary; general defense preparation engineering
IV	There are some wet spots, but there is no wet streaks and bleeding soil or fine sand and no leakage of sediment; the average leakage capacity of the whole work is no more than 2L/($m^2 \cdot d$); the average leakage capacity of every waterproof area of 100m^2 is no more than 4L/($m^2 \cdot d$).	buildings that have no special requirement for waterproofing

The main waterproof methods for underground works are waterproof concrete structure waterproofing, sheet waterproofing, coating waterproofing and so on.

根据地下防水工程的特点及环境要求，坚持多道设防、刚柔相济、扬长避短、综合防治的作法是十分必要的。刚性防水材料从普通防水混凝土向高性能、外加剂纤维抗裂以及聚合物水泥混凝土方向发展；柔性防水材料从普通纸胎沥青油毡向聚酯胎、玻纤胎高聚物改性沥青以及合成高分子片材方向发展；防水涂料和密封防水材料也从沥青基向高聚物改性沥青、高分子以及聚合物无机涂料方向发展。新材料、新技术、新工艺的推广促使我国地下防水应用技术水平有新的飞跃和提高。

表8.8 明挖法地下工程防水设防

工程部位		主体						施工缝					后浇带				变形缝、诱导缝					
防水措施		防水混凝土	防水砂浆	防水卷材	防水涂料	塑料防水板	金属板	遇水膨胀止水条	中埋式止水带	外贴式止水带	外抹防水砂浆	外涂防水涂料	膨胀混凝土	遇水膨胀止水条	外贴式止水带	防水嵌缝材料	中埋式止水带	可卸式止水带	防水嵌缝材料	外贴防水卷材	外涂防水涂料	遇水膨胀止水条
防水等级	一级	应选	应选1~2种					应选2种					应选	应选2种			应选	应选2种				
	二级	应选	应选1种					应选1~2种					应选	应选1~2种			应选	应选1~2种				
	三级	应选	宜选1种					宜选1~2种					应选	宜选1~2种			应选	宜选1~2种				

Chapter 8 Waterproofing Construction

The waterproof methods should meet the requirement of the underground works and the condition of the environment and it is necessary to follow the principles of muti-waterproof methods, combination of rigid and flexible waterproofing, making good use of the advantages of one waterproof method and trying to avoid its disadvantages and adopting comprehensive waterproof methods. The rigid waterproof materials have been developed from ordinary waterproof concrete to high performance, crack resistance with admixture fiber concrete and polymer cement. The flexible waterproof materials have been developed from ordinary paper bituminous felt to polyester tire, glass fiber polymer modified bituminous and synthetic polymer sheet. Waterproof coatings and waterproof sealing materials have also been developed from bituminous base to polymer modified bitumen, polymer and polymer inorganic coatings. The popularization of new materials and technology promote the rapid development of the applied technology in waterproofing of underground works in China.

Table 8.8 Waterproof Measures of Underground Works in Open Excavation Conditions

Parts of the Work	Main Building Structures						Construction Joins						Post Cast Band				Deformation Joints and Inducing Joint					
Waterproof grade	selectable waterproofing applications	waterproof concrete	waterproof mortar	waterproof sheet membrane	waterproof coating	plastic waterproof board	steel plate	swell-type (hydrophilic strip) waterstop	half embedded waterstop	water bar applied to the external	waterproof mortar plastered on the external	waterproof coating plastered on the external	expansion concrete	hydrophilic expansion sealing strip	water bar applied to the external	inlay sealing waterproof material	half embedded waterstop	water bar applied to the external	demountable waterstop	inlay sealing waterproof material	waterproof sheet membrane applied to the external	waterproof coating plastered on the external	hydrophilic expansion sealing strip
I		required	one or two selections of the above to be adopted					two selections of the above to be adopted						required	two selections of the above to be adopted			required	two selections of the above to be adopted			
II		required	one selection of the above to be adopted					one or two selections of the above to be adopted						required	one or two selections of the above to be adopted			required	one or two selections of the above to be adopted			
III		required	one selection of above recommended					one or two selections of the above to be recommended						required	one or two selections of the above to be recommended			required	one or two selections of the above to be recommended			

续表

工程部位		主体	施工缝		后浇带		变形缝、诱导缝	
防水等级	四级	宜选	—	宜选1种	应选	宜选1种	应选	宜选1种

注：明挖法是指敞口开挖基坑，再在基坑中修建地下工程，最后用土石等回填的施工方法。

表8.9 暗挖法地下工程防水设防

工程部位		主体				内衬砌施工缝				内衬砌变形缝、诱导缝				
防水措施		复合式衬砌	离壁式衬砌、衬套	贴壁式衬砌	喷射混凝土	外贴式止水带	遇水膨胀止水条	防水嵌缝材料	中埋式止水带	外涂防水涂料	中埋式止水带	外贴式止水带	可卸式止水带	防水嵌缝材料	遇水膨胀止水条
防水等级	一级	应选1种			—	应选2种			应选		应选2种				
	二级	应选1种				应选1~2种			应选		应选1~2种				
	三级	—	应选1种			宜选1~2种			应选		宜选1种				
	四级		应选1种			宜选1种			应选		宜选1种				

注：暗挖法是指不挖开地面，采用从施工通道在地下开挖、支护、衬砌的方法修建隧道等地下工程的施工方法。

Chapter 8 Waterproofing Construction

continued

Parts of the Work		Main Building Structures	Construction Joins	Post Cast Band	Deformation Joints and Inducing Joint
Waterproof grade	IV	selective	—	one selection of the above to be recommended	required	one selection of the above to be recommended	required	one selection of the above to be recommended

Note: Open excavation is a construction method to excavate the foundation trench and construct the underground works, and then backfill the trench with soil, rock and so on.

Table 8.9 Waterproof Measures of Underground Works in Underground Excavation Conditions

Parts of the Work		Main Part				Inside Lining Construction Joint						Inside Lining Deformation Joint and Inducing Joint				
Waterproof measures		combined lining	detached lining with blocks or sleeves	adhesive lining	shot concrete lining	water bar applied to the external	hydrophilic expansion sealing strip	waterproof material inlay sealing	half embedded waterstop	waterproof coating plastered on the external	half embedded water bar	water bar plastered on the external	demountable waterstop	waterproof material inlay sealing	hydrophilic expansion sealing strip
Waterproof grade	I	one selection of the above to be adopted				two selections of the above to be adopted					required	two selections of the above to be adopted			
	II	one selection of the above to be adopted				one or two selections of the above to be adopted					required	one or two selections of the above to be adopted			
	III			one selection of the above to be adopted	—	one or two selections of the above to be recommended					required	one selection of the above to be recommended			
	IV			one selection of the above to be adopted		one selection of the above to be recommended					required	one selection of the above to be recommended			

Note: Underground excavation is a construction method to construct tunnels and other underground works with underground trench, timbering and lining.

8.2.2 普通防水混凝土防水层施工

因混凝土自身的密实性而具有一定防水能力的混凝土或钢筋混凝土结构形式称之为混凝土结构自防水。它兼具承重、围护功能,且可满足一定的耐冻融和耐侵蚀要求。随着混凝土工业化、商品化生产和与其配套的先进运输及浇捣设备的发展,它已成为地下防水工程首选的一种主要结构形式,广泛适用于一般工业与民用建筑地下工程的建(构)筑物。例如地下室、地下停车场、水池、水塔、地下转运站、桥墩、码头、水坝等。混凝土结构自防水不适用于以下情况:允许裂缝开展宽度大于 0.2 mm 的结构、遭受剧烈振动或冲击的结构、环境温度高于 80 ℃ 的结构,以及可致耐蚀系数小于 0.8 的侵蚀性介质中使用的结构。

防水混凝土分为普通防水混凝土防水和外加剂防水混凝土。普通防水混凝土是在普通混凝土骨料级配的基础上,调整配合比,控制水灰比、水泥用量、灰砂比和坍落度来提高混凝土的密实性,从而抑制混凝土中的孔隙,达到防水的目的。外加剂防水混凝土是加入适量外加剂(减水剂、防水剂),改善混凝土内部组织结构,增加混凝土的密实性,提高混凝土的抗渗能力。本节仅介绍普通防水混凝土防水层施工。

1. 普通防水混凝土质量要求

防水混凝土适用于抗渗等级不低于 P6 的地下混凝土结构。不适用于环境温度高于 80 ℃ 的地下工程。处于侵蚀性介质中,防水混凝土的耐侵蚀性要求应符合现行国家标准《工业建筑防腐蚀设计标准》(GB/T 50046—2018)和《混凝土结构耐久性设计标准》(GB/T 50476—2019)的有关规定。

(1)水泥的选择应符合下列规定:

①宜采用普通硅酸盐水泥或硅酸盐水泥,采用其他品种水泥时应经试验确定;

②在受侵蚀性介质作用时,应按介质的性质选用相应的水泥品种;

③不得使用过期或受潮结块的水泥,并不得将不同品种或强度等级的水泥混合使用。

(2)砂、石的选择应符合下列规定:

①砂宜选用中粗砂,含泥量不应大于 3.0%,泥块含量不宜大于 1.0%;

②不宜使用海砂;在没有使用河砂的条件时,应对海砂进行处理后才能使用,且控制氯离子含量不得大于 0.06%;

Chapter 8　Waterproofing Construction

8.2.2　The Construction of Waterproof Layer with Ordinary Waterproof Concrete

Concrete or concrete structure that has a certain waterproof ability due to the concrete's compactness is called self-waterproofing of concrete structure. It has load-bearing and enclosing functions and can meet the requirements of freeze-thaw resistance and corrosion resistance. With the development of concrete industrialization, commercialized production, advanced transportation and casting equipment, it has become the first choice of waterproofing for underground works, which is widely used in general industrial and civil engineering. For example, concrete has been widely used in the construction of basement, underground parking, water tank, water tower, underground transfer station, bridge pier, dock and dam. The self-waterproofing of the concrete structure is not suitable for the following situations: the structure that has cracks larger than 0.2mm, the structure that suffers from severe vibrations or shocks, the structure that is subject to ambient temperature higher than 80℃ and the structure that can be used in corrosive medium with corrosion resistance factors of less than 0.8.

Waterproof concrete can be classified into ordinary waterproof concrete and waterproof concrete with admixture. The waterproofing of the ordinary waterproof concrete is achieved by adjusting the mix ratio, controlling of water-cement ratio, cement content, cement-sand ratio and the slump to improve the compactness of concrete and restrain the voids occurring in concrete. The waterproof concrete with admixture is a kind of concrete with certain amount of admixture (water reducing agent and water repellent agent) to adjust the internal structure of concrete, to increase the compactness of concrete and improve the anti-seepage ability of concrete. In this part, we mainly focus on the construction of waterproof layer with ordinary waterproof concrete.

1. The Quality Requirements for Ordinary Waterproof Concrete

Waterproof concrete is suitable for underground concrete structures with anti-seepage grade not less than P6, and it is not suitable for underground works with ambient temperatures above 80℃. When it is used in the corrosive medium, the corrosion resistance requirements of waterproof concrete should meet the current national standards in *Standard for Anticorrosion Design of Industrid Construction* (GB 50046—2018) and *Standard of Design of Concrete Structure Durability* (GB/T 50476—2019).

(1) Theselection of cement shall meet the following requirements:

①It is advisable to use ordinary silicate cement or silicate cement and when use other types of cement, the cement should be tested before using;

②When the cement is affected by corrosive medium, the cement should be selected according to the properties of the medium;

③Using of expired cement and agglomerated cement due to damping shall not be permitted and also the blended cement mixed with different types and strength grades of cement are not permitted.

(2) The selection of sand and stone should meet the following requirements:

①It's better to use medium sized coarse sand, and the ratio of mud should not be greater than 3%, and the ratio of mud block should not be greater than 1%;

②It's better not to use sea sand. If we have to use sea sand because there is no river sand, the sea sand should be subject to further treatment and the ratio of chlorine ion must not be greater than 0.06%;

③碎石或卵石的粒径宜为 5~40 mm,含泥量不应大于 1.0%,泥块含量不应大于 0.5%;

④对长期处于潮湿环境的重要结构混凝土用砂、石,应进行碱活性检验。

(3)矿物掺合料的选择应符合下列规定:

①粉煤灰的级别不应低于二级,烧失量不应大于 5%;

②硅粉的比表面积不应小于 15 000 m^2/kg,SiO_2 含量不应小于 85%;

③粒化高炉矿渣粉的品质要求应符合现行国家标准《用于水泥、砂浆和混凝土中的粒化高炉矿渣粉》(GB/T 18046—2017)的有关规定。

(4)混凝土拌合用水应符合现行行业标准《混凝土用水标准》(JGJ63—2006)的有关规定。

(5)外加剂的选择应符合下列规定:

①外加剂的品种和用量应经试验确定,所用外加剂应符合现行国家标准《混凝土外加剂应用技术规范》(GB 50119—2013)的质量规定;

②掺加引气剂或引气型减水剂的混凝土,其含气量宜控制在 3%~5%;

③考虑外加剂对硬化混凝土收缩性能的影响;

④严禁使用对人体产生危害、对环境产生污染的外加剂。

(6)防水混凝土的配合比应经试验确定,并应符合下列规定:

①试配要求的抗渗水压值应比设计值提高 0.2 MPa;

②混凝土胶凝材料总量不宜小于 320 kg/m^3,其中水泥用量不宜少于 260 kg/m^3;粉煤灰掺量宜为胶凝材料总量的 20%~30%,硅粉的掺量宜为胶凝材料总量的 2%~5%;

③水胶比不得大于 0.50,有侵蚀性介质时水胶比不宜大于 0.45;

④砂率宜为 35%~40%,泵送时可增加到 45%;

⑤灰砂比宜为 1∶1.5~1∶2.5;

⑥混凝土拌合物的氯离子含量不应超过胶凝材料总量的 0.1%;混凝土中各类材料的总碱量即 Na_2O 当量不得大于 3 kg/m^3。

(7)防水混凝土采用预拌混凝土时,入泵坍落度宜控制在 120~140 mm,坍落度每小时损失不应大于 20 mm,坍落度总损失值不应大于 40 mm。

③The diameter of gravel or pebble should be 5—40mm, and the ratio of mud should not be greater than 1%, and the ratio of mud block should not be greater than 0.5%;

④ The sand and stone for concrete used in important structure and long-lasting wet environment should be tested for alkali activity.

(3) The selection of mineral admixtures shall meet the following requirements:

①The fly ash grade should not be less than level two, and the amount of burnout should not be greater than 5%;

②The specific surface area of silica fume should not be less than 15000 m^2/kg, and the ratios of SiO_2 should not be less than 85%;

③The quality of granular blast furnace slag should meet the requirements in current national standard *Ground Granulated Blast Furnace Slag Used for Cement, Mortar and Concrete* (GB/T 18046—2017);

(4) The mixing water used for concrete mixes should meet the requirements in current national standard *Standard of Water for Concrete* (JGJ 63—2006);

(5) The selection of admixtures should meet the following requirements:

①The type and the amount of admixture should be determined by test, and the admixture should meet the requirements in current national standard *Code for Utility Technical of Concrete Admixture* (GB 50119—2013);

②For concrete with air entraining agent or air entraining water reducing agent, the ratio of water should be 3%—5%;

③We should consider the effect of admixture on the shrinkage performance of hardened concrete;

④It is forbidden to use adhesives that are harmful to human beings or will cause pollution to the environment.

(6) The mix ratio of waterproof concrete shall be determined by test and shall meet the following requirements:

①The anti-seepage pressure value should be increased by 0.2MPa compared with the design value;

②The total amount of cementing materials should not be less than 320kg/m^3, in which the amount of cement should not be less than 260kg/m^3, the amount of flyash should be 20%—30% and the silica fume should be 2%—5% of the total amount of cementing materials;

③The water-cement ratio should not be greater than 0.50, and the water-cement ratio should not be greater than 0.45 when used in corrosive medium;

④The sand ratio should be 35%—40%, and when it is pumped, the sand ratio should be increased to 45%;

⑤The cement-sand ratio should be 1 : 1.5—1 : 2.5;

⑥The ratio of chlorine ion should not exceed 0.1% of the total amount of cementing materials, and the total amount of alkali (Na_2O) in all kinds of materials must not be greater than 3kg/m^3.

(7) When using ready-mixed waterproof concrete, the pumping slump should be controlled within the range of 120—140mm, and the loss of the slump per hour should not be greater than 20mm, and the total loss of the slump should not be greater than 40mm.

2. 细部防水构造做法

(1)穿墙管。给排水管、电缆管和供暖管穿过地下室外墙,应做好防水处理。

穿墙管埋设方式有两种:一种是直埋(图 8.8),一种是加套管(图 8.9)。无论采用何种方式,必须与墙外防水层相结合,严密封堵。为了保证防水施工和管道的安装方便,穿墙管位置应距离内墙角或凸出部位 25 cm。如果几根穿墙管并列,管与管的间距应大于 30 cm。数根穿墙管集中时,应设穿墙盒,做法见图 8.10。穿管盒的埋设和施工较为复杂,应注意以下几点:预留洞四周边埋角钢框;封口钢板打孔穿管,穿管与封口钢板焊接要严密;封口钢板与边框角钢焊接严密;穿墙盒内填充松散物质,如发泡聚氨酯或沥青玛蹄脂等,亦有防水功能。

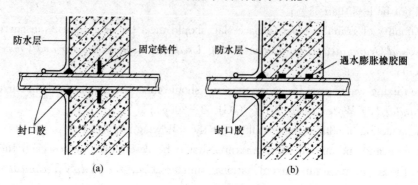

图 8.8 穿墙管直埋式

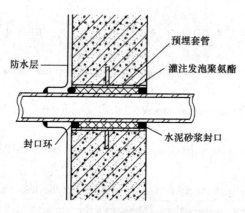

图 8.9 穿墙管加套管的埋设方式

2. Waterproofing Construction in Details

(1) Thru-wall pipes.

Waterproof treatment is very important in the parts that the water supply and drainage pipes, cable pipes and heat supply pipes penetrate the outer walls of the basement.

There are two alternative options to lay thru-wall pipes: one is to directly embed through the walls (Fig. 8.8) and the other is to provide inside sleeves for pipes penetrating the wall (Fig. 8.9). No matter which approach is adopted, the pipes or sleeves must be connected with the positive-side of waterproofing membrane and the perimeter joint should be sealed tightly. For the convenience of waterproofing operations and laying of pipes, the thru-wall pipes shall be kept 25cm away from the inner corner of the wall or any projections. If several thru-wall pipes are laid parallelly, the spacing between them should be greater than 30cm. When a number of thru-wall pipes are laid together, a wall box should be used, and the usage of wall box can be shown in Fig. 8.10. The laying and construction of wall box is complicated and we should pay attention to the following aspects: embed steel angle frame around the reversed holes of the pipes; weld the pipe and sealing steel plate together; weld sealing steel plate and steel angle frame together tightly; fill the wall box with loose substances, such as foaming polyurethane or bituminous horseshoe.

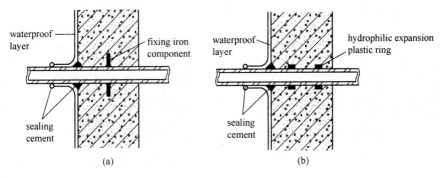

Fig. 8.8 Direct Embed Thru-wall Pipes

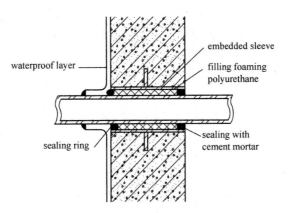

Fig. 8.9 Embed Thru-wall Pipe with Sleeve

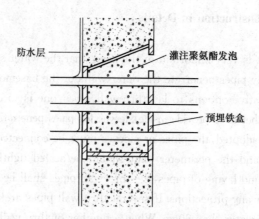

图 8.10 设穿墙盒的埋设方式

(2)预埋件。地下室内墙上或地板上,埋置铁件用来固定、安装设备。预埋铁件(图 8.11、图 8.12)用吊挂或专用工具固定,预埋件往往与结构钢筋接触,导致水沿铁件渗入室内。为此预留洞、槽均应作防水处理。预埋件受外力作用较大,为防止扰动周围混凝土,破坏防水层,预埋件端至墙外表面厚度不得小于 25 cm。如达不到 25 cm 应局部加厚,特殊工程需要做内防水,防水层一定与预埋件紧密结合,封闭严实。

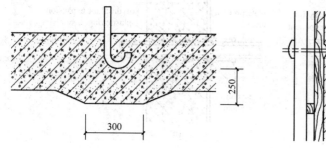

图 8.11 地下室底板上预埋螺栓图

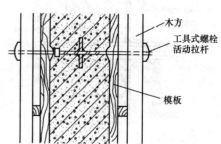

图 8.12 地下室墙体采用工具式螺栓

(3)施工缝。大面积浇筑混凝土一次完成有困难,须分两次或三次浇筑完。两次浇筑相隔几天或数天,前后两次浇筑的混凝土之间形成的缝即施工缝,此缝完全不是设计所需要的,由于混凝土的收缩,导致渗水通道,所以应对施工缝进行防水处理。

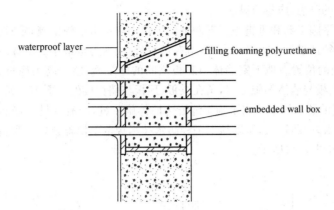

Fig. 8.10 Embed Thru-wall Pipe with Wall Box

(2) Embedded items. There are some metal items embedded in the basement walls or floor slabs used to fix or lay equipment. As shown in Fig. 8.11 and Fig. 8.12, the metal items are fixed by suspension or specific devices. Generally, these ancillary components connect with the reinforcement to give water easy seeping inside the basement along the components. Therefore, the reinforcing waterproofing should be made to the reserved holes or grooves for fixing these items. The embedded components endure great force from the outside, so in order to avoid damaging waterproof layer by influencing the concrete adjacent to the components, the space between the end of the component to the bottom of basement slab or the outer surface of the wall envelop should not be less than 25cm. If the space can't meet this requirement, the partial of basement envelop elements should be thickened around the items. If it is impractical for the partial envelops thickened to 25cm, the negative waterproofing must be applied, and the waterproof layer must be tightly integrated with the embedded items and tightly sealed.

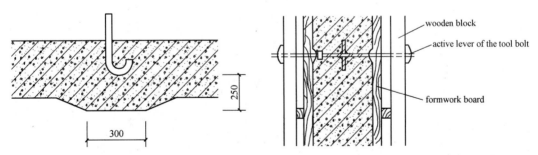

Fig. 8.11 Embedded Bolt in the Floor Slab   Fig. 8.12 Embedded Tool Bolt in the Wall

(3) Construction joints.

If it is proved impracticable that the placement of concrete in large area cannot be completed at one time, the second or third placement should be provided. It may be several days between the placements of the concrete and the joints between the placements are construction joints. Because of the shrinkage of the concrete, it may leak water in the joints, so waterproofing should be strengthened in the construction joints.

施工缝是渗水的隐患,应尽量减少。

施工缝分为水平施工缝和垂直施工缝两种。工程中多用水平施工缝,垂直施工缝尽量利用变形缝。留施工缝必须征求设计人员的同意,留在弯矩最小、剪力也最小的位置。

a. 水平施工缝的位置。地下室墙体与底板之间的施工缝,留在高出底板表面 30 cm 的墙体上。地下室顶板、拱板与墙体的施工缝,留在拱板、顶板与墙交接处之下 15~30 cm 处。

b. 水平施工缝的防水构造。水平施工缝皆为墙体施工缝,因有双排立筋和连接箍筋的影响,表面不可能平整光滑,凹凸较大,所以地下工程防水技术规范不推荐企口状和台阶状,只用平面的交接施工缝,构造如图 8.13 所示。

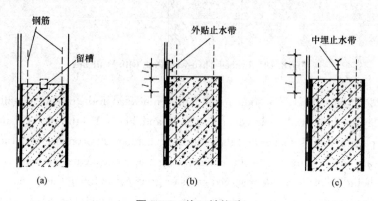

图 8.13 施工缝构造
(a)施工缝中设置遇水膨胀止水条;(b)外贴止水带;(c)中埋止水带

施工缝后浇混凝土之前,清理前期混凝土表面是非常重要的,因两次浇捣相差时间较长,在表面存留很多杂物和尘土细砂,清理不干净就成为隔离层,成为渗水通道。清理时必须用水冲洗干净。再铺 30~50 mm 厚的 1∶1 水泥砂浆或者刷涂界面剂,然后及时浇筑混凝土。

Chapter 8 Waterproofing Construction

Construction joints have high risk of leaking water, so there should be as less construction joints as possible.

The construction joints can be horizontal or vertical and in construction, the horizontal joints are more common because the vertical construction joints mainly coincide with the deformation joints. The set of the construction joints should get the permission from the designer and the location of the construction joints should be placed in the position where bending moment and shear force are minimum.

a. The location of the horizontal construction joints: The initial horizontal construction joints in basement walls should be placed 30cm above the surfaces of the basement slabs.

The construction joints between the walls and the arch cover or top plate of the basement should be set on the walls 15 to 30cm lower the conjunction of them.

b. Waterproofing in the horizontal construction joints: the horizontal construction joints are construction joints on the walls, and since there are double row ribs and the impact of connection hoops, the surface of the joint is uneven with bumps. Therefore, the code of practice for waterproofing of underground works do not recommend employing of the tongue-and-groove and step-shaped joints, so only the flat form of construction joints is considered as a preferred application. The structures of construction joints are shown in Fig. 8. 13.

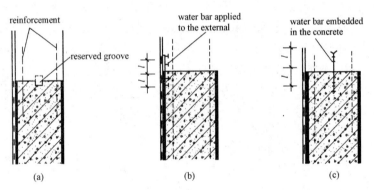

Fig. 8. 13 The Structure of Construction Joints
(a)hydrophilic expansion sealing strip set in the construction joints;
(b)water bar applied to the exteranl; (c)water bar embedded in the concrete

Before the latter casting concrete on the construction joints, it is necessary to clean the surface of the concrete(it is important to clean up the previously cast concrete surfaces). Because it may be a long time between the castings, there are debris, dust or fine sand on the surface of the concrete. If the surface is not cleaned, there may be holes between the two casting and it will leak water. So the surface must be washed thoroughly, and then lay mortar (water-cement ratio 1 : 1) with 30 to 50mm thickness or brush the interface agent. Finally, cast the concrete in time.

使用遇水膨胀止水条要特别注意防水。由于需先预留沟槽且受钢筋影响,操作不方便,很难填实,如果后浇混凝土未浇之前逢雨就会膨胀,这样将失去止水的作用。另外清理施工缝表面杂物时,止水之后应立即浇捣混凝土,不能留有膨胀的时间。

中埋止水带宜用"一"字形,但要求墙体厚度不小于 30 cm,它的止水作用,不如外贴式止水带好,外贴止水带拒水于墙外,使水不能进入施工缝。中埋止水带,水已进入施工缝中,可以绕过止水带进入室内。为此建议多用外贴止水带。

(4)变形缝及止水带。变形缝的构造比较复杂,施工难度较大,地下室发生渗漏常常在此部位,修补堵漏也很困难。变形缝两侧由于建筑沉降不等,产生沉降差,因沉降差导致止水带拉伸变形、防水层拉裂、嵌缝材料揭开等现象多有发生。建议沉降差不要超过 3 cm。

变形缝的宽度由结构设计决定。建筑越高,变形缝越宽,一般宽为 20~30 cm。变形缝处混凝土的厚度不小于 30 cm。

止水带是地下工程沉降缝必用的防水配件,它的功能:其一,可以阻止大部分地下水沿沉降缝进入室内;其二,当缝两侧建筑沉降不一致时,止水带可以变形,继续起阻水作用;其三,一旦发生沉降缝中渗水,止水带可以成为衬托,便于堵漏修补。

制作止水带的材料有橡胶止水带、塑料止水带、铜板止水带和橡胶加钢边止水带。目前我国多用橡胶止水带。止水带形状有多种,如图 8.14 所示。埋置止水带的形式如图 8.15 所示。

(5)后浇带。后浇带顾名思义是底板留出一条宽缝,若干天后再行浇捣混凝土,填实补平。

混凝土底板未达到龄期之前,产生大量水化热,引起收缩,如果底板较长,在收缩过程中会发生中间部位断裂。所以预先在底板中间部位留出 70~100 cm 宽的缝。40 d 左右后浇带两侧的混凝土达到了龄期,停止了收缩后,再作后浇带。两条后浇带相距一般为 30~60 m。

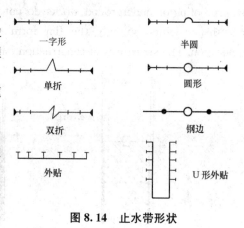

图 8.14 止水带形状

When using hydrophilic expansion sealing strip, we should pay special attention to waterproofing. Because of the reserved grooves and the influence of reinforcement, it is hard to operate and the sealing can't be finished compactly. If it rains before the latter casting of concrete, the strip will expand and it will lose the ability of waterproofing. Furthermore, after clearing the surface of the construction joint, the concrete must be cast immediately so that the strip is prevented from expansion.

The embedded waterstop usually takes the shape of "—", and the thickness of the walls for the embedded waterstop should be at least 30cm, but its performance of waterproofing is not as good as the waterstop applied to the external. For waterstop applied to the external, it can prevent the water from the walls, so the water can't enter the construction joints. For embedded waterstop, the water can enter the construction joints and it may also enter room, so waterstop applied to the external is preferred.

(4) Deformation joints and waterstop. The structure of deformation joint is complicated, so the construction of it is difficult. In basement, the leakage often takes place at this part and the repairing is difficult. Because of the differential settlement occurring between two sides of the joint that lead to the waterstop stretched and deformed, the waterproofing layer split, and the sealant may crack, so the differential settlement between the structural elements separated by the joints should not exceed 3cm.

The width of deformation joint is determined by the design of the structure. The higher the structure is, the wider the deformation joint will be. Generally, the width of the deformation joint is 20 to 30mm, and the concrete on that part is not less than 30cm.

Waterstop is the necessary waterproof accessory used in settlement joints in basement. First, it can prevent most water from entering the room; second, when the settlement besides the joint is differential, it can be deformed and will still function well; third, once the water leak into the deformation joint, the waterstop can be used as aid in repairing.

There are all kinds of waterstop according to the different materials: rubber waterstop, plastic waterstop, copper plate waterstop and rubber with steel edge waterstop and so on. Rubber waterstop is widely used in China. The waterstop may take different shapes, as shown in Fig. 8.14. The forms for embedding waterstop are shown in Fig. 8.15.

(5) Post-cast strip. Post-cast strip, as its name implies, is a strip on concrete slab which is left open and after several days, filled with concrete and is compacted and levelled. Before the concrete of slab reaches its full-age, a large amount of hydration heat is produced and it will cause shrinkage. If the total length of slab is large, it may break at the middle part during the shrinkage in process.

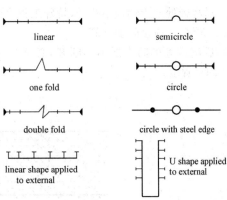

Fig. 8.14 The Shapes of Waterstop

So, it is necessary to reserve a gap of 70 to 100mm at the middle part of the bottom slab. About 40 days later, when the concrete reaches its age and the shrinkage stops, the casting of concrete can be finished. Generally, the distance between two post-cast strips is 30 to 60cm.

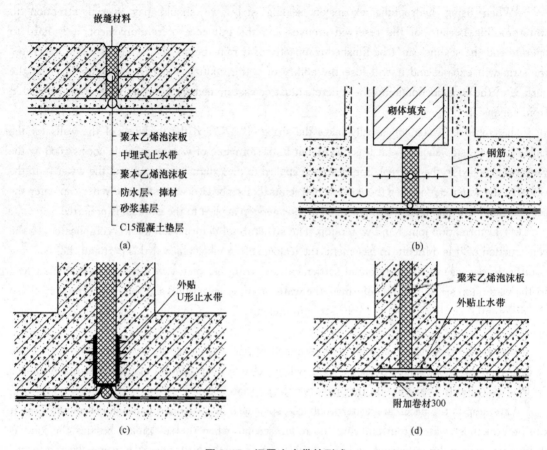

图 8.15 埋置止水带的形式

后浇带处底板钢筋不断开,特殊工程也可以断开,但两侧钢筋伸出,搭接长度应不小于主筋直径的 45 倍,还应设附加钢筋。

后浇带处的防水层不得断开,必须是一个整体,并采取设附加层和外贴止水带的措施,如图 8.16 所示。

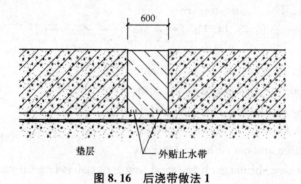

图 8.16 后浇带做法 1

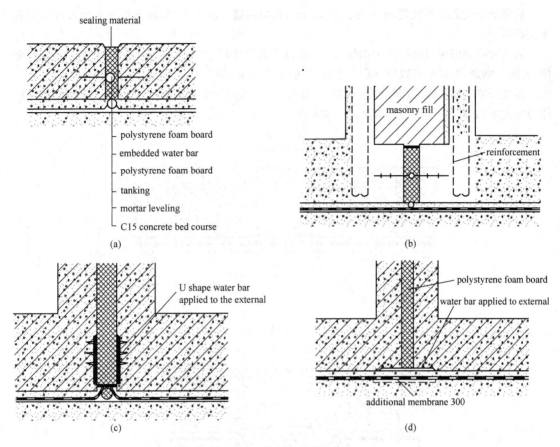

Fig. 8. 15　Forms for Embedding Waterstop

The reinforcement on the bottom slab of the post-cast strip will not be cut off, but for some special projects, the reinforcement can be cut off and the ends of the reinforcement should outstretch on both sides. When connecting the reinforcement, the length of the overlap should be at least 45 times of the diameter of the main reinforcement and additional reinforcement is also required.

The waterproof layer on the post-cast strip must be integrated and waterproof methods such as additional layer and waterstop applied to the external are also necessary, as shown in Fig. 8. 16.

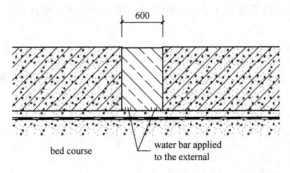

Fig. 8. 16　Construction Method for Post-cast Strip (I)

后浇带宽度宜窄不宜宽,最好不大于 70 cm,以防浇捣混凝土之前,地下水向上压力过大将防水层破坏。

后浇带两侧底板(建筑)产生沉降差,后浇带下方防水层会受拉伸或撕裂,为此,局部加厚垫层,并附加钢筋,沉降差可以使垫层产生斜坡,而不会断裂,如图 8.17 所示。

后浇带防水还可以采用超前止水方式(图 8.18)。其做法是将底板局部加厚,并设止水带,宜用外贴式止水带。由于底板局部加厚一般不超过 25 cm,不宜设中埋止水带。

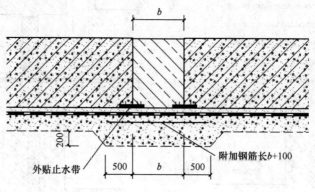

图 8.17 后浇带做法 2

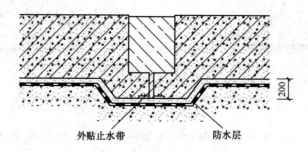

图 8.18 后浇带做法 3

后浇带两侧底板的立断面,可以做成企口,也可做成平面。

浇捣后浇带的混凝土之前,应清理掉落缝中杂物,因底板很厚,钢筋又密,清理杂物较困难。应认真做好清理工作。

后浇带的混凝土宜用膨胀混凝土,亦可用普通混凝土,但强度等级不能低于两侧混凝土。

The width of the post-cast strip shall be as narrow as possible and it is normally no more than 70cm. Otherwise, before casting the concrete, the pressure of the ground water may be too strong that it may destroy the waterproof layer.

The settlement of the bottom slabs besides the post-cast strip may be different, so the waterproof layer beneath the post-cast strip may be stretched or cracked; therefore, in this part, the bed course should be thicker and addition reinforcement is needed. In this case, the differential settlement may only lead to slope on this part and cracking may be avoided (Fig. 8.17).

We may also adopt the waterproof method of advanced water stopping on this part, as shown in Fig. 8.18. The construction method is to thicken the bottom slab partly and apply the waterstop to the external. Since typically the partial enhanced thickness of the bottom slabs doesn't exceed 25cm, the embedded waterstop is not suitable for this situation.

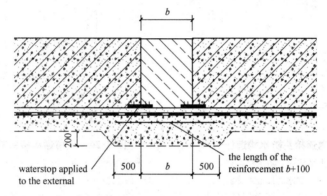

Fig. 8.17 Construction Method for Post-cast Strip (II)

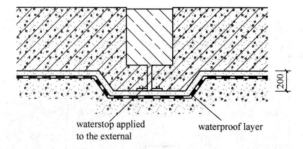

Fig. 8.18 Construction Method for Post-cast Strip (III)

The vertical section of the bottom slab besides the post-cast area can be T-G or flat. Before casting concrete on this area, remove the debris in the gaps to make it completely clean. The bottom slab is thick and the reinforcement is dense, so it is hard but very necessary to clean it well.

It's better to use expansive concrete on this area, and ordinary concrete is also acceptable, but the strength grade should not be lower than the strength grade of the concrete besides the post-cast strip.

(6)预留通道接头防水构造。预留通道接缝处的最大沉降差值不得大于 30 mm。留通道接头应采取复合防水构造形式,如图 8.19~图 8.21 所示。预留通道接头的防水施工应符合以下规定:中埋式止水带,遇水膨胀橡胶条,嵌缝材料、可卸式止水带的施工应符合本节变形缝施工中的有关规定;预留通道先施工部位的混凝土、中埋式止水带、与防水相关的预埋件等应及时保护,确保端部表面混凝土和中埋式止水带清洁,埋件不锈蚀。

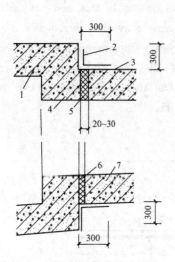

图 8.19 预留通道接头防水构造(一)
1—先浇混凝土结构;2—防水涂料;
3—填缝材料;4—遇水膨胀止水条;
5—嵌缝材料;6—背衬材料;7—后浇混凝土结构

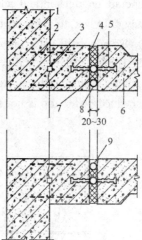

图 8.20 预留通道接头防水构造(二)
1—先浇混凝土结构;2—连接钢筋;3—遇水膨胀止水条;
4—填缝材料;5—中止式止水带;6—后浇混凝土结构;
7—遇水膨胀橡胶条;8—嵌缝材料;9—背衬材料

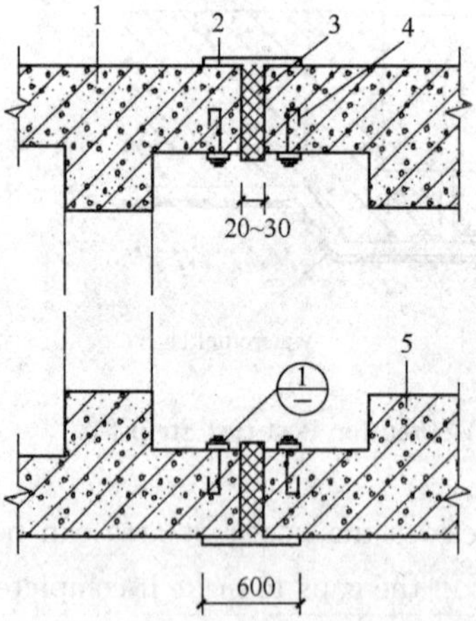

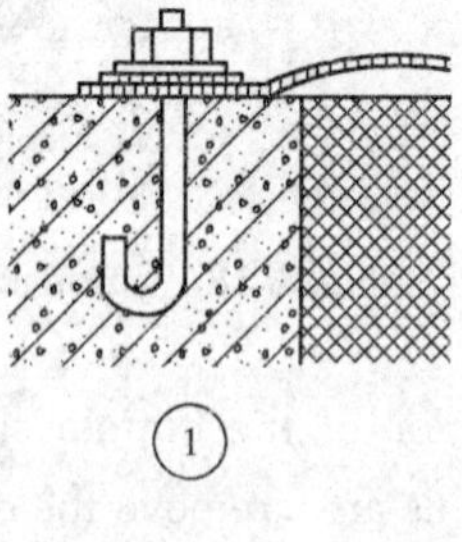

图 8.21 预留通道接头防水构造(三)
1—先浇混凝土结构;2—防水涂料;3—填缝材料;
4—可卸式止水带;5—后浇混凝土结构

(6) Waterproof structure at the joints of reserved passageways. At the joints of reserved passageways, the maximum differential settlement between two passageway structures shall not exceed 30mm. Multiple waterproofing methods should be adopted at the joints, as shown in Fig. 8.19—Fig. 8.21. The applications of waterproofing at this part should meet the following requirement: for the embedded waterstop, hydrophilic waterstop, joint sealant and removable waterstop, the requirement for the laying is the same as that of construction joints as described in this chapter. The previously cast concrete of the reserved passageways, section-center embedded waterstop and other embedded items at the joints should be protected in time to make them free from concrete stain and make the embedded items free from rusting.

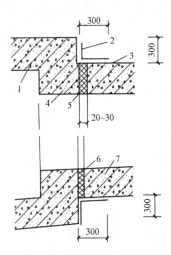

Fig. 8.19 Waterproof Structure at the Joints of the Reserved Passageway (I)

1—pre-cast concrete; 2—waterproof coating; 3—filling material; 4—hydrophilic expansion sealing strip; 5—caulking material; 6—backing material; 7—post-cast concrete

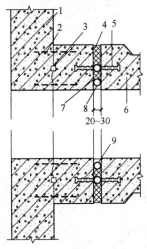

Fig. 8.20 Waterproof Structure at the Joints of the Reserved Passageway (II)

1—pre-cast concrete; 2—connecting reinforcement; 3—hydrophilic expansion sealing strip; 4—filling material; 5—embedded waterstop; 6—post-cast concrete; 7—hydrophilic expansion sealing strip; 8—caulking material; 9—backing material

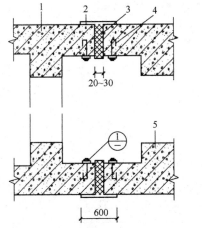
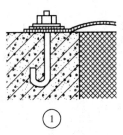

Fig. 8.21 Waterproof Structure at the Joints of the Reserved Channels (III)

1—pre-cast concrete; 2—waterproof coating; 3—filling material; 4—removable waterstop; 5—post-cast concrete

(7)桩头防水构造。桩头防水构造形式如图8.22和图8.23所示。桩头防水施工应符合以下规定：破桩后如发现渗漏水，应先采取措施将渗漏水止住；采用其他防水材料进行防水时，基面应符合防水层施工的要求；采取措施对遇水膨胀止水条进行保护。

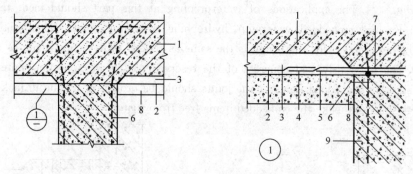

图 8.22 桩头防水构造（一）
1—结构底板；2—底板防水层；3—细石混凝土保护层；4—聚合物水泥防水砂浆；
5—水泥基渗透结晶型防水涂料；6—桩基受力筋；7—遇水膨胀止水条；
8—混凝土垫层；9—桩基混凝土

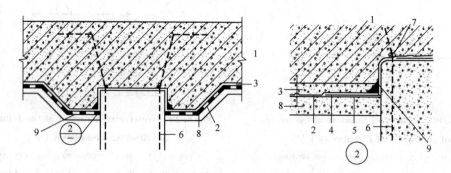

图 8.23 桩头防水构造（二）
1—构造底板；2—底板防水层；3—细石混凝土保护层；4—聚合物水泥防水砂浆；
5—水泥基渗透结合型防水涂料；6—桩基受力筋；7—遇水膨胀止水条；
8—混凝土垫层；9—桩基混凝土

(7) The waterproof structure of pile head is shown in Fig. 8.22, Fig. 8.23. The waterproof construction of pile head should meet the following requirement: after removing the pile head, if there is leakage besides the pile, take some measures to stop the water. When using other materials for waterproofing, the base level should meet the requirement for waterproof layer construction; take some measures to protect the hydrophilic expansion sealing strip.

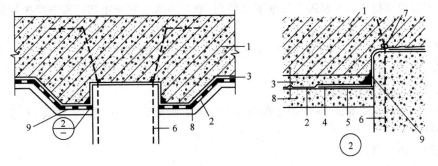

Fig. 8.22 Waterproof Structure of the Pile Head (I)

1—bottom slab; 2—waterproof layer of the bottom slab;
3—fine aggregate stone concrete protective covering;
4—polymeric cement waterproof mortar;
5—cement-based capillary crystalline waterproof coating;
6—tensile reinforcement of the pile foundation;
7—hydrophilic expansion sealing strip;
8—concrete bed course; 9—concrete of the pilefoundation

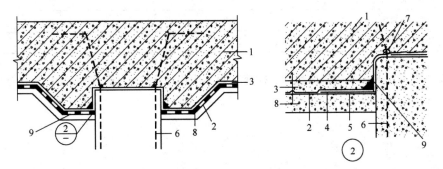

Fig. 8.23 Waterproof Structure of the Pile Head (II)

1— bottom slab; 2— waterproof layer of the bottom slab; 3—fine aggregate stone concrete protective covering;
4—polymeric cement waterproof mortar; 5—cement-based capillary crystalline waterproof coating;
6—tensile reinforcement of the pile foundation; 7—hydrophilic expansion sealing strip;
8—concrete bed course; 9—concrete of the pile foundation

3. 施工准备

(1)作业准备。钢筋、模板上道工序完成,施工的后一步骤是办理隐检、预检手续。注意检查固定模板的铁丝、螺栓是否穿过混凝土墙,如必须穿过时,应采取止水措施。特别是管道或预埋件穿过处是否已做好防水处理。木模板提前浇水湿润,并将落在模板内的杂物清理干净。

根据施工方案,做好技术交底。材料需经检验,由试验室试配提出混凝土配合比,试配的抗渗等级应按设计要求提高 0.2 MPa。如地下水位高,地下防水工程施工期间需继续做好降水,排水的工作。

(2)施工工具。混凝土搅拌运输车、车泵或拖式泵、布料机、搅拌机、翻斗车、磅秤、手推车、振捣器、试模、串桶、漏斗等。

4. 工艺流程及操作要点

(1)工艺流程:施工准备 → 混凝土配制、搅拌 → 混凝土运输 → 混凝土浇筑 → 混凝土养护。
(2)操作要点。
①混凝土的配制。

a. 严格按照经试配选定的施工配合比计算原材料用量。准确称量每种材料用量,按石子—水泥—砂子的顺序投入搅拌机。

b. 所用各种材料的品种、规格和用量,每工作班检查不应少于两次。每盘混凝土各组成材料计量结果的偏差应符合表 8.10 的规定。

表 8.10 混凝土组成材料计量结果的允许偏差(%)

混凝土组成材料	每盘计量	累计计量
水泥、掺合料	±2	±1
粗、细骨料	±3	±2
水、外加剂	±2	±1

注:累计计量仅适用于微机控制计量的搅拌站。

c. 防水混凝土必须采用机械搅拌。搅拌时间不应小于 120 s。掺外加剂时,应根据外加剂的技术要求确定搅拌时间。

3. Pre-construction Preparation

(1) Preparation for construction work.

In the construction process of reinforcement and formwork, the latter step of construction is based on the inspection of concealed work and pre-inspection of the previous work. Check whether the wire and bolt that fix the formwork penetrate the concrete wall; if they have to penetrate the concrete wall, take some waterstop measures. Check the parts where the pipes or the embedded component penetrate the wall to make sure they have been waterproofed. Wooden formwork should be wet and be cleaned to remove the debris in it.

The practice requirements should be interpreted and explained to the relevant operatives based on the building method statement. The material needs testing and the concrete mix ratio is based on the testing of the laboratory. The impermeability level of the concrete should be 0.2MPa higher than the requirement in the construction plan. If the ground water is high, the reducing and discharging of water is still necessary in the process of waterproof construction.

(2) Construction tools

The tools we may use in the construction are concrete mixer carrier, concrete pump truck, concrete pump trailer, spreader, mixer, dumper, platform scale, trolley, vibrator, testing formwork, chute, funnel and so on.

4. Construction Process and Main Operating Skills

(1) Construction process: pre-construction preparation → concrete preparation and mixing → concrete transportation → concrete placement → concrete curing

(2) Main operating skills.

①Concrete preparation.

a. The amount of materials used in the construction should be exactly calculated and prepared as the ratio determined in the test, and then put them into mixer in the order of gravel, finst and then cement, and then sand.

b. The type, specification and amount of the materials should be checked at least twice every working shift. The allowable amount derivation of the constituent materials mixed in each batch of concrete should meet the requirements shown in Table 8.10.

Table 8.10 Allowable Amount Derivation of the Constituent Materials of Concrete(%)

Constituent Materials	Amount in Every Batch	Amount in Total
cement, mineral admixture	±2	±1
coarse and fine aggregates	±3	±2
water, admixture	±2	±1
Note: total amount derivation is only applicable in the mixing station where compute is used in controlling the amount		

c. Waterproof concrete must be mixed by machine and the mixing time should be no less than 120s. When adding mixture, the mixing time is determined by the technical requirement of the mixture.

d. 采用集中搅拌或商品混凝土时,亦应符合上述规定,确保防水混凝土质量。

②混凝土运输。运输过程中应采取措施防止混凝土拌合物产生离析,以及坍落度和含气量的损失,同时要防止漏浆。防水混凝土拌合物在常温下应为半小时以内运至现场;运送距离较远或气温较高时,可掺入缓凝型减水剂,缓凝时间宜为6~8 h。

防水混凝土拌合物在运输后如出现离析,则必须进行二次搅拌。当坍落度损失后不能满足施工要求时,应加入原水灰比的水泥浆或二次掺加减水剂进行搅拌,严禁直接加水搅拌。

③混凝土浇筑。浇筑前,应清除模板内的积水、木屑、钢丝、铁钉等杂物,并以水湿润模板。使用钢模应保持其表面清洁无浮浆。

混凝土浇筑应分层,每层厚度不宜超过30~40 cm,相邻两层浇筑时间间隔不应超过2 h,夏季可适当缩短。

混凝土在浇筑地点须检查坍落度,每工作班至少检查两次。普通防水混凝土坍落度不宜大于50 mm,且实测坍落度与要求坍落度之间的偏差应符合表8.11的规定。

表 8.11 混凝土坍落度允许偏差

要求坍落度/mm	允许偏差/mm
≤40	±10
50~90	±15
≥100	±20

泵送混凝土在交货地点的入泵坍落度,每工作班至少检查两次。混凝土入泵时坍落度允许偏差应符合表8.12的规定。

表 8.12 混凝土入泵时坍落度允许偏差

所需坍落度/mm	允许偏差/mm
≤100	±20
>100	±30

泵送混凝土拌合物在运输后出现离析,必须进行二次搅拌。当坍落度损失后不能满足施工要求时,应加入原水胶比的水泥浆或掺加同品种的减水剂进行搅拌,严禁直接加水。

Chapter 8 Waterproofing Construction

d. When concentrated mixing concrete or commercial concrete is used, the above requirement should also be met to ensure the quality of the waterproof concrete.

②Concrete transportation.

In concrete transportation, some measures should be taken to prevent the concrete mixture from segregation and the loss of slump and air content, as well as to prevent the leakage of slurry. Waterproof concrete mixture should be transported to the construction site within half an hour after mixing at room temperature; if the distance is far or the temperature is high, retarding type water reducing agent can be added and the retarding time should better be 6 to 8 hours.

If the waterproof concrete mixture is segregated, it needs to be mixed again. When the loss of slurry can't meet the construction requirement, add mortar with the same water-cement ratio or water reducing agent and mix it. It is forbidden to mix with water directly.

③Concrete placement.

Before concrete placement, we should remove the water and the debris in the formwork, such as the wood chips, steel wire, nails, etc and wet the formwork. The surface of the steel formwork should be clean and there should be no laitance on it.

Concrete placement should be divided into several layers and the thickness of each layer should be 30—40mm. The time interval in placing the adjacent layers should not exceed 2h and in summer, the time interval may be shorter.

The concrete slump should be inspected at the placement site at least twice every working shift. ordinary waterproof concrete slump should not be greater than 50mm and the allowable slump derivation in entering pump should meet the requirements shown in Table 8.11.

Table 8.11 Concrete Allowable Slump Derivation

required slump(mm)	allowable derivation(mm)
≤40	±10
50—90	±15
≥100	±20

The slump of pumped concrete at the delivery point should be check at least twice every vorking shift. The allowable slump deviation of concrete during pumping should meet the requirements shown in Table 8.12.

Table 8.12 Concrete Allowable Slump Derivation in Entering Pump

Required Slump(mm)	Allowable Derivation(mm)
≤100	±20
>100	±30

If the concrete mixture is segregated after its delivery, it needs to be mixed again. When the slump loss of concrete cannot satisfy the workability requirement, add mortar with the same water-cement ratio or water reducer to the unworkable concrete and mix it once more. It is forbidden to mix with water directly.

④混凝土振捣。防水混凝土必须采用高频机械振捣,振捣时间宜为10～30 s,以混凝土泛浆和不冒气泡为准。要依次振捣密实,应避免漏振、欠振和超振。掺加引气剂或引气型减水剂时,应采用高频插入式振捣器振捣密实。

⑤混凝土养护。防水混凝土的养护对其抗渗性能影响极大,特别是早期湿润养护更为重要,一般在混凝土进入终凝(浇筑后4～6 h)即应覆盖,浇水湿润养护不少于14 d。因为在湿润条件下,混凝土内部水分蒸发缓慢,不致形成早期失水,有利于水泥水化,特别是浇筑后的前14 d,水泥硬化速度快,强度增长几乎可达28 d标准强度的80%,由于水泥充分水化,其生成物将毛细孔堵塞,切断毛细通路,并使水泥石结晶致密,混凝土强度和抗渗性均能很快提高;14 d以后,水泥水化速度逐渐变慢,强度增长亦趋缓慢,虽然继续养护依然有益,但对质量的影响不如早期大,所以应注意前14 d的养护。

防水混凝土不宜用电热法养护,该方法会使混凝土内形成连通毛细管网路,且因易产生干缩裂缝致使混凝土不能致密而降低抗渗性;又因这种方法不易控制混凝土内部温度均匀,更难控制混凝土内部与外部之间的温差,因此很容易使混凝土产生温差裂缝,从而降低混凝土质量,也对混凝土抗渗性不利。

防水混凝土也不宜用蒸汽养护。因为蒸汽养护会使混凝土内部毛细孔在蒸汽压力下大大扩张,导致混凝土抗渗性下降。

5. 成品保护

(1)为保护钢筋、模板尺寸位置正确,不得踩踏钢筋,并不得碰撞、改动模板、钢筋。
(2)在拆模或吊运其他物件时,不得碰坏施工缝处企口及止水带。
(3)保护好穿墙管、电线管、电门盒、预埋件等,振捣时勿挤偏或使预埋件挤入混凝土内。

6. 质量控制
(1)主控项目。
①防水混凝土的原材料、配合比及坍落度必须符合设计要求。检验方法:检查产品合格证、产品性能检测报告、计量措施和材料进场检验报告。

④ Concrete vibration

Waterproof concrete must be vibrated with high-frequency vibrator, and the vibrating time should be 10 to 30s. The exact vibrating time can be determined by the observation of the appearance of laitance and no bubbles in concrete. The vibration should be carried out in order to avoid the case of no vibration, lack vibration and over vibration in some parts. When adding air entraining agent or air entraining water reducing agent, the concrete should be vibrated compactly with high-frequency vibrator.

⑤ Curing concrete

Curing waterproof concrete is very important for its impermeability, and the early wet curing is more important. The concrete should be covered when it comes to the final setting time (4 to 6 hour after placement of concrete), and the wet curing period should be more than 14 days. When the concrete is wet, the water evaporation inside the concrete is slow, and it is beneficial to the cement hydration. In 14 days after placement, the cement hardening speed is fast, and it strength can reach 80% of the standard strength reached in 28 days after placing of concrete; in this period, the cement is fully hydrated and its production will clog the pores in it, and the cement stone crystallization will dense, so the strength and impermeability of the concrete can be improved quickly. 14 days later, the hydration speed and the increase of the strength will slow down. The curing is also beneficial but it is not as important as it is in the first 14 days. So, we should pay special attention to the curing of the concrete in the first 14 days.

Electric curing of concrete is not suitable for waterproof concrete because it may produce network of capillary pores in concrete and it may also reduce its impermeability because of the drying shrinkage cracks and the lack of density. Furthermore, in this curing method, it is not easy to control the internal temperature of the concrete and temperature difference inside and outside of the concrete, so it is easy to cause cracks in concrete and this will reduce the quality of the concrete and influence the impermeability of it.

Steam curing is also unsuitable for waterproof concrete because it will expand the pores in the concrete because of the steam pressure, which will lead to the decline of its impermeability.

5. Finished Product Protection

(1) To ensure the correct location and size of the reinforcement and formwork, do not step on the reinforcement, and do not bump into or change the location of the formwork or reinforcement.

(2) Do not bump into the grooved mouth and waterstop at the construction joints in striking off the formwork or laying of other items.

(3) Protect the thru-wall pipes, wire pipes, switch box, embedded components and other items. In vibrating the concrete, do not crush them to make them oblique or embedded into the concrete.

6. Quality Standard

(1) For key quality control assurance.

① The materials, mix ratio and slump of waterproof concrete should meet the design requirements. The inspection methods include the checking of the product certification, product performance test report, measurement methods and material entry inspection report.

②防水混凝土的抗压强度和抗渗性能必须符合设计要求。检验方法：检查混凝土抗压强度、抗渗性能检验报告。

③防水混凝土结构的变形缝、施工缝、后浇带、穿墙管、埋设件等设置和构造必须符合设计要求。检验方法：观察检查和检查隐蔽工程验收记录。

(2) 一般项目。

①防水混凝土结构表面应坚实、平整，不得有露筋、蜂窝等缺陷；埋设件位置应准确。检验方法：观察检查。

②防水混凝土结构表面的裂缝宽度不应大于 0.2 mm，且不得贯通。检验方法：用刻度放大镜检查。

③防水混凝土结构厚度不应小于 250 mm，其允许偏差应为 +8 mm、-5 mm；主体结构迎水面钢筋保护层厚度不应小于 50 mm，其允许偏差为 ±5 mm。检验方法：尺量检查和检查隐蔽工程验收记录。

8.2.3 外防外贴卷材防水层施工

卷材防水层是依靠结构的刚度由多层卷材铺贴而成的，要求结构层坚固、形式简单，粘贴卷材的基层面要平整干燥。卷材防水层适用于受侵蚀性介质作用或受振动作用的地下工程。

卷材防水层应采用高聚物改性沥青防水卷材和合成高分子防水卷材。所选用的基层处理剂、胶粘剂、密封材料等配套材料，均应与铺贴卷材材性相容，卷材防水层应铺设在主体结构的迎水面。

地下防水工程一般把卷材防水层设在建筑结构的外侧，称为外防水；受压力水的作用紧压在结构上，防水效果好。外防水有两种施工方法，即外防外贴法和外防内贴法。

②The compressive strength and impermeability of waterproof concrete should meet the design requirements. The inspection methods include checking the testing report of the compressive strength and impermeability.

③The set and structure of the deformation joints, construction joints, post-cast strip, thru-wall pipes, embedded components and other items should meet the design requirement of the waterproof concrete structure. The inspection methods include the observation and checking the record of the inspection and acceptance of the concealed work.

(2) For general quality control assurance.

①The surface of waterproof concrete structure should be compact and smooth, and there should be no exposure of reinforcement and honeycomb on it. The location of the embedded components should be exact. The checking method is observation.

②The crack on the surface of waterproof concrete structure should not be greater than 0.2mm, and the cracks should not be connected. The checking method is tape measure surveying and inspect of conceal work accepta records with magnifying glass.

③The thickness of waterproof concrete structure should be at least 250mm and the allowable derivation is $+8mm$, $-5mm$. The concrete cover to the positive-side of RC structure members should be at least 50mm, and the allowable derivation is $\pm 5mm$. The inspection methods include tape measure surveying and checking the acceptance record of the concealed work.

8.2.3 Positive Waterproofing with Sheet Waterproof Layer Applied to External

Sheet waterproof layer depends on the rigidity of the structure and it is composed of several layers of sheet son it. The structural layer should be solid and simple, and the substrate for the sheet should be smooth and dry. This kind of waterproof layer is suitable for the underground projects subjected to the influence of corrosion or vibration.

Sheet waterproof layer is composed of polymer modified bituminous waterproofing sheet membrane and synthetic polymer waterproofing sheet membrane. The other materials such as substrate treatment agent, adhesive agent and sealing material used in the construction should be compatible with the sheet. Sheet waterproof layer should be laid in the upstream face of the main structure.

In underground waterproofing constructions, the waterproof sheet membrane is normally laid on the outside of the underground building envelops, which is called positive waterproofing. This kind of waterproof layer has good waterproof performance because it is compacted to the structure by the confined groundwater. There are two construction methods for positive waterproofing with sheet waterproof layer. One is to lay waterproofing sheet membrane to the external and the other is to apply waterproofing sheet membrane to the internal of the structure.

外防外贴法(图 8.24)是将立面卷材防水层直接铺设在需防水结构的外墙外表面。适用于防水结构层高大于 3 m 的地下结构防水工程。外防内贴法(图 8.25)是浇筑混凝土垫层后,在垫层上将永久保护墙全部砌好,将卷材防水层铺贴在永久保护墙和垫层上。适用于防水结构层高小于 3 m 的地下结构防水工程。本节仅介绍外防外贴法。

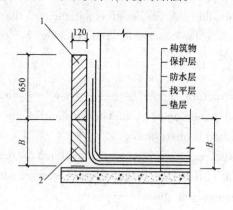

图 8.24 外防外贴法
1—临时保护墙;2—永久保护墙

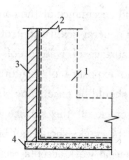

图 8.25 外防内贴法
1—待施工的构筑物;2—防水层;
3—保护层;4—垫层

1. 施工准备:

(1)技术准备:熟悉图纸,掌握细部构造的防水技术要求;编制施工方案或作业指导书,在经批准后向操作人员进行施工技术和安全交底。

(2)机具准备:手推车、压辊、喷灯、喷枪、热熔工具、安全帽等。

(3)材料准备:防水卷材及配套的底胶应有材料质量证明,并按要求见证取样送检,外观质量和物理性能均应满足相关规范要求。

2. 施工工艺

(1)工艺流程:基层清理、修补→基层验收→喷(涂)基层处理剂→特殊部位增强处理→弹铺贴线、试铺→卷材铺贴→收头、节点密封→检查修补→验收→保护层施工。

In the construction of positive waterproofing with waterproofing sheet membrane applied to the external (Fig. 8.24), the vertical sheet is applied directly to the surface of the external wall of the structure, and this kind of waterproof method is applicable to underground waterproof project with waterproof layer higher than 3m. The positive waterproofing with waterproofing sheet membrane applied to the internal (Fig. 8.25) is to pour the blinding concrete, and then on it lay the permanent protective brick veneer, followed by applying waterproofing sheet membrane onto the blinding and the veneer. This kind of waterproof method is applicable to underground waterproof project with waterproof layer lower than 3m. In this part, we mainly introduce the construction method of the sheet waterproof layer applied to the external of the structure.

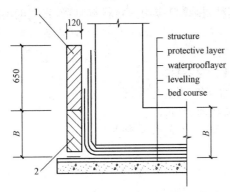

Fig. 8.24 Sheet Waterproof Layer Applied to the External
1—temporary protective wall;
2—permanent protective brick veneer

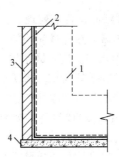

Fig. 8.25 Sheet Waterproof Layer Applied to the Internal
1—to be built structure; 2—waterproof layer;
3—protective covering; 4—bed course

1. Construction Preparation

① Technical preparation: be familiar with the drawings; master the waterproof technical requirements for the specific parts; prepare the construction plan or work instruction; communicate with the operators about the technology and safety requirements after approval.

②Preparation of tools: trolley, compression roller, gas torch, spray gun, hot melt tool, helmet and so on.

③Preparation of materials: There should be certification for the quality of waterproofing sheet membrane and substrate treatment agent. The selection and inspection of sample should follow the requirements and the appearance and physical properties should meet the relevant specifications.

2. Construction Technology

(1) Construction process: clean and mend the substrate → inspect and accept the substrate → spray or brush substrate treatment agent → further treatment for some parts with complex structure → set out lines with ink for the sheet and make trial laying → bind and seal the joints and ends → check and repair → accept the work → construct protective covering.

(2)操作要点。

①基层要求和处理:清理干净基层表面的异物和砂浆疙瘩等,局部孔洞、蜂窝、裂纹应修补严密,保持基层表面干燥(将 1 m² 卷材在白天铺放在找平层上,3~4 h 后,掀起卷材检查无水印)、平整、洁净、均匀一致,无起砂、脱皮等现象。

②冷粘法铺贴卷材应符合下列规定:
 a. 胶粘剂涂刷应均匀,不露底,不堆积。
 b. 根据胶粘剂的性能,应控制胶结剂涂刷与卷材铺贴的间隔时间。
 c. 铺贴时不得用力拉伸卷材,排除卷材下面的空气,辊压粘结牢固。
 d. 铺贴卷材应平整、顺直,搭接尺寸准确,不得有扭曲、皱折。
 e. 卷材接缝部位应采用专用粘结剂或胶结带满粘,接缝口应用密封材料封严,其宽度不应小于 10 mm。

③热熔法铺贴应符合下列规定:
 a. 火焰加热器加热卷材应均匀,不得加热不足或烧穿卷材。
 b. 卷材表面热熔后应立即滚铺,排除卷材下面的空气,并粘结牢固。
 c. 铺贴卷材应平整、顺直,搭接尺寸准确,不得有扭曲、皱折。
 d. 卷材接缝部位应溢出热熔的改性沥青胶料,并粘结牢固,封闭严密。

④自粘法铺巾卷材应符合下列规定:
 a. 铺贴卷材时,应将有黏性的一面朝向主体结构。
 b. 外墙、顶板铺贴时,应排除卷材下面的空气,并粘结牢固。
 c. 铺贴卷材应平整、顺直,搭接尺寸准确,不得有扭曲、皱折。
 d. 立面卷材铺贴完成后,应将卷材端头固定,并应用密封材料封严。
 e. 低温施工时,宜对卷材和基面采用热风适当加热,然后铺贴卷材料。

⑤卷材防水层完工并经验收合格后应及时做保护层。保护层应符合下列规定:

Chapter 8 Waterproofing Construction

(2) Main operating skills.

①Substrate requirements and treatment: clean the substrate surface to remove the debris and mortar bumps. If there are holes, honeycomb and cracks on the surface, they should be repaired. Keep the surface dry, smooth, clean and uniform. The way to check whether the surface is dry or not is to lay sheet of $1m^2$ on the leveling in daylight, and check whether there is watermark under the sheet 3 to 4 hours after laying. There should be no laitance or sand dust on the surface of the substrate.

②The laying of sheet with cold-melt application method should meet the following requirements:

a. The brushing of the adhesive agent should be even and flat, and the adhesive agent should cover all the surface of the substrate.

b. According to the performance of the adhesive agent, the time span between the brushing of adhesive agent and laying of sheet should be controlled.

c. It is forbidden to pull the sheet with great force, and the rolling of sheet should be compact to squeeze out the air and be compressed firmly to the substrate.

d. The sheet should be even and straight in laying, and the overlap length should be accurate. There should be no distortion and wrinkle on the overlap.

e. The joints between sheets should be fully bonded to the substrate with special adhesive agent or cement belt; joint application of sealing material seal mouth; its width should not be less than 10mm.

③ The laying of sheets with hot-melt application method should meet the following requirements:

a. The heating of the sheet with gas torch should be uniform. Lack of heating or over heating to burn through the sheet is forbidden.

b. Roll the sheet on the substrate immediately after the hot melt of its surface. Squeeze out the air under the sheet and bond it firmly to the substrate.

c. The sheet should be even and straight in laying, and the overlap length should be accurate. There should be no distortion and wrinkle on the overlap.

d. The hot-melt polymer modified bituminous adhesive agent should overflow at the joints and the joints should be bonded and sealed compactly.

④The laying of sheet with self-adhesive method should meet the following requirements:

a. In laying the sheet, the sticky side should face the main structure.

b. In laying the sheet on the exterior wall or the top plate, squeeze out the air under the sheet and bond it firmly.

c. The sheet should be even and straight in laying, and the overlap length should be accurate. There should be no distortion and wrinkle on the overlap.

d. After finalizing the laying of vertical sheet, the ends should be fixed and be sealed compactly with sealing material.

e. When the temperature is low, the sheet and the substrate can be heated with hot air first and then lay the sheet.

⑤ Make the protect layer in time after finalizing the sheet waterproof layer and being accepted. The protect layer should meet the following requirements:

a. 顶板的细石混凝土保护层与防水层之间宜设置隔离层。细石混凝土保护层厚度：机械回填时不宜小于 70 mm，人工回填时不宜小于 50 mm。

b. 底板的细石混凝土保护层厚度不应小于 50 mm。

c. 侧墙宜采用软质保护材料或铺抹 20 mm 厚 1∶2.5 水泥砂浆。

3. 质量控制

卷材防水层分项工程检验批的抽检数量，应按铺贴面积每 100 m^2 抽查 1 处，每处 10 m^2，且不得少于 3 处。卷材防水层应采用高聚物改性沥青防水卷材和合成高分子防水卷材。所选用的基层处理剂、胶粘剂、密封材料等均应与铺贴的卷材相匹配。防水卷材的搭接宽度应符合表 8.13 的要求。铺贴双层卷材时，上下两层和相邻两幅卷材的接缝应错开 1/3～1/2 幅宽，且两层卷材不得相互垂直铺贴。

表 8.13　防水卷材的搭接宽度

卷材品种	搭接宽度(mm)
弹性体改性沥青防水卷材	100
改性沥青聚乙烯胎防水卷材	100
自粘聚合物改性沥青防水卷材	80
三元乙丙橡胶防水卷材	100/60（胶粘剂/胶结带）
聚氯乙烯防水卷材	60/80（单面焊/双面焊）
	100（胶结剂）
聚乙烯丙纶复合防水卷材	100（粘结料）
高分子自粘胶膜防水卷材	70/80（自粘胶/胶结带）

(1) 主控项目。

① 卷材防水层所用卷材及其配套材料必须符合设计要求。检验方法：检查产品合格证、产品性能检测报告和材料进场检验报告。

② 卷材防水层在转角处、变形缝、施工缝、穿墙管等部位做法必须符合设计要求。检验方法：观察检查和检查隐蔽工程验收记录。

(2) 一般项目。

① 卷材防水层的搭接缝应粘贴或焊接牢固，密封严密，不得有扭曲、皱折、翘边和起泡等缺陷。检验方法：观察检查。

a. It is better to set up separation layer between the fine aggregate stone concrete protective covering and waterproof layer of the top plate. The thickness of the fine aggregate stone concrete should be at least 70mm in mechanical backfill and at least 50mm in manual backfill;

b. The thickness of the fine aggregate stone protective covering on the top plate should be at least 50mm;

c. Side walls should be protected by soft protective materials or mortar (cement : sand=1 : 2.5) of 20mm thick.

3. Safety Standard

The number of samples for the inspection of sheet waterproof layer should be at least three parts of 10m² every 100m². Polymer modifier bituminous waterproofing sheet membrane and synthetic polymer waterproofing sheet membrane should be used in waterproof layer. The substrate treatment agent, adhesive agent, sealing materials and so on should match with the sheet. The overlap width of the waterproofing sheet membrane should meet the requirement in Table 8.13. When there are two layers of waterproofing sheet membrane, the horizontal or vertical adjacent sheets should be staggered 1/3 to 1/2 of the its width and the vertical sheets should not be perpendicular to each other.

Table 8.13 **The Overlap Width of the Sheets**

Types of Sheet	Overlap Width(mm)
elastomer modified bituminous waterproofing sheet membrane	100
modified bituminous polyethylene tire waterproofing sheet membrane	100
self-adhering polymer modified bituminous waterproofing sheet membrane	80
EPDM rubber waterproofing sheet membrane	100/60(adhesive agent/cement belt)
polyvinyl chloride waterproofing sheet membrane	60/80(single side welding/double side welding)
	100(cement agent)
polyethylene polypropylene fiber composite waterproofing sheet membrane	100(adhesive agent)
polymer self-adhesive film waterproofing sheet membrane	70/80(self-adhesive agent/cement belt)

(1) For key quality control assurance.

①The sheet and the matching materials used in waterproof layer should meet the design requirement, and the inspection methods include the inspection of the product quality certificate, product performance test report and material entry inspection report.

②The construction of waterproof layer at specific parts should meet the design requirements, such as at the corner, deformation joints, construction joints, thru-wall pipes, etc. The inspection methods include observation and the checking of the acceptance of the concealed work.

(2) For general quality control assurance.

①The overlap joints of the waterproofing sheet membranes should be bonded together and sealed firmly. There should be no defects like twisting, wrinkles, warping and blisters. The inspection method is observation.

②采用外防外贴法铺贴卷材防水层时,立面卷材接槎的搭接宽度,高聚物改性沥青类卷材应为 150 mm,合成高分子类卷材应为 100 mm,且上层卷材应盖过下层卷材。检验方法:观察和尺量检查。

③侧墙卷材防水层的保护层与防水层应结合紧密、保护层厚度应符合设计要求。检验方法:观察和尺量检查。

④卷材搭接宽度的允许偏差应为 －10 mm。检验方法:观察和尺量检查。

4. 成品保护

(1)卷材运输和保管时应平放且不得高于 4 层,不得横放、斜放,应避免日晒、雨淋、受潮。

(2)操作人员应穿平底鞋施工;立面铺贴卷材时应注意保护好已铺卷材不受损坏,架子或梯子两端应用橡胶包裹,防止打滑和压破卷材。

(3)卷材层施工完毕,不允许上人踩踏,以免损伤防水层。并及时进行保护层施工。且有专人检查,发现破坏点后应及时标记、修补。

(4)外防外贴法墙角留槎的卷材要防止断裂和损伤,及时砌筑保护墙。外防内贴防水层,在地下防水结构施工前贴在永久性保护墙上,在防水层铺完后,按设计和规范要求及时做好保护层。

8.2.4 防水砂浆防水层施工

水泥砂浆防水层是在混凝土或砌砖的基层上用多层抹面的水泥砂浆等构成的防水层,它是利用抹压均匀、密实,并交替施工构成坚硬封闭的整体,具有较高的抗渗能力(2.5～3.0 MPa,30 d 无渗漏),以达到阻止压力水渗透的作用。水泥砂浆防水层适用于地下工程主体结构的迎水面或背水面。不适用于受持续振动或环境温度高于 80 ℃ 的地下工程。

②When the waterproof layer is constructed by applying sheets to the external, the overlap width of the vertical sheets should be 150mm for polymer modified bituminous sheet and 100mm for synthetic polymer sheets, and the sheet on the upper should cover the sheet on the lower layer. The inspection methods include observation and tape measure surveying.

③The veneer attached waterproofing sheet membrane layers and their protective coverings should be tightly combined, and the thickness of the coverings should satisfy the design requirements. Inspection method: observation and tape measure surveying.

④The allowable derivation of overlap width should be ±10mm, and the inspection methods include observation and tape measure surveying.

4. Finished Product Protection

(1) In the transportation and storage of sheets, they should not be piled to four layers, and horizontal and oblique placement is forbidden; what's more, they should avoid sunshine, rain and moisture.

(2) The roofers must wear flat shoes, and when laying the vertical sheet, try to protect the laid sheet from being destroyed. The ends of scaffold and ladders should be rapped with rubber, so that it will not slip on the sheet or destroy the sheet.

(3) After the laying of the sheet membrane is completed, no one is allowed to step on it to avoid damaging the waterproof layer, and the protective covering should be applied on it in time. Special person shall be assigned to check the formed condition, and the identified damage spots should be marked and repaired promptly.

(4) If there is joint on the sheet applied to the external of the corner, we need to construct protective walls in time to prevent the sheet from cracking or being destroyed. For the sheets used in waterproof layer applied to the internal, they should be applied to the permanent protective wall before underground waterproof construction, and after the construction of waterproof layer, protective covering should be constructed according to the design requirement.

8.2.4 Construction of Waterproof Layer with Waterproof Mortar

Cement/mortar waterproof layer is the waterproof treatment on the substrate of the concrete or bricks by pasting several layers of cement/mortar. Since the mortar is compressed evenly and compactly and several layers are constructed alternatively, an enclosed structure will be formed, so it has high impermeability (2.5 to 3.0MPa, 30 days with no leakage) to prevent the penetration of pressure water. The cement/mortar waterproof layer is applicable for the positive and negative side of the main structural envelopes for underground works. It is not suitable for the underground work subject to continuous vibration or ambient temperature above 80℃.

1. 施工准备：

(1)材料准备。

水泥砂浆防水层应采用聚合物水泥防水砂浆、掺外加剂或掺合料的防水砂浆。

水泥应使用普通硅酸盐水泥、硅酸盐水泥或特种水泥，不得使用过期或受潮结块的水泥；砂宜采用中砂，含泥量不应大于1%，硫化物和硫酸盐含量不得大于1%；用于拌制水泥砂浆的水应采用不含有害物质的洁净水；聚合物乳液的外观为均匀液体，无杂质、无沉淀、不分层。外加剂的技术性能应符合国家或行业有关标准的质量要求。

(2)机械工具：砂浆搅拌机、运料手推车、铁锹、铁抹子、米靠尺、楔形塞尺、水准仪、磅秤、小锤等。

(3)作业条件：主体结构验收合格，已办好验收手续。地下防水施工期间做好降水、排水工作，直至防水工程全部完工。降水、排水措施应按施工方案进行。地下室门窗口、预留孔洞、管道井出口等细部处理完毕。基层的混凝土和砌筑砂浆强度应不低于设计值的80%。基层表面应坚实、平整、洁净，并充分湿润，无积水。施工前，应有专业施工员进行技术交底、人员组织及现场安全文明管理工作。

2. 工艺流程及操作要点

(1)工艺流程：基层处理→刷水泥素浆(第一层)→抹底层砂浆(第二层)→刷水泥素浆(第三层)→抹面层砂浆(第四层)→刷水泥素浆(第五层)→养护。

(2)操作要点

①基层处理：基层表面应平整、坚实、清洁，并应充分湿润，无明水；基层表面的孔洞、缝隙应采用与防水层相同的水泥砂浆填塞并抹平；施工前应将埋设件、穿墙管预留凹槽内嵌填密封材料后，再进行水泥砂浆防水层施工。基层混凝土和砌筑砂浆的强度不应低于设计值的80%。

②刷素水泥浆：分两次抹压，基层浇水润湿后，先均匀刮抹1层素浆作为结合层，随后再抹1层素浆找平层，厚度要均匀。抹完后，用湿毛刷或排笔蘸水在素浆层表面依次均匀水平涂刷一遍，以堵塞和填平毛细孔道，增加不透水性。

③抹底层砂浆：在素浆初凝时进行，底灰厚度在5~10 mm。

Chapter 8 Waterproofing Construction

1. Construction Preparation

(1) Material preparation.

In cement/mortar waterproof layer, polymer cement waterproof mortar or waterproof mortar with admixture or addition shall be used. The cement should be ordinary portland cement, portland cement or special cement, and we can't use sticky cement or expired cement. Medium sfied sand shall be used and the mud content should not be higher than 1%, and the content of sulfide and sulfate should not be higher than 1%. The water for mixing cement/mortar shall be clean water without harmful substance. The appearance of polymer emulsion should be homogeneous, pure with no precipitation and stratification. The technical properties of admixtures should meet the relevant standards in China or in this industry.

(2) Mechanical tools.

Mortar mixer, trolley, shovel, iron trowel, batter stick by meter, wedge tucking ruler, level gauge, pound scale, hammer and so on.

(3) Operating Conditions

The main structure should be qualified and accepted. In the construction of underground waterproofing, precipitation and drainage is important and the measures for precipitation and drainage should meet the requirements of construction plan. The waterproof treatment should be finished at the specific parts such as openings for door and window, reserved holes, outlets of tunnels and so on. The strength of the concrete and masonry mortar of the substrate should be at least 80% of the designed value. The surface of the concrete base should be solid, smooth, clean, fully wet and with no stagnant water. Before construction, professional workers should come to the site to explain, organize and manage the process.

2. Construction Process and Main Operating Skills

(1) Construction process: treat the substrate →brush cement slurry (the first layer) →paste bottom mortar (second layer) →brush cement slurry (third layer) →brush surface mortar (the fourth layer) →brush cement slurry (the fifth layer) →curing.

(2) Main operating skills.

①Base treatment: The surface of the concrete base should be solid, smooth, clean, fully wet and with no stagnant water. The holes or gaps on the surface should be filled with the same mortar as used in the waterproof layer and the mortar levelling should be smoothed. The preserved grooves for the embedded components and thru-wall pipes should be filled with sealing material and then come to the construction of cement/mortar waterproof layer. The strength of the concrete and masonry mortar of the base should be at least 80% of the design value.

②Brush cement slurry: The cement slurry should be brushed twice. After the base is fully wet, brush a layer of cement slurry as binder course, and then brush a layer of cement slurry to make leveling, and the thickness of the slurry should be even. After that, use wet brush or broad brush to brush the surface of the slurry horizontally and evenly to block or fill the pores, so that it can strengthen the impermeability.

③Paste the base mortar: when the cement slurry begins to set, paste the base mortar with the thickness of 5—10mm.

④刷素水泥浆：抹完底层砂浆 1~2 d 后，再刷素水泥浆。

⑤面层砂浆：刷完素水泥浆后紧接着抹面层砂浆，配合比同底层砂浆，厚度在 5~10 mm，抹灰同第一层垂直，先用木抹子搓平，后用铁抹子压实抹光；

⑥刷水泥素浆：抹完底层砂浆 1 d 后，再刷素水泥浆。

⑦养护：水泥砂浆终凝后应及时进行养护，养护温度不宜低于 5 ℃并保持湿润，养护时间不得少于 14 d。聚合物水泥防水砂浆未达到硬化状态时，不得浇水养护或直接受雨水冲刷，硬化后应采用干湿交替的养护方法。潮湿环境中，可在自然条件下养护。

⑧接槎及阴阳角做法：每层宜连续施工，必须留施工缝时应采用阶梯坡形槎，每层施工缝错开 100 mm 左右，且离开阴阳角处不得小于 200 mm。所有阴角要做成半径 50 mm 的圆弧形，阳角做成半径为 10 mm 的圆弧形。

3. 质量标准

(1)主控项目。

①水泥砂浆防水层的原材料及配合比必须符合设计要求。检验方法：检查出厂合格证、质量检验报告、计量措施和现场抽样试验报告。

②水泥砂浆防水层各层之间必须结合牢固，无空鼓现象。检验方法：观察和用小锤轻击检查。

(2)一般项目。

①水泥砂浆防水层表面应密实、平整，不得有裂纹、起砂、麻面等缺陷；阴阳角处应做成圆弧形。检验方法：观察检查。

②水泥砂浆防水层施工缝留槎位置应正确，接槎应按层次顺序操作，层层搭接紧密。检验方法：观察检查和检查隐蔽工程验收记录。

③水泥砂浆防水层的平均厚度应符合设计要求，最小厚度不得小于设计值的 85%。检验方法：观察和尺量检查。

4. 成品保护

(1)抹灰层凝结硬化前应防止快干、水冲、撞击、振动和挤压，以保证防水层有足够强度。

(2)要保护好预埋件、预留孔洞等，不要随意抹死。

④Brush cement slurry: after one or two days after the pasting of base mortar, brush a layer of cement slurry.

⑤Brush surface mortar: Paste the surface mortar immediately after the brushing of cement slurry and the mix ratio should be the same with the mortar for the base mortar, with the thickness of 5—10mm. The mortar of this layer should be vertical to the first layer, and the mortar should be pasted with wooden trowel and then with trowel to make it compact and even.

⑥Brush cement slurry: brush cement slurry one day after the pasting of base mortar.

⑦Curing: cure the cement/mortar immediately after its final setting and the curing temperature should not be lower than 5℃ and keep it wet. The curing should be more than 14 days. When the polymer cement waterproof mortar has not been hardened, it should not be cured with casting water or directly scoured by rain. After hardened, it should be cured with wet and dry method alternatively. In wet condition, it can be cured naturally.

⑧The joints and corners: At the joints and the inner or outer corners, each layer should be constructed continuously. If there must be construction joints, the joints should be left in the shape of stairs and every joint should be about 100mm away from the other and the joints should be at least 200mm away from the inner or outer corner. The inner corner should be made into arc with a radius of 50mm, and the outer corner with a radius of 10mm.

3. Quality Standard

(1) For key quality control assurance.

①The materials and composing ratio of the cement/mortar waterproof layer should meet the design requirements. The inspection methods include checking the factory certificate, the quality inspection report, measurement methods and the test report for site sampling.

②The cement/mortar waterproof layer should be bonded to each other firmly without hollow between them. The inspection methods include observation and knocking it slightly with a hammer.

(2) For general quality control assurance.

①The surface of the cement/mortar waterproof layer should be compact and smooth. There should be no defects as cracking, laitance, sand dust, etc. The inner and outer corner should be made to the shape of arc. The inspection method is observation.

②The location of the construction joints of cement/mortar waterproof layer should be correct, and the joints should be constructed in order and every layer should be bond to each other firmly. The inspection methods include observation and checking the acceptance record of the concealed work.

③The average thickness of the cement/mortar waterproof layer should meet the design requirement and the minimum thickness shall not be less than 85% of the design value. The inspection methods include observation and tape measure surveying.

4. Finished Products Protection

(1) Before the hardening of the mortar, it should avoid drying quickly, being scoured, being stroke, being vibrated or crushed, so that it can ensure the strength of the waterproof layer.

(2) The embedded components and the preserved holes should be protected and do not paste mortar on it to make it compact.

8.2.5 涂料防水层施工

1. 涂料防水层施工的一般要求

涂料防水层包括有机防水涂料和无机防水涂料。有机防水涂料应采用反应型、水乳型、聚合物水泥等涂料；无机防水涂料应采用掺外加剂、掺合料的水泥基防水涂料或水泥基渗透结晶型防水涂料。

涂料防水层适用于受侵蚀性介质作用或受振动作用的地下工程；有机防水涂料宜用于主体结构的迎水面，无机防水涂料宜用于主体结构的迎水面或背水面。有机防水涂料基面应干燥。当基面较潮湿时，应涂刷湿固化型胶结剂或潮湿界面隔离剂；无机防水涂料施工前，基面应充分润湿，但不得有明水。防水涂料厚度应根据防水等级按表8.14选用。

表8.14 防水涂料厚度(mm)

防水等级	设防道数	有机涂料			无机涂料	
		反应型	水乳型	聚合物水泥	水泥基	水泥基渗透结晶型
Ⅰ级	三道及以上	1.2~2	1.2~1.5	1.5~2	1.5~2	≥0.8
Ⅱ级	二道设防	1.2~2	1.2~1.5	1.5~2	1.5~2	≥0.8
Ⅲ级	一道设防	—	—	≥2	≥2	—
	复合设防	—	—	≥1.5	≥1.5	—

2. 施工要点

(1)工艺流程：基层清理、修补→基层验收→刷界面处理剂→特殊部位增强处理→多次涂刷防水涂料至规定厚度→收头、节点密封→检查修补→保护层施工→验收。

(2)施工要点。

①多组分涂料应按配合比准确计量，搅拌均匀，并应根据有效时间确定每次配制的用量。

②涂料应分层涂刷或喷涂，涂层应均匀，涂刷应待前遍涂层干燥成膜后进行；每遍涂刷时应交替改变涂层的涂刷方向，同层涂膜的先后搭压宽度宜为 30~50 mm。

③涂料防水层的甩槎处接缝宽度不应小于 100 mm，接涂前应将其甩槎表面处理干净。

8.2.5 The Construction of Coating Waterproof Layer

1. General Requirements for the Construction of Coating Waterproof Layer

Coating waterproof layer can be classified into organic waterproof coating and inorganic waterproof layer. The materials for organic waterproof coating are reactive coating, water emulsion coating, polymer cement and so on; the materials for inorganic waterproof coating are cement-based waterproof coating with admixture or addition or cement-based permeable crystalline waterproof coating.

This kind of waterproof layer is suitable for the underground projects subjected to the influence of corrosion or vibration. The organic waterproof coating application is suitable for the positive side of the main structure and the inorganic waterproof coating is suitable for the positive side or negative side. The base for the organic waterproof coating should be dry and when the base is wet, brush wet curing binder or partitioning agent for wet surface. Before the construction of inorganic waterproof coating, the base should be wet fully without stagnant water. The thickness of the waterproof coating should be based on the waterproof levels, as shown in Table 8.14.

Table 8.14 The Thickness of Waterproof Coating (mm)

Waterproof Level	Layers of Waterproof membrane	Organic Coating			Inorganic Coating	
		reactive	water emulsion	polymer cement	cement base	cement-based permeable crystalline
I	three and above	1.2—2	1.2—1.5	1.5—2	1.5—2	≥0.8
II	two	1.2—2	1.2—1.5	1.5—2	1.5—2	≥0.8
III	one	—	—	≥2	≥2	—
	compound	—	—	≥1.5	≥1.5	—

2. The Main Operating Skills

(1) Construction process: clean and mend the substrate → accept the substrate → brush surface treatment agent → strengthen waterproofing for some parts with complex structure → brush waterproof coating repeatedly to the required thickness → strengthen waterproofing and seal the joints and ends → check and repair → make protective covering → make acceptance.

(2) Main operating skills.

①The coating is prepared in different batches, and in each batch, the amount of each kind of material should be measured exactly according the required mix ratio and stir well to mix them fully. The amount of coating prepared each time is based on the valid time of each kind of material.

②The coating should be brushed or sprayed in layers, and each layer should be even and the brushing of the latter layer should wait until the coating dries on the previous layer. In brushing the coating in different layers, the brushing direction should be changed alternately, and the overlap width for the brushing in the same layer should be 30—50mm.

③The overlap width at the joints of the coating waterproof layer should be no less than 100mm, and before brushing coating on the joints, the surface of the joints should be cleaned.

④采用有机防水涂料时,基层阴阳角处应做成圆弧;在转角处、变形缝、施工缝、穿墙管等部位应增加胎体增强材料和增涂防水涂料,宽度不应小于 50 mm。

⑤胎体增强材料的搭接宽度不应小于 100 mm,上下两层和相邻两幅胎体的接缝应错开 1/3 幅宽,且上下两层胎体不得相互垂直铺贴。

⑥涂料防水层完工并经验收合格后应及时做保护层。保护层应符合下列规定:顶板的细石混凝土保护层与防水层之间宜设置隔离层。细石混凝土保护层厚度:机械回填时不宜小于 70 mm;人工回填时不宜小于 50 mm;底板的细石混凝土保护层厚度不应小于 50 mm;侧墙宜采用软质保护材料或铺抹 20 mm 厚 1∶2.5 水泥砂浆。

8.3 厨房及厕浴间防水

厨房、厕浴间的防水涉及千家万户,也是百姓住房最关切、最敏感的问题之一。房屋渗漏现象严重,而厨房、厕浴间的渗漏首当其冲。楼上漏水,楼下受害,造成了邻里不和,直接影响到社会的和谐,所以厨房、厕浴间的防水与屋面、地下室建筑防水问题同等重要,而且更有直观感受。其防水的特点大致如下:

(1)一般公用的厨房、厕浴间面积稍大一些,但住宅的厨房、厕浴间面积均较小,而且构造比较复杂。

(2)管道多(指上下水、冷热水、暖气管),平立面连接拐角多,卫生设施多(指坐桶、便池、面盆、淋浴或洗澡盆),及排水地漏等或厨房间的所需管道、排水口、橱柜等均集中在独立的房间里。

(3)日常洗浴,房间带水,环境特殊。

(4)防水施工作业面小,费时、费工、费料,必须精心作业。

(5)防水层尽量整体防水、无接缝。

(6)不受大自然气候的影响,温度变化不大,对材料的延伸率要求不高。

④ When organic waterproof materials are used, the inner and outer corner of the substrate should be in the shape of arc. Matrix reinforcing material and waterproof material should be strengthened at some specific parts, such as at the corner, deformation joints, construction joints, thru-wall pipes and so on, the width of the matrix reinforcing material and waterproof material should be no less than 50 mm.

⑤ The overlap width of the matrix reinforcing material should be at least 100mm, and the joints of the adjacent matrix (vertical or horizontal) should be staggered 1/3 of their width. The vertically adjacent matrix should not be perpendicular to each other.

⑥ Make protective covering in time after the coating waterproof layer is finished and accepted. The protective covering should meet the following requirements: the separation layer should be set between the fine concrete protective covering and waterproof layer; the thickness of fine concrete protective covering should be no less than 70mm in mechanical backfill and no less than 50mm in manual backfill; the thickness of fine aggregate stone concrete protective covering of the bottom plate should be no less than 50mm; side walls should be protected by soft protective materials or 20mm thick mortar (cement : sand =1 : 2.5).

8.3　Waterproofing in Kitchen, Toilet and Bathroom

The leakage of water in houses, especially in kitchen, toilet and bathroom is a serious problem because it may cause damage to the rooms downstairs and arise quarrels between neighbors, which will destroy the harmony of the neighborhood. Therefore, the waterproofing in kitchen, toilet and bathroom is as important as in roof and basement. The waterproofing in kitchen, toilet and bathroom has the following characteristics:

(1) Generally, the areas of communal kitchen, toilet and bathroom are large, but in personal residence, the areas of kitchen, toilet and bathroom are small and the structures of them are complicated.

(2) There are many pipes in kitchen, toilet and bathroom, such as water supply and drainage pipes, hot water pipe and cold water pipe, and heating pipes. There are many corners at the connection of vertical and horizontal surfaces. There are many sanitary facilities, such as commode, urinal, basin, shower or bath basin. Furthermore, the floor drain or the necessary pipes, outlets or the cabinets are all located in these rooms.

(3) The kitchen, toilet and bathroom are all in wet environment because of daily washing or bathing.

(4) The operation areas for waterproofing are small, and waterproofing takes lots of time, work and materials and the waterproofing work should be operated carefully.

(5) The waterproof layer should better be an overall part in waterproofing and it's better to avoid joints in the waterproof layer of kitchen, toilet and bathroom.

(6) The waterproofing in kitchen, toilet and bathroom is not influenced by the climate and is not subject to great change in room temperature, so the requirement of the elongation of the materials is low.

基于以上特点，在作业上采用无接缝涂膜防水做法，施工最灵便。

8.3.1 厨房、厕浴间防水材料

厨房、厕浴间防水材料，可结合使用功能选用本课程提出的合成高分子防水涂料、聚合物水泥防水涂料、水泥基渗透结晶型防水材料、界面渗透型防水材料与涂料复合、聚乙烯丙纶防水卷材与聚合物水泥粘结料复合等多种选择，并对应其施工做法。

用于厨房、厕浴间的防水涂料主要有：合成高分子类、高聚物改性沥青类、沥青类和水泥基类。常用的防水涂料主要品种有：聚氨酯类（焦油聚氨酯类，沥青聚氨酯类）、高聚物改性沥青类以及聚合物水泥类。本节仅介绍聚氨酯防水施工。

聚氨酯防水涂料为双组份化学反应固化型防水涂料，甲组是以聚醚树脂和二异氰酸酯等经聚合反应制成的聚氨酯预聚体，乙组由硫化剂、催化剂、树脂等多种助剂精制而成。甲、乙组料按一定配合比例混合搅拌均匀，涂刮在基面上，经固化后形成整体而富有弹性的防水涂膜。

8.3.2 聚氨酯防水涂料施工

1. 施工准备

(1) 主要机具。电动搅拌器、油漆刷、拌料桶、滚动刷、小型油漆桶、小抹子、塑料刮板、油工铲刀、铁皮小刮板、墩布、橡胶刮板、笤帚、磅秤等。

(2) 作业条件。

①从事防水施工的队伍均有资质证书，主要操作人员均持证上岗。

②基层表面平整、密实、粘结、牢固、无起砂、开裂、空鼓等缺陷

③基层泛水坡度宜在2%以上，不得积水，管根、阴阳角等部位应抹成圆弧或钝角。

④与基层相连的管件、卫生洁具、地漏、排水口等应在防水层施工前安装牢固，预留管或管道未安装完不得进行防水层施工。

⑤厕浴间光线不足时，应有低压照明和通风设施，施工现象严禁烟火。

Based on the above features, the waterproofing in kitchen, toilet and bathroom should adopt the method of coating waterproofing without joints. , and this is the most convenient method.

8.3.1 Waterproof Materials for Kitchen, Toilet and Bathroom

The choice of waterproof materials for kitchen, toilet and bathroom should be based on the performance of the materials and the following materials can be applied, such as the synthetic polymer waterproof coating, polymer cement waterproof coating, cement-based capillary crystalline waterproof coating, interface permeability combination of polyethylene polypropylene waterproofing sheet membrane.

The main waterproof coatings for kitchen, toilet and bathroom are synthetic polymer material, polymer modified bituminous, general bituminous and cementitious material. The main types of waterproof coatings are polyurethane (tar polyurethane, bituminous polyurethane), polymer modified bitumen and polymer cement. In this part, we mainly focus on the waterproofing construction with polyurethane.

Polyurethane waterproof coating is a two-part chemical reaction curing waterproof coating. The first component for polyurethane waterproof coating is prepolymer of polyurethane in polymerization of polyether resin and diisocyanate. The second component is composed of several additives, such as vulcanizing agent, catalyst and resin. The first and second components are mixed together according to the required ratio, and then paste them on to the surface of the substrate. After curing, it will form a whole and elastic waterproof coating.

8.3.2 The Construction of Polyurethane Waterproof Coating

1. Construction Preparation

(1) The main tools.

Motor agitator, paint brush, bucket for mixing materials, rotary broom, small paint bucket, small trowel, plastic scrapper, stripping knife, iron sheet small scrapper, mop, rubber scrapper, broom, platform scale and so on.

(2) Operating Conditions.

①The team for waterproofing construction should have the qualification certificate, and the main workers for the construction should have the working certification.

②The surface of the substrate should be smooth, compact, sticky and solid, and there should be no defects like laitance, sand dust, hollow, etc.

③The flashing slope of the substrate should be more than 2% and there should be no stagnant water on the surface. The shape of the root of pipes and inner and outer corners should be arc or in obtuse angle.

④The pipes, sanitary facilities, drains and outlets embedded to the substrate should be laid and fixed before the construction of waterproof layer. Before the installation of preserved pipes or the laying of pipes, it is forbidden to make waterproof layer.

⑤When the light is insufficient, there should be low pressure lighting and ventilation facilities. Fires and smoking are strictly prohibited on the construction site.

2. 工艺流程及操作要点

(1)工艺流程：清理基层 → 涂刷底胶 → 细部附加层施工 → 第一遍涂膜防水层 → 第二遍涂膜防水层 → 第三遍涂膜防水层 → 第一次蓄水试验 → 保护层、饰面层施工 → 第二次蓄水试验 → 工程质量验收。

(2)操作要点。

①清理基层：基层表面必须认真清扫干净。如有油污，应用钢丝刷和砂纸刷掉；表面必须平整，凹陷处用 1∶3 水泥砂浆找平。

②涂刷底胶：将聚氨酯甲、乙两组分和二甲苯按 1∶1.5∶2（质量比）配合搅拌均匀，即可使用。用滚刷均匀地涂刷基层，不得露底，待涂层干固后方可进行下一道涂膜。

③细部附加层施工：应用甲、乙两组分按 1∶1.5 的比例混合搅拌均匀后，在地漏、管根、阴阳角和出入口等容易发生渗漏的薄弱部位涂刷。

④第一遍涂膜施工：将聚氨酯甲、乙两组分和二甲苯按 1∶1.5∶0.2（质量比）配合搅拌均匀，用橡胶刮板在基层表面均匀涂刮，厚度一致，涂刮量以 $0.8 \sim 1 \ kg/m^2$ 为宜。

⑤第二遍涂膜施工：在第一遍涂膜固化后，再进行第二遍聚氨酯涂刮。对平面的涂刮方向应与第一遍刮涂方向相垂直，涂刮量与第一遍相同。

⑥第三遍涂膜和粘砂粒施工：第二遍涂膜固化后，进行第三遍聚氨酯涂刮，达到设计厚度。在最后一遍涂膜施工完毕尚未固化时，在其表面应均匀地撒上少量干净的粗砂，以增加与即将覆盖的水泥砂浆保护层之间的粘结。厨房、厕浴间防水层经多遍涂刷，单组分聚氨酯涂膜总厚度应大于等于 1.5 mm。

⑦当涂膜固化完全并经蓄水试验验收合格才可进行保护层、饰面层施工。

3. 质量控制

(1)厨房、厕浴间防水工程使用的涂膜、刚性防水材料、聚乙烯丙纶卷材及其粘结材料、配套材料的质量、品种、配合比等均应符合设计要求和国家现行有关标准的规定。施工单位应提供材料检测报告、材料进入现场的复验报告及其他存档资料。

(2)涂膜防水层与预埋管件、表面坡度等细部做法，应符合设计要求和国家现行有关标准的规定，不得有渗漏现象（蓄水 24 h）。

Chapter 8 Waterproofing Construction

2. Construction Process and Main Operating Skills

(1) Construction process clean the substrate surface → brush treatment agent → make additional layer for detailed parts →brush coating for the first time →brush coating for the second time→ brush coating for the third time →carry out the first leakage test →make the protective covering and the decorative layer →carry out the second leakage test → make acceptance.

(2)Main Operating Skills.

①Clean the substrate surface: the substrate surface must be cleaned totally and carefully, and if there is any grease stain, it must be removed by wire brush or sand paper. The surface of the substrate must be even, and if there is sunken spot, it should be filled and leveled with mortar (cement : sand =1 : 3)

②Brush treatment agent: mix the two components of polyurethane and xylene at the weight ratio of 1 : 1.5 : 2 and stir them together well. Brush the treatment agent evenly with rotary broom on the substrate and cover all the surface of the substrate. Only after the agent is dry, can the brushing of coating begin.

③Make additional layer for detailed parts: mix the two components of polyurethane with the ratio of 1:1.5 and stir them well together, and then brush them on the parts that is easily subject to leakage, such as at the drains, the root of pipes, the inner and outer corner, and the outlets.

④Brush coating for the first time: mix the two components of polyurethane and xylene at the weight ratio of 1 : 1.5 : 2 and stir them together well and then spread them evenly on the substrate with rubber scraper. The thickness of the coating waterproof layer should be uniform and the amount of coating should better be 0.8—1kg/m².

⑤Brush coating for the second time: after the curing of the coating of the first waterproof layer, brush coating with polyurethane for the second time. The brushing direction of polyurethane on the second layer should be perpendicular with the direction of the first layer, and the amount of coating should be the same as the first coating.

⑥Brush coating for the third time and stick sand grains: after the curing of coating, brush the polyurethane waterproof coating with scrapper for the third time to make the waterproof layer to the required thickness. After brushing the last coating and before the curing of it, a small amount of clean coarse sand should be evenly sprinkled on its surface to increase the bond with the cement/mortar protective covering to be made. The waterproof layer in kitchen, toilet and bathroom should be constructed in different layers and the thickness of each polyurethane coating should be at least 1.5mm.

3. Quality Control

(1)The quality, types, mix ratio, etc. of the materials used in waterproofing in kitchen, toilet and bathroom, such as the coating, rigid waterproof material, polyethylene polypropylene sheet and the adhesive material and supporting material should meet the design requirements and the current national standards. The construction unit shall provide material inspection reports, re-inspection reports of materials entering the site and other archived materials.

(2)The construction of detailed parts such as coating waterproof layer, embedded components and slope of the surface should meet the design requirements and current national standards. There should be no leakage in these parts (water storage for 24h).

(3)找平层含水率低于9%时,且经检查合格后,方可进行防水工程施工。

(4)涂膜防水层应均匀一致,厚度应符合设计要求,不得有开裂、脱落、气泡、孔洞及收头不严密等缺陷。

(5)底胶和涂料附加层的涂刷方法、搭接收头,应符合施工规范要求,黏结牢固、紧密,接缝封严,无空鼓。

(6)涂膜防水层不起泡、不流淌、平整无凹凸,颜色亮度一致,与管件、洁具等接风严密,收头圆滑。

4. 成品保护

(1)操作人员应严格保护已做好的涂膜防水层,并及时做好保护层。在做保护层以前,非防水施工人员不得进入施工现场,以免损坏防水层。

(2)地漏要防止杂物堵塞,确保排水畅通。

(3)施工时,不允许涂膜材料污染已做好饰面的墙壁、卫生洁具、门窗等。

(4)材料必须密封储存于阴凉干燥处,严禁与水接触。存放材料地点和施工现场必须通风良好。

(5)存料、施工现场严禁烟火。

习 题

1. 简述热熔法高聚物改性沥青卷材防水层施工的施工工艺。
2. 简述合成高分子防水卷材屋面的工艺流程及主要操作工艺。
3. 简述涂膜防水屋面的质量控制标准。
4. 简述地下防水工程的防水等级及主要防水形式。
5. 举例说明普通防水混凝土防水层施工中,细部防水构造的做法。
6. 简述厨房、厕浴间聚氨酯防水涂料施工的施工工艺。

(3) When the moisture content of the levelling is less than 9% and the construction is qualified in inspection, the waterproofing construction can begin.

(4) The coating waterproof layer should be even and the thickness of the waterproof layer should meet the design requirement. There should be no defects on the waterproof layer, such as cracks, sand dust, hollow, voids and improper sealing of ends.

(5) The brushing method of the base course treatment agent and additional layer of the coating and the lapping or ends sealing should meet the design requirement. The binding between the different layers should be firm and compact and the sealing of the ends should be firm and there should be no hollows on the binding.

(6) There should be no defects on the coating waterproof layer, such as hollows or bleeding. The surface of the coating waterproof layer should be smooth and even, and the color and brightness of the coating should be uniform. The joints between the coating waterproof layer and the components or sanitary facilities should be sealed firmly and the ends of the coating should be sealed firmly and smoothly.

4. Finished Products Protection

(1) The constructors should protect the finished coating waterproof layer and make protective covering in time. Before the finishing of the protective covering, only the constructors in charge of waterproofing construction can enter the site.

(2) The floor drain should not be blocked by debris to ensure smooth drainage.

(3) In construction, coating materials should be prevented from staining the wall, sanitary facilities, doors and windows that have been finished.

(4) The coating materials must be sealed and stored in cool and dry environment and they are strictly prohibited from contacting with water. Good ventilation is important in the places storing coating materials and in construction site.

(5) There should be no fires and smoking on the construction site or on the places storing coating materials.

Exercises

1. Briefly describe the construction process of laying polymer modified bituminous waterproofing sheet membrane in hot-melt application method.

2. Briefly describe the construction process and main operating skills in constructing synthetic polymer waterproofing sheet applied roof.

3. Brief describe the quality control standard in constructing roof with waterproof coating.

4. Briefly describe the waterproof grade of underground works and the main waterproof methods.

5. Give examples to illustrate the waterproofing construction in details in the construction of waterproof layer with ordinary waterproof concrete.

6. Briefly describe the construction process of polyurethane waterproof coating in kitchen, toilet and bathroom.

第9章 装饰装修工程施工

9.1 抹灰工程

9.1.1 一般抹灰基本知识

1. 抹灰工程的概念

将水泥、砂、石灰膏、水等一系列材料拌合起来,直接涂抹在建筑物的表面,形成连续均匀抹灰层的作法叫抹灰工程。

2. 抹灰工程的分类

根据使用要求及装饰效果的不同,抹灰工程可分为一般抹灰、装饰抹灰和特种砂浆抹灰。

(1)一般抹灰:通常是指使用石灰砂浆、水泥砂浆、水泥混合砂浆、聚合物水泥砂浆、膨胀珍珠岩水泥砂浆和麻刀灰、纸筋灰、石灰膏等材料的抹灰叫一般抹灰。根据工序和质量的要求的不同,一般抹灰又分为高级抹灰和普通抹灰两个级别。

(2)装饰抹灰:按照不同施工方法和不同面层材料形成不同装饰效果的抹灰。可分为以下两类:水泥石灰类装饰抹灰和水泥石粒类装饰抹灰。

(3)特种抹灰:指特种功能要求的抹灰,即在普通砂浆中添加特种性能材料的抹灰,如保温隔热砂浆,耐酸、耐碱和防水砂浆等。

3. 抹灰工程的组成

为了使抹灰层与基层粘结牢固,防止起鼓开裂,并使抹灰层的表面平整,抹灰应分层涂抹。抹灰一般分为底层、中层、面层。底层为粘结层,主要起粘结兼初步找平作用;中层主要起找平的作用;面层的作用是美化装饰。当饰面用其他装饰材料时,只有底层、中层抹灰。

抹灰应采用分层分遍涂抹,应注意控制每遍厚度。如果一次涂抹太厚,由于自重和内外收缩快慢不一,易出现干裂,起鼓和脱落。

Chapter 9　Finishing and Decoration Work Construction

9.1　Plastering

9.1.1　Basic Knowledge of Plastering

1. Concept of Plastering

A series of materials, such as cement, sand, lime paste and water are mixed and directly applied to the surface of the building to form a continuous and uniform backing coat, which is called plastering.

2. Classification of Plastering

According to the different operating requirements and decorative effects, plastering can be divided into general plastering, decorative plastering and special mortar plastering.

(1) General plastering: It usually refers to plastering with lime mortar, cement mortar, cement mixed mortar, polymer cement mortar, expanded perlite cement mortar, hemp cut lime mortar, paper strip mixed with lime mortar, lime paste and other materials. According to the different requirements of process and quality, general plastering is divided into two levels: high grade plastering and ordinary plastering.

(2) Decorative plastering: Due to different construction methods and different surface materials, plastering will have different decorative effects. It can be divided into cement lime decorative plastering and cement stone decorative plastering.

(3) Special mortar plastering: It refers to the plastering for special functions. That is to say, materials with special performance will be added to the normal mortar in plastering, such as thermal insulation mortar, acid, alkali-resistant and waterproof mortar, etc.

3. Composition of Plastering

In order to make the plaster coat bond with the base firmly, to prevent bulging and cracking, and to make the surface of the backing coat smooth, plastering should be applied coat by coat. Plastering is generally divided into first coat, second coat, finish coat. The first coat is the bonding coat, which mainly plays the role of bonding and preliminary leveling; the second coat mainly plays the role of leveling; the function of the finish coat is to beautify decoration. When other decorative materials are used for the surface, only the first coat and the second coat are needed.

Plastering shall be applied sequentially in several coats over a base, and the thickness of each coat should be inspected. If the coating is too thick with one plastering, it is easy to crack, bulge and peel due to the difference of self-weight and internal and external contraction speed.

水泥砂浆和水泥混合砂浆的抹灰层,应等第一层抹灰层凝结后,方可涂抹下一层;石灰砂浆抹灰层,应等第一层七至八成干后,方可涂抹下一层。

9.1.2 一般抹灰的施工工艺

1. 工艺流程

基层清理→浇水湿润、吊垂直、套方、找规矩、抹灰饼→抹水泥踢脚或墙裙→做护角→抹水泥窗台→墙面充筋→抹底灰→修补预留孔洞,电箱槽、盒等→抹罩面灰。

2. 操作工艺

(1)基层清理:

砖砌体:应清除表面杂物,残留灰浆、舌头灰、尘土等。

混凝土基体:表面凿毛或在表面洒水润湿后涂刷1∶1水泥砂浆(加适量胶粘剂或界面剂)。

加气混凝土基体:应在湿润后边涂刷界面剂,边抹强度不大于 M5 的水泥混合砂浆。

(2)浇水湿润。一般在抹灰前一天,用软管或胶皮管或喷壶顺墙自上而下浇水湿润,每天宜浇两次。

(3)吊垂直、套方、找规矩、做灰饼。根据设计图纸要求的抹灰质量,根据基层表面平整垂直情况,用一面墙做基准,吊垂直、套方、找规矩,确定抹灰厚度,抹灰厚度不应小于 7 mm。当墙面凹度较大时应分层衬平。每层厚度不大于 7～9 mm。操作时应先抹上灰饼,再抹下灰饼。抹灰饼时应根据室内抹灰要求确定灰饼的正确位置,再用靠尺板找好垂直与平整。灰饼宜用 1∶3 水泥砂浆抹成 5 cm 见方形状。

房间面积较大时应先在地上弹出十字中心线,然后按基层面平整度弹出墙角线,随后在距墙阴角 100 mm 处吊垂线并弹出铅垂线,再按地上弹出的墙角线往墙上翻引弹出阴角两面墙上的墙面抹灰层厚度控制线,以此做灰饼,然后根据灰饼充筋。

In the backing coat of cement mortar or cement mixed mortar, the latter coat should not be applied before the solidification of the prior coat; in the backing coat of lime mortar, the latter coat should not be applied before the prior coat reaches 70%—80% of its solidification.

9.1.2 Construction Technology of General Plastering

1. Process Flow

Clean the base→ wet the base, check verticality with plumb bob, square room plan, square wall corner and make dots→ make cement skirting or dado→ make mortar corner guard→ make cement window sill→ make screed on wall→ apply first coat→ patch existing holes, fill the gaps coming from installation of conduits, distribution board and embedded electric outlet boxes, etc. → apply finish coat.

2. Operation Process

(1) Clean base.

Brick masonry: Removing debris, residual mortar, excess mortar, dust and other substances on the surface.

Concrete substratum: Roughen surfaces of set concrete, wet surface and then apply cement-sand slurry (cement : sand=1 : 1) with appropriate amount of adhesive or interface agent.

Aerated concrete substratum: The interface agent shall be applied after wetting, and the cement with mixed mortar shall be applied at the same time. The strength of the cement with mixed mortar should be no more than M5.

(2) Wet base. Generally, the day before plastering, wet the wall with hose, rubber hose or spray pot from top to bottom, twice a day.

(3) Check verticality with plumb bob, square room plan, square wall corner and make dots.

According to the requirement in the design drawings for the plastering and the flatness and verticality of the base surface, a wall is used as reference to check verticality with plumb bob, square room plan, square wall corner, and determine the plastering thickness. The plastering thickness should be at least 7mm. When the wall concavity is large, it should be plastered to flat by coats. The thickness of each coat is not more than 7—9mm. The dots on the top should be made first and then comes to the dots on the bottom. In the construction, the correct position of the dots should be determined according to the indoor plastering requirements, and then the verticality and flatness should be determined with the guiding rule. The dot should be made with cement mortar (cement : sand=1 : 3) to the volume of $5cm^3$.

When the room is large, we should mark the cross-center line with ink on the ground first and then mark the corner line according to the flatness of the base surface. Then, hang plumb bob at the point 100mm away from the inner corner of the wall to provide vertical line and mark the plumb line. After that, the control line of the thickness of the backing coat on the two walls of the inner corner shall be marked according the marked corner line on the ground. The dot can be made according to the control line, and then screed shall be made according to the dots.

(4)抹水泥踢脚(或墙裙)。根据已抹好的灰饼充筋(此筋可以冲的宽一些,8～10 cm 为宜,因此筋即为抹踢脚或墙裙的依据,同时也作为墙面抹灰的依据),底层抹 1∶3 水泥砂浆,抹好后用大杠刮平,木抹搓毛,常温第二天用 1∶2.5 水泥砂浆抹面层并压光,抹踢脚或墙裙厚度应符合设计要求,无设计要求时凸出墙面 5～7 mm 为宜。凡凸出抹灰墙面的踢脚或墙裙上口必须保证光洁顺直,踢脚或墙面抹好将靠尺贴在大面与上口平,然后用小抹子将上口抹平压光,凸出墙面的棱角要做成钝角,不得出现毛茬和飞棱。

(5)做护角。墙、柱间的阳角应在墙、柱面抹灰前用 1∶2 水泥砂浆做护角,其高度自地面以上 2 m。然后将墙、柱的阳角处浇水湿润。第一步在阳角正面立上八字靠尺,靠尺突出阳角侧面,突出厚度与成活抹灰面持平。然后在阳角侧面,依靠尺边抹水泥砂浆,并用铁抹子将其抹平,按护角宽度(不小于 5 cm)将多余的水泥砂浆铲除。

第二步待水泥砂浆稍干后,将八字靠尺移至到抹好的护角面上(八字坡向外)。在阳角的正面,依靠尺边抹水泥砂浆,并用铁抹子将其抹平,按护角宽度将多余的水泥砂浆铲除。抹完后去掉八字靠尺,用素水泥浆涂刷护角尖角处,并用捋角器自上而下捋一遍,使形成钝角。

(6)抹水泥窗台。先将窗台基层清理干净,松动的砖要重新补砌好。砖缝划深,用水润透,然后用 1∶2∶3 豆石混凝土铺实,厚度宜大于 2.5 cm,次日刷胶粘性素水泥一遍,随后抹 1∶2.5 水泥砂浆面层,待表面达到初凝后,浇水养护 2～3 天,窗台板下口抹灰要平直,没有毛刺。

(7)墙面充筋。当灰饼砂浆达到七八成干时,即可用与抹灰层相同砂浆充筋,充筋根数应根据房间的宽度和高度确定,一般标筋宽度为 5 cm。两筋间距不大于 1.5 m。当墙面高度小于 3.5 m 时宜做立筋。大于 3.5 m 时宜做横筋,做横向冲筋时灰饼的间距不宜大于 2 m。

Chapter 9   Finishing and Decoration Work Construction

(4) Making cement skirting (or dado). According to the finished dots and screed (The screed can be wider, and the width of 8—10cm is appropriate. It is the basis for making skirting or dado and the wall surface at the same time), the bottom coat is plastered with 1 : 3 cement mortar. After plastering, it should be scraped with strike-off board and roughened by wood float. In room temperature, it should be plastered with cement mortar (cement : sand =1 : 2.5) and compacted in the next day. The thickness of the skirting or dado should meet the design requirement, and it is advisable to protrude 5—7mm from the wall surface when there is no design requirement. The top of the skirting or dado protruding from the wall must keep clean and smooth. The guiding rule shall be put on the side of the wall to make the surface of the skirting or dado at the same level with it, and then a small spatula should be used to smooth and compact the top of the skirting or dado. The parts protruding the wall surface should be made to obtuse angle, and there should be no rough place and sharp angle on it.

(5) Make mortar corner guard.

The outer corner between walls and columns should be protected by cement mortar (cement : sand = 1 : 2) before plastering. The height of protecting area should be 2m above the ground. The corner between walls and columns should be wet. The first step is to set up the splayed guiding rule on the front side of the outer corner. The edge of the guiding rule protrudes out at the outer corner to an extent of the finished plaster thickness. Then, on the side of the outer corner, the cement mortar is applied along the guiding rule and then smoothed by iron spatula, and the excess cement mortar should be removed according to the width of the corner guard (not less than 5cm).

In the second step, when the cement mortar is slightly dry, move the guiding rule to the finished surface of corner guard (the splay of the rule is outward). In front of the outer corner, the cement mortar is applied along the guiding rule and then smoothed by iron spatula. The excess cement mortar is removed according to the width of the corner guard. After applying cement mortar, remove the rule, brush the sharp corner of the corner guard with cement slurry, and use the corner trowel to shape it from top to bottom, so as to make it to obtuse angle.

(6) Make cement window sill. Firstly, the base of the window sill should be cleaned, and the loose bricks should be repaired. Brick joints should be deepened and wet thoroughly, and then apply pisolite concrete (cement : sand : gravel = 1 : 2 : 3) to the base, and the thickness of pisolite concrete should be more than 2.5cm. The adhesive plain cement should be brushed on it once the next day, and then 1 : 2.5 cement mortar is applied on the surface. After the surface reaches the initial setting, it should be wet with water to cure for 2—3 days. Plastering of the bottom of the window sill board should be straight and there should be no burrs on it.

(7) Make screeds on wall. When the mortar of dot reaches 70%—80% of its solidification, the same mortar as the backing coat can be used to make screed. The number of screeds should be determined by the width and height of the room. Generally, the width of the standard screed is 5cm. The spacing between the two screeds is not more than 1.5m. When the room height is less than 3.5m, it is better to make vertical pillars, and when the wall is higher than 3.5m, it is advisable to make transverse pillars. When the screeds are transverse, the distance between dots should not be more than 2m.

(8)抹底灰。一般情况下充筋完成 2 h 左右可开始抹底灰为宜,抹前应先抹一层薄灰,要求将基体抹严,抹时用力压实使砂浆挤入细小缝隙内,接着分层装档、抹与充筋平,用木杠刮找平整,用木抹子搓毛。然后全面检查底子灰是否平整,阴阳角是否方直、整洁,管道后与阴角交接处、墙顶板交接处是否光滑平整、顺直,并用托线板检查墙面垂直与平整情况。散热器后边的墙面抹灰,应在散热器安装前进行,抹灰面接槎应平顺,地面踢脚板或墙裙,管道背后应及时清理干净,做到活完底清。

(9)修抹预留孔洞,配电箱、槽、盒。当底灰抹平后,要随即由专人把预留孔洞,配电箱、槽、盒周边 5 cm 宽的石灰砂刮掉,并清除干净,用大毛刷沾水沿周边刷水湿润,然后用 1∶1∶4 水泥混合砂浆,把洞口、箱、槽、盒周边压抹平整、光滑。

(10)抹罩面灰。应在底灰六七成干时开始抹罩面灰(抹时如底灰过干应浇水湿润),罩面灰两遍成活,厚度约 2 mm,操作时最好两人同时配合进行,一人先刮一遍薄灰,另一人随即抹平。依先上后下的顺序进行,然后赶实压光,压时要掌握火候,既不要出现水纹,也不可压活,压好后随即用毛刷蘸水将罩面灰污染处清理干净。施工时整面墙不宜甩破活,如遇有预留施工洞时,可甩下整面墙待抹为宜。

9.1.3 装饰抹灰的施工工艺

1. 水刷石施工工艺流程

堵门窗口缝→基层处理→浇水湿润墙面→吊垂直、套方、找规矩、抹灰饼、充筋→分层抹底层砂浆→分格弹线、粘分格条→做滴水线条→抹面层石渣浆→修整、赶实压光、喷刷→起分格条、勾缝→养护。

(8) Apply first coat. Generally, it is advisable to apply coat in 2 hours after finishing screed. Before plastering, a thin coat of plaster should be applied to cover the base completely. When plastering, the mortar should be squeezed into small cracks by hard compaction. Then, apply plaster between the screeds in coats with its top slightly lower than that of the screeds, and level the top plaster with wood strike-off board and roughen it with a wood float shall be used to roughen it. Then, check whether the plaster of the bottom coat is flat, whether the inner and outer corners are square and neat, whether the junction between the pipeline and inner corner and the junction between the wall top plates is smooth and straight, and check the verticality and flatness of the wall surface using the plumb rule. The plastering behind the radiator should be carried out before the installation of the radiator. The joint of plastering surfaces should be smooth, and the skirting or dado, and the back of the pipeline should be cleaned in time, so as to keep clean thoroughly after finishing the work.

(9) Patch existing holes, fill the gaps coming from installation of conduits, distribution board and embedded electric outlet boxes, etc. When the first coat is completed, scrape the limestone sands for the width of 5 cm around the existing holes, reserved conduit and distribution board by special constructor and clean it completely. The area should be wet with a large brush, and then the mixed mortar (cement : sand : gravel=1 : 1 : 4) is used to compact and smooth the surrounding area of the holes, board, conduits and outlet boxes.

(10) Apply finish coat. When the first coat comes to 60%—70% of its solidification, the finish coat should be applied (if the first coat is too dry, it should be wet). The finish coat should be applied twice, and the thickness should be about 2mm. It is better for two people to cooperate in the operation. One person should apply the thin plaster at first, and the other person should smoothen it. Plaster applying should start from the upper to the lower, then followed by compacting and smoothing the topping forward in sequence. The compacting work should be performed with right extent, avoid making water mark on the topping or delaying the pressing. After compacting, clean up the contaminated area of the finish coat with a wet brush. The plastering for each wall unit should be finished at once in progress, and any interruptive plastering work is not permitted. If there are preformed construction holes, it is better to leave the finish coat to be done in the future.

9.1.3 Construction Technology of Decorative Plastering

1. Construction Process of Plastering with Pebble-dash Finish

Caulk the gaps besides doors and windows→ clean the base→ wet the wall→ check verticality with plumb bob, square room plan, square wall corner and make dots and screeds→ apply bottom mortar by coats→ mark lines with ink for division and stick dividing strips→ make drip line→ apply small pebbles or crushed stone slurry coat on the surface→ trim, scrap, compact it and brush cement slurry on it→ remove dividing strips and make joint pointing→ maintain.

2. 水刷石施工操作工艺

(1)堵门窗口缝。抹灰前检查门窗口位置是否符合设计要求,安装牢固,四周缝按设计及规范要求填塞完成后,再用1:3水泥砂浆塞实抹严。

(2)基层清理。混凝土墙基层处理:用钢钻子将混凝土墙面均匀凿出麻面,并将板面酥松部分剔除干净,用钢丝刷将粉尘刷掉,用清水冲洗干净,然后浇水湿润。

清洗处理:用10%的火碱水将混凝土表面油污及污垢清刷除净,然后用清水冲洗晾干,采用涂刷素水泥浆或混凝土界面剂等处理方法均可。如采用混凝土界面剂施工时,应按所使用产品要求使用。

砖墙基层处理:抹灰前需将基层上的尘土、污垢、灰尘、残留砂浆、舌头灰等清除干净。

(3)浇水湿润。基层处理完后,要认真浇水湿润,浇水时应将墙面清扫干净,浇透浇均匀。

(4)吊垂直、套方、找规矩、做灰饼、充筋。根据建筑高度确定放线方法,高层建筑可利用墙大角、门窗口两边,用经纬仪打直线找垂直。多层建筑时,可从顶层用大线坠吊垂直,绷铁丝找规矩,横向水平线可依据楼层标高或施工+50 cm线为水平基准线交圈控制,然后按抹灰操作层抹灰饼,做灰饼时应注意横竖交圈,以便操作。每层抹灰时则以灰饼做基准充筋,使其保证横平竖直。

(5)分层抹底层砂浆。

混凝土墙:先刷一道胶粘性素水泥浆,然后用1:3水泥砂浆分层装档抹与筋平,然后用木杠刮平,木抹子搓毛或花纹。

砖墙抹1:3水泥砂浆,在常温时可用1:0.5:4混合砂浆打底,抹灰时以充筋为准,控制抹灰层厚度,分层分遍装档与充筋抹平,用木杠刮平,然后木抹子搓毛或花纹。底层灰完成24 h后应浇水养护。抹头遍灰时,应用力将砂浆挤入砖缝内使其粘结牢固。

Chapter 9 Finishing and Decoration Work Construction

2. Operation Process of Plastering with Pebble Finish

(1) Caulk the gaps besides doors and windows. Before plastering, check whether the position of doors and windows meet the design requirements and being installed firmly. Check whether the gaps besides them are filled according to the design and specifications, and then plaster the gaps with cement mortars (cement : sand=1 : 3) tightly.

(2) Clean the base.

Base treatment of concrete wall: roughen the concrete wall surface evenly with a steel drill, remove the loose part of the board surface, brush off the dust with steel wire brush, wash it with clear water, and then wet it.

Cleaning: The oil and dirt on the surface of the base should be cleaned with 10% caustic soda water, and then washed with clean water and wait till the surface is dry. Cement slurry or concrete interface agent can be brushed on it. If concrete interface agent is used in construction, it should be used according to the usage direction.

Base treatment of brick wall: before plastering, the dust, dirt, residual mortar and excess mortar on the base shall be removed.

(3) Wet the wall. After the base treatment, we should wet the wall carefully. When watering, we should clean the walls and wet them thoroughly and evenly.

(4) Check verticality with plumb bob, square room plan, square wall corner and make dots and screeds. The method of setting out is determined by the height of the building. For high-rise buildings, we can use theodolite to make straight lines to determine verticality on both sides of large angle of the walls, doors and windows. In multi-story buildings, we can hang the line with a heavy plumb bob to determine the verticality and tighten the tension wire to ensure flatness and verticality. The transverse horizontal line can be controlled by crossing circles according to the floor elevation or construction elevation plus 50cm as the horizontal datum line. Then, make dot on the plastering coat. In making the cake, we should pay attention to connection of the horizontal-vertical crossing for convenient operation. When each coat is plastered, the dot can be used as the reference for screed, which can make the screed horizontal and vertical.

(5) Apply bottom mortar by coats. For concrete walls: First, brush the adhesive cement slurry, then apply cement mortar (cement : sand =1 : 3) by coats to the wall between the screeds to it on the same level with the screed, and use the wood strike-off board to scrape it and wood float to roughen it or make decorative veins on it.

For brick walls: First, plaster cement mortar (cement : sand =1 : 3), or use mixed mortar (cement : lime powder : sand=1 : 0.5 : 4) as the primer at room temperature. When plastering is carried out, the screed should be regarded as the standard to control the thickness of it. The plastering between the screeds will be applied by coats and by times to reach the level of the screed and then scrape it with a wood strike-off board, and roughen it or make decorative veins on it with wood float. The plaster of the bottom coat should be cured with water after 24 hours. When the first plastering is applied, the mortar shall be squeezed into the joints of bricks to make them firmly bonded.

(6)弹线分格、粘分格条。根据图纸要求弹线分格、粘分格条,分格条宜采用红松制作,粘前应用水充分浸透(一般应浸 24 h 以上),粘时在条两侧用素水泥浆抹成 45°八字坡形,粘分格条时注意竖条应粘在所弹立线的同一侧,防止左右乱粘,出现分格不均匀,条粘好后待底层灰呈七八成干后可抹面层灰。

(7)做滴水线。在抹檐口、窗台、窗眉、阳台、雨篷、压顶和突出墙面的腰线以及装饰凸线等时,应将其上面作成向外的流水坡度,严禁出现倒坡。下面做滴水线(槽),窗台上面的抹灰层应深入窗框下坎裁口内,堵密实。流水坡度及滴水线(槽)距外表面不小于 40 cm,滴水线深度和宽度一般不小于 10 mm,应保证其坡度方向正确。抹滴水线(槽)应先抹立面,再抹顶面,最后抹底面。分格条在其面层灰抹好后即可拆除。采用"隔夜"拆条法时须待面层砂浆达到适当强度后方可拆除。滴水线做法同水泥砂浆抹灰做法。

(8)抹面层石渣浆。待底层灰六七成干时首先将墙面润湿涂刷一层胶粘性素水泥浆,然后开始用钢抹子抹面层石渣浆。自下往上分两遍与分格条抹平,并及时用靠尺或小杠检查平整度(抹石渣层高于分格条 1 mm 为宜),有坑凹处要及时填补,边抹边拍打揉平。

(9)修整、赶实压光、喷刷。将抹好在分格条块内的石渣浆面层拍平压实,并将内部的水泥浆挤压出来,压实后尽量保证石渣大面朝上,再用铁抹子溜光压实,反复 3~4 遍。拍压时特别要注意阴阳角部位石渣饱满,以免出现黑边。待面层初凝时(指捺无痕),用水刷子刷不掉石粒为宜。

然后开始刷洗面层水泥浆,喷刷分两遍进行,第一遍先用毛刷蘸水刷掉面层水泥浆,露出石粒,第二遍紧随其后用喷雾器将四周相邻部位喷湿,然后自上而下顺序喷水冲洗,喷头一般距墙面 10~20 cm,喷刷要均匀,使石子露出表面 1~2 mm 为宜。

(6) Mark lines with ink for division and stick dividing strips.

Marking lines with ink for division and sticking dividing strips should be carried out in accordance with the drawings. Korean pine should be used to make the dividing strips. Before sticking, it should be fully soaked (generally for more than 24 hours). When the sticking is carried out, cement slurry should be wiped on both sides of the strips to form a 45 degree slope in shape of "八". When sticking the dividing strips, the vertical strips should be stuck on the same side of the marked lines to prevent the disorderly sticking of the left and right sides, resulting in uneven division. After the strips are stuck, the surface will be plastered when the plaster of the bottom coat reaches 70%—80% of its solidification.

(7) Make drip line.

When the eaves, window sills, eyebrows, balconies, awnings, blocking course and protruding waistlines of walls and decorative convex lines are plastered, on the surfaces of their upper part, there should be outward water slope to prevent the inverted slope. When the drip line (groove) is on the lower part, the backing coat on the window sill should be deepened in the cut of the lower sill of the window frame, and it should be sealed. Water slope and drip line (groove) should not be less than 40cm away from the outer surface. The depth and width of drip line are generally not less than 10mm, and its sloping direction should be correct. When the drip line (groove) is plastered, the operation sequence is as follows: first, the elevation, then the top and finally the bottom. Dividing strips can be removed after plastering the surface. When using "overnight" removing method, the surface mortar must reach the appropriate strength before it can be demolished. The plastering of the drip line is the same as the cement mortar plastering method.

(8) Apply small pebbles or crushed stone slurry coat on the surface. When the plaster of the bottom coat reaches 60%—70% dry, the wall should be wet first and brushed with a coat of adhesive cement slurry, and then the steel spatula shall be used to plaster the surface with crushed stone slurry. The plastering will be carried out from bottom to top twice to be at the same level with dividing strip. The flatness should be checked in time with the guiding rule or small bars (it is advisable that the coat of crushed stone ballast slurry is 1mm higher than dividing strip). The pits should be filled in time and the crushed stone ballast slurry should be pressed tightly to make it even.

(9) Trim, scrap, compact it and brush cement slurry on it.

The coat of crushed stone ballast slurry between the dividing strips should be patted and compacted, and the internal cement mortar should be squeezed. After compaction, the large facet of the crushed stone ballasts should better face up, and then the surface should be smoothed and compacted with iron spatula, repeating it for 3—4 times. In patting and compacting, make sure there are enough crushed stone ballasts in the inner and outer corners and the crushed stone ballasts cover all the base. When the surcoat is initially set (The concrete surface doesn't get obvious finger mark when pressing with fingers), brush it with a water brush until pebble can't be brushed off.

Then the coat of crushed stone ballast slurry shall be brushed and wet twice. For the first time, the cement slurry of the surcoat shall be brushed with water to expose the gravel particles. In the second time, spray water on the adjacent parts to wet them, and then spray water to wash the surface from top to bottom. The sprinkler is usually 10—20cm away from the wall, and the spraying should be the same, and pebble should be exposed to the surface for 1—2mm.

最后用水壶从上往下将石渣表面冲洗干净,冲洗时不宜过快,同时注意避开大风天,以避免造成墙面污染发花。若使用白水泥砂浆做水刷石墙面时,在最后喷刷时,可用草酸稀释液冲洗一遍,再用清水洗一遍,使墙面更显洁净、美观。

(10)起分格条、勾缝。喷刷完成后,待墙面水分控干后,小心将分格条取出,然后根据要求用线抹子将分格缝溜平抹顺直。

(11)养护。待面层达到一定强度后,可喷水养护防止脱水、收缩造成空鼓、开裂。勾缝 3 d 后洒水养护,养护时间不少于 4 d。

(12)阳台、雨罩、门窗璇脸部位做法。门窗璇脸、窗台、阳台、雨罩等部位水刷石施工时,应先做小面,后做大面,刷石喷水应由外往里喷刷,最后用水壶冲洗,以保证大面的清洁美观。檐口、窗台、璇脸、阳台、雨罩等底面应做滴水槽,滴水线(槽)应做成上宽 7 mm,下宽 10 mm,深 10 mm 的木条,便于抹灰时木条容易取出,保持棱角不受损坏。滴水线距外皮不应小于 4 cm,且应顺直。当大面积墙面做水刷石一天不能完成时,在继续施工冲刷新活前,应将前面做的刷石用水淋湿,以防喷刷时沾上水泥浆,防止对原墙面造成污染。施工槎应留在分格缝上。

3. 外墙斩假石抹灰工程施工工艺流程

基层处理→浇水湿润墙面→吊垂直、套方、找规矩、抹灰饼、充筋→抹底层砂浆→分格弹线、粘分格条→抹面层水泥石子浆→斩剁面层→养护。

4. 外墙斩假石抹灰工程施工操作工艺

(1)掌握斩剁时间,在常温下经 3 d 左右或面层达到设计强度 60%～70% 时即可进行,大面积施工应先试剁,以石子不脱落为宜。

(2)斩剁前应先弹顺线,并离开顺线适当距离按线操作,以避免剁纹跑斜。

(3)斩剁应自上而下进行,首先将四周边缘和棱角部位仔细剁好,再剁中间大面。若有分格,每剁一行应随时将上面和竖向分格条取出,并及时将分块内的缝隙、小孔用水泥浆修补平整。

Chapter 9 Finishing and Decoration Work Construction

Finally, the surface of the crushed stone ballasts shall be washed from top to bottom with a water kettle, and it should not be carried out too fast. The operation should avoid the days with strong wind because it may cause pollution on the wall. If white cement mortar is used on the walls of pebble, oxalic acid diluent can be used to wash them once at last, and then clean water will be used to wash them again to clean and beautify the wall.

(10) Remove dividing strips and make joint pointing. Removing the dividing strips carefully after the wall is dry, and then use a trowel to make the joints smooth and straight according to the requirements.

(11) Maintain. When the surcoat reaches certain strength, it can be cured by spraying water to prevent dehydration and shrinkage from causing hollowing and cracking. 3 days after joint pointing, water will be sprinkled on the wall for maintenance. The maintenance shall not be less than 4 days.

(12) The plastering of pebble on balconies, rain covers and upper frames of doors and windows. When plastering pebble on upper frames of doors and windows, window sill, balconies, rain covers and other specific parts, small facets should be plastered first and large facets should be plastered later. Water should be sprayed on the pebble from the owner side to inner side, and finally wash it with kettle to ensure the cleanliness and beauty of large facets. The drip lines should be made on the under surface of eaves, window sills, upper frames of doors and windows, balconies, rain covers and so on. Drip grooves (lines) should be made by wooden strips with the width of 7mm on the top, the width of 10mm on the bottom and the depth of 10mm so that they can be removed easily, and the edges and corners will not be damaged easily. The drip lines should be not less than 4cm away from the outer face and should be straight. When the plastering of the pebble on the large area of the wall can't be finished in a day, before continuing the work on the next day, the pebble made before should be wet to prevent the cement mortar from splashing the wall when spraying and avoid pollution on the wall. The construction joint should be reserved on the dividing joint.

3. Construction Process of Plastering with Bush-hammered Finish on the Exterior Wall

Clean the base→ wet the wall→ check verticality with plumb bob, square room plan, square wall corner, make dots and screeds→ apply bottom mortar→ mark lines with ink for division and stick dividing strips→ apply cement mortar with gravel on the surface→ chop the finishing coat→ maintain.

4. The Operation Process of the Plastering with Bush-hammered Finish on the Exterior wall

(1) The time of chopping should be controlled well, and the chopping work can be carried out at room temperature after about 3 days of plastering cement mortar with gravel or when the surcoat reaches 60%—70% of the design strength. A chopping test should be done before large-scale construction. If gravels do not fall off in chopping, it meets the requirement.

(2) Before chopping, it is necessary to mark the control line. The chopping should be operated at an appropriate distance away from the veins in order to avoid the deviation of the them.

(3) Chopping should be carried out from top to bottom. First, the edges and corners should be carefully chopped, and then the middle large surface should be chopped. If there are dividing strips, the upper and vertical dividing strips should be removed when each row is chopped, and the cracks and holes should be repaired and smoothed with cement mortar in time.

(4)斩剁时宜先轻剁一遍,再盖着前一遍的剁纹剁出深痕,操作时用力应均匀,移动速度应一致,不得出现漏剁。

(5)柱子、墙角边棱斩剁时,应先横剁出边缘横斩纹或留出窄小边条(边宽3～4 cm)不剁。剁边缘时应使用锐利的小剁斧轻剁,以防止掉边掉角,影响质量。

(6)用细斧斩剁墙面饰花时,斧纹应随剁花走势而变化,严禁出现横平竖直的剁斧纹,花饰周围的平面上应剁成垂直纹,边缘应剁成横平竖直的围边。

(7)用细斧剁一般墙面时,各格块体中间部分应剁成垂直纹,纹路相应平行,上下各行之间要均匀一致。

(8)斩剁完成后面层要用硬毛刷顺剁纹刷净灰尘,分格缝按设计要求做归正。

(9)斩剁深度一般以剁掉石渣的1/3比较适宜,这样可使剁出的假石成品美观大方。

9.2　饰面板(砖)工程

9.2.1　贴面砖施工工艺

1. 施工工艺

基层处理→吊垂直、套方、找规矩→贴灰饼→抹底层砂浆→弹线分格→排砖→浸砖→镶贴面砖→面砖勾缝及擦缝。

2. 操作工艺

(1)基体为混凝土墙面时的操作方法。

①基层处理:将凸出墙面的混凝土剔平,对大钢模施工混凝土墙面应凿毛,并用钢丝刷满刷一遍,清除干净,然后浇湿润;对于基体混凝土表面很光滑的,可采取"毛化处理"较高的强度(用手掰不动为止)。

(4) When the chopping is carried out, it is advisable to chop once lightly and then chop the deep marks along the previous chopping trace. The strength should be equal in operation, and the moving speed should be the same, and no missing chopping should occur.

(5) When chopping the edges of columns and corners, the transverse veins on the edges should be chopped or the narrow regula (3—4cm wide) should be left. When chopping edges, a sharp small chopping axe should be used to chop the edges lightly, so as to prevent the edges and corners from falling off and affecting the quality.

(6) When decorative figures are chopped with fine axe, the veins should follow the shapes of the figures. It is strictly forbidden to leave horizontal and vertical veins on the figures. The planes around the figures should be chopped into vertical veins, and the edge should be chopped into horizontal and vertical veins.

(7) When ordinary wall surface is chopped with fine axe, the middle part of each block should be chopped into vertical veins, and the veins should be parallel and uniform between the upper and lower rows.

(8) After chopping, the surface should be cleaned up with hard brush along the chopping veins, and the dividing joints should be corrected according to the design requirements.

(9) The chopping should stop when 1/3 stone ballasts are chopped off, which can make the stone plaster beautiful and generous.

9.2 Tile or Board Cladding Finish

9.2.1 Construction Technology of Tile Cladding Finish

1. Construction Process

Treat the base→ check verticality with plumb bob, square room plan, square wall corner→ making dots→ apply the bottom mortar →mark lines with ink for division→ lay tiles → soak tiles→ stick tiles→ make joint pointing and wipe tiles.

2. Operation Process

(1) Operation method when the base is concrete wall.

①Treat the base: The concrete protruding from the wall surface should be trimmed to make the wall flat. The concrete wall constructed in large steel formwork should be roughened, brushed with wire brush completely, cleaned and wet; If the surface of the concrete wall is very smooth, roughening treatment can be taken on it to increase its strength until it will not be broken off with hands.

②吊垂直、套方、找规矩、贴灰饼、冲筋:高层建筑物应在四大角和门窗口边用经纬仪打垂直线找直;多层建筑物,可从顶层开始用特制的大线坠绷低碳钢丝吊垂直,然后根据面砖的规格尺寸分层设点、做灰饼,间距1.6 m。横向水平线以楼层为水平基准线交圈控制,竖向垂直线以四周大角和通天柱或墙垛子为基准线控制,应全部是整砖。阳角处要双面排直。每层打底时,应以此灰饼作为基准点进行冲筋,使其底层灰做到横平竖直。同时要注意找好突出檐口、腰线、窗台、雨篷等饰面的流水坡度和滴水线(槽)。

③抹底层砂浆:先刷一道掺水重10%的界面剂胶水泥素浆,打底应分层分遍进行抹底层砂浆(常温时采用配合比为1∶3水泥砂浆),第一遍厚度宜为5 mm,抹后用木抹子搓平;待第一遍六至七成干时,即可抹第二遍,厚度为8~12 mm,随即用木杠刮平、木抹子搓毛,终凝后洒水养护。砂浆总厚不得超过20 mm,否则应作加强处理。

④弹线分格:待基层灰六至七成干时,即可按图纸要求进行分段分格弹线,同时亦可进行面层贴标准点的工作,以控制面层出墙尺寸及垂直、平整。

⑤排砖:根据大样图及墙面尺寸进行横竖向排砖,以保证面砖缝隙均匀,符合设计图纸要求,注意大墙面、通天柱子和垛子要排整砖,以及在同一墙面上的横竖排列,均不得有一行以上的非整砖。非整砖行应排在次要部位,如窗间墙或阴角处等。但亦要注意一致和对称。如遇有突出的卡件,应用整砖套割吻合,不得用非整砖随意拼凑镶贴。面砖接缝的宽度不应小于5 mm,不得采用密缝。

⑥选砖、浸泡:釉面砖和外墙面砖镶贴前,应挑选颜色、规格一致的砖;浸泡砖时,将面砖清扫干净,放入净水中浸泡2 h以上,取出待表面晾干或擦干净后方可使用。

⑦粘贴面砖:粘贴应自上而下进行。高层建筑采取措施后,可分段进行。在每一分段或分块内的面砖,均为自下而上镶贴。从最下一层砖下皮的位置线先稳好固定靠尺,以此托住第一皮面砖。

②Check verticality, square wall plan, square wall corner, and make dots and screed: for high-rise buildings, the odolites should be used at the four corners and the openings for doors and windows to determine the verticality; for multi-story buildings, low-carbon steel wire with plumb bob can be used to determine the verticality from the top, and then controlling points and dots can be made by coats according to the specifications and sizes of tile cladding, and the spacing of them is 1.6m. The transverse horizontal line is controlled by the intersection of the base level line of floors, and the vertical line is controlled by the large angel around and the sky pillar or the wall pier as the datum line, and the tiles on them should be complete ones. The tiles on outer corners should be straight on both sides. In applying each coat, the dots should be used as controlling point for screeds, and the plaster of the bottom coat should be horizontal and vertical. At the same time, attention should be paid to finding the gradient for water flow and drip line (groove) of the protruding eave, the waist line, the window sill, the canopy and other facings.

③Apply bottom mortar: First, brush a coat of interface agent cement paste with 10% water in weight, and then apply the bottom mortar by coats and by times (mix ratio of 1 : 3 cement mortar is used at normal temperature). The thickness of the first coat should be 5mm. After plastering, a wood float is used to rub it flat; when the first coat is 60%—70% dry, the second coat with a thickness of about 8—12mm can be applied. Then the wood strike-off board is used for scraping and the wood float is used to roughen it. After the final setting, water is sprinkled on it to cure it. The total thickness of the mortar should not exceed 20mm; otherwise, it should be strengthened.

④Mark lines with ink for division: When the first coat is 60%—70% dry, the division line can be marked with ink in each segment or block according to the requirements of the drawings. At the same time, the work of plastering controlling points on the surcoat can be carried out to control the size protruding from the wall, verticality and flatness of the surcoat.

⑤Lay tiles: According to the detailed drawings and wall dimensions, the tiles should be laid horizontally and vertically to ensure uniform joints of the tile cladding and meet the requirements of the design drawings. Tiles on the large facet of the wall, the superhigh pillar and the wall pier should be complete, and there should be not more than one row of non-complete tiles on the same horizontal and vertical wall surface. Rows of non-complete tiles should be placed in secondary parts, such as walls between windows or inner corners; however, we should pay attention to their consistency and symmetry. If there are some protruding components, the complete tile should be cut to match their shapes, and don't use the non-complete tiles to match the protruding shapes casually. The width of the joint between the tile cladding shall not be less than 5mm, and close joints shall not be used.

⑥Select and soak tiles: Before the glazed tiles and exterior wall tiles are used, they should be classified according to colors and specifications; when the tiles are soaked, the tile cladding should be cleaned and soaked in clean water for more than 2 hours. They can only be used after being dried or wiped clean.

⑦Stick tiles: The tiles should be sticked from the top to the bottom of walls. For high-rise buildings, special measure should be taken to divide the wall into different sections. The tiles in each section or block should be sticked from bottom to top. The guiding rule should be set and fixed at the bottom line of the tile on the bottom coat, so it can support the first course of tiles.

在面砖背面宜采用1∶0.2∶2＝水泥∶白灰膏∶砂的混合砂浆镶贴，砂浆厚度为6～10 mm，贴上后用灰铲柄轻轻敲打，使之附线，再用钢片开刀调整竖缝，并用小杠通过标准点调整平面和垂直度。另外一种做法是，用1∶1水泥砂浆加水泥重20%的界面剂胶，在砖背面抹3～4 mm厚粘贴即可。但此种做法其基层灰必须抹得平整，而且砂子必须用窗纱筛后使用。不得采用有机物作主要粘结材料。另外也可用胶粉来粘贴面砖，其厚度为2～3 mm，在此种做法中其基层灰必须更平整。如要求釉面砖拉缝镶贴时，面砖之间的水平缝宽度用米厘条控制，米厘条用贴砖用砂浆与中层灰临时镶贴，米厘条贴在已镶贴好的面砖上口，为保证其平整，可临时加垫小木楔。女儿墙压顶、窗台、腰线等部位平面也要镶贴面砖时，除流水坡度符合设计要求外，应采取顶面砖压立面面砖的做法，预防向内渗水，引起空裂；同时还应采取立面中最低一排面砖必须压底平面面砖，并低出底平面面砖3～5 mm的做法，让其超滴水线(槽)的作用，防止尿檐，引起空裂。

⑧面砖勾缝与擦缝：面砖铺贴拉缝时，用1∶1水泥砂浆勾缝或采用勾缝胶，先勾水平缝再勾竖缝，勾好后要求凹进面砖外表面2～3 mm。若横竖缝为干挤缝，或小于3 mm者，应用白水泥配颜料进行擦缝处理。面砖缝子勾完后，用布或棉丝蘸稀盐酸擦洗干净。

(2)基体为砖墙面时的操作方法。

①基层处理：抹灰前，墙面必须清扫干净，浇水湿润。

②吊垂直、套方、找规矩：大墙面和四角、门窗口边弹线找规矩，必须由顶层到底一次进行，弹出垂直线，并决定面砖出墙尺寸，分层设点、做灰饼(间距为1.6 m)。横线则以楼层为水平基线交圈控制，竖向线则以四周大角和通天垛、柱子为基准线控制。每层打底时则以此灰饼作为基准点进行冲筋，使其底层灰做到横平竖直。同时要注意找好突出檐口、腰线、窗台、雨篷等饰面的流水坡度。

The mixed mortar (cement : white plaster : sand=1 : 0.2 : 2) should be used at the back of the tiles to stick them to the bottom mortar and the thickness of the mixed mortar is 6—10mm. After sticking, use the handle of shovel to knock the tile slightly and make it attach to the base. Then the steel knife is used to adjust the vertical joint, and the small bar is used to adjust the plane and verticality with reference to the controlling point. There is another method of sticking tiles, that is to add interface agent with 20% cement to cement mortar (cement : sand=1 : 1), and brush it to the back of tiles with thickness of 3—4mm to stick it to the bottom mortar; in this method, the bottom mortar must be smoothed, and the sand can only be used after selecting by window screening. Organic materials should not be used as the primary bonding material. In addition, adhesive powder with thickness of 2—3mm can also be used to stick the tiles; and in this method, the bottom mortar must be smoother. If it is required to set aligned joints for glazed tiles, the width of the horizontal joint between the tiles should be controlled by a millimeter strip, which is temporarily pasted to the second coat by mortar. The millimeter strip should be put on the upper line of the finished tiles and in order to ensure its flatness, the small wooden wedge can be temporarily added as the pad. When the capping of parapet, sill, waist line and other parts are also pasted with tile cladding, the water slope should meet the waterproof design requirements, and the top tile cladding should press the vertical facing tile to prevent inward leakage and cracking; at the same time, the lowest row of the tile cladding on the vertical face must press the facing tiles of the bottom, and the tile cladding of the bottom should be 3—5mm higher, which can work as drip lines (groove) to prevent water leakage and cracking.

⑧Make joint pointing and wipe the tiles: When the tiles are sticked and the joints are set, joint pointing should be finished with cement mortar (cement : sand=1 : 1) or sealant. The horizontal joints shall be filled up first, and then comes to the vertical joints. After finishing, the thickness of the joints should be 2—3mm lower than the exterior face of the tiles. If the horizontal and vertical joints are dry-extruded joints, or the width is less than 3mm, joints wiping should be done with the white cement mixed with the pigment. After joint pointing is finished, cloth or mercerized cotton dipped dilute hydrochloric acid can be used to make it clean.

(2) Operating method when the base is brick wall.

①Treat the base: Before sticking tiles, the wall must be cleaned and wet.

②Check verticality with plumb bob, square room plan, square wall corners: The flatness and verticality of large facet, four corners and edges of doors and windows shall be ensured with ink lines marked on them. Ink lines must be marked from the top coat to the bottom once a time. The vertical line should be marked first to determine the protruding size of the tile cladding, to set controlling points by coats, and set dots (The spacing is 1.6m). The horizontal line is determined by the horizontal baseline of the floor and connect with it smoothly, and the vertical line is determined by the baseline of the four corners, the exceedingly high piers and the columns. When the base of each coat start plastering, dots are used as reference points for the screed, and the plaster of bottom should be horizontal and vertical. At the same time, attention should be paid to finding the water slope of the protruding eave, the waist line, the sill, the awning and other cladding.

抹底层砂浆：先把墙面浇水湿润，然后用1∶3水泥砂浆刮5～6 mm厚，紧跟着用同强度等级的灰与所冲的筋抹平，随即用木杠刮平，木抹搓毛，隔天浇水养护。其他同基层为混凝土墙面做法。

9.2.2 大理石、磨光花岗岩饰面施工工艺

1. 工艺流程

(1)薄型小规格块材(边长小于40 cm)工艺流程：基层处理→吊垂直、套方、找规矩、贴灰饼→抹底层砂浆→弹线分格→石材刷防护剂→排块材→镶贴块材→表面勾缝与擦缝。

(2)普通型大规格块材(边长大于40 cm)工艺流程：施工准备(钻孔、剔槽)→穿铜丝或镀锌铅丝与块材固定→绑扎、固定钢丝网→吊垂直、找规矩、弹线→石材刷防护剂→安装石材→分层灌浆→擦缝。

2. 大理石、磨光花岗岩饰面操作工艺

(1)薄型小规格块材(一般厚度10 mm以下)：边长小于40 cm，可采用粘贴方法。

进行基层处理和吊垂直、套方、找规矩，其他可参见镶贴面砖施工要点有关部分。要注意同一墙面不得有一排以上的非整材，并应将其镶贴在较隐蔽的部位。

在基层湿润的情况下，先刷胶界面剂素水泥浆一道，随刷随打底；底灰采用1∶3水泥砂浆，厚度约12 mm，分二遍操作，第一遍约5 mm，第二遍约7 mm，待底灰压实刮平后，将底子灰表面划毛。

石材表面处理：石材表面充分干燥(含水率应小于8%)后，用石材防护剂进行石材六面体防护处理，此工序必须在无污染的环境下进行，将石材平放于木枋上，用羊毛刷蘸上防护剂，均匀涂刷于石材表面，涂刷必须到位，第一遍涂刷完间隔24 h后用同样的方法涂刷第二遍石材防护剂，如采用水泥或胶粘剂固定，间隔48 h后对石材粘结面用专用胶泥进行拉毛处理，拉毛胶泥凝固硬化后方可使用。

Apply bottom mortar: First, water the wall surface to wet, and then paste cement mortar (cement : sand =1 : 3) for 5—6mm in thickness. Then plaster of same strength should be applied to make it as the same thickness of the screeds. It should be smoothened with wood float immediately after it is finished and wood strike-off board is used to roughen it. The wall shall be wet for maintenance every other day.

The operating methods of other parts are the same as the methods when the base is concrete wall.

9.2.2 Construction Technology of Cladding with Marble or Polished Granite

1. Construction Process

(1) Construction process of cladding with thin small-size block (side length is less than 40cm): treat the base→ check verticality with plumb bob, square room plan, square wall corner, and make dots→ apply bottom mortar →mark lines with ink for division→ brush protective agent for stone material → lay blocks→ stick blocks→ make joint pointing and joints wiping.

(2) Construction process of cladding with ordinary large-size block (Side length is greater than 40cm): make preparation (drilling, grooving) → pass copper wire or galvanized lead wire through the blocks and fixing with blocks→ bind and fix steel wire mesh →check verticality, square wall corner, and mark lines with ink→ apply protective agent to stone material → install blocks→ grout by coats→ make joints wiping.

2. Construction Process of Cladding with Marble or Polished Granite

(1) For thin small-size block (The thickness is usually less than 10mm): When the side length is less than 40cm, sticking can be used to install it.

The construction methods in the base treatment, checking verticality with plumb bob, squaring room plan, squaring room corner, etc. are the same as the methods in sticking tiles. There should be no more than one row of non-complete blocks on the same wall, and the non-complete blocks should be applied to the concealed parts.

When the first coat is wet, first, brush the interfacial agent cement slurry on the first coat and make it even; cement mortar (cement : sand =1 : 3) should be used as the first coat with the thickness of 12mm. It should be finished at two times. For the first time, the cement mortar is about 5mm thick, and for the second time, it is about 7mm. After the first coat is compacted and flattened, the surface of the it should be roughened.

Treatment of the surface of the marble or polished granite: After the surface of marble or polished granite is fully dry (moisture content should be less than 8%), the stone hexahedron protective treatment is carried out with stone protective agent in a non-polluting environment. The marble or polished granite is placed on the wooden batten, and then the worker applies the protective agent evenly on the surface of the it with wool brush. The brushing should be finished fully. After 24 hours of the first brushing, the worker can use the same method to apply the protective agent for the second time. If it is fixed with cement or adhesive agent, 48 hours later, special puddle can be used on the bonding surface of stone to improve adhesion, and the marble or polished granite can only be used after solidification and hardening of the puddle.

待底子灰凝固后便可进行分块弹线,随即将已湿润的块材抹上厚度为 2~3 mm 的素水泥浆,内掺水泥重 20%的界面剂进行镶贴,用木槌轻敲,用靠尺找平找直。

(2)大规格块材:边长大于 40 cm,镶贴高度超过 1 m 时,可采用如下安装方法。

①钻孔、剔槽:安装前先将饰面板按照设计要求用台钻打眼,事先应钉木架使钻头直对板材上端面,在每块板的上、下两个面打眼,孔位打在距板宽的两端 1/4 处,每个面各打两个眼,孔径为 5 mm,深度为 12 mm,孔位距石板背面以 8 mm 为宜。如大理石、磨光花岗岩,板材宽度较大时,可以增加孔数。钻孔后用云石机轻轻剔一道槽,深 5 mm 左右,连同孔眼形成象鼻眼,以备埋卧铜丝之用。若饰面板规格较大,如下端不好拴绑镀锌钢丝或铜丝时,亦可在未镶贴饰面的一侧,采用手提轻便小薄砂轮,按规定在板高的 1/4 处上、下各开一槽,(槽长为 3~4 cm,槽深约 12 mm 与饰面板背面打通,竖槽一般居中,亦可偏外,但以不损坏外饰面和不反碱为宜),可将镀锌铅丝或铜丝卧入槽内,便可拴绑与钢筋网固定。此法亦可直接在镶贴现场做。

②穿铜丝或镀锌铅丝:把备好的铜丝或镀锌铅丝剪成长 20 cm 左右,一端用木楔粘环氧树脂将铜丝或镀锌铅丝进孔内固定牢固,另一端将铜丝或镀锌铅丝顺孔槽弯曲并卧入槽内,使大理石或磨光花岗石板上、下端面没有铜丝或镀锌铅丝突出,以便和相邻石板接缝严密。

③绑扎钢筋:首先剔出墙上的预埋筋,把墙面镶贴大理石的部位清扫十净。先绑扎一道竖向 6 mm 钢筋,并把绑好的竖筋用预埋筋弯压于墙面。横向钢筋为绑扎大理石或磨光花岗石板所用,如板材高度为 60 cm 时,第一道横筋在地面以上 10 cm 处与主筋绑牢,用作绑扎第一层板材的下口固定铜丝或镀锌铅丝。第二道横筋绑在 50 cm 水平线上 7~8 cm,比石板上口低 2~3 cm 处,用于绑扎第一层石板上上口固定铜丝或镀锌铅丝,再往上每 60 cm 绑一道横筋即可。

After the solidification of the bottom plaster, marking of division lines can be carried out, and then the wet block is coated with the cement slurry with a thickness of 2—3mm and with the interfacial agent with 20% cement to plaster it to the bottom plaster. Then it will be knocked slightly with a wooden hammer, and the guiding rule is used to make it horizontal and vertical.

(2) For large-size block: When the side length is more than 40cm and the plastering height is more than 1m, the following installation method can be used:

①Drilling and grooving: Before plastering, holes should be drilled on the cladding board with a bench drill based on the design requirements. A wooden frame should be nailed in advance so that the drill bit is directly opposite to the upper end face of the board. The hole should be drilled on the top and bottom of each board at 1/4 away from the ends of it, and two holes should be drilled at each side with a diameter of 5mm and a depth of 12mm. The hole should be 8mm away from the back of the board. When the width of some boards such as marble and polished granite is large, we can increase the number of holes. After drilling, carefully make a groove with a depth of about 5mm with cutting tool, and combined with the drilled holes, they can be used for burying copper wire. When the size of the cladding board is large and the bottom is not suitable for binding galvanized steel wire or copper wire, we can use a portable small thin grinding wheel to make two grooves on the upper and lower parts of the board at 1/4 of its height, and the groove is 3—4mm long and 12mm deep, and the grooves should get through the back of the cladding board. The vertical groove is generally in the middle or sometimes to one side, and it should not damage the exterior face of the board or cause alkali on it. The galvanized lead wire or copper wire can be buried in the groove, and it can be tied and fixed with steel mesh. This method can also be done directly on the site.

②Passing copper wire or galvanized lead wire through blocks: Cut the prepared copper wire or galvanized lead wire into sections with the length of about 20cm. Wood wedge with epoxy resin adhesive is used to stick copper wire or galvanized lead wire to the hole and fix them firmly on the one end, and the other end of the copper wire or galvanized lead wire will bend and bury in the groove, so that no copper wire or galvanized lead wire protrudes from the upper and lower surfaces of marble or polished granite board to ensure the tight joint between the adjacent stone blocks.

③Binding steel bar: First of all, find the embedded bar on the wall, and clean up the part of the wall for pasting marble. A vertical bar with a diameter of 6mm is tied up first, and it should be pressed by the embedded bar to bend to the wall. The transverse bar is used for binding marble or polished granite boards. When the height of the block is 60cm, the first transverse bar is tied to the main bar at 10cm above the ground, which is used to tie the steel wire or galvanized lead wire at the bottom of the first coat of the block. The second bar is tied at 7—8cm above the 50cm horizontal line, which is 2—3cm lower than the top of the block. It is used to tie the steel wire or galvanized lead wire at the top of the first coat of the block. After that, set transverse bar every 60cm.

④弹线：首先将要贴大理石或磨光花岗石的墙面、柱面和门窗套用大线坠从上至下找出垂直。应考虑大理石或磨光花岗石板材厚度、灌注砂浆的空隙和钢筋网所占尺寸，一般大理石、磨光花岗石外皮距结构面的厚度应以 5~7 cm 为宜。找出垂直后，在地面上顺墙弹出大理石或磨光花岗石等外廓尺寸线。此线即为第一层大理石或花岗岩等的安装基准线。编好号的大理石或花岗岩板等在弹好的基准线上画出就位线，每块留 1 mm 缝隙（如设计要求拉开缝，则按设计规定留出缝隙）。

⑤石材表面处理：石材表面充分干燥（含水率应小于 8%）后，用石材防护剂进行石材六面体防护处理，此工序必须在无污染的环境下进行，将石材平放于木方上，用羊毛刷蘸上防护剂，均匀涂刷于石材表面，涂刷必须到位，第一遍涂刷完间隔 24 h 后用同样的方法涂刷第二遍石材防护剂，如采用水泥或胶粘剂固定，间隔 48 h 后对石材粘接面用专用胶泥进行拉毛处理，拉毛胶泥凝固硬化后方可使用。

⑥基层准备：清理预做饰面石材的结构表面，同时进行吊直、套方、找规矩，弹出垂直线水平线。并根据设计图纸和实际需要弹出安装石材的位置线和分块线。

⑦安装大理石或磨光花岗石：按部位取石板并舒直铜丝或镀锌铅丝，将石板就位，石板上口外仰，右手伸入石板背面，把石板下口铜丝或镀锌铅丝绑扎在横筋上。绑时不要太紧可留余量，只要把铜丝或镀锌铅丝和横筋拴牢即可，把石板竖起，便可绑大理石或磨光花岗石板上口铜丝或镀锌铅丝，并用木楔子垫稳，块材与基层间的缝隙一般为 30~50 mm。用靠尺板检查调整木楔，再拴紧铜丝或镀锌铅丝，依次向另一方进行。柱面可按顺时针方向安装，一般先从正面开始。第一层安装完毕再用靠尺板找垂直，水平尺找平整，方尺找阴阳角方正，在安装石板时如发现石板规格不准确或石板之间的空隙不符，应用铅皮垫牢，使石板之间缝隙均匀一致，并保持第一层石板上口的平直。找完垂直、平直、方正后，用碗调制熟石膏，把调成粥状的石膏贴在大理石或磨光花岗石板上下之间，使这二层石板结成一整体，木楔处亦可粘贴石膏，再用靠尺检查有无变形，等石膏硬化后方可灌浆。（如设计有嵌缝塑料软管者，应在灌浆前塞放好）。

④Marking lines with ink: First of all, determine the verticality of the parts that install marble or polished granite, such as the wall, column, doors and windows, by hanging a plumb bob from top to the bottom. The thickness of marble or polished granite boards, the spacing of grouting mortar and the size of steel mesh should be considered. The exterior coat of general marble or polished granite should be 5—7cm away from the structural surface. After determining the verticality, mark the outline lines of marble or polished granite on the ground along the wall and this line is the installation baseline of the marble or granite on the first coat. The alignment line should be drawn on the marble or granite boards with ID numbers along with the marked baseline, leaving 1mm spacing in each board (if the design requires the spacing, it will be set according to the design regulations).

⑤Surface treatment of marble or polished granite: After the surface is fully dry (the moisture content should be less than 8%), the stone hexahedron protective treatment is carried out with stone protective agent in a non-polluting environment. The stone is placed on the timber, and then the protective agent is applied evenly on the surface of the stone with the wool brush. The brushing must be applied fully and 24 hours after the first brushing, the workers can use the same method to apply protective agent for the second time. If it is fixed with cement or adhesive agent, 48 hours later, special puddle can be used on the bonding surface of stone to improve adhesion, and the stone can only be used after solidification and hardening of the puddle.

⑥Base preparation: The structural surface for pasting marble or polished granite shall be cleaned. At the same time, checking verticality with plumb bob, squaring room plan, squaring wall corner, marking the horizontal line and the vertical line with ink can be carried out. Then the position line and block line of the installation of marble or polished granite shall be marked according to the design drawings and actual needs.

⑦Installation of marble or polished granite: Take the block for different parts and get them ready, straighten the copper wire or galvanized lead wire. Put the block in place and the top edge of it should incline outward. The worker can extend the right hand into the back of the block, and tie the copper wire or the galvanized lead wire at the bottom edge of it to the transverse bar. The binding should not be too tight, only if the copper wire or the galvanized lead wire can attach to the transverse bar firmly. Then the blocks can be erected, and the copper wire or galvanized lead wire at the top edge of the marble or polished granite can be attached to the transverse bar. In binding, the wooden wedge can be used to make it steady. The gap between the block and the base is generally 30—50mm. The guiding rule is used to check and adjust the wooden wedge, and the copper wire or galvanized lead wire shall be tightened from one side to the other. The installation of marble or polished granite on the column should carry out clockwise and it usually starts from the front. After the first coat is installed, the guiding rule is used to determine the verticality; the leveling rule is used to check flatness; the square rule is used to square the inner and outer corners. If the specification of the blocks is not accurate or the gaps between the blocks don't meet the requirement, the lead protection sheet is used as a pad to ensure the gaps' uniformity and the flatness of the top edge of the first coat of the blocks. After checking the verticality, flatness, squareness and straightness, the plaster like porridge is prepared on a bowl and pasted between the marble or polished granite slabs, so that the two coats of blocks can be integrated and the wooden wedge can also be pasted with the plaster. Then check whether there is any deformation with a guiding rule, and grouting can carry out after the plaster hardening (if there are caulking plastic hose, it should be plugged before grouting).

⑧灌浆：把配合比为1∶2.5水泥砂浆放入半截大桶加水调成粥状,用铁簸箕舀浆徐徐倒入,注意不要碰大理石,边灌边用橡皮锤轻轻敲击石板面使灌入砂浆排气。第一层浇灌高度为15 cm,不能超过石板高度的1/3;第一层灌浆很重要,因要锚固石板的下口铜丝又要固定饰面板,所以要轻轻操作,防止碰撞和猛灌。如发生石板外移错动,应立即拆除重新安装。

⑨擦缝：全部石板安装完毕后,清除所有石膏和余浆痕迹,用麻布擦洗干净,并按石板颜色调制色浆嵌缝,边嵌边擦干净,使缝隙密实、均匀、干净、颜色一致。

9.2.3 墙面干挂石材施工工艺

1. 工艺流程

清理结构表面→结构上弹出垂直线→大角挂两竖直钢丝→临时固定上层墙板→钻孔插入膨胀螺栓→镶不锈钢固定件→镶顶层墙板→挂水平位置线、支底层板托架→放置底层板用其定位→调节与临时固定→嵌板缝密封胶→饰面板刷二层罩面剂→灌 M20 水泥砂浆→设排水管→结构钻孔并插固定螺栓→镶不锈钢固定件→用胶粘剂灌下层墙板上孔→插入连接钢针→将胶粘剂灌入上层墙板的下孔内。

2. 操作工艺

(1)石材表面处理:石材表面充分干燥(含水率应小于8%)后,用石材护理剂进行石材六面体防护处理,此工序必须在无污染的环境下进行,将石材平放于木方上,用羊毛刷蘸上防护剂,均匀涂刷于石材表面,涂刷必须到位,第一遍涂刷完间隔24 h后用同样的方法涂刷第二遍石材防护剂,间隔48 h后方可使用。

⑧Grouting: The cement mortar (cement : sand =1 : 2.5) will be put into a barrel, and water will be added to make it to mushy state. It will be slowly poured with a dustpan, and the operation should avoid the marble. When the grouting is carried out, a rubber hammer will be used to strike the surface of the block slightly to squeeze the air of the mortar. The grouting height of the first coat is 15cm, which cannot exceed 1/3 of the height of the block. The grouting of the first coat is very important. Because the copper wire at the bottom edge of the block should bind and fix the block, it is necessary to operate carefully to prevent collision and hasty grouting. If the blocks are moved or misplaced, they should be removed and reinstalled immediately.

⑨Joints wiping: After all the blocks are installed, remove all traces of plaster and residual slurry and wipe them cleanly with duster. Prepare color paste according to the color of the block to caulk the joints, and wipe the edges of the joints to make them clean. The joints should be dense, uniform, clean and the color should be consistent.

9.2.3 Construction Technology of Cladding with Dry-hanging Building Stone

1. Construction Process

Clean the surface of the structure →mark the vertical line with ink on the structure → hang two vertical steel wires on the large angel→ fix the upper wall plate temporarily→ perforate holes and insert expansion bolts into them→ install the stainless steel fixing members → install the top wall slate → hang the horizontal position line and install the bracket for the bottom cladding slate→ place the bottom slate for positioning → adjust and fix temporarily → caulk the joints with sealant → brush two-coat covering agent on the facing board→ grout M20 cement mortar → set drainpipe → perforate holes on the structure and insert fixing bolts → install the stainless steel fixing members → pour adhesive agent into the upper hole of the bottom wall plate→ insert the connecting steel needles → pour the adhesive agent into the lower hole of the upper wall slate.

2. Operation Process

(1) Surface treatment of dry-hanging building stone: After the surface of dry-hanging building stone is fully dry (Moisture content should be less than 8%), the stone surface protective treatment is carried out with stone protective agent in a non-polluting environment. The dry-hanging building stone is placed on the timber and then the worker applies the protective agent evenly on the surface of the stone with the wool brush. The brushing must be applied fully, and 24 hours after the first brushing, the worker can use the same method to apply the protective agent for the second time. The stone can be used 48 hours after the surface treatment.

(2)石材准备:首先用比色法对石材的颜色进行挑选分类;安装在同一面的石材颜色应一致,并根据设计尺寸和图纸要求,将专用模具固定在台钻上,进行石材打孔,为保证位置准确垂直,要钉一个定型石材托架,使石板放在托架上,要打孔的小面与钻头垂直,使孔成型后准确无误,孔深为 22~23 mm,孔径为 7~8 mm,钻头为 5~6 mm。随后在石材背面刷不饱和树脂胶,主要采用一布二胶的做法,布为无碱、无捻 24 目的玻璃丝布,石板在刷头遍胶前,先把编号写在石板上,并将石板上的浮灰及杂污清除干净,如锯锈、铁抹子锈,用钢丝刷、粗砂子将其除掉再刷胶,胶要随用随配,防止固化后造成浪费。要注意边角地方一定要刷好。特别是打孔部位是个薄弱区域,必须刷到。布要铺满,刷完头遍胶,在铺贴玻璃纤维网格布时要从一边用刷子赶平,铺平后再刷二遍胶,刷子沾胶不要过多,防止流到石材小面给嵌缝带来困难,出现质量问题。

(3)基层准备:清理预做饰面石材的结构表面,同时进行吊直、套方、找规矩;弹出垂直线水平线。并根据设计图纸和实际需要弹出安装石材的位置线和分块线。

(4)挂线:按设计图纸要求,石材安装前要事先用经纬仪打出大角两个面的竖向控制线,最好弹在离大角 20 cm 的位置上,以便随时检查垂直挂线的准确性,保证顺利安装。竖向挂线宜用直径 1.0~1.2 mm 的钢丝为好,下边沉铁随高度而定,一般 40 m 以下高度沉铁重量为 8~10 kg,上端挂在专用的挂线角钢架上,角钢架用膨胀螺栓固定在建筑大角的顶端,一定要挂在牢固、准确、不易碰动的地方,并要注意保护和经常检查。并在控制线的上、下作出标记。

(5)支底层饰面板托架:把预先加工好的支托按上平线支在将要安装的底层石板上面。支托要支承牢固,相互之间要连接好,也可和架子接在一起,支架安好后,顺支托方向铺通长的 50 mm 厚木板,木板上口要在同一水平面上,以保证石材上下面处在同一水平面上。

Chapter 9 Finishing and Decoration Work Construction

(2) Preparation of dry-hanging building stone: First, the stone is selected and classified by the color with colorimetric method; the color of the stones installed on the same face should be consistent. According to the design size and drawing requirements, special mold is fixed to the bench drill for stone drilling. To ensure the accurate position and verticality of the hole, a modular stone bracket should be made so that the stone can be placed on the bracket. The small surface for drilling should be perpendicular to the drilt bit to make the hole accurate in size. The hole is 22—23mm in depth and its diameter is 7—8mm, and diameter of the drill bit is 5—6mm. Then, the unsaturated resin glue is brushed on the back of the dry-hanging building stone, mainly using the method of one cloth and two glues. The cloth is made of twistless glass fiber cloth without alkali with 24holes in 25.4mm^2. Before brushing adhesive on the dry-hanging building stone for the first time, the ID number of it should be written on it and the dust and dirt on it should be cleaned. If there is rust of sawing or iron trowel, it should be removed by wire brush or coarse sand, and then the adhesive is brushed. The adhesive should be prepared when it is needed to prevent wasting because of solidification. In brushing, pay attention to the edges and corners, especially the parts for hole, and it should be carried out carefully. After the brushing of the first adhesive is finished, the cloth should be laid fully on it. In the laying of the cloth, a brush will be used to scrape it from one side to the other to make it flat, and then the second adhesive will be brushed. The brush should not dip too much adhesive to prevent the adhesive from flowing to the small surface of the building stone, which may bring difficulties to the caulking of joints and causing quality problems.

(3) Base preparation: The structural surface for cladding with dry-hanging building stone shall be cleaned. At the same time, the works of checking verticality with plumb bob, squaring room plan, squaring wall corner, marking the vertical and horizontal line with ink should be carried out. Then the position line and block line of installing the dry-hanging building stone shall be marked according to the design drawings and actual needs.

(4) Hanging the line: According to the requirements of the design drawings, before installing the dry-hanging building stone, the vertical control lines on the two sides of large angle should be determined by theodolite and marked with ink 20cm away from the large angle, so as to check the accuracy of the vertical hanging line at any time and ensure the successful installation. The vertical hanging line is made by steel wire with diameter of 1.0—1.2mm and a sinking iron at the bottom. The weight of the sinking iron depends on the height of the wall. Generally, if the height of the wall is less than 40m, the weight of the sinking iron is 8—10kg. The upper end of the steel wire is hung on the special steel angle frame and the angle steel frame is fixed to the top of the large angle of the building with expansion bolts, and the hanging line must be placed firmly and accurately and it should not be touched easily. Pay attention to its protection and regular inspection. Do not forget to make corresponding marks at the top and bottom of the control line.

(5)Installing the bracket for the bottom facing board: install the prepared corbel to the facing board of the stone. The corbels should be firm and they should connect well to each other, and they can also be connected with the bracket. After the brackets are installed, the 50mm thick planks should be laid along the direction of the corbels, and the top edge of the planks should be on the same level and they should be aligned with each other, so as to ensure that the upper and lower sides of the stones are on the same level.

(6)在围护结构上打孔、下膨胀螺栓:在结构表面弹好水平线,按设计图纸及石材料钻孔位置,准确的弹在围护结构墙上并做好标记,然后按点打孔,打孔可使用冲击钻,上 $\phi 12.5$ 的冲击钻头,打孔时先用尖錾子在预先弹好的点上凿一个点,然后用钻打孔,孔深在 $60\sim 80$ mm,若遇结构里的钢筋时,可以将孔位在水平方向移动或往上抬高,要连接铁件时利用可调余量调回。成孔要求与结构表面垂直,成孔后把孔内的灰粉用小勾勺掏出,安放膨胀螺栓,宜将本层所需的膨胀螺栓全部安装就位。

(7)上连接铁件:用设计规定的不锈钢螺栓固定角钢和平钢板。调整平钢板的位置,使平钢板的小孔正好与石板的插入孔对正,固定平钢板,用力矩扳子拧紧。

(8)底层石材安装:把侧面的连接铁件安好,便可把底层面板靠角上的一块就位。方法是用夹具暂时固定,先将石材侧孔抹胶,调整铁件,插固定钢针,调整面板固定。依次按顺序安装底层面板,待底层面板全部就位后,检查一下各板水平是否在一条线上,如有高低不平的要进行调整;低的可用木楔垫平;高的可轻轻适当退出点木楔,退出面板上口在一条水平线上为止;先调整好面板的水平与垂直度,再检查板缝,板缝宽应按设计要求,板缝均匀,将板缝嵌紧被衬条,嵌缝高度要高于 25 cm。其后用 1∶2.5 的用白水泥配制的砂浆,灌于底层面板内 20 cm 高,砂浆表面上设排水管。

(9)石板上孔抹胶及插连接钢针:把 1∶1.5 的白水泥环氧树脂倒入固化剂、促进剂,用小棒将配好的胶抹入孔中,再把长 40 mm 的 $\phi 4$ 连接钢针通过平板上的小孔插入直至面板孔,上钢针前检查其有无伤痕,长度是否满足要求,钢针安装要保证垂直。

(10)调整固定:面板暂时固定后,调整水平度,如板面上口不平,可在板底的一端下口的连接平钢板上垫一相应的双股铜丝垫,若铜丝粗,可用小锤砸扁,若高,可把另一端下口用以上方法垫一下。调整垂直度,并调整面板上口的不锈钢连接件的距墙空隙,直至面板垂直。

Chapter 9 Finishing and Decoration Work Construction

(6) Perforating holes and inserting expansion bolts on the enclosure structure: Tap the horizontal ink line on the surface of the structure, and mark the positions of the holes on enclosure structure with ink line according to the design drawings and position of holes on the dry-hanging building stone. Then perforate the holes according to the marks on the enclosure structure. Hammer drill with the bit of $\phi 12.5$ can be used. In perforating, first, use a sharp chisel to drill a point at the place marked in advance, and then perforate the hole to the depth of 60—80mm. If the perforating meets the steel bar in the structure, the position of the hole can be moved horizontally or raised upwards, and the position can be adjusted in connecting to the iron members with their adjustable space. The holes should be perpendicular to the surface of the structure. After the holes are perforated, the ash in the hole should be scooped out with a small hooking spoon and the required expansion bolts should be installed in place. It is better to insert all the expansion bolts to the holes in the same coat.

(7) Connecting upper iron members: Angle steel and flat steel plates are fixed with stainless steel bolts according to the design specification. The position of the flat steel plate is adjusted so that the holes of the flat steel plate are aligned with the insertion holes of the dry-hanging building stone. Then, fix the flat steel plate firmly and tighten it with the torque wrench.

(8) Installation of dry-hanging building stone slate on the bottom: If the connecting iron members on the side are installed, dry-hanging building stone slate on the bottom can be positioned at the corner. In installation, fix the bottom slate temporarily by clippers. Firstly, brush adhesive to the side holes of the slate, and then adjust the iron members. Then steel needles are inserted and fixed, and the bottom slate shall be adjusted and fixed. The bottom slates shall be installed sequentially, and after all the bottom slates are installed in place, check whether the slates are in the same horizontal line. If some slates are uneven, they need to be adjusted. For the slates lower than the horizontal line, wood wedge can be used to make it flat, and for the slates higher than the horizontal line, remove some wood wedges to make them on the same horizontal line with others. Adjust the horizontal and vertical condition of the slates, and then check the joints between the slates. The width of joints should meet the design requirement, and width of the joints should be the same. The joint should be tight and covered by the furring, and the height of the joint should be higher than 25cm. Then, grout the mortar (white cement : sand = 1 : 2.5) to the bottom slate to be 20cm high and install drainage pipes on the surface of the mortar.

(9) Applying adhesive to the upper holes on the slate and inserting connecting steel needles: Add curing agent and accelerant into white cement epoxy resin (1 : 1.5) and mix them to adhesive, and then use a small stick to apply the prepared adhesive to the hole. Then the 40mm connecting steel needles with the diameter of 40mm are inserted into small holes of the slate and get to the holes on the surface of the wall. Before the steel needle is used, check whether it has any scars and whether the length meets the requirement to ensure the verticality of it.

(10) Adjusting and fixing: After the slates are temporarily fixed, the levelness shall be adjusted. If the top edges of the plates are uneven, the double-truss copper wire pad can be placed on the connecting flat steel plate at the bottom of slate. If the copper wire is thick, it can be smashed with a small hammer, and if it is higher on one side, the above method can be adopted at the other side on the bottom of the slate. In adjusting the verticality, the spacing between the stainless-steel connecting members at the top of the slates and the wall shall be adjusted to ensure the verticality of the slates.

(11)顶部面板安装:顶部最后一层面板除了一般石材安装要求外,安装调整后,在结构与石板缝隙里吊一通长的 20 mm 厚木条,木条上平为石板上口下去 250 mm,吊点可设在连接铁件上,可采用铅丝吊木条,木条吊好后,在石板与墙面之间的空隙里塞放聚苯板,聚苯板条要略宽于空隙,以便填塞严实,防止灌浆时漏浆,造成蜂窝、孔洞等,灌浆至石板口下 20 mm 作为压顶盖板之用。

(12)贴防污条、嵌缝:沿面板边缘贴防污条,应选用 4 cm 左右的纸带型不干胶带,边沿要贴齐、贴严,在大理石板间缝隙处嵌弹性泡沫填充(棒)条,填充(棒)条也可用 8 mm 厚的高连发泡片剪成 10 mm 宽的条,填充(棒)条嵌好后离装修面 5 mm,最后在填充(棒)条外用嵌缝枪把中性硅胶打入缝内,打胶时用力要均,走枪要稳而慢。如胶面不太平顺,可用不锈钢小勺刮平,小勺要随用随擦干净,嵌底层石板缝时,要注意不要堵塞流水管。根据石板颜色可在胶中加适量矿物质颜料。

(13)清理大理石、花岗石表面,刷罩面剂:把大理石、花岗石表面的防污条掀掉,用棉丝将石板擦净,若有胶或其他粘结牢固的杂物,可用开刀轻轻铲除,用棉丝蘸丙酮擦至干净。在刷罩面剂的施工前,应掌握和了解天气趋势,阴雨天和 4 级以上风天不得施工,防止污染漆膜;冬、雨季可在避风条件好的室内操作,刷在板块面上。罩面剂按配合比在刷前半小时对好,注意区别底漆和面漆,最好分阶段操作。配制罩面剂要搅匀,防止成膜时不均,涂刷要用羊毛刷,沾漆不宜过多,防止流挂,尽量少回刷,以免有刷痕,要求无气泡、不漏刷,刷的要平整有光泽。

也可参考金属饰面板安装工艺中的固定骨架的方法,来进行大理石、花岗石饰面板等干挂工艺的结构连接法的施工,尤其是室内干挂饰面板安装工艺。

(11) Installation of the slates on the top: The last coat of slates on the top should meet the general installation requirements. In addition, the 20mm thick planks should be suspended in the gap between the structure surface and the slates and they should be aligned with each other, and the top edge of the wooden plank is 250mm below the top edge of the slate. The hanging point can be set on the connecting iron members, and lead wire can be used to suspend the plank. After the suspension of the wooden plank, insert polystyrene strip into the gap between the structure surface and the slates, and the polystyrene strip should be slightly wider than the gap in order to fill tightly and prevent grouting leakage, resulting in honeycombs, holes and so on. The grouting level is 20mm below the top edge of the slates to leave position for the skin plate.

(12) Sticking the anti-fouling strip and caulking with gel: The anti-fouling strip shall be stuck along the edge of the slate. The paper adhesive tapes with the width of 4cm are mainly used. The edges of the tape should be stuck regularly and tightly. The elastic foam filling strip (bar) should be placed in the gaps between the marble slates, and the filling (bar) strip can be made by cutting strips with the width of 10mm by high foaming sheets with the thickness of 8mm. After the installation of the filling strip, it should be about 5mm away from the structural surface. Finally, the neutral silica gel is injected into the joint with a caulking gun outside the filling strip (bar). In the injection of the gel, the strength should be equal, and it should be controlled steadily and slowly. If the surface of the gel is not smooth, small stainless-steel spoon can be used to flat it. The small spoon should be wiped clean immediately after it is used. When caulking the joints between the bottom slates, make sure not block the drain pipes. Mineral pigments can be added to the gel to adjust its color to be similar to the color of the slate.

(13) Cleaning the surface of marble and granite and brushing coating agent: The anti-fouling strips on the surface of marble and granite shall be removed, and the slate shall be cleaned with cotton yarn. If there is gel or other bonded debris, it will be removed gently with a knife and wiped clean with cotton yarn dipped with acetone. Before brushing the coating agent, check the weather forecast. In cloudy and rainy days or if the wind is stronger than moderate breeze, it is forbidden to do the work to prevent the pollution of the coat film. In winter and rainy seasons, it can be done indoors under good shelter conditions, and the coating agent can be brushed on the surface of the slate. Coating agent is prepared according to the mix ratio half an hour before brushing. The difference between primer and top coating should be noticed, and it is best to operate in different stages. Coating agent should be mixed completely to prevent uneven film formation. The wool brush will be used in brushing, and do not dip too much agent once to prevent sagging. Move the brush forward and try not to move it backward to avoid brushing marks. There should be no bubbles and no missing space in the brushing. The brushing should be smooth and glossy.

The method of fixing framework in the installation metal facing board can also be used in the structural connection method of dry hanging construction process, such as marble and granite facing slate, especially in the installation process of indoor dry hanging facing slates.

9.3 涂饰工程

9.3.1 混凝土及抹灰表面施涂油漆涂料施工

本节适用于工业与民用建筑中室内混凝土表面及泥砂浆、混合砂浆抹灰表面施涂油性涂料工程。

1. 材料

(1)涂料：各色油性调和漆(酯胶调和漆、酚醛调和漆、醇酸调和漆等)，或各色无光调和漆等。

(2)填充料：大白粉、滑石粉、石膏粉、光油、清油、地板黄、红土子、黑烟子、立德粉、羧甲基纤维素、聚醋酸乙烯乳液等。

(3)稀释剂：汽油、煤油、松香水、酒精、醇酸稀料等与油漆性能相应配套的稀料。

2. 工具

高凳子、脚手板、半截大桶、小油桶、铜丝箩、橡皮刮板、钢皮刮板、笤帚、腻子槽、开刀、刷子、排笔、砂纸、棉丝、擦布等。

3. 工艺流程

基层处理→修补腻子→磨砂纸→第一遍满刮腻子→磨砂纸→第二遍满刮腻子→磨砂纸→弹分色线→刷第一道涂料→补腻子磨砂纸→刷第二道涂料→磨砂纸→刷第三道涂料→磨砂纸→刷第四道涂料。

4. 操作工艺

(1)基层处理：将墙面上的灰渣等杂物清理干净，用笤帚将墙面及浮土等扫净。

(2)修补腻子：用石膏腻子将墙面、门窗口角等磕碰破损处、麻面、风裂、接槎缝隙等分别找平补好，干燥后用砂纸将凸出处磨平。

(3)第一遍满刮腻子：满刮遍腻子干燥后，用砂纸将腻子残渣、斑迹等打磨平、磨光，然后将墙面清扫干净，腻子配合比为聚醋酸乙烯乳液(即白乳胶)：滑石粉或大白粉：2%羧甲基纤维素溶液=1:5:35(质量比)，以上为适用于室内的腻子。如厨房、厕所、浴室等应采用室外工程的乳胶防水腻子，这种腻子耐水性能较好，其配合比为聚醋酸乙烯乳液(即白乳胶)：水泥：水=1:5:1(质量比)。

Chapter 9　Finishing and Decoration Work Construction

9.3　Coating Construction

9.3.1　Applying Oil Paint on the Surface of Concrete and Plaster

The technology in this part is adoptable to the construction of applying oil paint on the surface of concrete and plaster in mortar or mixed mortar in the indoor works of industrial and civil buildings.

1. Materials

(1) Paint: oily blending paints of various colors (ester gum mixed paint, phenolic aldehyde paint, alkyd mixed paint, etc.), or various matt blending paints and so on.

(2) Filler: white powder, talcum powder, plaster powder, top oil, boiled oil, a kind of yellow filling powder, a kind of red filling powder, a kind of black filling powder, lithopone, Carboxymethyl Cellulose, polyvinyl acetate emulsion and so on.

(3) Diluter: gasoline, kerosene, rosin water, alcohol, alkyd thinner and other diluters that match the performance of the paint.

2. Tools

High stools, scaffolding boards, half vats, small oil drums, copper wire basket, rubber scraper, steel scraper, broom, putty groove, knife, brush, broad brush comprising a row of pen-shaped brushes, sandpaper, cotton, duster and so on.

3. Process Flow

Base treatment → repairing putty → sand-papering → applying the first coat of putty fully → sand-papering → applying the second coat of putty fully → sand-papering → tapping ink line for color separation → brushing the first coat of oil paint → repairing putty and sand-papering → brushing the second coat of oil paint → sand-papering → brushing the third coat of oil paint → sand-papering → brushing the fourth coat of oil paint.

4. Operation Process

(1) Base treatment: Clean up the ash and other debris on the wall, and use a broom to sweep the dust and make the surface of the wall clean.

(2) Repair putty: Use gypsum putty to level and repair the damaged areas on the wall surface, corners of door and window, matte finish, cracks and joints. After drying, use sandpaper to polish the protruding parts.

(3) Applying the first coat of putty fully: After the putty is applied fully and dried, the putty residue and stains should be sanded and polished with sandpaper, and then the wall surface should be cleaned up. The components in the puffy should follow the mix ratio: polyvinyl acetate emulsion (white latex) : talcum powder or calcium carbonate : 2% carboxymethyl cellulose solution = 1 : 5 : 35 (in weight). The above ratio is suitable for putty in indoor works; in the places such as kitchen, toilet and bathroom, we should use latex waterproof putty as in the outdoor works. This kind of putty has good water resistance performance, and the mix ratio for the components follows: polyvinyl acetate emulsion (white latex) : cement : water = 1 : 5 : 1 (in weight).

(4)第二遍腻子:涂刷高级涂料要满刮第二遍腻子。腻子配合比和操作方法同第一遍腻子。待腻子干透后个别地方再复补腻子,个别大的孔洞可复补腻子,彻底干透后,用1号砂纸打磨平整,清扫干净。

(5)弹分色线:如墙面设有分色线,应在涂刷前弹线,先涂刷浅色涂料,后涂刷深色涂料。

(6)涂刷第一遍油漆涂料:第一遍可涂刷铅油,它是遮盖力较强的涂料,是罩面涂料基层的底漆。铅油的稠度以盖底、不流淌、不显刷痕为宜,涂饰每面墙面的顺序应从上而下,从左到右,不得乱涂刷,以防漏涂或涂刷过厚,涂刷不均匀等。第一遍涂料干燥后个别缺陷或漏刮腻子处要复补,待腻子干透后打磨砂纸,把小疙瘩、腻子渣、斑迹等磨平、磨光、并清扫干净。

(7)涂刷第二遍涂料:涂刷操作方法同第一遍涂料。(如墙面为中级涂料,此遍可涂铅油;如墙面为高级涂料,此遍可涂调和漆),待涂料干燥后,可用较细的砂纸把墙面打磨光滑,清扫干净,同时用潮布将墙面擦抹一遍。

(8)涂刷第三遍涂料:用调和漆涂刷,如墙面为中级涂料,此道工序可作罩面,即最后一遍涂料,其涂刷顺序同上。由于调和漆粘度较大,涂刷时应多刷多理,以达到涂膜饱满、厚薄均匀一致、不流不坠。

(9)涂刷第四遍涂料:用醇酸磁漆涂料,如墙面为高级涂料,此道涂料为罩面涂料,即最后一遍涂料。如最后一遍涂料改为无光调和漆时,可将第二遍铅油改为有光调和漆,其余做法相同。

9.4 楼地面工程

9.4.1 水泥混凝土面层施工

1. 材料

(1)水泥:宜采用硅酸盐水泥、普通硅酸盐水泥或矿渣硅酸盐水泥,其强度等级应在32.5级以上。

(2)砂:应选用水洗粗砂,含泥量不大于3%。

(4) Applying the second coat of putty: When the high-grade paint is used, the second coat of putty should be applied fully. The mix ratio and operation method of putty are the same as the first coat of putty. After the putty dries, the putty will be brushed again in some places. Some large holes can be brushed again. After the puffy dries thoroughly, it should be polished with No. 1 sandpaper and then be cleaned up.

(5) Marking line with ink for color separation: If there is color separation line on the surface, the line should be marked before painting. Light color paint shall be applied first, and then dark paint shall be applied.

(6) Brushing the first coat of ink paint: The lead oil should be applied first, which is strong hiding paint and primer for the first coat of the finishing coating. The lead oil should be mixed to the condition that can cover the bottom and do not cause flowing and leave brushing marks. The brushing of the ink paint on each wall should be carried out from top to bottom and from left to right, and it should be carried out step by step in case of missing some parts or brushing too much or causing other mistakes. After the drying of the oil paint, if there are some defects or some parts missing putty, they should be applied by putty again, and after the putty dries, it should be polished by sandpaper, and the small flaws, putty slag and stains should be smoothed, polished and then they should be cleaned.

(7) Brushing the second coat of oil paint: The brushing method is the same as the first brushing (If mediate grade paint is used, the lead oil can be used; if the high-grade paint is used, the blending paint can be used). After the paint dries, fine sandpaper can be used to polish the wall and then clean the surface, and a piece of wet cloth can be used to wipe the wall.

(8) Brushing the third coat of oil paint: If we use mediate grade paint on the wall, we can use blending paint as the third coat and it can be used as the finishing, which means the last coating, and it can be finished as the same operation process as the previous coats. Since the viscosity of the blending paint is high, it should be brushed more times to achieve full coating, uniform thickness, and there should be no flowing and falling down.

(9) Brushing the fourth coat of oil paint: If we use high-grade paint on the wall, alkyd resin enamel can be used as the finishing, which means the last coating. If the last paint is matt blending paint, the second coat of the lead oil should be light blending paint, and the other construction methods are the same.

9.4　Flooring

9.4.1　The Construction of Cement Concrete Coat

1. Material

(1) Cement: Portland cement, ordinary Portland cement or slag Portland cement can be used, and their strength grade should be above 32.5.

(2) Sand: Washed coarse sand should be used, and its mud content is no more than 3%.

(3)粗骨料:水泥混凝土采用的粗骨料最大粒径不大于面层厚度的 2/3,细石混凝土面层采用的石子粒径不应大于 15 mm。

2. 机具设备

(1)根据施工条件,应合理选用适当的机具设备和辅助用具,以能达到设计要求为基本原则,兼顾进度、经济要求。

(2)常用机具设备有:混凝土搅拌机、平板振捣器、手推车、计量器、筛子、木耙、铁锹、小线、钢尺、胶皮管、木拍板、刮杠、木抹子、铁抹子等。

3. 工艺流程

检验水泥、砂子、石子质量→配合比实验→技术交底→准备机具设备→基底处理→找标高→贴饼冲筋→搅拌→铺设混凝土面层→振捣→撒面找平→压光养护→检查验收。

4. 操作工艺

(1)基层处理:把沾在基层上的浮浆、落地灰等用錾子或钢丝刷清理掉,再用扫帚将浮土清扫干净;如有油污,应用 5%～10%浓度火碱水溶液清洗。湿润后,刷素水泥浆或界面处理剂,随刷随铺设混凝土,避免间隔时间过长风干形成空鼓。

(2)找标高:根据水平标准线和设计厚度,在四周墙、柱上弹出面层的上平标高控制线。

(3)按线拉水平线抹找平墩(60 mm×60 mm 见方,与面层完成面同高,用同种混凝土),间距双向不大于 2 m。有坡度要求的房间应按设计坡度要求拉线,抹出坡度墩。

(4)面积较大的房间为保证房间地面平整度,还要做冲筋,以做好的灰饼为标准抹条形冲筋,高度与灰饼同高,形成控制标高的"田"字格,用刮尺刮平,作为混凝土面层厚度控制的标准。当天抹灰墩、冲筋,当天应该抹完灰,不应该隔夜。

(5)搅拌:混凝土的配合比应根据设计要求通过试验确定。投料必须严格过磅,精确控制配合比。每盘投料顺序为石子—水泥—砂—水。应严格控制用水量,搅拌要均匀,搅拌时间不少于 90 s,坍落度一般不应大于 30 mm。按照规定留制试块。

Chapter 9 Finishing and Decoration Work Construction

(3) Coarse aggregate: The diameter of the largest coarse aggregate in cement concrete is no more than 2/3 of the thickness of the coat, and the diameter of the crushed stone used in the fine stone concrete for coat should not exceed 15mm.

2. Machines and Equipment

(1) According to the construction conditions, suitable equipment and auxiliary equipment should be reasonably selected to meet the design requirements. At the same time, schedule and economic requirements should be considered.

(2) Commonly used machines and equipment: concrete mixers, plate vibrators, trolleys, gauges, sieves, wooden rakes, shovels, small lines, steel rulers, hoses, wooden clappers, scraping bars, wood float, iron trowels, etc.

3. Process Flow

Inspection the quality of cement, sand, stone → mix ratio experiment → technological explanation → preparation of machine and equipment → base treatment → elevation locating → making dot and screeding → stirring → laying the concrete coat → vibrating → sprinkling and leveling → compacting and maintenance → checking and accepting.

4. Operation Process

(1) Base treatment: The laitance, dusts and other debris on the base shall be cleaned with a chisel or wire brush, and then the dirt shall be cleaned with a broom; if there are oil stains, the sodium hydroxide solution with a 5%—10% concentration can be used to wash them off. After wetting, the cement slurry or interface finishing agent shall be brushed in the process of pouring concrete to avoid drying and bubbling.

(2) Elevation locating: According to the horizontal standard line and the design thickness of the coat, the elevation control line of the top of the coat should be tapped with ink on the walls or columns around it.

(3) The leveling pier (with the size of 60mm×60mm, and with the same height and the same concrete as the finished surface of the coat) shall be made according to the horizontal line, and the spacing between piers should be no more than 2m in both vertical and horizontal directions. For rooms with slope requirements, the line shall be drawn according to the design slope requirements, and levelling piers with slope shall be made.

(4) In large rooms, it is necessary to set screeds, which can ensure the flatness of the floor. The screed strips can be made based on the dots (leveling pier) and their height should be the same and they can form the layout in the shape of "田" to control the elevation, and scrapper is used to make it flat. Their height can be used as the standard in controlling the thickness of concrete coat. The levelling pier and the screed should be made in one day, and the work can't be left for the next day.

(5) Stirring: The mix ratio of concrete should be determined in experiment according to design requirements. The materials must be strictly weighed, and its mix ratio should be precisely controlled. The sequence for putting the materials into the mixing drum is as follows: stone—cement—sand—water. The amount of water should be strictly controlled. The stirring should be complete, and the stirring time should be no less than 90s. The slump should be no more than 30mm. The test block shall be reserved as required.

(6)铺设:铺设前应将基底湿润,并在基底上刷一道素水泥浆或界面结合剂,将搅拌均匀的混凝土,从房间内退着往外铺设。在振捣或滚压时低洼处应用混凝土补平。

(7)振捣:用铁锹铺混凝土,厚度略高于找平墩,随即用平板振捣器振捣。厚度超过 200 mm 时,应采用插入式振捣器,其移动距离不大于作用半径的 1.5 倍,做到不漏振,确保混凝土密实。振捣以混凝土表面出现泌水现象为宜。或者用 30 kg 重滚纵横滚压密实,表面出浆即可。

(8)撒面找平:混凝土振捣密实后,以墙柱上的水平控制线和找平墩为标志,检查平整度,高的铲掉,凹处补平。撒一层干拌水泥砂(水泥:砂=1:1),用水平刮杠刮平。有坡度要求的,应按设计要求的坡度施工。

(9)压光:当面层灰面吸水后,用木抹子用力搓打、抹平,将干拌水泥砂拌和料与混凝土浆混合,使面层达到紧密接合。第一遍抹压:用铁抹子轻轻抹压一遍直到出浆为止;第二遍抹压:当面层砂浆初凝后(上人有脚印但不下陷),用铁抹子把凹坑、砂眼填实抹平,注意不得漏压;第三遍抹压:当面层砂浆终凝前(上人有轻微脚印),用铁抹子用力抹压。把所有抹纹压平压光,达到面层表面密实光洁。

(10)养护:应在施工完成后 24 h 左右覆盖和洒水养护,每天不少于 2 次,严禁上人,养护期不得少于 7 天。

(11)冬季施工时,环境温度不应低于 5 ℃。如果在负温下施工时,所掺抗冻剂必须经过试验室试验合格后方可使用。不宜采用氯盐、氨等作为抗冻剂,不得不使用时掺量必须严格按照规范规定的控制量和配合比通知单的要求加入。

Chapter 9   Finishing and Decoration Work Construction

(6) Laying: Before laying, the base should be wet, and cement slurry or interface bonding agent should be brushed to the base. The fully mixed concrete is poured from the inner side to the outer side of the room. Concrete shall be used to level the low-lying areas during vibration or rolling.

(7) Vibrating: Concrete is laid with shovel, and its thickness is slightly higher than the leveling pier. The plate vibrator is used for vibrating. When the thickness of the concrete exceeds 200mm, the internal vibrator should be used, and the movement distance is not more than 1.5 times of the action radius of the vibrator so as to avoid missing some parts in vibrating and to ensure the compactness of the concrete. It is better to vibrate the concrete surface to the condition of bleeding, or use a roller weighed 30kg to roll on it vertically and horizontally to compact it until the grout appears on the face.

(8) Sprinkling and leveling: After the concrete is vibrated and compacted, check the flatness of it as referring to the horizontal control line on the wall and column and the leveling pier. If the concrete is higher, it should be removed by shovel, and if it is lower, lay some concrete to make it flat. Then sprinkle a coat of dry mixed cement sand (cement : sand = 1 : 1) and scrape it with a level scraper. If there is slope requirement, the construction should meet the slope requirement in the design.

(9) Compacting: After the plastering surface absorbs water, it should be rubbed and smoothed with a wood float, and the dry-mixed cement sand and the concrete slurry are blended together to make the coat tightly joint. In the first time of compacting, an iron trowel is used to compact it gently until slurry appears on the face; in the second time, after the mortar of the surface is initially set (when someone steps on the surface, the surface has footprints but it does not sink), the pit and the sand hole should be filled and smoothed by the iron trowel, and the surface should be compacted completely; in the third time, when the mortar of the surface is finally set (when someone steps on the surface, the surface has few footprint), a trowel shall be used to rub and compact it. All the working traces should be smoothed and compacted to make the surface dense and smooth.

(10) Maintenance: About 24 hours after the construction, the coat should be covered and wet at least twice a day. No one is allowed to step on it, and the maintenance period should be at least 7 days.

(11) In winter, the room temperature should not be lower than 5℃. If the temperature is below 0℃, the antifreeze tested qualified in the laboratory can be used. It is not advisable to use chlorine salt, ammonia, etc. as antifreeze. When we have to use these materials, the mixing amount must be strictly controlled in accordance with the requirements and the proportion notice in the specification.

9.5 门窗工程

9.5.1 钢门窗安装施工

1. 材料

钢门窗:钢门窗厂生产的合格的钢门窗,型号品种符合设计要求。

水泥、砂:水泥32.5级以上,砂为中砂或粗砂。

玻璃、油灰:按设计要求的玻璃。

焊条:符合要求的电焊条。

进场前应先对钢门窗进行验收,不合格的不准进场。运到现场的钢门窗应分类堆放,不能参差挤压,以免变形。堆放场地应干燥,并有防雨、排水措施。搬运时轻拿轻放,严禁扔摔。

2. 机具设备

电钻、电焊机、手锤、螺丝刀、活扳手、钢卷尺、水平尺、线坠。

3. 工艺流程

划线定位→钢门窗就位→钢门窗固定→以五金配件安装位置线为准,用线坠或经纬仪将顶层分出的门窗边线标划到各楼层相应位置。

4. 操作工艺

(1)画线定位。根据图纸中门窗的安装位置、尺寸和标高,以门窗中线为准向两边量出门窗边线。从各楼层室内+50 cm水平线量出门窗的水平安装线。依据门窗的边线和水平安装线做好各楼层门窗的安装标记。

Chapter 9　Finishing and Decoration Work Construction

9.5　Doors and Windows

9.5.1　Installation of Steel Doors and Windows

1. Material

Steel doors and windows: Qualified steel doors and windows manufactured by the steel doors and windows factories shall be used, and their model and specification shall meet the design requirements.

Cement, sand: The cement grade should be 32.5 or above, and the sand is medium sand or coarse sand.

Glass, putty: The glass should meet the design requirement.

Welding rod: The welding rod should meet the design requirement.

Steel doors and windows should be checked and accepted before entering the site, and non-qualified ones are not allowed to enter the construction site. Steel doors and windows delivered to the site should be stored by classification and size to avoid extrusion and deformation. The storage area should be dry, and there should be some measures to prevent rain and keep drainage. They should be moved carefully, and it is forbidden to throw them and drop them suddenly.

2. Machinery and Equipment

Electric drill, electric welder, hand hammer, screw driver, adjustable spanner, steel tape, level bar, plumb.

3. Process Flow

Positioning by marking lines→ preparation of steel doors and windows→ fixing steel doors and windows→ marking the positions of the doors and windows on each floor according the positions of them on the top floor by plumb bob or theodolite as referring to the installation lines of hardware fittings.

4. Operation Process

(1) Positioning by marking lines. According to the installation position, size and elevation of doors and windows in the drawings, the side lines of doors and windows can be marked by measurement from the center line of doors and windows to the left and right. The horizontal installation line of the doors and windows on each floor is the indoor horizontal line plus 50cm in each floor. According to the side lines of doors and windows and the horizontal installation line, the installation positions can be marked on each floor.

(2)钢门窗就位。按图纸中要求的型号、规格及开启方向等,将所需要的钢门窗搬运到安装地点,并垫靠稳当。将钢门窗立于图纸要求的安装位置,用木楔临时固定,将其铁脚插入预留孔中,然后根据门窗边线、水平线及距外墙皮的尺寸进行支垫,并用托线板靠吊垂直。钢门窗就位时,应保证钢门窗上框距过梁要有 20 mm 缝隙。框左右缝宽一致,距外墙皮尺寸符合图纸要求。

(3)钢门窗固定。钢门窗就位后,校正其水平和正、侧面垂直,然后将上框铁脚与过梁预埋件焊牢,将框两侧铁脚插入预留孔内,用水把预留孔内湿润,用1∶2较硬的水泥砂浆或 C20 细石混凝土将其填实后抹平。终凝前不得碰动框扇。三天后取出四周木楔,用1∶2水泥砂浆把框与墙之间的缝隙填实,与框同平面抹平。当为钢大门时,应将合页焊到墙中的预埋件上。要求每侧预埋件必须在同一垂直线上,两侧对应的预埋件必须在同一水平位置上。

(4)五金配件的安装。检查窗扇开启是否灵活,关闭是否严密,如有问题必须调整后再安装。在开关零件的螺孔处配置合适的螺钉,将螺钉拧紧。当拧不进去时,检查孔内是否有多余物。若有,将其剔除后再拧紧螺丝。当螺钉与螺孔位置不吻合时,可略挪动位置,重新攻丝后再安装。钢门锁的安装按说明书及施工图要求进行,安好后锁应开关灵活。

9.5.2　铝合金门窗安装施工

1. 材料

(1)铝合金门窗的规格、型号应符合设计要求,五金配件配套齐全,并具有出厂合格证、材质检验报告书并加盖厂家印章。

(2)防腐材料、填缝材料、密封材料、防锈漆、水泥、砂、连接板等应符合设计要求和有关标准的规定。

(2) Preparation of the steel doors and windows. According to the model, specifications and opening direction required in the drawing, the steel doors and windows are transported to the installation site and laid stably on the site. Steel doors and windows are placed on the required position and temporarily fixed with wooden wedges, and their iron feet should be inserted into the reserved holes. Then, they can be adjusted with pad according to the side lines of the door and window, and the horizontal line and the distance from the outer surface of the wall and the plumb bob should be used to determine verticality. When the steel doors and windows are in place, a gap of 20mm between the upper frame of the steel doors or windows and the lintels shall be left. The width of the joints beside the left and right frames shall be the same, and the distance from the outer surface of the wall shall meet the requirements in the drawing.

(3) Fixing of steel doors and windows. After the steel doors and windows are in place, correct their levelness and verticality of the front and two sides, and then weld the iron feet of the upper frame and the embedded parts of the lintel together firmly; the iron feet on both sides of the frame should be inserted into the reserved holes, and wet the reserved holes with water, fill them with 1 : 2 hard cement mortar or C20 fine stone concrete, and then smoothen them. The sash should not be touched before final setting. Three days later, the wooden wedges shall be taken out, and the gap between the frame and the wall shall be filled with cement mortar(cement : sand=1 : 2) to be on the same level with the frame. If it is steel door, the hinge shall be welded to the embedded parts in the wall. The embedded parts on each side of the door must be on the same vertical line, and the corresponding embedded parts on the wall must be on the same horizontal position.

(4) Installing hardware fittings. Check whether the sash is flexible and whether it can be closed tightly. If there is any problem, it must be adjusted and re-installed. The appropriate screw should be fixed and tightened at the screw hole of the switch components. If the screw cannot enter into the screw hole, check whether there is any excess substance in the hole, and it shall be removed if there is something in it, and the screws shall be tightened. When the screw and the screw hole do not match, their position can be slightly moved, and then installation will start again after tapping. The installation of the steel door lock shall be carried out according to the instructions and the construction drawing requirements. After installation, the lock shall be switched flexibly.

9.5.2 Installation of Aluminum Alloy Doors and Windows

1. Material

(1) The specifications and models of aluminum alloy doors and windows should meet the design requirements; the hardware fittings should be fully prepared, and the quality certificate and quality inspection report with the manufacturer's seal are also required.

(2) Anti-corrosion materials, filling materials, sealing materials, anti-rust paint, cement, sand, connecting plate, etc. shall meet the design requirements and relevant standards.

(3)进场前应对铝合金门窗进行验收检查,不合格者不准进场。运到现场的铝合金门窗应分型号、规格堆放整齐,并存放于仓库内。搬运时轻拿轻放,严禁扔摔。

2. 机具设备

电钻、电焊机、水准仪、电锤、活扳手、钢卷尺、水平尺、线坠、螺丝刀。

3. 工艺流程

划线定位→铝合金窗披水安装→防腐处理→铝合金门窗的安装就位→门窗的固定→门窗框与墙体间缝隙的处理→门窗扇及门窗玻璃的安装→安装五金配件。

4. 操作工艺

(1)划线定位。根据设计图纸中门窗的安装位置、尺寸和标高,依据门窗中线向两边量出门窗边线。若为多层或高层建筑时,以顶层门窗边线为准,用线坠或经纬仪将门窗边线下引,并在各层门窗口处划线标记,对个别不直的口边应剔凿处理。门窗的水平位置应以楼层室内+50 cm的水平线为准向上测量出窗下皮标高,弹线找直。每一层必须保持窗下皮标高一致。

(2)铝合金窗披水安装。按施工图纸要求将披水固定在铝合金窗上,且要保证位置正确、安装牢固。

(3)防腐处理。门窗框四周外表面的防腐处理设计有要求时,按设计要求处理。如果设计没有要求时,可涂刷防腐涂料或粘贴塑料薄膜进行保护,避免水泥砂浆直接与铝合金门窗表面接触,产生电化学反应,腐蚀铝合金门窗。安装铝合金门窗时,如果采用连接铁件固定,则连接铁件,固定件等安装用金属零件最好用不锈钢件。否则必须进行防腐处理,以免产生电化学反应,腐蚀铝合金门窗。

(4)铝合金门窗的安装就位。根据划好的门窗定位线,安装铝合金门窗框。并及时调整好门窗框的水平、垂直及对角线长度等符合质量标准,然后用木楔临时固定。

(3) Aluminum alloy doors and windows should be checked and accepted before entering the site, and doors and windows that are not qualified are forbidden to enter the site. Aluminum alloy doors and windows on the construction site should be classified according to their models and specifications, piled up neatly, and stored in the warehouse. They should be moved carefully and it is forbidden to throw and drop them suddenly.

2. Machinery and equipment

Electric drill, electric welder, level gauge, electric hammer, adjustable spanner, steel tape, level bar, plumb, screwdriver.

3. Process Flow

Positioning by marking lines → installation the upstand of aluminum alloy windows → anti-corrosion treatment → installation of aluminum alloy doors and windows → fixing of aluminum alloy doors and windows → the treatment of the gap between the door or window frame and the wall → installation of the sashes and glasses of doors and windows → installation of hardware fittings.

4. Operation Process

(1) Positioning by marking. According to the installation position, size and elevation of doors and windows in the drawings, the side lines of doors and windows can be marked by measurement from the center line of doors and windows to the left and right. If it is multi-story or high-rise building, the side lines of doors and windows on the top floor are referred as standard, and they can be extended with a plumb or theodolite to the lower floors and the position lines of door and window of each floor can be marked. If the edges are not straight, they should be treated properly to make it straight. The horizontal installation line of the doors and windows on each floor is the indoor horizontal line plus 50cm in each floor and it can be used as the elevation of the bottom of the window and it should be marked with ink to make it straight. The elevation of the bottom of the window should be the same on each floor.

(2) Installation the upstand of aluminum alloy windows. The upstand should be fixed to the aluminum alloy window according to the drawing and the fixing position should be accurate, and the installation should be firm.

(3) Anti-corrosion treatment. If there is requirement for the anti-corrosion treatment of frames of doors and windows, the anti-corrosion treatment should meet the design requirement. If there is no special requirement, the frames can be coated with anti-corrosion coating or plastic film to avoid its direct contact with cement mortar, resulting in electrochemical reaction and corrosion of aluminum alloy doors and windows. If we use connecting iron members to fix the aluminum alloy doors and windows, the connecting iron members and other iron fittings should better be stainless steel. Otherwise, anti-corrosion treatment must be carried out on them to avoid electrochemical reaction and corrosion of aluminum alloy doors and windows.

(4) Installation of aluminum alloy doors and windows. The frames of aluminum alloy door and window are installed according to the marked position line of the doors and windows. The horizontal, vertical and diagonal length of the doors and windows should be adjusted in time to meet the quality standards, and then use wooden wedges to fix them temporarily.

(5)铝合金门窗的固定。

当墙体上预埋有铁件时,可直接把铝合金门窗的铁脚与墙体上的预埋铁件焊牢,焊接处需做防锈处理。当墙体上没有预埋铁件时,可用金属膨胀螺栓或塑料膨胀螺栓将铝合金门窗的铁脚固定到墙上。

当墙体上没有预埋铁件时,也可用电钻在墙上打80 mm深、直径为6 mm的孔,用L型80 mm×50 mm的6 mm钢筋。在长的一端粘涂108胶水泥浆,然后打入孔中。待108胶水泥浆终凝后,再将铝合金门窗的铁脚与埋置的6 mm钢筋焊牢。

(6)门窗框与墙体间缝隙的处理。铝合金门窗安装固定后,应先进行隐蔽工程验收,合格后及时按设计要求处理门窗框与墙体之间的缝隙。如果设计未要求时,可采用弹性保温材料或玻璃棉毡条分层填塞缝隙,外表面留5~8 mm深槽口填嵌嵌缝油膏或密封胶。

(7)门窗扇及门窗玻璃的安装。门窗扇和门窗玻璃应在洞口墙体表面装饰完工验收后安装。推拉门窗在门窗框安装固定后,将配好玻璃的门窗扇整体安入框内滑槽,调整好与扇的缝隙即可。平开门窗在框与扇格架组装上墙、安装固定好后再安玻璃,即先调整好框与扇的缝隙,再将玻璃安入扇并调整好位置,最后镶嵌密封条及密封胶。地弹簧门应在门框及地弹簧主机入地安装固定后再安门扇。先将玻璃嵌入门扇格架并一起入框就位,调整好框扇缝隙,最后填嵌门扇玻璃的密封条及密封胶。

(8)安装五金配件。五金配件与门窗连接用镀锌螺钉。安装的五金配件应结实牢固,使用灵活。

(5) Fixing of aluminum alloy door and window.

When there are embedded iron members on the wall, the iron feet of the aluminum alloy doors and windows can be directly welded to the embedded iron members on the wall, and the welding position should take anti-rust treatment. When there is no embedded iron member on the wall, the iron feet of the aluminum alloy doors and windows can be fixed to the wall by metal expansion bolts or plastic expansion bolts.

When there is no embedded iron member on the wall, we can use the electric drill to make a hole with the depth of 80mm and the diameter of 6mm on the wall. Apply polyvinyl formal adhesive on the long edge of an L-shaped steel bar (80mm×50mm with the length of 6mm) and insert it into the hole. After the final set of polyvinyl formal adhesive, the iron feet of the aluminum alloy doors and windows should be welded firmly to the steel bars.

(6) The treatment of the gap between the door or window frame and the wall. After installation of aluminum alloy doors and windows, acceptance of concealed works should be carried out first, and the gap between door or window frame and the wall should be properly treated according to the design requirements. If there is no design requirement, elastic insulation material or glass-cotton felt can be used for caulking the gaps in different coats, and leave a groove with the depth of 5—8mm deep on the surface for caulking with ointment or sealant.

(7) Installation of the sashes and glasses of doors and windows. The sashes and glasses of doors and windows shall be installed after the completion of the decoration of the wall surface around the openings. For sliding doors and windows, after the frames are fixed, their sashes with glasses should be placed into the slots in the frames, and the gap between the sash and frame should be adjusted. For side hung doors and windows, after the frames and the sash frameworks are installed and fixed to the wall, the glass can be installed. In this process, the gap between the sash and frame should be adjusted first, and then the glass should be installed in the sash and its position should be adjusted; finally, the sealing strip and sealant shall be set. For ground spring door, the door leaf shall be installed and fixed after the door frame and the main frame of the ground spring are installed and fixed. The glass of the door or window will be embedded into the framework of the sash, and they will be installed into the frame as a whole and the gap between the frame and the sash will be adjusted. Finally, the sealing strip and sealant will be set to seal the glasses.

(8) Installation of hardware fittings. Hardware fittings are connected with doors and windows by galvanized screws. The installed hardware fittings should be firm and flexible.

9.6 吊顶工程

9.6.1 木龙骨吊顶施工

1. 工艺流程

弹线→安装吊顶紧固件→木龙骨防火、防腐处理→划分龙骨分档线→固定边龙骨→龙骨架的拼装→分片吊装→龙骨架与吊点固定→龙骨架分片间的连接→龙骨架的整体调平→吊顶骨架质量检验→安装罩面板→安装压条、面层刷涂料。

2. 操作工艺

(1) 弹线。

弹标高线：根据楼层+500 mm 标高水平线，顺墙高量至顶棚设计标高，沿墙和柱的四周弹顶棚标高水平线。根据吊顶标高线，检查吊顶以上部位的设备、管道、灯具对吊顶是否有影响。

弹吊顶造型位置线：有叠级造型的吊顶，依据标高线按设计造型在四面墙上角部弹出造型断面线，然后在墙面上弹出每级造型的标高控制线。检查叠级造型的构造尺寸是否满足设计要求，管道、设备等是否对造型有影响。

在顶板上弹出龙骨吊点位置线和管道、设备、灯具吊点位置线。

(2) 安装吊顶紧固件。无预埋的吊顶，可用金属胀铆螺栓或射钉将角钢块固定于楼板底(或梁底)作为安设吊杆的连接件。

小面积轻型的木龙骨装饰吊顶，可用胀铆螺栓固定方木(截面约为 40 mm×50 mm)，吊顶骨架直接与方木固定或采用木吊杆。

(3) 木龙骨防腐防火处理。

防腐处理：按规定选材并实施在构造上的防潮处理，同时涂刷防腐防虫药剂。

9.6 Suspended Ceiling

9.6.1 Construction of Wooden Keel Suspended Ceiling

1. Process Flow

Marking lines with ink → installing the fastening elements of the suspended ceiling → anti-corrosion and fire prevention treatment of the wooded keel → marking the keel division lines → fixing the side keels → assembling the keel framework → hoisting the keel slices one by one → fixing the keel framework with the hoisting point → the connection of different keel slices → the leveling of the whole keel framework → quality inspection of the keel framework → installation of the skin plate → installation of batten and brushing paint on the surface.

2. Operation Process

(1) Marking lines with ink.

Marking elevation lines with ink: Mark the horizontal elevation line of the floor as the elevation of the floor plus 500mm, and the designed elevation of the ceiling should be measured along the wall, and the elevation line of the ceiling should be marked with ink along the wall and column around the ceiling. According to the elevation line of the suspended ceiling, check whether the equipment, pipes and luminaire above the suspended ceiling influence it.

Marking the position line of the suspended ceiling: As for the ceiling with overlapping shape, the position of the suspended ceiling should be designed according to the elevation line, and the section line of the suspended ceiling should be marked on the upper corners of the walls, and then mark the elevation control line of each shape on the wall. Check whether the structural dimension of the overlapping shape meets the design requirements and the pipeline and equipment have an impact on the shape.

Mark the position lines of hoisting points for the keel and for the pipe, equipment, luminaire on the top plate.

(2) Installing the fastening elements of the suspended ceiling.

When there are no embedded parts on the suspended ceiling, the angle steel block can be fixed to the bottom of the floor slab (or the bottom of the beam) with the metal expansion rivet bolt or the nail as the connecting parts for installing the suspension rod.

As for small-area light wooden keel decorative suspended ceiling, wooden batten can be fixed with expansion rivet bolt (the section is about 40mm×50mm), and the framework of the suspended ceiling can be fixed directly to the timber or be fixed to it with the wooden suspension rod.

(3) Anti-corrosion and fireproofing treatment of the wooden keel.

Anti-corrosion treatment: select proper wooden keel according to the requirement, make damp-proof treatment on them, and apply anti-corrosion and insect-proof agents on them.

防火处理：将防火涂料涂刷或喷于木材表面，或把木材置于防火涂料槽内浸渍。防火涂料视其性质分为油质防火涂料（内掺防火剂）与氯乙烯防火涂料、可赛银（酪素）防火涂料、硅酸盐防火涂料，施工可按设计要求选择使用。

(4)划分龙骨分档线。沿已弹好的顶棚标高水平线，划好龙骨的分档位置线。

(5)固定边龙骨。沿标高线在四周墙（柱）面固定边龙骨方法主要有两种：沿吊顶标高线以上10 mm处在建筑结构表面打孔，孔距500～800 mm，在孔内打入木楔，将边龙骨钉固于木楔上；混凝土墙、柱面，可用水泥钉通过木龙骨上钻孔将边龙骨钉固于混凝土墙、柱面。

(6)龙骨架的拼装。为方便安装，木龙骨吊装前可先在地面进行分片拼接。

分片选择：确定吊顶骨架面上需要分片或可以分片的位置和尺寸，根据分片的平面尺寸选取龙骨纵横型材。

拼接：先拼接组合大片的龙骨骨架，再拼接小片的局部骨架。拼接组合的面积不可过大以便于吊装。

成品选择：对于截面为 25 mm×30 mm 的木龙骨，可选用市售成品凹方型材；如为确保吊顶质量木现场制作，应在方木上按中心线距 300 mm 开凿深 15 mm、宽 25 mm 的凹槽。

骨架拼接按凹槽对凹槽的方法咬口拼联，拼口处涂胶并用圆钉固定。可采用化学胶，如酚醛树脂胶、尿醛树脂胶和聚醋酸乙烯乳液等。

(7)分片吊装。将拼接组合好的木龙骨架托起，至吊顶标高位置。对于顶底低于 3 m 的吊顶骨架，可用定位杆作临时支撑；吊顶高度超过 3 m 时，可用钢丝在吊点上作临时固定。

根据吊顶标高线拉出纵横水平基准线，作为吊顶的平面基准。

将吊顶龙骨架向下略做移位，使之与基准线平齐。待整片龙骨架调正调平后，即将其靠墙部分与沿墙龙骨钉接。

Fireproofing treatment: apply or spray fireproof coatings on the surfaces of wooden keel, or impregnate the wooden keels in tank with fireproof paint. Fireproofing coatings are divided into oil fireproofing coatings (mixed with fire retardant), vinyl chloride fireproofing coatings, casein fireproofing coatings, silicate fireproofing coatings. They can be selected and used according to the design requirements.

(4) Marking the dividing line of the keel. Mark the dividing position line of the keel along the marked horizontal elevation line of the ceiling.

(5) Fixing the side keels. There are two main methods for fixing the side keels along the horizontal elevation line on the surrounding walls (columns): for the first method, perforate holes on the surface of the structure at 10mm above the elevation line of the suspended ceiling, and the spacing between holes is 500—800mm; insert wooden wedges into the holes, and then nail the side keels onto the wooden wedges. For the second method, the side keels can be nailed to concrete walls or columns by drilling holes in the wooden keel with the cement nail.

(6) Splicing the keel framework.

For convenient installation, the wooden keel can be spliced and connected before hoisting.

The selection and division of keel sections: Determine the position and size of the sections that need to be divided and can be divided, and select proper vertical and horizontal keels for the section.

Splicing: We should splice the section of big keel framework and then splice the section of small keel framework. For convenient hoisting, the area of the spliced section should not be too large.

Selection of made-up keels: For wooden keel with the section of 25mm×30mm, we can select the grooved wooden keels on the market. To ensure the quality of suspended ceiling, they can be made on the site. The grooves with the depth of 15mm and the width of 25mm should be made on the wooden batten and the spacing between the center lines of grooves should be 300mm.

The connection of keels follows the rule of connecting grooves with grooves, and the joints are coated with adhesive and are fixed with round-head nails. Chemical adhesives such as phenol formaldehyde resin adhesive, urea resin adhesive and polyvinyl acetate emulsion can be used.

(7) Hoisting the keel frameworks by section.

The spliced frameworks of the wooden keels should be hoisted to the elevation position of the suspended ceiling. If the height of the suspended ceiling is lower than 3m, the frameworks can be supported temporarily by positioning rod; if the height of the suspended ceiling is more than 3m, steel wire can be used for temporary fixing of the frameworks on the hoisting point.

The vertical and horizontal reference lines are determined by the elevation line of the suspended ceiling and they can be used as reference plane of the suspended ceiling.

The keel framework of the suspended ceiling can be moved down slightly to make it at the same level with the reference line. After the whole keel framework is adjusted and leveled, its parts against the wall will be nailed to the keel along the wall.

(8)龙骨架与吊点固定。固定做法有多种,视选用的吊杆及上部吊点构造而定:以 $\phi 6$ mm 钢筋吊杆与吊点的预埋钢筋焊接;利用扁钢与吊点角钢以 M6 螺栓连接;利用角钢作吊杆与上部吊点角钢连接等。

吊杆与龙骨架的连接,根据吊杆材料可分别采用绑扎、勾挂及钉固等,如扁钢及角钢杆件与木龙骨可用两个木螺钉固定。

(9)龙骨架分片间的连接。分片龙骨架在同一平面对接时,将其端头对正,再用短方木进行加固,将方木钉于龙骨架对接处的侧面或顶面均可。重要部位的龙骨接长,应采用铁件进行连接紧固。

(10)龙骨的整体调平。在吊顶面下拉出十字或对角交叉的标高线,检查吊顶骨架的整体平整度。

骨架底平面出现下凸的部分,要重新拉紧吊杆;有上凹现象的部位,可用木方杆件顶撑,尺寸准确后将方木两端固定。

各个吊杆的下部端头均按准确尺寸截平,不得伸出骨架的底部平面。

(11)安装罩面板。在木骨架底面安装顶棚罩面板,罩面板固定方式分为圆钉钉固法、木螺钉拧固法、胶结粘固法三种方式。

圆钉钉固法:用于石膏板、胶合板、纤维板的罩面板安装以及灰板条吊顶和 PVC 吊顶。

①固定罩面板的钉距为 200 mm。装饰石膏板,钉子与板边距离应不小于 15 mm,钉子间距宜为 150~170 mm,与板面垂直。钉帽嵌入石膏板深度宜为 0.5~1.0 mm,并应涂刷防锈涂料。钉眼用腻子找平,再用与板面颜色相同的色浆涂刷。

②软质纤维装饰吸声板,钉距为 80~120 mm,钉长为 20~30 mm,钉帽进入板面 0.5 mm,钉眼用油性腻子抹平。

(8) Fixing the keel framework and hoisting point.

There are many kinds of fixing methods, and the choice of the methods depends on the types of the suspension rod and upper structure of the hoisting point. For example, weld the reinforcement suspension rod ($\phi 6$ mm) with the embedded steel bar at the hoisting point; connect the flat steel bar and the angle steel bar at the hoisting point by M6 bolts; connect the angle steel bar (used as the suspension rod) with the upper angle steel at the suspension point.

The connecting methods of the suspension rod and the keel formwork are determined by the materials of the suspension rod, and there are several methods, such as tying, hooking and nailing and so on. For example, for suspension rod made from flat steel bar and angle steel bar, the wooden keel can be fixed with two wooden screws.

(9) The connection between different sections of keel frameworks. When different sections of keel frameworks are jointed on the same level, their ends should be aligned, and short timber should be nailed to connection parts of the framework for reinforcement. If the connection of the keel frameworks is on the important parts, the joints should be fastened by iron members.

(10) Leveling of the whole keel framework. Make the crossed or diagonally crossed elevation line under the suspended ceiling plane, and check the overall flatness of the framework of the suspended ceiling.

When the bottom of the framework protrudes the plane, the suspension rod should be tightened; when there is concave part, timber can be used to support it, and the two ends of the wood block will be fixed after the frameworks are all on the same plane.

The lower end of each suspension rod should be cut based on the same precise length and no rod is allowed to protrude from the bottom plane of the framework.

(11) Installation of the skin plate.

Install skin plate on the bottom plane of wooden keel framework. The fixing methods of the skin plate can be classified into three: the round-head nail fixing method, the wooded screw fixing method, and the cementing method.

Round-head nail fixing method: This method is used for the installation of skin plate made of gypsum board, plywood, and fiber board as well as board lath suspended ceiling and PVC suspended ceiling.

① The spacing between nails fixing the skin plate is 200mm. For decorative gypsum board, the nail should be no less than 15mm away from the edge of the plate, and the spacing between the nails should be 150—170mm and the nails should be vertical to the plate. The depth of the nail cap embedded in gypsum board should be 0.5—1.0mm, and the nail cap should be treated with antirust coating. The nail hole shall be leveled with putty and brushed with paste in the same color of the plate.

② As for soft fiber decorative sound-absorbing board, the spacing between nails is 80—120mm, and the length of the nail is 20—30mm. The depth of the nail cap embedded in the surface of the board should be 0.5mm, and the nail hole shall be levelled with oily putty.

③硬质纤维装饰吸声板,板材应用水浸透,自然晾干后安装,采用圆钉固定;对于大块板材,应使板的长边垂直于横向次龙骨,即沿着纵向次龙骨铺设。

④塑料装饰罩面板,一般用20~25 mm宽的木条,制成500 mm的正方形木格,用小圆钉钉,再用20 mm宽的塑料压条或铝压条或塑料小花固定板面。

⑤灰板条铺设,板与板之间应留8~10 mm的缝,板与板接缝应留3~5 mm,板与板接缝应错开,一般间距为500 mm左右。

木螺钉固定法:用于塑料板、石膏板、石棉板、珍珠岩装饰吸声板以及灰板条吊顶。在安装前罩面板四边按螺钉间距先钻孔,安装程序与方法基本上同圆钉钉固法。珍珠岩装饰吸声板螺钉应深入板面1~2 mm,用同色珍珠岩砂混合的粘结腻子补平板面,封盖钉眼。

胶结粘固法:用于钙塑板,安装前板材应选配修整,使厚度、尺寸、边楞齐整一致。每块罩面板粘贴前进行预装,然后在预装部位龙骨框底面刷胶,同时在罩面板四周刷胶,刷胶宽度为10~15 mm,经5~10 min后,将罩面板压粘在预装部位。

每间顶棚先由中间行开始,然后向两侧分行逐块粘贴,胶粘剂按设计规定。

(12)安装压条。木骨架罩面板顶棚,设计要求采用压条作法时,待一间罩面板全部安装后,先进行压条位置弹线,按线进行压条安装。其固定方法可同罩面板,钉固间距为300 mm,也可用胶结料粘贴。

9.6.2 轻钢龙骨吊顶

1. 工艺流程

测量放线定位→吊件加工与固定→固定吊顶边部骨架材料→安装主龙骨→安装次龙骨→双层骨架构造的横撑龙骨安装→吊顶龙骨质量检验→安装罩面板→安装压条(或嵌缝)→面层刷涂料。

③ As for rigid fiber decorative sound-absorbing board, the board should be installed only after being soaked with water and dried naturally, and the board can be fixed with round-head nails; for large boards, the long edge of the board should be perpendicular to the transverse secondary keel, that is to say, it will be laid along the longitudinal secondary keel.

④ The plastic decorative skin plate is generally made of square wooden lattice with wooden strips of 20—25mm wide and 500mm long. The plate can be nailed to the framework with small round-head nails and fixed with 20mm wide plastic batten, aluminum batten or plastic flower.

⑤ For board lath suspended ceiling, joints with the width of 8—10mm shall be left between board lath and the connection joints with the width of 3—5mm should be left. The joints between the lathes shall be staggered and the spacing is about 500mm.

Wooden screw fixing method: The method is used for the suspended ceiling of plastic board, gypsum board, asbestos board, perlite decorative sound-absorbing board and board lath. Holes shall be drilled at the four sides of the skin plate before installation with reference to the screw spacing. The installation procedure and method are basically the same as the fixing method with round-head nail. The screws for the perlite decorative sound-absorbing board should be embedded 1—2mm into the board surface and the nail holes should be leveled and covered with putty mixed with the same color perlite sand.

Adhesive bonding method: This method is used for calcified plastic board. Before installation, the boards should be selected and treated to make them with the same thickness, size and edge length. The plastic boards should be assembled before bonding, and then the adhesive is applied to the bottom surface of the framework of the to-be-installed keel and all the sides of the board. The brushing width on the board is 10—15mm and then the board is pressed and adhered to the to-be-installed parts after 5—10min.

Installation of the skin plate of each room begins from the middle row, and then comes to the two sides in rows, and the adhesive is selected according to design requirement.

(12) Installation of batten. When batten is required in the design for the wooden skin plate, the position line of the batten should be marked with mk after the skin plate is finished in one room, and the batten is installed according to the position line. The fixing method can be the same as the skin plate and the distance of the fixing nails is 300mm, and the batten can also be fixed with adhesive.

9.6.2 Light Steel Keel Suspended Ceiling

1. Process Flow

Surveying and positioning by marking→ making and fixing of suspension components→ fixing the framework material on the edge of the suspended ceiling→ installing the main keel→ installing the secondary keel→ installation of lateral brace keel of double-coat framework→ quality inspection of the keel → installing skin plate → installing batten (or caulking) → brushing surface coating.

2. 操作工艺

(1)测量放线定位。

在结构基层上,按设计要求弹线,确定龙骨及吊点位置。主龙骨端部或接长部位要增设吊点。较大面积的吊顶、龙骨和吊点间距应进行单独设计和验算。

确定吊顶标高。在墙面和柱面上,按吊顶高度弹出标高线。要求弹线清楚,位置准确,水平允许偏差控制在±5 mm。

(2)吊件加工与固定

吊点间距当无设计规定时,一般应小于1.2 m,吊杆应通直,距主龙骨端部距离不得超过300 mm。当吊杆与设备相遇时,应调整吊点构造或增设吊杆。

龙骨与结构连接固定有三种方法:

①在吊点位置钉入带孔射钉,用镀锌钢丝连接固定。射钉在混凝土基体上的最佳射入深度22~32 mm(不包括混凝土表面的涂敷层),一般取27~32 mm(仅在混凝土强度特高或基体厚度较小时才取下限值)。

②在吊点位置预埋膨胀管螺栓,再用吊杆连接固定。

③在吊点位置预留吊钩或埋件,将吊杆直接与预留吊钩或预埋件焊接连接,再用吊杆连接固定龙骨。采用吊杆时,吊杆端头螺纹部分应预留长度不小于30 mm的调节量。

(3)固定吊顶边部骨架材料。

吊顶边部的支承骨架应按设计的要求加以固定。

无附加荷载的轻便吊顶,用L形轻钢龙骨或角铝型材等,可用水泥钉按400~600 mm的钉距与墙、柱面固定。

有附加荷载的吊顶,或有一定承重要求的吊顶边部构造,需按900~1 000 mm的间距预埋防腐木砖,将吊顶边部支承材料与木砖固定。吊顶边部支承材料底面应与吊顶标高基准线平(罩面板钉装时应减去板材厚度)且必须牢固可靠。

2. Operation Process

(1) Surveying and positioning by marking.

According to the requirement of the design, determine the position of the keel and the hoisting points and mark the lines with ink on the structure. There should be hoisting points at the ends of the main keel and at the connecting points of the keels. As for suspended ceiling with large area, the spacing of keels and hanging points should be designed and checked separately.

Determine the elevation of the suspended ceiling. Mark the elevation line on the surrounding walls and columns according to the height of the suspended ceiling. The lines should be clearly visible, and the position should be accurate, and the horizontal allowable deviation is ±5mm.

(2) Making and fixing of suspension components.

When there is no design requirement for the spacing of hanging points, it should be less than 1.2m. The suspension rods should be aligned and they should be no more than 300mm away from the end of the main keel. When the suspension rod collides with other equipment, the hanging point should be adjusted or the additional suspension rod should be installed.

There are 3 ways to connect and fix the keel with the structure.

① The studs with holes are driven into the hanging point, and the keels are connected and fixed to them with galvanized steel wire. The suitable depth of the stud embedded into the concrete base is 22—32mm (excluding the coating of the concrete surface), and it is generally 27—32mm (The lower limit value is only being considered when the concrete strength is extremely high or the thickness of the base is small)

② The expansion pipe bolts should be embedded at the hanging point, and the keels can be connected and fixed to them with suspension rods.

③ Reserve hooks or embed components at the hanging point, and weld the suspension rod directly to the reserved hook or embedded components, and then connect and fix the keel with the suspension rod.

When the suspension rod is used, the thread at the end of it should leave at least 30mm for adjusting it.

(3) Fixing the framework at the edge of the suspended ceiling.

The supporting framework at the edge of the suspended ceiling shall be fixed according to the design requirement.

Light suspended ceiling without additional load can be made of L-shaped light steel keel, angle aluminum section bar and so on, and they can be fixed to the wall and columns with cement nails at a spacing of 400—600mm

For the suspended ceilings with additional loads or supporting structure with load requirement at the edge of suspended ceiling, the anti-corrosion wood blocks shall be embedded at an interval of 900—1000mm, and the supporting materials of suspended ceiling edge shall be fixed with the wood blocks. The bottom surface of the supporting material of the ceiling edge should be level with the ceiling elevation datum line (the thickness of the plate should be deducted when the cover panel is nailed) and they must be firm and reliable.

(4)安装主龙骨。

轻钢龙骨吊顶骨架施工,应先高后低。主龙骨间距一般为 1 000 mm。离墙边第一根主龙骨距离不超过 200 mm(排列最后距离超过 200 mm 应增加一根),相邻接头与吊杆位置要错开。吊杆规格轻型宜用 $\phi 6$ mm,重型(上人)用 $\phi 8$ mm,如吊顶荷载较大,需经结构计算,选定吊杆断面。

主龙骨与吊杆(或镀锌铁丝)连接固定。与吊杆固定时,应用双螺帽在螺杆穿过部位上下固定。轻钢龙骨系列的重型大龙骨 U、C 形,以及轻钢或铝合金 T 形龙骨吊顶中的主龙骨,悬吊方式按设计进行。与吊杆连接的龙骨安装有三种方法:

①有附加荷载的吊顶承载龙骨,采用承载龙骨吊件与钢筋吊杆下端套丝部位连接,拧紧螺母卡稳卡牢。

②附加荷载的 C 形轻钢龙骨单层构造的吊顶主龙骨,采用轻型吊件与吊杆连接,可利用吊件上的弹簧钢片夹固吊杆,下端勾住 C 形龙骨槽口两侧。

③轻便吊顶的 T 形主龙骨,可以采用其配套的 T 形龙骨吊件,上部连接吊杆,下端夹住 T 形龙骨,也可直接将镀锌钢丝吊杆穿过龙骨上的孔眼勾挂绑扎。

安装调平主龙骨:主龙骨安装就位后,以一个房间为单位进行调平。调平方法可采用木方按主龙骨间距钉圆钉,将龙骨卡住先作临时固定,按房间的十字和对角拉线,根据拉线进行龙骨的调平调直。根据吊件品种,拧动螺母或通过弹簧钢片,或调整钢丝,准确后再行固定。使用镀锌铁丝作吊杆者宜采取临时支撑措施,可设置方木,上端顶住吊顶基体底面,下端顶稳主龙骨,待安装吊顶板前再行拆除。在每个房间和中间部位,用吊杆螺栓进行上下调节,预先给予 5~20 mm 起拱量,水平度全部调好后,逐个拧紧吊杆螺帽。如吊顶需要开孔,先在开孔的部位划出开孔的位置,将龙骨加固好,再用钢锯切断龙骨和石膏板,保持稳固牢靠。

Chapter 9　Finishing and Decoration Work Construction

(4) Installing the main keel.

For the framework of light steel keel suspended ceiling, the construction should begin from the higher parts and then go to the lower parts. The spacing of the main keels is generally 1000mm. The distance from the first main keel to the wall is no more than 200mm (If the distance of final main keel to the wall is more than 200mm, an additional keel shall be added), and the adjacent joints and the position of the suspension rod should be staggered. For light keel, the diameter of the suspension rod is 6mm and for heavy keel which can support the load of people, the diameter of the suspension rod is 8m. If the load on the ceiling is large, the cross section of the suspension rod should be selected by calculation.

The main keel is fixed by attaching to the suspension rod (or galvanized iron wire). When it is fixed with suspension rod, two nuts are used to fix the upper and lower end of the screw. As for the U-shaped and C-shaped heavy keel in the light steel keel series and the main keel in the light steel or aluminum alloy T-shaped keel suspended ceiling, their hoisting methods should follow the design requirement. There are 3 ways to connect the keel and the suspension rod:

①As for the load-bearing keel of the suspended ceiling with additional load, the suspension components can be connected to the thread at the lower end of the reinforced suspension rod, and the nuts should be tightened to hold them tightly.

②As for the main keel of the suspended ceiling without additional load in C-shaped light steel keel single-coat structure, the light suspension components can be connected to the suspension rod. In this way, the spring steel sheet on the suspension component can clamp the suspension rod, and the lower end hooks the two sides of the groove of the C-shaped keel.

③As for the T-shaped main keel of the light suspended ceiling, the matching T-shaped suspension components can be used in its connection. The upper part of the suspension component is connected with the suspension rod, and the lower end of it clamps the T-shaped keel. In the other way, the galvanized steel wire suspension rod can pass through the holes on the keel to hook and bind it directly.

Installing and leveling the main keels: After the main keels are installed in place, they should be leveled room by room. The following leveling method can be adopted. The timbers are nailed with round-head nails according to the spacing of the main keels, and the keels are locked for temporary fixing at first. According to the cross and diagonal lines of the room, the keels are leveled and straightened. According to the types of suspension rods, the keels can be leveled by twisting the nut or adjusting the spring steel plate or the steel wire, and after that, they can be fixed. When the suspension rod is made of galvanized iron wire, temporary supporting measures are preferred. For example, we can set timber and the upper part of it props the under surface of the suspended ceiling base, and its lower part props the main keel. Before the ceiling plate is installed, the timber shall be removed. In each room and the joints between them, adjust the main keels by adjusting the bolts on the suspension rod. At first, spring it for 5—20mm, and after the leveling of the whole keels, the nuts of the suspension rod should be tightened one by one. If there is opening on the suspended ceiling, the position of the opening should be marked first and the keel should be strengthened, and finally, the keel and the gypsum board is cut off with a hacksaw to make the opening. In the process, we should keep the keel firm and stable.

(5)安装次龙骨。

双层构造的吊顶骨架,次龙骨(中龙骨及小龙骨)紧贴承载主龙骨安装,通长布置,利用配套的挂件与主龙骨连接,在吊顶平面上与主龙骨相垂直。次龙骨的中距由设计确定,并因吊顶装饰板采用封闭式安装或是离缝及密缝安装等不同的尺寸关系而异。

单层吊顶骨架,其次龙骨即为横撑龙骨。主龙骨与次龙骨处于同一水平面,主龙骨通长设置,横撑(次)龙骨按主龙骨间距分段截取,与主龙骨丁字连接。

以 C 形轻钢龙骨组装的单层构造吊顶骨架,在吊顶平面上的主、次 C 形龙骨垂直交接点,应采用其配套的挂插件(支托),挂插件一方面插入次龙骨内托住主龙骨段,另一方面勾挂住主龙骨,将二者连接。

T 形轻金属龙骨组装的单层构造吊顶骨架,其主、次龙骨的连接通常是 T 形龙骨侧面开有圆孔和方孔,圆孔用于悬吊,方孔则用于次龙骨的凸头直接插入。对于不带孔眼的 T 形龙骨连接方法有三种:

①在次龙骨段的端头剪出连接耳(或称连接脚),折弯 90°与主龙骨用拉铆钉、抽芯铆钉或自攻螺钉进行丁字连接。

②在主龙骨上打出长方孔,将次龙骨的连接耳插入方孔。

③采用角形铝合金块(或称角码),将主次龙骨分别用抽芯铆钉或自攻螺钉固定连接。小面积轻型吊顶,其纵、横 T 形龙骨均用镀锌钢丝分股悬挂,调平调直,只需将次龙骨搭置于主龙骨的翼缘上,再搁置安装吊顶板。

每根次龙骨用两只卡夹固定,校正主龙骨平正后再将所有的卡夹依次全部夹紧。

(6)双层骨架构造的横撑龙骨安装。

U 形、C 形轻钢龙骨的双层吊顶骨架在相对湿度较大的地区,必须设置横撑龙骨。

以轻钢 U 形(或 C 形)龙骨为承载龙骨,以 T 形金属龙骨作覆面龙骨的双层吊顶骨架,一般需设置横撑龙骨。吊顶饰面板作明式安装时,则必须设置横撑龙骨。

(5) Installing the secondary keel.

For double-coat suspended ceiling framework, the secondary keels (middle keel and small keel) should be close to the load-bearing main keel, and they should be aligned with each other. The secondary keels are connected with the main keels by using the matching suspension components, and they are perpendicular to the main keels on the plane of the suspended ceiling. The distance between the secondary keels is determined by the design and it varies a lot based on the different installation methods of the skin plate, such as enclosed installation, installation with joints or with sealed joints.

The secondary keel of the single-coat suspended ceiling framework is the lateral brace keel. The main keel and the secondary keel are on the same level, and the main keels should be aligned with each other; the lateral brace keels (secondary keels) are set according to the spacing of the main keels in different section and they are connected with the main keel in the shape of "T".

As for the single-coat suspended ceiling framework assembled by C-shaped light steel keel, at the vertical joints of the main keel and the secondary C-shaped keel on the same level, the matching plug-in components should be used. One end of the plug-in component should be inserted into the secondary keel and the other end of it should hook the main keel to connect them together.

As for single-coat suspended ceiling framework assembled by T-shaped light metal keel, the connection of its main and secondary keels usually depends on round and square holes on the edges of the T-shaped keel. The round holes are used for hanging, and the square holes are used for the direct insertion of the convex end of the secondary keel. There are three connection methods for T-shaped keels without holes:

①Cut the engaging lug (or engaging foot) at the end of the secondary keel and bend it 90° to connect it with the main keel in the shape of "T" with pulling rivets, self-plugging rivets or self-tapping screws.

②Cut square holes on the main keel, and insert the engaging lug on the secondary keel into the square hole.

③Angle aluminum alloy blocks (or angle code) are used to fix and connect the main and secondary keels with self-plugging rivets or self-tapping screws respectively. As for the small light suspended ceiling, its longitudinal and transverse T-shaped keels are all hung by galvanized steel wire truss, and they will be leveled and straightened. Then, the secondary keels are connected to the edge of the main keel, and then the suspended ceiling plate is placed and installed.

Each keel is fixed to the main keel with two clips and after the main keels are leveled and straightened, all the clips should be tightened one by one.

(6) Installation of lateral brace keel with double-coat framework.

In the area with relatively high humidity, for double-coat suspended ceiling framework with U-shaped or C-shaped light steel keel, lateral brace keel must be set.

For double-coat suspended ceiling framework with U-shaped (or C-shaped) keel as the load-bearing keel and T-shaped metal keel as the cladding keel, it is generally required to set the lateral brace keel. When the skin plate is installed on the surface, the lateral brace keel must be set.

C形轻钢吊顶龙骨的横撑龙骨由C形次龙骨截取,与纵向的次龙骨的T字交接处,采用其配套的龙骨支托(挂插件)将二者连接固定。

双层骨架的T形龙骨覆面层的T形横撑龙骨安装,根据其龙骨材料的品种类型确定,与上述单层构造的横撑龙骨安装做法相同。

(7)安装罩面板。

①石膏板罩面安装:

a. 应从吊顶的一边角开始,逐块排列推进。石膏板用镀锌3.5 mm×25 mm自攻螺钉固定在龙骨上,钉头应嵌入石膏板内为0.5~1 mm,钉距为150~170 mm,钉距板边15 mm。板与板之间和板与墙之间应留缝,一般为3~5 mm。

采用双层石膏板时,其长短边与第一层石膏板的长短边均应错开一个龙骨间距以上位置,且第二层板也应如第一层一样错缝铺钉,采用3.5 mm×35 mm自攻螺钉固定在龙骨上,螺钉应适当错位。

b. 纸面石膏板应在自由状态下进行安装,并应从板的中间向板的四周固定,纸包边长应沿着次龙骨平行铺设,纸包边宜为10~15 mm,切割边宜为15~20 mm,铺设板时应错缝。

c. 装饰石膏板可采用粘结安装法:对U、C形轻钢龙骨,可采用胶粘剂将装饰石膏板直接粘贴在龙骨上。胶粘剂应涂刷均匀,不得漏刷,粘贴牢固。胶粘剂未完全固化前板材不得有强烈振动。

d. 吸声穿孔石膏板与U形(或C形)轻钢龙骨配合使用,龙骨吊装找平后,在每4块板的交角点和板中心,用塑料小花以自攻螺钉固定在龙骨上。采用胶粘剂将吸声穿孔石膏板直接粘贴在龙骨上。安装时,应注意使吸声穿孔石膏板背面的箭头方向和白线方向一致。

e. 嵌式装饰石膏板可采用企口暗缝咬接安装法。将石膏板加工成企口暗缝的形式,龙骨的两条肢插入暗缝,靠两条肢将板托住。安装宜由吊顶中间向两边对称进行,墙面与吊顶接缝应交圈一致;安装过程中,接插企口用力要轻,避免硬插硬撬而造成企口处开裂。

②装饰吸声罩面板安装:

The lateral brace keel of C-shaped light steel keel of the suspended ceiling is made from cutting the C-shaped secondary keel into sections. At the T-shaped joints between the lateral brace keel and the longitudinal secondary keel, the matching bracket (plug-in component) should be used to connect and fix the two parts.

The installation of T-shaped lateral brace keel on the T-shaped keel cladding of the double-coat framework is determined by the types of the keel material, and its installation method is the same as the installation method of lateral brace keel in single-coat framework, which has been explained in the previous part.

(7) Installing of the skin plate.

① Installation of gypsum board:

a. The installation of gypsum board should begin from one corner of the ceiling and then come to the other part one by one. The gypsum board is fixed to the keel with galvanized self-tapping screws at the size of 3.5mm×25mm (diameter×length). The depth of the nail head embedded in the gypsum board should be about 0.5—1mm; the distance between nails is 150—170mm, and the nail is 15mm away from the edge of the board. Joints should be reserved between the boards and between the board and the wall, and the width of the joint is generally 3—5mm.

When double-coat gypsum board is used, the long and short edge of the first coat should stager at least the distance as the spacing between the keels with the long and short edge of the keel framework, and the second coat should be constructed as the first coat. The 3.5mm×35mm self-tapping screws are fixed to the keels, and the screws should stagger some distance with each other.

b. The gypsum board with kraft paper should be installed in the free state, and should be fixed from the middle of the board to the edge of it. The edge with kraft paper should be parallelly laid along the secondary keel and its length is usually 10—15mm; the length of the cutting edge should be 15—20mm. The joints between the boards should be staggered.

c. The adhesive bonding method for decorative gypsum board: as for U-shaped and C-shaped light steel keels, the decorative gypsum board can be directly bonded to the keel by adhesive. The adhesive should be applied on the boards evenly and fully, and bond the boards firmly. There should be no strong vibration on the boards before the complete solidification of the adhesive.

d. When sound-absorbing perforated plaster boards and the U-shaped (or C-shaped) light steel keels are used at the same time, after the keels are installed and levelled, the gypsum boards are fixed to the keel with plastic flowers by self-tapping screws and nails at the intersection point of each of four boards and at the center of the boards. The sound-absorbing perforated gypsum board can be directly bonded to the keel by adhesive. In bonding, please pay attention to the direction of the arrow on the back of the board and make it the same with the white line.

e. The inserted decorative gypsum board can be installed by the tongue-and-groove concealed joint method. The gypsum board is made into the shape of tongue-and-groove joint, and the two limbs of the keel are inserted into the concealed joints to support the board. The installation should be carried out symmetrically from the middle of the suspended ceiling to the two sides. The wall and the suspended ceiling should be connected closely and uniformly. In the installation, don't push too hard to connect them, avoiding making cracks by inserting or prying too hard.

② Installation of decorative sound-absorbing skin plate:

矿棉装饰吸声板在房间内湿度过大时不宜安装。安装前应先排板,安装时,吸声板上不得放置其他材料,防止板材受压变形。

a. 暗龙骨吊顶安装法。将龙骨吊平、矿棉板周边开槽,然后将龙骨的肢插到暗槽内,靠肢将板托住。房间内温度过高时不宜安装。

b. 复合平贴法。其构造为龙骨＋石膏板＋吸声饰面板。龙骨可采用上人龙骨或不上人龙骨,将石膏板固定在龙骨上,然后将装饰吸声板背面用胶布贴几处,用专用钉固定。

c. 复合插贴法。其构造为龙骨＋石膏板＋吸声板。吸声板背面双面胶布贴几个点,将板平贴在石膏板上,用打钉器将"冂"形钉固定在吸声板开榫处,吸声板之间用插件连接、对齐图案。粘贴法要求石膏板基层非常平整,粘贴时,可采用粘贴矿棉装饰吸声板的874型建筑胶粘剂。珍珠岩装饰吸声板的安装,可在龙骨上钻孔,将板用螺钉与龙骨固定。先在板的四角用塑料小花钉牢,再在小花之间沿板边按等距离加钉固定。

③塑料装饰罩面板安装:与轻钢龙骨固定时,可采用自攻螺钉,也可根据不同材料采用相应的胶粘剂粘贴在龙骨上。

④水泥加压板安装:宜采用胶粘剂和自攻螺钉粘、钉结合的方法固定。纤维增强水泥平板与龙骨固定时,应钻孔,钻头直径应比螺钉直径小 0.5～1.0 mm,固定时钉帽必须压入板面 1～2 mm,螺钉与板边距离宜为 8～15 mm,板周边钉距宜为 150～170 mm,板中钉距不得大于 200 mm。钉帽需作防锈处理,并用油性腻子嵌平。两张板接缝与龙骨之间,宜放一条 50 mm×3 mm 的再生橡胶垫条;纤维增强硅酸钙板加工打孔时,不得用铣子冲孔,应用手电钻钻孔,钻孔时宜在板下垫一木块。

⑤铝合金条板吊顶安装:

Chapter 9 Finishing and Decoration Work Construction

Mineral wool decorative sound-absorbing plate should not be installed when the humidity in the room is too high. Before installation, the plates should be arranged properly to make the figures on it in the correct order. At installation, it is forbidden to place any material on the sound-absorbing plate in case it may cause deformation.

a. Installation of decorative sound-absorbing skin plate for concealed keel suspended ceiling. In the installation, the keels should be installed and levelled, and grooves should be made on the edges of the mineral wool plate. Then the keel's limbs are inserted into the concealed grooves to support the plate. It is not suitable to install the plate when the temperature in the room is too high.

b. Composite flat pasting method: The structure for this method is the keel, the gypsum board and the sound-absorbing skin plate. The keels can be load-bearing or non-load bearing. The gypsum boards are fixed to the keels, and then some places on the back of the decorative sound-absorbing plate should be pasted with adhesive tape, and then they can be fixed to the gypsum boards by special nails.

c. Compound inserting and pasting method: The structure for this method is the keel, the gypsum board and the sound-absorbing plate. Some places on the back of the sound-absorbing plate should be pasted with double-sided adhesive tape and then the plate can be flatly attached to the gypsum board, and the nails in the shape of "⌐¬" are fixed to the tenoning of the sound-absorbing plate by the nailing machine. The sound-absorbing plates are connected by plugs, and the figures on them should align perfectly. In this pasting method, the base of the gypsum board should be very flat, and we can use the same adhesive (Mode 874) as used in pasting the mineral wool decorative sound-absorption plate. In the installation of the perlite decorative sound-absorbing plate, holes are drilled on the keel, and the plate and the keel are fixed with screws. First, the plate should be nailed to the gypsum board on the four corners of it with plastic flowers, and then the nails are added to fix it at equal distance between the plastic flowers along the edge of the plate.

③Installation of plastic decorative skin plate: When the plate is fixed to light steel keel, self-tapping screws can be used, and the matching adhesive can be applied to the keel according to different materials.

④Installation of cement pressure plate: The cement pressure plate can be fixed in pasting and nailing with adhesive and self-tapping screws. When the fiber reinforced cement plate is fixed to the keel, holes should be drilled on it, and the diameter of the drill should be 0.5—1.0mm smaller than the diameter of the screw. When fixing, the nail cap must be pressed into the plate surface for 1—2mm. The distance from the nails to the edge of the plate shall be 8—15mm; the distance between the nails on the edges of the plate shall be 150—170mm, and the distance between the nails in the plate shall not be greater than 200mm. The nail caps should take antirust treatment, and the places for the nails should be levelled with oily putt. Between the joints of the boards and the keel, recycled rubber cushion strip (50mm×3mm) should be placed; when the fiber reinforced calcium silicate plate is made and holes are drilled on it, it is not allowed to use a punch to make holes, and an electric hand drill should be used to drill holes. In drilling the holes, timber should be placed under the plate.

⑤Installation of aluminum alloy batten suspended ceiling:

a. 全面检查中心线,复核龙骨标高线和龙骨布置线,复核龙骨是否调平调直,以保证板面平整。

b. 卡固法条板的安装:适用于板厚为 0.8 mm 以下、板宽在 100 mm 以下的条板。条板安装应从一个方向依次安装,如果龙骨本身兼卡具,只要将条板托起后,先将条板的一端用力压入卡脚,再顺势将其余部分压入卡脚内。

c. 螺钉固定铝合金条板吊顶:适用于板宽超过 100 mm、板厚超过 1mm 的扣板式的铝合金条板材。

采用自攻螺钉固定,自攻螺钉头在安装后完全隐蔽在吊顶内。条板切割时,除控制好切割的角度,同时要对切口部位用锉刀修平,将毛边及不妥处修整好,再用相同颜色的胶粘剂(可用硅胶)将接口部位进行密合。

⑥铝合金方形板吊顶安装:铝合金块板与轻钢龙骨骨架的安装,可采用吊钩悬挂式或自攻螺钉固定式,也可采用铜丝扎结。用自攻螺钉固定时,应先用手电钻打出孔位后再上螺钉。安装时按照弹好的布置线,从一个方向开始依次安装,吊钩先与龙骨连接固定,再钩住板块侧边的小孔。铝合金板在安装时应轻拿轻放,保护板面不受碰伤或刮伤。

(8)嵌缝。吊顶石膏板铺设完成后,即进行嵌缝处理。

嵌缝的填充材料:老粉(双飞粉)、石膏、水泥及配套专用嵌缝腻子。常见的材料一般配以水、胶,也可根据设计的要求把水与胶水搅拌均匀之后使用。专用嵌缝腻子不用加胶水,只根据说明加适量的水搅拌均之后即可使用。

嵌缝的程序为:螺钉的防锈处理→板缝清扫干净→腻子嵌缝密实→干燥养护→第二道嵌缝腻子→贴盖缝带(品种有专用纤维纸带,玻璃纤维网格带等。盖缝带可适当湿润处理,防止腻子或胶干缩收缩时产生应力造成撕裂)→干燥→满批腻子。

a. Thoroughly inspect the center line, re-check the elevation line and the layout line of the keel, and check whether the keel is leveled and straightened to ensure the flatness of the surface.

b. The installation of batten in clamping method: this method is suitable for battens with the thickness below 0.8mm and the width below 100mm. The installation of the batten should start from one point and then battens can be installed one by one in the same direction. If there are clamping apparatus on the keel, the battens can be lifted and clamp one end of it into the clamping feet on the keel, and then the rest of it can be pressed to the clamping feet.

c. Installation of aluminum alloy batten suspended ceiling with screw: it is suitable for buckle aluminum alloy plate with the width more than 100mm and the thickness more than 1mm.

When it is fixed with self-tapping screws, the self-tapping screw head should completely conceal in the suspended ceiling after installation. When the batten is cut, the angle of the cutting should be controlled. At the same time, the incision should be flattened with a file, and the raw edges and the parts with defects should be trimmed to make them flat. Then the joints should be sealed with the adhesive of the same color (silica gel can be used).

⑥Installation of aluminum alloy square plate suspended ceiling: The aluminum alloy plate can be fixed to the framework of light steel keel by hanging with hooks or fixing with self-tapping screws, or tying with copper wire. When fixing with self-tapping screws, the hole should be drilled with an electric hand drill first, and then the screw should be used. At installation, it should start from one point and then the plates will be installed one by one in turn based on the tapped layout line. The hook should be fixed to the keel first and then it will hook the small hole on the side of the plate. The aluminum alloy plate shall be handled with care during installation to prevent the plate surface from being hit or scratched.

(8) Caulking.

After the gypsum boards of suspended ceiling are laid, the caulking treatment should be carried out immediately.

Filling materials for caulking: heavy calcium powder (talc), gypsum, cement and the matching caulking putty. The filling materials are generally mixed with water and adhesive, or they can be used after being completely stirred with water and adhesive according to the design requirements. It's no need to mix adhesive in the special caulking putty and according to the instruction, certain amount of water is mixed and stirred completely and then the special caulking putty can be used.

The following is the process of caulking: the antirust treatment of the screws → cleaning the joints of the plates → caulking compactly with putty → dry curing → caulking with putty for the second time → pasting the cover strip for the joint (there are special fiber paper strip, glass fiber grid strip and so on, which can be properly wet to prevent being tore by the stress generated in the shrink of the putty or adhesive) → drying → applying the putty fully.

习 题

1. 抹灰工程分为几类？抹灰由哪几层组成？
2. 简述一般抹灰的施工工艺。
3. 简述大理石饰面施工工艺流程。
4. 简述混凝土表面施涂油漆涂料操作工艺。
5. 简述铝合金门窗安装施工操作工艺。
6. 简述轻钢龙骨吊顶操作工艺。

Chapter 9 Finishing and Decoration Work Construction

Exercises

1. What are the classifications of plastering and what are the different coats of plastering?
2. Briefly describe the construction technology of general plastering.
3. Briefly describe the operation process of facing with marble.
4. Briefly describe the operation process of applying oil paint on the surface of concrete.
5. Briefly describe the operation process of installation of aluminum alloy doors and windows.
6. Briefly describe the operation process of light steel keel suspended ceiling.